Microbiome in Plant Health and Disease

Vivek Kumar • Ram Prasad • Manoj Kumar
Devendra K. Choudhary
Editors

Microbiome in Plant Health and Disease

Challenges and Opportunities

Springer

Editors
Vivek Kumar
Himalayan School of Biosciences
Swami Rama Himalayan University
Dehradun, Uttarakhand, India

Ram Prasad
School of Environmental Science & Engg
Sun Yat-sen University
Guangzhou, Guangdong, China

Manoj Kumar
Centre for Life Sciences
Central University of Jharkhand
Ranchi, Jharkhand, India

Devendra K. Choudhary
Amity Institute of Microbial Technology
Amity University
Noida, Uttar Pradesh, India

ISBN 978-981-13-8494-3 ISBN 978-981-13-8495-0 (eBook)
https://doi.org/10.1007/978-981-13-8495-0

This Springer imprint is published by the registered company Springer Nature Singapore Pte Ltd.
The registered company address is: 152 Beach Road, #21-01/04 Gateway East, Singapore 189721, Singapore

Preface

Microbiome diversity in association with plant roots is gargantuan; the microbial number includes tens of thousands of various species. This multifaceted plant accompanying microbial population, also referred to as the integrated part of the plant, is imperative for plant health. Current progresses in plant-microbe synergism research disclosed that plants have capabilities to shape their rhizospheric microbial community. This has been substantiated by the information that dissimilar plant species host particular microbial populations even when grown in the same climatic conditions and soil. Accumulation of not good or unwanted microbiome communities in host plant leads to negative effect on plant health, on the other hand, association of beneficial and constructive microbiome community in plant leads to positive effect on the host health.

Pathogenic microbiome could lead to adverse effects on plant health. Interactions between pathogens and plants are regularly considered as conflict between the two organisms, this ignores the significance of beneficial microbiome. But the prevalence and proliferation of virtuous microbiome can appreciably influence infection progression. Plants survive in close connection with microbial community that thrives the habitat in which the plants grow. The soil microbial population structure exemplifies the largest reservoir of biodiversity known till date. The rhizospheric zone of plant is the confined zone of soil around roots, which is manipulated by root exudates, can harbor up to 10^{11} microbial cells per gram of root and numerous other prokaryotic species. The combined genome of this microbial population is much higher than that of host plant and is therefore also insinuated as the host plant's additional genome.

We understand that the human system also benefits from beneficial probiotic bacteria. Similarly, host plants also have the dependence on specific favorable microbiome which are also recognized as plant growth strengtheners, biostimulators, phytomodulators, biofertilizers, bioinoculants, phytostimulators, biopesticides, biocontrol agents, etc., which are eventually advantageous to plant health. The exact mechanism through which these microbiome become associated with the plant system is unknown. This association depends upon the type of plant species and its age, though it also depends on diverse ecological factors.

A lot of investigation has been done in the past but not much has been achieved to understand the mutualistic interaction between microbiome and their host plants. The plant microbiome are one of the most significant reasons for plant health, its sustainability and productivity, furthermore, this research area has grasped wide attention and consideration, in recent years. Microbiome associated with host plants also play a decisive and essential role in general biogeochemical cycles. Plant-associated beneficial microbiome also help hosts to surmount pathogenic microbes, encourage growth, and inhabit space that would otherwise be reachable to host pathogens. Besides these beneficial aspects, good microbiome also stimulate various stress resistance, and ultimately persuade plant growth promotion through nutrients mobilization, uptake, and transport. Consequently, the plant microbiome proves to be a noteworthy and substantial determinant for host health and production.

To understand complex communication pathways regulation within the plant and their associated microbiome involves manifold functions of microbiome and plant root excretions and their influences on genome expression and translation. The universal and all-inclusive approach to understanding any organismic structure and function is to apprehend the organism in its entireness. The root associated or the endophytic microbiome and their functional inputs are undoubtedly essential for all plants on the sphere. It should be apparent that how the associated microbiome influence or are being influenced by host plant will definitely vary across species, as well as by several genetic and environmental factors. Studies on the plant microbiome need to authenticate the microbial population, this will help us in understand the change or fluctuations in microbiome community correlated with environmental habitats. Further development is required for functional examination that exploits metagenomics and metatranscriptomics techniques. This screening will be a front runner for us to understand plant attributes and behaviors based on microbiome knowledge. This will help us comprehend when, where, and how this "additional genome" also functions as an intact organ system of the host plant.

In this book, we discuss evidences related to associated plant microbiome, whether rhizospheric, phyllospheric, or endophytic, playing a significant role in plant health or disease formation. *Microbiome in Plant Health and Disease: Challenges and Opportunities* is focused on, but not limited to: microbiome colonization, their role in plant growth, development, nutrient recycling, mycorrhizae, and an overview of phytospheric microbiome in sustainable agro-eco-system.

It is believed that the enthusiasm, eagerness, and remarkable opportunities presented in this work about our latest perception of the challenges and relationships that bring about learning plant microbiome mutualistic approach will encourage and inspire readers to push the field forward to new frontiers.

Dehradun, Uttarakhand, India — Vivek Kumar
Guangzhou, Guangdong, China — Ram Prasad
Ranchi, Jharkhand, India — Manoj Kumar
Noida, Uttar Pradesh, India — Devendra K. Choudhary

Contents

Editors and Contributors

About the Editors

Vivek Kumar is working as an Associate Professor at the Himalayan School of Biosciences, Swami Rama Himalayan University, Jolly Grant, Dehradun, India, and is leading plant-microbe interaction, bioremediation, and microbial ecology research group. He has obtained his masters and doctoral degrees from CCS Haryana Agricultural University, Hisar, India. Serving as Editor and Reviewer of several reputed international journals, he has more than hundred publications to his credit. He has authored a practical manual book and edited several books with Springer Nature. He had also served as Microbiologist in Public Authority of Agricultural Affairs and Fish Resources, Kuwait, and has been credited with first time reporting of Pink Rot inflorescence disease of Date palm in Kuwait caused by *Serratia marcescens*.

Ram Prasad is associated with Amity Institute of Microbial Technology, Amity University, Uttar Pradesh, India, since 2005. His research interests include applied microbiology, plant-microbe interaction, sustainable agriculture, and nanobiotechnology. Dr. Prasad has more than 150 publications to his credit, including research papers, review articles, book chapters, 5 patents issued or pending, and several edited or authored books. He has been awarded the Young Scientist Award and Prof. J.S. Datta Munshi Gold Medal by the International Society for Ecological Communications; FSAB fellowship by the Society for Applied Biotechnology; the American Cancer Society

UICC International Fellowship for Beginning Investigators, USA; Outstanding Scientist Award in the field of Microbiology by Venus International Foundation; BRICPL Science Investigator Award; and Research Excellence Award. He serves as an editorial board member for *Frontiers in Microbiology*, *Frontiers in Nutrition*, and *Academia Journal of Biotechnology*. Previously, he served as a Visiting Assistant Professor at Whiting School of Engineering, Department of Mechanical Engineering at Johns Hopkins University, USA, and presently works as a Research Associate Professor at the School of Environmental Sciences and Engineering, Sun Yat-Sen University, Guangzhou, China.

Manoj Kumar is currently working as an Associate Professor, Life Sciences, at the Central University of Jharkhand, heading the department since 2018 onwards. Dr. Kumar is having long research experience in plant developmental biology with doctorate and postdoctoral research experiences from prestigious organizations (India and abroad). He has opted academic career with a passion in fundamental sciences (plant developmental biology) where he excelled numerous collaborative research activities at national and international levels. He has published valuable research inputs in accredited journals and edited books. He has also guided several students at master, M.Phil., doctorate, and post-doctorate levels. He has been a principle investigator of several projects of national funding agencies – DBT, DRDO, etc.

He is associated as Editor and Reviewer with several research journals and offering consultancy to several biotechnology firms at international level.

Currently, he is the principal investigator of DBT-BUILDER program and leading an elite research group at multidisciplinary level. He is exploring academic world with a vision of empowering young generation with fact-finding approach destined for rural India.

Devendra K. Choudhary shows his presence as an Associate Professor at Amity Institute of Microbial Technology. He is an active researcher and operated major projects sponsored by DBT, DST, and SERB, New Delhi, India, as principal investigator and co-investigator. He has published quorum of research/review articles along with several book chapters for reputed journals, edited several books through Springer, and filed three patents to India Patent Office, New Delhi. He has supervised several research scholars for their doctoral program. In addition, he is a recipient of NASI-life membership and accorded with Indian National Science Academy (INSA) visiting and summer research fellowship.

Contributors

Rasheed A. Adeleke Unit for Environmental Science and Management, North-West University, Potchefstroom Campus, Potchefstroom, South Africa

Microbiology and Environmental Biotechnology Research Group, Agricultural Research Council - Institute for Soil, Climate and Water, Pretoria, South Africa

Maqshoof Ahmad Department of Soil Science, University College of Agriculture and Environmental Sciences, The Islamia University of Bahawalpur, Bahawalpur, Pakistan

Qasim Ali College of Agriculture, Bahauddin Zakariya University, Layyah, Pakistan

Galdino Andrade Microbial Ecology Laboratory, Department of Microbiology, State University of Londrina, Londrina, Paraná, Brazil

Elena Andriūnaitė Institute of Horticulture, Lithuanian Research Centre for Agriculture and Forestry, Babtai, Kaunas reg., Lithuania

Sana Ashraf Institute of Soil and Environmental Sciences, University of Agriculture, Faisalabad, Pakistan

João Lúcio Azevedo Departamento de Genética, Escola Superior de Agricultura "Luiz de Queiroz", Universidade de São Paulo, Piracicaba, São Paulo, Brasil

Anabela Marisa Azul Centre for Functional Ecology (CFE), Department of Life Sciences, University of Coimbra, Coimbra, Portugal

Danas Baniulis Institute of Horticulture, Lithuanian Research Centre for Agriculture and Forestry, Babtai, Kaunas reg., Lithuania

Paula Baptista CIMO, School of Agriculture – Polytechnic Institute of Bragança, Bragança, Portugal

Anelise Beneduzi Departament of Agricultural Research and Diagnosis, Secretariat of Agriculture, Livestock and Irrigation of Rio Grande do Sul, Porto Alegre, Rio Grande do Sul, Brazil

Lorena Carro Departamento de Microbiología y Genética, Universidad de Salamanca, Salamanca, Spain

Andreas Lazaros Chryssafidis Agroveterinary Sciences Center, State University of Santa Catarina, Lages, Santa Catarina, Brazil

Ghoson Mosbah Daba Chemistry of Natural and Microbial Products Department, Pharmaceutical Industries Researches Division, National Research Center, Giza, Egypt

Ahmed Mohamed Eid Department of Botany and Microbiology, Faculty of Science, Al-Azhar University, Nasr City, Cairo, Egypt

Waill Ahmed Elkhateeb Chemistry of Natural and Microbial Products Department, Pharmaceutical Industries Researches Division, National Research Center, Giza, Egypt

Janaina Emiliano Microbial Ecology Laboratory, Department of Microbiology, State University of Londrina, Londrina, Paraná, Brazil

José David Flores-Félix Departamento de Microbiología y Genética, Universidad de Salamanca, Salamanca, Spain

Amr Fouda Department of Botany and Microbiology, Faculty of Science, Al-Azhar University, Nasr City, Cairo, Egypt

Camille Eichelberger Granada Universidade do Vale do Taquari – UNIVATES, Lajeado, Rio Grande do Sul, Brazil

Saad El-Din Hassan Department of Botany and Microbiology, Faculty of Science, Al-Azhar University, Nasr City, Cairo, Egypt

Muhammad Ijaz College of Agriculture, Bahauddin Zakariya University, Layyah, Pakistan

Irum Iqrar Department of Biotechnology, Quaid-i-Azam University, Islamabad, Pakistan

Mohamed A. Ismail Department of Botany and Microbiology, Faculty of Science, Al-Azhar University, Nasr City, Cairo, Egypt

Jackson Freitas Brilhante São José Departament of Agricultural Research and Diagnosis, Secretariat of Agriculture, Livestock and Irrigation of Rio Grande do Sul, Porto Alegre, Rio Grande do Sul, Brazil

P. Jishma School of Biosciences, Mahatma Gandhi University, Kottayam, Kerala, India

Muhammad Kamran Department of Agronomy, University College of Agriculture, University of Sargodha, Sargodha, Pakistan

Vankayalapati Vijaya Kumar Trident Sugars Limited, Zaheerabad, Sangareddy, Telangana, India

Vivek Kumar Himalayan School of Biosciences, Swami Rama Himalayan University, Dehradun, Uttarakhand, India

Bruno Brito Lisboa Departament of Agricultural Research and Diagnosis, Secretariat of Agriculture, Livestock and Irrigation of Rio Grande do Sul, Porto Alegre, Rio Grande do Sul, Brazil

Komal Mazhar Department of Forestry and Range Management, University of Agriculture Faisalabad, Faisalabad, Pakistan

Esther Menéndez Departamento de Microbiología y Genética, Universidad de Salamanca, Salamanca, Spain

Diogo Mina CIMO, School of Agriculture – Polytechnic Institute of Bragança, Bragança, Portugal

Fluvio Modolon Microbial Ecology Laboratory, Department of Microbiology, State University of Londrina, Londrina, Paraná, Brazil

Gholamreza Mohammadi Department of Plant Production and Genetic Engineering, College of Agriculture and Natural Resources, Razi University, Kermanshah, Iran

Kalaivani K. Nadarajah School of Environmental and Natural Resources Science, Faculty of Science and Technology, Universiti Kebangsaan Malaysia, Bangi, Selangor, Malaysia

Sajid Mahmood Nadeem University of Agriculture Faisalabad, Burewala-Vehari, Pakistan

Mahmoud Nasr Sanitary Engineering Department, Faculty of Engineering, Alexandria University, Alexandria, Egypt

Miguel O. P. Navarro Microbial Ecology Laboratory, Department of Microbiology, State University of Londrina, Londrina, Paraná, Brazil

Bhavna Nunthkumar Microbiology and Environmental Biotechnology Research Group, Agricultural Research Council - Institute for Soil, Climate and Water, Pretoria, South Africa

Linda Obi Microbiology and Environmental Biotechnology Research Group, Agricultural Research Council - Institute for Soil, Climate and Water, Pretoria, South Africa

Department of Environmental Science, College of Agriculture and Environmental Science, Pretoria, South Africa

Andreia Mara Rotta de Oliveira Departament of Agricultural Research and Diagnosis, Secretariat of Agriculture, Livestock and Irrigation of Rio Grande do Sul, Porto Alegre, Rio Grande do Sul, Brazil

Luciane Maria Pereira Passaglia Department of Genetics, Universidade Federal do Rio Grande do Sul – UFRGS, Porto Alegre, Rio Grande do Sul, Brazil

Alvaro Peix Unidad Asociada Universidad de Salamanca- CSIC 'Interacción Planta-Microorganismo', Salamanca, Spain

IRNASA-CSIC, Salamanca, Spain

José Alberto Pereira CIMO, School of Agriculture – Polytechnic Institute of Bragança, Bragança, Portugal

Yelena Perevalova Department of Genetics, Universidade Federal do Rio Grande do Sul – UFRGS, Porto Alegre, Rio Grande do Sul, Brazil

Amanda C. M. Piva Microbial Ecology Laboratory, Department of Microbiology, State University of Londrina, Londrina, Paraná, Brazil

Maria Carolina Quecine Departamento de Genética, Escola Superior de Agricultura "Luiz de Queiroz", Universidade de São Paulo, Piracicaba, São Paulo, Brasil

E. K. Radhakrishnan School of Biosciences, Mahatma Gandhi University, Kottayam, Kerala, India

Ranjini Ramesh St Joseph's College, Autonomous, Bangalore, India

Martha-Helena Ramírez-Bahena Departamento de Microbiología y Genética, Universidad de Salamanca, Salamanca, Spain

Abdul Rehman College of Agriculture, Bahauddin Zakariya University, Layyah, Pakistan

Ashira Roopnarain Microbiology and Environmental Biotechnology Research Group, Agricultural Research Council - Institute for Soil, Climate and Water, Pretoria, South Africa

Salim S. Salim Department of Botany and Microbiology, Faculty of Science, Al-Azhar University, Nasr City, Cairo, Egypt

Zabta Khan Shinwari Department of Biotechnology, Quaid-i-Azam University, Islamabad, Pakistan

Ane S. Simionato Microbial Ecology Laboratory, Department of Microbiology, State University of Londrina, Londrina, Paraná, Brazil

Flávia R. Spago Microbial Ecology Laboratory, Department of Microbiology, State University of Londrina, Londrina, Paraná, Brazil

Vidmantas Stanys Institute of Horticulture, Lithuanian Research Centre for Agriculture and Forestry, Babtai, Kaunas reg., Lithuania

Inga Tamošiūnė Institute of Horticulture, Lithuanian Research Centre for Agriculture and Forestry, Babtai, Kaunas reg., Lithuania

Faouzia Tanveer Department of Biotechnology, Quaid-i-Azam University, Islamabad, Pakistan

Teresa Lino-Neto Biosystems and Integrative Sciences Institute (BioISI), Plant Functional Biology Center (CBFP), University of Minho, Braga, Portugal

Paul William Thomas Mycorrhizal Systems Ltd, Lancashire, UK

University of Stirling, Stirling, UK

Sami Ul-Allah College of Agriculture, Bahauddin Zakariya University, Layyah, Pakistan

Luciano Kayser Vargas Departament of Agricultural Research and Diagnosis, Secretariat of Agriculture, Livestock and Irrigation of Rio Grande do Sul, Porto Alegre, Rio Grande do Sul, Brazil

Encarna Velázquez Departamento de Microbiología y Genética, Universidad de Salamanca, Salamanca, Spain

Unidad Asociada Universidad de Salamanca- CSIC 'Interacción Planta-Microorganismo', Salamanca, Spain

Nikhi Verma Department of Medical Mycology, VP Chest Institute, University of Delhi, Delhi, India

Camila Gazolla Volpiano Department of Genetics, Universidade Federal do Rio Grande do Sul – UFRGS, Porto Alegre, Rio Grande do Sul, Brazil

Ting-Chi Wen The Engineering Research Center of Southwest Bio–Pharmaceutical Resources, Ministry of Education, Guizhou University, Guiyang, Guizhou Province, China

Zahir Ahmad Zahir Institute of Soil and Environmental Sciences, University of Agriculture, Faisalabad, Pakistan

Mohammad Javad Zarea Department of Agronomy and Plant breeding, Faculty of Agriculture, Ilam University, Ilam, Iran

Applications of Plant–Microbe Interactions in Agro-Ecosystems

1

Rasheed A. Adeleke, Bhavna Nunthkumar, Ashira Roopnarain, and Linda Obi

Abstract

The natural association between plants and microorganisms has historically been linked to improved plant growth, nutrition and health. Rhizospheric and phyllospheric microorganisms have received much attention due to their applications in improved nutrient acquisition, enhanced water sequestration, induced systemic resistance, competitive exclusion of plant pathogens and remediation of environmental pollutants. Such beneficial attributes have motivated the adoption of these plant–microbe interactions in agro-ecosystems to improve productivity. The application of commercially available plant beneficial microorganisms (CAPBM) in agro-ecosystems is largely due to their compatibility and complementarity with natural processes of nutrient cycling, plant protection and other related biological processes. While numerous studies have reported the huge potential of the use of plant beneficial microorganisms in agro-ecosystems, wide-scale com-

R. A. Adeleke (✉)
Unit for Environmental Science and Management, North-West University, Potchefstroom Campus, Potchefstroom, South Africa

Microbiology and Environmental Biotechnology Research Group, Agricultural Research Council - Institute for Soil, Climate and Water, Pretoria, South Africa
e-mail: Rasheed.Adeleke@nwu.ac.za

B. Nunthkumar · A. Roopnarain
Microbiology and Environmental Biotechnology Research Group, Agricultural Research Council - Institute for Soil, Climate and Water, Pretoria, South Africa
e-mail: NunthkumarB@arc.agric.za; RoopnarainA@arc.agric.za

L. Obi
Microbiology and Environmental Biotechnology Research Group, Agricultural Research Council - Institute for Soil, Climate and Water, Pretoria, South Africa

Department of Environmental Science, College of Agriculture and Environmental Science, Pretoria, South Africa
e-mail: ObiL@arc.agric.za

V. Kumar et al. (eds.), *Microbiome in Plant Health and Disease*,
https://doi.org/10.1007/978-981-13-8495-0_1

mercialization of microbial products are still lagging. Hurdles in the commercialization of CAPBM range from lack of awareness and regulatory framework to inaccurate product selection. The future prospects of the application of CAPBM will be determined by the adoption of new technologies that include multi-omics approach for improving the quality as well as applicability of these beneficial microorganisms in agro-ecosystems. Furthermore, government intervention is of utmost importance to ensure that the necessary regulatory framework is in place, thereby ensuring high quality of products. High-quality products will improve adoption rate, which would have downstream influences on job creation in the CAPBM and agricultural industries.

Keywords
Commercially available plant beneficial microorganisms · Biofertilizers · Applications · Agro-ecosystems

1.1 Introduction

As photosynthesizers, plants are primarily responsible for provision of energy for the entire ecosystem; hence, they are intimately linked to many activities in the ecosystem and do not usually exist in isolation. The type and nature of associations they form with microorganisms are crucial for their development, survival, diversity, abundance and ecology (Van Der Heijden et al. 2008). Therefore, plant–microbe associations will continue to shape and dictate the structure of the ecosystem. Over the years, scientists had capitalized on the benefits of such associations and borrowed ideas from natural ecosystems to improve practices in agro-ecosystems.

The agro-ecosystem is physiologically different from that of the immediate surrounding environments. It functions as a specialized niche due to its biological and physicochemical differences. These variations are caused by biological and chemical processes carried out by the residing macro- and microorganisms (Conway 1986). Microorganisms are ubiquitous in the environment and play key roles in agro-ecosystem functioning. The interaction of plants and microbes can be helpful or harmful to plant life, or they simply exist in mutual harmony. Pathogenic microbes can infect plants in a host–parasite relationship causing plant diseases (Gnanamanickam et al. 1999; Vurro et al. 2010). However, some commensal bacteria have been known to reside within a host plant for long periods with no damage caused to the host plant (Hardoim et al. 2008).

Although plant–microbe interactions could be positive or negative, applications in the agro-ecosystem are mainly positive with partial or full benefits to the entire ecosystem. Such relationships could be mutualistic, which involves the provision of shelter and/or nutrients by the plant for the microbes while, in turn, the microbes enhance plant growth and provide biological control against potential pathogens and predators using various strategies (Mendes et al. 2013). Important roles of microorganisms in nutrient cycling and plant protection have drawn interests of

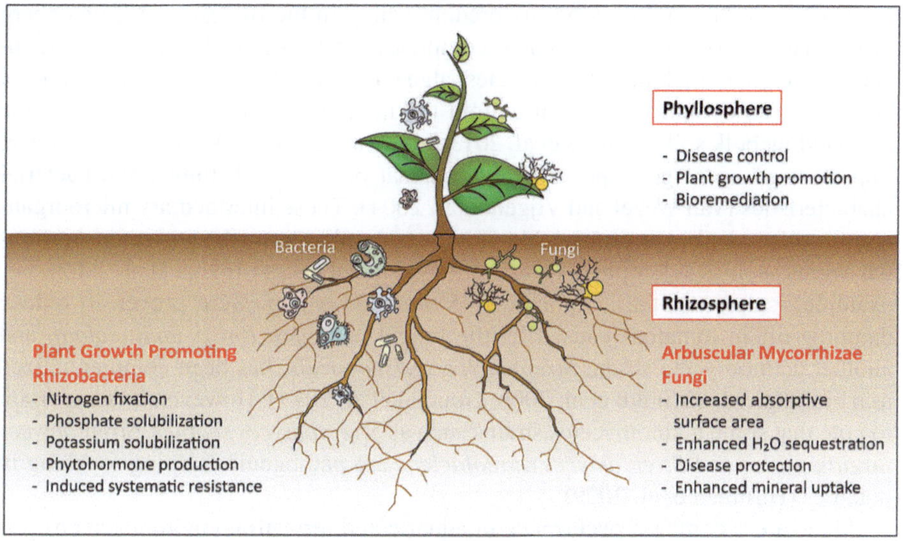

Fig. 1.1 Overview of applications of plant–microbe interactions in agro-ecosystems

researchers as they seek to replace chemical products with commercially available plant beneficial microorganisms (CAPBM) (also known as biofertilizer) for enhancing plant growth and for control of pathogens. Chemical fertilizers containing only nitrogen, phosphate and potassium, which saturate soils, can be replaced with species or a consortium of microorganisms that are able to transform unavailable substrates to an inorganic form that plants can assimilate (Sihi et al. 2017). Similarly, chemical pesticides can be replaced by CAPBM, which convey natural defences against pathogens (Mendes et al. 2013). Good crop yield and healthy soil microflora reflect the health of the agro-ecosystem. As such, this chapter focuses on plant–microbe interactions, which support agro-ecosystems, specifically, the effects of plant–microbe interactions on plant growth as well as plant and soil health in relation to the rhizosphere and phyllosphere (Fig. 1.1).

1.2 Plant–Microbe Interactions in the Rhizosphere

Although plants were the first eukaryotes to inhabit land around 450 million years ago, their survival was intrinsically linked to the symbiotic relationships they formed with microorganisms (Heckman et al. 2001). With no root structures, the earliest plants relied wholly on microorganisms for mineral nutrient acquisition. However, evolution of plants has resulted in the acquisition of vascular structures and root systems, which enables them to obtain mineral nutrients from their environment. Despite the increasing nutrient acquisition potential of plants over time, their symbiotic relationships with microorganisms are still integral to plant growth and development (Heckman et al. 2001; Kenrick and Strullu-Derrien 2014).

Plant–microbe symbiosis occurs predominantly in the rhizosphere (Lugtenberg 2015). The rhizosphere microbiome comprises a number of diverse microorganisms, which may include actinomycetes, algae, bacteria, fungi, viruses and archaea and is said to contain more than tenfold the number of microorganisms than the surrounding bulk soil (Mendes et al. 2013; Lugtenberg 2015). Actinomycetes are an important group of gram-positive bacteria that possess both fungal and bacterial characteristics (van Wezel and Vijgenboom 2004). These intermediary microorganisms as plant symbionts are capable of increasing plant growth and secrete antifungal agents such as siderophores and antifungal metabolites (Tokala et al. 2002). For example, isolates of the actinomycete *Streptomyces* spp. were proven to reduce damping-off in tomatoes caused by the fungal pathogen *Rhizoctonia solani,* and another actinomycete strain, *Streptomyces griseoviridis*, has been commercialized as a biofungicide (Minuto et al. 2006; Goudjal et al. 2014). However, it is important to note that some actinomycetes strains such as *Streptomyces scabies, Streptomyces turgidiscabies* and *Streptomyces aureofaciens* are pathogenic, causing scabbing in potatoes (Hiltunen et al. 2009).

Algae are a common occurrence in aquatic and terrestrial environments as primary producers of organic matter (Hristozkova et al. 2018). Cyanobacteria, blue-green algae and green algae are generally present in the rhizosphere. These algae play an important role in polysaccharide secretion and oxygen production, which contributes to soil aggregation and soil aeration, respectively (Hristozkova et al. 2018). In addition, algal exudates of *Ascophyllum nodosum* were used as biostimulants to improve cold tolerance in maize and drought tolerance in spinach (Xu and Leskovar 2015; Bradáčová et al. 2016).

Viruses present in the soil can cause plant diseases with deleterious effects on plant health. However, some plants exhibit increased tolerance to water and cold stress when infected with viruses such as plant mosaic and tobacco rattle viruses (Xu et al. 2008; Roossinck 2011).

Rhizospheric Fungi in Agro-Ecosystems

Fungi form one of the most diverse groups of eukaryotes and are an essential functional component of soil microbial communities. One of the most ubiquitous groups of microorganisms in the rhizosphere are the mycorrhizal fungi; they are present in harsh desert, thermal and arctic tundra soils as well as salt flats. Although there are many beneficial fungi in the rhizosphere, this chapter will focus primarily on mycorrhizal fungi. They are abundant in forest and agricultural soils (Gardes and Dahlberg 1996; Tian et al. 2006; Smith and Read 2008; Zabinski and Bunn 2014; Becerra et al. 2014). Different approaches have been advocated for the classification of this important group of fungi in the last two decades. One such approach is based on the trophic level, which consists of two major groups–ectotrophic and endotrophic mycorrhiza. Another classification approach is related to morphological and anatomical features of mycorrhizal fungi. In the latter approach, ectomycorrhiza, endomycorrhiza and ectendomycorrhiza (EM) are the recognized types. Perhaps, the most common form of classification is the seven popular categories that have been widely mentioned in literature. These are ectomycorrhiza, arbuscular mycorrhiza, ectendomycorrhiza, arbutoid, ericoid, monotropoid and orchid mycorrhiza (Brundrett 2004; Smith and Read 2008). These groups are broadly differentiated by the nutrient exchange compartments they form within the root. For instance, mycorrhizal fungi that penetrate the root and exhibit intracellular penetration of root cells are termed endomycorrhiza, while those that colonize outside of the root cells showing only intercellular penetration are termed ectomycorrhiza (Friberg 2001; Smith and Read 2008).

Ectomycorrhizal fungi form a mycorrhizal association characterized by means of their structural mycelial formation that does not penetrate but extends between the host root cells to form a Hartig net (Smith and Read 2008). Ectomycorrhizal fungal associations are formed by higher Basidiomycotina (*Agaricus, Amanita, Lactarius, Thelephora* and *Scleroderma*) and a few Ascomycotina (*Tuber and Terfezia*) and *Zygomycota* (Endogone) (Isaac 1991; Molina et al. 1992; Lodge 2000). The resilience and perhaps survival of trees from the families of Pinaceae, Fagaceae, Betulaceae and Myrtaceae are partly due to their association with ECM fungi (Muchovej 2004), which are important, particularly in environments where growth conditions are not ideal (Isaac 1991).

Unlike the AM fungal counterparts, some ECM fungi have been successfully cultivated in vitro, but they generally grow slowly (Horton and Bruns 2001; Finlay and Söderström 1992). This has been attributed to their ability to utilize simple sugars such as glucose, mannose and fructose, independent of their host plants, and is proof of the facultative nature of some ECM fungi. It has also been shown that ECM fungi can participate in the degradation of complex carbohydrates as shown by Entry et al. (1991). It was observed that the ECM fungus *Hysterangium setchellii* was able to provide an improved microenvironment for easy decomposition of lignin and cellulose.

Generally, ECM fungi have broad host ranges within restricted plant families and their hosts may be receptive to several ECM fungi (Molina et al. 1992). This means that a plant species can act as a host to a variety of ECM fungi and an ECM fungus can also colonize different suitable host plants (Isaac 1991; Molina et al. 1992). The advantage of this is the increased survival rate of new seedling species (dispersal of the seedlings) in new environments, as they can easily associate with a variety of ECM fungi. Examples of ECM fungi with broad host compatibility include *Amanita aspera*, *Boletus calopus*, *Tuber borchii*, *Ruber brumale*, *Tuber melanosporum*, *Choiromyces venosus* and *Pisolithus tinctorius* (Molina et al. 1992). This situation boosts the plants' access to more nutrients because of the ability of the individual ECM fungus involved in the association to source nutrients in the soil on behalf of the plant (Bruns et al. 2002). In contrast, some ECM fungi are host specific. These include those that can associate with one genus of host plant such as *Amanita diemii*, *Boletus loyo*, all *Suillus* spp. and *Tricholoma robustum*. Specificity can restrict the geographic scope of the ECM fungus; those that are too specific will be restricted to areas where the host is present.

Another important group of mycorrhizal fungi are the arbuscular mycorrhizal fungi. They are named after the finely branched structures they produce intracellularly, which are referred to as 'arbuscules.' This structure is present in most members of this group (Smith and Read 2008; Muchovej 2004). They are sometimes referred to as vesicular arbuscular mycorrhiza (VAM) but only in situations where the species produce vesicles (Brundrett 2002). Arbuscular mycorrhizal fungi are believed to be much older than land plants because of their various primitive characteristics such as simple spores, absence of sexual reproduction and their relationships with a wide variety of plants (Morton 1990; Brundrett 2002). They have been found to associate with primitive plants such as mosses and ferns as well as a wide range of angiosperms and gymnosperms. Arbuscular mycorrhizal fungi are obligate symbionts having no special enzymes to degrade simple or complex carbohydrates (Finlay and Söderström 1992; Brundrett 2002).

Arbuscular mycorrhizal fungi have been classified under the phylum Glomeromycota, which consists of several genera that coevolved with plants for over 400 million years (Walker and Schüßler 2004). Arbuscular mycorrhizal fungi colonize plant roots to scavenge organic carbon from root exudates while in return providing the plant with mineral nutrients. Members of the AM fungi family have identical genetic makeup that makes it difficult to identify individual species (Hosny et al. 1999; Pringle et al. 2000). They have either aseptate or rarely septate hypha with inter- and intracellular colonization of the cortical cells of the host plants and absence of Hartig net or mantle (Smith and Read 2008).

For symbiosis to occur between plants and AM fungi, communication is essential. Such communication is mediated by signalling molecules from both the plant and the fungal symbiont. The fungal symbiont senses roots in the vicinity by the detection of plant hormones released through root exudates. Plant hormones such as strigolactone have been found to initiate hyphal branching and speed up the metabolism of AMF (Bonfante and Desirò 2015). In order to colonize the root, fungi release lipochito-oligosaccharides (Myc factors), which also promote plant growth to increase root surface area (Lugtenberg 2015). Once the fungi have adhered to the surface of the root, the hyphae intrude the plant cell and branch out within the cell to form arbuscules (Sanders and Croll 2010; Lugtenberg 2015).

In ectendomycorrhizal associations, there is a special scenario in which the fungal hyphae form a Hartig net and a primitive sheath around the roots. The hyphae penetrate the cortical root cells, and there is a formation of reduced mantle, thereby possessing the characteristics of both ectomycorrhizal and endomycorrhizal fungi (Molina et al. 1992; Smith and Read 2008). Ectendomycorrhizal fungi have been shown to colonize seedlings in early stages over ECM, possibly due to the ability of some EM to break down complex polysaccharides for use as carbon source (Egger 1986; Caldwell et al. 2000; Yu et al. 2001). Yu et al. (2001) suggest that EM fungal association is favored, as the EM could possibly provide carbon to the developing seedling prior to autotrophy development and that the EM contributes less to the carbon drain of the young seedling than ECM.

Other endomycorrhizal groups include the arbutoid mycorrhizas such as *Leccinum* sp., which form a mycorrhizal association with two genera of Ericaceae

(*Arbutus* and *Arctostaphylos*) (Molina et al. 1992). The fungi colonize the plant root both intercellularly and intracellularly, but colonization is constrained to the epidermis and cortical cells (Molina et al. 1992). Mycorrhizal fungi responsible for this association can colonize other plants (Molina and Trappe 1982). Furthermore, there is another group known as the Monotropoid. These are mycorrhizal fungi associated with the plant family Monotropoideae, and they form a thick fungal sheath (Molina et al. 1992; Smith and Read 2008).

Ericoid mycorrhizal fungi (such as *Hymenoscyphus ericae*) are associated with host plants in the family Ericales. Their hyphae are septate and grow intercellularly, and their growth is, however, restricted to the epidermal cells (Molina et al. 1992). Lastly, there is Orchid mycorrhiza, in which the mycobionts responsible for this association belong to Basidiomycotina. This group is characterized by an intracellular colonization but an absence of Hartig net, fungal sheath and vesicles. An example of the group is *Rhizoctonia repens* (Smith and Read 2008).

1.2.1 Roles and Applications of Mycorrhizal Fungi in Agro-Ecosystems

1.2.1.1 Plant Growth Promotion

Ectomycorrhizal fungi play a major role in the sequestration of mineral nutrients on behalf of their symbiont plant. The ECM can improve host mineral uptake through enzymatic mobilization of organic nitrogen, phosphorus compounds or mineral weathering by organic acids (Landeweert et al. 2001). This group of fungi secretes enzymes, which are able to break down and utilize nitrogenous compounds such as glutamine, glutamate and alanine as well as other amino acids such as peptides and nucleic acids (Bücking et al. 2012). Some of the smaller organic molecules such as simple peptides and amino acids can also be directly absorbed by ECM hyphae, and such could be utilized for plant metabolism (Chalot and Brun 1998, Landeweert et al. 2001). Furthermore, the production of organic acids such as oxalic and citric acids by ECM fungi aids in the release of mineral ions from surrounding soils and solid lattices. The released organic acids also help in the mobilization of phosphorus, potassium, calcium and magnesium ions from mineral compounds such as apatite, biotite, iron ore, micas and feldspars (Wallander and Wickman 1999; Wallander 2000a, b; Adeleke et al. 2012; Adeleke 2014).

ECM fungi aid in extending the root surface area that further enhances the ability of ectomycorrhizal plants to obtain nutrients. ECM fungi have been known to form mycorrhizal mats (networks) just beneath the surface humus in forests. These networks are able to cover areas up to several meters and facilitate communication between ECM trees involving the transfer of nutrients and even defence priming against pathogens (Landeweert et al. 2001; Song et al. 2015). For example, studies of ECM network between a donor Douglas-fir (*Pseudotsuga menziesii*) and receiver ponderosa pine indicated that if the pine plant was subjected to defoliation stress, the Douglas-fir was able to transfer labile carbon to the pine directly through the mycorrhizal network in an attempt to alleviate the stress. When the Douglas-fir was further subjected to predation by herbivores, the defence responses of the pine were peaked at the very same time as that of the Douglas-fir, indicating the rapid transfer of defence signals to the neighbouring pine through the mycorrhizal network (Song et al. 2015).

The establishment of an ectomycorrhizal network can further increase mineral weathering and improve soil aggregation by the exudation of extracellular mucilage by fungal hyphae (Dunham et al. 2007; Smith and Read 2008). The stress ameliorating effects of the ECM fungal symbionts toward the plant hosts, including pollutant and heavy metal tolerance, increased mineral nutrient mobilization and uptake, as well as improved host plant stress responses. Hence, the ectomycorrhizal fungi can be considered for CAPBM formulation for forestry improvement. Further studies indicate capabilities of ECM in enhancing afforestation of heavy metal-polluted soils, as the fungi confer tolerance to the plants against pollutants through physical and chemical methods previously mentioned, allowing for improved plant growth in suboptimal conditions. Similar attributes of ECM fungi were reported by Bojarczuk and Kieliszewska-Rokicka (2010), where the growth of *Betula pendula* seedlings was enhanced in a 1:1 metal-contaminated soil supplemented with nonpolluted forest soil most likely due to improved mycorrhizal colonization.

Perhaps, the most important application of AM fungi is nutrient acquisition. To meet the growing phosphorus requirements, plants form symbiotic associations with AM fungi such as *Funneliformis mosseae, Rhizophagus irregularis* and *Gigaspora margarita*. Arbuscular mycorrhizal fungal hyphae branch out further from the phosphorus depletion zone, caused by direct phosphate uptake. Their smaller size allows them to explore surfaces, which plant roots cannot exploit, to obtain phosphates (Helgason and Fitter 2009) (Fig. 1.2). The fungal symbionts sequester phosphate from their surroundings using their extended hyphal network and transport it back to the plant root by translocation. The phosphate is then transported from the fungal arbuscules or coiled hyphae through an interfacial apoplast to the cortical plant cell through AM-induced phosphate transporters (Smith et al. 2011).

Phosphorus is required for many plant functions including vital nucleic acid and adenosine triphosphate (ATP) production (Shen et al. 2011). Not only does phosphorus deficiency limit plant growth but studies by Divito and Sadras (2014) have shown that a decrease in phosphorus availability to the plant causes a decrease in nodule formation and nitrogen fixation, largely due to lack of ATP metabolic actions in the associated plant cell. Hence, augmented phosphate uptake through mycorrhizal associations is a major contributor to enhanced plant growth.

An important application of AM fungi is their formulation into CAPBM, which are a sustainable alternative to chemical fertilizers (Abdullahi et al. 2015; Suhag 2016; Khan et al. 2017). For instance, inoculation of lettuce crops with the AM fungus *Rhizophagus intraradices* resulted in improvement of water use efficiency and salt tolerance – a favorable improvement under drought conditions (Aroca et al. 2008; Jahromi et al. 2008; Rouphael et al. 2015). Some mycorrhizal biofertilizers can further improve the host plant's systemic resistance against plant disease and predation by nematodes and insects (Schouteden et al. 2015; Song et al. 2015). In

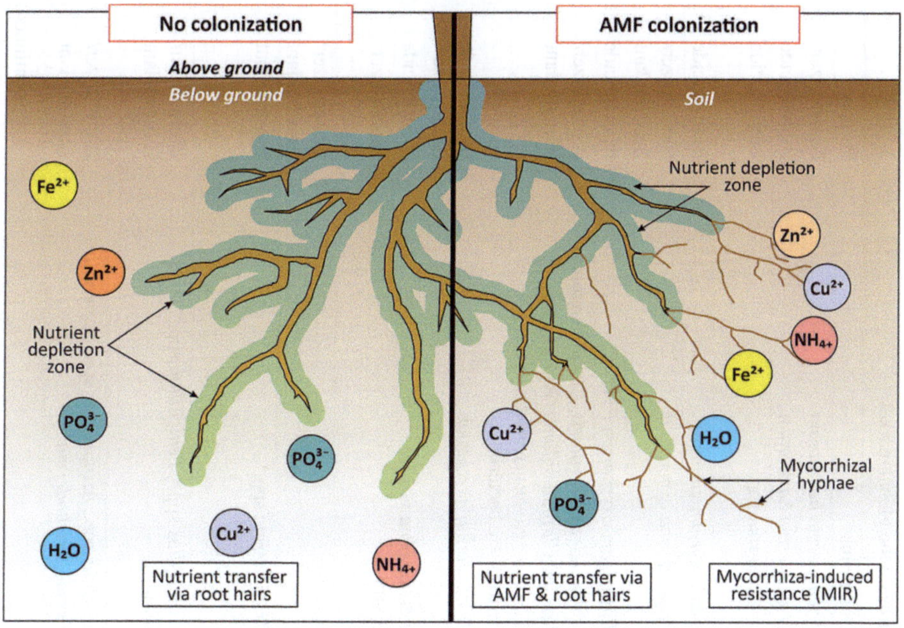

Fig. 1.2 Applications of fungi–plant interactions in agro-ecosystems

addition, AM fungi can also be used as biological control such as biofungicides and pest repellents.

Arbuscular mycorrhizal fungal inocula are usually applied on the seeds and seedlings to ensure early establishment of the mycorrhizal association (Baum et al. 2015; Malusà et al. 2016; Ercoli et al. 2017). Field trials of AMF biofertilizer have shown great potential; hence, numerous AMF biofertilizer products are commercially available (Table 1.1) (Berruti et al. 2016). Arbuscular mycorrhizal fungi biofertilizers are supplied as spores, colonized roots or AMF hyphae within a solid or liquid carrier material. The carrier material can influence the efficiency of the biofertilizer due to potential interactions between the material and different soil types (Baum et al. 2015; Rodrigues and Rodrigues 2017). Most commercially available AMF biofertilizers are 'broad spectrum' consisting of multiple species of AMF with *F. mosseae*, *R. irregularis* and *Glomus etunicatum* being the most common (Berruti et al. 2016).

1.2.1.2 Plant Protection

Drought Tolerance The AMFs extensive root structure is vital for plant protection against different environmental stresses such as drought (Fig. 1.2). The hyphal network extends further than the plant roots and is able to cover an exceptionally large surface area allowing the fungus to sequester scarce water trapped between soil particles and within bedrock micropores. Some fungal hyphae, which transport water to the plant, could measure up to 200 m per gram of soil (Van Der Heijden

Table 1.1 Biofertilizers containing plant beneficial fungi, which are commercially available on the global market

Region	Country	Company	Product	Organisms	Purpose
Europe	Sweden	**Bio-Innovation**	Myco-Rise	*Trichoderma harzianum*	Biofertilizer
			Bioderma-H	*Trichoderma harzianum*	Biofungicide
	Belgium	**Grondorts-mettingen de Cuester**	Biofungus	*Trichoderma* spp.	Biofungicide
	Germany	**Bayer-Cropscience**	Contans WG	*Coniothyrium minitans*	Biofungicide
	Hungary	**Bioved**	Trifender	*Trichoderma asperellum*	Biofertilizer/ Biostimulant
	France	**Natural Plant Protection**	Fusaclean	*Fusariuam oxysporum Fo47*	Biofungicide
	Italy	**SIAPA**	MycoApply DR	*Rhizophagus irregularis, Claroideoglomus luteum, Claroideoglomus etunicatum, Claroideoglomus claroideum* and 1% rhizosphere bacteria[a]	Biofertilizer/ Biostimulant
Asia Pacific	India	**Biotech Int. LTD**	Bioderma-H	*Trichoderma harzianum*	Biofungicide/ Biofertilizer
			Bioderma	*Trichoderma viridae*	Biofungicide
		Himalaya Pharmaceuticals Ltd.	Adhunik-VAM	AMF[a]	Biofertilizer
		PHMS Technocare Private Limited	Mycorrhizae Granular (PHMS MulShakti) GR	AMF[a]	Biofertilizer/ Biostimulant
	Thailand	**Probusiness Boonsongsuk Ltd**	Powdered organic fertilizer of Water Spraying Naga Brand	*Trichoderma* spp.[a]	Biostimulant
	Taiwan	**Yan Ten Biotech Corp**	Bio-Care Comprehensive Beneficial Effective Microorganism and PGPR	*Aspergillus* spp.[a], *Penicillium* spp.[a], other bacteria	Biofertilizer/ Biostimulant
	China	**Shandong Sukahan Bio-Technology Co., Ltd.**	BIO-GAIN ™ Microbial Fertilizer	*Bacillus* spp., *Azotobacter* spp., *Aspergillus oryzae, Trichoderma* spp.[a]	Biofertilizer/ Biostimulant

(continued)

	Australia	**Mycorrhizal Applications Int.**	MycoApply Defence	AMF[a], *Trichoderma* spp., *Bacillus* spp., other beneficial bacteria	Biofertilizer
	New Zealand	**Agrimm**	Abzorber Root Dip	*Trichoderma* spp. and AMF[a]	Biostimulant
	Australia	**Monsanto Ag**	RhizoBio	Six bacterial cultures and 18 species of endo- and ectomycorrhiza	Biostimulant
Americas	Iowa-USA	**Soil Technologies Corp.**	Plant Success – Granular Bio-Stimulant	*Rhizophagus irregularis, Funneliformis mosseae, Glomus aggregatum, Glomus etunicatum, Trichoderma koningii, Trichoderma harzianum,* other endomycorrhizae	Biofertilizer/ Biostimulant
	Columbia-USA	**Certis USA**	SoilGard	*Trichoderma virens*	Biofungicide
	California-USA	**JHBiotech Inc.**	Mycormax™	AMF[a]	Biostimulant
	Canada	**Bioworks**	RootShield Plus+ WP	*Trichoderma harzianum, Trichoderma virens*	Biofungicide
	Canada	**RiseHoP**	Myconnect Field Crops	*Rhizoglomus* sp.	Biofertilizer/ Biostimulant
		Sigma AgriScience	SigmaBio	*Rhizophagus irregularis, Funneliformis mosseae, Glomus etunicatum, Glomus aggregatum, Trichoderma harzianum, Trichoderma koningii, Bacillus* spp.	Biofertilizer/ Biostimulant
	Argentina	**Rizobacter**	Rizoderma	*Trichoderma harzanium*	Biofungicide
	Nicaragua	**Esagri**	MicoFert	*Rhizophagus irregularis*	Biostimulant
	Cuba	**National Institute of Tropical Viandas Research (INIVIT)**	EcoMic	*Glomus fasciculatum*	Biofertilizer/ Biostimulant
	Colombia		MycoBiol	*Glomus* spp., *Entrophospora colombiana, Acaulospora mellea*	Biofertilizer/ Biostimulant

(continued)

Table 1.1 (continued)

Region	Country	Company	Product	Organisms	Purpose
Africa[b]	South Africa	**Mycoroot**	Mycoroot™ SuperGro	AMF[a]	Biofertilizer
	South Africa	**Microbial Bio-fertilizers Int. (MBI)**	TRI-CURE WP	*Trichoderma harzianum*	Fungicide
	South Africa	**Zylem SA**	Nutri Life Platform	AMF[a], *Trichoderma harzianum*	Biofertilizer/ Biostimulant
	South Africa	**Biocult**	Biocult Mycorrhizae Powder	*Funneliformis mosseae, Rhizophagus intraradices, Glomus etunicatum, Scutellospora* spp., *Trichoderma harzianum*	Biofertilizer/ Biostimulant
	Kenya	**Dudutech Ltd**	Rhizatech	AMF[a]	Biofertilizer/ Biostimulant
	Kenya	**Juanco SPS LTD**	MYCORMAX WS	AMF[a]	Biostimulant

Adapted from Kaewchai et al. (2009), Uribe et al. (2010)

[a]Species not specified by producer [b]Many Northern African countries import their biofertilizers

The above table illustrates fungal biofertilizers available on the global market. Some biofertilizer producers choose not to specify the particular fungal strains. Single-species products such as Myco-Rise, Bioderma, SoilGard and Tricure WP have a specific role as biofertilizer, biofungicide or biostimulant. Products containing a 'cocktail' of fungi are more often claimed to have more than one purpose or are labelled as biostimulators. The most common mycorrhizal inoculants are *Rhizophagus irregularis, Funneliformis mosseae, Glomus aggregatum* and *Glomus etunicatum,* which feature in most mycorrhizal biofertilizers where the organisms are specified.

et al. 2008). Drought conditions are usually accompanied by low water retention in soils due to high levels of evaporation and loose soil structure. Hyphae of AMF increase soil structure by forming soil microaggregates through physical and chemical binding using AMF extracellular hyphae and exudates; often referred to as the 'sticky string bag' (Querejeta 2017). Studies by Hallett et al. (2009) have proven that AMF soils were capable of retaining more water than non-AMF soils even in the absence of plant roots, postulating that increased soil structure lead to enhanced root-soil hydraulic conduction.

The effects of drought are felt most severely by third world countries that rely on crop export and smallholder farms. For instance, in South Africa, the Western Cape has recently experienced a crop loss of up to 20% due to insufficient rains caused by the El Ninõ drought phenomenon (Chambers 2018). Planting crops with enhanced drought tolerance could reduce annual crop loss caused by drought and stabilize food production. In addition, the plant protection roles of AMF during drought condition could also be linked to the ability of the AMF to delay the onset of drought conditions through absorption of additional water absorbed by the fungus (Augé 2001, 2004; Augé et al. 2004; Querejeta 2017). Studies by Bitterlich et al. (2018) reported that transpiration rates of non-AMF plants were negatively impacted by drought conditions faster than AMF plants. This is another good evidence demonstrating the capability of drought-avoidance mechanisms associated with mycorrhizal plants.

Disease Resistance Apart from drought, plant diseases account for substantial global crop loss. Symbiont AMFs are good alternative to chemical fertilizers and biopesticides; they are able to stimulate and strengthen the host plant defence systems through the mycorrhiza-induced resistance (MIR), a form of induced systemic resistance (ISR) and disease priming (Fig. 1.2). Induced systemic resistance refers to the widespread expression of defence mechanisms in spatially separate parts of the plant, which are not under direct attack (Pieterse et al. 2014; Pieterse and Vanwees 2014). Arbuscular mycorrhizal fungi have been known to induce resistance in plants, as they have similar root invasion techniques as pathogens. Plants recognize microbial signature compounds such as bacterial flagellin or fungal hyphae, which are also referred to as pathogen- or microbial-associated molecular patterns (PAMPs). This plant recognition of PAMPs leads to PAMP-triggered immunity (PTI) (Zipfel 2008). The 'attack' on the roots of the plant by the AMF sends signal hormones to other parts of the plant to alert the plant about an invader. The defence mechanisms applied against AMF at the roots, that is, increased salicylic acid (SA) production, are then applied throughout other parts of the plant (Pieterse et al. 2014, Pieterse and Vanwees 2014).

Plant defence priming, which relies on the pre-conditioning of the host defence system by non-pathogenic AMF symbiont, potentiates the impacts of MIR. As the plant deploys some defences against the symbiont, there is an accumulation of defence signal molecules such as *MYC2* and mitogen-activated protein kinases

(MAPKs) within the plant (Pozo and Azcón-Aguilar 2007; Pieterse et al. 2014). With defence signals on "stand-by," the plant is in a primed state and able to respond to attack by pathogen or predator much faster and more efficiently than that in a non-primed state. Mycorrhiza-induced resistance relies on the jasmonic acid and ethylene signalling pathways, which respond by inducing resistance mechanisms against necrotizing pathogens and herbivorous insects (Jung et al. 2012).

Research by Song et al. (2011) has shown that priming maize plant defences with the AMF *Funneliformis mosseae* increases the plant's defences against the pathogen *Rhizoctonia solani*, the cause of sheath blight in maize plants. This resistance is conferred by enhanced production of plant antimicrobial 2,4-dihydroxy-7-methoxy-2H-1,4-benzoxazin-3(4 H)-one (DIMBOA) as a result of MIR (Song et al. 2011). Mycorrhizal inoculation has also been proven to potentiate resistance against *R. solani* in potatoes, tomatoes and soybean (Wyss et al. 1991; Yao et al. 2003; Song et al. 2015).

1.2.1.3 Phytoremediation

Microorganisms are the key drivers of bioremediation processes, but their participation in remediation is usually in association with other organisms in the environment. Fungi are the primary decomposers of organic matter, and such decomposition may occur through interactions with plants (plant–microbe interactions) or in combination with other microbes (microbe–microbe interactions) (Göltenboth et al. 2006, Smith and Read 2008). Therefore, fungi such as ECM fungi have crucial roles to play in phytoremediation. ECM fungi are able to degrade low- and high-molecular-weight aromatic pollutants (Cairney and Meharg 2002; Bello-Akinosho 2018). An example of these interactions is the study conducted by Sarand et al. (1999), where a mycorrhizal fungus, *Suillus bovinus,* individually and in conjunction with *Pseudomonas fluorescens* and *Pinus sylvestris*, was involved in the biodegradation of m-toluate, a low-molecular-weight aromatic environmental pollutant. Further plant–microbe interactions with the host plant *Pinus sylvestris* showed that the ECM fungi *Paxillus involutus* and *Suillus variegatus* improve the mineralization of short-chain 2,4-dichlorophenol and *Suillus bovinus* are able to improve the uptake of 3-chlorobenzoic acid (Meharg et al. 1997; Dittmann et al. 2002). It is generally accepted that biodegradation of hydrocarbons becomes less effective as the hydrocarbons increase in molecular weight and branching (Peixoto et al. 2011; Martin et al. 2014).

In contrast to studies validating the biodegradative capabilities of ECM on organic contaminants, Genney et al. (2004) reported no effect of the ECM fungi, in association with Scots pine, on the mineralization of PAH naphthalene, while Joner et al. (2006) confirmed these results as they report ECM/Scots pines to reduce the degradation of anthracene, anthraquinone, chrysene and dibenz[a,h]anthracene. A similar observation was reported in a study in which hybrid poplar plants colonized by *Pisolithus tinctorius* were used to remediate a diesel-contaminated site. The study showed that ECM-inoculated poplar plants had reduced efficiency in the removal of diesel from the site in comparison to an uninoculated control (Gunderson et al. 2007). However, despite this outcome, the ECM-colonized plants were reported to have greater whole-plant biomass and higher N and P concentrations in leaves than the uninoculated control. These results indicate that ECM conferred an amount of tolerance to poplar plants against the diesel contaminants. Such ability to tolerate PHC was also reported in studies relating to heavy metals. For instance, *Pisolithus tinctorium* improved growth of seedlings in a highly acidic mine soil containing heavy metals. Furthermore, in a laboratory study, the ability of *Pisolithus tinctorium* to withstand high concentrations of Al, Fe, Cu and Zn was demonstrated (Marx and Artman 1979; Tam 1995; Jentschke and Godbold 2000). With reference to these special attributes of ECM in the aforementioned examples, the stress-ameliorating effects of the ectomycorrhizal fungi on the host plant make it an ideal candidate for afforestation in polluted or disturbed soils (Bücking 2011).

1.2.2 Roles and Applications of Rhizobacteria in Agro-Ecosystems

Plants associate with a vast number of microorganisms, establishing relationships that drive many processes in the ecosystem, especially in the rhizosphere, which is a habitat for copious microorganisms including bacteria (Wu et al. 2009; Mendes et al. 2013). Rhizobacteria participate in several processes that are beneficial or detrimental to plants. Beneficial processes linked to the association between plants and rhizobacteria include improved nutrient acquisition, resistance to environmental stress and protection of plants against pathogens as well as improvement of metabolic and physiological processes in plants. Beneficial rhizobacteria involved in such processes are referred to as plant growth-promoting rhizobacteria (PGPR). On the other hand, some pathogenic bacteria are also found in the rhizosphere, which could have negative impact on plant health (Liu et al. 2010).

The PGPR are recognized non-pathogenic bacteria and have been categorized into two major types on the basis of their habitat, namely, extracellular PGPR (ePGPR) and intracellular PGPR (iPGPR). Extracellular PGPR exist in the rhizosphere or in spaces between the root cortex cells, while iPGPR are found within the root cells of plants. These PGPR are known to belong to various genera such as *Serratia, Bacillus, Allorhizobium, Paenibacillus, Actinomyces, Clostridium, Bradyrhizobium, Azoarcus, Azotobacter, Mesorhizobium, Enterobacter, Arthrobacter, Pseudomonas, Gluconacetobacter, Azospirillum, Caulobacter, Chromobacterium, Burkholderia, Micrococcus, Agrobacterium, Erwina, Flavobacterium* and *Rhizobium* (Bhattacharyya and Jha 2012; Gouda et al. 2017).

On the basis of the PGPR abilities to stimulate and improve plant growth, processes involved in such activities can be classified as direct or indirect methods. Direct plant growth-promoting methods include the ability of PGPR to enhance the availability of macronutrients such as nitrogen, phosphate and potassium in the rhizosphere to plants (Babu et al. 2015). This can be achieved through biological nitrogen-fixing abilities of PGPR, phosphate and potassium-solubilizing capabilities of the PGPR and production of phytohormones such as indole acetic acid, cytokinins, and gibberellic acid by PGPR. Indirect plant growth-promoting methods deal with the capabilities of PGPR to inhibit lethal effects of rhizosphere-inhabiting plant pathogens on plants. These microorganisms improve plant productivity and development through the activation of useful plant enzymes to enhance physiological processes in plants and amplification of plants' ability to resist diseases and environmental stress (Wang et al. 2016). Production of siderophores, ethylene and antibiotics forms part of the indirect plant growth-promoting methods.

1.2.2.1 Plant Growth Promotion

Biological Nitrogen Fixation Nitrogen is a vital macronutrient in plant growth and development, but it is not accessible to most plants. Atmospheric nitrogen gas is transformed to ammonia by PGPR through nitrogen fixation. This form of nitrogen (ammonia) is utilizable by plants for productivity (Ahemad and Kibret 2014). Rhizobacteria that are capable of fixing atmospheric nitrogen include *Rhizobium etli, Rhizobium trifolii, Bradyrhizobium* sp., *Sinorhizobium meliloti* (Shamseldin 2013), *Azotobacter agilis, Azotobacter chroococcum, Azotobacter vinelandii* and *Klebsiella pneumoniae* (Kennedy and Bishop 2004). Biological nitrogen fixation forms a renewable source of nitrogen to crops and plant species. Complex enzymes, such as *nitrogenases,* catalyze the process of biological nitrogen fixation, and some genes responsible for the nitrogen-fixing activities include *nifH, nifD* and *nifK* (Souza et al. 2015).

Phosphate Solubilization Phosphorus is an essential macronutrient in metabolic and physiological processes in plants such as photosynthesis, biological oxidation and cell division (Gupta et al. 2012). It is also an important nutrient for plant growth and productivity. Unavailability of soluble phosphorus to plants limits their ability to perform these crucial functions, hence the need for soluble forms of phosphate (Sharma et al. 2013). Application of PGPR that are capable of solubilizing insoluble phosphate by discharging organic acids increases the availability of this element to plants thereby improving soil fertility and crop productivity (Souza et al. 2015). Some of the rhizobacteria that are capable of solubilizing insoluble P include *Serratia phosphoticum, Serratia marcescens, Thiobacillus ferrooxidans, Rhodococcus, Pseudomonas putida, Pseudomonas striata, Pseudomonas fluorescens, Pseudomonas calcis, Alcaligens* sp., *Citrobacter* sp., *Rhizobium meliloti, Gordonia* sp., *Phyllobacterium* sp., *Bacillus cereus, Bacillus subtilis, Bacillus mycoides, Bacillus pumilus, Arthrobacter* sp., *Xanthomonas* sp., *Enterobacter* sp., *Chryseobacterium* sp., *Azotobacter* sp., *Pantoea* sp., *Klebsiella* sp., *Achromobacter* sp., *Aerobacter aerogenes* and *Erwinia* sp. (Gupta et al. 2012; Sharma et al. 2013).

Potassium Solubilization Potassium is another macronutrient essential for plant growth and development. Deficiency of potassium in the rhizosphere has been associated with reduced root growth and crop productivity. Generally, there is low concentration of soluble potassium in the rhizosphere, and potassium has the capacity to form insoluble complexes when applied as an inorganic fertilizer (Gupta et al. 2015). Insoluble potassium is made available by PGPR through the production of inorganic acids thus improving agricultural sustainability and aiding in the production of crops (Liu et al. 2012). Although many PGPRs have been linked to potassium solubilization, *Bacillus circulans, Bacillus edaphicus* and *Bacillus mucilaginosus* have been reported to be the most efficient potassium solubilizers (Meena et al. 2016). Other potassium solubilizers also comprise *Paenibacillus glucanolyticus, Burkholderia, Paenibacillus mucilaginosus* and *Enterobacter hormaechei* (Etesami et al. 2017).

Production of Phytohormones Various biochemical and physiological processes in plants are regulated by phytohormones. These hormones play a crucial role in controlling the response of plants to environmental stresses. The action of these hormones in response to environmental stimuli may be at the site of production of the hormones or in close proximity to the site (Fahad et al. 2015). In addition to the capability to stimulate plant growth and development, phytohormones may also act in defence of the plants against pathogens (da Costa et al. 2014). Such phytohormones include cytokinins, indole acetic acids (IAA) and gibberellins. Cytokinins are produced by PGPRs such as *Paenibacillus polymyxa,* and *Pseudomonas fluorescens, Enterobacter cloacae* and *Azospirillum brasilense* produce phytohormones like IAA, while gibberellins are produced by various bacteria including *Acetobacter diazotropicus, Bacillus pumilus* and *Bacillus cereus* (Kaymak 2010).

Production of Siderophores Siderophores are low-molecular-mass molecules that have strong attraction for iron – an essential nutrient for plant growth. Iron is involved in biological processes such as photosynthesis, respiration and chlorophyll production. It occurs in the form of insoluble complexes such as hydroxides and oxyhydroxides, which have low bioavailability. Siderophores solubilize iron in the rhizosphere by chelation thereby allowing its extraction from the environment. Secretion of siderophores by PGPR is their iron uptake strategy, which also benefits other organisms in the rhizosphere, especially plants (Ahemad and Kibret 2014; Ahmed and Holmström 2015). Plant growth-promoting rhizobacteria such as *Pseudomonas putida* and *Pseudomonas fluorescens* are known siderophore producers, which also protect plants against phytopathogens such as *Erwinia carotovora* and *Fusarium* sp. Other siderophore producers include *Pseudomonas cepacia, Enterobacter cloacae, Bacillus subtilis* and *Rhizobium meliloti* (Sayyed et al. 2013).

Aminocyclopropane-1-carboxylate (ACC) Deaminase Production Aminocyclopropane-1-carboxylate is a precursor of ethylene, and high concentrations of ethylene could hinder plant growth or even cause death. Production of ACC deaminase by PGPR enhances plant growth and productivity by reducing the quantity of ethylene in plants. These enzymes catalyze the metabolism of ACC to ammonia and α-ketobutyrate (Glick 2014). Absolute reduction of ACC in the environment could also protect plants against stress such as drought through the induction of water absorption from deep soil (Zhang et al. 2015).

1.2.2.2 Plant Protection

Antibiotic Production The synthesis of antibiotics to eradicate or decrease the growth of plant pathogens is a beneficial biocontrol characteristic of PGPR (Beneduzi et al. 2012; Gupta et al. 2015). For example, *Bacillus amyloliquefaciens* and *Bacillus subtilis* are known to produce an extensive range of antifungal and antibacterial antibiotics. *Pseudomonas aeruginosa* and *Pseudomonas fluorescens* are also antibiotic producers. These PGPRs repress the growth of plant pathogens by secreting an extracellular metabolite that hinders the growth of phytopathogens even at low concentration. Such metabolites include subtilin, sublancin, chlorotetain, rhizoctinins and surfactins (Goswami et al. 2016).

1.3 Plant–Microbe Interactions in the Phyllosphere

Unlike the microbiology of the rhizosphere, the phyllosphere and its associated microbiome have received less attention. However, in recent years, more investigations have been conducted on the phyllosphere microbiome (Vorholt 2012; Kirschner 2015; Remus-Emsermann and Schlechter 2018). The phyllosphere is generally considered a hostile environment for microorganisms due to fluctuating temperature and humidity, exposure to elevated levels of solar irradiation and heterogeneous availability of nutrients. Despite these challenges, the phyllosphere is colonized by numerous microbial communities comprising bacteria, archaea, fungi, protozoa and yeast (Andrews and Harris 2000; Whipps et al. 2008). The colonization of the phyllosphere is primarily through immigration of microorganisms from seeds, soil, water, air and animals (Vorholt 2012). Studies that have been conducted on phyllosphere microbiology have primarily focused on the understanding of plant–pathogen interactions due to the economic importance of such associations. Examples of these studies include analysis of pathogen colonization, spread, survival and mechanisms of pathogenicity (Wilson et al. 1999; Hirano and Upper 2000; Brandl et al. 2001). Nevertheless, of late, beneficial phyllosphere microorganisms have also been investigated with particular reference to involvement of the microbiota in plant protection and the promotion of plant growth (Rastogi et al. 2013; Senthilkumar and Krishnamoorthy 2017; Mashiane et al. 2017, 2018; Durand et al. 2018).

Leaf surfaces are a major entry point of microorganisms due to their large surface area and perhaps could be one of the largest microbial niches in the world. The cumulative, terrestrial leaf surface area is estimated to be about 6.4×10^8 km^2, and the colonization of such surfaces by bacteria contributes to a global phyllosphere bacterial population of approximately 10^{26} cells (Lindow and Brandl 2003). A majority of these bacteria are commensals and mutualists and could provide various ecosystem functions to the plant including competitive exclusion of pathogens, phytoremediation of toxic pollutants and cycling of important elements (Rastogi et al. 2013; Kembel et al. 2014; Bello-Akinosho et al. 2016, 2017) (Fig. 1.3). Some phyllosphere bacteria such as *Methylobacterium* species may be considered as plant growth-promoting bacteria (PGPB), which are sometimes referred to as 'plant probiotics' (Kwak et al. 2014). Phyllosphere bacteria may improve plant health by

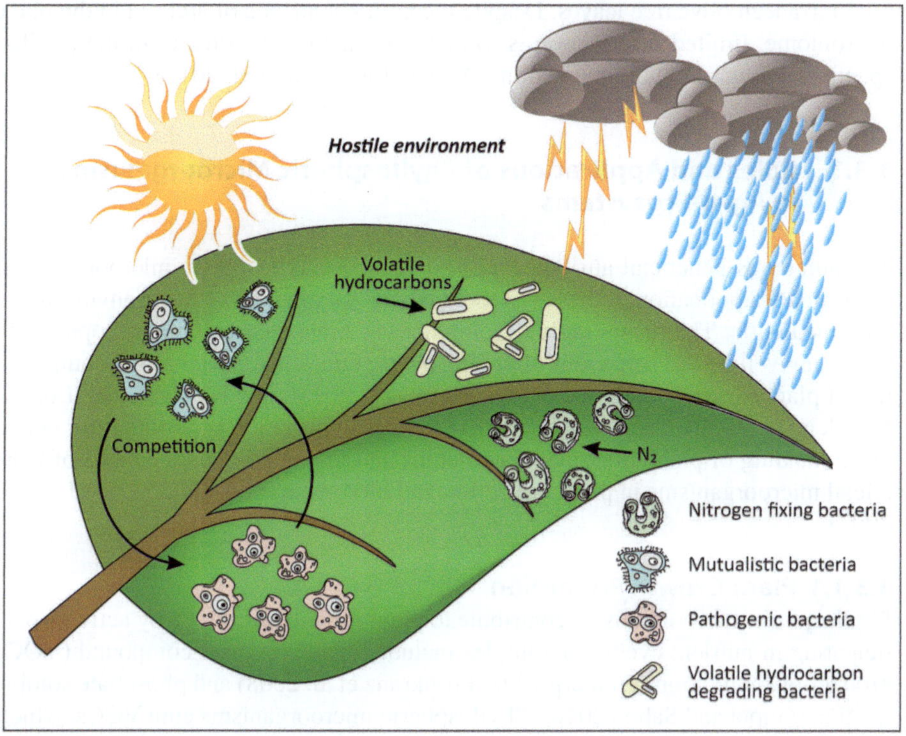

Fig. 1.3 Applications of phyllosphere–bacteria interactions in agro-ecosystems

increasing tolerance of the plant to stresses including frost-induced injury and drought stress (Lindow and Leveau 2002; Kumar et al. 2017). Furthermore, phyllosphere bacteria can influence plant growth using various mechanisms including the production of plant growth-promoting hormones, regulation of plant metabolic pathways and participation in increasing nutrient availability (Berlec 2012).

Although bacteria predominate the phyllosphere, other microorganisms such as fungi play key roles in plant eco-physiological functions. Similar to the roles played by bacteria in the phyllosphere, fungi are able to participate in competitive exclusion of pathogens as well as leaf litter degradation thereby contributing to soil organic matter with associated positive effects on plant growth (Fokkema and Van der Meulen 1976; Osono 2006; Voříšková and Baldrian 2013). Unlike phyllospheric bacteria and fungi, minimal information is available on applications of phyllospheric archaea. Studies on plant-associated archaea initially reported that these microorganisms are more widely distributed in the rhizosphere than in the phyllosphere (Knief et al. 2011). This is due to the elevated oxygen levels and environmental instability at the phyllosphere interface. However, in perennial plants, large numbers of archaea have been evidenced within phyllospheric plant tissues, referred to as the endosphere. For instance, Müller et al. (2016) attributed the archaeal population to constitute as much as 36% of the total endophytic microbial community of

Mediterranean olive tree leaves. Despite the high abundance of archaea in the plant microbiome, limited information is available on their functionality in the phyllosphere and archaea–plant interactions (Moissl-Eichinger et al. 2017).

1.3.1 Roles and Applications of Phyllospheric Microorganisms in Agro-Ecosystems

The numerous beneficial attributes associated with phyllospheric microorganisms motivate the application of these microorganisms for plant growth and environmental remediation. The recent rapid development of molecular and microscopic tools aid in unravelling the behaviour of phyllospheric microbes as well as the complexities of plant surfaces that these organisms inhabit (Singh and Kothari 2017; López-Mondéjar et al. 2017; Subudhi et al. 2018). Such information is pertinent for better understanding of plant–microbe relationships and maximizing applications of beneficial microorganisms in plant protection and bioremediation (Lindow and Leveau 2002).

1.3.1.1 Plant Growth Promotion

Phyllosphere microorganisms contribute to plant growth promotion by actively participating in nutrient cycling. Examples include volatile organic compound (VOC) production, N fixation (Freiberg 1998; Fürnkranz et al. 2008) and phosphate solubilization (Gupta and Sahoo 2010). Phyllospheric microorganisms emit VOCs, which have the potential to enhance plant growth and aid in plant stress resistance (Junker et al. 2011). Plants also emit VOCs that play a significant role in microbiome structure of the phyllosphere. This is primarily due to VOC potential as a source of carbon (Sy et al. 2005) and their antimicrobial effects (Junker et al. 2011). On the other hand, phyllosphere microorganisms inhabit the interface between the plant and the atmosphere; hence, these microorganisms are able to influence VOC production in the plant. This signifies the bidirectional relationship between plant VOC production and phyllosphere microbiota (Farré-Armengol et al. 2016).

1.3.1.2 Plant Protection

Plant pathogenic microorganisms negatively influence plant health thereby forming a major threat to global food security and ecosystem stability. Albeit agro-chemicals have been frequently utilized for plant pathogen control, this control method is gradually becoming less attractive due to their growing costs as well as a multitude of associated adverse environmental effects. Biological control of plant pathogens, a more attractive alternative to chemical control, has revolutionized the agricultural sector due to the study of the ecology of pathogens and antagonists (Compant et al. 2005). One such example is the biological control of fire blight disease of apple and pear using non-pathogenic phyllospheric bacteria. For infection to occur, the causal agent of fire blight, *Erwinia amylovora*, needs to increase its population size on the stigmatic surface of blossoms. The stigma is also colonized by other bacterial epiphytes that may have the ability to interact with and suppress the growth of *E.*

amylovora. These antagonistic bacteria such as *Pseudomonas fluorescens* A506 have been identified and are commercially available as biological control products for fire blight (Johnson and Stockwell 1998). They are sprayed onto the blossoms early in the season to enable pre-emptive competitive exclusion of the pathogen (Lindow and Leveau 2002). Recently, numerous studies have been conducted on the discovery, isolation and efficacy of antagonistic microorganisms for the control of plant diseases (Müller et al. 2016; Sartori et al. 2017).

Phyllospheric bacteria may also be used to prevent frost-induced injury. Ice nucleation-active (INA) bacteria are epiphytic bacterial species such as *Pseudomonas syringae* that contribute to frost injury of cold-sensitive plants (Lindow and Leveau 2002; Rostami et al. 2018). Frost injury is achieved by the presence of ice nucleation-active proteins (INAP) on the outer bacterial cell wall of INA bacteria, which facilitate inter- or intracellular ice formation in plant tissue. Ice formation in plant tissue is prevented in the absence of INA bacteria due to the ability of the plant to 'supercool' to temperatures below 0° to −12 °C without ice formation. This supercooling ability is reduced, in the presence of INA bacteria, where ice formation in the inter- or intracellular space of plants is evidenced at temperatures as high as 0 to −2 °C (Rostami et al. 2018). Furthermore, the temperature at which freezing occurs increases with the population size of the INA bacteria. Hence, the pre-emptive, competitive exclusion of INA bacteria with naturally occurring non-INA, phyllospheric bacteria has been proposed and tested as a method of frost control (Lindow and Leveau 2002). Initially, antagonistic bacterial products that were developed primarily targeted INA bacteria (e.g. Frostban®). However, subsequent products such as Blightban® were developed with the intention of alleviating frost injury and pathogen invasion. This was achieved by developing a dual-purpose product containing a bacterial antagonist that competitively excludes both INA bacteria and *E. amylovora*, the causative agent of fire blight of apples and pears (Skirvin et al. 2000; Lindow and Leveau 2002).

1.3.1.3 Bioremediation

Phytoremediation is a form of bioremediation in which higher plants are used in pollutant degradation or removal (Sorkhoh et al. 2010). Phytoremediation is usually achievable when pollutants are water soluble, but contributions from rhizospheric microorganisms are required when remediating soil containing water-insoluble pollutants (Ali et al. 2012). When microbes are involved, the contribution of plants is indirect, secreting root exudates that promote proliferation of pollutant-degrading, rhizospheric microbial populations (Ali et al. 2012). Although numerous studies have been conducted on the utilization of rhizospheric microorganisms for phytoremediation purposes (Radwan et al. 1995; Al-Awadhi et al. 2009; Bello-Akinosho et al. 2016), this technology is primarily targeted at decontamination of soils but not widely applicable for remediation of volatile hydrocarbons.

Phyllospheric microorganisms are well suited for the remediation of volatile/atmospheric hydrocarbon pollutants. Their abilities to remediate aliphatic and aromatic hydrocarbons (toluene, xylene and phenanthrene) have been reported in numerous studies (Sandhu et al. 2007; Scheublin et al. 2014; Sangthong et al. 2016).

Ali and colleagues (2012) showed, in their study, that the leaves of legumes harboured up to 9×10^7 cells/g of oil-utilizing bacteria. The potential of such phyllospheric microorganisms is very important for mitigation of the harmful effects of air pollution on the ecosystem. Environmentalists have now adopted an approach of utilizing these microorganisms, in association with their host plants, in bioremediation processes through their introduction (bioaugmentation) or enhancement of their growth (biostimulation) in polluted environments. One such study focused on the utilization of the toluene-degrading, phyllosphere bacterium, *Pseudomonas putida* TVA8 for the removal of airborne toluene. The bacterium was inoculated on *Azalea indica* leaves and resulted in the rapid decrease in toluene levels in the air in comparison to uninoculated plants (De Kempeneer et al. 2004).

1.4 Challenges in the Application of Plant–Microbe Interactions for the Benefit of Agro-Ecosystems

The use of plant–microbe interactions is essential for agricultural development and sustainability. However, some challenges are encountered during production, application or management of microbial formulations that contain plant growth-promoting microorganisms (PGPM). Such challenges include the following:

Climate Variation/Change About 50% of agricultural losses have been attributed to abiotic factors such as rise/fall in temperatures and reduced precipitation (Tyagi et al. 2014). These factors could also affect viability and adaptability of CAPBM. In addition, there is a possibility of compatibility challenges where products developed in a specific region may not meet the needs of plants in another region due to varying environmental conditions. This problem could be resolved by confirming compatibility of imported products before application.

Quality Control Variations in the content and efficacy of microbial formulations is a barrier to controlling the quality of CAPBMs. Inconsistency in the regulations of various countries concerning the safety of microbial strains in the environment and production of sub-standard products is another drawback for the use of CAPBMs to support agro-ecosystems. Local regulatory authorities should adopt regional, continental or other international standards and legislations to ensure consistency and proper quality management of CAPBM (Ochieng 2015).

Replication of Technology and Biofertilizer Storage One of the challenges of the application of CAPBM is difficulty in matching the results obtained in the laboratory with field results. For example, it has been reported that repeated subculture of ECM fungi on agar media, like other fungi, over a long period of time affects their natural ability to colonize host plants (Thomson et al. 1994). Fortunately, such problem could be overcome by inoculation onto and re-isolation from a compatible host (Thomson et al. 1994). Marx and Daniel (1976) showed that viability could be retained for a period of 1–3 years by storing ECM mycelia in sterile water at

5 °C. Other PGPMs exhibit similar loss of viability and biological activities during attempted replication of laboratory experiments in the field. In addition, improper storage of CAPBM or use of unsuitable carrier materials could affect the biological activities of associated microorganisms and may even result in decreased viability of PGPMs. Technological improvement of CAPBM is a potential solution to prolonging product shelf life.

Selectivity of PGPMs Commercially available plant beneficial microorganisms are generally specific in their functions. Some selective functions include nitrogen fixation, auxin and siderophore production, as well as phosphate and potassium solubilization. Application of unsuitable PGPMs could cause reductions in their efficiency when applied in the field unlike broad-spectrum chemical fertilizers, which are formulated in accordance to the basic requirements of plants to enhance productivity (Timmusk et al. 2017). Specificity is also a challenge when implementing ECM fungi as a CAPBM. Ectomycorrhizal fungal species have host-specific associations, which limit their beneficial effects to certain plant species, and as such these ECM-based products cannot be applied as biofertilizer on all ectomycorrhizal plants (Smith and Read 2008). Although, some species of ECM fungi such as *P. tinctorius* have a broad plant host range; hence, they have greater effectivity in field application (Martin et al. 2002; Brearley 2011). Further research including fungal and host plant genomic analyses could be undertaken to reveal mechanisms of fungi–host symbiosis aiding in the development of superior CAPBM products (Acioli-Santos et al. 2011).

Lack of Awareness Presently, adoption of CAPBM is not extensive due to lack of awareness and regulatory framework as well as the specificity of the available products. Awareness may be improved through government intervention, innovative policies and training programmes. This would aid farmers in the informed adoption of the CAPBM product, thereby ensuring wider usage. Subsequent increase in demand for CAPBM products would result in increased production, hence the need for proficiency of both supplier and user as well as investment into the production technology.

1.5 Conclusion

Globally, as climate change becomes inevitable, atmospheric CO_2 concentrations are expected to increase continuously during the twenty-first century. This will be accompanied by increase in temperature and alterations in precipitation patterns. Most of these climate-changing parameters will have major effects on the agro-ecosystem including plants, microorganisms and other members of this system. Proper understanding of this challenge requires more investigations about the potential applications of plant–microbe interactions, especially for adaptation and mitigation purposes. This will require more than laboratory experiments; hence, pot and field trial experiments should be prioritized. Extensive commercial adoption of plant–microbe interactions will require special approaches to be undertaken (Timmusk et al. 2017). Perhaps, integrated -omics approaches and metabolic modelling hold the key to unravel the fundamentals of plant–microbe interactions in the future.

References

Abdullahi R, Lihan S, Edward R (2015) Effect of arbuscular mycorrhizal fungi and poultry manure on growth and nutrients uptake by maize under field condition. Int J Agric Innov Res 4:158–163. ISSN (online): 2319-1473

Acioli-Santos B, Vieira HEE, Lima CE, Maia LC (2011) The molecular ectomycorrhizal fungus essence in association: a review of differentially expressed fungal genes during symbiosis formation. In: Diversity and biotechnology of ectomycorrhizae. Springer, Berlin/Heidelberg, pp 87–121. https://doi.org/10.1007/978-3-642-15196-5_5

Adeleke RA (2014) Getting rid of the unwanted: highlights of developments and challenges of biobeneficiation of iron ore minerals—a review. J Industrial Microbiol Biotechnol 41(12):1731–1741. https://doi.org/10.1007/s10295-014-1514-4

Adeleke RA, Cloete TE, Bertrand A, Khasa DP (2012) Iron ore weathering potentials of ectomycorrhizal plants. Mycorrhiza 22(7):535–544. https://doi.org/10.1007/s00572-012-0431-5

Ahemad M, Kibret M (2014) Mechanisms and applications of plant growth promoting rhizobacteria: current perspective. JKSUS 26(1):1–20. https://doi.org/10.1016/j.jksus.2013.05.001

Ahmed E, Holmström SJ (2015) Siderophore production by microorganisms isolated from a podzol soil profile. Geomicrobiol J 32(5):397–411. https://doi.org/10.1080/01490451.2014.925011

Ali N, Sorkhoh N, Salamah S, Eliyas M, Radwan S (2012) The potential of epiphytic hydrocarbon-utilizing bacteria on legume leaves for attenuation of atmospheric hydrocarbon pollutants. J Environ Manag 93(1):113–120

Aroca R, Vernieri P, Ruiz-Lozano JM (2008) Mycorrhizal and non-mycorrhizal Lactuca sativa plants exhibit contrasting responses to exogenous ABA during drought stress and recovery. J Exp Bot 59:2029–2041. https://doi.org/10.1093/jxb/ern057

Augé RM (2001) Water relations, drought and vesicular-arbuscular mycorrhizal symbiosis. Mycorrhiza 11(1):3–42. https://doi.org/10.1007/s005720100097

Augé RM (2004) Arbuscular mycorrhizae and soil/plant water relations. Can J Soil Sci 84(4):373–381. https://doi.org/10.4141/S04-002

Augé RM, Sylvia DM, Park S, Buttery BR, Saxton AM, Moore JL, Cho K (2004) Partitioning mycorrhizal influence on water relations of Phaseolus vulgaris into soil and plant components. Can J Bot 82(4):503–514. https://doi.org/10.1139/b04-020

Al-Awadhi H, El-Nemr I, Mahmoud H, Sorkhoh NA, Radwan SS (2009) Plant-associated bacteria as tools for the phytoremediation of oily nitrogen-poor soils. Int J Phytoremediation 11(1):11–27. https://doi.org/10.1080/15226510802363261

Andrews JH, Harris RF (2000) The ecology and biogeography of microorganisms on plant surfaces. Annu Rev Phytopathol 38(1):145–180. https://doi.org/10.1146/annurev.phyto.38.1.145

Babu AN, Jogaiah S, Ito SI, Nagaraj AK, Tran LSP (2015) Improvement of growth, fruit weight and early blight disease protection of tomato plants by rhizosphere bacteria is correlated with their beneficial traits and induced biosynthesis of antioxidant peroxidase and polyphenol oxidase. Plant Sci 231:62–73

Baum C, El-Tohamy W, Gruda N (2015) Increasing the productivity and product quality of vegetable crops using arbuscular mycorrhizal fungi: a review. Sci Hort 187:131–141. https://doi.org/10.1016/j.scienta.2015.03.002

Becerra A, Bartoloni N, Cofré N, Soteras F, Cabello M (2014) Arbuscular mycorrhizal fungi in saline soils: vertical distribution at different soil depth. Braz J Microbiol 45(2):585–594

Bello-Akinosho M (2018) Biodegradation of polycyclic aromatic hydrocarbons in contaminated soils using a tripartite association of ectomycorrhizal fungi, pine plant and their rhizosphere bacteria. Philosophiae Doctor (Microbiology). University of Pretoria, Pretoria, South Africa

Bello-Akinosho M, Makofane R, Adeleke R, Thantsha M, Pillay M, Chirima GJ (2016) Potential of polycyclic aromatic hydrocarbon-degrading bacterial isolates to contribute to soil fertility. Biomed Res Int 2016. https://doi.org/10.1155/2016/5798593

Bello-Akinosho M, Adeleke R, Thantsha MS, Maila M (2017) *Pseudomonas* sp. (strain 10–1B): a potential inoculum candidate for green and sustainable remediation. Remed J 27(3):75–79. https://doi.org/10.1002/rem.21521

Beneduzi A, Ambrosini A, Passaglia LM (2012) Plant growth-promoting rhizobacteria (PGPR): their potential as antagonists and biocontrol agents. Genet Mol Biol 35(4):1044–1051

Berlec A (2012) Novel techniques and findings in the study of plant microbiota: search for plant probiotics. Plant Sci 193:96–102. https://doi.org/10.1016/j.plantsci.2012.05.010

Berruti A, Lumini E, Balestrini R, Bianciotto V (2016) Arbuscular mycorrhizal fungi as natural biofertilizers: Let's benefit from past successes. Front Microbiol 6:1559. https://doi.org/10.3389/fmicb.2015.01559

Bhattacharyya PN, Jha DK (2012) Plant growth-promoting rhizobacteria (PGPR): emergence in agriculture. World J Microbiol Biotechnol 28(4):1327–1350. https://doi.org/10.1007/s11274-011-0979-9

Bitterlich M, Franken P, Graefe J (2018) Arbuscular mycorrhiza improves substrate hydraulic conductivity in the plant available moisture range under root growth exclusion. Front Plant Sci 9:301. https://doi.org/10.3389/fpls.2018.00301

Bojarczuk K, Kieliszewska-Rokicka B (2010) Effect of ectomycorrhiza on Cu and Pb accumulation in leaves and roots of silver birch (*Betula pendula* Roth.) seedlings grown in metal-contaminated soil. Water Air Soil Pollut 207(1–4):227–240. https://doi.org/10.1007/s11270-009-0131-8

Bonfante P, Desirò A (2015) Arbuscular mycorrhizas: the lives of beneficial fungi and their plant hosts. Principles of plant-microbe interactions. Springer, Cham, pp 235–245. https://doi.org/10.1007/978-3-319-08575-3_25

Brandl H, Bosshard R, Wegmann M (2001) Computer-munching microbes: metal leaching from electronic scrap by bacteria and fungi. Hydrometallurgy 59(2–3):319–326. https://doi.org/10.1016/S0304-386X(00)00188-2

Bradáčová K, Weber NF, Morad-Talab N, Asim M, Imran M, Weinmann M, Neumann G (2016) Micronutrients (Zn/Mn), seaweed extracts, and plant growth-promoting bacteria as cold-stress protectants in maize. Chem Biol Technol Agric 3(1):19. https://doi.org/10.1186/s40538-016-0069-1

Brearley FQ (2011) The importance of ectomycorrhizas for the growth of dipterocarps and the efficacy of ectomycorrhizal inoculation schemes. In: Diversity and biotechnology of ectomycorrhizae. Springer, Berlin/Heidelberg, pp 3–17. https://doi.org/10.1007/978-3-642-15196-5_1

Brundrett M (2002) Coevolution of roots and mycorrhizas of land plants. New Phytol 154:275–304. https://doi.org/10.1046/j.1469-8137.2002.00397.x

Brundrett M (2004) Diversity and classification of mycorrhizal associations. Biol Rev 79(3):473–495. https://doi.org/10.1017/S1464793103006316

Bruns TD, Bidartondo MI, Taylor LD (2002) Host specificity in ectomycorrhizal communities: what do the exceptions tell us? Integr Comp Biol 42(2):352–359. https://doi.org/10.1093/icb/42.2.352

Bücking H (2011) Ectomycoremediation: an eco-friendly technique for the remediation of polluted sites in: diversity and biotechnology of ectomycorrhizae. Springer, Berlin, Heidelberg, pp 209–229. https://link.springer.com/book/10.1007/978-3-642-15196-5

Bücking H, Liepold E, Ambilwade P (2012) The role of the Mycorrhizal symbiosis in nutrient uptake of plants and the regulatory mechanisms underlying these transport processes in: Plant Science, Intech Open. https://doi.org/10.5772/52570

Cairney JW, Meharg AA (2002) Interactions between ectomycorrhizal fungi and soil saprotrophs: implications for decomposition of organic matter in soils and degradation of organic pollutants in the rhizosphere. Can J Bot 80(8):803–809. https://doi.org/10.1139/b02-072

Caldwell BA, Jumpponen A, Trappe JM (2000) Utilization of major detrital substrates by dark-septate, root endophytes. Mycologia 92(2):230–232. https://doi.org/10.2307/3761555

Chalot M, Brun A (1998) Physiology of organic nitrogen acquisition by ectomycorrhizal fungi and ectomycorrhizas. FEMS Microbiol Rev 22(1):21–44. https://doi.org/10.1111/j.1574-6976.1998.tb00359.x

Chambers, D (2018) Drought has cut Western cape farm production by 20%, says finance MEC. Sunday Times, 7 March 2018. https://www.timeslive.co.za/news/south-africa/2018-03-07-drought-has-cut-western-cape-farm-production-by-20-says-finance-mec/

Conway GR (1986) Agroecosystem analysis for research and development. Winrock International Institute for Agricultural Development Bangkok: 19–35

Compant S, Reiter B, Sessitsch A, Nowak J, Clément C, Barka EA (2005) Endophytic colonization of *Vitis vinifera* L. by plant growth-promoting bacterium Burkholderia sp. strain PsJN. Appl Environ Microbiol 71(4):1685–1693. https://doi.org/10.1128/AEM.71.4.1685-1693.2005

da Costa PB, Granada CE, Ambrosini A, Moreira F, de Souza R, dos Passos JFM, Arruda L, Passaglia LM (2014) A model to explain plant growth promotion traits: a multivariate analysis of 2,211 bacterial isolates. PLoS One 9(12):e116020. https://doi.org/10.1371/journal.pone.0116020

De Kempeneer L, Sercu B, Vanbrabant W, Van Langenhove H, Verstraete W (2004) Bioaugmentation of the phyllosphere for the removal of toluene from indoor air. Appl Microbiol Biotechnol 64(2):284–288. https://doi.org/10.1007/s00253-003-1415-3

Dittmann J, Heyser W, Bücking H (2002) Biodegradation of aromatic compounds by white rot and ectomycorrhizal fungal species and the accumulation of chlorinated benzoic acid in ectomycorrhizal pine seedlings.Chemosphere 49(3):297–306. https://doi.org/10.1016/S0045-6535(02)00323-5

Divito GA, Sadras VO (2014) How do phosphorus, potassium and Sulphur affect plant growth and biological nitrogen fixation in crop and pasture legumes? A meta-analysis. Field Crops Res 156:161–171. https://doi.org/10.1016/j.fcr.2013.11.004

Dunham SM, Larsson KH, Spatafora JW (2007) Species richness and community composition of mat-forming ectomycorrhizal fungi in old-and second-growth Douglas-fir forests of the HJ Andrews Experimental Forest, Oregon, USA. Mycorrhiza 17(8): 633–645. https://doi.org/10.1007/s00572-007-0141-6

Durand N, Monger HC, Canti MG, Verrecchia EP (2018) Calcium carbonate features. In: Interpretation of micromorphological features of soils and regoliths. Elsevier, pp 205–258. https://doi.org/10.1016/B978-0-444-63522-8.00009-7

Egger KN (1986) Substrate hydrolysis patterns of post-fire ascomycetes (Pezizales). Mycologia 78(5):771–780

Entry JA, Donnelly PK, Cromack K (1991) Influence of ectomycorrhizal mat soils on lignin and cellulose degradation. Biol Fertil Soils 11(1):75–78. https://doi.org/10.1007/BF00335839

Ercoli L, Schüßler A, Arduini I, Pellegrino E (2017) Strong increase of durum wheat iron and zinc content by field-inoculation with arbuscular mycorrhizal fungi at different soil nitrogen availabilities. Plant Soil 419:153–167. https://doi.org/10.1007/s11104-017-3319-5

Etesami H, Emami S, Alikhani HA (2017) Potassium solubilizing bacteria (KSB): mechanisms, promotion of plant growth, and future prospects: a review. J Soil Sci Plant Nutr 17(4):897–911. https://doi.org/10.4067/S0718-95162017000400005

Fahad S, Hussain S, Bano A, Saud S, Hassan S, Shan D, Tabassum MA (2015) Potential role of phytohormones and plant growth-promoting rhizobacteria in abiotic stresses: consequences for changing environment. Environ Sci Pollut R 22(7):4907–4921. https://doi.org/10.1007/s11356-014-3754-2

Farré-Armengol G, Filella I, Llusia J, Peñuelas J (2016) Bidirectional interaction between phyllospheric microbiotas and plant volatile emissions. Trends Plant Sci 21(10):854–860. https://doi.org/10.1016/j.tplants.2016.06.005

Finlay R, Söderström B (1992) Mycorrhiza and carbon flow to the soil in mycorrhizal functioning. Chapman & Hall, New York, pp 134–160. https://doi.org/10.1007/978-3-319-08575-3_3
Fokkema NJ, Van der Meulen F (1976) Antagonism of yeastlike phyllosphere fungi against Septoria nodorum on wheat leaves. Neth J Plant Pathol 82(1):13–16. https://doi.org/10.1007/BF01977343
Freiberg E (1998) Microclimatic parameters influencing nitrogen fixation in the phyllosphere in a Costa Rican premontane rain forest. Oecologia 117(1–2):9–18. https://doi.org/10.1007/s004420050625
Friberg S (2001) Distribution and diversity of arbuscular mycorrhizal fungi in traditional agriculture on the Niger inland delta, Mali, West Africa. CBM: s Skriftserie 3:53–80
Fürnkranz M, Wanek W, Richter A, Abell G, Rasche F, Sessitsch A (2008) Nitrogen fixation by phyllosphere bacteria associated with higher plants and their colonizing epiphytes of a tropical lowland rainforest of Costa Rica. ISME J 2(5):561. https://doi.org/10.1038/ismej.2008.14
Gardes M, Dahlberg A (1996) Mycorrhizal diversity in arctic and alpine tundra: an open question. New Phytol 133(1):147–157
Genney DR, Alexander IJ, Killham K, Meharg AA (2004) Degradation of the polycyclic aromatic hydrocarbon (PAH) fluorene is retarded in a Scots pine ectomycorrhizosphere. New Phytol 163(3):641–649. https://doi.org/10.1111/j.1469-8137.2004.01131.x
Glick BR (2014) Bacteria with ACC deaminase can promote plant growth and help to feed the world. Microbiol Res 169(1):30–39. https://doi.org/10.1016/j.micres.2013.09.009
Gnanamanickam SS, Priyadarisini VB, Narayanan NN, Vasudevan P, Kavitha S (1999) An overview of bacterial blight disease of rice and strategies for its management. Curr Sci 77:1435–1444. https://www.jstor.org/stable/24105230
Göltenboth F, Langenberger G, Widmann P (2006) Tropical lowland evergreen rainforest. In: Ecology of insular Southeast Asia, pp 297–383
Goswami D, Thakker JN, Dhandhukia PC (2016) Portraying mechanics of plant growth promoting rhizobacteria (PGPR): a review. Cogent Food Agric 2(1):1127500. https://doi.org/10.1080/23311932.2015.1127500
Gouda S, Kerry RG, Das G, Paramithiotis S, Shin HS, Patra JK (2017) Revitalization of plant growth promoting rhizobacteria for sustainable development in agriculture. Microbiol Res 206:131–140. https://doi.org/10.1016/j.micres.2017.08.016
Goudjal Y, Toumatia O, Yekkour A, Sabaou N, Mathieu F, Zitouni A (2014) Biocontrol of Rhizoctonia solani damping-off and promotion of tomato plant growth by endophytic actinomycetes isolated from native plants of Algerian Sahara. Microbiol Res 169(1):59–65. https://doi.org/10.1016/j.micres.2013.06.014
Gunderson JJ, Knight JD, Van Rees KCJ (2007) Impact of ectomycorrhizal colonization of hybrid poplar on the remediation of diesel-contaminated soil. J Environ Qual 36(4):927–934. https://doi.org/10.2134/jeq2006.0260
Gupta N, Sahoo D (2010) Evaluation of in vitro solubilization potential of phosphate solubilising Streptomyces isolated from phyllosphere of Heritiera fomes (mangrove). Afr J Microbial Res 4(3):136–142
Gupta M, Kiran S, Gulati A, Singh B, Tewari R (2012) Isolation and identification of phosphate solubilizing bacteria able to enhance the growth and aloin-A biosynthesis of Aloe barbadensis Miller. Microbiol Res 167(6):358–363. https://doi.org/10.1016/j.micres.2012.02.004
Gupta G, Parihar SS, Ahirwar NK, Snehi SK, Singh V (2015) Plant growth promoting rhizobacteria (PGPR): current and future prospects for development of sustainable agriculture. J Microb Biochem Technol 7(2):096–102. https://doi.org/10.4172/1948-5948.1000188
Hallett PD, Feeney DS, Bengough AG, Rillig MC, Scrimgeour CM, Young IM (2009) Disentangling the impact of AM fungi versus roots on soil structure and water transport. Plant Soil 314:183–196. https://doi.org/10.1007/s11104-008-9717-y
Hardoim PR, Van Overbeek LS, Van Elsas JD (2008) Properties of bacterial endophytes and their proposed role in plant growth. Trends Microbiol 16:463–471. https://doi.org/10.1016/j.tim.2008.07.008

Heckman DS, Geiser DM, Eidell BR, Stauffer RL, Kardos NL, Hedges SB (2001) Molecular evidence for the early colonization of land by fungi and plants. Science 293:1129–1133. https://doi.org/10.1126/science.1061457
Helgason T, Fitter AH (2009) Natural selection and the evolutionary ecology of the arbuscular mycorrhizal fungi (Phylum Glomeromycota). J Exp Bot 60:2465–2480. https://doi.org/10.1093/jxb/erp144
Hiltunen LH, Ojanperä T, Kortemaa H, Richter E, Lehtonen MJ, Valkonen JPT (2009) Interactions and biocontrol of pathogenic Streptomyces strains co-occurring in potato scab lesions. J Appl Microbiol 106(1):199–212. https://doi.org/10.1111/j.1365-2672.2008.03992.x
Hirano SS, Upper CD (2000) Bacteria in the leaf ecosystem with emphasis onPseudomonas syringae—a pathogen, ice nucleus, and epiphyte. Microbiol Mol Biol Rev 64(3):624–653. https://doi.org/10.1128/MMBR.64.3.624-653.2000
Horton TR, Bruns TD (2001) The molecular revolution in ectomycorrhizal ecology: peeking into the black-box. Mol Ecol 10(8):1855–1871. https://doi.org/10.1046/j.0962-1083.2001.01333.x
Hosny M, Van Tuinen D, Jacquin F, Füller P, Zhao B, Gianinazzi-Pearson V, Franken P (1999) Arbuscular mycorrhizal fungi and bacteria: how to construct prokaryotic DNA-free genomic libraries from the Glomales. FEMS Microbiol Lett 170(2):425–430. https://doi.org/10.1111/j.1574-6968.1999.tb13404.x
Hristozkova M, Gigova L, Geneva M, Stancheva I, Velikova V, Marinova G (2018) Influence of mycorrhizal Fungi and microalgae dual inoculation on basil plants performance. Gesunde Pflanzen 70(2):99–107. https://doi.org/10.1007/s10343-018-0420-5
Isaac S (1991) Fungal-plant interactions. Springer, Dordrecht
Jahromi F, Aroca R, Porcel R, Ruiz-Lozano JM (2008) Influence of salinity on the in vitro development of Glomus intraradices and on the in vivo physiological and molecular responses of mycorrhizal lettuce plants. Microb Ecol 55:45. https://doi.org/10.1007/s00248-007-9249-7
Jentschke G, Godbold DL (2000) Metal toxicity and ectomycorrhizas. Physiol Plant 109(2):107–116. https://doi.org/10.1034/j.1399-3054.2000.100201.x
Johnson KB, Stockwell VO (1998) Management of fire blight: a case study in microbial ecology. Annu Rev Phytopathol 36(1):227–248. https://doi.org/10.1146/annurev.phyto.36.1.227
Joner EJ, Leyval C, Colpaert JV (2006) Ectomycorrhizas impede phytoremediation of polycyclic aromatic hydrocarbons (PAHs) both within and beyond the rhizosphere. Environ Pollut 142(1):34–38. https://doi.org/10.1016/j.envpol.2005.09.007
Jung SC, Martinez-Medina A, Lopez-Raez JA, Pozo MJ (2012) Mycorrhiza-induced resistance and priming of plant defenses. J Chem Ecol 38:651–664. https://doi.org/10.1007/s10886-012-0134-6
Junker RR, Loewel C, Gross R, Dötter S, Keller A, Blüthgen N (2011) Composition of epiphytic bacterial communities differs on petals and leaves. Plant Biol 13(6):918–924. https://doi.org/10.1111/j.1438-8677.2011.00454.x
Kaewchai S, Soytong K, Hyde K (2009) Mycofungicides and fungal biofertilizers. Fungal Divers 38:25–50
Kaymak HC (2010) Potential of PGPR in agricultural innovations. In: plant growth and health promoting bacteria. Springer, Berlin/Heidelberg, pp 45–79. https://doi.org/10.1007/978-3-642-13612-2_3
Kembel SW, O'Connor TK, Arnold HK, Hubbell SP, Wright SJ, Green JL (2014) Relationships between phyllosphere bacterial communities and plant functional traits in a neotropical forest. Proc Natl Acad Sci 111(38):13715–13720. https://doi.org/10.1073/pnas.1216057111
Kennedy C, Bishop P (2004) Genetics of nitrogen fixation and related aspects of metabolism in species of Azotobacter: history and current status. In: Genetics and regulation of nitrogen fixation in free-living bacteria. Springer, Dordrecht, pp 27–52. https://doi.org/10.1007/1-4020-2179-8_2
Kenrick P, Strullu-Derrien C (2014) The origin and early evolution of roots. Plant Physiol:114. https://doi.org/10.1104/pp.114.244517
Khan MMA, Haque E, Paul NC, Khaleque MA, Al-Garni SMS, Rahman M, Islam MT (2017) Enhancement of growth and grain yield of Rice in nutrient deficient soils by Rice probiotic Bacteria. Rice Sci 24:264–273. https://doi.org/10.1016/j.rsci.2017.02.002

Kirschner R (2015) Fungi on the leaf—a contribution towards a review of phyllosphere microbiology from the mycological perspective. In: Biodiversity and ecology of fungi, lichens and mosses—Kerner von Marilaun workshop, pp 426–441

Knief C, Delmotte N, Vorholt JA (2011) Bacterial adaptation to life in association with plants–a proteomic perspective from culture to in situ conditions. Proteomics 11(15):3086–3105. https://doi.org/10.3389/fpls.2014.00216

Kumar AS, Sridar R, Uthandi S (2017) Mitigation of drought in rice by a phyllosphere bacterium Bacillus altitudinis FD48. Afr J Microbiol Res 11(45):1614–1625

Kwak MJ, Jeong H, Madhaiyan M, Lee Y, Sa TM, Oh TK, Kim JF (2014) Genome information of Methylobacterium oryzae, a plant-probiotic methylotroph in the phyllosphere. PLoS One 9(9):e106704

Landeweert R, Hoffland E, Finlay RD, Kuyper TW, van Breemen N (2001) Linking plants to rocks: ectomycorrhizal fungi mobilize nutrients from minerals. Trends Ecol Evol 16(5):248–254. https://doi.org/10.1016/S0169-5347(01)02122-X

Lindow SE, Leveau JH (2002) Phyllosphere microbiology. Curr Opin Biotechnol 13(3):238–243. https://doi.org/10.1016/S0958-1669(02)00313-0

Lindow SE, Brandl MT (2003) Microbiology of the phyllosphere. Appl Environ Microbiol 69(4):1875–1883. https://doi.org/10.1128/AEM.69.4.1875-1883.2003

Liu YH, Huang CJ, Chen CY (2010) Identification and transcriptional analysis of genes involved in Bacillus cereus-induced systemic resistance. Lilium. Biol Plantarum 54(4):697–702. https://doi.org/10.1007/s10535-010-0123-y

Liu D, Lian B, Dong H (2012) Isolation of Paenibacillus sp. and assessment of its potential for enhancing mineral weathering. Geomicrobiol J 29(5):413–421. https://doi.org/10.1080/01490451.2011.576602

Lodge DJ (2000) Ecto- or arbuscular mycorrhizas – which are best? New Phytol 146(3):353–354. https://doi.org/10.1046/j.1469-8137.2000.00668.x

López-Mondéjar R, Kostovčík M, Lladó S, Carro L, García-Fraile P (2017) Exploring the plant microbiome through multi-omics approaches. In: Probiotics in agroecosystem. Springer, Singapore, pp 233–268. https://doi.org/10.1007/978-981-10-4059-7_13

Lugtenberg B (2015) Life of microbes in the rhizosphere. In: Principles of plant-microbe interactions: microbes for sustainable agriculture. Springer, Cham, pp 7–15. https://www.springer.com/gp/book/9783319085746

Malusà E, Pinzari F, Canfora L (2016) Efficacy of biofertilizers: challenges to improve crop production. In: Microbial inoculants in sustainable agricultural productivity. Springer, New Delhi, pp 17–40. https://doi.org/10.1007/978-81-322-2644-4_2

Martin F, Díez J, Dell B, Delaruelle C (2002) Phylogeography of the ectomycorrhizal Pisolithus species as inferred from nuclear ribosomal DNA ITS sequences. New Phytol 153(2):345–357. https://doi.org/10.1046/j.0028-646X.2001.00313.x

Martin BC, George SJ, Price CA, Ryan MH, Tibbett M (2014) The role of root exuded low molecular weight organic anions in facilitating petroleum hydrocarbon degradation: current knowledge and future directions. Sci Tot Environ 472:642–653. https://doi.org/10.1016/j.scitotenv.2013.11.050

Marx DH, Artman JD (1979) Pisolithus tinctorius ectomycorrhizae improve survival and growth of pine seedlings on acid coal spoils in Kentucky and Virginia. Reclam Rev 2:23–31

Marx DH, Daniel WJ (1976) Maintaining cultures of ectomycorrhizal and plant pathogenic fungi in sterile water cold storage. Can J Microbiol 22(3):338–341

Mashiane RA, Ezeokoli OT, Adeleke RA, Bezuidenhout CC (2017) Metagenomic analyses of bacterial endophytes associated with the phyllosphere of a Bt maize cultivar and its isogenic parental line from South Africa. World J Microbiol Biotechnol 33(4):80. https://doi.org/10.1007/s11274-017-2249-y

Mashiane RA, Adeleke RA, Bezuidenhout CC, Chirima GJ (2018) Community composition and functions of endophytic bacteria of Bt maize. S Afr J Sci 114(7–8):88–97. https://doi.org/10.17159/sajs.2018/20170018

Meena VS, Maurya BR, Verma JP, Meena RS (eds) (2016) Potassium solubilizing microorganisms for sustainable agriculture. Springer

Meharg AA, Cairney JW, Maguire N (1997) Mineralization of 2, 4-dichlorophenol by ectomycorrhizal fungi in axenic culture and in symbiosis with pine. Chemosphere 34(12):2495–2504. https://doi.org/10.1016/S0045-6535(97)00005-2

Mendes R, Garbeva P, Raaijmakers JM (2013) The rhizosphere microbiome: significance of plant beneficial, plant pathogenic, and human pathogenic microorganisms. FEMS Microbiol Rev 37:634–663. https://doi.org/10.1111/1574-6976.12028

Minuto A, Spadaro D, Garibaldi A, Gullino ML (2006) Control of soilborne pathogens of tomato using a commercial formulation of Streptomyces griseoviridis and solarization. Crop Protect 25(5):468–475. https://doi.org/10.1016/j.cropro.2005.08.001

Moissl-Eichinger C, Pausan M, Taffner J, Berg G, Bang C, Schmitz RA (2017) Archaea are interactive components of complex microbiomes. Trends Microbiol 26(1):70–85

Molina R, Trappe J (1982) Lack of mycorrhizal specificity in the ericaceous hosts Arbutus menziessi and *Arctostaphylos uva-ursi*. New Phytol 90:495–509. https://doi.org/10.1111/j.1469-8137.1982.tb04482.x

Molina R, Massicotte H, Trappe JM (1992) Specificity phenomena in mycorrhizal symbioses: community-ecological consequences and practical implications. In: Mycorrhizal functioning: an integrative plant-fungal process, pp 357–423

Morton JB (1990) Evolutionary relationships among arbuscular mycorrhizal Fungi in the Endogonaceae. Mycologia 82(2):192–207. https://doi.org/10.1080/00275514.1990.12025865

Muchovej RM (2004) Importance of mycorrhizae for agricultural crops. Institute of Food and Agricultural Sciences (IFAS), University of Florida. SS-AGR-170, Gainesville, FL

Müller T, Behrendt U, Ruppel S, von der Waydbrink G, Müller ME (2016) Fluorescent pseudomonads in the phyllosphere of wheat: potential antagonists against fungal phytopathogens. Curr Microbiol 72(4):383–389

Ochieng JRA (2015) Towards a regulatory framework for increased and sustainable use of biofertilizers in Kenya, MA thesis, Centre for Advanced Studies in Environmental Law and Policy (CASELAP), University of Nairobi

Osono T (2006) Role of phyllosphere fungi of forest trees in the development of decomposer fungal communities and decomposition processes of leaf litter. Can J Microbiol 52(8):701–716. https://doi.org/10.1139/w06-023

Peixoto RS, Vermelho AB, Rosado AS (2011) Petroleum-degrading enzymes: bioremediation and new prospects. Enzyme Res 2011. https://doi.org/10.4061/2011/475193

Pieterse CMJ, Vanwees SCM (2014) Induced disease resistance. In: Lugtenberg B (ed) Principles of plant-microbe interactions: microbes for sustainable agriculture, Cham, Springer, pp 123–134

Pieterse CM, Zamioudis C, Berendsen RL, Weller DM, Van Wees SC, Bakker PA (2014) Induced systemic resistance by beneficial microbes. Annu Rev Phytopathol 52. https://doi.org/10.1146/annurev-phyto-082712-102340

Pozo MJ, Azcón-Aguilar C (2007) Unraveling mycorrhiza-induced resistance. Curr Opin Plant Biol 10:393–398. https://doi.org/10.1016/j.pbi.2007.05.004

Pringle A, Moncalvo JM, Vilgalys R (2000) High levels of variation in ribosomal DNA sequences within and among spores of a natural population of the arbuscular mycorrhizal fungus Acaulospora colossica. Mycologia 92(2):259–268. https://doi.org/10.2307/3761559

Querejeta J (2017) Soil water retention and availability as influenced by mycorrhizal symbiosis: consequences for individual plants, communities, and ecosystems. Mycorrhizal Mediation of Soil. Elsevier, pp 299–317. https://doi.org/10.1016/B978-0-12-804312-7.00017-6

Radwan SS, Sorkhoh NA, Fardoun F, Al-Hasan RH (1995) Soil management enhancing hydrocarbon biodegradation in the polluted Kuwaiti desert. Appl Microbiol Biotechnol 44(1–2):265–270. https://doi.org/10.1007/BF00164513

Rastogi G, Coaker GL, Leveau JH (2013) New insights into the structure and function of phyllosphere microbiota through high-throughput molecular approaches. FEMS Microbiol Lett 348(1):1–10. https://doi.org/10.1111/1574-6968.12225

Remus-Emsermann MN, Schlechter RO (2018) Phyllosphere microbiology: at the interface between microbial individuals and the plant host. New Phytol 218(4):1327–1333. https://doi.org/10.1111/nph.15054

Rodrigues K & Rodrigues B (2017) Development of carrier based in vitro produced arbuscular mycorrhizal (AM) fungal inocula for organic agriculture. AOAS. 1(1): 26-37. http://irgu.unigoa.ac.in/drs/handle/unigoa/4909

Rostami M, Hasanzadeh N, Khodaygan P, Riahi-Madvar A (2018) Ice nucleation active bacteria from pistachio in Kerman Province, Iran. JPP 100(1):51–58

Roossinck MJ (2011) The good viruses: viral mutualistic symbioses. Nat Rev Microbiol 9:99. https://doi.org/10.1038/nrmicro2491

Rouphael Y, Franken P, Schneider C, Schwarz D, Giovannetti M, Agnolucci M, De Pascale S, Bonini P, Colla G (2015) Arbuscular mycorrhizal fungi act as biostimulants in horticultural crops. Sci Hort 196:91–108. https://doi.org/10.1016/j.scienta.2015.09.002

Sanders IR, Croll D (2010) Arbuscular mycorrhiza: the challenge to understand the genetics of the fungal partner. Annu Rev Genet 44:271–292. https://doi.org/10.1146/annurev-genet-102108-134239

Sandhu A, Halverson LJ, Beattie GA (2007) Bacterial degradation of airborne phenol in the phyllosphere. Environ Microbiol 9:383–392

Sangthong S, Suksabye P, Thiravetyan P (2016) Air-borne xylene degradation by Bougainvillea buttiana and the role of epiphytic bacteria in the degradation. Ecotoxicol Environ Saf 126:273–280. https://doi.org/10.1016/j.ecoenv.2015.12.017

Sarand I, Timonen S, Koivula T, Peltola R, Haahtela K, Sen R, Romantschuk M (1999) Tolerance and biodegradation of m-toluate by Scots pine, a mycorrhizal fungus and fluorescent pseudomonads individually and under associative conditions. J Appl Microbiol 86(5):817–826. https://doi.org/10.1046/j.1365-2672.1999.00731.x

Sartori M, Nesci A, García J, Passone MA, Montemarani A, Etcheverry M (2017) Efficacy of epiphytic bacteria to prevent northern leaf blight caused by Exserohilum turcicum in maize. Rev Argent Microbiol 49(1):75–82

Sayyed RZ, Chincholkar SB, Reddy MS, Gangurde NS, Patel PR (2013) Siderophore producing PGPR for crop nutrition and phytopathogen suppression. In: Bacteria in agrobiology: disease management. Springer, Berlin/Heidelberg, pp 449–471. https://doi.org/10.1007/978-3-642-33639-3_17

Scheublin TR, Deusch S, Moreno-Forero SK, Müller JA, van der Meer JR, Leveau JH (2014) Transcriptional profiling of G ram-positive a rthrobacter in the phyllosphere: induction of pollutant degradation genes by natural plant phenolic compounds. Environ Microbiol 16(7):2212–2225. https://doi.org/10.1111/1462-2920.12375

Schouteden N, De Waele D, Panis B, Vos CM (2015) Arbuscular mycorrhizal fungi for the biocontrol of plant-parasitic nematodes: a review of the mechanisms involved. Front Microbiol 6(1280). https://doi.org/10.3389/fmicb.2015.01280

Senthilkumar M, Krishnamoorthy R (2017) Isolation and characterization of tomato leaf Phyllosphere Methylobacterium and their effect on plant growth. Int J Curr Microbiol App Sci 6(11):2121–2136

Shamseldin A (2013) The role of different genes involved in symbiotic nitrogen fixation—review. Globl J Biotechnol Bioch 8(4):84–94. https://doi.org/10.5829/idosi.gjbb.2013.8.4.82103

Sharma SB, Sayyed RZ, Trivedi MH, Gobi TA (2013) Phosphate solubilizing microbes: sustainable approach for managing phosphorus deficiency in agricultural soils. Springer Plus 2(587):1–14. https://doi.org/10.1186/2193-1801-2-587

Shen J, Yuan L, Zhang J, Li H, Bai Z, Chen X, Zhang W, Zhang F (2011) Phosphorus dynamics: from soil to plant. Plant Physiol 156(3):111. https://doi.org/10.1104/pp.111.175232

Sihi D, Dari B, Sharma DK, Pathak H, Nain L, Sharma OP (2017) Evaluation of soil health in organic vs. conventional farming of basmati rice in North India. J Plant Nutr Soil Sci 180:389–406. https://doi.org/10.1002/jpln.201700128

Singh R, Kothari R (2017) The omics era and host microbiomes. In: Understanding host-microbiome interactions-an omics approach. Springer, Singapore, pp 3–12. https://doi.org/10.1007/978-981-10-5050-3_1

Skirvin RM, Kohler E, Steiner H, Ayers D, Laughnan A, Norton MA, Warmund M (2000) The use of genetically engineered bacteria to control frost on strawberries and potatoes. Whatever happened to all of that research? Sci Hortic 84(1–2):179–189. https://doi.org/10.1016/S0304-4238(99)00097-7

Smith SE, Read D (2008) Mycorrhizal Symbiosis, 3rd edn. Academic, London

Smith SE, Jakobsen I, Grønlund M, Smith FA (2011) Roles of arbuscular mycorrhizas in plant phosphorus (P) nutrition: interactions between pathways of P uptake in arbuscular mycorrhizal (AM) roots have important implications for understanding and manipulating plant P acquisition. Plant Physiol 156(3):111. https://doi.org/10.1104/pp.111.174581

Song YY, Cao M, Xie LJ, Liang XT, Zeng RS, Su YJ, Huang JH, Wang RL, Luo SM (2011) Induction of DIMBOA accumulation and systemic defense responses as a mechanism of enhanced resistance of mycorrhizal corn (Zea mays L.) to sheath blight. Mycorrhiza 21:721–731. https://doi.org/10.1007/s00572-011-0380-4

Song Y, Chen D, Lu K, Sun Z, Zeng R (2015) Enhanced tomato disease resistance primed by arbuscular mycorrhizal fungus. Front Plant Sci 6:786. https://doi.org/10.3389/fpls.2015.00786

Sorkhoh NA, Ali N, Al-Awadhi H, Dashti N, Al-Mailem DM, Eliyas M, Radwan SS (2010) Phytoremediation of mercury in pristine and crude oil contaminated soils: contributions of rhizobacteria and their host plants to mercury removal. Ecotoxicol Environ Saf 73(8):1998–2003. https://doi.org/10.1016/j.ecoenv.2010.08.033

Souza R, Ambrosini A, Passaglia LM (2015) Plant growth-promoting bacteria as inoculants in agricultural soils. Genet Mol Biol 38(4):401–419. https://doi.org/10.1590/S1415-475738420150053

Subudhi E, Sahoo RK, Dey S, Das A, Sahoo K (2018) Unraveling plant-endophyte interactions: an omics insight. In: Endophytes and secondary metabolites. Springer, Cham, pp 1–19. https://doi.org/10.1007/978-3-319-76900-4_2-1

Suhag M (2016) Potential of biofertilizers to replace chemical fertilizers. Int Adv Res J Sci Eng Technol 3:163–167. ISSN (Online): 2393-8021

Sy A, Timmers AC, Knief C, Vorholt JA (2005) Methylotrophic metabolism is advantageous for Methylobacterium extorquens during colonization of *Medicago truncatula* under competitive conditions. Appl Environ Microb 71(11):7245–7252

Tam PC (1995) Heavy metal tolerance by ectomycorrhizal fungi and metal amelioration by Pisolithus tinctorius. Mycorrhiza 5(3):181–187. https://doi.org/10.1007/BF00203335

Thomson BD, Grove TS, Malajczuk N, Hardy GSJ (1994) The effectiveness of ectomycorrhizal fungi in increasing the growth of *Eucalyptus globulus Labill.* In relation to root colonization and hyphal development in soil. New Phytol 126(3):517–524

Tian C, Shi Z, Chen Z, Feng G (2006) Arbuscular mycorrhizal associations in the Gurbantunggut Desert. Chin Sci Bull 51(1):140–146. https://doi.org/10.1007/s11434-006-8218-8

Timmusk S, Behers L, Muthoni J, Muraya A, Aronsson AC (2017) Perspectives and challenges of microbial application for crop improvement. Front Plant Sci 8:49. https://doi.org/10.3389/fpls.2017.00049

Tokala RK, Strap JL, Jung CM, Crawford DL, Salove MH, Deobald LA, Morra MJ (2002) Novel plant-microbe rhizosphere interaction involving Streptomyces lydicus WYEC108 and the pea plant (Pisum sativum). Appl Environ Microbiol 68(5):2161–2171. https://doi.org/10.1128/AEM.68.5.2161-2171.2002

Tyagi S, Singh R, Javeria S (2014) Effect of climate change on plant-microbe interaction: an overview. Eur J Mol Biotech (3):149–156

Uribe D, Sánchez-Nieves J, Vanegas J (2010) Role of microbial biofertilizers in the development of a sustainable agriculture in the tropics. In: Dion P (ed) Soil biology and agriculture in the tropics. Springer, Berlin, pp 235–250. https://doi.org/10.1007/978-3-642-05076-3_11

van Der Heijden MG, Bardgett RD, Van Straalen NM (2008) The unseen majority: soil microbes as drivers of plant diversity and productivity in terrestrial ecosystems. Ecol Lett 11:296–310. https://doi.org/10.1111/j.1461-0248.2007.01139.x

van Wezel GP, Vijgenboom E (2004) Novel aspects of signaling in Streptomyces development. Adv Appl Microbiol 56:65–88. https://doi.org/10.1016/S0065-2164(04)56002-1

Vorholt JA (2012) Microbial life in the phyllosphere. Nat Rev Microbiol 10(12):828. https://doi.org/10.1038/nrmicro2910

Voříšková J, Baldrian P (2013) Fungal community on decomposing leaf litter undergoes rapid successional changes. ISME J 7(3):477. https://doi.org/10.1038/ismej.2012.116

Vurro M, Bonciani B, Vannacci G (2010) Emerging infectious diseases of crop plants in developing countries: impact on agriculture and socio-economic consequences. Food Secur 2:113–132. https://doi.org/10.1007/s12571-010-0062-7

Walker C, Schüßler A (2004) Nomenclatural clarifications and new taxa in the Glomeromycota. Mycol Res 108(9):981–982. https://doi.org/10.1017/S0953756204231173

Wallander H (2000a) Uptake of P from apatite by Pinus sylvestris seedlings colonised by different ectomycorrhizal fungi. Plant Soil 218(1–2):249–256. https://doi.org/10.1023/A:1014936217105

Wallander H (2000b) Use of strontium isotopes and foliar K content to estimate weathering of biotite induced by pine seedlings colonised by ectomycorrhizal fungi from two different soils. Plant Soil 222(1–2):215–229. https://doi.org/10.1023/A:1004756221985

Wallander H, Wickman T (1999) Biotite and microcline as potassium sources in ectomycorrhizal and non-mycorrhizal Pinus sylvestris seedlings. Mycorrhiza 9(1):25–32. https://doi.org/10.1007/s005720050259

Wang Q, Dodd IC, Belimov AA, Jiang F (2016) Rhizosphere bacteria containing 1-aminocyclopropane-1-carboxylate deaminase increase growth and photosynthesis of pea plants under salt stress by limiting Na+ accumulation. Funct Plant Biol 43(2):161–172. https://doi.org/10.1071/FP15200

Whipps J, Hand P, Pink D, Bending GD (2008) Phyllosphere microbiology with special reference to diversity and plant genotype. J Appl Microbial 105(6):1744–1755. https://doi.org/10.1111/j.1365-2672.2008.03906.x

Wilson M, Hirano SS, Lindow SE (1999) Location and survival of leaf-associated bacteria in relation to pathogenicity and potential for growth within the leaf. Appl Environ Microbiol 65(4):1435–1443

Wu CH, Bernard SM, Andersen GL, Chen W (2009) Developing microbe–plant interactions for applications in plant-growth promotion and disease control, production of useful compounds, remediation and carbon sequestration. Microb Biotech 2(4):428–440. https://doi.org/10.1111/j.1751-7915.2009.00109.x

Wyss P, Boller T, Wiemken A (1991) Phytoalexin response is elicited by a pathogen (Rhizoctonia solani) but not by a mycorrhizal fungus (Glomus mosseae) in soybean roots. Exp 47:395–399. https://doi.org/10.1007/BF01972082

Xu C, Leskovar DI (2015) Effects of A. nodosum seaweed extracts on spinach growth, physiology and nutrition value under drought stress. Sci Hort 183:39–47. https://doi.org/10.1016/j.scienta.2014.12.004

Xu P, Chen F, Mannas JP, Feldman T, Sumner LW, Roossinck MJ (2008) Virus infection improves drought tolerance. New Phytol 180(4):911–921. https://doi.org/10.1111/j.1469-8137.2008.02627.x

Yao M, Desilets H, Charles M, Boulanger R, Tweddell R (2003) Effect of mycorrhization on the accumulation of rishitin and solavetivone in potato plantlets challenged with Rhizoctonia solani. Mycorrhiza 13:333–336. https://doi.org/10.1007/s00572-003-0267-0

Yu TE, Egger KN, Peterson LR (2001) Ectendomycorrhizal associations–characteristics and functions. Mycorrhiza 11(4):167–177. https://doi.org/10.1007/s005720100110

Zabinski CA, Bunn RA (2014) Function of mycorrhizae in extreme environments. In: Mycorrhizal Fungi: use in sustainable agriculture and land restoration. Springer, Berlin, Heidelberg, pp 201–214. https://doi.org/10.1007/978-3-662-45370-4_12

Zhang F, Zhang J, Chen L, Shi X, Lui Z, Li C (2015) Heterologous expression of ACC deaminase from Trichoderma asperellum improves the growth performance of Arabidopsis thaliana under normal and salt stress conditions. Plant Physiol Biochem 94:41–47. https://doi.org/10.1016/j.plaphy.2015.05.007

Zipfel C (2008) Pattern-recognition receptors in plant innate immunity. Curr Opin Immunol 20:10–16. https://doi.org/10.1016/j.coi.2007.11.003

Exploring the Phyllosphere Bacterial Community for Improving Tree Crop Protection

2

Diogo Mina, José Alberto Pereira, Teresa Lino-Neto, and Paula Baptista

Abstract

Plants are able to interact with plentiful bacteria resulting in a number of positive or negative outcomes for plant health. The ecological balance between pathogens and beneficial bacteria could be strategically disturbed and manipulated for improving host plant protection. As bacterial communities present in the phyllosphere of herbaceous plants have been largely studied, a number of biocontrol agents for controlling host diseases are already identified and used with promising results. A few studies on the use of phyllosphere biocontrol agents on woody crop tree plants have revealed encouraging results toward a future where plant disease control could be attained without the application of chemical compounds. In addition to the use of biocontrol agents, disease suppression can be achieved by the manipulation of microbial communities through plant management practices. In this review, an overview of the available knowledge on phyllosphere bacterial communities of woody tree crop species is provided, giving special emphasis to the structural differences of bacterial communities living on and within important tree crop species. Studies and challenges on the application and/or manipulation of these bacteria under *in planta* conditions are discussed, disclosing new sustainable ways for dealing with woody crop diseases.

Keywords

Woody plants · Bacteria · Microbiome · Plant disease · Biological control

D. Mina · J. A. Pereira · P. Baptista (✉)
CIMO, School of Agriculture – Polytechnic Institute of Bragança, Bragança, Portugal
e-mail: pbaptista@ipb.pt

T. Lino-Neto
Biosystems and Integrative Sciences Institute (BioISI), Plant Functional Biology Center (CBFP), University of Minho, Braga, Portugal

V. Kumar et al. (eds.), *Microbiome in Plant Health and Disease*,
https://doi.org/10.1007/978-981-13-8495-0_2

2.1 Introduction

Microbial communities on or around plants have already been described to play a pivotal role in plant growth and health (Vorholt 2012). Such action has been mostly recognized for root-associated microorganisms (Sahu et al. 2018), while the microbial community associated with aerial parts of plants has been less studied and characterized (Carvalho and Castillo 2018). However, the aerial part of plants (phyllosphere, in *lato* sensu) has been recognized as an important habitat for microorganisms (Roat and Saraf 2017). These microorganisms live either on the surface (usually referred as phylloplane) or inside (endosphere) the tissues of plant organs (Carvalho and Castillo 2018). Microorganisms inhabiting the phylloplane are generally referred as epiphytes, while the ones colonizing the endosphere are referred as endophytes (Newton et al. 2010). Accordingly, the phyllosphere microbiota comprises all microorganisms living on the surface and inside of all aboveground plant tissues (Lemanceau et al. 2017a). Either in phylloplane (Meyer and Leveau 2012) or in endosphere (Ibáñez et al. 2017) of most plant species, bacteria form an important part of microbial communities, surpassing by far other microbial groups in both abundance and diversity (Lindow and Brandl 2003). The load of bacteria in the leaf surface usually lies within the range of 10^6–10^7 cells/cm^2, up to 10^8 cells/g leaf fresh weight (Leach et al. 2017). The number of bacterial species in the phyllosphere of natural ecosystems is also enormous. Estimates of bacterial endophytes inhabiting the Brazilian Atlantic forest indicate the possible occurrence of 2–13 million species present in the aboveground plant parts, with almost 97% of these species being undescribed (Lambais et al. 2006).

Bacteria inhabiting the phyllosphere can interact with the host plant (Kembel et al. 2014) and with other microorganisms, including both beneficial and pathogenic microbes that share the same habitat (Müller and Ruppel 2014; Leach et al. 2017). These plant–bacteria and bacteria–microbe interactions significantly influence plant performance and defense against diseases and pests (Bulgarelli et al. 2013; Rastogi et al. 2013; Ciancio et al. 2016; O'Brien 2017). The importance of such interactions in promoting host plant defense against diseases was specially recognized in herbaceous plant species (El-Sayed et al. 2018; Rahman et al. 2018), whereas their role in woody plant protection against diseases has been less studied (Cazorla and Mercado-Blanco 2016). This would be key knowledge for developing new strategies for agricultural tree crop protection.

In this review, the diversity and structure of bacterial communities (both endophytic and epiphytic) inhabiting the phyllosphere of economically important agricultural woody tree crops will be highlighted. Both biotic and abiotic factors that contribute to the shaping of bacterial communities will be also addressed. The potential to explore phyllosphere-associated bacteria for protecting woody crops from diseases will be discussed, either through their use as biological control agents or through their management.

2.2 Diversity of Bacterial Communities in the Phyllosphere of Important Agricultural Woody Crop Trees

The structure and diversity of phyllosphere bacterial communities of agricultural woody crops have been primarily studied in economically important fruit trees such as citrus (*Citrus* sp.; Araújo et al. 2002; Passera et al. 2018), apple (*Malus pumila*; Yashiro et al. 2011; Yashiro and McManus 2012; He et al. 2012), banana (*Musa acuminata/balbisiana*; Thomas and Soly, 2009; Rossmann et al. 2012), chestnut (*Castanea sativa*; Valverde et al. 2017), coffee (*Coffea arabica/robusta*; Vega et al. 2005), olive (*Olea europaea*; Müller et al. 2015), and stone fruits (*Prunus dulcis, P. domestica, P. salicina, P. armeniaca, P. avium, P. cerasus* and *P. persica*; Jo et al. 2015) (Table 2.1).

The phyllosphere bacterial communities associated with such crops have been analyzed by using both culture-dependent and -independent molecular approaches such as PCR-SSCP fingerprinting, quantitative PCR, fluorescence in situ hybridization platforms, and/or high-throughput sequencing technologies. Although a broader spectrum of bacterial colonizers can be assessed by next-generation technologies than using cultural approaches (Pham et al. 2008), PCR limitations can bias diversity studies. For example, *primers* could display different affinities to templates, inhibitory compounds could be present in different environmental samples, and plant organelle-derived RNA sequences could interfere in microbial target amplifications (Müller and Ruppel 2014).

Combined culture-dependent and -independent approaches have revealed a high degree of bacterial diversity on the phyllosphere of seven fruit tree crops, spanning a total of 104 bacterial genera, belonging to 75 families and 12 phyla (Fig. 2.1a).

Globally, the bacterial communities of these fruit tree crops consisted predominantly of *Proteobacteria, Actinobacteria, Firmicutes* and *Bacteroidetes* but in different proportions according to the tree species (Fig. 2.1b). For instance, the phyllosphere of *Castanea* was found to be dominated by *Actinobacteria* (Valverde et al. 2017), while *Prunus* presented up to 90% of bacteria from *Proteobacteria* (Jo et al. 2015). From the seven surveyed tree species, the phyllosphere of both *Musa* and *Citrus* showed the highest proportion of bacteria from *Firmicutes* (Araújo et al. 2002; Thomas and Soly 2009; Rossmann et al. 2012; Passera et al. 2018). Other phyla were also specific from *Citrus* spp. and *Olea* spp. phyllospheres but at lower abundances. While *Planctomycetes* was found in both cultures, a number of phyla were specifically found on *Citrus* (*Fibrobacteres, Spirochaetes,* and *Tenericutes*) or *Olea* (*Acidobacteria, Verrucomicrobia,* and *Armatimonadetes*) phyllospheres (Araújo et al. 2002; Müller et al. 2015; Passera et al. 2018).

Differences between fruit tree species become more apparent when comparing bacterial communities at the class level (Fig. 2.1c). *Citrus* and *Olea* presented the highest number of classes (16), followed by *Malus* (12) and *Prunus* (6). While two bacterial classes were common among all the investigated fruit trees (i.e., *Actinobacteria* and *Gammaproteobacteria*), some classes were tree species-specific. Among the seven investigated tree species, the phyllosphere of *Olea* displayed the highest number of unique bacterial classes (8), followed by *Citrus* (6) and *Malus*

Table 2.1 Woody fruit crops surveyed for phyllospheric bacterial communities. For each tree species, the bacterial community surveyed (endophytic or epiphytic), plant organ, and methodological approach used are indicated

Plant host	Organ	Community	Method	Reference
Citrus (*Citrus* sp.)	Branches	Endophytes	Culture-dependent	Araújo et al. (2002)
			Culture-independent (PCR-DGGE analysis)	
	Leaves	Endophytes	Culture-independent (16S rRNA sequencing from ground leaf tissue)	Passera et al. (2018)
Apple (*Malus pumila*)	Leaves	Epiphytes	Culture-dependent	Yashiro et al. (2011)
			Culture-independent (DAPI and 16S rRNA gene cloning from leaves sonication extracts)	
	Leaves	Epiphytes	Culture-independent (16S rRNA gene cloning from leaves sonication extracts)	Yashiro and McManus (2012)
	Leaves	Epiphytes	Culture-dependent	He et al. (2012)
		Endophytes	Culture-independent (macroarray hybridization)	
Banana (*Musa acuminata*/*M. balbisiana*)	Branches	Endophytes	Culture-dependent	Thomas and Soly (2009)
	Fruit	Endophytes	Culture-dependent	Rossmann et al. (2012)
			Culture-independent (see reference)	
Chestnut (*Castanea sativa*)	Leaves	Epiphytes	Culture-dependent	Valverde et al. (2017)
Coffee (*Coffea arabica*/*C. robusta*)	Branches	Endophytes	Culture-dependent	Vega et al. (2005)
	Leaves			
	Fruits	Epiphytes		
Olive (*Olea europaea*)	Leaves	Endophytes	Culture-independent (Illumina sequencing and qPCR)	Müller et al. (2015)
Stone fruits (*Prunus dulcis*/*P. domestica*/*P. salicina*/*P. armeniaca*/*P. avium*/*P. cerasus*/*P. persica*)	Leaves	Epiphytes	Culture-independent (pyrosequencing)	Jo et al. (2015)

(1). Interestingly, three bacterial classes (i.e., *Cytophagia, Sphingobacteriia,* and *Rubrobacteria*) were exclusively found in the phyllosphere of *Malus* and *Prunus*, suggesting that these bacteria might represent the core microbiota of *Rosaceae* family plants.

Further analysis of bacterial community composition at the genus level in the phyllosphere of the studied host tree species indicates that *Bacillus, Pseudomonas, Pantoea, Micrococcus, Methylobacterium, Sphingomonas,* and *Enterobacter* are highly abundant and consistently found (data not showed). Therefore, these genera

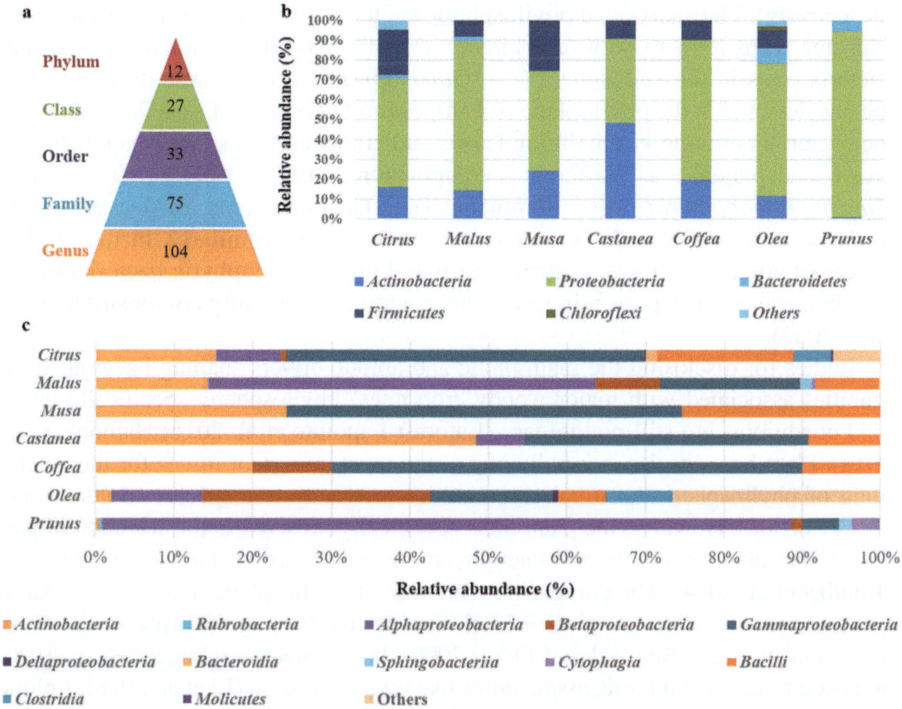

Fig. 2.1 Bacterial communities associated with the phyllosphere of woody tree crop species: *Citrus* (Araújo et al. 2002; Passera et al. 2018), *Malus* (Yashiro et al. 2011, Yashiro and McManus 2012; He et al. 2012), *Musa* (Thomas and Soly 2009; Rossmann et al. 2012), *Castanea* (Valverde et al. 2017), *Coffea* (Vega et al. 2005), *Olea* (Müller et al. 2015) and *Prunus* (Jo et al. 2015). (**a**) Number of distinct bacterial taxonomic groups detected across all tree crops; (**b**) Bacterial community composition, at the phylum level, for each tree crop. (**c**) Bacterial community composition, at the class level, for each tree crop

are likely to represent the core bacterial community of these fruit crops. The persistence of core members in apparently healthy trees suggests that they may be beneficial to the host. Indeed, the core microbiome is considered to encompass key microbial taxa that are critical for plant health. Evolutionary processes resulted in the selection and enrichment of microbiota carrying genes with essential functions for the fitness of holobiont (i.e., the plant plus all associated microbiota) (Lemanceau et al. 2017a). Besides core bacterial genera, surveys on the phyllosphere of fruit tree crops also detected bacterial genera specific to a particular tree species (Rossmann et al. 2012; Jo et al. 2015; Passera et al. 2018), reflecting the adaptation of bacteria to a specific environment (Lemanceau et al. 2017b).

Few studies have directly compared endophytic and epiphytic bacterial communities inhabiting the phyllosphere of woody crop trees. Despite the lack of studies comparing endo- and epiphytic bacterial communities within the same crop tree phyllosphere, Vega et al. (2005) found a higher number of bacterial species on the surface of Colombian coffee leaf than in internal leaf tissues (i.e., 18 vs. 8, respectively). The analysis of phyllospheric bacterial communities across the seven fruit tree crops (either endo- or epiphytic, or both) showed that the most abundant bacteria class in the endosphere (i.e., *Gammaproteobacteria*) was different from the one detected in the phylloplane (i.e., *Alphaproteobacteria*; Fig. 2.2a). Likewise, the bacterial community inhabiting leaves and branches, across the seven fruit tree crops, displayed a different composition (Fig. 2.2b). *Alpha-* and *Gammaproteobacteria* were dominant bacterial inhabitants of leaves, while *Actinobacteria* and *Bacilli* were common in branches. Similarly, in the phyllosphere of coffee seedlings, branches harbored greater endophytic bacterial diversity than leaves, both presenting a distinct bacterial community composition (Vega et al. 2005).

Studies for disclosing the main biotic and abiotic drivers shaping bacterial communities associated with major woody crop trees' phyllosphere, specifically under field conditions, are still preliminary (Laforest-Lapointe et al. 2016; Hamonts et al. 2018). Plant host species is usually one of the most important forces for the assembling of phyllospheric bacterial communities in woody tree species (Baldotto and Olivares 2008). However, the plant traits specifically involved in the selection of particular microbial epiphytic and endophytic colonizers are so far largely unknown (Kembel et al. 2014). The composition and size of phyllosphere-associated bacterial communities also depend on other biotic factors like host age (Carper et al. 2018), development stage (Redford and Fierer 2009), host genotype (Cregger et al. 2018), and occurrence of symbiotic associations like mycorrhization (Li et al. 2018). Abiotic factors are also known to influence phyllospheric bacterial community, such as geographical location (Finkel et al. 2011; Qvit-Raz et al. 2012) and climatic factors (Carper et al. 2018). A deeper understanding of bacterial communities in the

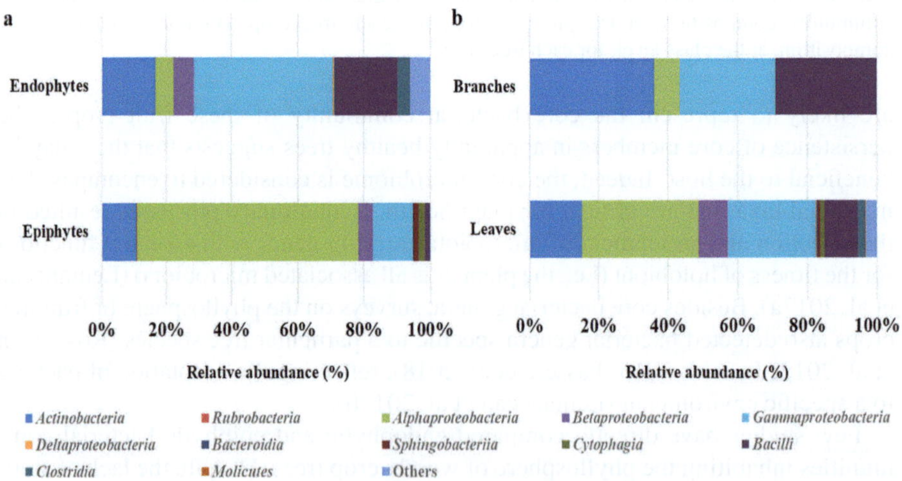

Fig. 2.2 Relative abundance of surveyed bacterial classes in the phyllosphere of the seven tree crops indicated in Table 2.1. (**a**) Bacterial classes detected within endophytic and epiphytic communities. (**b**) Bacterial classes detected on leaves and branches

phyllosphere of woody tree crops, as well as the drivers that shape their assembling, will offer new opportunities for controlling plant diseases and improve host plant health.

2.3 Exploiting Phyllosphere Bacterial Communities for Woody Tree Crop Protection

The use of bacterial isolates naturally adapted to crop species, resident microbiota and environment could provide an efficient approach to the biological control of plant diseases under field conditions (Ozaktan et al. 2012). The search for potential bacterial biological control agents, within the same host species as the pathogen, has begun more than 40 years ago (Wrather et al. 1973; McIntyre et al. 1973). However, up to the beginning of the twenty-first century, most of the investigation performed on woody tree crops mainly focused on apple and pear diseases (Utkhede 1987; Janisiewicz and Roitman 1988; Vanneste 1996; Pusey 2002). Furthermore, few studies have been illustrating the biocontrol of diseases in woody tree crops by using native phyllosphere-associated bacterial members. However, the antagonistic potential of phyllosphere microbiota has been explored against problematic pathogens over the last two decades, mainly through in vitro assays (e.g., Singh et al. 2004; Trivedi et al. 2011; Silva et al. 2012).

The control of woody crop diseases through the application of native phyllosphere-associated bacterial members presenting antagonistic activity, either in field or in greenhouse conditions, appears to be promising (Table 2.2). The level of disease suppression achieved by application of such bacteria ranged from 27% to 86%. Most research and development efforts have been focused on isolates of the genera *Pseudomonas* and *Bacillus*. *Pseudomonas* have been mostly effective for the biocontrol of bacterial diseases (e.g., *Erwinia* sp. and *Xanthomonas* sp.), while *Bacillus* have been mostly used for controlling fungal diseases (e.g., *Gnomoniopsis* sp., *Colletotrichum* sp., and *Cryphonectria* sp.). Accordingly, *Bacillus* strains are among the most exploited bacteria to be used as biocontrol agents against plant diseases (Bacon and Hinton 2002), in addition to their role in promoting plant growth (Pérez-Garcia et al. 2011). In recent years, there are also other bacteria that have received attention for the biocontrol of woody crop diseases, such as *Pantoea* sp. (Ozaktan et al. 2011; Gerami et al. 2013), *Serratia* sp. (Gerami et al. 2013), *Burkholderia* sp. (Silva and De Costa 2014), and *Alcaligenes* sp. (Abraham et al. 2013). These genera revealed to be effective in reducing the incidence and severity of important diseases that affect several hosts, mostly pear (Gerami et al. 2013) and apple (Pusey 2002; Ozaktan et al. 2011; Mikiciński et al. 2016), but also citrus (Kupper et al. 2011; Michavila et al. 2017), banana (Silva and De Costa 2014), mango (Yenjit et al. 2004), chestnut (Wilhelm et al. 1998; Pasche et al. 2016), avocado (Korsten et al. 1997), *Hevea* (Abraham et al. 2013), and pomegranate (Puneeth 2015). In particular, several *Pseudomonas* species (i.e., *P. graminis*, *P. agglomerans,* and *P. fluorescens*) were reported to be the most promising biocontrol agents against *Erwinia amylovora* on pear (Gerami et al. 2013) and apple

Table 2.2 Phyllospheric bacteria tested *in planta* for controlling diseases of woody tree crops. Efficiency of bacteria in preventing disease incidence (i) and severity (s) is shown

Microorganism	Host plant	Pathogen	Assay	Efficacy (%)	Reference
Pantoea vagans	Apple	*Erwinia amylovora*	Field	54% (i)	Ozaktan et al. (2011)
Pantoea agglomerans	Pear	*Erwinia amylovora*	Field	58–79% (i)	Gerami et al. (2013)
Serratia sp.	Pear	*Erwinia amylovora*	Field	27–58% (i)	Gerami et al. (2013)
Pseudomonas fluorescens	Apple	*Erwinia amylovora*	Field	27–36% (i)	Pusey (2002)
Pseudomonas fluorescens	Pear	*Erwinia amylovora*	Field	61–75% (i)	Gerami et al. (2013)
Pseudomonas protegens	Citrus	*Xanthomonas citri*	Greenhouse	78% (s)	Michavila et al. (2017)
Pseudomonas graminis	Apple	*Erwinia amylovora*	Greenhouse	86% (s)	Mikiciński et al. (2016)
			Field	73% (s)/40% (i)	
Burkholderia spinosa	Banana	*Colletotrichum musae*	Field	[a]	Silva and De Costa (2014)
Alcaligenes sp.	*Hevea*	*Phytophthora meadii*	Greenhouse	34–48% (s)	Abraham et al. (2013)
Bacillus licheniformis	Mango	*Colletotrichum gloeosporioides*	Greenhouse	50% (s)	Yenjit et al. (2004)
Bacillus subtilis	Chestnut	*Cryphonectria parasitica*	Greenhouse	71% (s)	Wilhelm et al. (1998)
Bacillus subtilis	Citrus	*Phyllosticta citricarpa*	Field	29% (s)	Kupper et al. (2011)
Bacillus subtilis	Pomegranate	*Xanthomonas axonopodis*	Field	78% (s)	Puneeth (2015)
Bacillus subtilis	Mango	*Colletotrichum gloeosporioides*	Greenhouse	44% (s)	Yenjit et al. (2004)
Bacillus subtilis	Avocado	*Pseudocercospora purpurea*	Field	[a]	Korsten et al. (1997)
Bacillus subtilis	Avocado	*Akaropeltopsis* sp.	Field	[a]	Korsten et al. (1997)

(continued)

Table 2.2 (continued)

Microorganism	Host plant	Pathogen	Assay	Efficacy (%)	Reference
Bacillus amyloliquefaciens	Chestnut	*Gnomoniopsis smithogylvyi*	Greenhouse	75% (i)	Pasche et al. (2016)
Bacterium fjat	Pomegranate	*Xanthomonas axonopodis*	Field	77% (s)	Puneeth (2015)

[a]Depending on the applied treatment, different values were obtained

(Mikiciński et al. 2016). *Bacillus subtilis* is a promising agent for controlling *Xanthomonas axonopodis* on pomegranate (Puneeth 2015).

Different mechanisms can be involved in the biological control of pathogens by these phyllospheric bacteria, although their effectiveness is still not totally understood. The antagonistic activity of *Pseudomonas* spp. toward pathogens is usually associated with the competition for nutrients (Cabrefiga et al. 2007) or the production of secondary metabolites such as siderophores (Duffy and Défago 1999; Sasirekha and Srividya 2016), antibiotics (e.g., pyoluteorin and phenazines), lytic enzymes (e.g., protease and cellulase), hydrogen cyanide (Weller 2007; Gerami et al. 2013; Zengerer et al. 2018), and antimicrobial volatile compounds (Hernández-León et al. 2015). Also, the mechanisms used by *Bacillus* strains to control plant pathogens have been attributed to the production of antibiotics and antimicrobial compounds, such as lipopeptides and lytic enzymes (Touré et al. 2004; Huang et al. 2012; Kumar et al. 2012), as well as to the induction of host plant defenses (Kloepper et al. 2004). Indeed, a strong antimicrobial effect against different phytopathogenic fungi and bacteria was already reported for lipopeptides, especially from iturin A, fengycin, and surfactin families (Touré et al. 2004; Ongena and Jacques 2008; Malfanova et al. 2012; Yuan et al. 2012). Sporulation of plant pathogens could be also compromised by certain *Bacillus* species, as previously reported for *B. subtilis, B. licheniformis,* and *B. cereus* that were able to reduce spore germination and germ-tube elongation in *Colletotrichum gloeosporioides* (Yenjit et al. 2004). *Bacillus* can also cause morphological abnormalities in the mycelium of pathogenic fungi (Chaurasia et al. 2005).

The genera *Pantoea, Enterobacter*, *Serratia,* and *Burkholderia* are also known to release antibiotics that are considered to be responsible for the antagonistic action against plant pathogens (Ishimaru et al. 1988; Subagio and Foster 2003; Buana et al. 2014). For example, *Burkholderia* was reported to be effective in inhibiting several fungal phytopathogens of oil palm through the production of antibiotics such as phenazine, pyrrolnitrin, pyoluteorin, and 2,4-DAPG (Subagio and Foster 2003; Buana et al. 2014). Expression analysis also revealed that the antagonistic effect of *Pantoea agglomerans* against fungal phytopathogens is related to up- and downregulation of genes, associated with fungal defense, virulence, and/or metabolic functions (Pandolfi et al. 2010). Furthermore, the suppression of phytopathogens by *Serratia* is related to a combination of mechanisms, including antibiosis (through the production of antimicrobial compounds), parasitism (through the release of extracellular cell wall-degrading enzymes), and competition (through siderophore release) (De Vleesschauwer and Höfte 2007). The biocontrol effect of *Alcaligenes* sp. toward several phytopathogens of different herbaceous crops has been also related to the production of siderophores (Sayyed and Patel 2011) and lytic enzymes, such as chitinase (Vaidya et al. 2001).

Beyond the use of biocontrol agents, disease suppression can also be achieved by the manipulation of phyllosphere microbial communities, in order to improve positive interactions with the host plant (Orozco-Mosqueda et al. 2018). In this "engineering" process, the microbial composition can be altered to maximize the benefits of the microbial social network for crop plants. To the best of our knowledge, such approach has not yet been explored in woody crop trees. Manipulation of microbiota is applied most extensively in humans for treatment of diseases (Young 2017; Larsen and Claassen 2018), and more recently in herbaceous crops for the control of root diseases (e.g., Gopal et al. 2013; Martínez-Hidalgo et al. 2018). This strategy, which is largely based on the transfer of natural microbial communities (by mixing disease suppressive soils with disease conducive ones), proved to be effective in the management of several root plant diseases, including rhizoctonia root in sugar beet, potato common scab, and tobacco black root rot (Gopal et al. 2013). Besides transfer of natural microbiome, soil inoculation with an artificial mixture of bacterial strains with desired functions has protected *Nicotiana attenuata* from sudden-wilt disease (Santhanam et al. 2015). There is also some evidence indicating that root microbiomes can be modulated by the phytohormone salicylic acid, whose role in the activation of defense responses is already well known (Lebeis et al. 2015). Similarly, the application of bioorganic fertilizers in banana nursery pots resulted in the manipulation of the rhizospheric microbial structure and subsequently decreased the incidence of Panama disease on banana (Xue et al. 2015). Despite all these successful approaches, more research is still required to better understand the impact of synthetic/natural microbial communities, as well as different cropping practices and abiotic parameters, on the microbiome structure and how microbiome shifts are translated to plant health (Müeller and Sachs 2015). In particular, microbiome engineering for improving woody tree crop health is a largely untapped area that deserves major research efforts.

2.4 Challenges for the Biocontrol of Woody Tree Crop Diseases

The efficiency of phyllospheric bacteria in controlling aerial diseases in woody tree crops, under field conditions, is often affected by several abiotic and biotic factors (Fig. 2.3). Specifically, some aspects related to the biocontrol agent (i.e., method of their application; Silva and De Costa 2014; Kupper et al. 2011; Ozaktan et al. 2011), pathogen (i.e., strain; Abraham et al. 2013), host plant (i.e., cultivar or organ; Gerami et al. 2013; Puneeth 2015), environment (i.e., weather conditions; Pusey 2002), and microbiome (Xue et al. 2015) have been described to play an important role in the

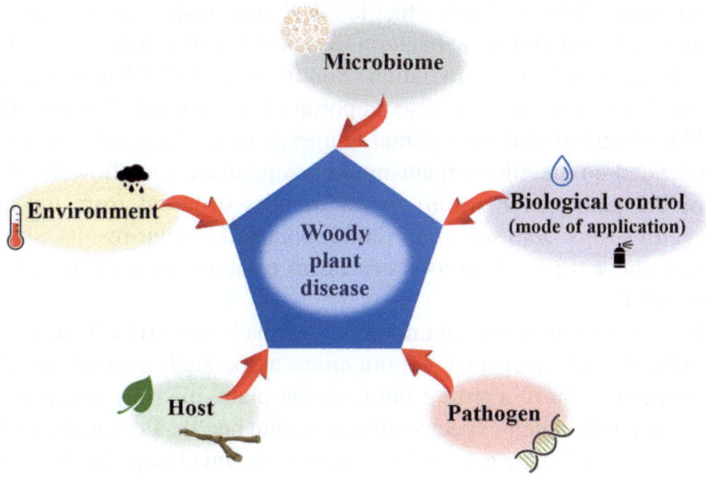

Fig. 2.3 Abiotic and biotic factors with recognized influence in the efficiency of biological control against aerial diseases of woody tree crops

efficiency of biocontrol agents against diseases affecting the phyllosphere of woody crop trees.

Concerning the method of biocontrol application, Silva and De Costa (2014) recognized that *Burkholderia spinosa* was more effective in suppressing the abundance of potential fungal pathogens (*Aspergillus* and *Fusarium*) on banana plants when applied as a foliar spray than as a soil drench. The application of plant regulators together with the bacterial agents was also reported to enhance their biocontrol ability. For example, when prohexadione-calcium was applied together with *Pantoea vagans* to the phyllosphere of pear tree, a higher biocontrol of *Erwinia amylovora* was achieved (Ozaktan et al. 2011). Similarly, complex microbial inoculums can improve biocontrol activity when compared with individual inoculums. For instance, Kupper et al. (2011) reported a synergistic effect of a complex mixture of several strains of *Bacillus subtilis* against *Guignardia citricarpa* on citrus tree.

Another limitation to effective bacterial biocontrol on woody crops is related to the specificity between the biocontrol agent and pathogenic strain. On *Hevea brasiliensis*, several strains of the same potential antagonist were reported to differently affect the same pathogenic agent (Abraham et al. 2013). In addition, depending on the host plant (e.g., type of cultivar) (Gerami et al. 2013) or target organ (Puneeth 2015), different effects were observed on the biocontrol of crop diseases.

The performance of a microbial control agent is widely influenced by the environment into which the antagonist is introduced. This is particularly important when biocontrol agents are applied on the aboveground parts of plants, especially for those diseases caused by airborne microorganisms. Indeed, the phyllosphere (in particular phylloplane) is a harsh environment for microorganisms to survive. Accordingly, the destructive influence of UV light was already reported to be a limiting factor for the application of potential biocontrol bacteria

(Jacobs and Sundin 2001). On the other hand, *Pseudomonas syringae pv. syringae*, a phytopathogenic and epiphytic bacteria associated with mango tree surfaces, was reported to be resistant to UV radiation (Sundin et al. 1996; Cazorla et al. 2008). Environmental temperatures are also important for bacterial thriving. Pusey and Curry (2004) observed that the optimal temperature for bacterial growth on apple flowers depended on the microorganism; the temperature that allowed a population increase of the pathogen *Erwinia amylovora* was different from the antagonist *Pantoea agglomerans*. Johnson et al. (2000) obtained similar results, showing that temperatures above 12 °C lead to a successful establishment of the antagonist *P. fluorescens* A506.

A further complication for obtaining the highest biocontrol efficiency is related to the complexity of microbial communities associated with plants. Microbe–microbe interactions have a strong influence on plant–microbe interactions (Kroll et al. 2017), disguising their expected effect on plant health. Taking this into account, the understanding of such a complex interaction (involving the host, pathogen, biocontrol agent, host microbiota, and environment) would be the major key to move forward to control woody plant diseases. Novel tools and technologies are being developed to provide deeper insights into the plant microbiome, as well as into microbe–microbe and microbe–plant interactions. In the first approach, the host core microbiome (i.e., microbial taxa consistently present in a healthy host) of phyllospheric woody tree crop should be identified and correlated with host health. This correlation could then be ascertained by employing, for example, metatranscriptomics and metaproteomics approaches that would infer the functional properties of the host core microbial community. This knowledge could help to fully understand the impact that core microbes have on woody crop tree health, revealing also strategies for microbiome engineering.

2.5 Conclusion

In the past 10 years, researchers have developed a much more in-depth and detailed understanding of how phyllosphere-associated bacteria can improve host plant health. However, such knowledge is so far higher for herbaceous crops than for woody crops. While more research work, both basic and applied, remains to be done, native phyllosphere-associated bacterial members are already being successfully used as biocontrol agents of some woody tree crop diseases, albeit on a small scale. Further studies are still required for enhancing the knowledge on the composition of microbial communities in the phyllosphere of woody trees, the factors shaping their assemblages, and their role/function in plant health. New approaches, such as omics technologies, can provide greater advances on all these aspects. Research efforts should also be carried on for elucidating the effects of inoculation with bacterial biocontrol agents (specific microbial strains, synthetic communities, or natural communities) on the management of woody crop diseases. Trials to identify efficient antagonists should be performed in conditions that mimic the host environment as much as possible, preferably using *in planta* assays. In this way, not only the antagonistic mechanisms that occur in natural environment would be unraveled, but the behavior of antagonistic (and pathogen) microorganisms would be evaluated (Pliego et al. 2006). Few efforts have been given on the manipulation of the microbiome to control woody crop diseases. Such approach could be further used to modulate intentionally the microbiome, recruiting disease antagonists, a process denoted as bioengineering. This could be also an interesting option for the management of woody crop diseases, in a more sustainable way.

Acknowledgments This work was partially funded by European Structural and Investment Funds in the FEDER component, through the Operational Competitiveness and Internationalization Programme (COMPETE 2020); and national funds, through the FCT – Portuguese Foundation for Science and Technology under the project POCI-01-0145-FEDER-031133. J.D. Mina thanks FCT, POPH-QREN and FSE for PhD grant SFRH/BD/105341/2014.

References

Abraham A, Philip S, Jacob CK, Jayachandran K (2013) Novel bacterial endophytes from *Hevea brasiliensis* as biocontrol agent against Phytophthora leaf fall disease. BioControl 58(5):675–684. https://doi.org/10.1007/s10526-013-9516-0

Araújo WL, Marcon J, Maccheroni W, van Elsas JD, van Vuurde JWL, Azevedo JL (2002) Diversity of endophytic bacterial populations and their interaction with *Xylella fastidiosa* in citrus plants. Appl Environ Microbiol 68(10):4906–4914. https://doi.org/10.1128/AEM.68.10.4906-4914.2002

Bacon CW, Hinton DM (2002) Endophytic and biological control potential of *Bacillus mojavensis* and related species. Biol Control 23:274–284. https://doi.org/10.1006/bcon.2001.101

Baldotto LE, Olivares FL (2008) Phylloepiphytic interaction between bacteria and different plant species in a tropical agricultural system. Can J Microbiol 54(11):918–931. https://doi.org/10.1139/w08-087

Buana RFN, Wahyudi AT, Toruan-Mathius N (2014) Control activity of potential antifungal-producing *Burkholderia* sp. in suppressing *Ganoderma boninense* growth in oil palm. AJAR 8:259–268. https://doi.org/10.3923/ajar.2014.259.268

Bulgarelli D, Schlaeppi K, Spaepen S, van Themaat EVL, Schulze-Lefert P (2013) Structure and functions of the bacterial microbiota of plants. Annu Rev Plant Biol 64:807–838. https://doi.org/10.1146/annurev-arplant-050312-120106

Cabrefiga J, Bonaterra A, Montesinos E (2007) Mechanisms of antagonism of *Pseudomonas fluorescens* EPS62e against *Erwinia amylovora*, the causal agent of fire blight. Int Microbiol 10(2):123–132

Carper DL, Carrell AA, Kueppers LM, Frank AC (2018) Bacterial endophyte communities in *Pinus flexilis* are structured by host age, tissue type, and environmental factors. Plant Soil 428(1–2):335–352. https://doi.org/10.1007/s11104-018-3682-x

Carvalho S, Castillo J (2018) Influence of light on plant–Phyllosphere interaction. Front Plant Sci 9. https://doi.org/10.3389/fpls.2018.01482

Cazorla FM, Mercado-Blanco J (2016) Biological control of tree and woody plant diseases: an impossible task? BioControl 61(3):233–242. https://doi.org/10.1007/s10526-016-9737-0

Cazorla FM, Codina JC, Abad C, Arrebola E, Torés JA, Murillo J, Pérez-García A, de Vicente A (2008) 62-kb plasmids harboring rulAB homologues confer UV-tolerance and epiphytic fitness to *Pseudomonas syringae* pv. *syringae* mango isolates. Microb Ecol 56(2):283–291. https://doi.org/10.1007/s00248-007-9346-7

Chaurasia B, Pandey A, Palni LM, Trivedi P, Kumar B, Colvin N (2005) Diffusible and volatile compounds produced by an antagonistic *Bacillus subtilis* strain cause structural deformations in pathogenic fungi *in vitro*. Microbiol Res 160(1):75–81. https://doi.org/10.1016/j.micres.2004.09.013

Ciancio A, Roccuzzo G, Ornat Longaron C (2016) Regulation of the citrus nematode *Tylenchulus semipenetrans* by a *Pasteuria* sp. endoparasite in a naturally infested soil. BioControl 61(3):337–347. https://doi.org/10.1007/s10526-015-9704-1

Cregger MA, Veach AM, Yang ZK, Crouch MJ, Vilgalys R, Tuskan GA, Schadt CW (2018) The Populus holobiont: dissecting the effects of plant niches and genotype on the microbiome. Microbiome 6:31. https://doi.org/10.1186/s40168-018-0413-8

De Vleesschauwer D, Höfte M (2007) Using *Serratia plymuthica* to control fungal pathogens of plants. CAB Rev 2(46). https://doi.org/10.1079/PAVSNNR20072046

Duffy BK, Défago G (1999) Environmental factors modulating antibiotic and siderophore biosynthesis by *Pseudomonas fluorescens* biocontrol strains. Appl Environ Microbiol 65:2429–2438

El-Sayed A, Akbar A, Iqrar I, Ali R, Norman D, Brennan M, Ali GS (2018) A glucanolytic *Pseudomonas* sp. associated with Smilax bona-nox L. displays strong activity against *Phytophthora parasitica*. Microbiol Res 207:140–152. https://doi.org/10.1016/j.micres.2017.11.018

Finkel OM, Burch AY, Lindow SE, Post AF, Belkin S (2011) Geographical location determines the population structure in phyllosphere microbial communities of a salt-excreting desert tree. Appl Environ Microbiol 77(21):7647–7655. https://doi.org/10.1128/AEM.05565-11

Gerami E, Hassanzadeh N, Abdollahi H, Ghasemi A, Heydari A (2013) Evaluation of some bacterial antagonists for biological control of fire blight disease. J Plant Pathol 95(1):127–134. https://doi.org/10.4454/JPP.V95I1.026

Gopal M, Gupta A, Thomas GV (2013) Bespoke microbiome therapy to manage plant diseases. Front Microbiol 4:355. https://doi.org/10.3389/fmicb.2013.00355

Hamonts K, Trivedi P, Garg A, Janitz C, Grinyer J, Holford P, Botha FC, Anderson IC, Singh BK (2018) Field study reveals core plant microbiota and relative importance of their drivers. Environ Microbiol 20(1):124–140. https://doi.org/10.1111/1462-2920.14031

He YH, Isono S, Shibuya M, Tsuji M, Purushothama CA, Tanaka K, Sano T (2012) Oligo-DNA custom macroarray for monitoring major pathogenic and non-pathogenic fungi and bacteria in the phyllosphere of apple trees. PLoS One 7(3):e34249. https://doi.org/10.1371/journal.pone.0034249

Hernández-León R, Rojas-Solís D, Contreras-Pérez M, Orozco-Mosqueda MC, Macías-Rodríguez LI, de la Cruz HR, Valencia-Cantero E, Santoyo G (2015) Characterization of the antifungal and plant growth-promoting effects of diffusible and volatile organic compounds produced by *Pseudomonas fluorescens* strains. Biol Control 81:83–92. https://doi.org/10.1016/j.biocontrol.2014.11.011

Huang CJ, Tsay JF, Chang SY, Yang HP, Wu WS, Chen CY (2012) Dimethyl disulfide is an induced systemic resistance elicitor produced by *Bacillus cereus* C1L. Pest Manag Sci 68(9):1306–1310. https://doi.org/10.1002/ps.3301

Ibáñez F, Tonelli ML, Muñoz V, Figueredo MS, Fabra A (2017) Bacterial endophytes of plants: diversity, invasion mechanisms and effects on the host. In: Maheshwari D (ed) Endophytes: biology and biotechnology. Sustainable development and biodiversity. Springer, Cham, pp 25–40

Ishimaru CA, Klos EJ, Brubaker RR (1988) Multiple antibiotic production by *Erwinia herbicola*. Phytopathology 78:746–750

Jacobs JL, Sundin GW (2001) Effect of solar UV-B radiation on a phyllosphere bacterial community. Appl Environ Microbiol 67(12):5488–5496. https://doi.org/10.1128/AEM.67.12.5488-5496.2001

Janisiewicz WJ, Roitman J (1988) Biological control of blue mold and gray mold on apple and pear with *Pseudomonas cepacian*. Phytopathology 78(12):1697–1700

Jo Y, Cho JK, Choi H, Chu H, Lian S, Cho WK (2015) Bacterial communities in the phylloplane of *Prunus* species. J Basic Microbiol 55(4):504–508. https://doi.org/10.1002/jobm.201400651

Johnson KB, Stockwell VO, Sawyer TL, Sugar D (2000) Assessment of environmental factors influencing growth and spread of *Pantoea agglomerans* on and among blossoms of pear and apple. Phytopathology 90:1285–1294. https://doi.org/10.1094/PHYTO.2000.90.11.1285

Kembel SW, O'Connor TK, Arnold HK, Hubbell SP, Wright SJ, Green JL (2014) Relationships between phyllosphere bacterial communities and plant functional traits in a neotropical forest. Proc Natl Acad Sci 111(38):13715–13720. https://doi.org/10.1073/pnas.1216057111

Kloepper JW, Ryu CM, Zhang SA (2004) Induced systemic resistance and promotion of plant growth by *Bacillus* spp. Phytopathology 94(11):1259–1266. https://doi.org/10.1094/PHYTO.2004.94.11.1259

Korsten L, De Villiers EE, Wehner FC, Kotzé JM (1997) Field sprays of *Bacillus subtilis* and fungicides for control of preharvest fruit diseases of avocado in South Africa. Plant Dis 81(5):455–459. https://doi.org/10.1094/PDIS.1997.81.5.455

Kroll S, Tagler M, Kemen E (2017) Genomic dissection of host–microbe and microbe–microbe interactions for advanced plant breeding. Curr Opin Plant Biol 90(11):1285–1294. https://doi.org/10.1094/PHYTO.2000.90.11.1285

Kumar P, Dubey RC, Maheshwari DK (2012) *Bacillus* strains isolated from rhizosphere showed plant growth promoting and antagonistic activity against phytopathogens. Microbiol Res 167(8):493–499. https://doi.org/10.1016/j.micres.2012.05.002

Kupper KC, Correa EB, Moretto C, Bettiol W, De Goes A (2011) Control of *Guignardia citricarpa* by *Bacillus subtilis* and *Trichoderma* spp. Rev Bras Frutic 33(4):1111–1118. https://doi.org/10.1590/S0100-29452011000400009

Laforest-Lapointe I, Messier C, Kembel SW (2016) Tree phyllosphere bacterial communities: exploring the magnitude of intra- and inter-individual variation among host species. PeerJ 4:e2367. https://doi.org/10.7717/peerj.2367

Lambais MR, Crowley DE, Cury JC, Büll RC, Rodrigues RR (2006) Bacterial diversity in tree canopies of the Atlantic forest. Science 312(5782):1917. https://doi.org/10.1126/science.1124696

Larsen OFA, Claassen E (2018) The mechanistic link between health and gut microbiota diversity. Sci Rep 2183(8):2045–2322. https://doi.org/10.1038/s41598-018-20141-6

Leach JE, Triplett LR, Argueso CT, Trivedi P (2017) Communication in the phytobiome. Cell 169(4):587–596. https://doi.org/10.1016/j.cell.2017.04.025

Lebeis SL, Paredes SH, Lundberg DS, Breakfield N, Gehring J, McDonald M, Malfatti S, Glavina del Rio T, Jones CD, Tringe SG, Dangl JL (2015) Salicylic acid modulates colonization of the root microbiome by specific bacterial taxa. Science 349(6250):860–864. https://doi.org/10.1126/science.aaa8764

Lemanceau P, Blouin M, Müller D, Moënne-Loccoz Y (2017a) Let the core microbiota be functional. Trends Plant Sci 22(7):583–595. https://doi.org/10.1016/j.tplants.2017.04.008

Lemanceau P, Barret M, Mazurier S, Mondy S, Pivato B, Fort T, Vacher C (2017b) Plant communication with associated microbiota in the spermosphere, rhizosphere and phyllosphere. In: G. Becard (eds) How plants communicate with their biotic environment. London: Academic. Adv Bot Res 82:101–133. https://doi.org/10.1016/bs.abr.2016.10.007

Li Q, Xiong C, Li X, Jin X, Huang W (2018) Ectomycorrhization of *Tricholoma matsutake* with *Quercus aquifolioides affects* the endophytic microbial community of host plant. J Basic Microbiol 58(3):238–246. https://doi.org/10.1002/jobm.201700506

Lindow SE, Brandl MT (2003) Microbiology of the phyllosphere. Appl Environ Microbiol 69(4):1875–1883. https://doi.org/10.1128/AEM.69.4.1875-1883.2003

Malfanova N, Franzil L, Lugtenberg B, Chebotar V, Ongena M (2012) Cyclic lipopeptide profile of the plant-beneficial endophytic bacterium *Bacillus subtilis* HC8. Arch Microbiol 194:893–899. https://doi.org/10.1007/s00203-012-0823-0

Martínez-Hidalgo P, Maymon M, Pule-Meulenberg F, Hirsch AM (2018) Engineering root microbiomes for healthier crops and soils using beneficial, environmentally safe bacteria. Can J Microbiol 18:1–14. https://doi.org/10.1139/cjm-2018-0315

McIntyre JL, Kuc J, Williams EB (1973) Protection of pears against fireblight by bacteria and bacterial sonicates. Phytopathology 63:872–877

Meyer KM, Leveau JH (2012) Microbiology of the phyllosphere: a playground for testing ecological concepts. Oecologia 168(3):621–629. https://doi.org/10.1007/s00442-011-2138-2

Michavila G, Adler C, De Gregorio PR, Lami MJ, Caram Di Santo MC, Zenoff AM, Cristobal RE, Vincent PA (2017) *Pseudomonas protegens* CS1 from the lemon phyllosphere as a candidate for citrus canker biocontrol agent. Plant Biol J 19(4):608–617. https://doi.org/10.1111/plb.12556

Mikiciński A, Sobiczewski P, Puławska J, Maciorowski R (2016) Control of fire blight (*Erwinia amylovora*) by a novel strain 49M of *Pseudomonas graminis* from the phyllosphere of apple (*Malus* spp.). Eur J Plant Pathol 145(2):265–276. https://doi.org/10.1007/s10658-015-0837-y

Müller T, Ruppel S (2014) Progress in cultivation-independent phyllosphere microbiology. FEMS Microbiol Ecol 87(1):2–17. https://doi.org/10.1111/1574-6941.12198

Müller UG, Sachs JL (2015) Engineering microbiomes to improve plant and animal health. Trends Microbiol 23(10):606–617. https://doi.org/10.1016/j.tim.2015.07.009

Müller H, Berg C, Landa BB, Auerbach A, Moissl-Eichinger C, Berg G (2015) Plant genotype-specific archaeal and bacterial endophytes but similar *Bacillus* antagonists colonize Mediterranean olive trees. Front Microbiol 6:138. https://doi.org/10.3389/fmicb.2015.00138

Newton AC, Gravouil C, Fountaine JM (2010) Managing the ecology of foliar pathogens: ecological tolerance in crops. Ann Appl Biol 157(3):343–359. https://doi.org/10.1111/j.1744-7348.2010.00437.x

O'Brien PA (2017) Biological control of plant diseases. Australasian Plant Pathol 46(4):293–304. https://doi.org/10.1007/s13313-017-0481-4

Ongena M, Jacques P (2008) *Bacillus* lipopeptides: versatile weapons for plant disease biocontrol. Trends Microbiol 16(3):115–125. https://doi.org/10.1016/j.tim.2007.12.009

Orozco-Mosqueda MC, Rocha-Granados MC, Glick BR (2018) Microbiome engineering to improve biocontrol and plant growth-promoting mechanisms. Microbiol Res 208:25–31. https://doi.org/10.1016/j.micres.2018.01.00

Ozaktan H, Akkopru A, Aslan E, Ilhan K, Koltuksuz T (2011) Integrated control of fire blight in a pear orchard in Turkey using prohexadione-ca and bacterial antagonists. Acta Hortic 896:441–446. https://doi.org/10.17660/actahortic.2011.896.64

Ozaktan H, Erdal M, Akkopru A, Aslan E (2012) Biological control of bacterial blight of walnut by antagonistic bacteria. J Plant Pathol 94(1):53–56

Pandolfi V, Jorge EC, Melo CMR, Albuquerque ACS, Carrer H (2010) Gene expression profile of the plant pathogen *Fusarium graminearum* under the antagonistic effect of *Pantoea agglomerans*. Genet Mol Res 9(3):1298–1311. https://doi.org/10.4238/vol9-3gmr828

Pasche S, Crovadore J, Pelleteret P, Jermini M, Mauch-Mani B, Oszako T, Lefort F (2016) Biological control of the latent pathogen *Gnomoniopsis smithogylvyi* in European chestnut grafting scions using *Bacillus amyloliquefaciens* and *Trichoderma atroviride*. Dendrobiology 75:113–122. https://doi.org/10.12657/denbio.075.011

Passera A, Alizadeh H, Azadvar M, Quaglino F, Alizadeh A, Casati P, Bianco PA (2018) Studies of microbiota dynamics reveals association of "*Candidatus* Liberibacter asiaticus" infection with citrus (*Citrus sinensis*) decline in south of Iran. Int J Mol Sci 19(6):1817. https://doi.org/10.3390/ijms19061817

Pérez-García A, Romero D, Vicente A (2011) Plant protection and growth stimulation by microorganisms: biotechnological applications of Bacilli in agriculture. Curr Opin Biotechnol 22(2):187–193. https://doi.org/10.1016/j.copbio.2010.12.003

Pham VD, Konstantinidis KT, Palden T, Delong EF (2008) Phylogenetic analyses of ribosomal DNA-containing bacterioplankton genome fragments from a 4000 m vertical profile in the North Pacific Subtropical Gyre. Environ Microbiol 10(9):2313–2330. https://doi.org/10.1111/j.1462-2920.2008.01657.x

Pliego C, De Weert S, Lamers GEM, Bloemberg G, Cazorla FM, Ramos C (2006) Ocupación diferencial de la rizosfera de aguacate por cepas de *Pseudomonas* spp. antagonistas frente a *Rosellinia necatrix*. In: Proceedings of the XIII Phytopathological Spanish Society Congress. Murcia, Spain, p 127

Puneeth ME (2015) Biocontrol of bacterial blight of pomegranate caused by *Xanthomonas axonopodis* pv. *punicae* (Hingorani and Singh) Vauterin *et al.* Department of plant pathology, University of agricultural sciences, Bengaluru

Pusey PL (2002) Biological control agents for fire blight of apple compared under conditions limiting natural dispersal. Plant Dis 86(6):639–644. https://doi.org/10.1094/PDIS.2002.86.6.639

Pusey PL, Curry EA (2004) Temperature and pomaceous flower age related to colonization by *Erwinia amylovora* and antagonists. Phytopathology 94(8):901–911. https://doi.org/10.1094/PHYTO.2004.94.8.901

Qvit-Raz N, Finkel OM, Al-Deeb TM, Malkawi HI, Hindiyeh MY, Jurkevitch E, Belkin S (2012) Biogeographical diversity of leaf-associated microbial communities from salt-secreting Tamarix trees of the Dead Sea region. Res Microbiol 163(2):142–150. https://doi.org/10.1016/j.resmic.2011.11.006

Rahman SFSA, Singh E, Pieterse CMJ, Schenk PM (2018) Emerging microbial biocontrol strategies for plant pathogens. Plant Sci 267:102–111. https://doi.org/10.1016/j.plantsci.2017.11.012

Rastogi G, Coaker GL, Leveau JH (2013) New insights into the structure and function of phyllosphere microbiota through high-throughput molecular approaches. FEMS Microbiol Lett 348(1):1–10. https://doi.org/10.1111/1574-6968.12225

Redford AJ, Fierer N (2009) Bacterial succession on the leaf surface: a novel system for studying successional dynamics. Microb Ecol 58(1):189–198. https://doi.org/10.1007/s00248-009-9495-y

Roat C, Saraf M (2017) Unravelling the interaction of plant and their phyllosphere microbiome. In: Singh R, Kothari R, Koringa P, Singh S (eds) Understanding host-microbiome interactions – an omics approach. Springer, Singapore, pp 157–172

Rossmann B, Müller H, Smalla K, Mpiira S, Tumuhairwe JB, Staver C, Berg G (2012) Banana-associated microbial communities in Uganda are highly diverse but dominated by *Enterobacteriaceae*. Appl Environ Microbiol 78(14):4933–4941. https://doi.org/10.1128/AEM.00772-12

Sahu PK, Singh DP, Prabha R, Meena KK, Abhilash PC (2018) Connecting microbial capabilities with the soil and plant health: options for agricultural sustainability. Ecol Indic. https://doi.org/10.1016/j.ecolind.2018.05.084

Santhanam R, Luu VT, Weinhold A, Goldberg A (2015) Native root-associated bacteria rescue a plant from a sudden-wilt disease that emerged during continuous cropping. Proc Natl Acad Sci U S A 112(36):E5013–E5020. https://doi.org/10.1073/pnas.1505765112

Sasirekha B, Srividya S (2016) Siderophore production by *Pseudomonas aeruginosa* FP6, a biocontrol strain for *Rhizoctonia solani* and *Colletotrichum gloeosporioides* causing diseases in chilli. AGNR 50(4):250–256. https://doi.org/10.1016/j.anres.2016.02.003

Sayyed RZ, Patel PR (2011) Biocontrol potential of siderophore producing heavy metal resistant *Alcaligenes* sp. and *Pseudomonas aeruginosa* RZS3 vis-a-vis organophosphorus fungicide. Indian J Microbiol 51(3):266–272. https://doi.org/10.1007/s12088-011-0170-x

Silva YMUKY, De Costa DM (2014) Potential of pre-harvest application of *Burkholderia spinosa* for biological control of epiphytic and pathogenic microorganisms on the phyllosphere of banana (*Musa* spp.). Trop Agric Res 25(4):443–454. https://doi.org/10.4038/tar.v25i4.8060

Silva HSA, Tozzi JPL, Terrasan CRF, Bettiol W (2012) Endophytic microorganisms from coffee tissues as plant growth promoters and biocontrol agents of coffee leaf rust. Biol Control 63(1):62–67. https://doi.org/10.1016/j.biocontrol.2012.06.005

Singh P, Piotrowski M, Kloppstech K, Gau AE (2004) Investigations on epiphytic living Pseudomonas species from *Malus domestica* with an antagonistic effect to *Venturia inaequalis* on isolated plant cuticle membranes. Environ Microbiol 6(11):1149–1158. https://doi.org/10.1111/j.1462-2920.2004.00622.x

Subagio A, Foster HL (2003) Implications of *Ganoderma* disease on loss in stand and yield production of oil palm in North Sumatra. In: Proceedings of MAPPS Conference, Kuala Lumpur, Malaysia

Sundin GW, Kidambi SP, Ullrich M, Bender CL (1996) Resistance to ultraviolet light in *Pseudomonas syringae*: sequence and functional analysis of the plasmid-encoded rulAB genes. Gene 177(1–2):77–81

Thomas P, Soly TA (2009) Endophytic bacteria associated with growing shoot tips of banana (*Musa* sp.) cv. Grand Naine and the affinity of endophytes to the host. Microb Ecol 58(4):952–964. https://doi.org/10.1007/s00248-009-9559-z

Touré Y, Ongena M, Jacques P, Guiro A, Thonart P (2004) Role of lipopeptides produced by *Bacillus subtilis* GA1 in the reduction of grey mould disease caused by *Botrytis cinerea* on apple. J Appl Microbiol 96(5):1151–1160. https://doi.org/10.1111/j.1365-2672.2004.02252.x

Trivedi P, Spann T, Wang N (2011) Isolation and characterization of beneficial bacteria associated with citrus roots in Florida. Microb Ecol 62(2):324–336. https://doi.org/10.1007/s00248-011-9822-y

Utkhede RS (1987) Chemical and biological control of crown and root rot of apple caused by *Phytophthora cactorum*. Can J Plant Pathol 9(4):295–300. https://doi.org/10.1080/07060668709501860

Vaidya RJ, Shah IM, Vyas PR, Chhatpar HS (2001) Production of chitinase and its optimization from a novel isolate *Alcaligenes xylosoxydans*: potential in antifungal biocontrol. World J Microbiol Biotechnol 17(7):691–696. https://doi.org/10.1023/A:1012927116756

Valverde A, González-Tirante M, Medina-Sierra M, Rivas R, Santa-Regina I, Igual JM (2017) Culturable bacterial diversity from the chestnut (*Castanea sativa* Mill.) phyllosphere and antagonism against the fungi causing the chestnut blight and ink diseases. AIMS Microbiol 3(2):293–314. https://doi.org/10.3934/microbiol.2017.2.293

Vanneste JL (1996) Honey bees and epiphytic bacteria to control fire blight, a bacterial disease of apple and pear. Biocontrol News Inf 17(4):67–78

Vega FE, Pava-Ripoll M, Posada F, Buyer JS (2005) Endophytic bacteria in *Coffea arabica* L. J Basic Microbiol 45(5):371–380. https://doi.org/10.1002/jobm.200410551

Vorholt JA (2012) Microbial life in the phyllosphere. Nat Rev Microbiol 10(12):828–840. https://doi.org/10.1038/nrmicro2910

Weller DM (2007) *Pseudomonas* biocontrol agents of soilborne pathogens: looking back over 30 years. Phytopathology 97(2):250–256. https://doi.org/10.1094/PHYTO-97-2-0250

Wilhelm E, Arthofer W, Schafleitner R, Krebs B (1998) *Bacillus subtilis*, an endophyte of chestnut (*Castanea sativa*), as antagonist against chestnut blight (*Cryphonectria parasitica*). Plant Cell Tiss Org 52(1–2):105–108

Wrather JA, Kuc J, Williams EB (1973) Protection of apple and pear fruit tissue against fireblight with nonpathogenic bacteria. Phytopathology 63:1075–1076

Xue C, Penton CR, Shen Z, Zhang R, Huang Q, Li R, Ruan Y, Shen Q (2015) Manipulating the banana rhizosphere microbiome for biological control of Panama disease. Nature 5(11124). https://doi.org/10.1038/srep11124

Yashiro E, McManus PS (2012) Effect of streptomycin treatment on bacterial community structure in the apple Phyllosphere. PLoS One 7(5):e37131. https://doi.org/10.1371/journal.pone.0037131

Yashiro E, Spear R, McManus P (2011) Culture-dependent and culture-independent assessment of bacteria in the apple phyllosphere. J Appl Microbiol 110(5):1284–1296. https://doi.org/10.1111/j.1365-2672.2011.04975.x

Yenjit P, Intanoo W, Chamswarng C, Siripanich J, Intana W (2004) Use of promising bacterial strains for controlling anthracnose on leaf and fruit of mango caused by *Colletotrichum gloeosporioides*. Walailak J Sci Tech 1(2):56–69. https://doi.org/10.2004/wjst.v1i2.186

Young (2017) The role of the microbiome in human health and disease: an introduction for clinicians. BMJ 356:j831. https://doi.org/10.1136/bmj.j831

Yuan J, Raza W, Huang Q, Shen Q (2012) The ultrasound-assisted extraction and identification of antifungal substances from *Bacillus amyloliquefaciens* strain NJN-6 suppressing *Fusarium oxysporum*. J Basic Microbiol 52:721–730. https://doi.org/10.1002/jobm.201100560

Zengerer V, Schmid M, Bieri M, Müller DC, Remus-Emsermann MNP, Ahrens CH, Pelludat C (2018) *Pseudomonas orientalis* F9: a potent antagonist against phytopathogens with phytotoxic effect in the apple flower. Front Microbiol 9:145. https://doi.org/10.3389/fmicb.2018.00145

Microbes: An Important Resource for Sustainable Agriculture

3

Kalaivani K. Nadarajah

Abstract

Microbes play an important role in the agricultural industry. While diseases caused by microbes pose a devastating effect to the industry, the application of beneficial microorganisms in agriculture has shown promise in addressing global issues such as disease reduction, yield and growth enhancement, and the reduction in the use of agrochemicals that contaminate the environment. In this chapter, we address (i) various beneficial plant-microbe interactions, (ii) explore the advances made in these beneficial relationships, and finally (iii) identify future directions of research to answer questions in the black box of knowledge pertaining to plant-microbe interaction for sustainable agriculture.

Keywords

Microbiome · Sustainable environment · Plant microbe interactions

3.1 Introduction

The agroecosystem is a complex interaction of many players, which includes above- and below-ground members. This makes the system difficult for control and manipulation by humans (Berendsen et al. 2012; Busby et al. 2017; De Vries and Wallenstein 2017). One subject that is gaining interest over the past decades is soil microbial diversity. Diversity of microbes and the development of several technologies including omics and big data analytical systems have resulted in the advancement of microbiome research. In the past decade, the importance of soil microbiome

K. K. Nadarajah (✉)
School of Environmental and Natural Resources Science, Faculty of Science and Technology, Universiti Kebangsaan Malaysia, Bangi, Selangor, Malaysia
e-mail: vani@ukm.edu.my

V. Kumar et al. (eds.), *Microbiome in Plant Health and Disease*,
https://doi.org/10.1007/978-981-13-8495-0_3

has been implicated in sustainable agriculture. Microbes are involved in plant nutrient uptake (Bonfante and Anca 2009; Elser et al. 2007; Ortiz et al. 2015; Friesen et al. 2011; Hiruma et al. 2016; van der Heijden et al. 1998), resistance toward pests and pathogens, response towards environmental stresses (Busby et al. 2016; Calvo et al. 2014; Mendes et al. 2011; Santhanam et al. 2015; Selosse et al. 2014; Vorholt et al. 2017; Zavala-Gonzalez et al. 2017), and phenology (Wagner et al. 2014). These minute organisms may be an untapped resource in addressing the issue of sustainable agriculture (Berendsen et al. 2012; Busby et al. 2017; de Vries and Wallenstein 2017).

One major contribution in microbiome research is the knowledge garnered from the Earth Microbiome Project. Through a large-scale sequencing project, a reference catalogue was developed for the microbiome in Earth (Thompson et al. 2017). The potential of metagenomics in deciphering plant-associated microbiome is likely to dwarf the genomic abilities of plants with the ability to improve host function. This has mainly been the reason behind the drive to incorporate microbes into agricultural systems to improve efficiency of the plant production systems (Bakker et al. 2012; Mueller and Sachs 2015; Schlaeppi and Bulgarelli 2015). Researchers have explored the terrains of plant microbiome function and structure in model and non-model plant systems in both natural and controlled environments. Some of the plant systems that have been studied are *Arabidopsis thaliana*, barley, soybean, corn, rice, wheat, and cottonwood trees (Aira et al. 2010; Bulgarelli et al. 2012; Delmotte et al. 2009; Donn et al. 2015; Edwards et al. 2015; Knief et al. 2012; Mendes et al. 2014; Rascovan et al. 2016). While many plant systems are being studied by researchers and academics, there is a disconnect in the flow of information from the research front into the hands of the farmers. The farmers need to be involved in the utilization of microorganisms for sustainable agriculture. Their field exposure coupled with the bench/laboratory-based knowledge may contribute to better application of this science in agriculture (Kavamura et al. 2013). In the recent years, research has been focused on the identification and application of single-microbe inoculation to improve crop nutrient uptake, stress management, and growth (Harman et al. 2004; Kandula et al. 2015). These applications have met with different success in field trials. This is largely due to the complexity of the microbial soil communities and the effect of the environment on the microbial population. However, there should be focus on understanding the efficacy of the single strains and to build a consortium that can function collectively in providing enhanced growth, better development, and resistance to both biotic and abiotic stresses. Most often, the research involving single microbial cultures (e.g., nitrogen fixing and mycorrhizae) has focused on the functional role of a particular microbial group and the associated plants (Andreote et al. 2009).

This chapter takes stock of all that has been done in soil microbiome and what has been derived on the role of microorganisms in facilitating sustainable agriculture. This chapter also addresses research priorities for the next decade and issues that need to be addressed in the utilization of soil microbiome in sustainable agriculture.

3.2 The Soil Microbiome

There are a plethora of microorganisms that inhabit diverse habitats from forest to agroecosystems (Robertson et al. 1994). One of the richest parts of the soil is the rhizosphere. The organisms within a rhizosphere are influenced by soil type, environmental conditions, plant type and developmental stages (Huang et al. 2014; van der Heijden and Schlaeppi 2015; İnceoğlu et al. 2011; Lebeis et al. 2015; Philippot et al. 2013; Turner et al. 2013; Raaijmakers et al. 2009). Soil microorganisms emit chemicals that interact with other microbes and plants in the vicinity. These compounds are perceived by the plant roots which in return excrete plant exudates that may differ according to plant species, ecotypes, and root type (Micallef et al. 2009; Uren 2000). Root exudates are made of sugars, amino acids, fatty acids, and others. These compounds are able to recruit both beneficial and detrimental organisms into the rhizosphere (Badri et al. 2009; Raaijmakers et al. 2009). However, these exudates can vary in their concentration and composition based on changes to the environment, soil, and developmental stages of the plant (De-la-Pena et al. 2010). While these root exudates act as a nutrient source for the microbial community, the amount and composition of this material will alter the microbial population of the soil (Badri et al. 2009; Micallef et al. 2009; Vandenkoornhuyse et al. 2007). Rhizodeposition is a process that comes at a very high energy and carbon expense to the plant. As such it is most likely that for such an investment it is highly likely that these deposits bring about a plethora of benefits such as secretion of growth hormones, disease prevention, and acquisition of nutrient from biochemically active root systems (Hamilton III and Frank 2001; Weisskopf et al. 2006). However, there are also situations where the number of soil microbes are found to decrease in the presence of certain exudates. For instance, invading plant species to an ecosystem have been known to attract pathogens to the native plants of the particular ecosystem as seen in the weed *Centaurea maculosa* and *Chromolaena odorata* on the native grass species. There have also been other reports where native populations of microbes are affected due to invasive species (Broz et al. 2007; Mangla and Callaway 2008; Stinson et al. 2006).

In his report, Zgadzaj et al. (2016) stated that there was a correlation between genetic determinants and the host microbial interactions. Here he reported that plants with mutation of the nodulation genes (*nfr5, nin,* and *lhk1*) led to alteration in the bacterial communities between the wild-type and the mutant varieties. Further, Bulgarelli et al. (2015) reported that some microbial families were dominant in barley such as *Rhizobiaceae*, *Flavobacteriaceae*, and *Comamonadaceae*. With the domestication of barley, these communities underwent some change. Meanwhile, Yeoh et al. (2017) concluded that while soil type was a major determinant in soil microbial diversity, plant genotypes also affected the composition of the soil microbial community. In his study, he identified common bacterial genera in a wide range of plant species, resulting in a hypothesis that a core root microbiome has co-evolved with terrestrial plants (Yeoh et al. 2017). Through this investigation, it was concluded that there are main bacterial taxa that live in association with crops, and in these taxa, there are some common clades that provide possible evolutionary perspectives (Chialva and Bonfante 2018). In addition to the bacterial taxa that have been reported in the above studies, fungi are also a crucial component of the soil microbiota that exerts crucial function in plant-associated relationships (Guttman et al. 2014). Mycorrhizal fungi have been outlined as an important member of the plant microbe association (Davison et al. 2015). These communities have been reviewed extensively by Porras-Alfaro and Bayman (2011) and Toju et al. (2013). In these reports, it was highlighted that roots act as compartments for Ascomycetes and Basidiomycetes, (Hacquard 2016) with the main orders being Pleosporales, Agaricales, Sordariales, Hypocreales, and Xylariales (Porras-Alfaro and Bayman 2011). While bacterial populations have been studied extensively, only a little is known with regards to fungal soil population. In recent years, studies have been conducted using molecular techniques to determine the fungal microbial population in model plant systems such as rice, (Wang et al. 2016) and wheat (Rascovan et al. 2016; Chialva and Bonfante 2018)

Further studies on soil microbiome provided information that a large portion of the rhizosphere and the surrounding soil has certain microbial-rich groups (Lundberg et al. 2012). Certain studies suggest that orders such as Sphingobacteriales do not discriminate against plant type, while other orders such as Sphingomonadales are specific about their nutrient source and hence are bound by the type of plants and roots that it inhabits. These findings led to further speculation on the involvement of host in the determination of rhizosphere microbiome (Philippot et al. 2013). An analysis of plant exudates and transcriptome profiling of the soil microbiome within the rhizosphere has provided information that plant exudates are developmentally regulated (Chaparro et al. 2013, 2014). These exudates are also responsible for recruiting soil microorganisms to the rhizosphere to fulfill specific needs of the plant (Philippot et al. 2013; Weinert et al. 2011). Therefore, based on these experiments, it is possible to allude that each plant type may have its own core microbes that allows the plant to optimize nutrient acquisition and address environmental stresses. Currently, there is improvement in technologies that are able to yield more data on the soil structure and microbial diversity. Through the use of metagenomics and metatranscriptomics, a holistic study of soil microbiome is made possible where niche organisms may be identified for any given environment and functional diversities may be determined for soil communities in a short duration. The various technological advancements have been reviewed and their potentials and drawbacks highlighted (Berlec 2012; Chaparro et al. 2012; Dini-Andreote and van Elsas 2013; Rincon-Florez et al. 2013; Turner et al. 2013). Therefore, it is likely that all soil microbiome studies will definitely incorporate molecular techniques alongside culture techniques to obtain the best representation possible of the microbiome. While there are hindrances listed by Dini-Andreote and van Elsas (2013) of the high-throughput sequencing systems in answering fundamental questions on diversity in both spatial and temporal levels, with the advent of new molecular platforms, these questions are gradually being answered.

Other than the microbial and plant exudates, it is impossible to discuss the rhizospheric microbial community without addressing the soil environment. Diverse soil types result in differences in density and diversity of microorganisms (Schloss and Handelsman 2006). The texture, soil nutrient content (C, N, P), and soil pH have largely controlled soil microbiome (Faoro et al. 2010; Frey et al. 2004; Fierer and Jackson 2006; Lauber et al. 2008; Rousk et al. 2010). Of these factors, pH is a crucial element, as it can determine the type of bacteria and fungi that may inhabit the soil (Fierer and Jackson 2006; Rousk et al. 2010). Sensitivity of cells to pH is a critical factor in determining the community structure, density, and diversity (Rousk et al. 2010). There is, however, research that stipulates nutrient content as a limiting factor (Faoro et al. 2010). Many factors converge in environmental control of soil microbial communities. For instance, if the pH is suitable, there could be nutrient deficiencies that limit the microorganisms. However, when the plant-microbe and environment work together, the deficiencies found in the soil can be addressed through the recruitment of microorganisms that are able to fix the nutrient deficiencies (Bonito et al. 2014; Broeckling et al. 2008; Chialva and Bonfante 2018; Vandenkoornhuyse et al. 2015).

3.3 Type of Root-Associated Microorganisms

In examining the plant-microbe associations, there are plant-associated bacteria and fungi that live in different trophic states resulting in beneficial interactions or detrimental effects on the plant host. The microbe-rich plain in the soil is referred to as the rhizosphere where bacteria, fungi, and endophytes forge relationships with plants in a particular ecosystem (Brader et al. 2014; Hardoim et al. 2013; Mercado-Blanco 2015; Ramond et al. 2013). In the following section, we will address the beneficial group of organisms that may be applied to assist with improved yield, growth, development and reduced disease incidences in the agricultural front.

3.4 Beneficial Interactions in Plant-Microbe Interactions

3.4.1 Nutrient Uptake

As frequently reported, the increase in world population and demand for food supply has imposed a strain on the agricultural industry to rise to meet market demands. Consequently, the Green Revolution was experienced in most parts of the world where to address the need for more food, the agricultural industry unremittingly used fertilizers and pesticides to increase the outputs. However, these applications led to detrimental effect to the environment such as the reduced soil microbial population, resulting in poor soil fertility. Hence, more eco-friendly solutions are needed to address soil fertility and plant yield. The beneficial mutualistic interactions between microbes are able to enhance stress tolerance, disease reduction, biodiversity enrichment, and improved growth and yield. Some of these interactions are directly related to acquisition of nutrients necessary not only for good plant growth but also for the maintenance of beneficial microbial population in the soil (Berg and Smalla 2009; Dobbelaere et al. 2003; Morrissey et al. 2014).

Yield and growth can be moderated through microorganisms through direct or indirect methods. Microbes can assist in nutrient acquisition through fixing, mineralization, or decomposition of material. They too are responsible for the reduction of pathogenic microorganisms in the soil through secretion of inhibitors. It is therefore crucial for microbes to be able to colonize plant habitats for efficient plant-microbe interaction (Kamilova et al. 2005). The process is initiated by the plant, which produces exudates that attracts the right kind of microbial populations to colonize the rhizospheric regions (Bais et al. 2006). We can cluster these interactions into three major groups, the first being the beneficial relationship that is built by symbionts like *Rhizobium* species that are a rich component of the rhizosphere. These organisms are involved in the fixing of nitrogen for utilization by the plant (Nadarajah 2016; Nadarajah 2017a, 2017b). The other interaction involves the function of fungi in the form of arbuscular mycorrhiza (AM), which is also able to recruit macro- and micronutrients to the plant (Harrison 1999). Finally, there is the group of microorganisms that are involved in the mineralization of organic matter making minerals such as N, P, and many other micronutrients available to the plant (Hayatsu et al. 2008). There is the nitrogen fixation symbiotically conducted by *Rhizobium* species in the root nodules of leguminous and other free-living soil bacteria like *Burkholderia*, *Stenotrophomonas,* and *Azospirillum* (Dobbelare et al. 2003). The amount of nitrogen level in the soil will determine the diversity and density of nitrogen-fixing organisms that exude chemicals that initiate the process of legume-rhizobia symbiosis (Stacey 2007). Compounds such as flavanols induce the *nod* genes and result in nodulation (Santi et al. 2013). Although individual nitrogen-fixing organisms are able to provide sufficient ammonium supply to the plants, a mixed culture (e.g. *Rhizobium* sp. and *Azotobacter* sp.) generally provides a higher level of nodulation and N_2 fixation. The mixed cultures are known to exude chemical compounds such as exopolysaccharides, daidzein, genistein, and luteolin, which are reported to induce the *nod* gene expression (Jones et al. 2008). These organisms have been reported to result in a plethora of positive interactions such as good shoot and root development, improved water utilization efficiency, improved nutrient uptake, and inhibition of pathogenic and non-beneficial interactions (de Bruijn 2015; Olivares et al. 2013; Santi et al. 2013).

Arbuscular mycorrhizal (AM) fungi form association with roots and assist in the adaptive strategy in stress modulation against both abiotic (drought, salinity, heavy metals, organic pollutants) and biotic (pathogen, insect) stresses. In addition to these, AM is efficient in nutrient acquisition and recycling, making it a useful member of the plant-soil interaction (Jeffries and Barea 2012; Smith and Read 2008; van der Heijden et al. 2015). In studying the mycorrhiza and plant interaction, it has been deemed that the plant provides carbon to these organisms in return for N_2 that is fixed from the atmosphere (Fellbaum et al. 2012). The colonization of soil by AMF changed the chemical composition of roots and soil. This results in physical and environmental modifications, which affects both soil diversity and structure (Barea et al. 2013). AMF is known to induce the defense mechanism in the plant through the increase in chemical defense compounds such as phytoalexins. In the interaction between AMF and plants, several types of flavonoids are exuded. These flavonoids are able to control root colonization, spore germination, and hyphal growth. The function of flavonoids such as strigolactone, glyceollin, coumestrol, and daidzein has been reported in legumes (Akiyama et al. 2005; Steinkellner et al. 2007). In non-leguminous plants however, sugars, carbohydrates, and strigolactone 5-deoxygol are reported to facilitate the AMF-plant interaction (Fang and Leger 2010; Kiers et al. 2011; Yoneyama et al. 2008). AM colonization with N_2 fixer improves plant growth (Nasto et al. 2014). However, (Larimer et al. 2014) experimentally showed that inoculation of AMF and N_2-fixing bacteria did not have an additive effect on growth. Vesicular-arbuscular mycorrhizae (VAM), on the other hand, are involved in phosphorus mobilization. Enhanced P intake improved nodulation and N_2 fixation that contributed towards good root and shoot development in legumes especially when co-inoculated with free-living organisms (Nadarajah 2017a; Nadarajah 2017b; Requena et al. 2001). In short, AMF inoculation of rhizosphere is able to improve symbiotic nitrogen fixation, improve phosphorus mobilization, enhance heavy metal remediation, inhibit pathogens, and improve overall soil health (Barea et al. 2013).

Phosphorus (P) mobilization is important as most P in agroecosystems is in immobilized and inorganic forms. Even when fertilizers are applied, only a small amount is available to the plant. P-mobilizing bacteria are able to turn the inorganic and immobilized forms into solubilized P (Yadav et al. 2014). These organisms not only solubilize but also mobilize P through enzymatic cleavage and translocate the P to the root system (Owen et al. 2015). Potassium (K) is another nutrient required in plant and soil health. However, this nutrient is found in minute amounts in the soil and is usually bound within phyllosilicate structures (Shelobolina et al. 2014). Certain organisms such as *Bradyrhizobium*, *Ralstonia solanacearum*, and *Nocardioides* sp. are able to oxidize iron and free the K for plant use. Further, certain acids like citric, oxalic, and succinic acids, when produced by certain *Bradyrhizobium* sp., are able to mobilize K from K-containing minerals (Sheng and He 2006). Similarly, fungi are also able to solubilize K from minerals through the production of citrate, malate, and oxalate (Meena et al. 2014; Sieverding et al. 2014). Improved mobilization was observed in degraded soils when AMF was inoculated into acidic soil samples (Clark et al. 1999). Two fungal species that have been used extensively in K mobilization are *Aspergillus terreus* and *Aspergillus niger* (Prajapati et al. 2012). Fungi and AMF have also been implicated in improving Cu, Zn, B, Mn, and Fe uptake from the soil. Siderophores produced by some of these organisms have been reported to assist in the mechanism of nutrient uptake and inhibition of soil pathogens (Djonović et al. 2006; Shoresh et al. 2010). Microorganisms from the genera *Azospirillum*, *Serratia*, *Streptomyces,* and *Trichoderma* have been extensively studied for their mode of interaction in antagonizing pathogens and regulating plant growth (De Vleeschauwer and Hofte 2007; Schrey and Tarkka 2008).

3.4.2 Disease and Pest Suppression

Two methods by which microorganisms are able to reduce disease incidence in plants are through priming and anti-quorum sensing (QS). Plants recruit beneficial microorganisms to their root systems through the classic production of plant root exudates. These beneficial organisms are able to prime the defense mechanism of the plant by enhancing the perception to pathogen-associated molecular patterns (PAMPs) by the host (Conrath 2011). Similar to the animal defense mechanism, parallel mechanisms have been drawn where there is the epigenetic inheritance in plants of traits for several subsequent generations. Pieterse et al. (2012) stipulated that this inheritance was due to DNA methylation and chromatin modification. Evidence of this epigenetic inheritance was reported by Slaughter et al. (2012) in his experiments concerning *Arabidopsis thaliana* which showed effective priming by an avirulent strain of *P. syringae* pv. tomato, which showed rapid accumulation of defense transcripts and activation of the SA signaling pathway. The following progenies also showed enhanced disease resistance against *P. syringae*. Their consequent treatment with a priming agent almost always resulted in a higher level of protection. This transgenerational priming is an effective system for defense against disease in plants that involve SA-mediated systemic resistance (Luna et al. 2012; Luna and Ton 2012; Pieterse et al. 2012; Slaughter et al. 2012).

Bacteria communicate from cell to cell using the quorum sensing ability. This process is controlled by signal molecules like N-acylhomoserine lactones (AHL), which act as autoinducer. QS has a lot to do with biofilm production and virulence determinants in bacterial species. However, the anti-quorum sensing ability also exists to interrupt QS and therefore results in lower levels of pathogenicity (Truchado et al. 2012). Contrary to QS, anti-QS has a role in reducing pathogenicity; therefore appearing to be a worthwhile approach to adopt in disease control (Alvarez et al. 2012). Various examples of using AHL production-related genes in the generation of transgenic plants have shown resistance against pathogens such as *Erwinia carotovora* (Dong et al. 2001). These genes expressed AHL lactonase which reduced QS and reduced disease incidence. In a nonpathogenic *Pseudomonas* sp., the AHL that was produced was able to suppress plant disease. As such, the potential of these AHL-degrading microbes as biocontrols should be further evaluated (Crépin et al. 2012a, 2012b).

Biopesticides and biocontrols have emerged as a means to address issues in the agricultural industry. These identified microorganisms are known to play multifaceted roles in disease and pest suppression for improved plant health and yield (Berg 2009).

3.4.3 Physical and Chemical Defense Mechanisms

Plants respond to pathogen infiltration through the production of ROS. While ROS helps circumvent the infection, high levels of this compound are detrimental to the host, and therefore, there is a need for scavengers such as superoxide dismutase

(SOD), catalase (CAT), and ascorbate peroxidase (APX) to navigate these levels *in planta* (Mittler 2002) via the Haber-Weiss or the Fenton reactions (Bowler et al. 1991; Asada and Takahashi 1987). In defense against pathogens, there are pathways that are activated such as the phenylpropanoid pathway which results in antimicrobial activity through the synthesis of phenolic compounds. These pathways are involved in activating SOD and peroxidase (POx) within host (Silva et al. 2004; Singhai et al. 2011).

At the microbial level, soil microflora in the rhizospheric regions are capable of augmenting antioxidant activities through the activation of ISR protective mechanisms in defense against pathogens. Soil pseudomonads have been reported to increase polyphenol oxidase (PPO) and phenylalanine ammonia lyase (PAL) in potatoes, protecting them against the potato scab disease. Diallo et al. (2011) reported that in tomatoes, the infection by *Pseudomonas syringae* has been inhibited through the application of *Serratia marcescens (*Singhai et al. 2011). *S. marcescens* induces the levels of PAL, POx, and lipoxygenase (LOX) within the host. More often than not the protection against pathogens is better induced in response to a consortium rather than a single inoculum. Singh et al. (2013) reported elevated PAL, POx, PPO, and SOD levels in response to consortium application (*Pseudomonas*, *Trichoderma*, and *Rhizobium*) in *Sclerotinia sclerotiorum*- and *Sclerotium rolfsii*-challenged environments. Taken together, these reports indicate the significant role played by these beneficial organisms in inducing tolerance in the plants (Jain et al. 2012).

Several beneficial microbes produce an array of secreted compounds that are able to elicit a defense response in the plants. Among these compounds are lipopolysaccharides, flagellin, surfactin, siderophores, and other antimicrobial substances. These compounds are able to inhibit the effect of pathogens. At the same time, some of these beneficial microorganisms are also able to create competition in the rhizosphere, which is able to inhibit non-beneficial relationships such as the competition for iron by forming LMW chelators (Bakker et al. 2007; Djonović et al. 2007; Meziane et al. 2005; Ran et al. 2005). Similar to bacteria, fungi also produce proteins that are able to enhance plant defense mechanism such as endochitinases. These proteins are able to induce the terpenoid pathway and result in the accumulation of phytoalexins and peroxidases (Djonović et al. 2006, 2007; Keswani et al. 2014). Fungal species such as *Trichoderma* are able to induce the expression of PR proteins that consequently activate the phenyl propanoid pathway resulting in the production of enzymes such as PAL and LOX (Harman and Shoresh 2007; Keswani et al. 2014) which eventually activates the SA signaling. Shoresh et al. (2010) reported the ability of *Trichoderma harzianum* to induce the expression of the *NPR1* gene which is central in the activation of SAR in the infection of *Arabidopsis* against *Pythium* sp. Similarly, Stein et al. (2008) showed that another fungus, *Piriformospora indica,* was able to inhibit soil pathogen through the induction of the jasmonate pathway within the hosts.

In addition to the chemical defenses mentioned above, plants also respond towards microbial infiltration through the production of physical barriers. Phenolic compounds and free radicals result in the lignification of plant cell walls. The cross-linking between the sugars and proteins in the wall results in a highly resistant wall structure that is impenetrable toward pathogens, insects, and herbivores (Boerjan et al. 2003; Davin and Lewis 2000; Hatfield and Vermerris 2001). Resistant cultivars exhibit rapid accumulation of lignin compared to susceptible varieties (Durrant and Dong 2004). Through lignification, the penetration of pathogen is restricted through (i) shielding against enzymatic hydrolysis of plant tissue, (ii) reduced fluids which restrict pathogen mobility and result in starvation, (iii) reduced cell wall damage due to chemical modification of the cell wall, (iv) inactivation of pathogen by the free radicals and phenolic compounds, and finally (v) utilization of fungal cell wall components in lignification.

While the contribution of the pathogen plant response in generating lignification has been studied extensively, little is known about the role of beneficial microorganisms in lignin deposition in plants. The effects of a microbial consortium of *Trichoderma*, *Rhizobium*, and *Pseudomonas* species on *S. rolfii* infection of chickpea resulted in a pronounced increase in lignin deposits in the treated plants. This therefore indicates a possible role of beneficial microbes in altering the physical structure within host to prevent proliferation of microbes (Singh et al. 2013). Mandal and Mitra (2007) in their study tested the effect of *Fusarium* and *Trichoderma* mycelium in lignin deposition. Results showed that *Fusarium* mycelial treatment elicited deposits in the plant host by several folds indicating that lignin-synthesizing enzymes were triggered by these treatments as part of the defense response in plants.

3.4.4 Hormones and Enzymes

Microbes are able to assist plants in synthesizing phytohormones such as indole-3-acetic acid, cytokinins, ethylene, and gibberellin which enables them to alter plant growth. ACC deaminase-producing bacteria remain an important group of organisms that are able to assist in stress modulation in plants (Glick 2005; Saleem et al. 2007). *Pseudomonas fluorescens* induces higher tryptophan levels in the host roots, thereby increasing root length and density (Kamilova et al. 2006). In addition, certain species of *Bacillus* have been reported to enhance photosynthesis and chlorophyll content in *A. thaliana* through abscisic and glucose signaling *in planta* (Contreras-Cornejo et al. 2009; Zhang et al. 2008).

3.5 The Mycobiome in Anthropogenic Soils

Sustainability in agriculture requires low input for reduced expense and low impact to the environment and humans. Intensified research is directed towards methods with reduced pesticide, fertilizer, herbicide, water and soil pollution (Douds et al. 2016; Sniegowski et al. 2011). Anthropogenic compounds have a detrimental effect on the diversity of soil microbial communities thus affecting the biochemical processes within the soil. Hussain et al. (2009) reported the negative and detrimental effects of pesticide on soil health. From his study, it was observed that there was a marked reduction in the abundance of nitrogen-fixing and phosphorus-solubilizing microorganisms in the ecosystem. Therefore, the trend in agriculture should look into organic methods versus conventional farming where the diversity, structure, and richness of the soil microbiota may be maintained or improved (Fließbach et al. 2007). While there are negative effects of these chemicals on the soil microbiota, there are studies that have shown that the application of beneficial microbes is able to improve or negate the negative effects of these chemicals on the soil (Imfeld and Vuilleumier 2012). Douds et al. (2016) in his study showed that the application of AMF increased soil health and improved plant growth and nutrient acquisition. Utilizing these microorganisms as biomixtures can be potentially effective in reducing the use of pesticides, herbicides, and various other chemicals (Ruiz-Hidalgo et al. 2016). The use of these biomixture and soil amendments is slowly gaining interest and wider application as the move towards organic products is rapidly adopted worldwide. Symbionts, composting, microbial inoculants, biochar, and other soil conditioners increase soil health and plant growth and development (Pagano and Covacevich 2011; Pagano and Jorio 2016; Viti et al. 2010). However, while we see reports on the use of microorganisms in improving soil conditions, it is important for us to study the biodiversity, the structure of soil, microbial density, and the affect and interaction of these microbes in influencing the above. Therefore, more studies should be conducted to isolate more beneficial microorganisms for different soils and plant systems, and there also needs to be a directed effort in studying the mechanism of action of these microbes and how they may collectively affect the soil health and agricultural produce (Prieto et al. 2016).

3.6 Redesigning Agroecosystems for Sustainability

3.6.1 Importance of Microbiome Information

The importance of microbiome in the management of host health in medicine has been significant over the past decade. Through these studies, the role played by the various organisms identified within the microbiome and its dynamics is slowly being unraveled (Lloyd-Price et al. 2017; Lozupone et al. 2012; Paramsothy et al. 2017). However, the general complexities in microbial interaction is also seen in plant soil microbiome. The research direction here is to identify the population dynamics, structure, relationship between host and microbe, and the factors that govern the population in the soil. Ideally, we hope that the black box of information contained in the soil microbiome will provide an insight into what is happening in the rhizosphere (Vorholt et al. 2017; Müller et al. 2016).

3.6.2 Microbiome Colonization

In environment, the microbial communities that colonize the roots and plant tissues are generally dependent on the plant species and the environment. There are core groups of organisms that may be present probably in different soil types and may be independent of the plant species. However, as the microbial population is determined by the root exudates, there is bound to be some differences in the species dynamics and density around the roots between plant systems. In the human gut microbiome, the natural population can move between alternate stable states where the unhealthy disease-causing microbiome can be reduced by the introduction of good microorganisms. However, these changes are temporary and are retained only as long as there is continuous introduction of these beneficial microorganisms. In soil systems however, a diseased soil microbiome will only be affected by the intruding beneficial organisms if these beneficial groups are able to efficiently colonize the root systems. If the colonization is not achieved, the soil is bound to revert to a diseased system again (Scheffer et al. 2001). However, it is easier and achievable to change the soil microbiota to one that is both inhibitory of non-beneficial organisms and encouraging of plant growth and health. Organisms that are root colonizers such as nitrogen fixers and mycorrhiza stand a higher chance of maintaining their population due to their stable colonization of host and soil and therefore persist and maintain a state of equilibrium (Edwards et al. 2015; Toju et al. 2018).

3.6.3 Order of Colonization

The order by which an organism colonizes an ecosystem will determine the efficiency of colonization. For instance, if a biocontrol agent was included prior to an infection, it is likely to inhibit disease incidence as the initial organism has the opportunity to propagate and produce antimicrobial products and or physical barriers that reduce the efficiency of colonization by subsequent organisms (Fukami 2015; Wei et al. 2014; Werner and Kiers 2015). The production of these antimicrobial products will exclude competition from other members of the rhizosphere, phyllosphere, and endosphere. In addition to influencing the population within these spheres, these organisms are also able to trigger the immune system of the host through the activation of JA/SA pathways (Pieterse et al. 2014). Based on the observation that time of colonization is important, it may be beneficial to study the efficacy of pretreatment of seed with core organisms that are not only able to colonize the soil but also able to induce the resistance of the plants from the point of germination. Further, in most diseases, disease incidence in early development stages has the most devastating effect. Therefore, it is important to look into methods that may be utilized to increase the level of protection and defense from the very beginning (Toju et al. 2018).

3.6.4 Core Organism Deployment

In identifying microbiomes for use in plant-microbe interactions, focus should be directed towards identifying species with specific functions, pathogen inhibitors, and growth and yield enhancers. As mentioned above colonizers are important to ensure the health of soil. These colonizers should be efficient recruiters of microorganisms that are beneficial to ensure a healthy equilibrium in any given soil system (Freilich et al. 2011). Most often the preferential recruitment involves native pools of communities that have the potential to be symbionts. An example of this symbiont interaction can be seen where nodulating bacteria recruit nitrogen-fixing organisms to the root system. The second method of early colonization involves the inhibition of pathogens and pests, where the colonizers produce antimicrobial products that can inhibit any other organism. In addition, the competition between initial organizers and entry of antagonistic late colonizers for resources can result in the inhibition of the new colonizers (Wei et al. 2014). The third group of core organisms are those that are mutualists or commensalists where the organisms improve growth and yield through the moderation of functions such as phytohormones (IAA, GA, etc.) (Cassan et al. 2009). The core microbiome is what is needed to ensure that all biological and chemical activities that are related to the plant-microbe interaction are executed effectively (Lundberg et al. 2012).

3.7 Future Research Direction

3.7.1 Development of Resources and Models

Plant model systems are a useful way of studying the mechanism of disease and/or symbiosis in the plant-microbe interaction. The information derived from these models may be utilized in understanding the mechanism of disease and symbiosis. Future directions in this area is further assisted with the extensive genome-based projects, the mutant collections, data repositories, and extensive laboratory- and field-based studies to better understand these relationships. In order for microbiomes to be utilized in achieving sustainable agriculture, resources should be established to fill the gaps in knowledge. For instance, for the longest time, the plant model system that has been used to study plant-microbe interactions is *A. thaliana*. However, as this plant is a dicot and a non-leguminous plant system, it does not provide information based on legume interactions and the basis for interactions in a monocot system. Hereafter, more than two decades of studying the tale cress, several other plant systems, databases, resources, and mutant lines have been included in the microbiome studies such as maize, sorghum, rice, and tomato (Peiffer et al. 2013; Stanton-Geddes et al. 2013; Tian et al. 2015). This is a critical step in establishing knowledge and resources necessary for studies in sustainable innovative agriculture (Busby et al. 2017).

3.7.2 Identifying Core Microbiomes

Through identifying models and resources, one expects these studies to identify core microorganisms that are necessary to create a healthy and rich soil environment. Therefore, there should always be an initiative towards identifying the core microbiome taxonomically and further to functionally analyze these taxonomically diverse groups. There are various molecular and genomic tools that will enable the identification of these core organisms right down to the species level. In addition, through the metagenome and metatranscriptome studies, we are likely to attribute functional roles for each taxa and thereafter determine their role and importance in the soil ecosystem. Identification of core microorganisms according to their functional roles will establish the interaction and provide their function in the plant-microbe interphase (Bulgarelli et al. 2013; Hacquard et al. 2015; Louca et al. 2016; Vandenkoornhuyse et al. 2015). By combining various platforms of omics research, we are able to zero in on the core, functionally important taxonomic groups based on (1) plant genotype, (2) soil type, (3) environment conditions, (4) artificial interference such as pesticides, herbicides, and fertilizers, and, finally, (5) the introduction of non-native organisms to the soil ecosystem. From this, we are able to decipher how the core populations differ in various conditions, taxonomic groups that persist in all stresses, and those that are most sensitive to changes. This data will also show us which group of organisms are most capable of producing significant positive effects on a particular plant genotype/species in any particular given environment (Bodenhausen et al. 2014; Horton et al. 2014; Peiffer et al. 2013; Wagner et al. 2014;) and the plant genes and functional traits that influence microbiome assembly. The plant exudates, plant cell structure, and the plant microflora are determining factors of the type of microorganisms that will be recruited to the rhizosphere (Berg et al. 2015; Lebeis et al. 2015; Ritpitakphong et al. 2016; Werner et al. 2015). These attributes of the plant host serves as recruiting factors for soil microorganisms, which subsequently determines the colonizing taxa (Busby et al. 2017; Chaston et al. 2014).

3.7.3 Microbiome Engineering and Resilience

Plant-associated microbiomes are a complex interaction that is still rather poorly understood. However, since plant microbiomes are important, it is recommended that synthetic microbial communities are developed to colonize plants and persist in the environment to bring benefit to the hosts. Diversity and species richness of core microbiomes are useful in the development of stable colonizers (Bai et al. 2015). Studies should also be directed to determine traits that will ensure that these cultures maintain their prevalence in the soil. Their ability to be resilient against biotic and abiotic stresses will further ensure their survivability in the soil ecosystem. A beneficial microbiome may be subject to competitive stress from the surrounding microbial population, and this may vary from farm to farm, across climates, and according to agricultural practices (Soman et al. 2017; DeAngelis et al. 2015).

Therefore, it is important to select for consortia that persist in a variety of heterogeneous ecosystems. The assembly and resilience of a synthetic microbiome in the soil is also dependent on the delivery system that is utilized. There are various methods of inoculation used such as air-, water-, carrier- and vector-based delivery system. While the synthetic community needs to be resilient, it is important that the community is not aggressive and therefore invasive of the environment (Schlaeppi and Bulgarelli 2015). Finally, we need to look into the development of communities that are robust enough to adapt to different crop species or to develop a more species-specific community that serves the plant host amicably (Nemergut et al. 2013). Identifying the link between microbes can help in designing efficient microbial communities that interact positively as keystone species that strongly impact the structure and function of the soil community (Agler et al. 2016; Busby et al. 2017). Lastly, Fig. 3.1 depicts the mapping of research to fill the gaps in soil microbiome research for sustainable agriculture.

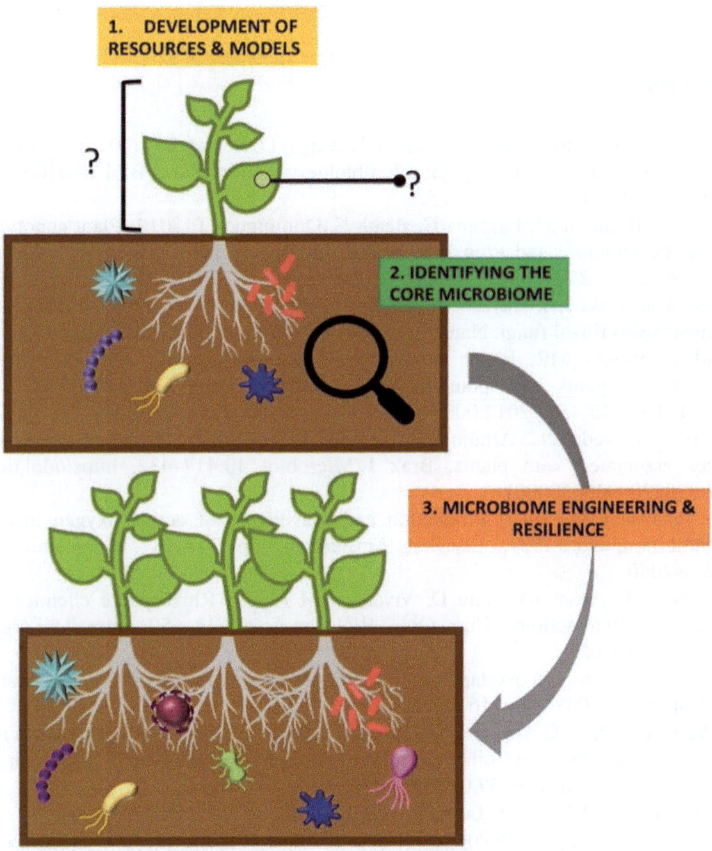

Fig. 3.1 Mapping research gaps in soil microbiome for sustainable agriculture

3.8 Conclusion

In conclusion, soil microbiomes have a large role to play in retaining soil health and contributing positively towards agricultural advancement and yield. However, while microorganisms are included into the design of modern-day agriculture, there remains many grey or unknown areas with regards to the functionality and the role of these organisms in disease suppression and yield enhancement. Some future directions in research involving soil microbiomes and their interactions with the plant host has been elaborated above. As these interactions are studied more carefully in diverse environments, against diverse hosts, and in varying climates and environmental conditions, we are more likely to get some clarity on these complex interaction. The advent of various techniques, resources, databases, and information further contributes towards the building of body of information in the role that microbes play in sustainable agriculture.

Acknowledgment The author wishes to acknowledge the technical and graphic contribution of Ms Ilakiya Sharanee Kumar.

References

Agler MT, Ruhe J, Kroll S, Morhenn C, Kim S-T, Weigel D, Kemen EM (2016) Microbial hub taxa link host and abiotic factors to plant microbiome variation. PLoS Biol 14:e1002352. https://doi.org/10.1371/journal.pbio.1002352

Aira M, Gómez-Brandón M, Lazcano C, Bååth E, Domínguez J (2010) Plant genotype strongly modifies the structure and growth of maize rhizosphere microbial communities. Soil Biol Biochem 42:2276–2281. https://doi.org/10.1016/j.soilbio.2010.08.029

Akiyama K, Matsuzaki K-i, Hayashi H (2005) Plant sesquiterpenes induce hyphal branching in arbuscular mycorrhizal fungi. Nature 435:824. https://doi.org/10.1038/nature03608

Alvarez MV, Moreira MR, Ponce A (2012) Antiquorum sensing and antimicrobial activity of natural agents with potential use in food. J Food Saf 32:379–387. https://doi.org/10.1111/j.1745-4565.2012.00390.x

Andreote FD, Azevedo JL, Araújo WL (2009) Assessing the diversity of bacterial communities associated with plants. Braz J Microbiol 40:417–432. https://doi.org/10.1590/S1517-83822009000300001

Asada K, Takahashi M (1987) Production and scavenging of active oxygen in chloroplasts Photoinhibition, vol 1987. Elsevier, Amsterdam, pp 227–287. https://doi.org/10.1104/pp.106.082040

Badri DV, Weir TL, van der Lelie D, Vivanco JM (2009) Rhizosphere chemical dialogues: plant–microbe interactions. Curr Opin Biotechnol 20:642–650. https://doi.org/10.1016/j.copbio.2009.09.014

Bai Y et al (2015) Functional overlap of the Arabidopsis leaf and root microbiota. Nature 528:364. https://doi.org/10.1038/nature16192

Bais HP, Weir TL, Perry LG, Gilroy S, Vivanco JM (2006) The role of root exudates in rhizosphere interactions with plants and other organisms. Annu Rev Plant Biol 57:233–266. https://doi.org/10.1146/annurev.arplant.57.032905.105159

Bakker PA, Pieterse CM, Van Loon L (2007) Induced systemic resistance by fluorescent Pseudomonas spp. Phytopathology 97:239–243. https://doi.org/10.1094/PHYTO-97-2-0239

Barea J et al. (2013) Arbuscular mycorrhizas and their significance in promoting soil-plant systems sustainability against environmental stresses. In: Beneficial plant-microbial interactions: ecology and applications pp 353–387. https://doi.org/10.1201/b15251-16

Berendsen RL, Pieterse CM, Bakker PA (2012) The rhizosphere microbiome and plant health. Trends Plant Sci 17:478–486. https://doi.org/10.1016/j.tplants.2012.04.001

Berg G (2009) Plant–microbe interactions promoting plant growth and health: perspectives for controlled use of microorganisms in agriculture. Appl Microbiol Biotechnol 84:11–18. https://doi.org/10.1007/s00253-009-2092-7

Berg G, Smalla K (2009) Plant species and soil type cooperatively shape the structure and function of microbial communities in the rhizosphere. FEMS Microbiol Ecol 68:1–13. https://doi.org/10.1111/j.1574-6941.2009.00654.x

Berg G, Rybakova D, Grube M, Köberl M (2015) The plant microbiome explored: implications for experimental botany. J Exp Bot 67:995–1002. https://doi.org/10.1093/jxb/erv466

Berlec A (2012) Novel techniques and findings in the study of plant microbiota: search for plant probiotics. Plant Sci 193:96–102. https://doi.org/10.1016/j.plantsci.2012.05.010

Bodenhausen N, Bortfeld-Miller M, Ackermann M, Vorholt JA (2014) A synthetic community approach reveals plant genotypes affecting the phyllosphere microbiota. PLoS Genet 10:e1004283. https://doi.org/10.1371/journal.pgen.1004283

Boerjan W, Ralph J, Baucher M (2003) Lignin biosynthesis. Annu Rev Plant Biol 54:519–546. https://doi.org/10.1146/annurev.arplant.54.031902.134938

Bonfante P, Anca I-A (2009) Plants, mycorrhizal fungi, and bacteria: a network of interactions. Annu Rev Microbiol 63:363–383. https://doi.org/10.1146/annurev.micro.091208.073504

Bonito G et al (2014) Plant host and soil origin influence fungal and bacterial assemblages in the roots of woody plants. Mol Ecol 23:3356–3370. https://doi.org/10.1111/mec.12821

Bowler C et al (1991) Manganese superoxide dismutase can reduce cellular damage mediated by oxygen radicals in transgenic plants. EMBO J 10:1723–1732

Brader G, Compant S, Mitter B, Trognitz F, Sessitsch A (2014) Metabolic potential of endophytic bacteria. Curr Opin Biotechnol 27:30–37. https://doi.org/10.1016/j.copbio.2013.09.012

Broeckling CD, Broz AK, Bergelson J, Manter DK, Vivanco JM (2008) Root exudates regulate soil fungal community composition and diversity. Appl Environ Microbiol 74:738–744. https://doi.org/10.1128/AEM.02188-07

Bulgarelli D et al (2012) Revealing structure and assembly cues for Arabidopsis root-inhabiting bacterial microbiota. Nature 488:91. https://doi.org/10.1038/nature11336

Bulgarelli D, Schlaeppi K, Spaepen S, van Themaat EVL, Schulze-Lefert P (2013) Structure and functions of the bacterial microbiota of plants. Annu Rev Plant Biol 64:807–838. https://doi.org/10.1146/annurev-arplant-050312-120106

Bulgarelli D et al (2015) Structure and function of the bacterial root microbiota in wild and domesticated barley. Cell Host Microbe 17:392–403. https://doi.org/10.1146/annurev-arplant-050312-120106

Busby PE et al (2017) Research priorities for harnessing plant microbiomes in sustainable agriculture. PLoS Biol 15:e2001793. https://doi.org/10.1371/journal.pbio.2001793

Calvo P, Nelson L, Kloepper JW (2014) Agricultural uses of plant biostimulants. Plant and Soil 383:3–41. https://doi.org/10.1007/s11104-014-2131-8

Cassan F, Perrig D, Sgroy V, Masciarelli O, Penna C, Luna V (2009) *Azospirillum brasilense* Az39 and *Bradyrhizobium japonicum* E109, inoculated singly or in combination, promote seed germination and early seedling growth in corn (*Zea mays* L.) and soybean (*Glycine max* L.). Eur J Soil Biol 45:28–35. https://doi.org/10.1016/j.ejsobi.2008.08.005

Chaparro JM, Sheflin AM, Manter DK, Vivanco JM (2012) Manipulating the soil microbiome to increase soil health and plant fertility. Biol Fertil Soils 48:489–499. https://doi.org/10.1007/s00374-012-0691-4

Chaparro JM, Badri DV, Bakker MG, Sugiyama A, Manter DK, Vivanco JM (2013) Root exudation of phytochemicals in Arabidopsis follows specific patterns that are developmentally programmed and correlate with soil microbial functions. PLoS One 8:e55731. https://doi.org/10.1371/journal.pone.0055731

Chaparro JM, Badri DV, Vivanco JM (2014) Rhizosphere microbiome assemblage is affected by plant development. ISME J 8:790. https://doi.org/10.1038/ismej.2013.196

Chaston JM, Newell PD, Douglas AE (2014) Metagenome-wide association of microbial determinants of host phenotype in *Drosophila melanogaster*. MBio 5:e01631–e01614. https://doi.org/10.1128/mBio.01631-14

Chialva M, Bonfante P (2018) Plant microbiota: from model plants to Mediterranean crops. Phytopathol Mediterr 57:123–145. https://doi.org/10.14601/Phytopathol_Mediterr-22052

Clark R, Zeto S, Zobel R (1999) Arbuscular mycorrhizal fungal isolate effectiveness on growth and root colonization of *Panicum virgatum* in acidic soil. Soil Biol Biochem 31:1757–1763. https://doi.org/10.1016/S0038-0717(99)00084-X

Conrath U (2011) Molecular aspects of defence priming. Trends Plant Sci 16:524–531. https://doi.org/10.1016/j.tplants.2011.06.004

Contreras-Cornejo HA, Macías-Rodríguez L, Cortés-Penagos C, López-Bucio J (2009) Trichoderma virens, a plant beneficial fungus, enhances biomass production and promotes lateral root growth through an auxin-dependent mechanism in Arabidopsis. Plant Physiol 149:1579–1592. https://doi.org/10.1104/pp.108.130369

Crépin A et al (2012a) Biological control of pathogen communication in the rhizosphere: a novel approach applied to potato soft rot due to *Pectobacterium atrosepticum*. Plant Soil 358:27–37. https://doi.org/10.1007/s11104-011-1030-5

Crépin A et al (2012b) N-acyl homoserine lactones in diverse Pectobacterium and Dickeya plant pathogens: diversity, abundance, and involvement in virulence. Sensors 12:3484–3497. https://doi.org/10.3390/s120303484

Davin LB, Lewis NG (2000) Dirigent proteins and dirigent sites explain the mystery of specificity of radical precursor coupling in lignan and lignin biosynthesis. Plant Physiol 123:453–462. https://doi.org/10.1104/pp.123.2.453

Davison J et al (2015) Global assessment of arbuscular mycorrhizal fungus diversity reveals very low endemism. Science 349:970–973. https://doi.org/10.1126/science.aab1161

De Bruijn FJ (2015) Biological nitrogen fixation. In: Principles of plant-microbe interactions. Springer, Cham, pp 215–224. https://doi.org/10.1007/978-3-319-08575-3

De Vries FT, Wallenstein MD (2017) Below-ground connections underlying above-ground food production: a framework for optimising ecological connections in the rhizosphere. J Ecol 105:913–920. https://doi.org/10.1111/1365-2745.12783

DeAngelis KM et al (2015) Long-term forest soil warming alters microbial communities in temperate forest soils. Front Microbiol 6:104. https://doi.org/10.3389/fmicb.2015.00104

De-la-Pena C et al (2010) Root secretion of defense-related proteins is development-dependent and correlated with flowering time. J Biol Chem:jbc. M110. 119040 https://doi.org/10.1074/jbc.M110.119040

Delmotte N et al (2009) Community proteogenomics reveals insights into the physiology of phyllosphere bacteria. Proc Natl Acad Sci 106:16428–16433. https://doi.org/10.1073/pnas.0905240106

Diallo S, Crepin A, Barbey C, Orange N, Burini JF, Latour X (2011) Mechanisms and recent advances in biological control mediated through the potato rhizosphere. FEMS Microbiol Ecol 75(3):351–364

Dini-Andreote F, van Elsas JD (2013) Back to the basics: the need for ecophysiological insights to enhance our understanding of microbial behaviour in the rhizosphere. Plant Soil 373:1–15. https://doi.org/10.1007/s11104-013-1687-z

Djonović S, Pozo MJ, Dangott LJ, Howell CR, Kenerley CM (2006) Sm1, a proteinaceous elicitor secreted by the biocontrol fungus Trichoderma virens induces plant defense responses and systemic resistance. Mol Plant-Microbe Interact 19:838–853. https://doi.org/10.1094/MPMI-19-0838

Djonović S, Vargas WA, Kolomiets MV, Horndeski M, Wiest A, Kenerley CM (2007) A proteinaceous elicitor Sm1 from the beneficial fungus *Trichoderma virens* is required for induced systemic resistance in maize. Plant Physiol 145:875–889. https://doi.org/10.1104/pp.107.103689

Dobbelaere S, Vanderleyden J, Okon Y (2003) Plant growth-promoting effects of diazotrophs in the rhizosphere. Crit Rev Plant Sci 22:107–149. https://doi.org/10.1080/713610853

Dong Y-H, Wang L-H, Xu J-L, Zhang H-B, Zhang X-F, Zhang L-H (2001) Quenching quorum-sensing-dependent bacterial infection by an N-acyl homoserine lactonase. Nature 411:813. https://doi.org/10.1038/35081101

Donn S, Kirkegaard JA, Perera G, Richardson AE, Watt M (2015) Evolution of bacterial communities in the wheat crop rhizosphere. Environ Microbiol 17:610–621. https://doi.org/10.1111/1462-2920.12452

Douds DD et al (2016) Utilization of inoculum of AM fungi produced on-farm increases the yield of *Solanum lycopersicum*: a summary of 7 years of field trials on a conventional vegetable farm with high soil phosphorus. Science Horticulture 207:89–96

Durrant WE, Dong X (2004) Systemic acquired resistance. Annu Rev Phytopathol 42:185–209. https://doi.org/10.1146/annurev.phyto.42.040803.14

Edwards J et al (2015) Structure, variation, and assembly of the root-associated microbiomes of rice. Proc Natl Acad Sci 112:E911–E920. https://doi.org/10.1073/pnas.1414592112

Elser JJ et al (2007) Global analysis of nitrogen and phosphorus limitation of primary producers in freshwater, marine and terrestrial ecosystems. Ecol Lett 10:1135–1142. https://doi.org/10.1111/j.1461-0248.2007.01113.x

Fang W, Leger RJS (2010) Mrt, a gene unique to fungi, encodes an oligosaccharide transporter and facilitates rhizosphere competency in *Metarhizium robertsii.* Plant Physiol https://doi.org/10.1104/pp.110.163014

Faoro H et al (2010) Influence of soil characteristics on the diversity of bacteria in the Southern Brazilian Atlantic Forest. Appl Environ Microbiol 76:4744–4749. https://doi.org/10.1128/AEM.03025-09

Fellbaum CR et al (2012) Carbon availability triggers fungal nitrogen uptake and transport in arbuscular mycorrhizal symbiosis. Proc Natl Acad Sci 109:2666–2671. https://doi.org/10.1073/pnas.1118650109

Fierer N, Jackson RB (2006) The diversity and biogeography of soil bacterial communities. Proc Natl Acad Sci 103:626–631. https://doi.org/10.1073/pnas.0507535103

Fließbach A et al (2007) Soil organic matter and biological soil quality indicators after 21 years of organic and conventional farming. Agriculture Ecosystem and Environment 118:273–284

Frey SD, Knorr M, Parrent JL, Simpson RT (2004) Chronic nitrogen enrichment affects the structure and function of the soil microbial community in temperate hardwood and pine forests. For Ecol Manag 196:159–171. https://doi.org/10.1016/j.foreco.2004.03.018

Friesen ML, Porter SS, Stark SC, von Wettberg EJ, Sachs JL, Martinez-Romero E (2011) Microbially mediated plant functional traits. Annu Rev Ecol Evol Syst 42. https://doi.org/10.1146/annurev-ecolsys-102710-145039

Fukami T (2015) Historical contingency in community assembly: integrating niches, species pools, and priority effects annual review of ecology. Evolution, and Systematics 46:1–23. https://doi.org/10.1146/annurev-ecolsys-110411-160340

Glick BR (2005) Modulation of plant ethylene levels by the bacterial enzyme ACC deaminase. FEMS Microbiol Lett 251:1–7. https://doi.org/10.1016/j.femsle.2005.07.030

Guttman DS, McHardy AC, Schulze-Lefert P (2014) Microbial genome-enabled insights into plant–microorganism interactions. Nat Rev Genet 15:797. https://doi.org/10.1038/nrg3748

Hacquard S (2016) Disentangling the factors shaping microbiota composition across the plant holobiont. New Phytol 209:454–457. https://doi.org/10.1111/nph.13760

Hacquard S, Garrido-Oter R, González A, Spaepen S, Ackermann G, Lebeis S, Schulze-Lefert P (2015) Microbiota and host nutrition across plant and animal kingdoms. Cell Host Microbe 17:603–616. https://doi.org/10.1016/j.chom.2015.04.009

Hamilton EW III, Frank DA (2001) Can plants stimulate soil microbes and their own nutrient supply? Evidence from a grazing tolerant grass. Ecology 82:2397–2402. https://doi.org/10.1890/0012-9658(2001)082[2397:CPSSMA]2.0.CO;2

Hardoim PR, Nazir R, Sessitsch A, Elhottová D, Korenblum E, van Overbeek LS, van Elsas JD (2013) The new species Enterobacter oryziphilus sp. nov. and Enterobacter oryzendophyticus sp. nov. are key inhabitants of the endosphere of rice. BMC Microbiol 13:164. https://doi.org/10.1186/1471-2180-13-164

Harman GE, Shoresh M (2007) The mechanisms and applications of symbiotic opportunistic plant symbionts. In: Novel biotechnologies for biocontrol agent enhancement and management. Springer, Dordrecht, pp 131–155. https://doi.org/10.1007/978-1-4020-5799-1_7

Harman GE, Howell CR, Viterbo A, Chet I, Lorito M (2004) Trichoderma species—opportunistic, avirulent plant symbionts. Nat Rev Microbiol 2:43. https://doi.org/10.1038/nrmicro797

Harrison MJ (1999) Molecular and cellular aspects of the arbuscular mycorrhizal symbiosis. Annu Rev Plant Biol 50:361–389. https://doi.org/10.1146/annurev.arplant.50.1.361

Hatfield R, Vermerris W (2001) Lignin formation in plants. The dilemma of linkage specificity. Plant Physiol 126:1351–1357. https://doi.org/10.1104/pp.126.4.1351

Hayatsu M, Tago K, Saito M (2008) Various players in the nitrogen cycle: diversity and functions of the microorganisms involved in nitrification and denitrification. Soil Sci Plant Nutr 54:33–45. https://doi.org/10.1111/j.1747-0765.2007.00195.x

Hiruma K et al (2016) Root endophyte *Colletotrichum tofieldiae* confers plant fitness benefits that are phosphate status dependent. Cell 165:464–474. https://doi.org/10.1016/j.cell.2016.02.028

Horton MW et al (2014) Genome-wide association study of *Arabidopsis thaliana* leaf microbial community. Nat Commun 5:5320. https://doi.org/10.1038/ncomms6320

Huang X-F, Chaparro JM, Reardon KF, Zhang R, Shen Q, Vivanco JM (2014) Rhizosphere interactions: root exudates, microbes, and microbial communities. Botany 92:267–275. https://doi.org/10.1139/cjb-2013-0225

Hussain S, Siddique T, Arshad M, Saleem M (2009) Bioremediation and phytoremediation of pesticides: recent advances. Crit Rev Environ Sci Technol 39:843–907. https://doi.org/10.1080/10643380801910090

Imfeld G, Vuilleumier S (2012) Measuring the effects of pesticides on bacterial communities in soil: a critical review. Eur J Soil Biol 49:22–30. https://doi.org/10.1016/j.ejsobi.2011.11.010

İnceoğlu Ö, Al-Soud WA, Salles JF, Semenov AV, van Elsas JD (2011) Comparative analysis of bacterial communities in a potato field as determined by pyrosequencing. PLoS One 6:e23321. https://doi.org/10.1371/journal.pone.0023321

Jain A, Singh S, Kumar Sarma B, Bahadur Singh H (2012) Microbial consortium–mediated reprogramming of defence network in pea to enhance tolerance against Sclerotinia sclerotiorum. J Appl Microbiol 112:537–550. https://doi.org/10.1111/j.1365-2672.2011.05220.x

Jeffries P, Barea J (2012) 4 Arbuscular Mycorrhiza: a key component of sustainable plant–soil ecosystems. In: Fungal associations. Springer, Berlin, pp 51–75. https://doi.org/10.1007/978-3-642-30826-0_4

Jones KM, Sharopova N, Lohar DP, Zhang JQ, VandenBosch KA, Walker GC (2008) Differential response of the plant *Medicago truncatula* to its symbiont Sinorhizobium meliloti or an exopolysaccharide-deficient mutant. Proc Natl Acad Sci 105:704–709. https://doi.org/10.1073/pnas.0709338105

Kamilova F, Validov S, Azarova T, Mulders I, Lugtenberg B (2005) Enrichment for enhanced competitive plant root tip colonizers selects for a new class of biocontrol bacteria. Environ Microbiol 7:1809–1817. https://doi.org/10.1111/j.1462-2920.2005.00889.x

Kamilova F, Kravchenko LV, Shaposhnikov AI, Makarova N, Lugtenberg B (2006) Effects of the tomato pathogen *Fusarium oxysporum* f. sp. radicis-lycopersici and of the biocontrol bacterium *Pseudomonas fluorescens* WCS365 on the composition of organic acids and sugars in tomato root exudate. Mol Plant-Microbe Interact 19:1121–1126. https://doi.org/10.1094/MPMI-19-1121

Kandula DR, Jones EE, Stewart A, McLean KL, Hampton JG (2015) Trichoderma species for biocontrol of soil-borne plant pathogens of pasture species. Biocontrol Sci Tech 25:1052–1069. https://doi.org/10.1080/09583157.2015.1028892

Kavamura VN et al (2013) Screening of Brazilian cacti rhizobacteria for plant growth promotion under drought. Microbiol Res 168:183–191. https://doi.org/10.1016/j.micres.2012.12.002

Keswani C, Mishra S, Sarma BK, Singh SP, Singh HB (2014) Unraveling the efficient applications of secondary metabolites of various Trichoderma spp. Appl Microbiol Biotechnol 98:533–544. https://doi.org/10.1007/s00253-013-5344-5

Kiers ET et al (2011) Reciprocal rewards stabilize cooperation in the mycorrhizal symbiosis. Science 333:880–882. https://doi.org/10.1126/science.1208473

Knief C et al (2012) Metaproteogenomic analysis of microbial communities in the phyllosphere and rhizosphere of rice. ISME J 6:1378. https://doi.org/10.1038/ismej.2011.192

Larimer AL, Clay K, Bever JD (2014) Synergism and context dependency of interactions between arbuscular mycorrhizal fungi and rhizobia with a prairie legume. Ecology 95:1045–1054

Lauber CL, Strickland MS, Bradford MA, Fierer N (2008) The influence of soil properties on the structure of bacterial and fungal communities across land-use types. Soil Biol Biochem 40:2407–2415. https://doi.org/10.1016/j.soilbio.2008.05.021

Lebeis SL et al (2015) Salicylic acid modulates colonization of the root microbiome by specific bacterial taxa. Science 349:860–864. https://doi.org/10.1126/science.aaa8764

Lloyd-Price J et al (2017) Strains, functions and dynamics in the expanded human microbiome project. Nature 550:61. https://doi.org/10.1038/nature23889

Louca S, Parfrey LW, Doebeli M (2016) Decoupling function and taxonomy in the global ocean microbiome. Science 353:1272–1277. https://doi.org/10.1126/science.aaf4507

Lozupone CA, Stombaugh JI, Gordon JI, Jansson JK, Knight R (2012) Diversity, stability and resilience of the human gut microbiota. Nature 489:220. https://doi.org/10.1038/nature11550

Luna E, Ton J (2012) The epigenetic machinery controlling transgenerational systemic acquired resistance. Plant Signal Behav 7:615–618. https://doi.org/10.4161/psb.20155

Luna E, Bruce TJ, Roberts MR, Flors V, Ton J (2012) Next-generation systemic acquired resistance. Plant Physiol 158:844–853. https://doi.org/10.1104/pp.111.187468

Lundberg DS et al (2012) Defining the core *Arabidopsis thaliana* root microbiome. Nature 488:86. https://doi.org/10.1038/nature11237

Mandal S, Mitra A (2007) Reinforcement of cell wall in roots of Lycopersicon esculentum through induction of phenolic compounds and lignin by elicitors. Physiol Mol Plant Pathol 71:201–209. https://doi.org/10.1016/j.pmpp.2008.02.003

Mangla S, Callaway RM (2008) Exotic invasive plant accumulates native soil pathogens which inhibit native plants. J Ecol 96:58–67. https://doi.org/10.1111/j.1365-2745.2007.01312.x

Meena VS, Maurya B, Verma JP (2014) Does a rhizospheric microorganism enhance K+ availability in agricultural soils? Microbiol Res 169:337–347. https://doi.org/10.1016/j.micres.2013.09.003

Mendes R et al (2011) Deciphering the rhizosphere microbiome for disease-suppressive bacteria. Science:1203980. https://doi.org/10.1126/science.1203980

Mendes LW, Kuramae EE, Navarrete AA, Van Veen JA, Tsai SM (2014) Taxonomical and functional microbial community selection in soybean rhizosphere. ISME J 8:1577. https://doi.org/10.1038/ismej.2014.17

Mercado-Blanco J (2015) Life of microbes inside the plant. In: Principles of plant-microbe interactions. Springer, Cham, pp 25–32. https://doi.org/10.1007/978-3-319-08575-3

Meziane H, Van Der Sluis I, Van Loon LC, Höfte M, Bakker PA (2005) Determinants of Pseudomonas putida WCS358 involved in inducing systemic resistance in plants. Mol Plant Pathol 6:177–185. https://doi.org/10.1111/j.1364-3703.2005.00276.x

Micallef SA, Shiaris MP, Colón-Carmona A (2009) Influence of Arabidopsis thaliana accessions on rhizobacterial communities and natural variation in root exudates. J Exp Bot 60:1729–1742. https://doi.org/10.1093/jxb/erp053

Mittler R (2002) Oxidative stress, antioxidants and stress tolerance. Trends Plant Sci 7:405–410

Morrissey EM, Gillespie JL, Morina JC, Franklin RB (2014) Salinity affects microbial activity and soil organic matter content in tidal wetlands. Glob Chang Biol 20:1351–1362. https://doi.org/10.1111/gcb.12431

Mueller UG, Sachs JL (2015) Engineering microbiomes to improve plant and animal health. Trends Microbiol 23:606–617. https://doi.org/10.1016/j.tim.2015.07.009

Müller DB, Vogel C, Bai Y, Vorholt JA (2016) The plant microbiota: systems-level insights and perspectives. Annu Rev Genet 50:211–234. https://doi.org/10.1146/annurev-genet-120215-034952

Nadarajah KK (2016) Rhizosphere interactions: life below ground. In: Plant-microbe interaction: an approach to sustainable agriculture. Springer, Puchong, pp 3–23. https://doi.org/10.1007/978-981-10-2854-0_1

Nadarajah KK (2017a) Induction of systemic resistance for disease suppression. In: Crop Improvement. Springer, pp 335–357. https://doi.org/10.1007/978-3-319-65079-1_15

Nadarajah KK (2017b) Rhizobium in Rice yield and growth enhancement. In: Rhizobium biology and biotechnology. Springer, Cham, pp 83–103

Nasto MK, Alvarez-Clare S, Lekberg Y, Sullivan BW, Townsend AR, Cleveland CC (2014) Interactions among nitrogen fixation and soil phosphorus acquisition strategies in lowland tropical rain forests. Ecol Lett 17:1282–1289. https://doi.org/10.1111/ele.12335

Nemergut DR et al (2013) Patterns and processes of microbial community assembly. Microbiol Mol Biol Rev 77:342–356. https://doi.org/10.1128/mmbr.00051-12

Olivares J, Bedmar EJ, Sanjuán J (2013) Biological nitrogen fixation in the context of global change. Mol Plant-Microbe Interact 26:486–494. https://doi.org/10.1094/MPMI-12-12-0293-CR

Ortiz N, Armada E, Duque E, Roldán A, Azcón R (2015) Contribution of arbuscular mycorrhizal fungi and/or bacteria to enhancing plant drought tolerance under natural soil conditions: effectiveness of autochthonous or allochthonous strains. J Plant Physiol 174:87–96. https://doi.org/10.1016/j.jplph.2014.08.019

Owen D, Williams A, Griffith G, Withers P (2015) Use of commercial bio-inoculants to increase agricultural production through improved phosphrous acquisition. Appl Soil Ecol 86:41–54. https://doi.org/10.1016/j.apsoil.2014.09.012

Paramsothy S et al (2017) Multidonor intensive faecal microbiota transplantation for active ulcerative colitis: a randomised placebo-controlled trial. Lancet 389:1218–1228. https://doi.org/10.1016/S0140-6736(17)30182-4

Peiffer JA et al (2013) Diversity and heritability of the maize rhizosphere microbiome under field conditions. Proceedings of the National Academy of Sciences:201302837. https://doi.org/10.1073/pnas.1302837110

Philippot L, Raaijmakers JM, Lemanceau P, Van Der Putten WH (2013) Going back to the roots: the microbial ecology of the rhizosphere. Nat Rev Microbiol 11:789. https://doi.org/10.1038/nrmicro3109

Pieterse CMJ et al (2012) Hormonal modulation of plant immunity. Annual Review of Cell Developmental Biology 28:489–521. https://doi.org/10.1146/annurev-cellbio-092910-154055

Pieterse CMJ, Pierik R, Van Wees SCM (2014) Different shades of JAZ during plant growth and defense. New Phytol 204:261–264. https://doi.org/10.1111/nph.13029

Porras-Alfaro A, Bayman P (2011) Hidden fungi, emergent properties: endophytes and microbiomes. Annu Rev Phytopathol 49:291–315. https://doi.org/10.1146/annurev-phyto-080508-081831

Prajapati K, Sharma M, Modi H (2012) Isolation of two potassium solubilizing fungi from ceramic industry soils. Life Sci Leaflets 5:71–75

Prieto I, Roldán A, Huygens D, del Mar Alguacil M, Navarro-Cano JA, Querejeta JI (2016) Species-specific roles of ectomycorrhizal fungi in facilitating interplant transfer of hydraulically redistributed water between *Pinus halepensis* saplings and seedlings. Plant Soil 406:15–27. https://doi.org/10.1007/s11104-016-2860-y

Raaijmakers JM, Paulitz TC, Steinberg C, Alabouvette C, Moënne-Loccoz Y (2009) The rhizosphere: a playground and battlefield for soilborne pathogens and beneficial microorganisms. Plant Soil 321:341–361. https://doi.org/10.1007/s11104-008-9568-6

Ramond JB, Tshabuse F, Bopda CW, Cowan DA, Tuffin MI (2013) Evidence of variability in the structure and recruitment of rhizospheric and endophytic bacterial communities associated with arable sweet sorghum (*Sorghum bicolor* (L) Moench). Plant Soil 372:265–278. https://doi.org/10.1007/s11104-013-1737-6

Ran L, Liu C, Wu G, Van Loon L, Bakker P (2005) Suppression of bacterial wilt in Eucalyptus urophylla by fluorescent Pseudomonas spp. in China. Biol Control 32:111–120. https://doi.org/10.1016/j.biocontrol.2004.08.007

Rascovan N et al (2016) Integrated analysis of root microbiomes of soybean and wheat from agricultural fields. Sci Rep 6:28084. https://doi.org/10.1038/srep28084

Requena N, Perez-Solis E, Azcón-Aguilar C, Jeffries P, Barea J-M (2001) Management of indigenous plant-microbe symbioses aids restoration of desertified ecosystems. Appl Environ Microbiol 67:495–498. https://doi.org/10.1128/AEM.67.2.495-498.2001

Rincon-Florez VA, Carvalhais LC, Schenk PM (2013) Culture-independent molecular tools for soil and rhizosphere microbiology. Diversity 5:581–612. https://doi.org/10.3390/d5030581

Ritpitakphong U, Falquet L, Vimoltust A, Berger A, Métraux JP, L'Haridon F (2016) The microbiome of the leaf surface of Arabidopsis protects against a fungal pathogen. New Phytol 210:1033–1043. https://doi.org/10.1111/nph.13808

Robertson GP, Gross KL, Caldwell M, Pearcy R (1994) Assessing the heterogeneity of belowground resources: quantifying pattern and scale. In: Exploitation of environmental heterogeneity by plants: ecophysiological processes above-and belowground, pp 237–253. https://doi.org/10.1016/B978-0-12-155070-7.50013-5

Rousk J et al (2010) Soil bacterial and fungal communities across a pH gradient in an arable soil. ISME J 4:1340. https://doi.org/10.1038/ismej.2010.58

Ruiz-Hidalgo K, Chin-Pampillo JS, Masís-Mora M, Carazo-Rojas E, Rodríguez-Rodríguez CE (2016) Optimization of a fungally bioaugmented biomixture for carbofuran removal in on-farm biopurification systems. Water Air Soil Pollut 227:3. https://doi.org/10.1007/s11270-015-2681-2

Saleem M, Arshad M, Hussain S, Bhatti AS (2007) Perspective of plant growth promoting Rhizobacteria (PGPR) containing ACC deaminase in stress agriculture. J Ind Microbiol Biotech 34:635–648. https://doi.org/10.1007/s10295-007-0240-6

Santhanam R, Weinhold A, Goldberg J, Oh Y, Baldwin IT (2015) Native root-associated bacteria rescue a plant from a sudden-wilt disease that emerged during continuous cropping. Proc Natl Acad Sci 112:E5013–E5020. https://doi.org/10.1073/pnas.1505765112

Santi C, Bogusz D, Franche C (2013) Biological nitrogen fixation in non-legume plants. Ann Bot 111:743–767. https://doi.org/10.1093/aob/mct048

Scheffer M, Carpenter S, Foley JA, Folke C, Walker B (2001) Catastrophic shifts in ecosystems. Nature 413:591. https://doi.org/10.1038/35098000

Schlaeppi K, Bulgarelli D (2015) The plant microbiome at work. Mol Plant-Microbe Interact 28:212–217. https://doi.org/10.1094/MPMI-10-14-0334-FI

Schloss PD, Handelsman J (2006) Toward a census of bacteria in soil. PLoS Comput Biol 2:e92. https://doi.org/10.1371/journal.pcbi.0020092

Schrey SD, Tarkka MT (2008) Friends and foes: streptomycetes as modulators of plant disease and symbiosis. Antonie Van Leeuwenhoek 94:11–19. https://doi.org/10.1007/s10482-008-9241-3

Selosse M-A, Bessis A, Pozo MJ (2014) Microbial priming of plant and animal immunity: symbionts as developmental signals. Trends Microbiol 22:607–613. https://doi.org/10.1016/j.tim.2014.07.003

Sheng XF, He LY (2006) Solubilization of potassium-bearing minerals by a wild-type strain of *Bacillus edaphicus* and its mutants and increased potassium uptake by wheat. Can J Microbiol 52:66–72. https://doi.org/10.1139/w05-117

Shoresh M, Harman GE, Mastouri F (2010) Induced systemic resistance and plant responses to fungal biocontrol agents. Annu Rev Phytopathol 48:21–43. https://doi.org/10.1146/annurev-phyto-073009-114450

Silva HSA, da Silva Romeiro R, Macagnan D, de Almeida Halfeld-Vieira B, Pereira MCB, Mounteer A (2004) Rhizobacterial induction of systemic resistance in tomato plants: non-specific protection and increase in enzyme activities Biol Control 29:288–295 https://doi.org/10.1016/S1049-9644(03)00163-4

Singh R, Soni SK, Kalra A (2013) Synergy between *Glomus fasciculatum* and a beneficial Pseudomonas in reducing root diseases and improving yield and forskolin content in *Coleus forskohlii* Briq. under organic field conditions. Mycorrhiza 23:35–44. https://doi.org/10.1007/s00572-012-0447-x

Singhai P, Sarma B, Srivastava J (2011) Biological management of common scab of potato through Pseudomonas species and vermicompost. Biol Control 57:150–157. https://doi.org/10.1016/j.biocontrol.2011.02.008

Slaughter A, Daniel X, Flors V, Luna E, Hohn B, Mauch-Mani B (2012) Descendants of primed Arabidopsis plants exhibit resistance to biotic stress. Plant Physiol 158:835–843. https://doi.org/10.1104/pp.111.191593

Smith S, Read D (2008) Mineral nutrition, toxic element accumulation and water relations of arbuscular mycorrhizal plants Mycorrhizal symbiosis 3:145–118 https://doi.org/10.1016/B978-012370526-6.50007-6

Sniegowski K, Bers K, Van Goetem K, Ryckeboer J, Jaeken P, Spanoghe P, Springael D (2011) Improvement of pesticide mineralization in on-farm biopurification systems by bioaugmentation with pesticide-primed soil. FEMS Microbiol Ecol 76:64–73. https://doi.org/10.1111/j.1574-6941.2010.01031.x

Soman C, Li D, Wander MM, Kent AD (2017) Long-term fertilizer and crop-rotation treatments differentially affect soil bacterial community structure. Plant Soil 413:145–159. https://doi.org/10.1007/s11104-016-3083-y

Stacey G (2007) The rhizobium-legume nitrogen-fixing symbiosis. In: Biology of the nitrogen cycle. Elsevier, Amsterdam, pp 147–163

Stanton-Geddes J et al (2013) Candidate genes and genetic architecture of symbiotic and agronomic traits revealed by whole-genome, sequence-based association genetics in *Medicago truncatula*. PLoS One 8:e65688. https://doi.org/10.1371/journal.pone.0065688

Stein E, Molitor A, Kogel K-H, Waller F (2008) Systemic resistance in Arabidopsis conferred by the mycorrhizal fungus *Piriformospora indica* requires jasmonic acid signaling and the cytoplasmic function of NPR1. Plant Cell Physiol 49:1747–1751. https://doi.org/10.1093/pcp/pcn147

Steinkellner S, Lendzemo V, Langer I, Schweiger P, Khaosaad T, Toussaint J-P, Vierheilig H (2007) Flavonoids and strigolactones in root exudates as signals in symbiotic and pathogenic plant-fungus interactions. Molecules 12:1290–1306. https://doi.org/10.3390/12071290

Thompson LR et al (2017) A communal catalogue reveals Earth's multiscale microbial diversity. Nature 551. https://doi.org/10.1038/nature24621

Tian B-Y, Cao Y, Zhang K-Q (2015) Metagenomic insights into communities, functions of endophytes, and their associates with infection by root-knot nematode, *Meloidogyne incognita*, in tomato roots. Sci Rep 5:17087. https://doi.org/10.1038/srep17087

Toju H, Yamamoto S, Sato H, Tanabe AS, Gilbert GS, Kadowaki K (2013) Community composition of root-associated fungi in a Quercus-dominated temperate forest: "codominance" of mycorrhizal and root-endophytic fungi. Ecol Evol 3:1281–1293. https://doi.org/10.1002/ece3.546

Toju H, Tanabe AS, Sato H (2018) Network hubs in root-associated fungal metacommunities. Microbiome 6:116. https://doi.org/10.1186/s40168-018-0497-1

Truchado P, Tomás-Barberán FA, Larrosa M, Allende A (2012) Food phytochemicals act as quorum sensing inhibitors reducing production and/or degrading autoinducers of *Yersinia enterocolitica* and *Erwinia carotovora*. Food Control 24:78–85. https://doi.org/10.1016/j.foodcont.2011.09.006

Turner TR et al (2013) Comparative metatranscriptomics reveals kingdom level changes in the rhizosphere microbiome of plants. ISME J 7:2248. https://doi.org/10.1038/ismej.2013.119

Uren NC (2000) Types, amounts, and possible functions of compounds released into the rhizosphere by soil-grown plants. In: The rhizosphere. CRC Press, Boca Raton, pp 35–56

van der Heijden MG, Schlaeppi K (2015) Root surface as a frontier for plant microbiome research. Proc Natl Acad Sci 112:2299–2300. https://doi.org/10.1073/pnas.1500709112

Van der Heijden MG, Boller T, Wiemken A, Sanders IR (1998) Different arbuscular mycorrhizal fungal species are potential determinants of plant community structure. Ecology 79:2082–2091. https://doi.org/10.1890/0012-9658(1998)079[2082:DAMFSA]2.0.CO;2

Van Der Heijden MG, Martin FM, Selosse MA, Sanders IR (2015) Mycorrhizal ecology and evolution: the past, the present, and the future. New Phytol 205:1406–1423. https://doi.org/10.1111/nph.13288

Vandenkoornhuyse P et al (2007) Active root-inhabiting microbes identified by rapid incorporation of plant-derived carbon into RNA. Proc Natl Acad Sci 104:16970–16975. https://doi.org/10.1073/pnas.0705902104

Vandenkoornhuyse P, Quaiser A, Duhamel M, Le Van A, Dufresne A (2015) The importance of the microbiome of the plant holobiont. New Phytol 206:1196–1206. https://doi.org/10.1111/nph.13312

Viti C et al (2010) Compost effect on plant growth-promoting rhizobacteria and mycorrhizal fungi population in maize cultivations. Compost Science & Utilization 18:273–281. https://doi.org/10.1080/1065657X.2010.10736966

Vorholt JA, Vogel C, Carlström CI, Mueller DB (2017) Establishing causality: opportunities of synthetic communities for plant microbiome research. Cell Host Microbe 22:142–155. https://doi.org/10.1016/j.chom.2017.07.004

Wagner MR, Lundberg DS, Coleman-Derr D, Tringe SG, Dangl JL, Mitchell-Olds T (2014) Natural soil microbes alter flowering phenology and the intensity of selection on flowering time in a wild Arabidopsis relative. Ecol Lett 17:717–726. https://doi.org/10.1111/ele.12276

Wang J et al (2016) Antifungal new oxepine-containing alkaloids and xanthones from the deep-sea-derived fungus *Aspergillus versicolor* SCSIO 05879. J Agric Food Chem 64:2910–2916. https://doi.org/10.1021/acs.jafc.6b00527

Wei J, van Loon JJ, Gols R, Menzel TR, Li N, Kang L, Dicke M (2014) Reciprocal crosstalk between jasmonate and salicylate defence-signalling pathways modulates plant volatile emission and herbivore host-selection behaviour. J Exp Bot 65:3289–3298. https://doi.org/10.1093/jxb/eru181

Weisskopf L et al (2006) White lupin has developed a complex strategy to limit microbial degradation of secreted citrate required for phosphate acquisition. Plant Cell Environ 29:919–927

Werner GD, Kiers ET (2015) Order of arrival structures arbuscular mycorrhizal colonization of plants. New Phytol 205:1515–1524. https://doi.org/10.1111/nph.13092

Yadav J, Verma JP, Jaiswal DK, Kumar A (2014) Evaluation of PGPR and different concentration of phosphorus level on plant growth, yield and nutrient content of rice (*Oryza sativa*). Ecol Eng 62:123–128. https://doi.org/10.1016/j.ecoleng.2013.10.013

Yeoh YK et al (2017) Evolutionary conservation of a core root microbiome across plant phyla along a tropical soil chronosequence. Nat Commun 8:215. https://doi.org/10.1038/s41467-017-00262-8

Yoneyama K et al (2008) Strigolactones, host recognition signals for root parasitic plants and arbuscular mycorrhizal fungi, from Fabaceae plants. New Phytol 179:484–494. https://doi.org/10.1111/j.1469-8137.2008.02462.x

Zavala-Gonzalez EA et al (2017) *Arabidopsis thaliana* root colonization by the nematophagous fungus *Pochonia chlamydosporia* is modulated by jasmonate signaling and leads to accelerated flowering and improved yield. New Phytol 213:351–364. https://doi.org/10.1111/nph.14106

Zgadzaj R, Garrido-Oter R, Jensen DB, Koprivova A, Schulze-Lefert P, Radutoiu S (2016) Root nodule symbiosis in Lotus japonicus drives the establishment of distinctive rhizosphere, root, and nodule bacterial communities. Proc Natl Acad Sci 113:E7996–E8005. https://doi.org/10.1073/pnas.1616564113

Zhang H, Xie X, Kim MS, Kornyeyev DA, Holaday S, Paré PW (2008) Soil bacteria augment Arabidopsis photosynthesis by decreasing glucose sensing and abscisic acid levels in planta. Plant J 56:264–273. https://doi.org/10.1111/j.1365-313x.2008.03593.x

Bacteria-Inducing Legume Nodules Involved in the Improvement of Plant Growth, Health and Nutrition

4

Encarna Velázquez, Lorena Carro, José David Flores-Félix, Esther Menéndez, Martha-Helena Ramírez-Bahena, and Alvaro Peix

Abstract

Bacteria-inducing legume nodules are known as rhizobia and belong to the class *Alphaproteobacteria* and *Betaproteobacteria*. They promote the growth and nutrition of their respective legume hosts through atmospheric nitrogen fixation which takes place in the nodules induced in their roots or stems. In addition, rhizobia have other plant growth-promoting mechanisms, mainly solubilization of phosphate and production of indoleacetic acid, ACC deaminase and siderophores. Some of these mechanisms have been reported for strains of rhizobia which are also able to promote the growth of several nonlegumes, such as cereals, oilseeds and vegetables. Less studied are the mechanisms that have the rhizobia to promote the plant health; however, these bacteria are able to exert biocontrol of some phytopathogens and to induce the plant resistance. In this chapter, we revised the available data about the ability of the legume nodule-inducing bacteria for improving the plant growth, health and nutrition of both legumes and nonlegumes. These data showed that rhizobia meet all the requirements of sustainable agriculture to be used as bio-inoculants allowing the total or partial replacement of chemicals used for fertilization or protection of crops.

E. Velázquez (✉)
Departamento de Microbiología y Genética, Universidad de Salamanca, Salamanca, Spain

Unidad Asociada Universidad de Salamanca- CSIC 'Interacción Planta-Microorganismo', Salamanca, Spain
e-mail: evp@usal.es

L. Carro · J. D. Flores-Félix · E. Menéndez · M.-H. Ramírez-Bahena
Departamento de Microbiología y Genética, Universidad de Salamanca, Salamanca, Spain

A. Peix
Unidad Asociada Universidad de Salamanca- CSIC 'Interacción Planta-Microorganismo', Salamanca, Spain

IRNASA-CSIC, Salamanca, Spain

V. Kumar et al. (eds.), *Microbiome in Plant Health and Disease*,
https://doi.org/10.1007/978-981-13-8495-0_4

Keywords
Bacteria · Root nodule · Plant growth · Sustainable agriculture

4.1 Introduction

Currently, two of the main challenges of global agriculture are the achievement of a sustainable crop production and the protection of natural environments. The increase of the world population requires an increase in food production but using agronomic practices that preserve the environment. In order to achieve these aims, the Food and Agriculture Organization of the United Nations (FAO) has proposed to declare the year 2020 as the International Year of Plant Health (IYPH 2020). Obtaining healthier plants implies the protection of the world plant resources from pests (https://www.ippc.int/en/iyph/). According to FAO expectations, healthier plants allow us to obtain higher crop yields avoiding diversity losses, to reduce the hunger and poverty and to achieve a safer trade, a higher economic development and a sustainable health. All of these aims are included in the goals of the Agenda 2030 for Sustainable Development, launched by the United Nations in September of the year 2015 (https://www.un.org/sustainabledevelopment/sustainable-development-goals/).

The increase in crop production that is necessary for the eradication of hunger and malnutrition in the world requires agronomic practices that are not just limited to the control of pests. The fertilization of crops, to date mainly based on the application of chemical fertilizers, is also essential to increase their productivity and get an adequate amount of food for the ever-growing world's population. Moreover, consumers currently also increasingly demand healthy and safe foods, which go beyond the obtaining of healthier plants themselves. It is hard to combine agronomic sustainable practices with the obtaining of safer and healthier plants because the current agronomic practices need to be changed. These changes involve the total or partial replacement of chemical fertilizers and pesticides by biofertilizers and biopesticides in order to protect the health of all living beings and also to preserve the environment.

Biofertilizers and biopesticides are mainly constituted by microorganisms which exert a positive effect on the growth, nutrition and health of the plants (Berg 2009; Berendsen et al. 2012; Abhilash et al. 2016; Vejan et al. 2016; Berg et al. 2017), and they are key factors for plant growth and protection (Berg et al. 2017). The plant microbiome, either rhizospheric or endospheric, is a determinant of the plant health, growth and nutrition (Berendsen et al. 2012; Gaiero et al. 2013; Santoyo et al. 2016; Berg et al. 2017). However, many species of bacteria present in the plant microbiome are opportunistic human pathogens (Mendes et al. 2013) and, despite some of them are plant growth promoters, they cannot be used as biofertilizers or biopesticides (Menéndez et al. 2016).

Within the plant beneficial bacteria that are also safe for human health, we can highlight the rhizobia, a diverse group of bacteria able to induce nodules in roots or stems of legumes where they carry out the nitrogen fixation (Velázquez et al. 2017b).

After their use as inoculants for more than one century, the rhizobia have proven to be non-pathogenic for humans, animals and plants. Moreover, they are able to improve plant growth and nutrition and to produce compounds, such as siderophores, involved in the biological control of plant pathogens (Gopalakrishnan et al., 2015; Vargas et al. 2017; Velázquez et al. 2017a).

In the present chapter, we revise the current knowledge about the plant growth mechanisms presented by strains belonging to different genera of rhizobia, as well as the effects of their inoculation on different plants from the point of view of their health, growth and nutrition.

4.2 Diversity of Bacteria-Inducing Legume Nodules

The existence of nodules in the roots of legumes was first reported in the seventeenth century by Malpighi, but it was at the end of the nineteenth century when Beijerinck (1888) isolated for the first time a bacterium from nodules of *Vicia*, which was initially named *Bacillus radicicola*. Later, this bacterium was renamed as *Rhizobium leguminosarum* (Frank 1889) and, until now, the bacteria nodulating legumes are generically called rhizobia. The rhizobia currently form a complex group of bacteria which belong to different phyla, classes, orders, families, and genera (Fig. 4.1) and are able to establish nitrogen-fixing symbioses with different legumes around the world.

The species nodulating legumes described before the year 2017 were recorded by Velázquez et al. (2017b) and those described from this year to date are listed in Table 4.1.

Most of rhizobia reported to date belong to the class *Alphaproteobacteria* within the phylum *Proteobacteria* and nodulate legumes from the subfamily Papilionoideae. They are distributed in several families of the order *Rhizobiales* (Velázquez et al. 2017b), and most of them belong to the genus *Rhizobium*, included in the family *Rhizobiaceae* (Conn, 1938) together with the old genera *Allorhizobium* (de Lajudie et al. 1998) and *Ensifer* (previously named *Sinorhizobium*) (Judicial Commission of the International Committee on Systematic of Prokaryotes, 2008) and the new genera *Neorhizobium* (Mousavi et al. 2014) and *Pararhizobium* (Mousavi et al. 2015).

All these mentioned genera contain species which present rapid growth on media containing mannitol as carbon source, whereas the genera *Bradyrhizobium* (Jordan, 1982) and *Azorhizobium* (Dreyfus et al. 1988) contain slow-growing species. They were included into the families *Bradyrhizobiaceae* (Garrity et al. 2005), whose correct name is *Nitrobacteraceae* (Validation list 107, 2016), and *Hyphomicrobiaceae* (Babudieri 1950; Skerman et al. 1980), respectively. Following the criteria of the growth rate in yeast mannitol agar (Vincent 1970), a new genus named *Mesorhizobium*, with an intermediate growth rate between the genera *Rhizobium* and *Bradyrhizobium*, was split from genus *Rhizobium* by Jarvis et al. (1997). The genus *Mesorhizobium* belongs to the family *Phyllobacteriaceae* (Mergaert and Swings 2005; Validation list No.107 2006).

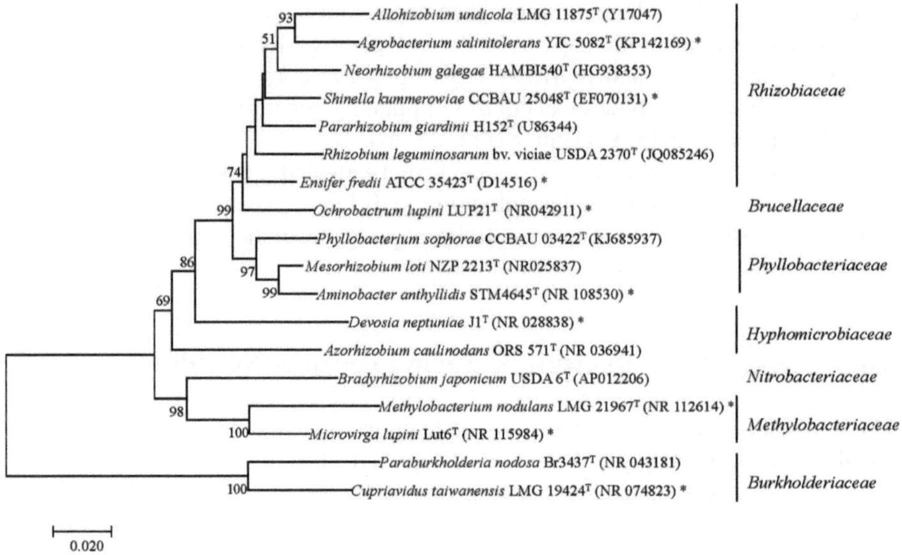

Fig. 4.1 Neighbour-joining phylogenetic tree based on nearly complete 16S rRNA gene sequences of type strains of 18 type species inducing legume nodules distributed in 7 families within the order *Rhizobiales*. The significance of each branch is indicated by a percentage of a bootstrap value calculated for 1000 subsets. Bar, 2 nt substitutions by 100 nt. Evolutionary analyses were conducted in MEGA7 software. Asterisks show the type strains of species that nodulate legumes, when the type species of the genus is not the one with this ability

Also, several species of non-classic rhizobial genera belonging to the *Alphaproteobacteria* have been reported as legume-nodulating bacteria (Velázquez et al. 2017b). Some of these genera belong to families that also contain classic rhizobia, such as *Phyllobacterium* (Valverde et al. 2005; Jiao et al. 2015) and *Aminobacter* (Maynaud et al. 2012) from the family *Phyllobacteriaceae*; *Shinella* (Lin et al. 2008) and *Agrobacterium* (Yan et al. 2017) from the family *Rhizobiaceae* and *Devosia* (Rivas et al. 2002) from the family *Hyphomicrobiaceae*. Other genera belong to families that classically did not include rhizobia, such as *Methylobacterium* (Sy et al. 2001) and *Microvirga* (Ardley et al. 2012; Radl et al. 2014) from the family *Methylobacteriaceae* and *Ochrobactrum* from the family *Brucellaceae* (Trujillo et al. 2005; Zurdo-Piñeiro et al. 2007).

Since the year 2000, several species belonging to two genera from *Betaproteobacteria* have also been reported as being able to induce nodules in several legumes (Velázquez et al. 2017b). When the first nodulating strains of *Betaproteobacteria* were reported, they were included in species of the genera *Burkholderia* and *Ralstonia* (Moulin et al. 2001; Chen et al. 2003), but they are currently included in the genera *Paraburkholderia* (Sawana et al. 2014; Dobritsa and Samadpour, 2016) and *Cupriavidus* (Vandamme and Coenye 2004) in both cases belonging to the family *Burkholderiaceae* (Velázquez et al. 2017b).

Table 4.1 Recently described species of rhizobia able to nodulate legumes

Species	Host legume or nodulated legumes	References
Order *Rhizobiales*, family *Rhizobiaceae*		
Genus *Rhizobium*		
R. hidalgonense	*Phaseolus vulgaris*	Yan et al. (2017a, b)
R. esperanzae	*Phaseolus vulgaris*	Cordeiro et al. (2017)
R. hedysari	*Hedysarum multijugum*	Xu et al. (2017)
Genus *Ensifer*		
E. shofinae	*Glycine max*	Chen et al. (2017)
Genus *Agrobacterium*		
A. salinitolerans	*Sesbania cannabina*	Yan et al. (2017)
Order *Rhizobiales*, family *Phyllobacteriaceae*		
Genus *Mesorhizobium*		
M. delmotii	*Anthyllis vulneraria*	Mohamad et al. (2017)
M. prunaredense	*Anthyllis vulneraria*	Mohamad et al. (2017)
M. helmanticense	*Lotus corniculatus*	Marcos-García et al. (2017)
M. zhangyense	*Thermopsis lanceolata*	Xu et al. (2018)
M. wenxiniae	*Cicer arietinum*	Zhang et al. (2018)
M. sanjuanii	*Lotus tenuis*	Sannazzaro et al. (2018)
Order *Rhizobiales*, family *Nitrobacteriaceae* ('*Bradyrhizobiaceae*')		
Genus *Bradyrhizobium*		
B. centrolobii	*Centrolobium paraense*	Michel et al. (2017)
B. macuxiense	*Centrolobium paraense*	Michel et al. (2017)
B. brasilense	*Vigna unguiculata*, *Macroptilium atropurpureum*	Martins da Costa et al. (2017)
B. sacchari	*Vigna unguiculata*, *Macroptilium atropurpureum*, *Cajanus cajan*	de Matos et al. (2017)
B. mercantei	*Deguelia costata*	Helene et al. (2017)
B. cajanii	*Cajanus cajan*	Araújo et al. (2017)
B. forestalis	*Inga* sp., *Swartzia* sp.	Martins da Costa et al. (2018)
B. algeriense	*Retama sphaerocarpa*	Ahnia et al. (2018)
B. ripae	*Indigofera rautanenii*, *Chamaecrista biensis*, *Vigna unguiculata*	Bünger et al. (2018)
B. shewense	*Erythrina brucei*	Aserse et al. (2017)
Order *Burkholderiales*, family *Burkholderiaceae*		
Genus *Paraburkholderia*		
P. piptadeniae	*Piptadenia gonoacantha*	Bournaud et al. (2017)
P. ribeironis	*Piptadenia gonoacantha*	Bournaud et al. (2017)

4.3 Plant Growth-Promoting Mechanisms

The bacteria able to induce legume nodules present direct and indirect mechanisms of plant growth promotion. The direct mechanisms include nitrogen fixation, phosphate solubilization and production of phytohormones and ACC deaminase, while the indirect mechanisms include production of siderophores, which can be also considered a direct mechanism because it enhances the Fe uptake by plants (García-Fraile et al. 2012; Suárez-Moreno et al. 2012; Laranjo et al. 2014; Das et al. 2017; Gopalakrishnan et al. 2015; Patil et al., 2017; Vargas et al. 2017; Velázquez et al. 2017a).

Nitrogen fixation was the first-studied plant growth-promoting mechanism of rhizobia since Hellriegel and Wilfarth at the end of the nineteenth century established that legume nodules are the responsible for nitrogen fixation (Hellriegel and Wilfarth 1888). From this time to date, many research works have focused on this ability in order to select the most effective rhizobial strains to be used as legume inoculants (Catroux et al. 2001; Checcucci et al. 2017). Nevertheless, the classic alpha rhizobia are specialized in the symbiotic nitrogen fixation with legumes (Remigi et al. 2016), and then, improvement of plant growth via nitrogen fixation is limited to these plants.

Phosphate solubilization is the second plant growth-promoting mechanism involved in nutrient mobilization presented by rhizobia (Rodríguez and Fraga, 1999; Thakur et al. 2014). Within them, the most active phosphate solubilizers in vitro are the species included into the genus *Mesorhizobium* (Peix et al. 2001; Rivas et al. 2006; Verma et al. 2013; Imen et al. 2015; Wdowiak-Wróbel and Małek 2016; Brígido et al. 2017), although this mechanism is also presented by strains of *Rhizobium* (Chabot et al. 1996; Antoun et al. 1998; Alikhani et al. 2007; Abril et al. 2007; Sridevi et al. 2007; Flores-Félix et al. 2013; Dahale et al. 2016; Othman and Tamimi 2016; Jiménez-Gómez et al. 2018), *Ensifer* (formerly *Sinorhizobium*) (Ormeño et al. 2007; Villar-Igea et al. 2007) and *Bradyrhizobium* (Boiero et al. 2007).

Also within the direct mechanisms, one of the most widely analysed is the production of the phytohormone indoleacetic acid (IAA), which is widely extended among rhizobial species nodulating legumes, such as those from the genus *Rhizobium* (Datta and Basu 2000; Bhattacharjee et al. 2012; García-Fraile et al. 2012; Kumar and Ram 2012; Flores-Félix et al. 2013; Jiménez-Gómez et al. 2018), *Allorhizobium* (Ghosh et al. 2015), *Ensifer* (formerly *Sinorhizobium*) (Bianco and Defez 2009; Dubey et al. 2010), *Mesorhizobium* (Wdowiak-Wróbel and Małek 2016; Vieira et al. 2017) and *Bradyrhizobium* (Boiero et al. 2007).

The production of aminocyclopropane-1-carboxylate (ACC) deaminase, responsible for the cleavage of the ethylene precursor ACC into ammonia and α-ketobutyrate, has been reported in different species of rhizobia from different genera (Nascimento et al. 2014, 2018), such as *Rhizobium* (Ma et al. 2003; Duan et al. 2009), *Allorhizobium* (Ghosh et al. 2015), *Ensifer* (formerly *Sinorhizobium*) (Ma et al. 2004; Kong et al. 2015), *Mesorhizobium* (Nascimento et al. 2012) and *Bradyrhizobium* (Rangel et al. 2017), and in nodulating species of *Methylobacterium* (Ekimova et al. 2018).

Different genera of rhizobia have been reported to produce siderophores (Carson et al. 2000; García-Fraile et al. 2012; Gopalakrishnan et al., 2015; Vargas et al. 2017; Velázquez et al. 2017), for example, *Rhizobium* (Patel et al. 1988; Carson et al. 1992; Wright et al. 2013; Jiménez-Gómez et al. 2018), *Mesorhizobium* (Berraho et al. 1997; Datta and Chakrabartty 2014; Wdowiak-Wróbel and Małek 2016; Brígido et al. 2017; Demissie et al. 2018), *Bradyrhizobium* (Nambiar and Sivaramakrishnan

1987; Lesueur et al. 1993; Abd-Alla 1998; Khandelwal et al. 2002; Boiero et al. 2007), *Allorhizobium* (Ghosh et al. 2015) and *Ensifer* (Lynch et al. 2001).

Some of these plant growth-promoting mechanisms have also been reported for several strains of *Paraburkholderia* and *Cupriavidus* from *Betaproteobacteria*, among which we must highlight the ability to fix nitrogen in symbiosis with several legumes (Remigi et al. 2016). Moreover, some species of *Paraburkholderia* have been shown to be able to produce indoleacetic acid, siderophores or ACC deaminase (Suárez-Moreno et al. 2012). Concretely, the species *Paraburkholderia tuberum* solubilizes phosphate and produces siderophores (Angus et al. 2013).

4.4 Growth Promotion of Legumes and Nonlegumes

The growth promotion of legumes by rhizobia via nitrogen fixation has been widely studied (Gopalakrishnan et al., 2015; Vargas et al. 2017; Velázquez et al. 2017b), and the inoculation with rhizobia of some legumes, such as soybean, has been performed for several decades in America with increases in the production, overall in South American countries (Leggett et al. 2017; Vargas et al. 2017). Moreover, increases in the production of other legumes, such as *Phaseolus vulgaris* and *Cajanus cajan,* have been obtained after the inoculation with *Rhizobium* and *Bradyrhizobium* strains, respectively (Mulas et al. 2011; Araújo et al. 2015; Koskey et al. 2017; Barros et al. 2018; Samago et al. 2018; Wolde-Meskel et al. 2018; Yanni et al. 2018).

It has also been reported that the co-inoculation of different rhizobial strains can improve the yield of legumes such as common bean (de Oliveira Longatti et al. 2013; Diez-Mendez et al. 2015; da Conceição Jesus et al. 2018). In the same line, the co-inoculation with rhizobia and other bacteria increased the nitrogen content on soybean (Subramanian et al. 2015) and improved the growth of chickpea (Verma et al. 2012; Yadav and Verma 2014; Prasanna et al. 2017), galega (Egamberdieva et al. 2010), lentil (Khanna and Sharma 2011), soybean (Hungria et al. 2013; Nimnoi et al. 2014; Htwe et al. 2018), peanut (Vicario et al. 2016) and mungbean (Kaur and Khanna 2016; Tarafder et al. 2016; Qureshi et al. 2011).

The co-inoculation of rhizobia and arbuscular mycorrhiza increases the nitrogen fixation in common bean (Tajini et al. 2011) and soybean (Meng et al. 2015), the nitrogen content on chickpea (Tavasolee et al. 2011) and pigeon pea (Bhattacharjee and Sharma 2012) and the productivity of pea (Shinde and Thakur 2016), cowpea (Haro et al. 2018), soybean (Hemmat Jou and Besalatpour 2018), *Styloshantes* (Crespo Flores et al. 2014), faba bean in alkaline soils (Abd-Alla et al. 2014; Hemid et al. 2014) and garden pea in acidic soils (Bai et al. 2017). Dual inoculations of rhizobia and arbuscular mycorrhiza also increase the grain protein content in chickpea under moderate water deficit (Oliveira et al. 2017).

In the last decades, also the study of the effect of rhizobial inoculation on the legume growth under different stresses is gaining interest, and several works have been performed in different legumes (Naveed et al. 2017). Drought and salt stresses are major limiting factors to plant productivity; nevertheless, inoculation with selected strains of rhizobia able to survive, grow and effectively nodulate legumes under these stress conditions can improve their productivity, quality and drought stress response (Faghire et al. 2012; Aamir et al. 2013; El-Akhal et al. 2013; Sharma et al. 2013; Bertrand et al. 2015; Staudinger et al. 2016; Yanni et al. 2016; Wang et al. 2016; Defez et al. 2017; Egamberdieva et al. 2017; Oliveira et al. 2017).

Several works also reported that the co-inoculation of different rhizobial strains (Ali et al. 2017; Ullah et al. 2017) and that of rhizobia and other bacteria or arbuscular mycorrhizal fungi can be a strategy to mitigate salt or drought stress (Ahmad et al. 2011a, b, 2012, 2013; Soliman et al. 2012; Martínez et al. 2015; Cerezini et al. 2016; Egamberdieva et al. 2016a, b; Ren et al. 2016; Zhu et al. 2016; da Piedade Melo et al. 2017; Fukami et al. 2017; Oliveira et al. 2017). Moreover, the co-inoculation of rhizobia and other bacteria can alleviate other stresses, such as copper stress (Challougui et al. 2015; Fatnassi et al. 2015).

Concerning the promotion of growth of nonlegumes, although the first works were carried out in the 1990s (Chabot et al. 1996; Yanni et al. 1997), most works have been carried out after the year 2000 (Velázquez et al. 2017a). Several of these works focused on the growth promotion of cereals by *Rhizobium* in rice (Yanni et al. 2001; Yanni and Dazzo 2010; Bhattacharjee et al. 2012; Granada et al. 2014), maize (Gutiérrez-Zamora and Martínez-Romero 2001; Shing et al. 2013) and wheat (Yanni et al. 2016) and by *Mesorhizobium* strains in barley (Peix et al. 2001). Although there are few reports to date, co-inoculation with rhizobia and other bacteria also increases the growth of some cereals, such as rice (Hasan et al. 2014; Tan et al. 2015).

Several works also showed that rhizobial inoculation also increased the growth of oil-containing plants such as canola and sunflowers, with high interest for human nutrition (McKevith 2005) and biodiesel production (Pimentel and Patzek 2005, Ge et al. 2017). The promotion of growth and the nitrogen uptake increase were reported by Alami et al. (2000) after the inoculation of a strain from the genus *Rhizobium* in sunflower plantlets. The inoculation with strains of the genus *Rhizobium* enhances the root growth of canola plants (Noel et al. 1996) and, under salinity stress conditions, treatments with different rhizobial strains increase the plant height and the dry weight of canola shoots and, moreover, the leaf area and relative water content (Saghafi et al. 2018).

In addition, the ability of rhizobia to promote the growth of fresh vegetables has been studied by several authors, dating also the first works in the 1990s (Chabot et al. 1996; Antoun et al. 1998). Nevertheless, most studies have been carried out in the recent years showing that *Rhizobium* strains are able to promote the growth and quality of tomato and pepper (García-Fraile et al. 2012), lettuce and carrots (Flores-Félix et al. 2013), strawberries (Flores-Félix et al. 2015, 2018), arugula (Rubio-Canalejas et al. 2016) and spinach (Jiménez-Gómez et al. 2018). The high potential of rhizobia to promote the growth of vegetables, together with the high safety level of these bacteria, highlights the need to perform more studies about the effect of different rhizobial species on the growth of other freshly consumed vegetables.

4.5 Biocontrol Mechanisms

The mechanisms of biocontrol presented by bacteria nodulating legumes have been less studied than those involved in plant growth promotion. Nevertheless, for some strains belonging to several rhizobial genera and species, different biocontrol mechanisms have been reported, including mycoparasitism, production of antibiotics and bacteriocins, antifungal metabolites, such as hydrocyanic acid (HCN), and phytoalexins, as well as the induction of systemic resistance in plants (Deshwal et al. 2003b; Das et al. 2017).

Concerning mycoparasitism, in 1978, it was reported that *Bradyrhizobium japonicum* colonized growing hyphal tips of *Phytophthora megasperma* being observed inside the hyphae. A decrease in the symptoms was observed with the application of *B. japonicum* to the soil after soybean planting suggesting that saprophytic soil rhizobia may reduce *Phytophthora* root rot by parasitizing hyphae of the fungus (Tu 1978). Also, Antoun et al. (1978) showed that strains of *Ensifer meliloti* (*Sinorhizobium meliloti*) were effective against *Fusarium oxysporum* in lucerne plants.

After this date, several works reported the in vitro inhibition of several fungi by strains of different rhizobial genera. For example, some *Bradyrhizobium* strains inhibit the mycelial growth and sclerotial formation and germination of *Sclerotium rolfsii* (Balasundaran and Sarbhoy 1988) and *Rhizoctonia solani* (Kelemu et al. 1995). In the same line, different fast-growing rhizobial strains are able to inhibit the growth of *Phytophtora cinnamomi* (Malajczuk et al. 1984), *Sclerotium rolfsii* (Balasundaran and Sarbhoy 1988), *Fusarium*, *Pythium* and *Rhizoctonia* (Ozkoc and Deliveli 2001).

In 1978, the production of bacteriocins by *Rhizobium trifolii* strains (currently *R. leguminosarum*) was reported, which were dominant in mixed cultures and were growing in peat, suggesting that they have advantages for competition (Schwinghamer and Brockwell 1978). Also, bacteriocin production by *Rhizobium japonicum* (currently *B. japonicum*) was also reported, although in this case, the producing strains were less competitive than the nonproducing ones (Gross and Vidaver 1978). More recently, the production of bacteriocins has been reported for other strains from *Rhizobium* (Hafeez et al. 2005; Ansari and Rao 2014), *Bradyrhizobium* (Hafeez et al. 2005) and several rhizobial strains nodulating mothbean, clusterbean and mungbean (Mondal et al. 2017). In addition, the genome of a bacteriocin-producing strain of *B. japonicum* has been sequenced, obtaining a better understanding of this molecule (Kohlmeier et al. 2015).

The production of peptide antibiotics active against other rhizobial strains, such as trifolitoxin, has also been reported for *R. leguminosarum* sv. *trifolii* (Triplett and Barta 1987) and *Rhizobium etli* (Robleto et al. 1997, 1998). The rhizobitoxine produced by *B. japonicum* (Minamisawa 1989) and *Bradyrhizobium elkanii* (Yuhashi et al. 2000) reduces the mycelial growth of *Macrophomina phaseolina* (Chakraborty and Purkayastha 1984). More recently, the analysis of the genetic region encoding a novel rhizobiocin produced by *R. leguminosarum* sv. *viciae* has been reported (Venter et al. 2001).

New genome sequences of rhizobia have shown the presence of bioclusters coding for secondary metabolites, such as the HCN, an antifungal metabolite produced by some rhizobia, although the abundance of strains producing this compound among rhizobia is low to date. For example, Antoun et al. (1998) reported the HCN production in three strains of *R. leguminosarum*, Arfaoui et al. (2006) in six strains of rhizobia-nodulating chickpea, Chandra et al. (2007) in a strain of *Mesorhizobium loti* and Priyanka and Wati (2017) in two strains of rhizobia isolated from *Vigna* nodules.

The production of siderophores, in addition to being a plant growth-promoting mechanism, is also a biocontrol mechanism because these compounds have high affinity for ferric iron-forming complexes which remove this ion from the rhizosphere preventing the growth and plant colonization by pathogenic microorganisms (Saha et al. 2016). Several types of siderophores are produced by different rhizobial species and genera (Das et al. 2017), and we recently found that *Rhizobium laguerreae* produced carboxylate-type siderophores (Jiménez-Gómez et al. 2018).

Induced systemic resistance is a plant defence mechanism against different types of pathogens which is elicited by several rhizobial strains alone (Elbadry et al. 2006) or combined with other bacteria (Dutta et al. 2008), endophytic fungi or arbuscular mycorrhiza (AM) (Martinuz et al. 2012; Gao et al. 2018a, b). The *Rhizobium etli* lipopolysaccharides have been shown to be agents inducing systemic resistance to infection by the cyst nematode *Globodera pallida* in potato roots (Reitz et al. 2000) and those of *R. leguminosarum* against the parasitic plant *Orobanche crenata* in pea (Mabrouk et al. 2016).

Rhizobial strains are also able to induce the production of some phytoalexins in plants treated with fungal pathogens, as occurred in the case of pea infected with *Fusarium solani* and inoculated with *R. leguminosarum* (Chakraborty and Chakraborty 1989), in the case of chickpea infected with *Fusarium oxysporum* and inoculated with rhizobia nodulating this legume (Arfaoui et al. 2007) and in the case of lucerne infected with *Phoma medicaginis* and treated with *Ensifer medicae* (*Sinorhizobium medicae*) and the AM *Funneliformis mosseae* (Gao et al. 2018a, b).

As occurred in the case of the plant growth-promoting mechanisms, those involved in the biocontrol of plant pathogens have been more studied in species of the classic rhizobial genera than in those of the new genus *Paraburkholderia*. Nevertheless, recent studies have been performed in legume-nodulating species of the genus *Paraburkholderia*, showing that three species of this genus showed antifungal activity (Eberl and Vandamme 2016). Therefore, also in this case, more studies should be performed to understand the biocontrol mechanisms in legume-nodulating species of this last genus.

4.6 Biocontrol of Phytopathogens from Legumes and Nonlegumes

Concerning the direct biocontrol of phytopathogens by rhizobia in plant assays, there are few studies to date (Das et al. 2017). Nevertheless, some studies showed the potential of rhizobial strains for the inhibition of some pathogenic fungi, such as *Macrophomina phaseolina* (Omar and Abd-Alla, 1998; Siddiqui et al. 2000; Arora et al. 2001; Deshwal et al. 2003a; Al-Ani et al. 2012), *Fusarium solani* (Omar and Abd-Alla, 1998; Rakib et al. 2012), *Fusarium oxysporum* (Arfaoui et al. 2006; Kumar et al. 2011), *Rhizoctonia solani* (Omar and Abd-Alla, 1998; Hemissi et al. 2011) and *Phytium* sp. (Bardin *et al.*, 2004) .

The co-inoculation of strains from *Rhizobium* and *Glomus* increased the biocontrol of the *Fusarium* wilt of chickpea (Singh et al. 2010) and the *Fusarium* root rot of *Phaseolus vulgaris* (Dar et al. 1997), also protecting *Vicia faba* plants against *Botrytis fabae* (Rabie 1998). The co-inoculations of *Rhizobium* and *Trichoderma* have been also shown to reduce the damping-off and root rot diseases in several legumes (Shaban and El-Bramawy 2011) and the incidence of collar rot disease caused by *Sclerotium rolfsii* in groundnut (Ganesan et al. 2007). In the same way, the co-inoculation of *Ensifer* (*Sinorhizobium*) and *Pseudomonas* significantly reduced *Fusarium* wilt in pigeon pea (Kumar et al. 2010).

Other studies showed a reduction in galling and nematode multiplication of Meloidogyne incognita in chickpea when the plants were inoculated with a strain of rhizobia nodulating this legume (Akhtar and Siddiqui 2008). The dual inoculation of *Rhizobium* and other *Pseudomonas* strains in lentils also controlled *Meloidogyne javanica* (Siddiqui et al. 2007). The co-inoculation of *Rhizobium* with *Pseudomonas* or *Bacillus* strains decreases the wilting of *Fusarium oxysporum* in lentils inoculated with this pathogen (Akhtar et al. 2010) and that of *Rhizobium* or *Bradyrhizobium* with *Bacillus* improved the bean root rot control in common bean and peanut, respectively (Estevez de Jensen et al. 2002; Yuttavanichakul et al. 2012).

The co-inoculation with rhizobia and arbuscular mycorrhiza could control soybean red crown rot in acidic soils (Gao et al. 2012). The co-inoculation of tomato with *Rhizobium etli* and the arbuscular mycorrhiza *Glomus intraradicis* leads to a 60% reduction in the galling by *Meloidogyne incognita* (Reimann et al. 2008). The tripartite inoculation of *Rhizobium* with *Glomus* and *Pseudomonas* also controlled the root rot disease in chickpea caused by *Meloidogyne incognita* and *M. phaseolina* (Akhtar and Siddiqui 2008).

The co-inoculation with *Rhizobium* and *Trichoderma* of faba bean plants has been shown to reduce 57%, on average, the incidence of chocolate spot disease produced by *Botrytis fabae* and increasing 23%, on average, of the yield of faba bean (Saber et al. 2009). Moreover, the dual inoculation of these microorganisms reduced the stem rot incidence promoting the growth of the groundnut (Ganesan et al. 2007), as well as the incidence of the damping-off and root rot in several legumes such as *Vicia, Cicer* and *Lupinus* (Shaban et al. 2011).

Although all these studies showed that rhizobia are promising bacteria to control different plant pathogens through different mechanisms, this ability has been poorly studied to date. Therefore, also taking into account the ability of these bacteria to improve the plant growth of legumes and nonlegumes and, especially, their safety as biofertilizers for human health, the effects of rhizobia on plant health should be further studied.

4.7 Conclusions

Bacteria-inducing legume nodules, commonly called rhizobia, are mainly known to produce beneficial effects on legumes via atmospheric nitrogen fixation. However, they are also able to promote the growth of other economically valuable crops, such as cereals, oleaginous plants or horticultural crops through other plant growth-promoting mechanisms, such as solubilization of phosphate and production of indoleacetic acid, among others. Since this group of bacteria is considered safe for human, animal and plant health and for the environment, they are good candidates for the formulation of biofertilizers. The ability of rhizobia to produce compounds involved in biocontrol and to induce systemic resistance in plants also makes them good candidates as biocontrollers, although research in this field is still limited. Thus, further studies are necessary to be performed in order to include rhizobia in the formulation of biopesticides.

Acknowledgments The authors would like to thank their numerous collaborators and students involved in this research over the years. Funding was provided by "Ministerio de Economía, Ciencia, Industria y Competitividad (MINECO)" through the Projects AGL2010-17380 and AGL2013-48098-P, by "Junta de Castilla y León" through the project SA058U16 and by the "Diputación de Salamanca (local Government)" through the Projects V113/463AC06 and 18VB2I/463AC06.

References

Aamir M, Aslam A, Khan MY, Jamshaid MU, Ahmad M, Asghar HN, Zahir ZA (2013) Co-inoculation with rhizobium and plant growth promoting rhizobacteria (PGPR) for inducing salinity tolerance in mung bean under field condition of semi-arid climate. Asian J Agri Biol 1:7–12

Abd-Alla MH (1998) Growth and siderophore production in vitro of *Bradyrhizobium* (Lupin) strains under iron limitation. Eur J Soil Biol 34:99–104

Abd-Alla MH, El-Enany AWE, Nafady NA, Khalaf DM, Morsy FM (2014) Synergistic interaction of *Rhizobium leguminosarum* bv. viciae and arbuscular mycorrhizal fungi as a plant growth promoting biofertilizers for faba bean (*Vicia faba* L.) in alkaline soil. Microbiol Res 169:49–58

Abhilash PC, Dubey RK, Tripathi V, Gupta VK, Singh HB (2016) Plant growth-promoting microorganisms for environmental sustainability. Trends Biotechnol 34:847–850

Abril A, Zurdo-Piñeiro JL, Peix A, Rivas R, Velázquez E (2007) Solubilization of phosphate by a strain of *Rhizobium leguminosarum* bv. Trifolii isolated from *Phaseolus vulgaris* in El Chaco Arido soil (Argentina). In: Velázquez E, Rodríguez-Barrueco C (eds) First international meeting on microbial phosphate solubilization. Developments in plant and soil sciences, vol 102. Springer, Dordrecht, pp 135–138

Ahmad M, Zahir ZA, Asghar HN, Asghar M (2011a) Inducing salt tolerance in mung bean through coinoculation with rhizobia and plant-growth-promoting rhizobacteria containing 1-aminocyclopropane-1-carboxylate deaminase. Can J Microbiol 57:578–589

Ahmad M, Zahir ZA, Asghar HN, Asghar M (2011b) The combined application of rhizobial strains and plant growth promoting rhizobacteria improves growth and productivity of mung bean (*Vigna radiata* L.) under salt-stressed conditions. Ann Microbiol 62:1321–1330

Ahmad M, Zahir ZA, Khalid M, Nazli F, Arshad M (2013) Efficacy of *Rhizobium* and *Pseudomonas* strains to improve physiology, ionic balance and quality of mung bean under salt-affected conditions on farmer's fields. Plant Physiol Biochem 63:170–176

Ahnia H, Bourebaba Y, Durán D, Boulila F, Palacios JM, Rey L, Ruiz-Argüeso T, Boulila A, Imperial J (2018) *Bradyrhizobium algeriense* sp. nov., a novel species isolated from effective nodules of *Retama sphaerocarpa* from Northeastern Algeria. Syst Appl Microbiol 41:333–339

Akhtar MS, Siddiqui ZA (2008) Biocontrol of a root-rot disease complex of chickpea by *Glomus intraradices*, *Rhizobium* sp. and *Pseudomonas striata*. Crop Prot 27:410–417

Akhtar MS, Shakeel U, Siddiqui ZA (2010) Biocontrol of *Fusarium* wilt by *Bacillus pumilus*, *Pseudomonas*, *Alcaligenes* and *Rhizobium* sp. on lentil. Turk J Biol 34:1–7

Alami Y, Achouak W, Marol C, Heulin T (2000) Rhizosphere soil aggregation and plant growth promotion of sunflowers by an exopolysaccharide-producing *Rhizobium* sp. strain isolated from sunflower roots. Appl Environ Microbiol 66:3393–3398

Al-Ani RA, Adhab MA, Mahdi MH, Abood HM (2012) *Rhizobium japonicum* as a biocontrol agent of soybean root rot disease caused by *Fusarium solani* and *Macrophomina phaseolina*. Plant Protect Sci 48:149–155

Ali Q, Zahir ZA, Asghar HN, Jamil A (2017) Inoculation with rhizobial consortium for improving the growth, yield and quality of maize under salt-stressed conditions. Pak J Agric Sci 54:97–105

Alikhani HA, Saleh-Rastin N, Antoun H (2007) Phosphate solubilization activity of rhizobia native to Iranian soils. In: Velázquez E, Rodríguez-Barrueco C (eds) First international meeting on microbial phosphate solubilization. Developments in plant and soil sciences, vol 102. Springer, Dordrecht, pp 35–41

Angus AA, Lee A, Lum MR, Shehayeb M, Hessabi R, Fujishige NA, Yerrapragada S, Kano S, Song N, Yang P, Estrada de los Santos P, de Faria SM, Dakora FD, Weinstock G, Hirsch AM (2013) Nodulation and effective nitrogen fixation of *Macroptilium atropurpureum* (siratro) by *Burkholderia tuberum*, a nodulating and plant growth promoting beta-proteobacterium, are influenced by environmental factors. Plant Soil 369:543–562

Ansari PG, Rao DLN (2014) Soybean rhizobia in Indian soils: populations, host specificity and competitiveness. Proc Natl Acad Sci, India Section B: Biol Sci 84:457–464

Antoun H, Bordeleau LM, Gagnon C (1978) Antagonisme entre *Rhizobium meliloti* at *Fusarium oxysporum* en relation avec lefficacite symbiotique. Can J Plant Sci 58:75–78

Antoun H, Beauchamp CJ, Goussard N, Chabot R, Lalande R (1998) Potential of *Rhizobium* and *Bradyrhizobium* species as plant growth promoting rhizobacteria on non-legumes: effect on radishes (*Raphanus sativus* L.). Plant Soil 204:57–67

Araújo J, Díaz-Alcántara CA, Velázquez E, Urbano B, González-Andrés F (2015) *Bradyrhizobium yuanmingense* related strains form nitrogen-fixing symbiosis with *Cajanus cajan* L. in Dominican Republic and are efficient biofertilizers to replace N fertilization. Sci Hortic 192:421–428

Araújo J, Flores-Félix JD, Igual JM, Peix A, González-Andrés F, Díaz-Alcántara CA, Velázquez E (2017) *Bradyrhizobium cajani* sp. nov. isolated from nodules of *Cajanus cajan*. Int J Syst Evol Microbiol 67:2236–2241

Ardley JK, Parker MA, De Meyer SE, Trengove RD, O'Hara GW, Reeve WG, Yates RJ, Dilworth MJ, Willems A, Howieson JG (2012) *Microvirga lupini* sp. nov., *Microvirga lotononidis* sp. nov. and *Microvirga zambiensis* sp. nov. are alphaproteobacterial root-nodule bacteria that specifically nodulate and fix nitrogen with geographically and taxonomically separate legume hosts. Int J Syst Evol Microbiol 62:2579–2588

Arfaoui A, Sifi B, Boudabous A, Hadrami IE, Chérif M (2006) Identification of *Rhizobium* isolates possessing antagonistic activity against *Fusarium oxysporum* f. sp. ciceris, the causal agent of *Fusarium* wilt of chickpea. J Plant Pathol 88:67–75

Arfaoui A, El Hadrami A, Mabrouk Y, Sifi B, Boudabous A, El Hadrami I, Daayf F, Chérif M (2007) Treatment of chickpea with *Rhizobium* isolates enhances the expression of phenylpropanoid defense-related genes in response to infection by *Fusarium oxysporum* f. sp. ciceris. Plant Physiol Biochem 45:470–479

Arora NK, Kang SC, Maheshwari DK (2001) Isolation of siderophore-producing strains of *Rhizobium meliloti* and their biocontrol potential against *Macrophomina phaseolina* that causes charcoal rot of groundnut. Curr Sci 81:673–677

Aserse AA, Woyke T, Kyrpides NC, Whitman WB, Lindström K (2017) Draft genome sequences of *Bradyrhizobium shewense* sp. nov. ERR11^T and *Bradyrhizobium yuanmingense* CCBAU 10071^T. Stand Genomic Sci 12:74

Babudieri B (1950) Natura delle cosidette "S-formen" delle leptospire. Loro identificazione con *Hyphomicrobium vulgare* Stutzer e Hartleb. Studio di quest. Ultimo germe. R.C. 1st Supplement Sanita Roma 13:580–591

Bai B, Suri VK, Kumar A, Choudhary AK (2017) Tripartite symbiosis of *Pisum–Glomus–Rhizobium* leads to enhanced productivity, nitrogen and phosphorus economy, quality, and biofortification in garden pea in a Himalayan acid alfisol. J Plant Nutr 40:600–613

Balasundaran V, Sarbhoy A (1988) Inhibition of plant pathogenic fungi by *Rhizobium japonicum*. Indian Phytopathol 41:128–130

Bardin SD, Huang H-C, Pinto J, Amundsen EJ, Erickson RS (2004) Biological control of *Pythium* damping-off of pea and sugar beet by *Rhizobium leguminosarum* bv. Viceae. Can J Bot 82:291–296

Barros LRN, Barbosa de Oliveira L, Barros Magalhães W, Oliveira Médici L, Pimentel C (2018) Interaction of biological nitrogen fixation with sowing nitrogen fertilization on common bean in the two seasons of cultivation in Brazil. J Plant Nutr 41:774–781

Beijerinck MW (1888) Cultur des *Bacillus radicicola* aus den Knöllchen. Bot Ztg 46:740–750

Berendsen RL, Pieterse CMJ, Bakker PAHM (2012) The rhizosphere microbiome and plant health. Trends Plant Sci 17:478–486

Berg G (2009) Plant–microbe interactions promoting plant growth and health: perspectives for controlled use of microorganisms in agriculture. Appl Microbiol Biotechnol 84:11–18

Berg G, Köberl M, Rybakova D, Müller H, Grosch R, Smalla K (2017) Plant microbial diversity is suggested as the key to future biocontrol and health trends. FEMS Microbiol Ecol 93. https://doi.org/10.1093/femsec/fix050

Berraho EL, Lesueur D, Diem HG, Sasson A (1997) Iron requirement and siderophore production in *Rhizobium ciceri* during growth on an iron-deficient medium. World J Microbiol Biotechnol 13:501–510

Bertrand A, Dhont C, Bipfubus M, Chalifour FP, Drouin P, Beauchamp CJ (2015) Improving salt stress responses of the symbiosis in alfalfa using salt-tolerant cultivar and rhizobial strain. Appl Soil Ecol 87:108–117

Bhattacharjee S, Sharma GD (2012) Effect of dual inoculation of arbuscular mycorrhiza and rhizobium on the chlorophyll, nitrogen and phosphorus contents of pigeon pea (*Cajanus cajan* L.). Adv Microbiol 2:561–564

Bianco C, Defez R (2009) *Medicago truncatula* improves salt tolerance when nodulated by an indole-3-acetic acid-overproducing *Sinorhizobium meliloti* strain. J Exp Bot 60:3097–3107

Boiero L, Perrig D, Masciarelli O, Penna C, Cassán F, Luna V (2007) Phytohormone production by three strains of *Bradyrhizobium japonicum* and possible physiological and technological implications. Appl Microbiol Biotechnol 74:874–880

Bournaud C, Moulin L, Cnockaert M, Faria S, Prin Y, Severac D, Vandamme P (2017) *Paraburkholderia piptadeniae* sp. nov. and *Paraburkholderia ribeironis* sp. nov., two root-nodulating symbiotic species of *Piptadenia gonoacantha* in Brazil. Int J Syst Evol Microbiol 67:432–440

Brígido C, Glick BR, Oliveira S (2017) Survey of plant growth-promoting mechanisms in native Portuguese chickpea *Mesorhizobium* isolates. Microb Ecol May 73:900–915

Bünger W, Grönemeyer JL, Sarkar A, Reinhold-Hurek B (2018) *Bradyrhizobium ripae* sp. nov., a nitrogen-fixing symbiont isolated from nodules of wild legumes in Namibia. Int J Syst Evol Microbiol. https://doi.org/10.1099/ijsem.0.002955

Carson KC, Dilworth MJ, Glenn AR (1992) Siderophore production and iron transport in *Rhizobium leguminosarum* bv. viciae MNF710. J Plant Nutr 15:2203–2220

Carson KC, Meyer JM, Dilworth MJ (2000) Hydroxamate siderophores of root nodule bacteria. Soil Biol Biochem 32:11–21

Catroux G, Hartmann A, Revellin C (2001) Trends in rhizobial inoculant production and use. Plant Soil 230:21–30

Cerezini P, Harumi Kuwano B, Barbosa dos Santos M, Terassi F, Hungria M, Nogueira MA (2016) Strategies to promote early nodulation in soybean under drought. Field Crops Res 196:160–167

Chabot R, Antoun H, Cescas MP (1996) Growth promotion of maize and lettuce by phosphate-solubilizing *Rhizobium leguminosarum* biovar. phaseoli. Plant Soil 184:311–321

Chakraborty U, Chakraborty BN (1989) Interaction of *Rhizobium leguminosarum* and *Fusarium solani* f. sp. pisi on pea affecting disease development and phytoalexin production. Can J Bot 67:1698–1701

Chakraborty U, Purkayastha RP (1984) Role of rhizobitoxine in protecting soybean roots from *Macrophomina phaseolina* infection. Can J Microbiol 30:285–289

Challougui I, Chibou FM, Saadani O, Jebara M, Jebara SH (2015) Impact of dual inoculation with *Rhizobium* and PGPR on growth and antioxidant status of *Vicia faba* L. under copper stress. Compt Rend Biol 338:241–254

Chandra S, Choure K, Dubey RC, Maheshwari DK (2007) Rhizosphere competent *Mesorhizobium loti* MP6 induces root hair curling, inhibits *Sclerotinia sclerotiorum* and enhances growth of Indian mustard (*Brassica campestris*). Braz J Microbiol 38:124–130

Checcucci A, DiCenzo GC, Bazzicalupo M, Mengoni A (2017) Trade, diplomacy, and warfare: the quest for elite rhizobia inoculant strains. Front Microbiol 8:2207

Chen WM, James EK, Prescott AR, Kierans M, Sprent JI (2003) Nodulation of *Mimosa* spp. by the beta-proteobacterium *Ralstonia taiwanensis*. Mol Plant-Microbe Interact 16:1051–1061

Chen WH, Yang SH, Li ZH, Zhang XX, Sui XH, Wang ET, Chen WX, Chen WF (2017) *Ensifer shofinae* sp. nov., a novel rhizobial species isolated from root nodules of soybean (*Glycine max*). Syst Appl Microbiol 40:144–149

Conn HJ (1938) Taxonomic relationships of certain non-sporeforming rods in soil. J Bacteriol 36:320–321

Cordeiro AB, Ribeiro RA, Helene LCF, Hungria M (2017) *Rhizobium esperanzae* sp. nov., a N2-fixing root symbiont of *Phaseolus vulgaris* from Mexican soils. Int J Syst Evol Microbiol 67:3937–3945

Crespo Flores G, Ramírez JF, González PJ, Hernández I (2014) Co-inoculation of *Rhizobium* strains and one of the arbuscular mycorrhizal fungus on *Stylosanthes guianensis* cv. CIAT-184. Cuban J Agric Sci 48:297–300

da Conceição Jesus E, de Almeida Leite R, do Amaral Bastos R, da Silva Aragão OO, Araújo AD (2018) Co-inoculation of *Bradyrhizobium* stimulates the symbiosis efficiency of *Rhizobium* with common bean. Plant Soil 425:201–215

da Piedade Melo A, Lopes Olivares F, Oliveira Médici L, Torres-Neto A, Barros Dobbss L, Pasqualoto Canellas L (2017) Mixed rhizobia and *Herbaspirillum seropedicae* inoculations with humic acid-like substances improve water-stress recovery in common beans. Chem Biol Techn Agric 4:6

Dahale SK, Prashanthi SK, Krishnaraj PU (2016) *Rhizobium* mutant deficient in mineral phosphate solubilization activity shows reduced nodulation and plant growth in green gram. Proc Natl Acad Sci, India Section B: Biol Sci 86:723–734

Dar GH, Zargar MY, Beigh GM (1997) Biocontrol of *Fusarium* root rot in the common bean (*Phaseolus vulgaris* L.) by using symbiotic *Glomus mosseae* and *Rhizobium leguminosarum*. Microb Ecol 34:74–80

Das K, Prasanna R, Saxena AK (2017) Rhizobia: a potential biocontrol agent for soilborne fungal pathogens. Folia Microbiol 62:425–435

Datta C, Basu PS (2000) Indole acetic acid production by a *Rhizobium* species from root nodules of a leguminous shrub, *Cajanus cajan*. Microbiol Res 155:123–127

Datta B, Chakrabartty PK (2014) Siderophore biosynthesis genes of *Rhizobium* sp. isolated from *Cicer arietinum* L. 3 Biotech 4:391–401

de Lajudie P, Laurent-Fulele WA, Torck U, Coopman R, Collins MD, Kersters K, Dreyfus B, Gillis M (1998) *Allorhizobium undicola* gen. nov., sp. nov., nitrogen-fixing bacteria that efficiently nodulate *Neptunia natans* in Senegal. Int J Syst Bacteriol 48:1277–1290

de Matos GF, Zilli JE, de Araújo JLS, Parma MM, Melo IS, Radl V, Baldani JI, Rouws LFM (2017) *Bradyrhizobium sacchari* sp. nov., a legume nodulating bacterium isolated from sugarcane roots. Arch Microbiol 199:1251–1258

de Oliveira Longatti SM, Marra LM, de Souza Moreira FM (2013) Evaluation of plant growth-promoting traits of *Burkholderia* and *Rhizobium* strains isolated from Amazon soils for their co-inoculation in common bean. Afr J Microbiol Res 7:948–959

Thakur D, Kaushal R, Shyam V (2014) Phosphate solubilising microorganisms: role in phosphorus nutrition of crop plants-a review. Agric Rev 35:159–171

Defez R, Andreozzi A, Dickinson M, Charlton A, Tadini L, Pesaresi P, Bianco C (2017) Improved drought stress response in alfalfa plants nodulated by an IAA over-producing *Rhizobium* strain. Front Microbiol 8:2466

Demissie N, Degefu T, Ergena A, Ojiewo C (2018) Phenotypic characteristics of rhizobial and non-rhizobial isolates recovered from root nodules of chickpea (*Cicer arietinum* L.) grown in Ethiopia. Afr J Microbiol Res 12:73–85

Deshwal VK, Dubey RC, Maheshwari DK (2003a) Isolation of plant growth-promoting strains of *Bradyrhizobium* (*Arachis*) sp. with biocontrol potential against *Macrophomina phaseolina* causing charcoal rot of peanut. Curr Sci 84:443–448

Deshwal V, Pandey P, Kang S, Maheshwari D (2003b) Rhizobia as a biological control agent against soil borne plant pathogenic fungi. Indian J Exp Biol 41:1160–1164

Diez-Mendez A, Menéndez E, García-Fraile P, Celador-Lera L, Rivas R, Mateos PF (2015) *Rhizobium cellulosilyticum* as a co-inoculant enhances *Phaseolus vulgaris* grain yield under greenhouse conditions. Symbiosis 67:135–141

Dobritsa AP, Samadpour M (2016) Transfer of eleven species of the genus *Burkholderia* to the genus *Paraburkholderia* and proposal of *Caballeronia* gen. nov. to accommodate twelve species of the genera *Burkholderia* and *Paraburkholderia*. Int J Syst Evol Microbiol 66:2836–2846

Dreyfus B, Garcia JL, Gillis M (1988) Characterization of *Azorhizobium caulinodans* gen. nov., sp. nov., a stem-nodulating nitrogen-fixing bacterium isolated from *Sesbania rostrata*. Int J Syst Bacteriol 38:89–98

Duan J, Müller KM, Charles TC, Vesely S, Glick BR (2009) 1-aminocyclopropane-1-carboxylate (ACC) deaminase genes in rhizobia from southern Saskatchewan. Microb Ecol 57:423–436

Dubey RC, Maheshwari DK, Kumar H, Choure K (2010) Assessment of diversity and plant growth promoting attributes of rhizobia isolated from *Cajanus cajan* L. African J Biotechnol 9:8619–8629

Dutta S, Mishra AK, Kumar BSD (2008) Induction of systemic resistance against fusarial wilt in pigeon pea through interaction of plant growth promoting rhizobacteria and rhizobia. Soil Biol Biochem 40:452–461

Eberl L, Vandamme P (2016) Members of the genus *Burkholderia*: good and bad guys. F1000 Res 5:1007

Egamberdieva D, Berg G, Lindström K, Räsänen LA (2010) Co-inoculation of *Pseudomonas* spp. with *Rhizobium* improves growth and symbiotic performance of fodder galega (*Galega orientalis* Lam.). Eur J Soil Biol 46:269–272

Egamberdieva D, Jabborova D, Berg G (2016a) Synergistic interactions between *Bradyrhizobium japonicum* and the endophyte *Stenotrophomonas rhizophila* and their effects on growth, and nodulation of soybean under salt stress. Plant Soil 405:35–45

Egamberdieva D, Li L, Lindström K, Räsänen LA (2016b) A synergistic interaction between salt-tolerant *Pseudomonas* and *Mesorhizobium* strains improves growth and symbiotic performance of liquorice (*Glycyrrhiza uralensis* Fish.) under salt stress. Appl Microbiol Biotechnol 100:2829–2841

Egamberdieva D, Reckling M, Wirtha S (2017) Biochar-based *Bradyrhizobium* inoculum improves growth of lupin (*Lupinus angustifolius* L.) under drought stress. Eur J Soil Biol 78:38–42

Ekimova GA, Fedorov DN, Tani A, Doronina NV, Trotsenko YA (2018) Distribution of 1-aminocyclopropane-1-carboxylate deaminase and D-cysteine desulfhydrase genes among type species of the genus *Methylobacterium*. Antonie Van Leeuwenhoek. https://doi.org/10.1007/s10482-018-1061-5

El-Akhal MR, Rincón A, Coba de la Peña T, Lucas MM, El Mourabit N, Barrijal S, Pueyo JJ (2013) Effects of salt stress and rhizobial inoculation on growth and nitrogen fixation of three peanut cultivars. Plant Biol (Stuttg) 15:415–421

Elbadry M, Taha RM, Eldougdoug KA, Gamal-Eldin H (2006) Induction of systemic resistance in faba bean (*Vicia faba* L.) to bean yellow mosaic potyvirus (BYMV) via seed bacterization with plant growth promoting rhizobacteria. J Plant Dis Protect 113:247–251

Estevez de Jensen C, Percich JA, Graham PH (2002) Integrated management strategies of bean root rot with *Bacillus subtilis* and *Rhizobium* in Minnesota. Field Crops Res 74:107–115

Faghire M, Mandri B, Oufdou K, Bargaz A, Ghoulam C, Ramírez-Bahena MH, Velázquez E, Peix A (2012) Identification at the species and symbiovar levels of strains nodulating *Phaseolus vulgaris* in saline soils of the Marrakech region (Morocco) and analysis of the *otsA* gene putatively involved in osmotolerance. Syst Appl Microbiol 35:156–164

Fatnassi IC, Chiboub M, Saadani O, Jebara M, Jebara SH (2015) Phytostabilization of moderate copper contaminated soils using co-inoculation of *Vicia faba* with plant growth promoting bacteria. J Basic Microbiol 55:303–311

Flores-Félix JD, Menéndez E, Rivera LP, Marcos-García M, Martínez-Hidalgo P, Mateos PF, Martínez-Molina E, Velázquez E, García-Fraile P, Rivas R (2013) Use of *Rhizobium leguminosarum* as a potential biofertilizer for *Lactuca sativa* and *Daucus carota* crops. J Plant Nutr Soil Sci 176:876–882

Flores-Félix JD, Marcos-García M, Silva LR, Menéndez E, Martínez-Molina E, Mateos PF, Velázquez E, García-Fraile P, Andrade P, Rivas R (2015) *Rhizobium* as plant probiotic for strawberry production under microcosm conditions. Symbiosis 67:25–32

Flores-Félix JD, Velázquez E, García-Fraile P, González-Andrés F, Silva LR, Rivas R (2018) *Rhizobium* and *Phyllobacterium* bacterial inoculants increase bioactive compounds and quality of strawberries cultivated in field conditions. Food Res Int 111:416–422

Frank B (1889) Ueber die Pilzsymbiose der Leguminosen. Bet Dtsch Bot Ges 7:332–346

Fukami J, de la Osa C, Ollero FJ, Megías M, Hungria M (2017) Co-inoculation of maize with *Azospirillum brasilense* and *Rhizobium tropici* as a strategy to mitigate salinity stress. Funct Plant Biol 45:328–339

Gaiero JR, McCall CA, Thompson KA, Day NJ, Best AS, Dunfield KE (2013) Inside the root microbiome: bacterial root endophytes and plant growth promotion. Am J Bot 100:1738–1750

Ganesan S, Kuppusamy RG, Sekar R (2007) Integrated management of stem rot disease (*Sclerotium rolfsii*) of groundnut (*Arachis hypogaea* L.) using *Rhizobium* and *Trichoderma harzianum* (ITCC-4572). Turk J Agric For 31:103–108

Gao X, Lu X, Wu M, Zhang H, Pan R, Tian J, Li S, Liao H (2012) Co-inoculation with rhizobia and AMF inhibited soybean red crown rot: from field study to plant defense-related gene expression analysis. PLoS One 7:e33977

Gao P, Guo Y, Li Y, Duan T (2018a) Effects of dual inoculation of AMF and rhizobium on alfalfa (*Medicago sativa*) root rot caused by *Microdochium tabacinum*. Australas Plant Pathol 47:195–203

Gao P, Li Y, Guo Y, Duan T (2018b) Co-inoculation of lucerne (*Medicago sativa*) with an AM fungus and a *Rhizobium* reduces occurrence of spring black stem and leaf spot caused by *Phoma medicaginis*. Crop Pasture Sci 69:933–943

García-Fraile P, Carro L, Robledo M, Ramírez-Bahena MH, Flores-Félix JD, Fernández MT, Mateos PF, Rivas R, Igual JM, Martínez-Molina E, Peix A, Velázquez E (2012) *Rhizobium* promotes non-legumes growth and quality in several production steps: towards a biofertilization of edible raw vegetables healthy for humans. PLoS One 7:e38122

Garrity GM, Bell JA, Lilburn T (2005) Family VII. Bradyrhizobiaceae fam. nov. In: Brenner DJ, Krieg NR, Staley JT, Garrity GM (eds) Bergey's manual of systematic bacteriology, second edition, vol. 2 (the Proteobacteria), part C (the Alpha-, Beta-, Delta-, and Epsilonproteobacteria). Springer, New York, pp 438–443

Ge JC, Yoon SK, Choi NJ (2017) Using canola oil biodiesel as an alternative fuel in diesel engines: a review. Appl Sci 7:881

Ghosh PK, De TK, Maiti TK (2015) Production and metabolism of indole acetic acid in root nodules and symbiont (*Rhizobium undicola*) isolated from root nodule of aquatic medicinal legume *Neptunia oleracea* Lour. J Bot 2015. ID 575067

Gopalakrishnan S, Sathya A, Vijayabharathi R, Varshney RK, Laxmipathi Gowda CL, Krishnamurthy L (2015) Plant growth promoting rhizobia: challenges and opportunities. 3 Biotech 5:355–377

Granada CE, Arruda L, Brito Lisboa B, Passaglia LMP, Vargas LK (2014) Diversity of native rhizobia isolated in South Brazil and their growth promotion effect on white clover (*Trifolium repens*) and rice (*Oryza sativa*) plants. Biol Fertility Soils 50:123–132

Gross DC, Vidaver AK (1978) Bacteriocin-like substances produced by *Rhizobium japonicum* and other slow-growing rhizobia. Appl Environ Microbiol 36:936–943

Gutiérrez-Zamora ML, Martínez-Romero E (2001) Natural endophytic association between *Rhizobium etli* and maize (*Zea mays* L.). J Biotechnol 91:117–126

Hafeez FY, Naeem FI, Naeem R, Zaidi AH, Malik KA (2005) Symbiotic effectiveness and bacteriocin production by *Rhizobium leguminosarum* bv. *viciae* isolated from agriculture soils in Faisalabad. Environ Exp Botany 54:142–147

Haro H, Sanon KB, Le Roux C, Duponnois R, Traoré AS (2018) Improvement of cowpea productivity by rhizobial and mycorrhizal inoculation in Burkina Faso. Symbiosis 74:107–120

Hasan M, Bano A, Hassan SG, Iqbal J, Awan U, Rong-ji D, Khan KA (2014) Enhancement of rice growth and production of growth-promoting phytohormones by inoculation with *Rhizobium* and other rhizobacteria. World Appl Sci J 31:1734–1743

Helene LCF, Delamuta JRM, Ribeiro RA, Hungria M (2017) *Bradyrhizobium mercantei* sp. nov., a nitrogen-fixing symbiont isolated from nodules of *Deguelia costata* (syn. *Lonchocarpus costatus*). Int J Syst Evol Microbiol 67:1827–1834

Hellriegel and Wilfarth H (1888) Untersuchungen über die stickstoffnahrung der gramineen und leguminosen. Beilageheft zu der Zeitschrift des Vereins Rübenzucker-Industrie Deutschen Reiches 1–234

Hemid M, Abdel-Waha AA, El-Enany E, Allam N, David N, Khalaf M, Morsy FM (2014) Synergistic interaction of *Rhizobium leguminosarum* bv. viciae and arbuscular mycorrhizal fungi as a plant growth promoting biofertilizers for faba bean (*Vicia faba* L.) in alkaline soil. Microbiol Res 169:49–58

Hemissi I, Mabrouk Y, Abdi N, Bouraoui M, Saidi M, Sifi B (2011) Effects of some *Rhizobium* strains on chickpea growth and biological control of *Rhizoctonia solani*. Afr J Microbiol Res 5:4080–4090

Hemmat Jou MH, Besalatpour AA (2018) Interactive effects of co-inoculation of *Bradyrhizobium japonicum* strains and mycorrhiza species on soybean growth and nutrient contents in plant. J Plant Nutr 41:10–18

Htwe AZ, Moh SM, Moe K, Yamakawa T (2018) Effects of co-inoculation of *Bradyrhizobium japonicum* SAY3-7 and *Streptomyces griseoflavus* P4 on plant growth, nodulation, nitrogen fixation, nutrient uptake, and yield of soybean in a field condition. Soil Sci Plant Nutr 64:222–229

Hungria M, Nogueira MA, Silva Araújo R (2013) Co-inoculation of soybeans and common beans with rhizobia and azospirilla: strategies to improve sustainability. Biol Fertil Soils 49:791–801

Imen H, Neila A, Adnane B, Manel B, Mabrouk Y, Saidi M, Bouaziz S (2015) Inoculation with phosphate solubilizing *Mesorhizobium* strains improves the performance of chickpea (*Cicer arietinum* L.) under phosphorus deficiency. J Plant Nutr 38:1656–1671

Jarvis BDW, van Berkum P, Chen WX, Nour SM, Fernandez MP, Cleyet-Marel JC, Gillis M (1997) Transfer of *Rhizobium loti, Rhizobium huakuii, Rhizobium ciceri, Rhizobium mediterraneum,* and *Rhizobium tianshanense* to *Mesorhizobium* gen. nov. Int J Syst Bacteriol 47:895–898

Jiao YS, Yan H, Ji ZJ, Liu YH, Sui XH, Zhang XX, Wang ET, Chen WX, Chen WF (2015) *Phyllobacterium sophorae* sp. nov., a symbiotic bacterium isolated from root nodules of *Sophora flavescens*. Int J Syst Evol Microbiol 65:399–406

Jiménez-Gómez A, Flores-Félix JD, García-Fraile P, Mateos PF, Menéndez E, Velázquez E, Rivas R (2018) Probiotic activities of *Rhizobium laguerreae* on growth and quality of spinach. Sci Rep 8:295

Jordan DC (1982) Transfer of *Rhizobium japonicum* Buchanan 1980 to *Bradyrhizobium* gen. nov., a genus of slow-growing, root nodule bacteria from leguminous plants. Int J Syst Bacteriol 32:136–139

Kaur S, Khanna V (2016) Evaluation of synergistic potential of plant growth promoting rhizobacteria with *Rhizobium* in mungbean (*Vigna radiata* L.). J Appl Nat Sci 8:995–998

Kelemu S, Thomas RJ, Moreno CX, Ocampo GI (1995) Strains of *Bradyrhizobium* from tropical forage legumes inhibit *Rhizoctonia solani* AG-1 in vitro. Australas Plant Pathol 24:168–172

Khandelwal S, Manwar AV, Chaudhari BL, Chincholkar SB (2002) Siderophoregenic Bradyrhizobia boost yield of soybean. Appl Biochem Biotechnol 102:155–168

Khanna V, Sharma P (2011) Potential for enhancing lentil (*Lens culinaris*) productivity by co-inoculation with PSB, plant growth-promoting rhizobacteria and *Rhizobium*. Indian J Agric Sci 81:932–934

Kohlmeier MG, Yudistira H, Zhang XL, Fristensky B, Levin DB, Sparling R, Oresnik IJ (2015) Draft genome sequence of the bacteriocin-producing *Bradyrhizobium japonicum* strain FN1. Genome Announc 3:e00812–e00815

Kong Z, Glick BR, Duan J, Ding S, Tian J, McConkey BJ, Wei G (2015) Effects of 1-aminocyclopr opane-1-carboxylate (ACC) deaminase-overproducing *Sinorhizobium meliloti* on plant growth and copper tolerance of *Medicago lupulina*. Plant Soil 391:383–398

Koskey G, Mburu SW, Njeru EM, Kimiti JM, Ombori O, Maingi JM (2017) Potential of native rhizobia in enhancing nitrogen fixation and yields of climbing beans (*Phaseolus vulgaris* L.) in contrasting environments of Eastern Kenya. Front Plant Sci 8:443

Kumar PR, Ram MR (2012) Production of indole acetic acid by *Rhizobium* isolates from *Vigna trilobata* (L) Verdc. African J Microbiol Res 6:5536–5541

Kumar H, Bajpai VK, Dubey RC, Maheshwari DK, Kang SC (2010) Wilt disease management and enhancement of growth and yield of *Cajanus cajan* (L) var. Manak by bacterial combinations amended with chemical fertilizer. Crop Prot 29:591–598

Kumar H, Dubey RC, Maheshwari DK (2011) Effect of plant growth promoting rhizobia on seed germination, growth promotion and suppression of *Fusarium* wilt of fenugreek (*Trigonella foenum-graecum* L.). Crop Prot 30:1396–1403

Laranjo M, Alexandre A, Oliveira S (2014) Legume growth-promoting rhizobia: an overview on the *Mesorhizobium* genus. Microbiol Res 169:2–17

Leggett M, Diaz-Zorita M, Koivunen M, Bowman R, Pesek R, Stevenson C, Leister T (2017) Soybean response to inoculation with *Bradyrhizobium japonicum* in the United States and Argentina. Agron J 109:1031–1038

Lesueur D, Diem HG, Meyer JM (1993) Iron requirement and siderophore production in *Bradyrhizobium* strains isolated from *Acacia mangium*. J Appl Bacteriol 74:675–682

Lin DX, Wang ET, Tang H, Han TX, He YR, Guan SH, Chen WX (2008) *Shinella kummerowiae* sp. nov., a symbiotic bacterium isolated from root nodules of the herbal legume *Kummerowia stipulacea*. Int J Syst Evol Microbiol 58:1409–1413

Lynch D, O'Brien J, Welch T, Clarke P, Ócuiv P, Crosa JH, O'Connell M (2001) Genetic organization of the region encoding regulation, biosynthesis, and transport of rhizobactin 1021, a siderophore produced by *Sinorhizobium meliloti*. J Bacteriol 183:2576–2585

Ma W, Sebestianova SB, Sebestian J, Burd GI, Guinel FC, Glick BR (2003) Prevalence of 1-aminocyclopropane-1-carboxylate deaminase in *Rhizobium* spp. Antonie Van Leeuwenhoek 83:285–291

Ma W, Charles TC, Glick BR (2004) Expression of an exogenous 1-aminocyclopropane-1-carboxylate deaminase gene in *Sinorhizobium meliloti* increases its ability to nodulate alfalfa. Appl Environ Microbiol 70:5891–5897

Mabrouk Y, Mejri S, Belhadj O (2016) Biochemical mechanisms of induced resistance by rhizobial lipopolysaccharide in pea against crenate broomrape. Braz J Bot 39:107–114

Malajczuk N, Pearce M, Litchfield RT (1984) Interactions between *Phytophthora cinnamomi* and *Rhizobium* isolates. Trans Br Mycol Soc 82:491–500

Marcos-García M, Menéndez E, Ramírez-Bahena MH, Mateos PF, Peix Á, Velazquez E, Rivas R (2017) *Mesorhizobium helmanticense* sp. nov., isolated from *Lotus corniculatus* nodules. Int J Syst Evol Microbiol 67:2301–2305

Martínez R, Espejo A, Sierra M, Ortiz-Bernad I, Correa D, Bedmar E, López-Jurado M, Porres JM (2015) Co-inoculation of *Halomonas maura* and *Ensifer meliloti* to improve alfalfa yield in saline soils. Appl Soil Ecol 87:81–86

Martins da Costa E, Azarias Guimarães A, Pereira Vicentin R, de Almeida Ribeiro PR, Ribas Leão AC, Balsanelli E, Lebbe L, Aerts M, Willems A, de Souza Moreira FM (2017) *Bradyrhizobium brasilense* sp. nov., a symbiotic nitrogen-fixing bacterium isolated from Brazilian tropical soils. Arch Microbiol 199:1211–1221

Martins da Costa E, Azarias Guimarães A, Soares de Carvalho T, Louzada Rodrigues T, de Almeida Ribeiro PR, Lebbe L, Willems A, de Souza Moreira FM (2018) *Bradyrhizobium forestalis* sp. nov., an efficient nitrogen-fixing bacterium isolated from nodules of forest legume species in the Amazon. Arch Microbiol 200:743–752

Martinuz A, Schouten A, Menjivar RD, Sikora RA (2012) Effectiveness of systemic resistance toward *Aphis gossypii* (Hom., Aphididae) as induced by combined applications of the endophytes *Fusarium oxysporum* Fo162 and *Rhizobium etli* G12. Biol Control 62:206–212

Maynaud G, Willems A, Soussou S, Vidal C, Mauré L, Moulin L, Cleyet-Marel JC, Brunel B (2012) Molecular and phenotypic characterization of strains nodulating *Anthyllis vulneraria* in mine tailings, and proposal of *Aminobacter anthyllidis* sp. nov., the first definition of *Aminobacter* as legume-nodulating bacteria. Syst Appl Microbiol 35:65–72

McKevith B (2005) Nutritional aspects of oilseeds. Nutr Bull 30:13–26

Mendes R, Garbeva P, Raaijmakers JM (2013) The rhizosphere microbiome: significance of plant beneficial, plant pathogenic, and human pathogenic microorganisms. FEMS Microbiol Rev 37:634–663

Menéndez E, Ramirez-Bahena MH, Peix A, Tejedor C, Mulas R, González-Andrés F, Martínez-Molina E, Velázquez E (2016) Analysis of cultivable endophytic bacteria in roots of maize in a soil from León province in mainland Spain. In: González-Andrés F, James E (eds) Biological nitrogen fixation and beneficial plant-microbe interaction. Springer, Cham, pp 45–53

Meng L, Zhang A, Wang F, Han X, Wang D, Li S (2015) Arbuscular mycorrhizal fungi and *Rhizobium* facilitate nitrogen uptake and transfer in soybean/maize intercropping system. Front Plant Sci 6:339

Michel DC, Passos SR, Simões-Araújo JL, Baraúna AC, da Silva K, Parma MM, Melo IS, De Meyer SE, O'Hara G, Zilli JE (2017) *Bradyrhizobium centrolobii* and *Bradyrhizobium macuxiense* sp. nov. isolated from *Centrolobium paraense* grown in soil of Amazonia, Brazil. Arch Microbiol 199:657–664

Minamisawa K (1989) Comparison of extracellular polysaccharide composition, rhizobitoxine production, and hydrogenase phenotype among various strains of *Bradyrhizobium japonicum*. Plant Cell Physiol 30:877–884

Mohamad R, Willems A, Le Quéré A, Maynaud G, Pervent M, Bonabaud M, Dubois E, Cleyet-Marel JC, Brunel B (2017) *Mesorhizobium delmotii* and *Mesorhizobium prunaredense* are two new species containing rhizobial strains within the symbiovar anthyllidis. Syst Appl Microbiol 40:135–143

Mondal HK, Mehta S, Kaur H, Gera R (2017) Characterization of abiotic stress tolerant rhizobia as PGPR of mothbean, clusterbean and mungbean grown in hyper-arid zone of Rajasthan. Int J Bio-Res & Stress Manag 8:309–315

Moulin L, Munive A, Dreyfus B, Boivin-Masson C (2001) Nodulation of legumes by members of the beta-subclass of Proteobacteria. Nature 411:948–950. Erratum in: Nature 412:926

Mousavi SA, Österman J, Wahlberg N, Nesme X, Lavire C, Vial L, Paulin L, De Lajudie P, Lindström K (2014) Phylogeny of the *Rhizobium-Allorhizobium-Agrobacterium* clade supports the delineation of *Neorhizobium* gen. nov. Syst Appl Microbiol 37:208–215

Mousavi SA, Willems A, Nesme X, de Lajudie P, Lindström K (2015) Revised phylogeny of *Rhizobiaceae*: proposal of the delineation of *Pararhizobium* gen. nov., and 13 new species combinations. Syst Appl Microbiol 38:84–90

Mulas D, García-Fraile P, Carro L, Ramírez-Bahena MH, Casquero P, Velázquez E, González-Andrés F (2011) Distribution and efficiency of *Rhizobium leguminosarum* strains nodulating *Phaseolus vulgaris* in Northern Spanish soils: selection of native strains that replace conventional N fertilization. Soil Biol Biochem 43:2283–2293

Nambiar PTC, Sivaramakrishnan S (1987) Detection and assay of siderophores in cowpea rhizobia (*Bradyrhizobium*) using radioactive Fe (59Fe). Lett Appl Microbiol 4:37–40

Nascimento FX, Brígido C, Glick BR, Oliveira S (2012) ACC deaminase genes are conserved among *Mesorhizobium* species able to nodulate the same host plant. FEMS Microbiol Lett 336:26–37

Nascimento FX, Rossi MJ, Soares CR, McConkey BJ, Glick BR (2014) New insights into 1-ami nocyclopropane-1-carboxylate (ACC) deaminase phylogeny, evolution and ecological significance. PLoS One 9:e99168

Nascimento FX, Rossi MJ, Glick BR (2018) Ethylene and 1-Aminocyclopropane-1-carboxylate (ACC) in plant-bacterial interactions. Front Plant Sci 9:114

Naveed M, Hussain MB, Mehboob I, Zahir ZA (2017) Rhizobial amelioration of drought stress in legumes. In: Zaidi A, Khan M, Musarrat J (eds) Microbes for legume improvement. Springer, Cham, pp 341–365

Nimnoi P, Pongsilp N, Lumyong S (2014) Co-inoculation of soybean (*Glycine max*) with actinomycetes and *Bradyrhizobium japonicum* enhances plant growth, nitrogenase activity and plant nutrition. J Plant Nutr 37:432–446

Noel TC, Sheng C, Yost CK, Pharis RP, Hynes MF (1996) *Rhizobium leguminosarum* as a plant growth-promoting rhizobacterium: direct growth promotion of canola and lettuce. Can J Microbiol 42:279–283

Oliveira RS, Carvalho P, Marques G, Ferreira L, Nunes M, Rocha I, Ma Y, Carvalho MF, Vosátka M, Freitas H (2017) Increased protein content of chickpea (*Cicer arietinum* L.) inoculated with arbuscular mycorrhizal fungi and nitrogen-fixing bacteria under water deficit conditions. Sci Food Agric 97:4379–4385

Omar SA, Abd-Alla MH (1998) Biocontrol of fungal root rot diseases of crop plants by the use of rhizobia and bradyrhizobia. Folia Microbiol 43:431–437

Ormeño E, Torres R, Mayo J, Rivas R, Peix A, Velázquez E, Zúñiga D (2007) *Phaseolus lunatus* is nodulated by a phosphate solubilizing strain of *Sinorhizobium meliloti* in a Peruvian soil. In: Velázquez E, Rodríguez-Barrueco C (eds) First international meeting on microbial phosphate Solubilization. Springer, Heidelberg, pp 143–147

Othman H, Tamimi SM (2016) Characterization of rhizobia nodulating faba bean plants isolated from soils of Jordan for plant growth promoting activities and N2 fixation potential. Int J Adv Res Biol Sci 3:20–27

Ozkoc I, Deliveli MH (2001) In vitro inhibition of the mycelial growth of some root rot fungi by *Rhizobium leguminosarum* biovar phaseoli isolates. Turk J Biol 25:435–445

Patel HN, Chakraborty RN, Desai SB (1988) Isolation and partial characterization of phenolate siderophore from *Rhizobium leguminosarum* IARI 102. FEMS Microbiol Lett 56:131–134

Patil A, Kale A, Ajane G, Sheikh R, Patil S (2017) Plant growth-promoting *Rhizobium*: mechanisms and biotechnological prospective. In: Hansen A, Choudhary D, Agrawal P, Varma A (eds) *Rhizobium* biology and biotechnology. Soil Biology, vol 50. Springer, Cham, pp 105–134

Peix A, Rivas-Boyero AA, Mateos PF, Rodríguez-Barrueco C, Martınez-Molina E, Velázquez E (2001) Growth promotion of chickpea and barley by a phosphate solubilizing strain of *Mesorhizobium mediterraneum* under growth chamber conditions. Soil Biol Biochem 33:103–110

Pimentel D, Patzek TW (2005) Ethanol production using corn, switchgrass, and wood; biodiesel production using soybean and sunflower. Nat Resour 14:65–76

Prasanna R, Ramakrishnan B, Simranjit K, Ranjan K, Kanchan A, Hossain F, Nain L (2017) Cyanobacterial and rhizobial inoculation modulates the plant physiological attributes and nodule microbial communities of chickpea. Arch Microbiol 199:1311–1323

Priyanka M, Wati L (2017) Screening of rhizobial isolates from *Vigna radiata* for plant growth promoting traits. Res Crops 18:190–195

Qureshi MA, Shakir MA, Iqbal A, Akhtar N, Khan A (2011) Co-inoculation of phosphate solubilizing bacteria and rhizobia for improving growth and yield of mungbean (*Vigna radiata* L.). J Anim Plant Sci 21:491–497

Rabie GH (1998) Induction of fungal disease resistance in Vicia faba by dual inoculation with *Rhizobium leguminosarum* and vesicular-arbuscular mycorrhizal fungi. Mycopathologia 141:159–166

Radl V, Simões-Araújo JL, Leite J, Passos SR, Martins LM, Xavier GR, Rumjanek NG, Baldani JI, Zilli JE (2014) *Microvirga vignae* sp. nov., a root nodule symbiotic bacterium isolated from cowpea grown in semi-arid Brazil. Int J Syst Evol Microbiol 64:725–730

Rangel WM, de Oliveira Longatti SM, Ferreira PAA, Bonaldi DS, Guimarães AA, Thijs S, Weyens N, Vangronsveld J, Moreira FMS (2017) Leguminosae native nodulating bacteria from a gold mine as-contaminated soil: multi-resistance to trace elements, and possible role in plant growth and mineral nutrition. Int J Phytoremediation 19:925–936

Reimann S, Hauschild R, Hildebrandt U, Sikora RA (2008) Interrelationships between *Rhizobium etli* G12 and *Glomus intraradices* and multitrophic effects in the biological control of the root-knot nematode *Meloidogyne incognita* on tomato. J Plant Dis Protect 115:108–113

Reitz M, Rudolph K, Schröder I, Hoffmann-Hergarten S, Hallmann J, Sikora RA (2000) Lipopolysaccharides of *Rhizobium etli* strain G12 act in potato roots as an inducing agent of systemic resistance to infection by the cyst nematode *Globodera pallida*. Appl Environ Microbiol 66:3515–3518

Remigi P, Zhu J, Young JPW, Masson-Boivin C (2016) Symbiosis within symbiosis: evolving nitrogen-fixing legume symbionts. Trends Microbiol 24:63–75

Ren CG, Bai YJ, Kong CC, Bian B, Xie ZH (2016) Synergistic interactions between salt-tolerant rhizobia and arbuscular mycorrhizal fungi on salinity tolerance of *Sesbania cannabina* plants. J Plant Growth Regul 35:1098–1107

Rivas R, Velázquez E, Willems A, Vizcaíno N, Subba-Rao NS, Mateos PF, Gillis M, Dazzo FB, Martínez-Molina E (2002) A new species of *Devosia* that forms a unique nitrogen-fixing root-nodule symbiosis with the aquatic legume *Neptunia natans* (L.f.) druce. Appl Environ Microbiol 68:5217–5222

Rivas R, Peix A, Mateos PF, Trujillo ME, Martínez-Molina E, Velázquez E (2006) Biodiversity of populations of phosphate solubilizing rhizobia that nodulates chickpea in different Spanish soils. Plant Soil 287:23–33

Robleto EA, Scupham AJ, Triplett EW (1997) Trifolitoxin production in *Rhizobium etli* strain CE3 increases competitiveness for rhizosphere growth and root nodulation of *Phaseolus vulgaris* in soil. Mol Plant-Microbe Interact 10:228–233

Robleto EA, Borneman J, Triplett EW (1998) Effects of bacterial antibiotic production on rhizosphere microbial communities from a culture independent perspective. Appl Env Microbiol 64:5020–5022

Rodríguez H, Fraga R (1999) Phosphate solubilizing bacteria and their role in plant growth promotion. Biotech Adv 17:319–339

Rubio-Canalejas A, Celador-Lera L, Cruz-González X, Menéndez E, Rivas R (2016) *Rhizobium* as potential biofertilizer of *Eruca Sativa*. In: González-Andrés F, James E (eds) Biological nitrogen fixation and beneficial plant-microbe interaction. Springer, Heidelberg, pp 213–220

Saber WIA, Abd El-Hai KM, Ghoneem KM (2009) Synergistic effect of *Trichoderma* and *Rhizobium* on both biocontrol of chocolate spot disease and induction of nodulation, physiological activities and productivity of *Vicia faba*. Res J Microbiol 4:286–300

Saghafi D, Ghorbanpour M, Lajayer BA (2018) Efficiency of *Rhizobium* strains as plant growth promoting rhizobacteria on morpho-physiological properties of *Brassica napus* L. under salinity stress. J Soil Sci Plant Nutr 18:253–268

Saha M, Sarkar S, Sarkar B, Kumar B, Bhattacharjee SS, Tribedi P (2016) Microbial siderophores and their potential applications: a review. Environ Sci Pollution Res 23:3984–3999

Samago TY, Anniye EW, Dakora FD (2018) Grain yield of common bean (*Phaseolus vulgaris* L.) varieties is markedly increased by rhizobial inoculation and phosphorus application in Ethiopia. Symbiosis 75:245–255

Sannazzaro AI, Torres Tejerizo G, Fontana MF, Cumpa Velásquez LM, Hansen LH, Pistorio M, Estrella MJ (2018) *Mesorhizobium sanjuanii* sp. nov., isolated from nodules of *Lotus tenuis* in the saline-alkaline lowlands of flooding Pampa, Argentina. Int J Syst Evol Microbiol 68:2936–2942

Santoyo G, Moreno-Hagelsieb G, Orozco-Mosqueda Mdel C, Glick BR (2016) Plant growth-promoting bacterial endophytes. Microbiol Res 183:92–99

Sawana A, Adeolu M, Gupta RS (2014) Molecular signatures and phylogenomic analysis of the genus *Burkholderia*: proposal for division of this genus into the emended genus *Burkholderia* containing pathogenic organisms and a new genus *Paraburkholderia* gen. nov. harboring environmental species. Front Genet 5:429

Schwinghamer EA, Brockwell J (1978) Competitive advantage of bacteriocin and phage-producing strains of *Rhizobium trifolii* in mixed culture. Soil Biol Biochem 10:383–387

Shaban WI, El-Bramawy MA (2011) Impact of dual inoculation with *Rhizobium* and *Trichoderma* on damping off, root rot diseases and plant growth parameters of some legumes field crop under greenhouse conditions. Int Res J Agric Sci Soil Sci 1:98–108

Sharma SR, Rao NK, Gokhale TS, Ismail S (2013) Isolation and characterization of salt-tolerant rhizobia native to the desert soils of United Arab Emirates. Emirates J Food Agric 25:102–108

Shinde BP, Thakur J (2016) The effect of co-inoculation of pea plants with arbuscular mycorrhizal fungi and rhizobium on the nodulation, growth and productivity. Int J Bioassays 10:4954–4957

Siddiqui IA, Ehteshamul-Haque S, Zaki MJ, Ghaffar A (2000) Greenhouse evaluation of rhizobia as biocontrol agent of root-infecting fungi in okra. Acta Agrobot 53:13–22

Siddiqui ZA, Baghel G, Akhtar MS (2007) Biocontrol of *Meloidogyne javanica* by *Rhizobium* and plant growth-promoting rhizobacteria on lentil. World J Microbiol Biotechnol 23:435–441

Singh PK, Singh M, Vyas D (2010) Biocontrol of fusarium wilt of chickpea using arbuscular mycorrhizal fungi and *Rhizobium leguminosarum* biovar. Caryologia 63:349–353

Skerman VBD, McGowan V, Sneath PHA (1980) Approved lists of bacterial names. Int J Syst Bacteriol 30:225–420

Soliman AS, Shanan NT, Massoud ON, Swelim DM (2012) Improving salinity tolerance of *Acacia saligna* (Labill.) plant by arbuscular mycorrhizal fungi and *Rhizobium* inoculation. African J Biotechnol 11:1259–1266

Sridevi M, Mallaiah KV, Yadav NCS (2007) Phosphate solubilization by *Rhizobium* isolates from *Crotalaria* species. J Plant Sci 2:635–639

Staudinger C, Mehmeti-Tershani V, Gil-Quintana E, Gonzalez EM, Hofhansl F, Bachmann G, Wienkoop S (2016) Evidence for a rhizobia-induced drought stress response strategy in *Medicago truncatula*. J Proteome 136:202–213

Suárez-Moreno ZR, Caballero-Mellado J, Coutinho BG, Mendonça-Previato L, James EK, Venturi V (2012) Common features of environmental and potentially beneficial plant-associated *Burkholderia*. Microb Ecol 63:249–266

Subramanian P, Ramasamy KK, Sundaram KS, Sa T (2015) Endophytic bacteria improve nodule function and plant nitrogen in soybean on co-inoculation with *Bradyrhizobium japonicum* MN110. Plant Growth Regul 76:327–332

Sy A, Giraud E, Jourand P, Garcia N, Willems A, de Lajudie P, Prin Y, Neyra M, Gillis M, Boivin-Masson C, Dreyfus B (2001) Methylotrophic *Methylobacterium* bacteria nodulate and fix nitrogen in symbiosis with legumes. J Bacteriol 183:214–220

Tajini F, Trabelsi M, Drevon JJ (2011) Co-inoculation with *Glomus intraradices* and *Rhizobium tropici* CIAT899 increases P use efficiency for N_2 fixation in the common bean (*Phaseolus vulgaris* L.) under P deficiency in hydroaeroponic culture. Symbiosis 53:123

Tan KZ, Radziah O, Halimi MS, Khairuddin AR, Shamsuddin ZH (2015) Assessment of plant growth-promoting rhizobacteria (PGPR) and rhizobia as multi-strain biofertilizer on growth and N_2 fixation of rice plant. Austr J Crop Sci 9:1257–1264

Tarafder HK, Dey A, Dasgupta S (2016) Co-inoculation of phosphate solubilizing bacteria and rhizobia for improving growth and yield of mungbean (*Vigna radiata* L.). Asian J Soil Sci 11:207–212

Tavasolee A, Aliasgharzad N, SalehiJouzani G, Mardi M, Asgharzadeh A (2011) Interactive effects of arbuscular mycorrhizal fungi and rhizobial strains on chickpea growth and nutrient content in plant. Afr J Biotechnol 10:7585–7591

Thakur D, Kaushal R, Shyam V (2014) Phosphate solubilising microorganisms: role in phosphorus nutrition of crop plants-a review. Agric Rev 35:159–171

Triplett EW, Barta TM (1987) Trifolitoxin production and nodulation are necessary for the expression of superior nodulation competitiveness by *Rhizobium leguminosarum* bv. Trifolii strain T24 on clover. Plant Physiol 85:335–342

Trujillo ME, Willems A, Abril A, Planchuelo AM, Rivas R, Ludeña D, Mateos PF, Martínez-Molina E, Velázquez E (2005) Nodulation of Lupinus albus by strains of Ochrobactrum lupini sp. nov. Appl Environ Microbiol 71:1318–1327

Tu JC (1978) Protection of soybean from severe *Phytophthora* root rot by *Rhizobium*. Physiol Plant Pathol 12:233–240

Ullah S, Khan MY, Asghar HN, Akhtar MJ, Zahir ZA (2017) Differential response of single and co-inoculation of *Rhizobium leguminosarum* and *Mesorhizobium ciceri* for inducing water deficit stress tolerance in wheat. Ann Microbiol 67:739–749

Validation List no. 107 (2006) Int J Syst Evol Microbiol 56:1–6

Valverde A, Velázquez E, Fernández-Santos F, Vizcaíno N, Rivas R, Mateos PF, Martínez-Molina E, Igual JM, Willems A (2005) *Phyllobacterium trifolii* sp. nov., nodulating *Trifolium* and *Lupinus* in Spanish soils. Int J Syst Evol Microbiol 55:1985–1989

Vandamme P, Coenye T (2004) Taxonomy of the genus *Cupriavidus:* a tale of lost and found. Int J Syst Evol Microbiol 54:2285–2289

Vargas LK, Volpiano CG, Lisboa BB, Giongo A, Beneduzi A, Passaglia LMP (2017) Potential of rhizobia as plant growth-promoting rhizobacteria. In: Zaidi A, Khan M, Musarrat J (eds) Microbes for legume improvement. Springer, Cham, pp 153–174

Vejan P, Abdullah R, Khadiran T, Ismail S, Nasrulhaq Boyce A (2016) Role of plant growth promoting Rhizobacteria in agricultural sustainability-a review. Molecules 21:piiE573

Velázquez E, Carro L, Flores-Félix JD, Martínez-Hidalgo P, Menéndez E, Ramírez-Bahena MH, Mulas R, González-Andrés F, Martínez-Molina E, Peix A (2017a) The legume nodule microbiome: a source of plant growth-promoting bacteria. In: Kumar V, Kumar M, Sharma S, Prasad R (eds) Probiotics and plant health. Springer, Singapore, pp 41–70

Velázquez E, García-Fraile P, Ramírez-Bahena MH, Rivas R, Martínez-Molina E (2017b) Current status of the taxonomy of bacteria able to establish nitrogen-fixing legume symbiosis. In: Zaidi A, Khan M, Musarrat J (eds) Microbes for legume improvement. Springer, Cham, pp 1–43

Venter AP, Twelker S, Oresnik IJ, Hynes MF (2001) Analysis of the genetic region encoding a novel rhizobiocin from *Rhizobium leguminosarum* bv. Viciae strain 306. Can J Microbiol 47:495–502

Verma JP, Yadav J, Tiwari KN (2012) Enhancement of nodulation and yield of chickpea by co-inoculation of indigenous *Mesorhizobium* spp. and plant growth–promoting rhizobacteria in Eastern Uttar Pradesh. Comm Soil Sci Plant Anal 43:605–621

Verma JP, Yadav J, Tiwaric KN, Kumarb A (2013) Effect of indigenous *Mesorhizobium* spp. and plant growth promoting rhizobacteria on yields and nutrients uptake of chickpea (*Cicer arietinum* L.) under sustainable agriculture. Ecol Eng 51:282–286

Vicario JC, Primo ED, Dardanelli MS, Giordano W (2016) Promotion of peanut growth by co-inoculation with selected strains of *Bradyrhizobium* and *Azospirillum*. J Plant Growth Regul 35:413–419

Vieira JD, da Silva PRD, Stefenon VM (2017) In vitro growth and indoleacetic acid production by *Mesorhizobium loti* SEMIA806 and SEMIA816 under the influence of copper ions. Microbiol Res 8:57–58

Villar-Igea M, Velázquez E, Rivas R, Willems A, van Berkum P, Trujillo ME, Mateos PF, Gillis M, Martínez-Molina E (2007) Phosphate solubilizing rhizobia originating from *Medicago*, *Melilotus* and *Trigonella* grown in a Spanish soil. In: Velázquez E, Rodríguez-Barrueco C (eds) First international meeting on microbial phosphate Solubilization. Springer, Heidelberg, pp 149–156

Vincent JM (1970) The cultivation, isolation and maintenance of rhizobia. In: Vincent JM (ed) A manual for the practical study of root-nodule. Blackwell Scientific Publications, Oxford, pp 1–13

Wang Y, Zhang Z, Zhang P, Cao Y, Hu T, Yang P (2016) *Rhizobium* symbiosis contribution to short-term salt stress tolerance in alfalfa (*Medicago sativa* L.). Plant Soil 402:247–261

Wdowiak-Wróbel S, Małek W (2016) Properties of *Astragalus* sp. microsymbionts and their putative role in plant growth promotion. Arch Microbiol 198:793–801

Wolde-meskel E, Heerwaarden J, Abdulkadir B, Kassa S, Aliyi I, Degefu T, Wakwey K, Kanampiu F, Giller KE (2018) Additive yield response of chickpea (*Cicer arietinum* L.) to *Rhizobium* inoculation and phosphorus fertilizer across smallholder farms in Ethiopia. Agric Ecosyst Environ 261:144–152

Wright W, Little J, Liu F, Chakraborty R (2013) Isolation and structural identification of the trihydroxamate siderophore vicibactin and its degradative products from *Rhizobium leguminosarum* ATCC 14479 bv. trifolii. Bio Metals 26:271–283

Xu L, Shi J, Li C, Zhu S, Li B (2017) *Rhizobium hedysari* sp. nov., a novel species isolated from a root nodule of *Hedysarum multijugum* in China. Antonie Van Leeuwenhoek 110:479–488

Xu L, Zhang Y, Mohamad OA, Jiang C, Friman VP (2018) *Mesorhizobium zhangyense* sp. nov., isolated from wild *Thermopsis lanceolate* in northwestern China. Arch Microbiol 200:603–610

Yadav J, Verma JP (2014) Effect of seed inoculation with indigenous *Rhizobium* and plant growth promoting rhizobacteria on nutrients uptake and yields of chickpea (*Cicer arietinum* L.). Eur J Soil Biol 63:70–77

Yan J, Li Y, Yan H, Chen WF, Zhang X, Wang ET, Han XZ, Xie ZH (2017a) *Agrobacterium salinitolerans* sp. nov., a saline-alkaline-tolerant bacterium isolated from root nodule of *Sesbania cannabina*. Int J Syst Evol Microbiol 67:1906–1911

Yan J, Yan H, Liu LX, Chen WF, Zhang XX, Verástegui-Valdés MM, Wang ET, Han XZ (2017b) *Rhizobium hidalgonense* sp. nov., a nodule endophytic bacterium of *Phaseolus vulgaris* in acid soil. Arch Microbiol 199:97–104

Yanni YG, Dazzo FB (2010) Enhancement of rice production using endophytic strains of *Rhizobium leguminosarum* bv. trifolii in extensive field inoculation trials within the Egypt Nile delta. Plant Soil 336:129–142

Yanni YG, Rizk RY, Corich V, Squartini A, Ninke K, Philip-Hollingsworth S, Orgambide G, de Bruijn F, Stoltzfus J, Buckley D, Schmidt TM, Mateos PF, Ladha JK, Dazzo FB (1997) Natural endophytic association between *Rhizobium leguminosarum* bv. *trifolii* and rice roots and assessment of its potential to promote rice growth. In: Ladha JK, de Bruijn FJ, Malik KA (eds)

Opportunities for biological nitrogen fixation in rice and other non-legumes. Developments in plant and soil sciences, vol 75. Springer, Dordrecht, pp 99–114
Yanni YG, Rizk RY, Abd El-Fattah FK, Squartini A, Corich V, Giacomini A, de Bruijn F, Rademaker J, Maya-Flores J, Ostrom P, Vega-Hernandez M, Hollingsworth RI, Martínez-Molina E, Mateos P, Velázquez E, Wopereis J, Triplett E, Umali-García M, Anarna JA, Rolfe BG, Ladha JK, Hill J, Mujoo R, Ng PK, Dazzo FB (2001) The beneficial plant growth-promoting association of *Rhizobium leguminosarum* bv. trifolii with rice roots. Australian J Plant Physiol 28:845–870
Yanni YG, Dazzo FB, Squartini A, Zanardo M, Zidan MI, Elsadany AEY (2016) Assessment of the natural endophytic association between *Rhizobium* and wheat and its ability to increase wheat production in the Nile delta. Plant Soil 407:367–383
Yanni Y, Zidan M, Dazzo F, Rizk R, Mehesen A, Abdelfattah F, Elsadany A (2018) Enhanced symbiotic performance and productivity of drought stressed common bean after inoculation with tolerant native rhizobia in extensive fields. Agric Ecosyst Environ 232:119–128
Yuhashi K, Ichikawa N, Ezura H, Akao S, Minakawa Y, Nukui N, Yasuta T, Minamisawa K (2000) Rhizobitoxine production by *Bradyrhizobium elkanii* enhances nodulation and competitiveness on *Macroptilium atropurpureum*. Appl Environ Microbiol 66:2658–2663
Yuttavanichakul W, Lawongs P, Wongkaew S, Teaumroong N, Boonkerd N, Nomura N, Tittabutr P (2012) Improvement of peanut rhizobial inoculant by incorporation of plant growth promoting rhizobacteria (PGPR) as biocontrol against the seed borne fungus *Aspergillus niger*. Biol Control 63:87–97
Zhang J, Guo C, Chen W, de Lajudie P, Zhang Z, Shang Y, Wang ET (2018) *Mesorhizobium wenxiniae* sp. nov., isolated from chickpea (*Cicer arietinum* L.) in China. Int J Syst Evol Microbiol 68:1930–1936
Zhu RF, Tang F, Liu J, Liu FQ, Deng XY, Chen JS (2016) Co-inoculation of arbuscular mycorrhizae and nitrogen fixing bacteria enhance alfalfa yield under saline conditions. Pak J Bot 48:763–769
Zurdo-Piñeiro JL, Rivas R, Trujillo ME, Vizcaíno N, Carrasco JA, Chamber M, Palomares A, Mateos PF, Martínez-Molina E, Velázquez E (2007) *Ochrobactrum cytisi* sp. nov., isolated from nodules of *Cytisus scoparius* in Spain. Int J Syst Evol Microbiol 57:784–788

Applications of Beneficial Microbe in Arid and Semiarid Agroecosystem: IAA-Producing Bacteria

5

Mohammad Javad Zarea

Abstract

Plant harbors large population of various microorganisms due to release of root substance that favors bacterium proliferation. Volume of the soil that is influenced by the roots of plants is known as rhizosphere. Rhizosphere is occupied by distinct bacterial population that interrelate with each other and plant roots. There are several bacterial communities which influence the development and growth of plants through several mechanisms, including phytohormone production and nutrients mobilization. Interest in the application of bacteria of the plant growth encouraging rhizobia groups has been increased. Many authors have reviewed the important roles of PGPR in augmenting the yield and the growth of crops. Therefore, beneficial or application of plant growth enhancing rhizobacteria (PGER) in crop production is not the aim of this chapter. Arid and semiarid agricultural lands are characterized by drought stress; therefore, in such lands, plant does not grow well. Application of beneficial microbes in such lands helps in mitigation of drought as well as other unwanted abiotic and biotic stresses. The main goal of this chapter is to focus on the role of IAA-producing bacteria and their role in plant growth promotion as well as drought tolerance of crops.

Keywords

Microbes · Arid · Semiarid · Agroecosystem · Phytohormone

M. J. Zarea (✉)
Department of Agronomy and Plant breeding, Faculty of Agriculture, Ilam University, Ilam, Iran
e-mail: mj.zarea@ilam.ac.ir

V. Kumar et al. (eds.), *Microbiome in Plant Health and Disease*,
https://doi.org/10.1007/978-981-13-8495-0_5

5.1 Introduction

Plant growth-promoting rhizobacteria (PGPR) or plant growth-enhancing rhizobacteria (PGER) are capable of production of a commonly known phytohormone, indole acetic acid (IAA), also known as indole-3-acetic acid. Most of the bacteria, which produce IAA, induce a beneficial effect on plant growth. Production of IAA is a common characteristic among most microorganisms isolated from various plants (Patten and Glick 1996). It has been estimated that 80% of isolated microorganisms from different crops had the ability to produce auxins (Patten and Glick 1996, 2002). In another study, Joseph et al. (2007) stated that all the isolates of *Pseudomonas*, *Azotobacter*, and *Bacillus* isolated from chickpea rhizosphere were capable of producing IAA. IAA produced by microorganisms can influence rhizobacteria and plant interaction (Spaepen and Vanderleyden 2011). IAA can affect many plant developments and functions. IAA has a vital role in plant cell division and differentiation (Salisbury 1994). IAA positively influences seed germination, root growth and development, and root surface area and length. IAA increases root lateral and adventitious root formation as well. Physiological processes such as photosynthesis and pigment formation are affected by IAA. An important role of IAA in plants is the promotion of plant growth; this hormone allows rapid cell division of plant cells. The role of the PGPR *Azospirillum* in conferring drought tolerance to host plant has been attributed to IAA release (Fukami et al. 2016).

Rhizobacteria can improve drought tolerance of host plant through enhancing root biomass and density of root hairs. In the past, studies showed the role of rhizobacteria of the genus *Azospirillum* in improving root traits (Cassán and García de Salamone 2008; Lopes et al. 2011; Hungria et al. 2015; Saharan and Nehra 2011). Most of these modulations in root systems have been attributed to the IAA produced by the PGPR (Saharan and Nehra 2011). Developed root system plays a significant part in water acquisition by the plant under drought stress (Kumar et al. 2018). Canola inoculated with *A. lipoferum* had better performance in seed germination and in water potential improvement (Saeed and collaborators 2016). Improved growth of *A. brasilense*-inoculated *Arabidopsis* under drought stress was attributed to the enhanced levels of ABA (Cohen et al. 2015). Enhancement of drought tolerance in *A. brasilense*- or *Herbaspirillum seropedicae*-inoculated maize was also correlated with ABA and ethylene contents (Curá et al. 2017).

5.2 IAA and Environmental Stresses

IAA, as the most abundant auxin in plants alone or in combination with other plant hormones, regulates various aspects of plant development and growth (Cooke et al. 2002; Brumos et al. 2014). Auxin conjugates, free auxin, inactive methyl ester form of IAA and MeIAA, and inactive auxin precursors consist the pool auxin of plants (Bajguz and Piotrowska 2009; Korasick et al. 2013). Auxin has a significant role in development regulation and adaptation response of plants to stress (Park et al. 2007; Ludwig-Muller 2011; Liu et al. 2014). It has been assumed that a decrease in free auxin may be an adaptation mechanism response of plants to stress. IAA in the leaves of rice (Prakash and Prathapasenan 1990) and tomato (Albacete et al. 2008) has been reported to decline due to salt stress. In wheat roots, IAA was also found to diminish (Shakirova et al. 2003). However, alteration in the levels of IAA seems to be affected by the severity of stress induced (Pierik and Testerink 2014) and the kinds of abiotic stress imposed (Du et al. 2012, 2013). Based on the reported studies dealing with the role of auxin in plant response to stresses, two questions emerged: (1) the role of auxin in plants subjected to moderate stress and (2) the effect of IAA exogenous application in stress-induced plants. Lecube et al. (2014) described the IAA role in protecting plants against oxidative stress and improved drought tolerance of soybean plant, exogenously treated with 100 μM of IAA.

5.3 IAA and Drought Stress

Drought, like other environmental stresses, adversely affects plant growth and development. Drought stress increases reactive oxygen species (ROS) resulting in cell and organelle cell injury. Plants protect themselves from ROS by means of endogenous protective mechanisms. Protective mechanisms of plants in response to ROS include enzymatic and nonenzymatic systems. Indole acetic acid (IAA) is one of the vital plant hormones involved in several functions of plant growth and development like cell division and elongation, root formation, etc. The role of this naturally occurring plant growth hormone in drought resistance and tolerance is still rather antithetical. Water deficiency has been assumed to have a restriction effect on biosynthesis of IAA and decrease IAA content in plants (Pustovoitova and Zholkevich 1999). However, studies concerning the involvement of IAA in plant response to drought stress elucidated that drought stress is accompanied by IAA levels (Zholkevich and Pustovoitova 1993). Lecube et al. (2014) reported the role of IAA in protecting plants against oxidative stress in soybean. IAA indirectly increased heme oxygenase-1 (HO-1) through the modulation levels of nitric oxide (NO) (Lecube et al. 2014). HO-1 participates in the response of the plant to different stresses (Zhang et al. 2009; Gohya et al. 2006) like drought stress (Lecube et al. 2014). NO regulates the expression of HO-1 (Noriega et al. 2007). HO-1 guards the plant from oxidative stress induced by different stresses (Lecube et al. 2014; Santa-Cruz et al. 2010, 2017). Endogenous decreased level of H-1 results in promoting methylation. Moreover, Lecube et al. (2014) elucidated that exogenous application of IAA (100 μM) enhanced HO-1 activity by 75% in drought-induced soybean. Therefore, sustaining appropriate level of indigenous auxin is a vital element for plants to coordinate various cellular functions and processes.

5.4 Plant-Associated Bacteria

PGPR or bacteria associated with plant roots can activate plant pathogen resistance (Sharifi and Ryu, 2016). PGPR can help plants withstand abiotic stresses like famine (Lim and Kim 2013). Synthesis of auxin by different PGPR strains has been reported (Spaepen et al. 2007; Ahmed and Hasnain 2014; Júnior et al. 2011; Ali

2015). Root-associated bacteria can produce phytohormones (Bloemberg and Lugtenberg 2001; Bottini et al. 2004; Pirlak and Kose 2009). Symbiotic, free-living bacteria and other rhizobacteria have the ability to synthesize IAA (Tsavkelova et al. 2006; Costacurta and Vanderleyden 1995). Indole acetic acid has been reported in many soil bacteria such as *Azospirillum*, *Enterobacter*, *Azotobacter*, *Bacillus*, *Aeromonas*, *Burkholderia*, *Rhizobium*, and *Pseudomonas* (Dobbelaere et al. 1999; Halda-Alija 2003; Swain et al. 2007; Ahmad et al. 2008; Ghosh et al. 2003; Hariprasad and Niranjana 2009; Shoebitz et al. 2009). The most valuable influence of potential rhizobacteria on crop plants has been attributed to IAA synthesis. PGPR with negligible auxin production has no stimulating influence on plant growth (Singh et al. 2013). The mutant strain of *Azospirillum brasilense* FAJ0009 was shown to have no inducing effect on plants (Spaepen et al. 2014). It should be noted that the PGPR having the ability to synthesize high concentrations of auxin might cause an adverse effect on plants (Park et al. 2015). Elevated auxin in roots due to PGPR inoculation is attributed to the increased root growth (Poupin et al. 2016). It should be noted that some PGPR can utilize IAA as a nutritive substance. *Pseudomonas putida* 1290 has been reported to utilize or consume IAA, and therefore, this bacterium decreases levels of exogenous IAA, which leads to amelioration of the adverse effect (growth inhibition) of higher IAA (Leveau and Lindow 2005).

5.5 Bacterial IAA

Although the role of auxin in plant function has been to some extent elucidated, its effect on bacterial cell is not elucidated. However, it has been reported that bacterial IAA has been involved in overall plant growth and also may result in enhancing plant fitness (Patten and Glick 2002). Emerging hairs are one of the main parts of root colonized by the IAA-producing bacteria. IAA-producing bacteria, through the weakness of the plant defense system, can colonize roots easier than other bacteria. Some of the phytopathogen bacteria have the ability to produce IAA causing infectious diseases in plants such as blight diseases, gall diseases, and leafy gall. Table 5.1 shows the comparison of phytopathogenic phytostimulator bacteria.

Table 5.1 Comparison of the effect of auxin produced by phytopathogenic bacteria and phytostimulator on plants

Phytopathogenic bacteria	*Dickeya dadantii*	Blight diseases	Yang et al. (2007)
	Pantoea agglomerans	Gall disease	Chalupowicz et al. (2009)
	Rhodococcus fascians	Leafy gall	Vandeputte et al. (2005)
Phytostimulator	*Azospirillum brasilense*	Plant root development promotion	Dobbelaere et al. (1999)
	Pseudomonas putida	Plant root development promotion	Patten and Glick (2002)

Bacteria of the different genus, *Azospirillum* (Dobbelaere et al. 1999), *Pseudomonas* (Patten and Glick 2002), *Azotobacter* (Verma et al. 2001), *Bacillus* (Raddidi et al. 2008), Rhizobia (Hassen and Labuschagne 2010; Vega-Hernández et al. 2002), *Paenibacillus* (Phi et al. 2010), and *Methylobacterium* (Ivanova et al. 2001), have been shown to synthesize IAA in laboratory conditions. Table 5.2 shows bacteria of different genus with the ability to produce IAA. IAA-producing bacteria affect root growth and root hair and lateral formation. IAA of the bacteria also enhances both root length and area. Enhanced root area and length helps plants to increase soil water uptake and soil nutrient absorption (Vessey 2003).

5.6 IAA-Producing Bacteria May Improve Drought Tolerance

IAA produced by the PGPR plays a chief part in stimulation and root structure growth (Patten and Glick 2002). Rashad et al. (2006) attributed improved drought tolerance in sorghum to IAA produced by the PGPR *Bradyrhizobium* (*B. japonicum*) and *Rhizobium* (*R. leguminosarum*) strains. IAA-producing bacteria may improve tolerance of the plant to water scarcity stresses (Dimkpa et al., 2009; Vurukonda et al. 2016). Some PGPR have been demonstrated to improve the host plant drought tolerance by modifying ethylene production. These bacteria possess a specific enzyme known as 1-aminocyclopropane-1-carboxylate deaminase (ACC deaminase). This ACC displays a vital role in cleaving the precursor of plant hormone ethylene and 1-aminocyclopropane-1-carboxylate (ACC) into α-ketobutyrate and ammonia (Honma and Shimomura 1978) which results in the decline of ACC levels in plants (Glick et al. 1998, 2007). It has been assumed that some IAA produced and secreted by the bacteria is taken up by the host plant. This IAA uptake, in addition with plant endogenous IAA, may act as ethylene stimulator in the plant (Glick 2014). Therefore, the plant uptake of IAA, produced by bacteria, can persuade enzyme transcription of ACC synthase inside the root plant. ACC synthase has a vital role in catalyzing the formation of 1-aminocyclopropane-1-carboxylate (Fig. 5.1).

However, the role of IAA in drought-stressed plants is to some extent contradictory. Conducted experiments suggest that drought stress is accompanied by IAA levels (Zholkevich and Pustovoitova 1993). Recently, the role of IAA in protecting plants against oxidative stress in soybean has been elucidated (Lecube et al. 2014). Lecube et al. (2014) reported that IAA indirectly enhanced heme oxygenase-1 which participates in the response of the plant to different stresses (Zhang et al. 2009; Gohya et al. 2006) like drought stress (Lecube et al. 2014). Besides, exogenous application of IAA (100 μM) enhanced HO-1 activity by 75% in drought-induced soybean (Lecube et al. 2014). Therefore, sustaining appropriate level of indigenous auxin is an important factor for plants to coordinate various cellular functions and processes. From the obtained results presented above, several supposed pathways of IAA in plant response to stress can be suggested, as shown in Fig. 5.1. Although some studies indicated the protective role of IAA in plants against

Table 5.2 Bacteria of the different groups that exhibit ability of IAA production

Azotobacter chroococcum	Verma et al. (2001)
Azospirillum sp.	Yasmin et al. (2004)
Azomonas sp. *RJ4*	Sheng and Xia (2006)
Azospirillum amazonense	Rodrigues et al. (2008)
Mesorhizobium sp.	Wani et al. (2008)
Pseudomonas sp.	Poonguzhali et al. (2008)
Serratia marcescens	Selvakumar et al. (2008)
Enterobacter sp.	Kumar et al. (2008)
Burkholderia	Jiang et al. (2008)
Pseudomonas jessenii	Rajkumar and Freitas (2008)
Pseudomonas aeruginosa	Ganesan (2008)
Pseudomonas sp.	Rajkumar and Freitas (2008)
Azotobacter sp.	Ahmad et al. (2008)
Bradyrhizobium sp.	Wani et al. (2007a)
Rhizobium sp.	Wani et al. 2007b
Azotobacter chroococcum	Wani et al. (2007c)
Brevibacillus spp.	Vivas et al. (2006)
Xanthomonas sp. *RJ3*	Sheng and Xia (2006)
Bradyrhizobium japonicum	Shaharoona et al. (2008)
Pseudomonas fluorescens	Dey et al. (2004)
Bradyrhizobium	Antoun et al. (1998)
Rhizobium	Antoun et al. (1998)
Pseudomonas sp. *A3R3*	Ma et al. (2011)
Bacillus species PSB10	Wani and Khan (2010)
Paenibacillus polymyxa	Phi et al. (2010)
Rhizobium phaseoli	Zahir et al. (2010)
Rahnella aquatilis	Mehnaz et al. (2010)
Pseudomonas sp.	Tank and Saraf (2009)
Azospirillum	Fukami et al. (2016)
B. japonicum	Rashad et al. (2006)
R. leguminosarum	Rashad et al. (2006)
Rhizobium	Yanni et al. (2001)
Azotobacter	Zarrin and Sharon (2010)
Pseudomonas aeruginosa	Jay et al. (2013)
P. putida	Patten and Glick (2002)
P. fluorescens	Egamberdieva (2008)
P. aureantiaca TSAU22	Egamberdieva (2009)
P. extremorientalis TSAU6	
P. extremorientalis TSAU20	
Kocuria varians	Egamberdieva (2008)
Klebsiella	Chaiharn and Lumyong (2011)
Pseudomonas sp. *A3R3*	Ma et al. (2011)
Bacillus species PSB10	Wani and Khan (2010)
Paenibacillus polymyxa	Phi et al. (2010)
Rhizobium phaseoli	Zahir et al. (2010)

(continued)

Table 5.2 (continued)

Rahnella aquatilis	Mehnaz et al. (2010)
Pseudomonas sp.	Tank and Saraf (2009)
B. phytofirmans PsJN	Weilharter et al. (2011)
B. japonicum	Rashad et al. (2006)
R. leguminosarum	Rashad et al. (2006)
Azospirillum brasilense Sp245	Smets et al. (2004)
Aeromonas punctata PNS-1	Iqbal and Hasnain (2013)
Serratia marcescens 90–166	Shi et al. (2010)
Mycobacterium sp.	Tsavkelova et al. (2006)
B. megaterium (KBA-10)	Ekinci et al. (2014)
Pantoea agglomerans (RK-92)	Ekinci et al. (2014)
B. subtilis	Colo et al. (2014)
A. chroococcum	Colo et al. (2014)
Aeromonas punctata PNS-1	Iqbal and Hasnain (2013)
Enterobacter sp. I-3	Park et al. (2015)

stresses, studies in this field are to some extent contradictory. Tognetti et al. (2012) reported that auxin improved stress tolerance of plants through regulation of chloroplast structure and abundance of photosynthetic components. Auxin is affected by the reactive oxygen species (ROS) under stresses. ROS can decrease auxin signaling (Potters et al. 2007), causing a change in plant development and adaptation (Potters et al. 2007). According to a report by Iglesias et al. (2010), auxin perception impaired *Arabidopsis* showed enhanced tolerance to oxidative stress. Sharma et al. (2018) investigated the influence of external IAA on the various traits of corn and reported that drought stress reduced indigenous IAA content of rice panicles. These researchers reported that exogenous auxin was found to be useful in stabilizing the grain yield of rice and mitigating spikelet sterility under drought stresses. Kim et al. (2013) reported that overexpression of the gene AtYUC6 increased production of auxin and improved drought tolerance of transgenic *Arabidopsis*. Similarly, Shi et al. (2014) showed that *Arabidopsis* with higher endogenous IAA displayed better tolerance to drought stress as compared to plants with lower endogenous IAA level.

5.7 Conclusion

PGPR can affect plants via several mechanisms, for example, biocontrol of phytopathogenic bacteria or through inducing plant growth (Bashan and Holguin 1998). These bacteria indirectly enhance availability of the soil nutrient to plants through root architecture modulation (Navarro-Rodenas et al. 2016), enhance nitrogen content of plants via fixation of nitrogen (Kuan et al. 2016), affect soil phosphate mobilization (Mehta et al. 2015) and siderophore production, and mitigate the detrimental or inhibiting effects of abiotic or nonliving stress such as water scarcity (García et al. 2017) and salinity (Zarea et al. 2012). The application of plant growth-promoting bacterial strains has attracted attention of researchers all over the globe,

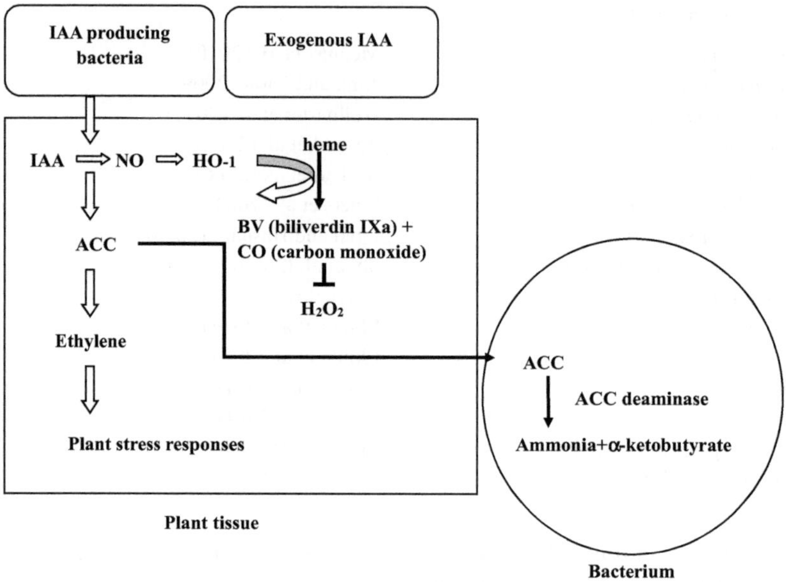

Fig. 5.1 A schematic model of possible mechanism of IAA action in plants in response to environmental stresses and possible mechanisms of IAA-producing bacteria and ACC deaminase-producing bacteria by which PGPR may improve or facilitate plant growth under stress. Plant indigenous IAA indirectly increased heme oxygenase-1 (HO-1) through the modulation levels of nitric oxide (NO) (Lecube et al. 2014). HO-1 participates in the response of the plant to different stresses (Zhang et al. 2009; Gohya et al. 2006) like drought stress (Lecube et al. 2014). NO regulates the expression of HO-1 (Noriega et al. 2007 Lecube et al. 2014; Santa-Cruz et al. 2010, 2017) and enhances both synthesis and activity of HO-1 (Lecube et al. 2014). Augmentation of the HO-1 activity results in biliverdin production. Biliverdin has been identified as potent antioxidant (Noriega et al. 2004) and could participate in the response of plants to drought stress (Lecube et al. 2014).

as an efficient means to mitigate the effect of drought on crop plants. On the other hand, it is imperative to take into account that PGPB strains might vary in their characteristics and in conferring drought tolerance, which justifies that most effective strains should be selected (García et al. 2017). From published reports, it has been observed that IAA-producing bacteria may prove as an important instrument in crop production especially in semiarid and arid regions of the planet. The phytohormone IAA has been reported to play a significant role in several productive mechanisms in the plant. In some studies, it has been reported that exogenous IAA leads to drought tolerance in the host plant. Therefore, considerations in using IAA-producing bacteria in crop production under arid and semiarid areas seem to be important. However, it should be noted that the influence of IAA on plants depends on the plant sensitivity to IAA and the quantity of IAA produced by efficient plant-associated bacterial strains and ultimately other phytohormone induction (Peck and Kende 1995).

References

Ahmad F, Ahmad I, Khan MS (2008) Screening of free-living rhizospheric bacteria for their multiple plant growth promoting activities. Microbiol Res 163:173–181

Ahmed A, Hasnain S (2014) Auxins as one of the factors of plant growth improvement by plant growth promoting rhizobacteria. Polish J Microbiol 63:261–266

Albacete A, Ghanem ME, Martinez-Andujar C, Acosta M, Sanchez-Bravo J, Martinez V et al (2008) Hormonal changes in relation to biomass partitioning and shoot growth impairment in salinized tomato (*Solanum lycopersicum* L.) plants. J Exp Bot 59:4119–4131

Ali B (2015) Bacterial auxin signaling: comparative study of growth induction in A. thaliana and Triticum aestivum. Turkish J Bot 39:1–9

Antoun H, Beauchamp CJ, Goussard N, Chabot R, Lalande R (1998) Potential of Rhizobium and Bradyrhizobium species as plant growth promoting rhizobacteria on non-legumes: effects on radishes (Raphanus sativus L.). Plant Soil 204:57–67

Bajguz A, Piotrowska A (2009) Conjugates of auxin and cytokinin. Phytochemistry 20:957–969

Bashan Y, Holguin G (1998) Proposal for the division of plant growth-promoting rhizobacteria into two classifications: biocontrol-PGPB (plant growth-promoting bacteria) and PGPB. Soil Biol Biochem 30:1225–1228

Bloemberg GV, Lugtenberg BJ (2001) Molecular basis of plant growth promotion and biocontrol by rhizobacteria. Curr Opin Plant Biol 4:343–350

Bottini R, Cassán F, Piccoli P (2004) Gibberellin production by bacteria and its involvement in plant growth promotion and yield increase. Appl Microbiol Biotechnol 65:497–503

Brumos J, Alonso JM, Stepanova AN (2014) Genetic aspects of auxin biosynthesis and its regulation. Physiol Plantarum 151:3–12

Cassán FD, García de Salamone IE (2008) Azospirillum sp. cell physiology, plant interactions and agronomic research in Argentina. In: I International workshop on Azospirillum: cell physiology, plant response and agronomic research in Argentine 2007

Chaiharn M, Lumyong S (2011) Screening and optimization of indole-3-acetic acid production and phosphate solubilization from Rhizobacteria aimed at improving plant growth. Curr Microbiol 62:173–181

Chalupowicz L, Barash I, Panijel M, Sessa G, Manulis-Sasson S (2009) Regulatory interactions between quorum-sensing, auxin, cytokinin, and the Hrp regulon in relation to gall formation and epiphytic fitness of *Pantoea agglomerans* pv. *gypsophilae*. Mol Plant-Microbe Interact 22:849–856

Cohen AC, Bottini R, Pontin M, Berli FJ, Moreno D, Boccanlandro H, Travaglia CN, Piccoli PN (2015) Azospirillum brasilense ameliorates the response of Arabidopsis thaliana to drought mainly via enhancement of ABA levels. Physiol Plant 153:79–90

Colo J, Hajnal-Jafari TI, Durić S, Stamenov D, Hamidović S (2014) Plant growth promotion rhizobacteria in onion production. Pol J Microbiol 63:83–88

Cooke TJ, Poli DB, Sztein AE, Cohen JD (2002) Evolutionary patterns in auxin action. Plant Mol Biol 49:319–338

Costacurta A, Vanderleyden J (1995) Synthesis of phytohormones by plant-associated bacteria. Crit Rev Microbiol 21:1–18

Curá A, Franz DR, Filosofía JE, Balestrasse KB, Burgueño LE (2017) Inoculation with Azospirillum sp. and Herbaspirillum sp. bacteria increases the tolerance of maize to drought stress. Microorganisms 5:41

Dey R, Pal KK, Bhatt DM, Chauhan SM (2004) Growth promotion and yield enhancement of peanut (Arachis hypogaea L.) by application of plant growth-promoting rhizobacteria. Microbiol Res 159:371–394

Dimkpa C, Weinand T, Asch F (2009) Plant-rhizobacteria interactions alleviateabiotic stress conditions. Plant Cell Environ 32:1682–1694

Dobbelaere S, Croonenborghs A, Thys A, Vande Broek A, Vanderleyden J (1999) Phytostimulatory effect of *Azospirillum brasilense* wild type and mutant strains altered in IAA production on wheat. Plant Soil 212:153–162

Du H, Wu N, Fu J, Wang S, Li X, Xiao J, Xiong L (2012) A GH3 family member, OsGH3-2, modulates auxin and abscisic acid levels and differentially affects drought and cold tolerance in rice. J Exp Bot 63:6467–6480

Du H, Liu H, Xiong L (2013) Endogenous auxin and jasmonic acid levels are differentially modulated by abiotic stresses in rice. Fron Plant Sci 4:397

Egamberdieva D (2008) Plant growth promoting properties of rhizobacteria isolated from wheat and pea grown in loamy sand soil. Turk J Biol 32:9–15

Egamberdieva D (2009) Alleviation of salt stress by plant growth regulators and IAA producing bacteria in wheat. Acta Physiol Plant 31:861–864

Ekinci M, Turan M, Yildirim E, Güneş A, Kotan R, Dursun A (2014) Effect of plant growth promoting rhizobacteria on growth, nutrient, organic acid, amino acid and hormone content of cauliflower (*Brassica oleracea* l. var. botrytis) transplants. Acta Sci Pol Hortorum Cultus 13:71–85

Fukami J, Nogueira MA, Araujo RS, Hungria M (2016) Accessing inoculation methods of maize and wheat with *Azospirillum brasilense*. AMB Express 6(1)

Ganesan V (2008) Rhizoremediation of cadmium soil using a cadmium-resistant plant growth-promoting rhizopseudomonad. Curr Microbiol 56:403–407

García JE, Maroniche G, Creus C, Suárez-Rodríguez R, Ramirez-Trujillo JA, Groppa MD (2017) In vitro PGPR properties and osmotic tolerance of different *Azospirillum* native strains and their effects on growth of maize under drought stress. Microbiol Res 202:21–29

Ghosh S, Penterman JN, Little RD, Chavez R, Click BR (2003) Three newly isolated plant growth-promoting bacilli facilitate the seedling growth of canola, *Brassica campestris*. Plant Physiol Biochem 41:277–281

Glick BR (2014) Bacteria with ACC deaminase can promote plant growth and help to feed the world. Microbiol Res 169:30–39

Glick BR, Penrose DM, Li J (1998) A model for the lowering of plant ethylene concentrations by plant growth promoting bacteria. J Theor Biol 190:63–68

Glick BR, Cheng Z, Czarny J, Duan J (2007) Promotion of plant growth by ACC deaminase-containing soil bacteria. Eur J Plant Pathol 119:329–339

Gohya T, Zhang X, Yoshida T, Migita C (2006) Spectroscopic characterization of a higher plant heme oxygenase isoform-1 from *Glycine max* (soybean): coordination structure of the heme complex and catabolism of heme. FEBS J 273:5384–5399

Halda-Alija L (2003) Identification of indole-3-acetic acid producing freshwater wetland rhizosphere bacteria associated with *Juncus effusus* L. Can J Microbiol 49:781–787

Hariprasad P, Niranjana SR (2009) Isolation and characterization of phosphate solubilizing rhizobacteria to improve plant health of tomato. Plant Soil 316:13–24

Hassen AI, Labuschagne N (2010) Root colonization and growth enhancement in wheat and tomato by rhizobacteria isolated from the rhizoplane of grasses. World J Microbiol Biotechnol 26:1837–1846

Honma M, Shimomura T (1978) Metabolism of 1-aminocyclopropane-1-carboxylic acid. Agric Biol Chem 43:1825–1831

Hungria M, Nogueira MA, Araujo RS (2015) Soybean seed co-inoculation with *Bradyrhizobium* spp. and *Azospirillum brasilense*: a new biotechnological tool to improve yield and sustainability. Am J Plant Sci 6:811–817

Iglesias MJ, Terrile MC, Bartoli CG, D'Ippólito S, Casalongué CA (2010) Auxin signaling participates in the adaptative response against oxidative stress and salinity by interacting with redox metabolism in Arabidopsis. Plant Molecular Biolog 74:215–222

Iqbal A, Hasnain S (2013) *Aeromonas punctata* PNS-1: a promising candidate to change the root morphogenesis of *A. thaliana* in MS and sand system. Acta Physiol Plant 35:657–665

Ivanova EG, Doronina NV, IuA T (2001) Aerobic methylobacteria are capable of synthesizing auxins. Mikrobiologiia 70:452–458

Jay PV, Janardan Y, Kavindra NT, Ashok K (2013) Effect of indigenous *Mesorhizobium* spp. and plant growth promoting rhizobacteria on yields and nutrients uptake of chickpea (*Cicer arietinum* L.) under sustainable agriculture. Ecol Eng 51:282–286

Jiang C, Sheng X, Qian M, Wang Q (2008) Isolation and characterization of a heavy metal-resistant *Burkholderia* sp. from heavy metal-contaminated paddy field soil and its potential in promoting plant growth and heavy metal accumulation in metal polluted soil. Chemosphere 72:157–164

Joseph B, Patra RR, Lawrence R (2007) Characterization of plant growth promoting rhizobacteria associated with chickpea (*Cicer arietinum*). Int J Plant Prod 2:141–152

Júnior RFG, Pedrinho EAN, Castellane TCL, de Macedo Lemos EG (2011) Auxin producing bacteria isolated from the roots of *Cattleya walkeriana*, an endangered Brazilian orchid, and their role in acclimatization. Rev Bras Ciênc Solo 35:729–737

Kim JI, Baek D, Park HC et al (2013) Overexpression of Arabidopsis YUCCA6 in potato results in high-auxin developmental phenotypes and enhanced resistance to water deficit. Mol Plant 6:337–349

Korasick DA, Enders TA, Strader LC (2013) Auxin biosynthesis and storage forms. J Exp Bot 64:2541–2555

Kuan KB, Othman R, Abdul Rahim K, Shamsuddin ZH (2016) Plant growth-promoting rhizobacteria inoculation to enhance vegetative growth, nitrogen fixation and nitrogen remobilization of maize under greenhouse conditions. PLoS One 11:e0152478

Kumar KV, Singh N, Behl HM, Srivastava S (2008) Influence of plant growth promoting bacteria and its mutant on heavy metal toxicity in *Brassica juncea* grown in fly ash amended soil. Chemosphere 72:678–683

Kumar M, Sharma S, Gupta S, Kumar V (2018) Mitigation of abiotic stresses in *Lycopersicon esculentum* by endophytic bacteria. Environ Sustain 1(1):71–81

Lecube ML, Noriega GO, Santa Cruz DM, Tomaro ML, Batlle A, Balestrasse KB (2014) Indole acetic acid is responsible for protection against oxidative stress caused by drought in soybean plants: the role of heme oxygenase induction. Redox Rep 19:242–250

Leveau JH, Lindow SE (2005) Utilization of the plant hormone indole-3-acetic acid for growth by *Pseudomonas putida* strain 1290. Appl Environ Microbiol 71:2365–2371

Lim JH, Kim SD (2013) Induction of drought stress resistance by multi-functional PGPR *Bacillus licheniformis* K11 in pepper. Plant Pathol 29:201–208

Liu L, Guo G, Wang Z, Ji H, Mu F, Li X (2014) Auxin in plant growth and stress responses. In: PhanTran L-S, Pal S (eds) Phytohormones: a window to metabolism, signaling and biotechnological applications. Springer, Berlin/Heidelberg, pp 1–35

Lopes MS, Araus JL, van Heerden PDR, Foyer CH (2011) Enhancing drought tolerance in C4 crops. J Exp Bot 62:3135–3153

Ludwig-Muller J (2011) Auxin conjugates: their role for plant development and in the evolution of land plants. J Exp Bot 62:1757–1773

Ma Y, Rajkumar M, Luo Y, Freitas H (2011) Inoculation of endophytic bacteria on host and non-host plants-effects on plant growth and Ni uptake. J Hazard Mater 195:230–237

Mehnaz S, Baig DN, Lazarovits G (2010) Genetic and phenotypic diversity of plant growth promoting rhizobacteria isolated from sugarcane plants growing in Pakistan. J Microbiol Biotechnol 20:1614–1623

Mehta P, Walia A, Kulshrestha S, Chauhan A, Shirkot CK (2015) Efficiency of plant growth-promoting P-solubilizing *Bacillus circulans* CB7 for enhancement of tomato growth under net house conditions. J Basic Microbiol 55:33–44

Navarro-Rodenas A, Berna LM, Lozano-Carrillo C, Andrino A, Morte A (2016) Beneficial native bacteria improve survival and mycorrhization of desert truffle mycorrhizal plants in nursery conditions. Mycorrhiza 26:769–779

Noriega GO, Balestrasse KB, Batlle A, Tomaro ML (2004) Heme oxygenase exerts a protective role against oxidative stress in soybean leaves. Biochem Biophys Res Commun 323:1003–1008

Noriega G, Yannarelli G, Balestrasse K, Batlle A, Tomaro M (2007) The effect of nitric oxide on heme oxygenase gene expression in soybean leaves. Planta 226:1155–1163

Park J-E, Park J-Y, Kim Y-S, Staswick PE, Jeon J, Yun J et al (2007) GH3-mediated auxin homeostasis links growth regulation with stress adaptation response in *Arabidopsis*. J Biol Chem 282:10036–10046

Park JM, Radhakrishnan R, Kang SM, Lee IJ (2015) IAA producing *Enterobacter* sp. I-3 as a potent bio-herbicide candidate for weed control: a special reference with lettuce growth inhibition. Indian J Microbiol 55:207–212

Patten CL, Glick BR (1996) Bacterial biosynthesis of indole-3- acetic acid. Can J Microbiol 42:207–220

Patten CL, Glick BR (2002) Role of *Pseudomonas putida* indoleacetic acid in development of the host plant root system. Appl Environ Microbiol 68:3795–3801

Peck SC, Kende H (1995) Sequential induction of the ethylene biosynthetic enzymes by indole-3-acetic acid in etiolated peas. Plant Mol Biol 28:293–301

Phi QT, Yu MP, Keyung JS, Choong MR, Seung HP, Jong GK, Sa YG (2010) Assessment of root-associated *Paenibacillus polymyxa* groups on growth promotion and induced systemic resistance in pepper. J Microbiol Biotechnol 20:1605–1613

Pierik R, Testerink C (2014) The art of being flexible: how to escape from shade, salt, and drought. Plant Physiol 166:5–22

Pirlak L, Kose M (2009) Effects of plant growth promoting rhizobacteria on yield and some fruit properties of strawberry. J Plant Nutr 32:1173–1184

Poonguzhali S, Madhaiyan M, Sa T (2008) Isolation and identification of phosphate solubilizing bacteria from Chinese cabbage and their effect on growth and phosphorus utilization of plants. J Microbiol Biotechnol 18:773–777

Potters G, Pasternak TP, Guise Y, Palme KJ, Jansen MAK (2007) Stress induced morphogenic responses: growing out of trouble? Trends Plant Sci 12:98–105

Poupin MJ, Greve M, Carmona V, Pinedo I (2016) A complex molecular interplay of auxin and ethylene signaling pathways is involved in *Arabidopsis* growth promotion by *Burkholderia phytofirmans* PsJN. Frontiers Plant Sci 7:492–502

Prakash L, Prathapasenan G (1990) NaCl-and gibberellic acid-induced changes in the content of auxin and the activities of cellulase and pectin lyase during leaf growth in rice (*Oryza sativa*). Ann Bot 65:251–257

Pustovoitova T, Zholkevich V (1999) Basic trends in the investigation of drought effects on physiological processes in plants. Fiziol Biokhim Kul't Rast 24:14–27

Raddidi N, Cherif A, Boudabous A, Daffonchio D (2008) Screening of plant growth promoting traits of *Bacillus thuringiensis*. Ann Microbiol 58:47–52

Rajkumar M, Freitas H (2008) Effects of inoculation of plant growth promoting bacteria on Ni uptake by Indian mustard. Bioresour Technol 99:3491–3498

Rashad MH, Ragab AA, Salem SM (2006) The influence of some *Bradyrhizobium* and *Rhizobium* strains as plant growth promoting rhizobacteria on the growth and yield of sorghum (*Sorghum bicolor* L.) plants under drought stress. In: WWJ H, Schenk MK, Bürkert A et al (eds) Plant nutrition: food security and sustainability of agro-ecosystems through basic and applied research. Fourteenth international plant nutrition colloquium, Hannover, Germany. Kluwer Academic Publishers, Dordrecht

Rodrigues EP, Rodrigues LS, de Oliveira ALM, Baldani VLD, Teixeira KRS, Urquiaga S, Reis VM (2008) *Azospirillum amazonense* inoculation: effects on growth, yield and N_2 fixation of rice (*Oryza sativa* L.). Plant Soil 302:249–261

Saeed M, Ilyas N, Mazhar R, Bibi F, Batool N (2016) Drought mitigation potential of *Azospirillum* inoculation in canola (*Brassica napus*). J Appl Bot Food Qual 278:270–278

Saharan BS, Nehra V (2011) Plant growth promoting rhizobacteria: a critical review. Life Sci Med Res 2011:1–30

Salisbury FB (1994) The role of plant hormones. In: Wilkinson RE (ed) Plant environment interactions. Dekker, New York

Santa-Cruz D, Pacienza N, Polizio A, Balestrasse K, Tomaro ML, Yannarelli G (2010) Nitric oxide synthase-like dependent NO production enhances heme oxygenase up-regulation in ultraviolet-B-irradiated soybean plants. Phytochemistry 71:1700–1707

Santa-Cruz D, Pacienza N, Zilli C, Pagano E, Balestrasse K, Yannarelli G (2017) Heme oxygenase up-regulation under ultraviolet-B radiation is not epigenetically restricted and involves specific stress-related transcriptions factors. Redox Biol 12:549–557

Selvakumar G, Mohan M, Kundu S, Gupta AD, Joshi P, Nazim S, Gupta HS (2008) Cold tolerance and plant growth promotion potential of *Serratia marcescens* strain SRM (MTCC8708) isolated from flowers of summer squash (*Cucurbita pepo*). Lett Appl Microbiol 46:171–175

Shaharoona B, Naveed M, Arshad M, Zahir ZA (2008) Fertilizer-dependent efficiency of Pseudomonads for improving growth, yield, and nutrient use efficiency of wheat (*Triticum aestivum* L.). Appl Microbiol Biotechnol 79:147–155

Shakirova FM, Sakhabutdinova AR, Bezrukova MV, Fatkhutdinova RA, Fatkhutdinova DR (2003) Changes in the hormonal status of wheat seedlings induced by salicylic acid and salinity. Plant Sci 164:317–322

Sharifi R, Ryu CM (2016) Are bacterial volatile compounds poisonous odors to a fungal pathogen *Botrytis cinerea*, alarm signals to *Arabidopsis* seedlings for eliciting induced resistance, or both? Front Microbiol 7(196):1–10

Sharma L, Dalal M, Verma RK, Kumar SVV, Yadav SK, Pushkar S, Kushwaha SR, Bhowmik A, Chinnusamy V (2018) Auxin protects spikelet fertility and grain yield under drought and heat stresses in rice. Environ Experim Bot 150:9–24

Sheng XF, Xia JJ (2006) Improvement of rape (*Brassica napus*) plant growth and cadmium uptake by cadmium-resistant bacteria. Chemosphere 64:1036–1042

Shi CL, Park HB, Lee JS, Ryu S, Ryu CM (2010) Inhibition of primary roots and stimulation of lateral root development in *A. thaliana* by the rhizobacterium *Serratia marcescens* 90-166 is through both auxin-dependent and -independent signaling pathways. Mol Cells 29:251–258

Shi H, Chen L, Ye T, Liu X, Ding K, Chan Z (2014) Modulation of auxin content in *Arabidopsis* confers improved drought stress resistance. Plant Physiol Biochem 82:209–217

Shoebitz M, Ribaudo CM, Pardo MA, Cantore ML, Ciampi L, Curá JA (2009) Plant growth promoting properties of a strain of *Enterobacter ludwigii* isolated from *Lolium perenne* rhizosphere. Soil Biol Biochem 41:1768–1774

Singh RK, Malik N, Singh S (2013) Improved nutrient use efficiency increases plant growth of rice with the use of IAA-overproducing strains of endophytic *Burkholderia cepacia* strain RRE25. Microb Ecol 66:375–384

Smets I, Bernaerts K, Cappuyns A, Ona O, Vanderleyden J, Prinsen E, Van Impe JF (2004) A prototype model for indole-3-acetic acid (IAA) production by *Azospirillum brasilense* sp 245. In: Shah SL, Mac-Gregor JF (Eds.) Proceedings of the 7th international conference on DYnamics and COntrol of Process Systems, Dycops 7, CDROM. IFAC Proceedings Volumes. 37:493–498

Spaepen S, Vanderleyden J (2011) Auxin and plant-microbe interactions. Cold Spring Harb Perspect Biol. https://doi.org/10.1101/cshperspect.a001438

Spaepen S, Vanderleyden J, Remans R (2007) Indole-3-acetic acid in microbial and microorganism plant signaling. FEMS Microbiol Rev 31:425–448

Spaepen S, Bossuyt S, Engelen K, Marchal K, Vanderleyden J (2014) Phenotypical and molecular responses of *A. thaliana* roots as a result of inoculation with the auxin producing bacterium *Azospirillum brasilense*. New Phytol 201:850–861

Swain MR, Naskar SK, Ray RC (2007) Indole-3-acetic acid production and effect on sprouting of yam (*Dioscorea rotundata* L.) minisetts by *Bacillus subtilis* isolated from culturable cow dung microflora. Pol J Microbiol 56:103–110

Tank N, Saraf M (2009) Enhancement of plant growth and decontamination of nickel-spiked soil using PGPR. J Basic Microbiol 49:195–204

Tognetti VB, Mühlenbock P, Van Breusegem F (2012) Stress homeostasis-the redox and auxin perspective. Plant Cell Environ 35:321–333

Tsavkelova EA, Klimova SY, Cherdyntseva TA, Netrusov AI (2006) Microbial producers of lant growth stimulators and their practical use: a review. Appl Biochem Microbiol 42:117–126

Vandeputte O, Oden S, Mol A, Vereecke D, Goethals K, El Jaziri M, Prinsen E (2005) Biosynthesis of auxin by the gram-positive phytopathogen *Rhodococcus fascians* is controlled by compounds specific to infected plant tissues. Appl Environ Microbiol 71:1169–1177

Vega-Hernández MC, León-Barrios M, Pérez-Galdona R (2002) Indole-3-acetic acid production from indole-3-acetonitrile in *Bradyrhizobium*. Soil Biol Biochem 34:665–668

Verma A, Kukreja K, Pathak DV, Suneja S, Narula N (2001) *In vitro* production of plant growth regulators (PGRs) by *Azorobacter chroococcum*. Indian J Microbiol 41:305–307

Vessey JK (2003) Plant growth promoting rhizobacteria as biofertilizers. Plant Soil 255:571–586

Vivas A, Biro B, Ruiz-Lozano JM, Barea JM, Azcon R (2006) Two bacterial strains isolated from a Zn-polluted soil enhance plant growth and mycorrhizal efficiency under Zn toxicity. Chemosphere 52:1523–1533

Vurukonda SSKP, Vardharajula S, Shrivastava M, SkZ A (2016) Enhancement of drought stress tolerance in crops by plant growth promoting rhizobacteria. Microbiol Res 184:13–24

Wani PA, Khan MS (2010) *Bacillus* species enhance growth parameters of chickpea (*Cicer arietinum* L.) in chromium stressed soils. Food Chem Toxicol 48:3262–3267

Wani PA, Khan MS, Zaidi A (2007a) Effect of metal tolerant plant growth promoting *Bradyrhizobium* sp. (vigna) on growth, symbiosis, seed yield and metal uptake by green gram plants. Chemosphere 70:36–45

Wani PA, Khan MS, Zaidi A (2007b) Co inoculation of nitrogen fixing and phosphate solubilizing bacteria to promote growth, yield and nutrient uptake in chickpea. Acta Agron Hung 55:315–323

Wani PA, Khan MS, Zaidi A (2007c) Synergistic effects of the inoculation with nitrogen fixing and phosphate solubilizing rhizobacteria on the performance of field grown chickpea. J Plant Nutr Soil Sci 170:283–287

Wani PA, Khan MS, Zaidi A (2008) Chromium-reducing and plant growth-promoting *Mesorhizobium* improves chickpea growth in chromium-amended soil. Biotechnol Lett 30:159–163

Weilharter A, Mitter B, Shin MV, Chain PS, Nowak J, Sessitsch A (2011) Complete genome sequence of the plant growth-promoting endophyte Burkholderia phytofirmans strain PsJN. J Bacteriol 193:3383–3384

Yang S, Zhang Q, Guo J, Charkowski AO, Glick BR, Ibekwe AM, Cooksey DA, Yang CH (2007) Global effect of indole-3-acetic acid biosynthesis on multiple virulence factors of *Erwinia chrysanthemi* 3937. Appl Environ Microbiol 73:1079–1088

Yanni YG, Rizk RY, Abd El-Fattah FK, Squartini A, Corich V, Giacomini A, de Bruijn F, Rademaker J, Maya-Flores J, Ostrom P, Vega-Hernandez M, Hollingsworth RI, Martinez-Molina E, Mateos P, Velazquez E, Wopereis J, Triplett E, Umali-Garcia M, Anarna JA, Rolfe BG, Ladha JK, Hill J, Mujoo R, Ng PK, Dazzo FB (2001) The beneficial plant growth promoting association of *Rhizobium leguminosarum* bv. *trifolii* with rice roots. Aust J Plant Physiol 28:845–870

Yasmin S, Rahman M, Hafeez FY (2004) Isolation, characterization and beneficial effects of rice associated plant growth promoting bacteria from Zanzibar soils. J Basic Microbiol 44:241–252

Zahir ZA, Shah MK, Naveed M, Akhter MJ (2010) Substrate dependent auxin production by *Rhizobium phaseoli* improves the growth and yield of *Vigna radiata* L. under salt stress conditions. J Microbiol Biotechnol 20:1288–1294

Zarea MJ, Hajinia S, Karimi N, Mohammadi Goltapeh E, Rejali F, Varma A (2012) Effect of *Piriformospora indica* and *Azospirillum* strains from saline or non-saline soil on mitigation of the effects of NaCl. Soil Biol Biochem 45:139–146

Zarrin K, Sharon LD (2010) Characterization of bacterial endophytes of sweet potato plants. Plant Soil 322:197–207

Zhang X, Ervin EH, Evanylo GK, Haering KC (2009) Impact of biosolids on hormone metabolism in drought-stressed tall fescue. Crop Sci 49:1893–1901

Zholkevich V, Pustovoitova T (1993) Growth and phytohormone content in *Cucumis sativus* L. leaves under water deficiency. Russ Plant Physiol 40:595–599

Role of Endophytes in Plant Health and Abiotic Stress Management

6

Ahmed Mohamed Eid, Salim S. Salim, Saad El-Din Hassan, Mohamed A. Ismail, and Amr Fouda

Abstract

Microbial endophytes are symbionts dwelling within plant tissues without appearance of disease symptoms on host plant and have been recently investigated for their plant growth-promoting properties and their beneficial functions associated with plant responses under abiotic stress conditions. This study focuses on the critical role of endophytic microbes in plant health and their stimulatory different mechanisms to tolerance against abiotic stress in plants. Endophytic microbial community can enhance plant growth through producing secondary active compounds which protect the plant from pathogens such as insect and fungi; also endophytes can produce extracellular enzymes which play critical roles in colonization of endophytes within the plant host. Microbial endophytes have the ability to act as plant growth-promoting agents through producing phytohormones and also enable plants to grow in contaminated soils through breakdown of hazardous compounds. Endophytes manage plant growth under adverse conditions such as salinity, drought, temperature, heavy metal stress, and nutrient stress through different mechanisms. This chapter may introduce new approaches for the use of endophytic inoculants to combat abiotic stresses in agricultural fields, which increases global crop production.

Keywords

Endophyte · Plant · Abiotic stress · Management

A. M. Eid · S. S. Salim · S. E.-D. Hassan · M. A. Ismail · A. Fouda (✉)
Department of Botany and Microbiology, Faculty of Science, Al-Azhar University, Nasr City, Cairo, Egypt

V. Kumar et al. (eds.), *Microbiome in Plant Health and Disease*,
https://doi.org/10.1007/978-981-13-8495-0_6

6.1 Introduction

Endophytes are microbial communities that currently dwell in the healthy plant tissues such as stems, roots, leaves, and seeds without affecting physiological plant functions and not causing any disease symptoms to the plant tissues. Under normal conditions, endophytes have important roles in host plant growth either by secondary metabolite or nutrient assimilation or by preventing induction of plant disease symptoms by different pathogens. Endophytic microbes including bacteria, actinomycetes, and fungi tend to form a network closely to their host plants and are additionally sheltered from unfavorable climatic and other unwanted change in the environment (Zhao et al. 2011; Passari et al. 2017).

Recent research work suggests that about 300,000 species of plants are present, the unmistakable dominant part of which contains endophytes (Smith et al. 2008). In fact, microbial endophytes especially bacteria and fungi have originated in most plant species that have been investigated. According to Partida-Martínez and Heil (2011), endophyte-free plant is an unusual state to what is typically obtained in nature; a plant without endophytes would be susceptible to environmental stress conditions and lose their ability to resist the pathogens (Timmusk et al. 2011).

The origin of endophytes is still not clear because of the multiplicity of the host's living environment and the complex association between the endophyte and its host plant. Two hypotheses explaining the origin of endophytes were exogenous and endogenous. It was believed that in the last decade, endophytes are gaged from the chloroplast and mitochondria of the plant, and so it has comparable genetic backgrounds to the host (Wen 2004); this is the endogenous hypothesis, while the latter believes that endophytes arrive from outside of the plant and insert into the host from root wound, induced channels, or surface (Li 2005); this is exogenous hypothesis.

Different parts of plants were used for isolation of microbial endophytes as meristem, scale primordia, resin ducts (Pirttilä et al. 2003), leaf segments with midrib and roots (Hata et al. 2002), leaf blade, stem, petiole, bark, and buds (Pirttilä et al. 2008).

Endophytic fungi insert through the hyphae and enter the kernels in the seeds of plant cells that come below vertical transmission. A variant was detected in horizontal and vertical transmission of the endophyte species invading the host plant cells (Tintjer et al. 2008). The procedures of endophytic microbial growth in plants and methods of propagation were paid more attention to know their role in transmission. The endophytic fungal species transmits horizontally by sexual spores or asexually between different plants in community or a population (Tadych et al. 2014).

The microbial community such as bacteria, algae, fungi, and actinomycetes colonizes the host plant roots (Saharan and Nehra 2011; Prashar et al. 2014). Among microbial population found in the rhizosphere, actinobacteria are considered the second most abundant microorganisms, and they comprise more than 30% of the total microorganisms in the soil (Glick 2014). Endophytes are transmitted between the soil rhizhosphere across the seeds. They spread quickly between endo-rhizosphere through the lateral root junction instigated through microbial phyto-pathogens or nematode (Chi et al. 2005). Also, bacterial endophytes can enter their host plant roots through spaces between root hairs and epidermal cells (Hardoim et al. 2008).

The most common endophytic fungi isolated and identified from numerous plants are *Alternaria infectoria*, *Aspergillus* sp., *Penicillium* sp., *Colletotrichum musae*, *Colletotrichum gloeosporioides*, *Nigrospora oryzae*, *Phomopsis* sp., *Nigrospora sphaerica*, *Guignardia* sp., *Cordana musae*, *Rhizoctonia* sp., species of *Phialocephala sphaeroides*, *Xylaria* (Wilson et al. 2004), *P. chrysogenum* Pc_25, *A. alternata* Aa_27, and Sterile hyphae Sh_26 (Fouda et al. 2015). On the other hand, endophytic *P. chrysogenum* Pc_25 was mediated biosynthesis of ZnO nanoparticles (Fouda et al. 2019a).

In addition, various endophytic bacterial strains were isolated from economically important plant species. Several of the novel endophytic bacterial species belong to the *Arthrobacter* spp., *Actinobacter* spp., *Aeromonas* spp., *Enterobacter* spp., *Agrobacterium* spp., *Alcaligenes* spp., *Bacillus* spp., *Flavobacterium* spp., *Azospirillum* spp., *Azotobacter* spp., *Pseudomonas* spp., *Burkholderia* spp., *Beijerinckia* spp., *Enterobacter* spp., *Flavobacterium* spp., *Erwinia* spp., *Rhizobium* spp., and *Serratia* spp. were characterized and identified (Gray and Smith 2005). In the last periods, other endophytic actinobacteria such as *Streptomyces*, *Amycolatopsis*, *Nocardia*, *Microbispora*, *Micromonospora*, and *Streptomyces capillispiralis* Ca-1 have been positively isolated from different plant species (Shi et al. 2009; Ruanpanun et al. 2010; Hassan et al. 2018).

6.2 Role of Endophytes in Plant Health

Great effort has been made to study the diversity of endophytic species in plants and their evolutionary biology, ecology, and their roles in defense mechanism against abiotic and biotic stress via production of different metabolites. Endophytic biotechnology can be expended for the efficient production of economically, agriculturally, and industrially significant plants and their crops. The reasonable application for different endophytic species associated with plants can help in improvement of the agricultural products, increasing metabolite productivity in different plants, as well as adjustment tolerance to numerous abiotic and biotic conditions (Wani et al. 2015).

Endophytic species have recently generated important new bioactive substances. It has been suggested that the relationship between different endophytic species and their host plant in the production of a great amount and diversity of biologically active molecules are related together, and this contrasted to epiphytes or soil-related microorganisms (Strobel 2003).

New biotechnology applications for endophytic species such as bioremediation and phytoremediation are gaining considerable impetus (Li et al. 2012a). Endophytes play critical roles in healthy plants through three different mechanisms known as biofertilization, phytostimulation, and biocontrol (Bloemberg and Lugtenberg 2001).

6.2.1 Endophytes as Source for Bioactive and Novel Compounds

Endophytic microbes release specialized biologically active compounds or metabolites without any observable damage to their host tissues (Liarzi et al. 2016). The bioactive compounds synthesized by different endophytic microbes that increase plant resistance against pathogenic microorganisms are too used in the pharmaceutical fields as anticancer, antimicrobial, antiviral, antidiabetic, and other biologically active compounds (Guo et al. 2008). Other biologically active compounds synthesized by endophytic microbes as alkaloids, terpenoids, steroids, peptides, poly-ketones, quinols, flavonoids, phenols, and insecticide azadirachtin are also investigated for their medical, agricultural, and industrial applications (Kusari et al. 2012a; Molina et al. 2012; Zinniel et al. 2002). Numerous of these bioactive compounds showed antioxidant, antimicrobial (antibacterial, antifungal, and antiviral activities), antineoplastic, antiproliferative, anti-leishmanial, cytotoxicity, and fuel production activities (Shankar Naik and Krishnamurthy 2010; Wang and Dai 2011). Examples of antifungal compounds produced by endophytes include cryptocandin, pestaloside, cryptocin, ecomycins, pestalopyrone, and pseudomycins (Yu, et al. 2010).

Naturally, different seasons, locations, environmental conditions, soil, age, and tissue of the host plant, all influence the endophyte biology and thus considerable variants in the synthesis of bioactive metabolites (Strobel and Daisy 2003). Also, cultivation conditions and separation methods can affect the type and variety of metabolites (Gunatilaka 2006). Aly et al. (2011) and Kusari et al. (2012b) reported that sampling, type, and size of the plant tissue used for isolation, composition, and culture conditions for media such as pH, incubation temperature, incubation period, agitation, and culture, all of these factors are controlling the productivity of bioactive compounds in the laboratory.

The productivity of bioactive compounds by endophytic microorganisms can be influenced both genetically and physicochemically (Kharwar et al. 2011). Separation and identification methods of bioactive metabolites from microbes, especially fungal and bacterial endophytic species, are fast growing, as can be detected from numbers of patents, reviews, and original articles available each year in the drug discovery field (Tejesvi and Pirttilä 2011). Fungal endophytic species are a native source for flavonoids, terpenoids, phenols, saponin, alkaloids, carbohydrates tannins, and nematode antagonistic compounds (Liu et al. 2016; Bogner et al. 2017). Endophytic actinomycetes are promising tool for synthesis of bioactive compounds, which can be used as therapeutic agents against different pathogens (Prashith-Kekuda 2016). On the other hand, endophytic bacteria have useful effects as enhancement of nitrogen fixation, phosphate solubilization, production of phytohormones, and reduction of ethylene biosynthesis as response to abiotic (pH, temperature, and osmotic pressure) and biotic (from fungi, bacteria, nematodes, and insects) stresses and have biocontrol activities. Supplementary than 300 endophytic bacteria and actinobacteria belonging to the genera *Streptomyces, Rhodococcus, Nocardiopsis, Microbacterium, Brevibacterium, Arthrobacter, Nocardia,*

Brachybacterium, Tsukamurella, Kocuria, Pseudonocardia, and *Nocardioides* were isolated from *Dracaena cochinchinensis* Lour. plant. Of these, 17 endophytic strains have antimicrobial, anthracyclines-producing activities, antifungal properties, and anticancer activities against Hep G2 and MCF-7 as cancer cells (Salam et al. 2017). Endophytic microorganisms have already a bulk for the discovery of biologically active compounds, but still innovative methods are demanded to natural product-based drug discovery.

6.2.2 Extracellular Enzyme Production

Endophytes play an important role in plant health via extracellular enzyme production which have been counted as the most significant and important mechanisms for endophyte colonization in plants. Among enzymes, extracellular enzymes or exoenzymes have industrial importance in different fields such as fermentation process, food, and other biotechnological applications. Microorganisms including fungi and bacteria produce different types of extracellular enzymes, which are oxidoreductases, hydrolases, transferases, and lyases (Traving et al. 2015). Extracellular enzymes breakdown numerous macromolecules such as lignin, sugar-based polymers, proteins, organic phosphate, and carbohydrates to micromolecules and are transported throughout the cells; they are continuously metabolized and help to instruct the host symbiosis process (Strong and Claus 2011; Wingender et al. 1999). Also, extracellular hydro-lyase enzymes increase plant responses to pathogenic infection (Leo et al. 2016).

Different extracellular enzymes such as β-1,3-glucanase, protease, lipase, and chitinase associated with endophytic microbes lyse the cell walls of pathogenic bacteria and fungi and hence can be used as biocontrol agents (Fouda et al. 2015, Wang et al. 2014).

On the other hand, improvement in endophytic microbial growth within host plant tissues and then reduction in the pathogens are influenced by production of various enzymes such as xylanases, cellulases, pectinases, lipases, proteases, phosphatases, amylase, and glucosidases (Kannan et al. 2015; Pereira et al. 2016; Khan et al. 2016; Ayob and Simarani 2016).

Chathurdevi and Gowrie (2016) reported that the endophytic fungi isolated from medicinal plants can support plant growth to overcome the adverse conditions through producing different extracellular enzymes. Also, approximately 50 endophytic fungal strains having amylase, laccase, cellulase, pectinase, lipase, and protease were isolated and identified according to Sunitha et al. (2013). In addition, bacterial endophytes have been investigated to produce 1-aminocyclopropane-1-carboxylate (ACC) deaminase, amylase, cellulases, esterase, pectinase, protease, lipase, asparaginase, phytase, and xylanase (Carrim et al. 2006; Gupta et al. 2013; Fouda et al. 2015; Akinsanya et al. 2016). Vijayalakshmi et al. (2016) isolated bacterial endophytes from medicinally significant plants producing various extracellular enzymes as cellulase, amylase, and protease.

6.2.3 Plant Growth-Promoting Activity

Endophytic microbes play a critical role in plants' adaptation to stress conditions and varying environments which can limit their development and growth. To contract with extreme environments, plants may form network with microorganisms in symbiotic relationship, which confer helpful effects on evolution to both partners and appropriateness (Rodriguez et al. 2009). Extreme environment conditions such as low water and nutrient availability, high radiation, strong winds, and low temperatures affect plant survival and organization (Convey 2011).

A practical strategy to reducing stress without cooperating plant growth is the use of plant growth-promoting microbe relationship. Widespread varieties of metabolite substances produced by endophytic fungi are used to improve plant growth (Waqas et al. 2015).

Endophytic microbes enhance plant growth through their ability to synthesize enzymes and various bioactive metabolites. Endophytic microorganisms, especially fungi such as *Sebacina vermifera, Piriformospora indica* and numerous species of *Colletotrichum* and *Penicillium,* are distinguished to have better plant growth-promoting effects under unfavorable conditions (Waller et al. 2005; Redman et al. 2011; Hamilton and Bauerle 2012). Plant growth-promoting microbes (PGPM) associated with many species of plants tend to have useful effects such as intensified plant growth and decreased sensitivity to diseases instigated by plant pathogenic viruses, fungi, bacteria, and nematodes. Main activities of PGPM are accompanied with plant hormone synthesis such as indole-3-acetic acid (IAA), cytokinins, gibberellins, siderophores, phosphate solubilization, nutrient uptake, and antagonism to phytopathogens. Also, PGPM can induce chemical or physical changes related to plant protection, a process signified as induced systemic resistance (ISR). PGPM have developed to be beneficial for plants constantly under many abiotic stresses. Several reports confirmed the critical roles of plant growth-promoting fungi (PGPF) in increasing tolerance against different stresses such as heat, drought, salinity, cold, and heavy metals (Khan et al. 2012). On the other hand, stresses like salinity and drought induce osmotic stress, which is conveyed through abscisic acid (ABA)-independent or ABA-dependent pathways (Cao et al. 2014), and low levels of ABA productions were achieved under fungal action (Jahromi et al. 2008; Khan et al. 2014). Miransari (2012) reported that plants needed low attempt to synthesize ABA and hence protect cell progress under stress, as the water equilibrium in plant was achieved by treatment with endophytic *Penicillium* spp.

Plant growth-promoting bacteria (PGPB) are proficient to enhance plant growth through independent or linked mechanisms for maintainable agriculture (Compant et al. 2010; Palacios et al. 2014). PGPB showed different responses against numerous stresses in plants (Kim et al. 2012), fighting against plant pathogens (Raaijmakers et al. 2009) and supplementary in the recovery of damaged cells or degraded constituents (de Bashan et al. 2012). Colonization of host plant tissues by endophytic bacterial species has been reported by Yang et al. (2016) and Tang et al. (2017), and its capability to promote growth, fix nitrogen, and repress phytopathogens with induced systemic resistance (ISR) of this pathogen was reported by Pieterse et al.

(2014), Puri et al. (2016), and Padda et al. (2016). Endophytic actinobacteria can improve plant growth via one or more plant growth-promoting mechanisms including nitrogen fixation, solubilization of inorganic nutrients, excretion of phytohormones, and siderophores (Dudeja et al. 2012).

Indoleacetic acid (IAA) is fundamental plant growth hormone used for the development and growth of shoot and root cells of plants; many microorganisms including plant growth-promoting rhizobacteria (PGPR) produce IAA (Hassan 2017). Soil microorganisms such as bacteria and fungi synthesize plant growth-promoting compounds such as gibberellins and IAA (Radhakrishnan et al. 2013; Limtong et al. 2014).

Several reports proved that endophytic actinobacteria synthesize plant growth regulators such as auxins, cytokinins, gibberellins (gibberellic acid), and IAA in vitro (Ghodhbane-Gtari et al. 2010; Fouda et al. 2019b).

Siderophores are small compounds with high-affinity iron chelators (soluble Fe^{3+}-binding agents) synthesized by microbes such as fungi, bacteria, and actinobacteria growing under low iron stress. Several endophytic microbes have been explored to synthesize siderophores, with a molecular weight ranging between 400 and 1500 daltons (Kannahi and Senbagam 2014). There are four types of siderophores synthesized by bacteria, viz., catecholate, salicylate, hydroxamate, and carboxylate. Endophytic actinobacteria including the genera *Pseudonocardia, Streptomyces, Nocardia, Actinopolyspora, Micromonospora, Salinispora, Actinomadura, and Kibdelosporangium* are recognized as siderophore producers (Gangwar et al. 2011; Kannahi and Senbagam 2014; Bhosale and Kadam 2015). Endophytic actinobacteria synthesize siderophores as an extramechanism, which act as plant growth regulators and in defense against pathogens (Rungin et al. 2012).

Moreover salicylic acid (SA), as a phytohormone, is a significant plant hormone concerned in many processes such as root initiation, seed germination, floral induction, stomatal closure, and increased tolerance of plant to abiotic and biotic stresses. Bacterial endophytes synthesize SA, which enhance the growth of plant seedlings under water stress and reduce the growth of plant pathogens such as fungi (Klessig et al. 2016).

6.2.4 Biocontrol Agents

Endophytic microorganisms are defined as a functional biocontrol agent, instead of chemical control. Endophytic fungi play a critical role in controlling insect herbivores not only in grasses but also in conifers (Parker 1995). Tefera and vidal (2009) reported that *Beauveria bassiana,* an endophytic fungi known as an entomopathogen, was used to control the borer insects in sorghum. Also, acute rotting caused by fungal pathogens in tomato fruits can be achieved during storage and shelf life. Different bacterial endophytic strains such as *Bacillus subtilis* isolated from *Speranskia tuberculata* (Bge.) Baill has an antagonistic effect in vitro against the pathogen *Botrytis cinerea,* which cause rotting of tomato fruits during storage (Wang et al. 2009). New endophytes, such as *Burkholderia pyrrocinia* JK-SH007 and *Bacillus cepacia,* were used in biocontrol study against poplar canker (Ren et al. 2011). New approaches in biocontrol studies induced genetically engineered gene expression into an endophytic microorganism to synthesize anti-pest proteins like lectins for insect control.

On the other hand, endophytic strains such as *Chaetomium globosum* YY-11 isolated from rape seedlings, *Enterobacter* sp., and *Bacillus subtilis* isolated from seedlings of rice were utilized for the expression of *Pinellia ternate* agglutinin (*PtA*) gene (Zhao et al. 2010). The previous endophytic recombinant fungal and bacterial strains which express the *PtA* gene were used successfully to control sap-sucking pests in numerous crop seedlings. Similarly, in another study, *Enterobacter cloacae,* as a recombinant endophytic bacterial strain expressing the *PtA* gene, was proved as a bio-insecticidal agent against the white backed plant hopper *Sogatella furcifera* (Zhang et al. 2011). The recombinant endophytic strains which easily dwell within several plants can be used as a new strategy to control different plant pests through expression of different anti-pest proteins. Also, Hassan et al. (2018) reported that copper nanoparticles synthesized using the endophyte *Streptomyces capillispiralis* Ca-1 have the ability to biocontrol *Culex pipiens* (Mosquito) and *Musca domestica* (housefly). On the other hand, copper oxide nanoparticles synthesized by two endophytic actinomycetes, *Streptomyces zaomyceticus* Oc-5 and *Streptomyces pseudogriseolus* Acv-11, isolated from *Oxalis corniculate L.* plant have antimicrobial activity against four phytopathogenic fungi, namely, *Phoma destructiva, Fusarium oxysporum, Alternaria alternata, and Curvularia lunata* (Hassan et al. 2019).

6.2.5 Bioremediation/Biodegradation Activity

Endophytic microorganisms have a powerful ability to enhance plant growth in contaminated soil through breakdown of hazard compounds. Bioremediation is defined as elimination of pollutants and hazardous wastes from contaminated environments by breakdown of these wastes using biological processes. This is due to the major microbial diversity. Mastretta et al. (2009) reported the ability of *Nicotiana tabacum* plants in bioremediation through inoculation of their seeds with endophytes. They have showed enhanced plant biomass production under Cadmium (Cd) as heavy metal stress, and the Cd concentration in plant tissue was higher compared to noninoculated plants. These results proved the useful effects of endophytic inoculated seeds on accumulation and assimilation of heavy metals.

To discover the role of endophytic microbes in the breakdown of contaminants such as plastics, different endophytic fungal strains were subjected to growth in agar and broth media containing polyester polyurethane (PUR) as a sole carbon source (Russell et al. 2011). Among fungal endophytic strains, two *Pestalotiopsis microspora* isolates have the ability to use PUR as the sole carbon source under aerobic and anaerobic conditions and using serine hydrolase enzyme for the degradation of PUR.

6.2.6 Induced Systemic Resistance (ISR)

Endophytic microorganisms increase plant resistance to pathogen through induce defense mechanisms, so-called induced systematic resistance (ISR) (Zamioudis and Pieterse 2012). At an initial stage, there is increasing evidence that interactions between endophytic microorganisms and their hosts stimulate immune response in host plants; this is similar to those happening against pathogens; later on, endophytic microorganisms colonize plants through escaping from defense responses as occurring in the bacterial genera *Bacillus* and *Pseudomonas* (Kloepper and Ryu 2006). Different bacterial factors such as antibiotics, salicylic acid, N-acyl-homoserine lactones, siderophores, jasmonic acid, lipopolysaccharides, and volatiles (e.g., acetoin) are responsible for induction of ISR (Bordiec et al. 2011). The defense mechanisms and protections of plants against herbivorous insects and pathogens were related to ISR. Although several endophytic bacteria have increased ISR via salicylic acid induction, ethylene (ET) and jasmonic acid (JA) as plant hormones have important regulatory roles in signaling pathways implicated in ISR induction (Pieterse et al. 2012). The endophytic bacterium *Pseudomonas fluorescens* 89B-61 was the first reported, explaining the ISR induction to protect cucumber plants against cucumber anthracnose (Kloepper and Ryu 2006). The resistance of potato plant against the pathogen *Pectobacterium atrosepticum* was increased in the presence of the endophyte *Methylobacterium* sp. IMBG290. The resistance manner was related to changes in composition of the native endophytic community. Changes in endophytic community were linked to disease resistance, which means the endophytic community has a critical role in disease repressions (Ardanov et al. 2011). Also, endophytic fungi have been involved in protection mechanisms via ISR induction but less than endophytic bacteria (Bae et al. 2011). The potentiality of endophytic fungi in producing metabolites has inhibitory activities against herbivores, and plant pathogens were recorded. These metabolites comprise steroids, alkaloids, peptides, terpenoids, flavonoids, polyketones, phenols, chlorinated compounds, and quinols (Higginbotham et al. 2013; Tejesvi et al. 2013). On the other hand, metabolites having antibacterial, antiviral, insecticidal, and antifungal activities were reported by fungal endophytes, which transmitted horizontally, forming local disease in their hosts (Gunatilaka 2006; Tejesvi et al. 2011).

6.3 Alleviation of Abiotic Stresses via Microbial Endophytes

Plant growth and development are restricted by different extreme conditions which include environmental stresses as well as stresses caused by living communities.

Plants can tolerate abiotic stress by two mechanisms: (i) plants can avoid negative effects of stress via activation of response systems directly after exposure to stress (Meena et al. 2017), and (ii) biochemical compounds are synthesized by endophytes and act as anti-stress agents (Schulz et al. 2002). The up- and downregulation for some stress-inducible genes in pepper plant were reduced after inoculation with the endophyte *Arthrobacter* sp. and *Bacillus* sp. when compared with gene expression in uninoculated plants. Assimilation of nutrients, such as magnesium, potassium, and calcium, plant biomass, growth parameters, and decreased sodium toxicity were significantly increased in cucumber plants under sodium chloride and drought stress after inoculation with *Phoma glomerata* and *Penicillium* sp. when compared with uninoculated plants (Waqas et al. 2012). Bailey et al. (2006) revealed that *Trichoderma* sp. isolated from *Theobroma cacao* increases tolerance in *cacao* plant against abiotic stress especially drought via gene expression change.

The resistance of tissue cultured *Kalmia latifolia* L. to drought stress was increased after seedling inoculation with *Streptomyces padanus* AOK-30 as endophytic actinomycetes as reported by Hasegawa et al. (2004).

Bae et al. (2009) reported that sugars and amino acids showed significant increase in endophyte-colonized plants due to drought stress. Increase in sugar and amino acid production (as indicators for increased osmolytic activity) is due to intricate symbiotic relationship in plants possessing a drought-tolerant phenotype (Shinozaki and Yamaguchi-Shinozaki 2007). Significantly producing biomass is one response to drought, temperature, and salt stress in endophyte-colonized plants than their non-colonized one (Redman et al. 2011). Zhang and Nan (2010) revealed that increased seedling growth as drought response was due to higher antioxidant activity. Also, Zhang and Nan (2007) showed that increase in biomass, proline concentrations, and relative water content as a result of endophyte colonization under low water conditions was investigated. Inoculation of wheat with *Burkholderia phytofirmans* PsJN increased CO_2 assimilation, photosynthetic rate, water use efficiency, and chlorophyll content under drought conditions (Naveed et al. 2014).

The following are examples for abiotic stresses that have a negative effect on plant growth and development, and alleviate via endophytes.

6.3.1 Drought Stress

Drought is one of the most important abiotic stresses that suppress plant growth, development, and productivity. Plants undergo drought conditions through either limiting water supply to the roots or very high transpiration (Anjum et al., 2011). It has been concluded that diurnal water stress normally occurs in most plant species during noon and afternoon hours in temperate climates, even though the soil water contents are normal. This temporary drought stress has a negative impact on the growth rate (Granier and Tardieu 1999). Drought causes reduced germination rates, membrane loss of its integrity, repression of photosynthesis, and increase in the productivity of reactive oxygen species (Greenberg et al. 2008). Furthermore, elevated drought and salinity were the main causes of osmotic stress to plants. While drought leads to osmotic stress, salinity shows both ionic or ion-toxicity, and osmotic stress impacts cells (Zhu 2002). The shoot system symptoms of osmotic stress caused by salinity interfere with that of drought stress including stunted growth and leaf senescence (Munns 2002).

Plants harboring endophytes (rice, tomato, dune grass, and panic grass) consumed significantly less water and had enhanced biomass than nonsymbiotic plants. Increased accumulation of solutes in tissues of endophyte-associated plants comparable with noninfected plants, or because of thicker cuticle formation, or by decreased leaf conductance and a slower transpiration stream may explain the drought tolerance phenomenon (Malinowski and Beleskey 2000). The ability of plant to tolerate water stress may be related to morphological and genetic adaptation and biochemical responses. However, the central response to water deficits is the increase in the biosynthesis of plant hormone ABA and/or reduction in ABA breakdown (Bray 2002). In plants suffering from drought, it is supposed that ABA behaves like the signal that manages the plant's resistance to water deficit, principally by controlling water loss and stomata closure (Zhang and Outlaw 2001). Also, other evidence proposes that ABA has a role in root branching, enhancing the plant water absorption capacity (De Smet et al. 2006).

ABA was defined using full scan mass spectrometry as a by-product of chemically enhanced growth cultures of *Azospirillum brasilense* Sp 245. Adding NaCl to the culture medium led to increased bacterial ABA production, and ABA levels were improved in *Arabidopsis thaliana* seedlings inoculated with *Azospirillum brasilense* Sp 245 (Cohen et al. 2008).

6.3.2 Salinity Stress

Soil salinization happens when water-soluble salts accumulate in the soil to a level that affected environmental health, agricultural production, and economics. In the first stages, salinity has a negative impact on the metabolism of soil organisms and hence decreases soil productivity, but in advanced stages, it destroys all vegetation and other organisms living in the soil, consequently transforming fertile and productive land into barren and desertified lands (Jones et al. 2012). A saline soil is known to have an electrical conductivity (EC) of the saturation extract (ECe) in the root zone more than 4 dS m^{-1} (nearly 40 mM NaCl) at 25 °C with exchangeable sodium of 15%. The yield of most crop plants is decreased at this ECe, and many crops showed reduced yield at lower ECes (Jamil et al., 2011).

It is a key factor contributing to reduced productivity of cultivated soils. Although accurate estimation is difficult, the salinized soil area is increased, and this phenomenon is particularly dense in irrigated soils. It is estimated that about 20% (45 million hectares) of irrigated land, which produces one-third of global food, is affected by salinity (Shrivastava and Kumar 2015). Soil salinity impacts an estimated one million hectares in the European Union, particularly in the Mediterranean countries, a major cause of desertification. In Spain, about 3% of irrigated land (3.5 million hectares) is severely affected, significantly reducing their agricultural potential, while another 15% are at high risk (Stolte et al. 2015). In the Mediterranean area, soil alkalization associated with land degradation may deteriorate at increasing rates in the coming decades due to the expected increase in irrigated regions and the increasing deficiency of good-quality water (Bowyer et al. 2009).

6.3.2.1 Effect of Soil Salinity on Plants

Salinity significantly affects agricultural crops, which reduces agricultural output and affects the physical and chemical properties of the soil and environmental balance of the region, as well as low economical findings and soil corrosions (Hu and Schmidhalter 2002).

Complex interactions led to salinity effects comprising biochemical, physiological, and morphological processes including vegetative growth, water uptake, seed germination, enzyme activity, seedling growth, protein synthesis, and mitosis of DNA and RNA (Akbarimoghaddam et al. 2011). It has deep impact on reproductive development by stamen filament elongation and inhibiting microsporogenesis, ovule abortion, and senescence of fertilized embryos and enhanced cell death in tissue types. Since many salts are also plant nutrients, increased salt concentrations in the soil can disturb the plant nutritional balance or interfere with the absorption of some nutrients (N, Ca, K, P, Fe, and Zn) leading to nutrient deficiency. Because P ions precipitate with Ca ions, P uptake is significantly reduced by soil salinity (Bano and Fatima 2009).

While K^+ has a main role in biochemical reactions, acting as a cofactor for various enzymes and in protein synthesis, high concentrations of K^+ mediates binding of tRNA to ribosomes. However, soil salinity leads to ion toxicity resulting from replacing K^+ by Na^+ in such reactions. Furthermore, Cl^- and Na^+ induced conformational modifications in proteins (Zhu 2002). Soil salinity imposes osmotic stress, which leads to loss of turgidity, cell dehydration, and, finally, death of cells. Osmotic stress and ion toxicity lead to metabolic imbalance, which in turn causes oxidative stress (Ashraf 2004).

Photosynthesis is negatively affected by salinity of the soil through reducing photosystem II capacity, chlorophyll content, leaf area, and stomatal conductance (Netondo et al. 2004). Moreover, salinity may impede the supply of hormones or photosynthetic assimilates to growing tissue (Ashraf 2004). The cell cycle is transiently arrested by salinity stress which causes reduction in the activity and expression of cyclins that result in fewer cells in the meristem, consequently limiting growth. In addition, the posttranslational inhibition during salinity stress causes reduction in the activity of cyclin-dependent kinase (Seckin et al. 2009).

6.3.2.2 Salinity Stress Alleviation by Microbial Endophytes

Salinity problem threatens more than 20% of agricultural soil (Zhu 2000), and by 2050, about 50% of important agricultural land will be affected by salinity stress (Munns and Tester 2008). Endophytic microbes can enhance growth properties and modulate metabolism and phytohormone signaling. In addition, endophytic microbes improve adaptation to abiotic and biotic stress. Endophytes represent a particular concern for improved crop adaptation to stress as they are relatively protected from the harsh soil environment under high salt, drought, or other stress conditions (Sturz et al. 2000).

The following are the major advantages of endophytes to minimize salinity impacts on plants.

Plant Antioxidant Status

Reactive oxygen species in plants are formed on the onset of salt and osmotic stress. Scavenging enzymes such as ascorbate peroxidase, catalase, and superoxide dismutase inhibit oxidation of DNA, membrane proteins, and lipids. Microorganisms use similar methods to deal with oxidative stress. Hamilton and colleagues in 2012 reported the fungal endophyte mediation of reactive oxygen species in plants (Hamilton et al. 2012). Previous studies have suggested the relationship between tolerance of plants to salt stress and the alleviation of antioxidant enzymes (Sekmen et al. 2007). Scavinging enzymes for ROS include glutathione reductases (GR), superoxide dismutases (SOD), catalases (CAT), dehydroascorbate reductases (DHAR), ascorbate or thiol-dependent peroxidases (APX), and monodehydroascorbate reductases (MDHAR), in addition to tocopherol and glutathione (Rouhier et al. 2008). These enzymes involved in the removal of ROS either directly (APX, SOD, CAT) or indirectly via regeneration of glutathione and ascorbate in the cell. On constant, when the nonsymbiotic plants *Leymus mollis* (dunegrass)subjected to 500 mmol l^{-1} NaCl solution becomes severely wilted, desiccated within 7 days and ultimately dead after 14 days (Rodriguez et al. 2008). Plants infected with *Fusarium culmorum* did not show the symptoms of wilt until it was subjected to 500 mmol l^{-1} NaCl for 14 days. The endophyte *Piriformospora indica* induces salt tolerance by improving the antioxidant status of barley (Baltruschat et al. 2008).

ACC Deaminase

Although endophytic microbes might produce ACC deaminase enzyme and do not benefit from it, the enzyme has a role in promoting plant growth and enhances plant stress tolerance through cleaving ethylene, which acts as a precursor for the synthesis of ACC (Glick 2014). ACC deaminase can reduce plant ethylene levels by cleaving its precursor ACC (1-aminocyclopropane-1-carboxylate) to 2-oxobutanoate and ammonia, inhibiting ethylene signaling (Glick et al. 1998). Ethylene is a significant plant hormone that contributes in seed germination, in response to several stresses, and it is the main regulator for bacterial colonization of plant tissues (Iniguez et al. 2005). Ethylene accumulation in plants as a stress response is commonly detrimental to plant health and growth (Czarny et al. 2006). In addition to stress relief, ACC deaminase enzyme supports bacterial endophyte colonization of the plant. *Burkholderia phytofirmans* PsJN lost the capacity of root elongation in canola plant seedlings when the gene of ACC deaminase was inactivated (sun et al. 2009). A previous study on cut flowers reported the ability of endophytic bacteria to colonize shoot. Moreover, the ACC deaminase enzyme delayed flower senescence (Ali et al. 2012).

Phytohormone Production

Endophytes capable of promoting plant growth considerably produce auxins, principally indole-3-acetic acid (IAA) (Witzel et al. 2012). Auxins act against ethylene and play a major role in promotion of root development and growth. So, endophytic management of auxin production might be a significant tool in awarding salt tolerance in halophytic plants. IAA production was found in (i) species of *Serratia, Bacillus, Vibrio, Brevundimonas, Oceanobacillus Exiguobacterium, Staphylococcus,* and *Halobacillus* isolated from four samples of halotolerant plants grown in coastal sandbank of China (Bian et al. 2011) and (ii) salinity-tolerant rhizobacteria (*Halomonas* sp., *Arthrobacter* sp., *Pseudomonas mendocina, Bacillus pumilus,* and *Nitrinicola lacisaponensis*) originating from extremely saline habitats (Tiwari et al. 2011). It was suggested that IAA, one of the auxins, increases the efficiency of colonization (Suzuki et al. 2003), probably via interference with the host defense system (Navarro et al. 2006), and the production of such compounds or other related compounds might be a significant property for colonization of plant by endophytes. The halophytic plant *Prosopis strombulifera* also produced ABA, gibberellins, and IAA (Piccoli et al. 2011). ABA is a vital hormone for plant development and growth, and plants increase their ABA levels in stressed conditions. The main role of ABA is to regulate water balance of plant and tolerance of osmotic stress (Tuteja 2007). Wheat plants growing in saline soil showed increased fitness when inoculated with rhizobacteria having IAA producing and salt-tolerant capacity (Tiwari et al. 2011). The function of phytohormones for enhancing salt tolerance has not been analyzed for root fungi either mycorrhizal or endophytic (Rupple et al. 2013).

Nitrogen Fixation

Benefits of endophytes include pathogen suppression, phytohormone production, nutrient supply, and nitrogen fixation; these mechanisms also contribute to the mitigating effects of endophytes when the host plant faces unfavorable ecological conditions (Rupple et al. 2013). Various root endophytes could fix nitrogen (e.g., *Azoarcus* spp., *Acetobacter diazotrophicus, and Herbaspirillum* spp.). Nitrogen fixation improves host plant fitness, mostly in poor nitrogen environments. Even if fixed nitrogen in single species is found in a low amount, it should be clarified whether fixed nitrogen is intended for the microbial demands and/or host plant demands. The endophytic strain *Paenibacillus* P22 isolated from poplar trees conferred the fixed nitrogen to the pool of total nitrogen of host plant, as well as induced changes in plant metabolism (Hardoim et al. 2015).

Compatible Solutes

Sequestration of Na^+ and Cl^- ions in the vacuole of plant cell causes osmotic pressure. To balance this pressure, metabolically compatible organic solutes must be accumulated (even at elevated concentrations) in organelles and cytosol. Sucrose, glycine betaine, and proline are the most prevalent accumulated solutes (Munns and Tester 2008).

Cumulating organic solutes consider a vital mechanism to counter osmotic pressure, and this was also found in halophytic plants (Flowers and Colmer 2008); proline amino acid was the topic of research to understand the increased salt tolerance in plants colonized with endophytes. Nevertheless, mycorrhizal fungi gave variable results and suggested that accumulation of proline is generally considered as the effect, but it is not the reason for salinity tolerance (Ruiz-Lozano et al. 2012). Osmosis can also be regulated by betaines and sugars. Increased levels of sugars and betaines in mycorrhizal plants suggested that they have a role in salinity tolerance (Manchanda and Garg 2011). The endophyte *Pseudomonas pseudoalcaligenes* showed an improvement in salinity tolerance of rice by stimulating the accumulation of glycine betaine-like compounds in high concentrations (Jha et al. 2011).

6.3.3 Temperature Stress

Extreme temperature adversely affects plant growth, and high temperature leads to significant damage to cellular proteins that are widely denaturated and aggregated, leading to cell death. On the other hand, low temperature causes impaired metabolism due to inhibition of enzyme reactions, interactions among macromolecules, changes in protein structure, and modulating the membrane properties (Andreas et al. 2012).

Only few reports have the detrimental effects of extreme temperatures, which are often related to water limitation. In this regard, *Burkholderia phytofirmans* enhance resistance of plants grown at low temperatures (Ait Barka et al. 2006). The grass *Dichanthelium lanuginosum* was able to survive, although soil temperatures ranged from 38 °C to 65 °C in the Yellowstone National Park due to *Curvularia protuberata* and its thermal tolerance mycovirus *Curvularia* (CThTV) (Redman et al. 2002). The fungal endophytes have increased wheat tolerance to temperature regarding grain weight and seed germination of the second generation (Hubbard et al. 2014).

High temperature, precipitation, and latitude can interact and influence the endophyte composition in plants. For example, in sweet root (*Osmorhiza depauperata*), the endophytes *Sinorhizobium meliloti* and *Agrobacterium tumefaciens* were more abundant in locations with higher precipitation and annual temperature, while *Paenibacillus* strains were more common at sites with lower precipitation and higher latitudes (Li et al. 2012b).

Matsouri et al. (2010) reported that enhanced tolerance of endophyte-colonized plants to temperature and salt stress arise from alterations in ratios of oxidized-to-reduced forms of ascorbate and glutathione as well as lipid peroxidation. Endophytes enhance the adaptation of plant with chilling temperatures. This results in reduced cellular damage, increased photosynthetic activity, and accumulation of various metabolites related to cold stress such as phenolic compounds, proline, and starch. Endophytes also have a positive effect on the metabolic balance, which also reduces the impact of drought stress on wheat growth in reduced watering conditions (Naveed et al. 2014).

6.3.4 Heavy Metal Stress

Toxicity by heavy metals is one of the most important abiotic stresses that cause the loss of about 25–80% of various cultivated crops. In acidic soils, low crop productivity and reduced soil fertility are principally due to manganese and aluminum toxicities along with nutrient deficiencies (K, Mg, P, and Ca) (Singh et al. 2011). Heavy metals are very toxic to roots of cultivated plants and cause poor development of the root system (Singh et al. 2011). Heavy metal toxicity has become a serious problem that restricts crop productivity in acidic soils, in addition to overlapping with many physiological and biochemical processes including nutrient uptake, protein and nitrogen metabolism, photosynthesis, and respiration (Zhang et al. 2009).

It is recognized that bacterial endophytes participate in immobilization and mobilization of the metal cations, which affect availability of cations to plants (Pandey et al. 2016). In Cd-stressed soil, the dark septate endophyte (DSE) *Exophiala pisciphila* associated with *Zea mays* root showed improved activity of the antioxidant enzymes (Wang et al. 2016). Three major genes contributing to detoxification, transport, and uptake of Cd have been identified as PCS and MTP upregulation and ZIP downregulation when plants were inoculated with DSE and subsequently exposed to high concentrations of Cd. Changes in the content of 1-aminocyclopropane-1-carboxylate (ACC) by *Gigaspora* and *Pseudomonas* can change heavy metal tolerance directly by manipulating levels of plant ethylene (Friesen et al. 2011).

6.3.5 Nutrient Stress

Light, mineral nutrients, carbon, and water are essential prerequisites of plants for development, reproduction, and optimal growth. Starvation and nutrient stress are important abiotic stresses that harm plants (Chaves and Oliveira 2004).

Endophytes can provide their host with micronutrients and macronutrients. Bacteria that have nitrogen-fixing capacity can metabolize root exudates of plants and, in turn, supply nitrogen for the synthesis of plant amino acids. Endophytes can promote growth of plant by gibberellins (GAs), phosphate solubilization, cytokinins, IAA, and production of siderophore and supply essential vitamins to the plant (Jha et al. 2011). Choi et al. (2008) have reported that solubilization of phosphate in wheat and rice was mediated by gibberellic acid produced by *Pseudomonas* sp. The uptake of mineral nutrients (particularly Zn) in wheat plant has been improved by *Azotobacter chroococcum* and *Piriformospora indica* (Abadi and Sepehri 2015). Studies proved the function of endophytes in biological degradation of litter of the host plants. Endophytes initially colonize plants, facilitating the action of saprophytic microbes by antagonism, in that way increasing decomposition of litter (Terekhova and Semenova 2005). Another study explained the ability of all endophytes to decompose the organic components including cellulose lignin and hemicelluloses that facilitate nutrient cycling (He et al. 2012).

6.4 Conclusion

About 300,000 species of plants in the world harbor one or more endophytes. Each endophyte has its own function that helps to improve plant growth and protect it from diverse biotic and abiotic stresses. This benefit does not involve host specificity, so we can use endophytes as inoculants to alleviate abiotic stresses arising from changeful environmental conditions. With increasing interest on environmental protection, food security, and sustainable agriculture, exploiting useful endophytes is urgent. Endophytes may also be a good tool for enhancement of yield and quality of the plant products by producing various kinds of pioneer biologically active metabolites which may be able to positively regulate plant physiological disorders. They can protect plants from pathogens and remediate toxic residues of insecticides, herbicides, and various heavy metals. In addition, it has quick responses in stimulation of immune defense of the host.

Endophytes can be used as alternative strategies to plants that adapted to many stresses like drought, salinity, temperature, nutrient stress, and heavy metals. Further studies of endophytes will provide a better understanding of their relationship with host plant and maximize its utilization as promoters of plant growth as well as its ability to protect the plant from many harmful factors.

References

Abadi VAJM, Sepehri M (2015) Effect of *Piriformospora indica* and *Azotobacter chroococcum* on mitigation of zinc deficiency stress in wheat (*Triticum aestivum* L). Symbiosis 69:9–19

Ait Barka E, Nowak J, Clément C (2006) Enhancement of chilling resistance of inoculated grapevine plantlets with a plant growth-promoting rhizobacterium, *Burkholderia phytofirmans* strain PsJN. Appl Environ Microbiol 72:7246–7252

Akbarimoghaddam H, Galavi M, Ghanbari A, Panjehkeh N (2011) Salinity effects on seed germination and seedling growth of bread wheat cultivars. Trakia J Sci 9(1):43–50

Akinsanya MA, Ting A, Goh JK, Lim SP (2016) Biodiversity, enzymatic and antimicrobial activities of bacterial endophytes in selected local medicinal plants. J Biomed Pharm Res 16:5(1)

Ali S, Charles TC, Glick BR (2012) Delay of flower senescence by bacterial endophytes expressing 1-aminocyclopropane-1-carboxylate deaminase. J Appl Microbiol 113:1139–1144. https://doi.org/10.1111/j.1365-2672.2012.05409.x

Aly AH, Debbab A, Proksch P (2011) Fungal endophytes: unique plant inhabitants with great promises. Appl J Microbiol Biotechnol 90:1829–1845

Andreas T, Christophe C, Essaïd AB (2012) Physiological and molecular changes in plants at low temperatures. Planta 235:1091–1105. https://doi.org/10.1007/s00425-012-1641-y

Anjum SA, Wang LC, Farooq M, Hussain M, Xue LL, Zou CM (2011) Brassinolide application improves the drought tolerance in maize through modulation of enzymatic antioxidants and leaf gas exchange. J Agron Crop Sci 197:177–185. https://doi.org/10.1111/j.1439-037X.2010.00459.x

Ardanov P, Ovcharenko L, Zaets I, Kozyrovska N, Pirttilä AM (2011) Endophytic bacteria enhancing growth and disease resistance of potato (Solanum tuberosum L.). Biol Control 56:43–49. https://doi.org/10.1016/j.biocontrol.2010.09.014

Ashraf M (2004) Some important physiological selection criteria for salt tolerance in plants. Flora 199:361–376

Ayob FW, Simarani K (2016) Endophytic filamentous fungi from a *Catharanthus roseus*: identification and its hydrolytic enzymes. Saudi Pharm J 24:273 278

Bae H, Sicher RC, Kim MS, Kim SH, Strem MD, Melnick RL, Bailey BA (2009) The beneficial endophyte *Trichoderma hamatum* isolate DIS 219b promotes growth and delays the onset of the drought response in *Theobroma cacao*. J Exp Bot 60(11):3279–3295

Bae H, Roberts DP, Lim H-S, Strem MD, Park S-C, Ryu C-M, Melnick RL, Bailey BA (2011) Endophytic Trichoderma isolates from tropical environments delay disease onset and induce resistance against *Phytophthora capsici* in hot pepper using multiple mechanisms. Mol Plant-Microbe Interact 24:336–351. https://doi.org/10.1094/MPMI-09-10-0221

Bailey BA, Bae H, Strem MD, Roberts DP, Thomas SE, Crozier J, Samuels GJ, Choi IY, Holmes KA (2006) Fungal and plant gene expression during the colonization of cacao seedlings by endophytic isolates of four Trichoderma species. Planta 224:1449–1464

Baltruschat H, Fodor J, Harrach BD, Niemczyk E, Barna B, Gullner G, Janeczko A, Kogel KH, Schäfer P, Schwarczinger I, Zuccaro A, Schäfer P, Schwarczinger I, Zuccaro A, Skoczowski A (2008) Salt tolerance of barley induced by the root endophyte *Piriformospora indica* is associated with a strong increase in antioxidants. New Phytol 180:501–510

Bano A, Fatima M (2009) Salt tolerance in *Zea mays* (L.) following inoculation with *Rhizobium* and *Pseudomonas*. Biol Fertil Soils 45:405–413

Bhosale HJ, Kadam TA (2015) Generic diversity and a comparative account on plant growth promoting characteristics of actinobacteria in roots and rhizosphere of *Saccharum officinarum*. Int J Curr Microbiol App Sci 4:230–244

Bian G, Zhang Y, Qin S, Xing K, Xie H, Jiang J (2011) Isolation and biodiversity of heavy metal tolerant endophytic bacteria from halotolerant plant species located in coastal shoal of Nantong. Acta Microbiol Sin 51(11):1538–1547

Bloemberg GV, Lugtenberg BJ (2001) Molecular basis of plant growth promotion and biocontrol by rhizobacteria. Curr Opin Plant Biol 4(4):343–350

Bogner CW, Kamdem RST, Sichtermann G, Mattheaus C, Heolscher D, Popp J, Proksch P, Grundler FMW, Schouten A (2017) Bioactive secondary metabolites with multiple activities from a fungal endophyte. Microb Biotechnol 10:175–188. https://doi.org/10.1111/1751-7915.12467

Bordiec S, Paquis S, Lacroix H, Dhondt S, Ait Barka E, Kauffmann S, Jeandet P, Mazeyrat-Gourbeyre F, Clement C, Baillieul F, Dorey S (2011) Comparative analysis of defence responses induced by the endo- phytic plant growth-promoting rhizobacterium *Burkholderia phytofirmans* strain PsJN and the non-host bacterium *Pseudomonas syringae* pv. pisi in grapevine cell suspensions. J Exp Bot 62:595–603. https://doi.org/10.1093/jxb/erq291

Bowyer C, Withana S, Fenn I, Bassi S, Lewis M, Cooper T, Benito P, Mudgal S (2009) Land Degradation and Desertification Policy Department Economic and Scientific Policy IP/A/ENVI/ST/2008-23. European Parliament, Brussels

Bray EA (2002) Abscisic acid regulation of gene expression during water-deficit stress in the era of the Arabidopsis genome. Plant Cell Environ 25:153–161

Cao MJ, Wang Z, Zhao Q, Mao JL, Speiser A, Wirtz M, Hell R, Zhu JK, Xiang CB (2014) Sulfate availability affects ABA levels and germination response to ABA and salt stress in Arabidopsis thaliana. Plant J 77:604–615

Carrim AJ, Barbosa EC, Vieira JD (2006) Enzymatic activity of endophytic bacterial isolates of Jacaranda decurrens Cham (Carobinha-do-campo). Braz Arch Biol Technol 49:353–359

Chathurdevi G, Gowrie SU (2016) Endophytic fungi isolated from medicinal plant—a source of potential bioactive metabolites. Int J Curr Pharm Res 8:50–56

Chaves MM, Oliveira MM (2004) Mechanisms underlying plant resilience to water deficits: prospects for water-saving agriculture. J Exp Bot 55:2365–2384

Chi F, Shen S, Cheng H, Jing Y, Yanni Y, Dazzo F (2005) Ascending migration of endophytic rhizobia, from roots to leaves, inside rice plants and assessment of benefits to rice growth physiology. Appl Environ Microbiol 71:7271–7278

Choi O, Kim J, Kim JG, Jeong Y, Moon JS, Park CS, Hwang I (2008) Pyrroloquinoline quinone is a plant growth promotion factor produced by *Pseudomonas fluorescens* B16. Plant Physiol 146:657–668

Cohen AC, Bottini R, Piccoli P (2008) *Azospirillum brasilense* Sp 245 produces ABA in chemically-defined culture medium and increases ABA content in *Arabidopsis* plants. Plant Growth Regul 54:97–103

Compant S, Clément C, Sessitsch A (2010) Plant growth-promoting bacteria in the rhizo- and endosphere of plants: their role, colonization, mechanisms involved and prospects for utilization. Soil Biol Biochem 42:669–678. https://doi.org/10.1016/j.soilbio.2009.11.024

Convey P (2011) Antarctic terrestrial biodiversity in a changing world. Polar Biol 34:1629e1641

Czarny JC, Grichko VP, Glick BR (2006) Genetic modulation of ethylene biosynthesis and signaling in plants. Biotechnol Adv 24(4):410–419. https://doi.org/10.1016/j.biotechadv.2006.01.003

De Bashan LE, Hernandez JP, Bashan Y (2012) The potential contribution of plant growth-promoting bacteria to reduce environmental degradation- a comprehensive evaluation. Appl Soil Ecol 61:171–189. https://doi.org/10.1016/j.apsoil.2011.09.003

De Smet I, Zhang H, Inzé D, Beeckman T (2006) A novel role for abscisic acid emerges from underground. Trends Plant Sci 11:434–439

Dudeja SS, Giri R, Saini R, Suneja-Madan P, Kothe E (2012) Interaction of endophytic microbes with legumes. J Basic Microbiol 52:248–260

Flowers TJ, Colmer TD (2008) Salinity tolerance in halophytes. New Phytol 179(4):945–963. https://doi.org/10.1111/j.1469-8137.2008.02531.x

Fouda AH, Hassan SE, Eid AM, Ewais EE (2015) Biotechnological applications of fungal endophytes associated with medicinal plant *Asclepias sinaica* (Bioss.). Ann Agric Sci 60:95–104

Fouda A, Abdel-Maksoud G, Abdel-Rahman MA, Salem SS, Hassan SE, El-Sadany MA (2019a) Eco-friendly approach utilizing green synthesized nanoparticles for paper conservation against microbes involved in biodeterioration of archaeological manuscript. Int Biodeterior Biodegrad 142:160–169

Fouda A, Hassan SE, Eid AM, El-Din Ewais E (2019b) The interaction between plants and bacterial endophytes under salinity stress. In: Jha S (ed) Endophytes and secondary metabolites. Springer, Cham, pp 1–18

Friesen M, Porter SS, Stark SC, von Wettberg EJ, Sachs JL, Martinez Romero E (2011) Microbially mediated plant functional traits. Annu Rev Ecol Evol Syst 42:23–46

Gangwar M, Dogra S, Sharma N (2011) Antagonistic bioactivity of endophytic actinobacteria isolated from medicinal plants. J Adv Lab Res Biol 2(4):1–4

Ghodhbane-Gtari F, Essoussi I, Chattaoui M, Chouaia B, Jaouani A, Daffonchio D, Boudabous A, Gtari M (2010) Isolation and characterization of non-Frankia actinobacteria from root nodules of *Alnus glutinosa, Casuarina glauca* and *Elaeagnus angustifolia.* Symbiosis 50:51–57

Glick BR (2014) Bacteria with ACC deaminase can promote plant growth and help to feed the world. Microbiol Res 169(1):30–39

Glick BR, Penrose DM, Li JP (1998) A model for the lowering of plant ethylene concentrations by plant growth-promoting bacteria. J Theor Biol 190:63–68. https://doi.org/10.1006/jtbi.1997.0532

Granier C, Tardieu F (1999) Water deficit and spatial pattern of leaf development. Variability in responses can be stimulated using a simple model of leaf development. Plant Physiol 119:609–619

Gray EJ, Smith DL (2005) Intracellular and extracellular PGPR: commonalities and distinctions in the plant–bacterium signaling processes. Soil Biol Biochem 37:395–412

Greenberg BM, Huang XD, Gerwing P, Yu XM, Chang P, Wu SS, Gerhardt K, Nykamp J, Lu X, Glick B (2008) Phytoremediation of salt impacted soils: greenhouse and the field trials of plant growth promoting rhizobacteria (PGPR) to improve plant growth and salt phyto-accumulation. In: Proceeding of the 33rd AMOP technical seminar on environmental contamination and response. Environment Canada, Ottawa, pp 627–637

Gunatilaka AAL (2006) Natural products from plant-associated micro-organisms: distribution, structural diversity, bioactivity, and implications of their occurrence. J Nat Prod 69:509–526. https://doi.org/10.1021/np058128

Guo B, Wang Y, Sun X, Tang K (2008) Bioactive natural products from endophytes: a review. Prikl Biokhim Mikrobiol 44:153–158

Gupta V, Trivedi N, Kumar M, Reddy CR, Jha B (2013) Purification and characterization of exo-b-agarase from an endophytic marine bacterium and its catalytic potential in bioconversion of red algal cell wall polysaccharides into galactans. Biomass Bioenergy 28:290–298

Hamilton CE, Bauerle TL (2012) A new currency for mutualism? Fungal endophytes alter antioxidant activity in hosts responding to drought. Fungal Divers 54:39–49

Hamilton CE, Gundel PE, Helander M, Saikkonen K (2012) Endophytic mediation of reactive oxygen species and antioxidant activity in plants: a review. Fungal Divers 54(1):1–10. https://doi.org/10.1007/s13225-012-0158-9

Hardoim PR, Overbeek LS, Elsas JD (2008) Properties of bacterial endophytes and their proposed role in plant growth. Trends Microbiol 16:463–471. https://doi.org/10.1016/j.tim.2008.07.008

Hardoim PR, van Overbeek LS, Berg G, Pirttilä AM, Compant S, Campisano A, Döring M, Sessitsch A (2015) The hidden world within plants: ecological and evolutionary considerations for defining functioning of microbial endophytes. Microbiol Mol Biol Rev 79(3):293–320. https://doi.org/10.1128/MMBR.00050-14

Hasegawa S, Meguro A, Nishimura T, Kunoh H (2004) Drought tolerance of tissue- cultured seedlings of mountain laurel (*Kalmia latifolia* L.) induced by an Endophytic actinomycete. Actinomycetologica 18:43–47

Hassan SE (2017) Plant growth-promoting activities for bacterial and fungal endophytes isolated from medicinal plant of *Teucrium polium* L. J Adv Res 8(6):687–695

Hassan SE, Salem SS, Fouda A, Awad MA, El-Gamal MS, Abdo AM (2018) New approach for antimicrobial activity and bio-control of various pathogens by biosynthesized copper nanoparticles using endophytic actinomycetes. J Radiat Res Appl Sci 11:262–270

Hassan SE, Fouda A, Radwan AA, Salem SS, Barghoth MG, Awad MA, El-Gamal MS, Abdo AM (2019) Endophytic actinomycetes *Streptomyces* spp. mediated biosynthesis of copper oxide nanoparticles as a promising tool for biotechnological applications. J Biol Inorg Chem 24(3): 377–393

Hata K, Atari R, Sone K (2002) Isolation of endophytic fungi from leaves of Pasania edulis and their within-leaf distributions. Mycoscience 43(5):369–373

He X, Han G, Lin Y, Tian X, Xiang C, Tian Q, Wang F, He Z (2012) Diversity and decomposition potential of endophytes in leaves of a *Cinnamomum camphora* plantation in China. Ecol Res 27:273–284

Higginbotham SJ, Arnold AE, Ibañez A, Spadafora C, Coley PD, Kursar TA (2013) Bioactivity of fungal endophytes as a function of endophyte taxonomy and the taxonomy and distribution of their host plants. PLoS One 8:e73192. https://doi.org/10.1371/journal.pone.0073192

Hu Y, Schmidhalter U (2002) Limitation of salt stress to plant growth. In: Hock B, Elstner CF (eds) Plant Toxicology. Marcel Dekker Inc, New York, pp 91–224

Hubbard M, Germida JJ, Vujanovic V (2014) Fungal endophytes enhance wheat heat and drought tolerance in terms of grain yield and second-generation seed viability. J Appl Microbiol 116:109–122

Iniguez AL, Dong YM, Carter HD, Ahmer BMM, Stone JM, Triplett EW (2005) Regulation of enteric endophytic bacterial colonization by plant defenses. Mol Plant-Microbe Interact 18:169–178. https://doi.org/10.1094/MPMI-18-0169

Jahromi F, Aroca R, Porcel R, Ruiz-Lozano JM (2008) Influence of salinity on the in vitro development of *Glomus intraradices* and on the *in vivo* physiological and molecular responses of mycorrhizal lettuce plants. Microb Ecol 55:45–53

Jamil A, Riaz S, Ashraf M, Foolad MR (2011) Gene expression profiling of plants under salt stress. Crit Rev Plant Sci 30(5):435–458

Jha Y, Subramanian RB, Patel S (2011) Combination of endophytic and rhizospheric plant growth promoting rhizobacteria in Oryza sativa shows higher accumulation of osmoprotectant against saline stress. Acta Physiol Plant 33:797–802. https://doi.org/10.1007/s11738-010-0604-9

Jones A, Panagos P, Barcelo S, Bouraoui F, Bosco C, Dewitte O, Gardi C, Hervás J, Hiederer R, Jeffery S (2012) The state of soil in Europe -a contribution of the JRC to the European Environment Agency's Environment State and Outlook R- SOER 2010

Kannahi M, Senbagam N (2014) Studies on siderophore production by microbial isolates obtained from rhizosphere soil and its antibacterial activity. J Chem Pharm Res 6(4):1142–1145

Kannan R, Damodaran T, Umamaheswari S (2015) Sodicity tolerant polyembryonic mango root stock plants: a putative role of endophytic bacteria. Afr J Biotechnol 14:350–359

Khan AL, Hamayun M, Hussain J, Kang SM, Lee IJ (2012) The newly isolated endophytic fungus *Paraconiothyrium* sp. LK1 produces ascotoxin. Molecules 17:1103–1112

Khan AL, Waqas M, Hussain J, Al-Harrasi A, Lee IJ (2014) Fungal endophyte *Penicillium janthinellum* LK5 can reduce cadmium toxicity in *Solanum lycopersicum* (Sitiens and Rhe). Biol Fertil Soils 50:75–85

Khan AL, Al-Harrasi A, Al-Rawahi A, Al-Farsi Z, Al-Mamari A, Waqas M, Asaf S, Elyassi A, Mabood F, Shin JH, Lee IJ (2016) Endophytic fungi from frankincense tree improves host growth and produces extracellular enzymes and indole acetic acid. PLoS One 11:e0158207

Kharwar RN, Mishra A, Gond SK, Stierle A, Stierle D (2011) Anticancer compounds derived from fungal endophytes: their importance and future challenges. Nat Prod Rep 28:1208–1228

Kim YC, Glick BR, Bashan Y, Ryu CM (2012) Enhancement of plant drought tolerance by microbes. In: Aroca R (ed) Plant responses to drought stress: from morphological to molecular features. Springer, Heidelberg, pp 383–413. https://doi.org/10.1007/978-3-642-32653-0_15

Klessig DF, Tian M, Choi HW (2016) Multiple targets of salicylic acid and its derivatives in plants and animals. Front Immunol 7:206. https://doi.org/10.3389/fimmu.2016.00206

Kloepper JW, Ryu CM (2006) Bacterial endophytes as elicitors of induced systemic resistance. In: Schulz BJE, Boyle CJC, Sieber TN (eds) Microbial root endophytes. Springer, Berlin, pp 33–52. https://doi.org/10.1007/3-540-33526-9_3

Kusari S, Verma VC, Lamshoeft M, Spiteller M (2012a) An endophytic fungus from *Azadirachta indica* A. Juss. That produces azadirachtin. World J Microbiol Biotechnol 28:1287–1294

Kusari S, Hertweck C, Spiteller M (2012b) Chemical ecology of endophytic fungi: origin of secondary metabolites. Chem Biol 19:792–798

Leo VV, Passari AK, Joshi JB, Mishra VK, Uthandi S, Ramesh N, Gupta VK, Saikia R, Sonawane VC, Singh BP (2016) A novel triculture system (CC3) for simultaneous enzyme production and hydrolysis of common grasses through submerged fermentation. Front Microbiol 7. https://doi.org/10.3389/fmicb.2016.00447

Li WK (2005) Endophytes and Natural Medicines. Chin J Nat Med 3(4):193–199

Li HY, Wei DQ, Shen M, Zhou ZP (2012a) Endophytes and their role in phytoremediation. Fungal Divers 54:11–18

Li L, Sinkko H, Montonen L, Wei G, Lindstrom K, Rasanen LA (2012b) Biogeography of symbiotic and other endophytic bacteria isolated from medicinal *Glycyrrhiza* species in China. FEMS Microbiol Ecol 79:46–68

Liarzi O, Bucki P, Braun Miyara S, Ezra D (2016) Bioactive volatiles from an endophytic *Daldinia* cf. *concentrica* isolate affect the viability of the plant parasitic nematode *Meloidogyne javanica*. PLoS One 11:e0168437. https://doi.org/10.1371/journal.pone.0168437

Limtong S, Kaewwichian R, Yongmanitchai W, Kawasaki H (2014) Diversity of culturable yeasts in phylloplane of sugarcane in Thailand and their capability to produce indole-3-acetic acid. World J Microbiol Biotechnol 30(6):1785–1796

Liu G, Lai D, Liu ZQ, Zhou L, Liu LZ (2016) Identification of nematicidal constituents of *Notopterygium incisum* rhizomes against *Bursaphelenchus xylophilus* and *Meloidogyne incognita*. Molecules 21:1276. https://doi.org/10.3390/molecules21101276

Malinowski CP, Beleskey DP (2000) Adaptations of endophyte-infected cool-season grasses to environmental stresses: mechanisms of drought and mineral stress tolerance. Crop Sci 40:923–940

Manchanda G, Garg N (2011) Alleviation of salt-induced ionic, osmotic and oxidative stresses in *Cajanus cajan* nodules by AM fungi inoculation. Plant Biosyt 145(1):88–97. https://doi.org/10.1080/11263504.2010.539851

Mastretta C, Taghavi S, van der Lelie D et al (2009) Endophytic bacteria from seeds of *Nicotiana tabacum* can reduce cadmium phytotoxicity. Int J Phytoremediation 11(3):251–267

Matsouri F, Björkman T, Harman GE (2010) Seed treatment with *Trichoderma harzianum* alleviates biotic, abiotic, and physiological stresses in germinating seeds and seedlings. Biol Control 100:1213–1221

Meena KK, Sorty AM, Bitla UM, Choudhary K, Gupta P, Pareek A, Singh DP, Prabha R, Sahu PK, Gupta VK, Singh HB, Krishanani KK, Minhas PS (2017) Abiotic stress responses and microbe-mediated mitigation in plants: the omics strategies. Front Plant Sci 8:172. https://doi.org/10.3389/fpls.2017.00172

Miransari M (2012) Role of phytohormone signaling during stress. In: Ahmad P, Prasad MNV (eds) Environmental adaptations and stress tolerance of plants in the era of climate change. Springer, New York, p 381_393

Molina G, Pimentel MR, Bertucci TCP, Pastore GM (2012) Application of fungal endophytes in biotechnological processes. Chem Eng Trans 27:289–294

Munns R (2002) Comparative physiology of salt and water stress. Plant Cell Environ 25:239–250. https://doi.org/10.1046/j.0016-8025.2001.00808.x

Munns R, Tester M (2008) Mechanisms of salinity tolerance. Annu Rev Plant Biol 59:651–681

Navarro L, Dunoyer P, Jay F, Arnold B, Dharmasiri N, Estelle M, Voinnet O, Jones JDG (2006) A plant miRNA contributes to antibacterial resistance by repressing auxin signaling. Science 312:436–439. https://doi.org/10.1126/science.1126088

Naveed M, Hussain MB, Zahir ZA, Mitter B, Sessitsch A (2014) Drought stress amelioration in wheat through inoculation with *Burkholderia phytofirmans* strain PsJN. Plant Growth Regul 73:121–131. https://doi.org/10.1007/s10725-013-9874-8

Netondo GW, Onyango JC, Beck E (2004) Sorghum and salinity: II. Gas exchange and chlorophyll fluorescence of sorghum under salt stress. Crop Sci 44:806–811

Padda KP, Puri A, Chanway CP (2016) Plant growth promotion and nitrogen fixation in canola by an endophytic strain of *Paenibacillus polymyxa* and its GFP-tagged derivative in a long-term study. Botany 94:1209–1217. https://doi.org/10.1139/cjb-2016-0075

Palacios OA, Bashan Y, de-Bashan LE (2014) Proven and potential involvement of vitamins in interactions of plants with plant growth promoting bacteria-an overview. Biol Fertil Soils 50:415–432. https://doi.org/10.1007/s00374-013-0894-3

Pandey PK, Singh S, Singh AK, Samanta R, Yadav RNS, Singh MC (2016) Inside the plant: Bacterial endophytes and abiotic stress alleviation. J Appl Nat Sci 8(4):1899–1904

Parker MA (1995) Plant fitness variation caused by different mutualist genotypes. Ecology 76(5):1525–1535

Partida-Martínez LP, Heil M (2011) The microbe-free plant: fact or artifact? Front Plant Sci 29(2):100. https://doi.org/10.3389/fpls.2011.00100.eCollection

Passari AK, Mishra VK, Singh G, Singh P, Kumar B, Gupta VK, Sarma RK, Saikia R, Donovan AO, Singh BP (2017) Insights into the functionality of endophytic actinobacteria with a focus on their biosynthetic potential and secondary metabolites production. Sci Rep 7:11809. https://doi.org/10.1038/s41598-017-12235-4

Pereira SIA, Monteiro C, Vega AL, Castro PML (2016) Endophytic culturable bacteria colonizing Lavandula dentate L. plants: isolation, characterization and evaluation of their plant growth-promoting activities. Ecol Eng 87:91–96

Piccoli P, Travaglia C, Cohen A, Sosa L, Cornejo P, Masuelli R, Bottini R (2011) An endophytic bacterium isolated from roots of the halophyte *Prosopis strombulifera* produces ABA, IAA, gibberellins A(1) and A(3) and jasmonic acid in chemically-defined culture medium. Plant Growth Regul 64(2):207–210. https://doi.org/10.1007/s10725-010-9536-z

Pieterse CM, Van der Does D, Zamioudis C, Leon-Reyes A, Van Wees SC (2012) Hormonal modulation of plant immunity. Annu Rev Cell Dev Biol 28:489–521. https://doi.org/10.1146/annurev-cellbio-092910-154055

Pieterse CM, Zamioudis C, Berendsen RL, Weller DM, Van Wees SC, Bakker PA (2014) Induced systemic resistance by beneficial microbes. Annu Rev Phytopathol 52:347–375

Pirttilä AM, Pospiech H, Laukkanen H, Myllyla R, Hohtola A (2003) Two endophytic fungi in different tissues of Scots pine buds (*Pinus sylvestris* L.). Microb Ecol 45(1):53–62

Pirttilä AM, Podolich O, Koskimäki JJ, Hohtola E, Hohtola A (2008) Role of origin and endophyte infection in browning of bud-derived tissue cultures of Scots pine (*Pinus sylvestris* L.). Plant Cell Tissue Organ Cult 95(1):47–55

Prashar P, Kapoor N, Sachdeva S (2014) Rhizosphere: its structure, bacterial diversity and significance. Rev Environ Sci Biotechnol 13(1):63–77

Prashith-Kekuda TR (2016) Isolation, characterization and antimicrobial potential of endophytic actinomycetes. Int J Curr Microbiol App Sci 5:100–116. https://doi.org/10.20546/ijcmas.2016.507.008

Puri A, Padda KP, Chanway CP (2016) Evidence of nitrogen fixation and growth promotion in canola (*Brassica napus* L.) by an endophytic diazotroph *Paenibacillus polymyxa* P2b-2R. Biol Fertil Soils 52:119–125. https://doi.org/10.1007/s00374-015-1051-y

Raaijmakers J, Paulitz TC, Steinberg C, Alabouvette C, Moënne-Loccoz Y (2009) The rhizosphere: a playground and battlefield for soil borne pathogens and beneficial microorganisms. Plant Soil 321:341–361. https://doi.org/10.1007/s11104-008-9568-6

Radhakrishnan R, shim k-b, lee b-w, Hwang c-d, Pae S-B, Park C-H, Kim S-U, Lee C-K, In YB (2013) IAA-producing Penicillium sp. NICS01 triggers plant growth and suppresses *Fusarium* sp.-induced oxidative stress in sesame (*Sesamum indicum* L.). J Microbiol Biotechnol 23:856–863

Redman RS, Sheehan KB, Stout RG, Rodriguez RJ, Henson JM (2002) Thermotolerance generated by plant/fungal symbiosis. Science 298:1581–1581

Redman RS, Kim YO, Woodward CJ, Greer C, Espino L, Doty SL, Rodriguez RJ (2011) Increased fitness of rice plants to abiotic stress via habitat adapted symbiosis: a strategy for mitigating impacts of climate change. PLoS One 6:e14823

Ren JH, Ye JR, Liu H, Xu XL, Wu XQ (2011) Isolation and characterization of a new *Burkholderia pyrrocinia* strain JK SH007 as a potential biocontrol agent. World J Microbiol Biotechnol 27(9):2203–2215

Rodriguez RJ, Henson J, Van Volkenburgh E, Hoy M, Wright L, Beckwith F, Kim YO, Redman RS (2008) Stress tolerance in plants via habitat-adapted symbiosis. ISME 2:404–416

Rodriguez RJ, White JFJR, Arnold AE, Redman RS (2009) Fungal endophytes: diversity and functional roles-Tansley review. New Phytol 182:314–330

Rouhier N, San Koh C, Gelhaye E, Corbier C, Favier F, Didierjean C, Jacquot JP (2008) Redox based anti- oxidant systems in plants: biochemical and structural analyses. Biochim Biophys Acta 1780:1249–1260

Ruanpanun P, Tangchitsomkid N, Hyde KD, Lumyong S (2010) Actinomycetes and fungi isolated from plant-parasitic nematode infested soils: screening of the effective biocontrol potential, indole-3-acetic acid and siderophore production. World J Microbiol Biotechnol 26:1569–1578

Ruiz-Lozano JM, Porcel R, Azcon C, Aroca R (2012) Regulation by arbuscular mycorrhizae of the integrated physiological response to salinity in plants: new challenges in physiological and molecular studies. J Exp Bot 63(11):4033–4044. https://doi.org/10.1093/jxb/ers126

Rungin S, Indanand C, Suttiviriya P, Kruasuwan W, Jaemsaeng R, Thamchaipenet A (2012) Plant growth enhancing effects by a siderophore producing endophytic streptomycete isolated from a Thai jasmine rice plant (*Oryza sativa* L. cv. KDML105). Antonie Van Leeuwenhoek 102:463–472

Rupple S, Franken P, Witzel K (2013) Properties of the halophyte microbiome and their implications for plant salt tolerance. Funct Plant Biol 40:940–951. https://doi.org/10.1071/FP12355

Russell JR, Huang J, Anand P, Kucera K, Sandoval AG, Dantzler KW, Hickman D, Jee J, Kimovec FM, Koppstein D, Marks DH, Mittermiller PA, Núñez SJ, Santiago M, Townes MA, Vishnevetsky M, Williams NE, Vargas MP, Boulanger LA, Bascom-Slack C, Strobel SA (2011) Biodegradation of polyester polyurethane by endophytic fungi. Appl Environ Microbiol 77(17):6076–6084

Saharan BS, Nehra V (2011) Plant growth promoting rhizobacteria: a critical review. Life Sci Med Res 21(1):30–37

Salam N, Khieu T, Liu M, Vu T, Chu-Ky S, Quach N, Phi Q, Rao MPN, Fontana A et al (2017) Endophytic actinobacteria associated with *Dracaena cochinchinensis* Lour.: isolation, diversity,

and their cytotoxic activities. Biomed Res Int 2017:1. https://doi.org/10.1155/2017/1308563. (Article ID 1308563
Schulz B, Boyle C, Draeger S, R€ommert AK, Krohn K (2002) Endophytic fungi: a source of novel biologically active secondary metabolites. Mycol Res 109:996–1004
Seckin B, Sekmen AH, Turkan I (2009) An enhancing effect of exogenous mannitol on the anti-oxidant enzyme activities in roots of wheat under salt stress. J Plant Growth Regul 28:12–20
Sekmen AH, Turkan I, Takio S (2007) Differential responses of antioxidative enzymes and lipid peroxidation to salt stress in salt-tolerant Plantago maritima and salt- sensitive Plantago media. Physiol Plant 131:399–411
Shankar Naik B, Krishnamurthy YL (2010) Endophytes: the real untapped high energy biofuel resource. Curr Sci 98(7):883
Shi Y, Lou K, Li C (2009) Promotion of plant growth by phytohormone-producing endophytic microbes of sugar beet. Biol Fertil Soils 45:645–653
Shinozaki K, Yamaguchi-Shinozaki K (2007) Gene networks involved in drought stress response and tolerance. J Exp Bot 58:221–227
Shrivastava P, Kumar R (2015) Soil salinity: a serious environmental issue and plant growth promoting bacteria as one of the tools for its alleviation. Saudi J Biol Sci 22:123–131
Singh D, Singh NP, Chauhan SK, Singh P (2011) Developing aluminium tolerant crop plants using biotechnological tools. Curr Sci 100(12):1807–1814
Smith SA, Tank DC, Boulanger LA, Bascom-Slack CA, Eisenman K, Kingery D, Babbs B, Fenn K, Greene JS, Hann BD, Keehner J, Kelley-Swift EG, Kembaiyan V, Lee SJ, Li P, Light DY, Lin EH, Ma C, Moore E, Schorn MA, Vekhter D, Nunez PV, Strobel GA, Donoghue MJ, Strobel SA (2008) Bioactive endophytes warrant intensified exploration and conservation. PLoS One 3(8):e3052
Stolte J, Tesfai M, Øygarden L, Kværnø S, Keizer J, Verheijen F, Panagos P, Ballabio C, Hessel R (2015) Soil threats in Europe: status, methods, drivers and effects on ecosystem services. A review report, deliverable 2.1 of the RECARE Project, vol EUR 27607. Office for Official Publications of the European Community, Luxembourg, pp 69–78
Strobel GA (2003) Endophytes as sources of bioactive products. Microbes Infect 5:535–544
Strobel G, Daisy B (2003) Bioprospecting for microbial endophytes and their natural products. Microbiol Mol Biol Rev 67:491–502
Strong PJ, Claus H (2011) Laccase: A review of its past and its future in bioremediation. Crit Rev Environ Sci Technol 41(4):373
Sturz AV, Christie BR, Nowak J (2000) Bacterial endophytes: potential role in developing sustainable systems of crop production. CRC Crit Rev Plant Sci 19:1–30
Sun Y, Cheng Z, Glick BR (2009) The presence of a 1-aminocyclopro- pane-1-carboxylate (ACC) deaminase deletion mutation alters the physiology of the endophytic plant growth-promoting bacterium *Burkholderia phytofirmans* PsJN. FEMS Microbiol Lett 296:131–136. https://doi.org/10.1111/j.1574-6968.2009.01625.x
Sunitha VH, Devi DN, Srinivas C (2013) Extracellular enzymatic activity of endophytic fungal strains isolated from medicinal plants. WJAS 9:1–9
Suzuki S, He YX, Oyaizu H (2003) Indole-3-acetic acid production in *Pseudomonas fluorescens* HP72 and its association with suppression of creeping bentgrass brown patch. Curr Microbiol 47:138–143. https://doi.org/10.1007/s00284-002-3968-2
Tadych M, Bergen M, White JF (2014) *Epichloë* spp. associated with grasses: new insights on life cycles, dissemination and evolution. Mycologia 106(2):181–201
Tang Q, Puri A, Padda KP, Chanway CP (2017) Biological nitrogen fixation and plant growth promotion of lodgepole pine by an endophytic diazotroph *Paenibacillus polymyxa* and its GFP-tagged derivative. Botany 95:611–619. https://doi.org/10.1139/cjb-2016-0300
Tefera T, Vidal S (2009) Effect of inoculation method and plant growth medium on endophytic colonization of sorghum by the entomopathogenic fungus *Beauveria bassiana*. BioControl 54(5):663–669

Tejesvi MV, Pirttilä AM (2011) Potential of tree endophytes as sources for new drug compounds. In: Pirttilä AM, Frank AC (eds) Endophytes of forest trees: biology and applications. Springer, New York, pp 295–312

Tejesvi MV, Kajula M, Mattila S, Pirttilä AM (2011) Bioactivity and genetic diversity of endophytic fungi in *Rhododendron tomentosum Harmaja*. Fungal Divers 47:97–107. https://doi.org/10.1007/s13225-010-0087-4

Tejesvi MV, Segura DR, Schnorr KM, Sandvang D, Mattila S, Olsen PB, Neve S, Kruse T, Kristensen HH, Pirttilä AM (2013) An antimicrobial peptide from endophytic *Fusarium tricinctum* of *Rhododendron tomentosum Harmaja*. Fungal Divers 60:153–159. https://doi.org/10.1007/s13225-013-0227-8

Terekhova VA, Semenova TA (2005) The structure of micromycete communities and their synecologic interactions with basidiomycetes during plant debris decomposition. Microbiology 74:91–96

Timmusk S, Paalme V, Pavlicek T, Bergquist J, Vangala A, Danilas T, Nevo E (2011) Bacterial distribution in the rhizosphere of wild barley under contrasting microclimates. PLoS One 6:e17968

Tintjer T, Leuchtmann A, Clay K (2008) Variation in horizontal and vertical transmission of the endophyte *Epichloë elymi* infecting the grass *Elymus hystrix*. New Phytol 179(1):236–246

Tiwari S, Singh P, Tiwari R, Meena KK, Yandigeri M, Singh DP, Arora DK (2011) Salt tolerant rhizobacteria mediated induced tolerance in wheat (*Triticum aestivum*) and chemical diversity in rhizosphere enhance plant growth. Biol Fertil Soils 47(8):907–916. https://doi.org/10.1007/s00374-011-0598-5

Traving SJ, Thygesen UH, Riemann L, Stedmon CA (2015) A model of extracellular enzymes in free-living microbes: which strategy pays off? Appl Environ Microbiol 81:7385–7393

Tuteja N (2007) Abscisic acid and abiotic stress signaling. Plant Signal Behav 2(3):135–138. https://doi.org/10.4161/psb.2.3.4156

Vijayalakshmi R, Kairunnisa K, Sivvaswamy SN, Dharan SS, Natarajan S (2016) Enzyme production and antimicrobial activity of endophytic bacteria isolated from medicinal plants. Indian J Sci Technol 9(14):23–32

Waller F, Achatz B, Baltruschat H, Fodor J, Becker K, Fischer M, Heier T, Hückelhoven R, Neumann C, von Wettstein D, Franken P, Kogel KH (2005) The endophytic fungus *Piriformospora indica* reprograms barley to salt stress tolerance, disease resistance, and higher yield. Proc Natl Acad Sci U S A 102:13386–13391

Wang Y, Dai CC (2011) Endophytes: a potential source for biosynthesis transformation and biodegradation. Ann Microbiol 61:207–215

Wang S, Hu T, Jiao Y, Wei J, Cao K (2009) Isolation and characterization of *Bacillus subtilis* EB-28, an endophytic bacterium strain displaying biocontrol activity against *Botrytis cineread* Pers. Front Agric China 3(3):247–252

Wang M, Xing Y, Wang J, Xu Y, Wang G (2014) The role of the chi1 gene from the endophytic bacteria *Serratia proteamaculans* 336x in the biological control of wheat take-all. Can J Microbiol 60(8):533–540

Wang JL, Li T, Liu GY, Smith JM, Zhao ZW (2016) Unravelling the role of dark septate endophyte (DSE) colonizing maize (*Zea mays*) under cadmium stress: physiological, cytological and genic aspects. Sci Rep 6:22028

Wani ZA, Ashraf N, Mohiuddin T, Riyaz-Ul-Hassan S (2015) Plant-endophyte symbiosis, an ecological perspective. Appl Microbiol Biotechnol 99:2955–2965

Waqas M, Khan AL, Kamran M, Hamayun M, Kang SM, Kim YH, Lee IJ (2012) Endophytic fungi produce gibberellins and indoleacetic acid and promotes host-plant growth during stress. Molecules 17:10754–10773

Waqas M, Khan AL, Hamayun M, Shahzad R, Kang SM, Kim JG et al (2015) Endophytic fungi promote plant growth and mitigate the adverse effects of stem rot: an example of *Penicillium citrinum* and *Aspergillus terreus*. J Plant Interact 10:280_287

Wen CY (2004) Recent advances and issues on the endophyte. Chinese J Ecol 23(2):86–91

Wilson BJ, Addy HD, Tsuneda A, Hambleton S, Currah RS (2004) *Phialocephala sphaeroides* sp. nov., a new species among the dark septate endophytes from a boreal wetland in Canada. Can J Bot 82:607–617. https://doi.org/10.1139/b04-030

Wingender J, Neu TR, Flemming HC (1999) Microbial extracellular polymeric substances: characterization, structure and function. Springer, Berlin/Heidelberg. https://doi.org/10.1007/978-3-642-60147-7

Witzel K, Gwinn-Giglio M, Nadendla S, Shefchek K, Ruppel S (2012) Genome sequence of Enterobacter radicincitans DSM16656T, a plant growth-promoting endophyte. J Bacteriol 194(19):5469. https://doi.org/10.1128/JB.01193-12

Yang H, Puri A, Padda KP, Chanway CP (2016) Effects of *Paenibacillus polymyxa* inoculation and different soil nitrogen treatments on lodgepole pine seedling growth. Can J For Res 46:816–821. https://doi.org/10.1139/cjfr-2015-0456

Yu H, Zhang L, Li L, Zheng C, Guo L, Li W, Sun P, Qin L (2010) Recent developments and future prospects of antimicrobial metabolites produced endophytes. Microbiol Res 165:437–449

Zamioudis C, Pieterse CMJ (2012) Modulation of host immunity by beneficial microbes. Mol Plant-Microbe Interact 25:139–150. https://doi.org/10.1094/MPMI-06-11-0179

Zhang YP, Nan ZB (2007) Growth and anti-oxidative systems changes in *Elymus dahuricus* is affected by *Neotyphodium* endophyte under contrasting water availability. J Agron Crop Sci 193:377–386

Zhang YP, Nan ZB (2010) Germination and seedling anti-oxidative enzymes of endophyte-infected populations of *Elymus dahuricus* under osmotic stress. Seed Sci Technol 38:522–527

Zhang SQ, Outlaw WH Jr (2001) Abscisic acid introduced into the transpiration stream accumulates in the guard-cell apoplast and causes stomatal closure. Plant Cell Environ 24:1045–1054

Zhang F, Zhang H, Wang G, Xu L, Shen Z (2009) Cadmium-induced accumulation of hydrogen peroxide in the leaf apoplast of *Phaseolus aureus* and *Vicia sativa* and the roles of different antioxidant enzymes. J Hazard Mater 168:76–84

Zhang X, Li J, Qi G, Wen K, Lu L, Zhao X (2011) Insecticidal effect of recombinant endophytic bacterium containing *Pinellia ternata* agglutinin against white backed plant hopper, *Sogatella furcifera*. Crop Prot 30(11):1478–1484

Zhao X, Qi G, Zhang X, Lan N, Ma X (2010) Controlling sapsucking insect pests with recombinant endophytes expressing plant lectin. Nat Precedings 21:21

Zhao K, Penttinen P, Guan T, Xiao J, Chen Q, Xu J et al (2011) The diversity and anti-microbial activity of endophytic actinomycetes isolated from medicinal plants in Panxi Plateau China. Curr Microbiol 62:182_190

Zhu JK (2000) Over expression of a delta-pyrroline-5- carboxylate synthetase gene and analysis of tolerance to water and salt stress in transgenic rice. Trends Plant Sci 6:66–72

Zhu JK (2002) Salt and drought stress signal transduction in plants. Annu Rev Plant Biol 53:247–273

Zinniel DK, Lambrecht P, Harris NB, Feng Z, Kuczmarski D, Higley P (2002) Isolation and characterization of endophytic colonizing bacteria from agronomic crops and prairie plants. Appl Environ Microbiol 68:2198–2208. https://doi.org/10.1128/AEM.68.5.2198-2208.2002

Affirmative Plant-Microbe Interfaces Toward Agroecosystem Sustainability

7

Qasim Ali, Sana Ashraf, Muhammad Kamran, and Muhammad Ijaz

Abstract

Soil microbes are the important part of every agroecosystem in the world. They live naturally in all soils and plant systems, in which they depict their dominant existence with regard through their number, vast diversity, and their multi-dynamic functional abilities. They carry out essential life- and soil-sustaining processes such as nutrient fixation, solubilization, recycling, decomposition, acquisition, mobilization, remediation, degradation, and sequestration. Natural balance in all these processes is the key determinant of soil fertility that is represented by diverse physical, chemical, and biological soil factors. Fertile soils are characterized by diverse microbes, and they guarantee sustainability in agroecology that results in better plant health and crop productivity. Functional capabilities of microbial communities present in soils and their interaction with plant parts have been critically explored and characterized in the last few decades. So, application of these beneficial plant-microbe interactions can be used to find out a substitute and/or supplement in the present agricultural systems that are extensively dependent on synthetic chemical and inorganic fertilizers. In this chapter, we provide the comprehensive details of soil plant-microbe interactions, their role in plant health, and sustainability of agroecosystems. Further, their potential roles that can be used to establish a sustainable soil ecological environment for optimum crop growth, better development, and maximum yield in the long run are briefly discussed.

Q. Ali (✉) · M. Ijaz
College of Agriculture, Bahauddin Zakariya University, Layyah, Pakistan
e-mail: ali_qasim@bzu.edu.pk

S. Ashraf
Institute of Soil and Environmental Sciences, University of Agriculture, Faisalabad, Pakistan

M. Kamran
Department of Agronomy, University College of Agriculture, University of Sargodha, Sargodha, Pakistan

V. Kumar et al. (eds.), *Microbiome in Plant Health and Disease*,
https://doi.org/10.1007/978-981-13-8495-0_7

7.1 Introduction

Agroecosystems in a changing climate are supposed to be the results of changes in the climate and atmosphere on reliability and integrity of agroecosystems. Globally, in changing climate, environmental CO_2 sequestration, elevated wind temperature, and rainfall pattern fluctuations are most crucial and prominent factors that influence agriculture production and agroecosystem adaptations. In precedent era, production of agriculture and agroecosystem was affected greatly due to huge and continuous change in the global climate, especially in the arid zones. Ultimately, climate change demands continuous adaptation of the cropping system by keeping in mind the needs and socioeconomic status of the farmers (Boiffin et al. 2001).

Globally, climate change not only changes the crop environment but also develops a behavioral change in adaptation of good and better agro-products for farmers. To fulfill food, feed, and fiber demands, there is a need to adopt modern and advanced agriculture practices for healthier and enhanced production. This adaptation has led to diversified field crops, a cropping pattern with ever-increasing dependence on petrochemical products and diminishing exercise of useful biotic interactions (Altieri 1999; Stoate et al. 2001). In general, agroecosystem's biodiversity is influenced by natural land destruction and intensive use of synthetic agrochemicals (Bianchi et al. 2006; Farwig et al. 2009), potentially threatening biological pest management and pollination pattern in crops. Thus, agroecosystems possess low biodiversity, and growing diversity is a complementary element to enhance sustainability and functioning of the agroecosystem (Gurr et al. 2003; Moonen and Bàrberi 2008). Abiotic factors such as air, soil, and environment are influenced by intensification of arable farming (Anonymous 2005; Le Roux et al. 2008).

Nowadays, modern and advanced agriculture practices are being adopted by the farmers. Different production systems are being practiced by the farmers to focus on the organic products that are environment friendly and economical and depend upon natural inputs with best agronomic practices to manage insect pest (Warner 2007). Both biodiversity and agroecosystem are interlinked, and understanding of both is complex. Someone needs to clarify the specific role of biodiversity and its advantages in the agroecosystem. Positive influence of biodiversity predominantly depends on interactions among biotic and abiotic components in any advanced and beneficial agriculture system. Among biotic factors, microorganisms (bacteria, virus, and fungi) are gaining nearly all attention of the researchers. These microorganisms live in the root zone called rhizosphere, where plant roots interact with herbivores and microbes in a mutual way (Barea et al. 2005; Bais et al. 2006).

In fact, the root zone acts as a trade zone among pathogens, neighboring plant roots, and plants for interactions and also hosts defensive microbes, which suppress severe diseases (Baetz and Martinoia 2014). Mainly, the rhizosphere is categorized into three zones, namely, endorhizosphere, rhizoplane, and ectorhizosphere. The first region, the endorhizosphere, is the first zone consisting of root cortical and endodermal tissue; rhizoplane, the second zone, comprises mucilage and root epidermis. The ectorhizosphere, the third zone, contains soil near the root (Badri and

Vivanco 2009). The rhizosphere contains approximately 10^{11} microbial cells/g of plant root (Egamberdieva et al. 2008) and over 30,000 species of prokaryotes. So, rhizosphere directly or indirectly affects the crop yield (Mendes et al. 2013). Even though cultural restrictions render us to underestimate the exact number of rhizospheric microorganisms, plants have the ability to maintain the protective layer of bacteria and fungi around their roots (Berendsen et al. 2012). Microbial action is complex and diverse in nature, and they act as a growth-promoting agent for almost all crops. However, single organism is not responsible for growth promotion effect on plants; rather, it is the result of collective impact caused by the number of interactions between all the organisms present in the rhizosphere of plants. All microorganisms in the rhizosphere are working together for beneficial influence in the plants. In the rhizosphere, plant-plant, plant-microbe, or microbe-microbe interactions are essential for a healthy, stable, and sustainable agroecosystem for better and more crop production (Broz et al. 2010; Pellegrino and Bedini 2014).

Plant growth-promoting rhizobacteria (PGPRs), AMF, and insidious plant species are the most common organisms used to study and identify better biotic relationship for sustainable plant production. PGPR are known to have a very important role in recruiting diverse bacterial species like *Azospirillum*, *Alcaligenes*, *Acinetobacter*, *Bacillus*, *Pseudomonas*, *Rhizobium*, and others. A crucial role of rhizobacteria has also been identified in phosphorus solubilization, organic fertilizers, plant development, and stress tolerance in plants during severe environmental conditions (Yang et al. 2009; Bhattacharyya and Jha 2012). Rhizobacteria as PGPR are classified into three major classes as biofertilizers, phytostimulators, and biopesticides (Bhardwaj et al. 2014; Pérez-Montaño et al. 2014). Biofertilizers in organic production are gaining much attention from the researchers and farmers to get better and good-quality crops as these enhance the uptake of essential minerals by host plant. In general, *Allorhizobium* spp., *Trichoderma* spp., *Pseudomonas fluorescens*, and *Rhizobium* spp. are extensively implemented in the field as biofertilizers (Badar and Qureshi 2012; Yadav et al. 2013). Similarly, natural plant growth substances are also helpful in growth promotion in plants. These phytostimulators produce hormones like IAA, GA3, and cytokinin to promote the growth of plants by altering growth and development mechanisms in plants even under stress conditions (Apine and Jadhav 2011; Duca et al. 2014). *Pseudomonas* spp., *Streptomyces* spp., *and Bacillus* spp. are examples of phytostimulators that promote plant growth by suppressing pathogen propagation (Radja et al. 2002; Bhattacharyya and Jha 2012).

In addition to biopesticides, biofertilizers, and phytostimulators, some natural growth-promoting bacteria improve tolerance in plants under various stress conditions. For example, *Achromobacter piechaudii*, *Paenibacillus polymyxa*, *Rhizobium tropici*, and *Achromobacter piechaudii* bestow tolerance in different vegetables. These plant growth-promoting substances probably accumulate abscisic acid and suppress ROS and ACC in host plants. Utilization of *Achromobacter piechaudii* and *B. subtilis* in crop production system enhances resistance against salinity in plants (Zhang et al. 2008; Yang et al. 2009). Globally, inclusion of natural growth-promoting substances is gaining more interest in agroecosystems to produce healthy and better quality plants by minimizing synthetic fertilizer use, so that plants may develop tolerance against changing environment and various stress conditions.

7.2 The Challenge of Microbiology in the Beginning of the Twenty-First Century

Worldwide, agriculture has to increase the food production by twofold by 2050 to sustain the world's increasing population and meanwhile has to reduce its dependence on inorganic composts and pesticides. To meet this demand, it is dire need of the day for researchers to exploit various beneficial plant-microbe interactions for the benefit of the farmer community for embellishment of the agroecosystem. This plant-microbe relationship is beneficial both by supplying significant nutrients to the plant roots and through secretions of microbes that make the soil structure better for root growth. Some of the most important microorganisms being used in the agriculture system to boost crop productivity include *Azospirillum*, *Bacillus*, *Mycorrhizae*, *Pseudomonas*, *Rhizobium*, *Streptomyces*, and *Trichoderma* species. Discovering the hidden efficient biota through advanced technologies and methods is helpful in finding new suites of beneficial microorganisms that can enhance crop production around the world.

In the second half of the twentieth century, the main focus of microbiology was on endeavoring to understand the role of important microorganisms under lab conditions. Challenges for the next decades are to study the role of microorganisms under complex, natural, and extreme conditions such as the rhizosphere and the phyllosphere.

A microbe seems largely to be driven by principles that also govern our own behavior: a microbe wants to survive also under hostile conditions, and if conditions are more favorable, it will eat and proliferate. In order to understand a microbe's behavior in a certain habitat, one should have to understand the gene expression mechanism, the essential role of that gene and interaction among gene products, and their dependency on biotic and abiotic factors and toward which behavior traits this leads. In theory, the use of the new technology often designated as "genomics" can answer these questions. However, before becoming overoptimistic, it is good to realize that even from the best-known organism *Escherichia coli* K-12, biochemical function has not yet been established for one-third of the 4288 ORFs. Presently, the DNA of several microbes and plants that are studied in microbe-plant model systems has been sequenced or will soon be sequenced. On the other hand, sequencing is still expensive, and the sequences of many interesting organisms will not soon be available. For example, although the opportunistic human pathogen *Pseudomonas aeruginosa* PAO-1 has been sequenced (5570 ORFs; Stover et al. 2000), it is unlikely that the genomic sequences of the many *Pseudomonas* strains, which are important plant pathogens, phytostimulators, bioremediators, or biocontrol agents, will soon be available. Based on comparative genomics between *E. coli* strains, it can be predicted that their genomes not only differ in a few "islands" but that two

Pseudomonas strains will differ in hundreds of segments larger than 50 bp. Also the genomes of fungi and plants are being sequenced in a process.

7.3 Microbe-Plant Interactions: General Concerns

In view of their effects on the plant, microorganisms interacting with plants can be categorized as pathogenic, saprophytic, and beneficial. Pathogens can affect leaves, stems, or roots. A fascinating novel research field is the interaction between waterborne microorganisms and underwater plants. Saprophytes, which live on dead plant material, will not be further treated in this overview, although they play a crucial role in important processes such as the cycling of elements and composting. Beneficial organisms are frequently employed as bioinoculants (Bloemberg and Lugtenberg 2001). They can be characterized following the purpose of their application: biofertilizers (e.g., rhizobia, which have been connected monetarily for over a century), phytostimulators (e.g., auxin-secreting, root-elongating *Azospirillum*), rhizoremediators (toxin degraders that utilize root exudates as their carbon supply), and biopesticides.

None of the several microbe-plant interactions is completely understood. Therefore, we have not chosen for summing up a huge number of facts per interesting microbe, but we will restrict ourselves to a number of newly discovered principles and mechanisms and discuss a limited number of interactions between microbes and plants. At this point, basically microorganisms in their interactions with plants, regardless of whether the organism is useful or pathogenic, often act using the similar mechanisms of action, although for various combinations and for various purposes. In addition, obviously, microorganisms in their association with plants employ a similar mechanism to that in their interactions with different eukaryotes, for instance, fungi and also humans (Chin-A-Woeng et al. 2000; Lugtenberg et al. 2001). It is, thus, good to realize that although one uses the term microbe-plant interactions, the reality is that in the rhizosphere and in the phyllosphere, microbes also interact with each other.

7.4 Beneficial Plant-Microbe Interactions

Rhizosphere is a chemically complex zone that has vibrant microbial communities (Haldar and Sengupta 2015). Soil physicochemical characteristics, crop growth, and microbial secretions, generally, are interlinked and depend on the rhizosphere (Lareen et al. 2016). Cultivated lands are in consistent adjustments because of abilities of microbial communities. Soil microorganism produces extracellular compounds with adhesive properties. These compounds form aggregates of nutrients and soil around the plant roots and facilitate more nutrient uptake by roots. The natural materials secreted by microorganisms equally act as a defensive layer against dehydration of the microbes alongside the soil structure. Microorganisms significantly contribute in maintaining a natural balance of soil minerals and organic matter. It is helpful to increase crop productivity by enhancing soil fertility (Kibblewhite et al. 2008). Microorganism involvement in recycling of vital nutrients like B, C, Fe, S, N, P, and K, and soil fertility improvement are controlled by the enzymatic actions released by the microbes (Johnston et al. 2009).

Microorganisms absorb essential minerals present in the vicinity of rhizosphere, which act as profitable natural organic substances that are gradually discharged to enhance the productivity of agroecosystem (Shahbaz et al. 2017; van der Wal and de Boer 2017). Accordingly, these microbes directly acclimatize and confiscate nutrients and guarantee nonstop and moderate supply. Microorganisms are helpful and play a vital role in nutrient cycling in the soil. Microorganisms are significant elements to give expansive scope of enzymes for OM decay (Wallenstein and Weintraub 2008).

Fractional and impartial crop residue degradation piles up soil organic matter in a beneficial way to improve soil structure and productivity (Castellano et al. 2015). Soil macro- and microorganisms play a crucial role in nutrient stream crosswise over various tropic levels in sustainable agroecosystems (Chen et al. 2003). In agricultural soils, continuous consumption of minerals by crop harvesting, nutrient leaching, and water evaporation causes a significant decrease in the amount of nitrogen and minerals, which, generally, supports crop production (Brussaard 2004). Microbial communities along with other soil flora and soil fauna play an important role in the fixation of biological nitrogen fixation, which is easily available to plant roots and further releases a lot of nitrogen after the decomposition of root residues left after crop harvesting (Barrios 2007). Thus, for soils that possess a large amount of nitrogen, microbes are supposed to be a crucial indicator of nitrogen supply by nitrogen cycling. Microorganisms maintain and improve soil fertility by adding a large amount of nitrogen.

In reality, significant research on molecular techniques promptly associated with investigating microorganism communities and characterizing community intensity is the basic tool as a biomarker for demonstrating and observing biological community as soil health worldwide (Trivedi et al. 2016). Soil-lost nutrients are recovered by biological decomposition and recycling of organic matter. Essential minerals are important to enhance soil fertility and plant growth. Biological degradation of soil fertility also depends on soil physical and chemical properties, mainly regulated by the microbes (Anderson 2003). Since microbial population generally undergoes dynamic change, they can specifically add to soil richness, increasing productivity, and they are an important indicator of soil health.

7.4.1 Plant Growth Promotion

Bacterial competencies for numerous characteristics such as N, P, and K mineralization, solubilization, fixation, and production of hydrocyanic acid (HCN) and siderophores make them efficient plant growth-promoting microbes (Meena et al. 2017a; Felestrino et al. 2017). Many experiments show the multitrait microbial metabolic functions that make them prominent bioinoculants for plant growth improvements, natural suppression of disease-causing agents, and bioremediation of polluted soils (Singh et al. 2016). Soils that are rich inhabitants of these microorganisms are useful and enhance soil fertility and productivity through phytohormone production, bioremediation, nitrogen fixation, and phosphorus solubilization.

The surge of environment-friendly approaches in agribusiness rises and advances utilization of microbial biofertilizers. Introduction of microbial formulations in the soil not only develops resistance against severe environments but also enhances crop productivity by aiding in positive and economical soil fertility (Bhardwaj et al. 2014). Application of the biofertilizer BioGro in rice cultivation decreased the dependency of chemical fertilizers by about 52% in Mekong Delta, Vietnam (Nguyen et al. 2017). In Vietnam and Australia, the outcomes depend on utilization of microbes for a long time, which ultimately enhances productivity, crop development, and yield attributes of various crops. The yield of sweet potato increased by combined application of composts and organic fertilizers in Uganda. Arbuscular mycorrhizal inoculation combined with NPK fertilizer enhanced the yield (12.8–20.1 T ha^{-1}) of the sweet potato cultivar NASPOT 11 when compared to the existing yield (4.5 T ha^{-1}) (Mukhongo et al. 2017). Similarly, application of organic fertilizers with cold water enhanced the growth and quality traits of funnel (El-Azim et al. 2017). The discoveries have bolstered the way that microorganisms can boost crop development as well as acquire qualitative and quantitative changes in producing quality content. Growth improvement as a result of application of PGPRs prevents the increase in greenhouse gas emission. Rice yield increase and greenhouse gas emission (N_2O, CH_4, and CO_2) decrease were observed in Indonesia by the addition of biofertilizer (biofertilizer Biotara and Biosure) in the alluvial soils (Hadi and Nur 2017). The efficiency of biofertilizer enhanced by adding *Trichoderma* up to 12.9%. This further enhanced the plant's defensive mechanism against stress and severe environment by developing ascorbic acid, β-carotene, and lycopene (cancer-preventing agents) levels in plant tissues (Khan et al. 2017). Models demonstrate that the inoculation with multitrait microbial species has potential use for the development and nutritional health of crops.

7.4.2 Production of Phytohormones

Phytohormones like indole acetic acid, gibberellins, cytokinins, and ethylene have a great potential to promote crop yield, and microbes play a crucial role and comprise intrinsic ability to deliver phytohormones (Gamalero and Glick 2015). Likewise, gibberellic acid (GA)-producing *Azospirillum brasilense* and *A. lipoferum* were accounted for shoot development, invert dwarfism, and improvement of root hair density in rice and maize (Baca and Elmerich 2007). GAs account for the promotion of germination and extension in plants (King and Evans 2003) and furthermore to control plant development by corrupting DELLA proteins (Pieterse et al. 2012). Cytokinins play an important role in cell division and plant defense mechanism against biotrophic plant disease-causing agents (Pieterse et al. 2012). Increase in the concentration of cytokinins due to endophytic microbes in the plants is largely reported (Ortiz Castro et al. 2008). Microorganisms additionally secrete ACC (1-aminocyclopropane-1-carboxylate) deaminase that lessens ethylene concentration in plants to eliminate the effects of stress (Glick 2014; Gamalero and Glick 2015). In such a way, phytohormone-producing microbial populations having the potential for plant development can be utilized as indicators in health management measures.

7.4.3 Soil Health Management

Good soil health is a prerequisite for the crop profitability, as physicochemical soil properties are also determined by good soil health. Soil is a natural pool for all essential living and nonliving organic substances. Plant growth and development also depend on soil health (Doran and Safley 1997). Microbes influence the physical, chemical, and growing conditions of the soil and give an awesome quality to the soil health. The inclination for zero tillage over conventional cultivation provides soil with good health and humidity preservation choices to enhance soil quality, since it builds the number of aerobic and facultative microorganisms meaningfully in comparison to deep cultivated lands. In zero tillage soils, organic C and N, mineralization of nutrients, and available P in soil water are relatively more than those in the other soils (Doran 1980). Outcome revealed the change in physical and yield traits because of improved organic action managed by microflora and fauna (Kaschuk et al. 2010). Thus, quantity and type of microorganisms in the rhizosphere determine the condition and quality of the soil health. Crop growth and productivity ultimately depend on microbes present in the soil (Singh et al. 2016).

Vermicompost in combination with microbial species like *Bacillus megaterium*, *B. subtilis*, *Chlorella* sp., *Glomus* sp., *Pantoea agglomerans*, and *Paenibacillus azotofixans* improves total dependability and natural carbon content in the earth's topsoil (Yilmaz and Sonmez 2017). Decomposed residues and bioformulation having *Trichoderma* improve soil fertility and health and bring about noteworthy decrease in brown spot, sheath blight, and bacterial leaf curve malady inferable from initiate systemic resistance and improvement in paddy yield (0.8–3.0 t ha^{-1}) (Simarmata et al. 2016).

Improved soil fertility, growth, and tomato yield resulted due to the application of biofertilizer containing *Bacillus* species (Tripti et al. 2017). Research indicates a clear direction and influence of microbe's utilization for plant development to get enormous and healthy food. Soil plays a fundamental role, which is influenced by living and nonliving agents (Asano and Wagai 2014). Some of fungi such as arbuscular mycorrhizal fungi make a symbiotic relationship with most of the land plants, which is much important for the development of soil macroaggregates (>250 μm) (Miller and Jastrow 2000). Thus, arbuscular mycorrhizal fungi (AMF) make a beneficial interaction with plant roots and have extensions of mycelium that release natural substance to rhizosphere to develop soil microaggregates (<250 μm) (Rillig and Mummey 2006).

Ultimately, these microorganisms play a vital role to improve soil fertility by improving soil aeration, soil temperature, soil porosity, and soil moisture. AMF secretes proteins called glomalin into the soil and influences the soil health positively. Higher glomalin content demonstrates more aggregate stability proposing that diverse plant and AMF species like *Gigaspora gigantea*, *Rhizophagus irregularis*, and *Septoglomus deserticola* (Leifheit et al. 2014; Kohler et al. 2016). In India, a survey was made to identify the impact of AMF on aggregate stability in semiarid vertisol in sorghum bicolor, and greater soil security and bigger soil aggregates were observed in inoculated soils (Bearden and Petersen 2000). Root type, rhizospheric microflora, and mycorrhizal affiliation affect the soil aggregation (Rillig et al. 2015).

7.4.4 Biocontrol Activity

The use of organic substances in the crop production system is the key to success to maintain the soil health and to get more and healthy foods from the field. Soil health can be maintained by enhancing microbe's activities. Inoculation of germs makes the soil to offer resistance against pathogens and diseases in the rhizosphere. *Bacillus* sp., *Enterobacter* sp., *Klebsiella* sp., and *Paenibacillus* sp. can be used against the pathogens *Colletotrichum falcatum* and *Macrophomina* sp. (Arthee and Marimuthu 2016). Report affirms that because of the production of salicylate and catechol-type siderophores alongside HCN, lipopeptide, iturin, and surfactin, microbial inoculants prevail to stifle phytopathogens. Microbes (*Bacillus thuringiensis*, *Beauveria bassiana*, and *Metarhizium anisopliae*) can be utilized in less fertile and unproductive soils to get good yields. Generally, these microbes act as natural biocontrol agents and show resistance against diseases and pathogens, for example, *Tuta absoluta* (tomato leaf mineworker) can be minimized by using *Pseudomonas fluorescens*, *Bacillus subtilis*, vermicompost, and humic manures (Mohamadi et al. 2017). Microorganism activity is important, as reduction in the net production rate of *T. absoluta* was observed by these biocontrol agents. So, the inoculant with arbuscular mycorrhizal (AM) growth influences the outflow of barrier-related qualities like β-1,3 glucanases, chitinases, and oxalate oxidase in wheat contaminated with *Fusarium oxysporum* (Sabbagh et al. 2017). Effective prevention of root knot nematode (*Meloidogyne incognita*) in Pusa Ruby cultivar of tomato was likewise seen by the use of microbial inoculants. The use of biofertilizer with adjusted measurement of NPK (125:50:100 kg ha^{-1}) brought about 74.8% decrease in nematode population in root knot (75.8%) and rise in plant dry weight (62–64%) (Patra et al. 2017). Biocontrol of *Ralstonia solanacearum* accounted for by the utilization of chitosan acquired from *Cunninghamella elegans* (Oliveira et al. 2017). In banana, treatment of *Bacillus amyloliquefaciens* NJN-6, which stifled the development of *Fusarium*- and *Ralstonia*-like pathogens under field conditions, likewise promoted plant growth and health (Fu et al. 2017). These investigations widen the role of PGPRs as biocontrol entities against pathogens and nematodes and as natural markers for disease and pest control.

7.4.5 Alleviation of Abiotic Stress

Organisms possess diverse characteristics (amino acids, genes, and metabolites) to overcome severe circumstances posed by nonliving factors. These abilities of microbes are helpful for crops to withstand severe environmental conditions (Bacilio et al. 2016; Meena et al. 2017a). Combined application of microbes (*Pseudomonas stutzeri*) with humic acid enhanced K^+/Na^+ and Ca^{2+}/Na^+ particle proportion and development in chime and bean stew pepper demonstrating alleviation of negative effects of salinity stress (Bacilio et al. 2016). Under abiotic stress, biochar-enriched *Bradyrhizobium* inoculations are very important for lupin, as it improves N and P uptake by enhancing root nodulation in arid regions (Egamberdieva et al. 2017).

7.4.6 Nutrient Acquirement

In the forest ecological system that is not managed properly, biological N-fixation is central to meet optimum plant nutrition requirement. Cyanobacteria add to the N-pool in soils of forest ecological system (Wang et al. 2010). In the paddy farming system, different N-fixing cyanobacteria accomplish N requirement of the growing crops by fixing ambient N_2. Likewise, it has been assessed that *Azolla-Anabaena* symbiosis contributes to about 580 kg N ha^{-1}. The cyanolichen *Peltigera aphthosa* considerably adds to N-fixation by employing vanadium nitrogenase enzyme rather than Mo-nitrogenase in Mo-deficient soils (Darnajoux et al. 2017). Hence, *P. aphthosa* is taken as a biological indicator that employs another nitrogenase pathway in Mo-deficient soils. Paddy soils containing abundant N-fixing cyanobacteria can decrease additional input of N through inorganic fertilizers. Many single-cell, filamentous, nonfilamentous, heterocystous, and nonheterocystous cyanobacterial classes are important N-fixing populations in the paddy soil. These species can be applied for extensive biological N-fixation to enhance N concentration in the soil and to meet plant N demand (Singh et al. 2016). Microalgae play an important role in paddy rice cultivation by adding N, other nutrients, plant hormones, and biomaterial to the soil (Dineshkumar et al. 2017). The use of inorganic fertilizers in paddy farming can be reduced by adding microalgae in the soil (Singh et al. 2011). Intercropping tree species can be used to enhance N-fixation, like hybrid walnut trees cropped with alfalfa showed 35% enhancement in light use efficacy and N-fixation by alfalfa (Querne et al. 2017). Likewise, combined inoculation of *Enterobacter* sp. and *Microbacterium arborescens* in wheat cultivation promoted N-fixation, IAA production, and P-solubilization (Kumar et al. 2017). Higher crop production was noted due to more plant uptake of essential growth nutrients. Bacteria attached to hyphae of mycorrhiza contribute to P-solubilization by the mycorrhizal species that, in response, support these bacteria to flourish by offering adhering place and growth nutrients (Kaiser et al. 2015). Mycorrhizae enhance the solubilization of P in the presence of PSB and promote the bioavailability of P to the cultivated crops (Taktek et al. 2017).

7.5 Nitrogen-Fixing Bacteria in Agricultural Soil

Concentration of N_2 in the atmosphere is about 4×10^{15} tons that is 20 times greater than that found in the underground rocks (Gallon and Chaplin 1987). These sources will remain nonbioavailable, until they are transformed to NH_3 nitrogen-fixing bacteria (Hernandez 2000). These bacteria can be symbiotic, free-living, or associative and live in interdependence on plants. The associative diazotrophs live within the root zone and also cross the intracellular spaces of the root and shoot (so-called as diazotrophic endophytes). Symbiotic N_2-fixing bacteria contribute a significant amount of N to plants as compared to free-living bacteria and promote plant growth (Dobbelaere et al. 2003). The rhizobia-legume symbiotic association is the most extensively investigated plant-microbe relationship (Sprent 2001). Legumes are greatly diverse including up to 19,000 species globally and provide N to plants by N-fixation through association with rhizobia. Hence, inoculation of legume plants with potential rhizobial species is of much concern for promoting the eco-friendly crop cultivation. Rhizobia existing within nodules of legume crops take atmospheric N_2 and transform it into plant-accessible N. The host plants develop nodules and provide O_2 and organic C to the bacteria; in turn, the bacteria provide N to the plants. Nonsymbiotic N-fixing bacteria live in the rhizosphere and can interchange fixed N with the plants for organic C. In the root zone, plant exudates that contain growth nutrients and organic C promote microbial populations. Investigations have revealed that root exudation is significantly correlated with soil inorganic N-pool (Hamilton and Frank 2001). Soil deposits of plant-derived N, comprising rhizodeposits (decaying roots), promote soil microbial population by providing nutrients (Høgh-Jensen and Schjoerring 2001). Crop remains also enhance the population of N-fixing bacteria in soil.

7.5.1 Symbioses with N_2-Fixing Cyanobacteria

Various N-fixing cyanobacteria develop symbioses with plants that are different from legume nodules. These dissimilarities are based on the capability of cyanobacteria to fix N_2 in free-living situation and to conduct the fixation on the cost of their own photosynthetic process. The processes of free-living N_2-fixation are employed in symbioses, but here it is organized by the hosts because of confinement of cyanobacteria in characteristic symbiotic structures. In the water fern *Azolla*, cyanobacteria are found inside the leaf voids; in the gymnosperm *Cycas*, inside the intercellular places of coralloid roots; and in the gymnosperm *Gunnera*, inside the glands at the sites of leaf petioles. When microbial progression is conducted in the absence of N, 10% of cells in the *Nostoc* or *Anabaena* filaments are developed into heterocysts that lack the reproductive potential. Heterocysts have dense walls that prevent the O_2 conduction into the cells. Due to the inhibition of photosynthesis, microaerobic environments are developed within the heterocysts, allowing nitrogenase synthesis. One more cellular variation in *Nostoc* includes the motile hormogonia filaments involving tiny cells with gas vacuoles. These cellular differences are also employed by cyanobacteria in symbioses with plants and to fix N. During the preinfection interactions, huge development of hormogonia that can move is encouraged in *Nostoc* that live within the *Gunnera* glands. Inside the glands, cyanobacteria spread vigorously and infect the plant cells and hence destruct cell walls, but they are reestablished after the entry of cyanobacteria. Inside the plant cytoplasm, the cyanobacteria are captured inside the symbiosome tissues, and consequently 80% of cells are transformed into heterocycts having N-fixing potential.

7.5.2 H_2-Consuming Bacteria Living in the Root Zone

It is documented that intercropping legumes with nonlegumes, or crop rotation including legumes and nonlegumes, can remarkably enhance the growth of nonlegume crops (Høgh-Jensen and Schjoerring 2001). It has also been demonstrated that the H_2 generated during N_2-fixation by legume nodules is accountable for growth-promoting effects of legume (Irvine et al. 2003). H_2 gas is a byproduct of the N_2-fixing nitrogenase enzyme. The majority of free-living diazotrophs and few symbiotic bacteria also generate the uptake hydrogenase (HUP) enzyme, which has the potential to oxidize H_2 and obtain chemical energy from it. However, most of the rhizobia used in crop production do not have uptake hydrogenase enzyme (Uratsu et al. 1982). Thus, the H_2 generated by the nitrogenase diffuses out from nodule into the soil, and this loss of H_2 is a drawback for the uptake hydrogenase enzyme (Dong and Layzell 2001). H_2 gas liberated from legume nodules can enhance the soil minerals to benefit the plant because soil microorganisms frequently oxidize the liberated H_2 and enhance the rhizobial biomass, O_2 utilization rate, and chemoautolithotrophic carbon dioxide fixation (Dong and Layzell 2001). Microbial oxidation of liberated H_2 alters the soil microbial population and consequently enhances plant growth. Significant quantity of energy is provided to rhizosphere microbes in the uptake hydrogenase legume field in the form of H_2, and subsequently, energy-enriched soils boost plant productivity in return. So, it can be believed that H_2 released during N_2-fixation in legumes promotes the microbial community in soil and yield of the succeeding crop.

7.6 Mycorrhizae

Mycorrhizae demonstrate the approach of plant-microbe symbiosis. They are developed by plants and the root-inhabiting fungi in which a part of the fungal partner (mycobiont) is within the root and the second part is outside of it. On the basis of anatomy of the intraradical part of mycobiont, the mycorrhizae are categorized as endomycorrhizae and ectomycorrhizae (EcM). The widespread group of endomycorrhizae is showed by arbuscular mycorrhiza (AM) developed by most of (75–90%) temporal plants. Division of the mycobiont into inter-radical and extraradical fragments reveals the elementary purpose of mycorrhizae as an interceder between the plant and the soil. This characteristic is of global significance, as only few plants have the potential to fulfill their mineral requirement and H_2O without the help of a mycobiont. In artificial circumstances, most plants can grow in the absence of a fungal partner, but under field circumstances, tough mineral competition takes place among different plant species. Thus, plants cannot survive independently and depend on the mycobionts. AMF associated with the earliest phylum *Glomeromycota* and have developed biotic connections with above 80% of terrestrial plant species like maize, rice, wheat, and soybean (Gutjahr and Parniske 2013). The habitation of roots by the AMF is accomplished by an interchange of chemical signals between fungi and plant. The main signal (strigolactones) is discharged by the host roots that influence sprouting and hyphal branching in AMF and also activate fungal metabolism (Zwanenburg and Pospisil 2013). In response to that signal, fungus discharge tetrameters and pentamers of N-acetylglucosamine and lipochitin-oligosaccharides (Gutjahr and Parniske 2013) and stimulate a signaling channel within the roots of the host. When the transmission system has been developed between the fungus and the plant, symbiotic exchange of minerals starts.

The association between plants and AMF is two-sided: AMF receives carbohydrates from the plant and, in return, supplies the plant with minerals, abiotic stress tolerance, and enhanced H_2O supply (Parniske 2008). This association transfers 4–20% of photosynthetically fixed C of plant to the AMF. The development of AFM relies on both the host plant root exudates and soil phosphorus (P) concentration (Tamasloukht et al. 2003). If the level of P is very high (10 mM), the development of the fungal hyphae is suppressed, and the AMF colonization is decreased. Some other aspects that also disturb AFM colonization include the practice of monoculture, tillage, advanced agricultural systems, and genetically modified crops. AM development includes different methods: (1) preinfection, (2) development of mycelium within the cells, (3) formation of symbiotic structures within the cells, and (4) formation of sporulating extraradical mycelium. It initiates from the sprouting of the fungal spores and branching of the growing germ tube caused by the root exudates. A speedy development of the germ tube from the spore to root is activated by functional chemotaxis and is completed by bonding of appressoria to the root surface. After bonding with root, the hyphae begin to develop from the appressorium into the cortex, inhabiting its outer and inner coatings but not crossing the pericycle. After forming a system of intercellular hyphae, fungus develops intracellular and extraradical structures. The completely grown arbuscules with high branches almost occupy the entire volume in infected cortical cells. A prominent distinction is a feature of the plant cells holding arbuscules. The key aspect of the arbuscules is to maintain the physical association with intercellular hyphae. However, arbuscules are short-lived structures, and every 4–7 days, they are consumed by plant cells, and the new arbuscule may be formed from the adjacent hyphae. The last growth stage in AM is the development of extraradical hyphae that are of vital significance for both partners. The extraradical hyphae are inconsistent in their morphology, with thick (10–20 mkM in diameter) first-order (runner) hyphae developing directly from the roots and thin (2–5 mkM) second-order (absorptive) hyphae branch developing from the runner hyphae. The enhanced network of extraradical hyphae (70–80 m per 1 m of root) enables the mycobiont to penetrate the soil within many millimeters around the roots vigorously taking the soil minerals. An essential aspect of the extraradical hyphae is their capability to enter the nearby roots and to develop an underground system of hyphae associating different members of plant species. During the growth of AM, the defense reactions are instigated within the root cortex such as amendments within the cell walls, production of phytoalexins, storage of callose, and of some pathogen-regulated proteins. During the glomus development, the strength of plant processes within roots is low; they are less sustained and extremely contrasted in time and space in comparison to pathogenesis.

7.7 Phosphate-Solubilizing Microorganisms (PSMs)

Phosphorus (P) is the second essential nutrient necessary for excellent plant development. It contributes to all metabolic systems like respiration, signal transduction, energy transfer, and photosynthesis (Anand et al. 2016). However, 95–99% soil P is not bioavailable; consequently, plants cannot take up the P. Plants take up P only in the form of monobasic ($H_2PO_4^-$) and dibasic (HPO_4^{-2}) ions. Dissolution and mineralization of P by phosphate-solubilizing bacteria is an imperative quality that can be attained by PGPR. Organic acids released by several bacteria inhabiting the soil enhance the bioavailability of inorganic P by increasing solubility (Sharma et al. 2013). Phosphate-solubilizing PGPR that belonged to the genera *Arthrobacter*, *Burkholderia*, *Beijerinckia*, *Bacillus*, *Enterobacter*, *Erwinia*, *Flavobacterium*, *Mesorhizobium*, *Microbacterium*, *Pseudomonas*, *Rhodococcus*, *Rhizobium*, and *Serratia* have drawn the attention of cultivators as soil inoculates increase plant productivity (Oteino et al. 2015). Although these microbes increase P-solubilization and soil mineral status, experiments on their application as a biofertilizer are deficient. Currently, P-fertilization is a main agronomic research area due to the high cost of fertilizers. Products containing high-quality rock phosphate (RP) are also expensive like food preservatives and fungicide, and reservoirs of top-grade phosphates are depleting swiftly and projected to be depleted in about 100 years. As a result, agronomic research is focused on low-quality RP (9–11% P_2O_5) as a source of fertilizer, as low-quality ore is accessible globally in huge quantities. PSMs are the main constituent of crop cultivation and improve the fertility of soil. PBRMs facilitate soil processes like decomposition, mineralization, and release of stored minerals. P-solubilizing potential can transform the insoluble phosphatic compounds into soluble compounds (Verma et al. 2017) in soil and enhance their accessibility to the edible crops. Bacterial strains from the genera *Bacillus*,*Enterobacter*, *Pseudomonas*, and *Rhizobium* including *Aspergillus* and *Penicillium* are highly potent P-solubilizers (Verma et al. 2014). *Enterobacter*, *Pseudomonas striata*, *B. sircalmous*, *B. polymyxa*, *B. subtilis*, *B. circulans*, and *Bacillus megaterium* could be stated as significant strains (Li et al. 2016). The PSMs include diverse classes of microbes that not only take up P from insoluble sources of phosphates, but they also discharge soluble phosphates in large amounts that are beyond their requirements (Meena et al. 2017b).

7.8 Potassium-Solubilizing Microorganisms (KSMs)

Potassium (K) is the third foremost important nutrient for plant growth. As more than 90% of K is available as insoluble rock and silicate ores, the level of soluble K is generally limited in soil (Parmar and Sindhu 2013). K-deficiency has become a main limitation in crop cultivation and results in poor crop production. It is indispensable to discover an alternative source of K for providing sufficient K in soils for sustaining crop cultivation (Kumar and Dubey 2012). The capability of PGPR to dissolve K through the production of organic acids has been extensively explored. K-solubilizing PGPR, like *Paenibacillus* sp., *Burkholderia* sp., *Pseudomonas* sp., *Bacillus mucilaginosus*, *B. ferrooxidans* sp., *Bacillus edaphicus*, and *Acidothiobacillus* sp., have been recognized as K solubilizers in the soil (Liu et al. 2012). Consequently, application of P-solubilizing PGPR as biofertilizers can decrease the requirement of fertilizers and promote sustainable crop production (Setiawati and Mutmainnah 2016). A diverse range of microorganisms are documented to be found in soil (particularly rhizosphere), which contribute to plant growth (Bahadur et al. 2017). A big gap exists between researchers and cultivators (Meena et al. 2016). The majority of the cultivators only practice urea as the N source and $(NH_4)_2HPO_4$ as the P source. However, very few apply K-fertilizer as muriate of potash for crop cultivation (Prakash and Verma 2016). Thus, accessible compounds of K decline in soil because plant uptake is in high concentration. Yet, crop residue contains more K-content in contrast to other elements, but cultivators do not add crop residues in the soil that is the main reason of K-deficiency in soil, which causes low crop yield (Nath et al. 2017). K-solubilizing bacteria (KSB) have the potential to discharge K from insoluble compounds (Nath et al. 2017). Investigators have found that KSB can promote plant development by controlling infectious agents and enhancing soil minerals and structure. A large population of KSMs found in the rhizosphere produce organic acids which in turn cause K-solubilization and enhance plant development (Velazquez et al. 2016). Thus, the implementation of KSMs is viewed as an inclusive strategy to improve the crop yield. This practice can also restore the mineral level of degraded agricultural soils (Bakhshandeh et al. 2017). Nevertheless, application of KSMs is restricted due to insufficient transfer of information from researchers to cultivators (World Bank 2008).

7.9 Zinc-Solubilizing Microorganisms (ZSMs)

Plants consume zinc (Zn) as a divalent cation (Zn^{+2}), but in soils of elevated pH, it is consumed as a monovalent cation ($ZnOH^+$). A high level of Zn in the soil results in improved crop performance and human health. Currently, Zn deficiency is common in both crops and humans (Bahadur et al. 2016). Plant tissues require Zn in low concentrations (5–100 mg kg^{-1}). The dominant cause of Zn scarcity in agricultural soil is reduced solubility of Zn (Gontia Mishra et al. 2016). Zn deficiency can be linked to elevated pH (>7.0) of soil. Zn dissolution declines with rise in pH, organic matter, and carbonate content, high Mg:Ca ratio, and immense accessibility of P and Fe (Li et al. 2016). When the soluble form of Zn ($ZnSO_4$) is introduced to agriculture soils, it is converted into a number of insoluble compounds like zinc hydroxide in soils of elevated pH, zinc carbonate in Ca-abundant alkali soils, and $Zn_3(PO_4)_2$ in neutral or alkali soils containing high concentration of P-containing fertilizers under reducing environments (Sarathambal et al. 2010). Plants cannot consume insoluble forms of Zn. The Zn scarcity can be controlled by introducing Zn-containing fertilizers in the soil, but inorganic fertilizers are expensive and have detrimental consequences on the ecosystem. Hence, to solve this problem, environment-friendly and inexpensive techniques are needed with time such as Zn-solubilization by ZSMs (Mishra et al. 2017). Rhizobacteria considerably solubilize insoluble Zn compounds (Krithika and Balachandar 2016). Zn-solubilizing bacteria (ZSB) increase the bioavailability of Zn by releasing the fixed form of Zn (Barbagelata and Mallarino 2012). Hence, application of ZSB for crop production is gaining popularity among farmers (Krithika and Balachandar 2016).

7.10 Organisms for Biological Control or Biopesticides

Different commercially available bioformulations are available in market, which contain numerous biological control agents that have the potential to protect plants from fungal diseases. The major applied organisms are the bacteria *Bacillus*, *Pseudomonas*, and *Streptomyces*, while fungal bioformulations contain *Trichoderma*, *Gliocladium*, and *Fusarium*. The mechanisms used by these biocontrol agents include (i) niche exclusion; (ii) competition for nutrients; (iii) production of chitinase by *Serratia marcescens*; (iv) release of AFMs (antifungal metabolites), like PHL (2,4-diacetylphloroglucinol); and (v) ISR (induced systemic resistance) that is triggered by certain nonpathogenic *Pseudomonas* rhizobacteria, which make the plant extremely reactive toward a range of pathogens including leaf pathogens (M'piga et al. 1997). Flagella, LPS (lipopolysaccharide), and siderophores have been implied as bacterial components involved in causing ISR (van Loon et al. 1998). ISR differs from SAR (systemic acquired resistance), which is instigated as a result of pathogenic infection. Instantaneous activation of both SAR and ISR caused supplementary effect on the intensity of induced protection of *Arabidopsis thaliana* against *P. syringae* pv. tomato (van Wees et al. 2000). Ethylene signaling plays a role in *P. fluorescens* WCS417r-mediated ISR functions. Introduction of the ACC deaminase gene, which encodes the cytoplasmic enzyme ACC deaminase that transforms ACC (1-aminocyclopropane-1-carboxylic acid, a precursor of ethylene) to ammonia and α-ketobutyrate, into *P. fluorescens* strain CHAO enhanced its potential to shield cucumber from *Pythium* damping off. It was proposed that ACC deaminase decreases the level of pathogen-induced plant ethylene and therefore eliminates the pathogen-induced inhibitory effect of ethylene on root elongation (Wang et al. 2000). Two *Pseudomonas* strains have the same strategy to kill eukaryotes. *P. aeruginosa* kills the nematode *C. elegans*, and this killing requires colonization as well as the synthesis of the phenazine-derivative pyocyanin (Mahajan et al. 1999), whereas *P. chlororaphis* kills fungi including *Fusarium*, and this killing is accompanied by colonization of hyphae (Lagopodi et al. 2002) and, if present, the root surface as well as by phenazine-1-carboxamide production (Chin-A-Woeng et al. 2000).

7.11 Phytostimulators

Some bacteria of the genus *Azospirillum* promote plant growth as free-living organisms. A polar flagellum is involved in the bonding of *Azospirillum* cells to plant roots. *Azospirillum* fixes atmospheric nitrogen, but the observed growth promotion may rather be related to plant growth promoters produced by the bacterium rather than by its nitrogen-fixing capacity. Actually, three kinds of elements have been identified in the supernatant fluids of *Azospirillum* cultures that enhance the plant growth, namely, auxins, cytokinins, and gibberellins. Among them, the auxin IAA (indole-3-acetic acid) is concentration-wise highly significant. Three pathways are known to convert tryptophan into IAA. Experiments conducted by inoculating *Azospirillum* mutants have shown amplified rooting that in turn promoted nutrient uptake (Steenhoudt and Vanderleyden 2000). Similar results were obtained with a plant growth-promoting *P. fluorescens* strain, which converts exogenous tryptophan into IAA. The strain promotes maturation of radish in greenhouse experiments, most likely due to the high concentration of tryptophan measured in radish exudate.

7.12 Probiotics for Plants

Currently, probiotic organisms are being used to sustain plant health. Plants stimulate the microbes in their rhizosphere to produce antibiotics as a defense against infections caused by soilborne pathogens. Infections caused by soilborne fungal pathogens adversely affect the crop yield. However, soil bacteria of the genus *Pseudomonas* are abundant in most soils and contribute to enhance plant growth, disease suppression, nutrient cycling, N-fixation, and bioremediation. They have the potential to respond suddenly to changes in physicochemical soil conditions. Pseudomonads have been extensively analyzed for their biocontrol capability against fungi. They suppress the diseases through the production of antibiosis, and different antimicrobial compounds have been recognized such as 2,4-diacetylphloroglucinol, phenazines, pyrrolnitrin, pyoluteorin, HCN, and biosurfactant antibiotics. Biochemistry-based characterization of the chemical compounds is conducted by using molecular techniques to understand the mechanisms of production and association with pathogens and plants and to analyze their activity in soil.

7.13 Bacterial Endophytes

An incessant apoplastic channel is ever present between plant root and shoot that is adequate for entry of microbes from the root cortex to xylem and then overall in the plant (Peterson et al. 1981). Therefore, most of the bacterial endophyte communities are developed during the colonizing process initiated by bacteria in the rhizosphere. Thus, the plants provide a diverse environment to microbial endophyte that is suitable for mutual association between plant and microbes (Stone et al. 2000). Complimentary association between plant and endophyte offers two most important beneficial effects, for example, plant growth promotion and disease suppression (Mathesius 2003). However, noncomplementary association between plant and endophyte has inhibitory allelopathic outcomes (Sturz and Christie 1996). It is necessary to find an important source of proficient endophytes in the soils and organic leftover of preceding crop plantings. Consequently, mutual association between rotation crops will take in a microbial consistency among the newly established crop and the indigenous soil microbial community. Accordingly, it has been suggested that the growth advantages from complementary cropping networks (legume rotations for residual N and improved soil structure) can be obtained by leftover residual endophytic populations that have the potential to enhance plant development and suppress disease progression (Sturz et al. 1998). These associations between crops in complementary rotations can be cultivar specific (Sturz et al. 2003). Therefore, selection standard adopted for crop rotation must comprise an assessment of the consequences of rhizobacteria and their related endoplant-competent bacterial populations when examining the long-term concerns of these crops to be selected.

7.14 Role of Microbes in Sustainable Agriculture

The intensive implementation of agrochemicals for crop production has resulted in environmental pollution due to accumulation of N and P in the soil and leaching to groundwater. Microbial-based formulations should be used for sustainable agriculture that will reduce the need for chemical fertilizers. Currently, PGPR are progressively employed for the inoculation of edible crops. Preparation of inoculants is a difficult process, and crops also behave differently with microbial inoculation. Old crops such as pea and soybean grow faster under N fertilization than after inoculation with rhizobial strains, whereas young crops such as clover and hairy vetch easily use symbiotically fixed N_2. During cultivation, plants dissipate the major part of their symbiotic capability, and the cultivated crops develop fast by taking inorganic fertilizers in contrast to suitable microbial symbionts. Consequently, special techniques are required to restore the symbiotic potential of cultured plants. It can be done by using the plant genes that deactivate symbioses with inefficient strains. The main requirement for enhancing the symbiotic potential is the coordination of genetic alterations in the plants and their microsymbionts. Two-factor analysis of variance in the productivity of symbiotic systems proposes that symbiotic potential is influenced by genotypes of both the partners. Hence, coordinated breeding must focus on establishing the specific combinations of partner's genotypes. Coordinated breeding aims at improving the plant associations for the particularly selected or genetically engineered potent strains. These associations are often evaluated on the account of competitiveness of implemented microbes (potential symbionts) that is demonstrated against the indigenous (mostly inefficient) symbionts. The rhizosphere microbial population consists of bacteria that are known as PGPR and promote plant growth through release of 1-aminocyclopropane-1-carboxylate deaminase, volatile compounds, phytohormones, and siderophores and also have potentials of disease suppression and antagonism to plant pathogens (Santoyo et al. 2012). The intensity of PGPR to promote agricultural performance has been precisely explained in literature but has not been rightly practiced mainly in developing states. Presently, agrochemicals are extensively applied in cropping system pesticides. As the human population continues to grow, now above 75,000 million people are occupying our earth, and it is estimated that the food requirement will become double by 2050 (Baus 2017). Therefore, it is crucial to understand the abiotic and biotic associations to best exploit the rhizosphere microbial population to promote agricultural yield. Microbiome-based bioinoculant with growth-promoting and biocontrol potentials can be generated by selecting hundreds of bacterial and/or fungal species. This bioinoculant can be directly applied to agriculture crops to enhance soil fertility and crop production. This methodology should avoid the application of those inorganic fertilizers that damage the environment and human and animal health or genetically modified organisms.

7.15 Impacts of Beneficial Plant-Microbe Interactions on Ecological and Agricultural Systems

Extensive exploration of plant-microbe interactions has changed the vision of plant as equipment that can transform inorganic compounds into organic substances with the help of sunlight. In spite of this concept, the plant behaves as a controller of symbiotic population, in which many essential host activities are transferred to the microsymbionts. It is mostly assessed that the function of favorable microbes toward plants is to enhance their nutritional status. Because of the widespread implementation of isotopic (15N, 14C) techniques, it has become easy to evaluate the metabolic exchange among the partners. However, plant-microbe associations are not merely limited to this exchange. Likewise, in defensive symbioses, the microbes do not contribute in supplying minerals to the host. At the plant community level, microbial-based nutrient supply is mostly conducted by enhancing competitiveness in symbiotic plants that can significantly alter the ecosystem structure.

7.16 Conclusion and Future Research Directions

Numerous bacteria and fungi inhabiting the root zone control different chemical processes that provide nutrients to plants, benefitting the plant growth and health. Thus, the importance of maintaining diverse microbial population in the rhizosphere should be considered for better future scenario. Regarding this, implementation of selected microbial consortia as plant inoculants for improving crop production has gained popularity. In order to keep healthy crop production and ecosystem, strategic research must be conducted to expand our existing understanding on microbial associations in the rhizosphere. New genetically modified and eco-friendly microbial inoculates can be applied to suppress the plant diseases and to promote plant growth. It is well documented that less than 5% of all soil bacteria, archaea, and fungi are culturable, and the remaining extensive microbial population cannot be cultured. There is a need to understand the activities of these nonculturable microbes in the rhizosphere and to explore the unidentified species of root-based soil microbes.

References

Altieri MA (1999) The ecological role of biodiversity in agroecosystems. Agric Ecosyst Environ 74:19–31

Anand K, Kumari B, Mallick MA (2016) Phosphate solubilizing microbes: an effective and alternative approach as bio-fertilizers. Int J Pharm Sci 8(2):37–40

Anderson TH (2003) Microbial eco-physiological indicators to asses soil quality. Agric Ecosyst Environ 98:285–293

Anonymous (2005) Millennium ecosystem assessment. Ecosystems and human well-being: biodiversity synthesis. World Resources Institute, Washington, DC

Apine OA, Jadhav JP (2011) Optimization of medium for indole-3-acetic acid production using Pantoea agglomerans strain PVM. J Appl Microbiol 110:1235–1244

Arthee R, Marimuthu P (2016) Isolation identification and screening of culturable endophytic bacterial community in sugarcane for antimicrobial activity on major fungal pathogens. Bioscan 11:2157–2167

Asano M, Wagai R (2014) Evidence of aggregate hierarchy at micro- to submicron scales in an allophanic Andisol. Geoderma 216:62–74

Baca BE, Elmerich C (2007) Microbial production of plant hormones. In: Associative and endophytic nitrogen-fixing bacteria and cyanobacterial associations. Springer, Dordrecht, pp 113–143

Bacilio M, Moreno M, Bashan Y (2016) Mitigation of negative effects of progressive soil salinity gradients by application of humic acids and inoculation with Pseudomonas stutzeri in a salt-tolerant and a salt-susceptible pepper. Appl Soil Ecol 107:394–404

Badar R, Qureshi SA (2012) Use of Trichoderma hamatum alone and in combination with rhizobial isolates as bio-fertilizer for improving the growth and strength of sunflower. J Basic Appl Sci Res 2:6307–6314

Badri DV, Vivanco JM (2009) Regulation and function of root exudates. Plant Cell Environ 32:666–681

Baetz U, Martinoia E (2014) Root exudates: the hidden part of plant defense. Trends Plant Sci 19:90–98

Bahadur I, Maurya BR, Kumar A, Meena VS, Raghuwanshi R (2016) Towards the soil sustainability and potassium-solubilizing microorganisms. In: Meena VS, Maurya BR, Verma JP, Meena RS (eds) Potassium solubilizing microorganisms for sustainable agriculture. Springer, New Delhi, pp 225–266. https://doi.org/10.1007/978-81-322-2776-218

Bahadur I, Maurya BR, Meena VS, Saha M, Kumar A, Aeron A (2017) Mineral release dynamics of tricalcium phosphate and waste muscovite by mineral-solubilizing rhizobacteria isolated from indo-gangetic plain of India. Geomicrobiol J 34(5):454–466

Bais HP, Weir TL, Perry LG, Gilroy S, Vivanco JM (2006) The role of root exudates in rhizosphere interactions with plants and other organisms. Annu Rev Plant Biol 57:233–266

Bakhshandeh E, Pirdashti H, Lendeh KS (2017) Phosphate and potassium-solubilizing bacteria effect on the growth of rice. Ecol Eng 103(1):164–169

Barbagelata PA, Mallarino AP (2012) Field correlation of potassium soil test methods based on dried and field-moist soil samples for corn and soybean. Nutr Manage Soil Plant Anal 77:318–327

Barea JM, Pozo MJ, Azcón R, Azcón-Aguilar C (2005) Microbial cooperation in the rhizosphere. J Exp Bot 56:1761–1778
Barrios E (2007) Soil biota, ecosystem services and land productivity. Ecol Econ 64:269–285
Baus D (2017) Overpopulation and the impact on the environment. CUNY Academic Works. http://academicworks.cuny.edu/gc_etds/1906
Bearden BN, Petersen L (2000) Influence of arbuscular mycorrhizal fungi on soil structure and aggregate stability of a vertisol. Plant Soil 218:173–183
Berendsen RL, Pieterse CMJ, Bakker PAHM (2012) The rhizosphere microbiome and plant health. Trends Plant Sci 17:478–486
Bhardwaj D, Ansari MW, Sahoo RK, Tuteja N (2014) Biofertilizers function as key player in sustainable agriculture by improving soil fertility, plant tolerance and crop productivity. Microb Cell Fact 13:66
Bhattacharyya PN, Jha DK (2012) Plant growth-promoting rhizobacteria (PGPR): emergence in agriculture. World J Microbiol Biotechnol 28:1327–1350
Bianchi FJJA, Booij CJH, Tscharntke T (2006) Sustainable pest regulation in agricultural landscapes: a review on landscape composition, biodiversity and natural pest control. Proc R Soc London Ser B 273:1715–1727
Bloemberg GV, Lugtenberg BJJ (2001) Molecular basis of plant growth promotion and biocontrol by rhizobacteria. Curr Opin Plant Biol 4:343–350
Boiffin J, Malezieux E, Picard D (2001) Cropping systems for the future. In: Nösberger J, Geiger HH, Struik PC (eds) Crop science: progress and prospects. CABI Publishing, New York, pp 261–279
Broz AK, Broeckling CD, De-la-Peña C, Lewis MR, Greene E, Callaway RM, Sumner LW, Vivanco JM (2010) Plant neighbor identity influences plant biochemistry and physiology related to defense. BMC Plant Biol 10:115
Brussaard CP (2004) Optimization of procedures for counting viruses by flow cytometry. Appl Environ Microbiol 70:1506–1513
Castellano MJ, Mueller KE, Olk DC, Sawyer JE, Six J (2015) Integrating plant litter quality, soil organic matter stabilization, and the carbon saturation concept. Glob Change Biol 21:3200–3209
Chen G, Zhu H, Zhang Y (2003) Soil microbial activities and carbon and nitrogen fixation. Res Microbiol 154:393–398
Chin-A-Woeng TFC, Bloemberg GV, Mulders IHM, Dekkers LC, Lugtenberg BJJ (2000) Root colonization by Phenazine-1-carboxamide-producing bacterium *Pseudomonas chlororaphis* PCL1391 is essential for biocontrol of tomato foot and root rot. Mol Plant Microbe Interact 13:1340–1345
Darnajoux R, Zhang X, McRose DL, Miadlikowska J, Lutzoni F, Kraepiel AM, Bellenger JP (2017) Biological nitrogen fixation by alternative nitrogenases in boreal cyanolichens: importance of molybdenum availability and implications for current biological nitrogen fixation estimates. New Phytol 213:680–689
Dineshkumar R, Kumaravel R, Gopalsamy J, Sikder MNA, Sampathkumar P (2017) Microalgae as bio-fertilizers for rice growth and seed yield productivity. Waste Biomass Valoriz:1–8
Dobbelaere S, Vanderleyden J, Okon Y (2003) Plant growth-promoting effects of diazotrophs in the rhizosphere. Crit Rev Plant Sci 22:107–149
Dong Z, Layzell DB (2001) H_2 oxidation, O_2 uptake and CO_2 fixation in hydrogen treated soils. Plant Soil 229:1–12
Doran JW (1980) Soil microbial and biochemical changes associated with reduced tillage. Soil Sci Soc Am J 44(4):765–771
Doran JW, Safley M (1997) Defining and assessing soil health and sustainable productivity. In: Pankhurst C, Doube BM, Gupta VVSR (eds) Biological indicators of soil health. CAB International, Wallingford, pp 1–28
Duca D, Lorv J, Patten CL, Rose D, Glick BR (2014) Indole-3-acetic acid in plant-microbe interactions. Antonie Van Leeuwenhoek 106:85–125

Egamberdieva D, Kamilova F, Validov S, Gafurova L, Kucharova Z, Lugtenberg B (2008) High incidence of plant growth-stimulating bacteria associated with the rhizosphere of wheat grown on salinated soil in Uzbekistan. Environ Microbiol 10:1–9

Egamberdieva D, Reckling M, Wirth S (2017) Biochar-based *Bradyrhizobium* inoculum improves growth of lupin (*Lupinus angustifolius* L.) under drought stress. Eur J Soil Biol 78:38–42

El-Azim A, WM K, Rania MR, Badawy MYM (2017) Effect of bio-fertilization and different licorice extracts on growth and productivity of *Foeniculum vulgare*. Mill Plant Middle East J 6:1–12

Farwig N, Bailey D, Bochud E, Herrmann JD, Kindler E, Reusser N, Schüepp C, Schmidt-Entling MH (2009) Isolation from forest reduces pollination, seed predation and insect scavenging in Swiss farmland. Landsc Ecol 24:919–927

Felestrino EB, Santiago IF, Silva F Ld, Rosa LH, Ribeiro SP, Moreira LM (2017) Plant Growth promoting bacteria associated with Langsdorffia hypogaea-rhizosphere-host biological interface: a neglected model of bacterial prospection. Front Microbiol. https://doi.org/10.3389/fmicb.2017.001722

Fu L, Penton CR, Ruan Y, Shen Z, Xue C, Li R, Shen Q (2017) Inducing the rhizosphere microbiome by biofertilizer application to suppress banana Fusarium wilt disease. Soil Biol Biochem 104:39–48

Gallon JR, Chaplin AE (1987) An introduction to nitrogen fixation. Cassel Educational Limited, London

Gamalero E, Glick BR (2015) Bacterial modulation of plant ethylene levels. Plant Physiol 169:13–22

Glick BR (2014) Bacteria with ACC deaminase can promote plant growth and help to feed the world. Microbiol Res 169:30–39

Gontia-Mishra I, Sapre S, Sharma A, Tiwari S (2016) Alleviation of mercury toxicity in wheat by the interaction of mercury-tolerant plant growth-promoting rhizobacteria. J Plant Growth Regul 35:1000–1012

Gurr GM, Wratten SD, Luna JM (2003) Multi-function agricultural biodiversity: pest management and other benefits. Basic Appl Ecol 4:107–116

Gutjahr C, Parniske M (2013) Cell and developmental biology of arbuscular mycorrhiza symbiosis. Annu Rev Cell Dev Biol 29:593–617

Hadi A, Nur HS (2017) The use of biofertilizers increased plant growth with no tradeoff effect on greenhouse gas emissions. J Wetlands Environ Manag 5:18–23

Haldar S, Sengupta S (2015) Plant-microbe cross-talk in the rhizosphere: insight and biotechnological potential. Open Microbiol 9:1–7. https://doi.org/10.2174/1874285801509010001

Hamilton EW, Frank DA (2001) Can plants stimulate soil microbes and their own nutrient supply? Evidence from a grazing tolerant grass. Ecology 82:2397–2402

Hernandez G (2000) Biological nitrogen fixation and sustainable agriculture. In: Fina TM, O'Brian MR, Layzell DB, Vessey JK, Newton W (eds) Nitrogen fixation, global perspectives. CABI Publishing, New York, pp 339–340

Høgh-Jensen H, Schjoerring JK (2001) Rhizodeposition of nitrogen by red clover, white clover and ryegrass. Soil Biol Biochem 33:439–448

Irvine P, Smith M, Dong Z (2003) Hydrogen fertilizer: bacteria or fungi? ISHS Acta Hortic 631:239–242

Johnston AE, Poulton PR, Coleman K (2009) Soil organic matter: its importance in sustainable agriculture and carbon dioxide fluxes. Adv Agron 101:1–57

Kaiser C, Kilburn MR, Clode PL, Fuchslueger L, Koranda M, Cliff JB, Solaiman ZM, Murphy DV (2015) Exploring the transfer of recent plant photosynthates to soil microbes: mycorrhizal pathway vs direct root exudation. New Phytol 205:1537–1551

Kaschuk G, Alberton O, Hungria M (2010) Three decades of soil microbial biomass studies in Brazilian ecosystems: lessons learned about soil quality and indications for improving sustainability. Soil Biol Biochem 42:1–13

Khan MY, Haque MM, Molla AH, Rahman MM, Alam MZ (2017) Antioxidant compounds and minerals in tomatoes by Trichoderma-enriched biofertilizer and their relationship with the soil environments. J Integr Agric 16:691–703

Kibblewhite MG, Ritz K, Swift MJ (2008) Soil health in agricultural systems. Philos Trans R Soc London Ser B 363:685–701. https://doi.org/10.1098/rstb.2007.2178

King RW, Evans LT (2003) Gibberellins and flowering of grasses and cereals: prizing open the lid of the "florigen" black box. Annu Rev Plant Biol 54:307–328

Kohler J, Roldán A, Campoy M, Caravaca F (2016) Unraveling the role of hyphal networks from arbuscular mycorrhizal fungi in aggregate stabilization of semiarid soils with different textures and carbonate contents. Plant Soil:1–9

Krithika S, Balachandar D (2016) Expression of zinc transporter genes in rice as influenced by zinc-solubilizing *Enterobacter cloacae* strain ZSB14. Front Plant Sci 7:446

Kumar P, Dubey RC (2012) Plant Growth Promoting Rhizobacteria for biocontrol of phytopathogens and yield enhancement of Phaseolus vulgaris. J Curr Perspect App Microbiol 1:6–38

Kumar A, Maurya BR, Raghuwanshi R, Meena VS, Islam MT (2017) Co-inoculation with Enterobacter and rhizobacteria on yield and nutrient uptake by wheat (*Triticum aestivum* L.) in the alluvial soil under Indo-gangetic plain of India. J Plant Growth Regul:1–10

Lagopodi AL, Ram AFJ, Lamers GEM, Punt JP, Van den Hondel CAMJJ, Lugtenberg BJJ, Bloemberg GV (2002) Novel aspects of tomato root colonization and infection by *Fusarium oxysporum* F. sp. radicislycopersici revealed by confocal laser scanning microscopic analysis using the green fluorescent protein as a marker. Mol Plant-Microbe Interact 15:172–179

Lareen A, Burton F, Schäfer P (2016) Plant root-microbe communication in shaping root microbiomes. Plant Mol Biol 90:575–587. https://doi.org/10.1007/s11103-015:0417-8

Le Roux X, Barbault R, Baudry J, Burel F, Doussan I, Garnier E, Herzog F, Lavorel S, Lifran R, Roger-Estrade J, Sarthou JP, Trommetter M (2008) Agriculture et biodiversité-Valoriser les synergies. Expertise scientifique collective, INRA, France

Leifheit EF, Veresoglou SD, Lehmann A, Morris EK, Rillig MC (2014) Multiple factors influence the role of arbuscular mycorrhizal fungi in soil aggregation a meta-analysis. Plant Soil 374:523–537

Li M, Ahammed GJ, Li C, Bao X, Yu J, Huang C, Yin H, Zhou J (2016) Brassinosteroid ameliorates zinc oxide nanoparticles-induced oxidative stress by improving antioxidant potential and redox homeostasis in tomato seedling. Front Plant Sci 7:615

Liu D, Lian B, Dong H (2012) Isolation of *Paeni bacillus* sp. and assessment of its potential for enhancing mineral weathering. Geomicrobiol J 29(5):413–421

Lugtenberg BJJ, Dekkers L, Bloemberg GV (2001) Molecular determinants of rhizosphere colonization by Pseudomonas. Annu Rev Phytopathol 39:461–490

M'piga PM, Bélanger RR, Paulitz TC, Benhamou N (1997) Increased resistance to *Fusarium oxysporum* f. sp. radicislycopersici in tomato plants treated with the endophytic bacterium Pseudomonas fluorescens strain 63-28. Physiol Mol Plant Pathol 50:301–320

Mahajan MS, Tan MW, Rahme LG, Ausubel FM (1999) Molecular mechanisms of bacterial virulence elucidated using a Pseudomonas aeruginosa Caenorhabditis elegans pathogenesis model. Cell 96:47–56

Mathesius U (2003) Conservation and divergence of signaling pathways between roots and soil microbes-the Rhizobium-legume symbiosis compared to the development of lateral roots, mycorrhizal interactions and nematode induced galls. Plant Soil 255:105–119

Meena RK, Singh RK, Singh NP, Meena SK, Meena VS (2016) Isolation of low temperature surviving plant growth-promoting rhizobacteria (PGPR) from pea (*Pisum sativum* L.) and documentation of their plant growth promoting traits. Biocatal Agric Biotechnol 4:806–811

Meena KK, Sorty AM, Bitla UM, Choudhary K, Gupta P, Pareek A, Singh DP, Prabha R, Sahu PK, Gupta VK, Singh HB (2017a) Abiotic stress responses and microbe-mediated mitigation in plants: the omics strategies. Front Plant Sci 8:172

Meena VS, Mishra PK, Bisht JK, Pattanayak A (2017b) Agriculturally important microbes for sustainable agriculture, Vol. 2: Applications in crop production and protection. Springer, Singapore. https://doi.org/10.1007/978-981-10-5343-6

Mendes R, Garbeva P, Raaijmakers JM (2013) The rhizosphere microbiome: significance of plant beneficial, plant pathogenic, and human pathogenic microorganisms. FEMS Microbiol Rev 37:634–663

Miller RM, Jastrow JD (2000) Mycorrhizal fungi influence soil structure. In: Arbuscular mycorrhizas: physiology and function. Springer, Dordrecht, pp 3–18

Mishra I, Sapre S, Kachare SA, Tiwari S (2017) Molecular diversity of 1-aminocyclopropane-1-carboxylate (ACC) deaminase producing PGPR from wheat (*Triticum aestivum* L.) rhizosphere. Plant Soil 414:213–227

Mohamadi P, Razmjou J, Naseri B, Hassanpour M (2017) Population growth parameters of *Tutaabsoluta* (*Lepidoptera: Gelechiidae*) on tomato plant using organic substrate and biofertilizers. J Insect Sci (Online) 17:36

Moonen AC, Bàrberi P (2008) Functional biodiversity: an agroecosystem approach. Agric Ecosyst Environ 127:7–21

Mukhongo RW, Tumuhairwe JB, Ebanyat P, AbdelGadir AH, Thuita M, Masso C (2017) Combined application of biofertilizers and inorganic nutrients improves sweet potato yields. Front Plant Sci 8

Nath D, Maurya BR, Meena VS (2017) Documentation of five potassium and phosphorus solubilizing bacteria for their K and P-solubilization ability from various minerals. Biocatal Agric Biotechnol 10:174–181

Nguyen TH, Phan TC, Choudhury AT, Rose MT, Deaker RJ, Kennedy IR (2017) BioGro: A plant growth-promoting biofertilizer validated by 15 years' research from laboratory selection to rice farmer's fields of the Mekong Delta. In: Agro- environmental sustainability. Springer, pp 237–254

Oliveira WS, Coelho IL, Oliveira JRS, Leite MCBS, Arnaud TMS, Stamford NP, Silva EVN, da Silva VSG, de Oliveira MW, de Oliveira DC, Oliveira TBA (2017) Biological control of the bacterial wilt *Ralstonia solanacearum* by bioprotector with fungi chitosan from *Cunninghamella elegans* on tomatoes. Afr J Agric Res 12:42–49

Ortíz-Castro R, Valencia-Cantero E, López-Bucio J (2008) Plant growth promotion by Bacillus megaterium involves cytokinin signalling. Plant Signal Behav 3:263–265

Oteino N, Lally RD, Kiwanuka S, Lloyd A, Ryan D, Germaine KJ, Dowling DN (2015) Plant growth promotion induced by phosphate solubilizing endophytic Pseudomonas isolates. Front Microbiol 6:745

Parmar P, Sindhu SS (2013) Potassium solubilization by rhizosphere bacteria: influence of nutritional and environmental conditions. J Microbiol Res 3:25–31

Parniske M (2008) Arbuscular mycorrhiza: the mother of plant root endosymbioses. Nat Rev Microbiol 6:763–775

Patra MK, Dash BK, Mahalik JK (2017) Effect of bio-fertilizers with graded doses of NPK on Tomato (cv. Pusa ruby) infected by *Meloidogyne incognita*. Ann Plant Prot Sci 25:190–194

Pellegrino E, Bedini S (2014) Enhancing ecosystem services in sustainable agriculture: biofertilization and biofortification of chickpea (*Cicer arietinum* L.) by arbuscular mycorrhizal fungi. Soil Biol Biochem 68:429–443

Pérez-Montaño F, Alías-Villegas C, Bellogín RA, del Cerro P, Espuny MR, Jiménez-Guerrero I, López-Baena FJ, Ollero FJ, Cubo T (2014) Plant growth promotion in cereal and leguminous agricultural important plants: from microorganism capacities to crop production. Microbiol Res 169:325–336

Peterson CA, Emanuel ME, Humphreys GB (1981) Pathway of movement of apoplastic fluorescent dye tracers through the endodermis at the site of secondary root formation in corn (*Zea mays*) and broad bean (*Vicia faba*). Can J Bot 59:618–625

Pieterse CM, Van der Does D, Zamioudis C, Leon-Reyes A, Van WSC (2012) Hormonal modulation of plant immunity. Annu Rev Cell Dev Biol 28:489–521

Prakash S, Verma JP (2016) Global perspective of potash for fertilizer production. In: Meena VS, Maurya BR, Verma JP, Meena RS (eds) Potassium solubilizing microorganisms for sustainable agriculture. Springer, New Delhi, pp 327–331. https://doi.org/10.1007/978-81-322-2776-2

Querne A, Battie-laclau P, Dufour L, Wery J, Dupraz C (2017) Effects of walnut trees on biological nitrogen fixation and yield of intercropped alfalfa in a Mediterranean agroforestry system. Eur J Agron 84:35–46

Radja CR, Nandakumar R, Kandan A, Suresh S, Bharathi M, Raguchander T, Samiyappan R (2002) *Pseudomonas fluorescens* based bio-formulation for the management of sheath blight disease and leaf folder insect in rice. Crop Prot 21:671–677

Rillig M, Mummey DL (2006) Mycorrhizas and soil structure. New Phytol 171:41–53

Rillig MC, Aguilar-Trigueros CA, Bergmann J, Verbruggen E, Veresoglou SD, Lehmann A (2015) Plant root and mycorrhizal fungal traits for understanding soil aggregation. New Phytol 205:1385–1388

Sabbagh SK, Poorabdollah A, Sirousmehr A, Gholamalizadeh-Ahangar A (2017) Biofertilizers and systemic acquired resistance in Fusarium infected wheat. J Agric Sci Technol 19:453–464

Santoyo G, Orozco-Mosqueda MD, Govindappa M (2012) Mechanisms of biocontrol and plant growth-promoting activity in soil bacterial species of Bacillus and Pseudomonas: a review. Biocontrol Sci Technol 22(8):855–872

Sarathambal C, Thangaraju M, Paulraj C, Gomathy M (2010) Assessing the zinc solubilization ability of *Gluconacetobacter diazotrophicus* in maize rhizosphere using labelled 65Zn compounds. Indian J Microbiol 50:103–S109

Setiawati TC, Mutmainnah L (2016) Solubilization of potassium containing mineral by microorganisms from sugarcane rhizosphere. Agric Sci Procedia 9:108–117

Shahbaz M, Kuzyakov Y, Sanaullah M, Heitkamp F, Zelenev V, Kumar A, Blagodatskaya E (2017) Microbial decomposition of soil organic matter is mediated by quality and quantity of crop residues: mechanisms and thresholds. Biol Fertil Soils 1:15

Sharma SB, Sayyed RZ, Trivedi MH, Gobi TA (2013) Phosphate solubilizing microbes: sustainable approach for managing phosphorus deficiency in agricultural soils. Springer Plus 2:587

Simarmata T, Turmuktini T, Fitriatin BN, Setiawati MR (2016) Application of bioameliorant and biofertilizers to increase the soil health and rice productivity. HAYATI J Biosci 23:181–184

Singh DP, Prabha R, Yandigeri MS, Arora DK (2011) Cyanobacteria-mediated phenylpropanoids and phytohormones in rice (*Oryza sativa*) enhance plant growth and stress tolerance. Antonie Van Leeuwenhoek 100:557–568

Singh DP, Singh HB, Prabha R (eds) (2016) Microbial inoculants in sustainable agricultural productivity. Vol. 1: Research perspectives. Springer, p 343

Sprent JI (2001) Nodulation in Legumes. Royal Botanic Gardens, Kew, UK. Staples, R. C. 2003. A novel gene for rust resistance. Trends Plant Sci 8:149–151

Steenhoudt O, Vanderleyden J (2000) *Azospirillum*, a free-living nitrogen-fixing bacterium closely associated with grasses: genetic, biochemical and ecological aspects. FEMS Microbiol Rev 24:487–506

Stoate C, Boatman ND, Borralho RJ, Carvalho CR, de Snoo GR, Eden P (2001) Ecological impacts of arable intensification in Europe. J Environ Manag 63:337–365

Stone JM, Heard JE, Asai T, Ausubel FM (2000) Simulation of fungal-mediated cell death by fumonisin B1 and selection of fumonisin B1–resistant (fbr) Arabidopsis mutants. Plant Cell 12(10):1811–1822

Stover CK, Pham XQ, Erwin AL, Mizoguchi SD, Warrener P, Hickey MJ, Brinkman FS, Hufnagle WO, Kowalik DJ, Lagrou M, Garber RL (2000) Complete genome sequence of *Pseudomonas aeruginosa* PAO1, an opportunistic pathogen. Nature 406:959–964

Sturz AV, Christie BR (1996) Endophytic bacteria of red clover as agents of allelopathic clover-maize syndromes. Soil Biol Biochem 28:583–588

Sturz AV, Christie BR, Matheson BG (1998) Associations of bacterial endophyte populations from red clover and potato crops with potential for beneficial allelopathy. Can J Microbiol 44:162–167

Sturz AV, Arsenault W, Christie BR (2003) Red clover–potato cultivar combinations for improved potato yield. Agron J 95(5):1089–1092

Taktek S, St-Arnaud M, Piché Y, Fortin JA, Antoun H (2017) Igneous phosphate rock solubilization by biofilm-forming mycorrhizobacteria and hyphobacteria associated with *Rhizoglomus irregulare* DAOM 197198. Mycorrhiza 27:13–22

Tamasloukht M, Séjalon-Delmas N, Kluever A, Jauneau A, Roux C, Bécard G, Franken P (2003) Root factors induce mitochondrial-related gene expression and fungal respiration during the developmental switch from asymbiosis to presymbiosis in the arbuscular mycorrhizal fungus *Gigaspora rosea*. Plant Physiol 131:1468–1478

Tripti KA, Usmani Z, Kumar V (2017) Biochar and flyash inoculated with plant growth promoting rhizobacteria act as potential biofertilizer for luxuriant growth and yield of tomato plant. J Environ Manag 190:20–27

Trivedi P, Delgado-Baquerizo M, Anderson IC, Singh BK (2016) Response of soil properties and microbial communities to agriculture: implications for primary productivity and soil health indicators. Front Plant Sci 7:990

Uratsu SL, Keyser HH, Weber DF, Lim ST (1982) Hydrogen uptake (HUP) activity of *Rhizobium japonicum* from major U.S. soybean production areas. Crop Sci 22:600–602

van der Wal A, de Boer W (2017) Dinner in the dark: illuminating drivers of soil organic matter decomposition. Soil Biol Biochem 105:45–48

Van Loon LC, Bakker PAHM, Pieterse CMJ (1998) Systemic resistance induced by rhizosphere bacteria. Annu Rev Phytopathol 36:453–483

Van Wees SCM, Swart de EAM, van Pelt JA, van Loon LC, Pieterse CMJ (2000) Enhancement of induced disease resistance by simultaneous activation of salicylate- and jasmonate dependent defense pathways in *Arabidopsis thaliana*. Proc Natl Acad Sci 97:8711–8716

Velazquez E, Silva LR, Ramírez-Bahena MH, Peix (2016) Diversity of potassium-solubilizing microorganisms and their interactions with plants. In: Meena VS, Maurya BR, Verma JP, Meena RS (eds) Potassium solubilizing microorganisms for sustainable agriculture. Springer, New Delhi, pp 99–110. https://doi.org/10.1007/978-81-322-2776-2

Verma JP, Yadav J, Tiwari KN, Jaiswal DK (2014) Evaluation of plant growth promoting activities of microbial strains and their effect on growth and yield of chickpea (*Cicer arietinum* L.) in India. Soil Biol Biochem 70:33–37

Verma R, Maurya BR, Meena VS, Dotaniya ML, Deewan P, Jajoria M (2017) Enhancing production potential of cabbage and improves soil fertility status of Indo-Gangetic Plain through application of bio-organics and mineral fertilizer. Int J Curr Microbiol Appl Sci 6(3):301–309

Wallenstein MD, Weintraub MN (2008) Emerging tools for measuring and modeling the in situ activity of soil extracellular enzymes. Soil Biol Biochem 40:2098–2106

Wang C, Knill E, Glick BR, Défago G (2000) Effect of transferring 1-aminocyclopropane-1-carboxylic acid (ACC) deaminase genes into *Pseudomonas fluorescens* strain CHAO and its gacA derivative CHA96 on their growth-promoting and disease-suppressive capacities. Can J Microbiol 46:898–907

Wang YP, Law RM, Pak B (2010) A global model of carbon, nitrogen and phosphorus cycles for the terrestrial biosphere. Bio Geosciences 7:2261–2282

Warner KD (2007) The quality of sustainability: agroecological partnerships and the geographic branding of California winegrapes. J Rural Stud 23:142–155

World Bank (2008) World development report: agriculture for development. Washington DC

Yadav SK, Dave A, Sarkar A, Singh HB, Sarma BK (2013) Co-inoculated biopriming with Trichoderma, Pseudomonas and Rhizobium improves crop growth in *Cicer arietinum* and Phaseolus vulgaris. Int J Agric Environ Biotechnol 6:255–259

Yang J, Kloepper JW, Ryu CM (2009) Rhizosphere bacteria help plants tolerate abiotic stress. Trends Plant Sci 14:1–4

Yilmaz E, Sonmez M (2017) The role of organic/biofertilizer amendment on aggregate stability and organic carbon content in different aggregate scales. Soil Tillage Res 168:118–124

Zhang H, Kim MS, Sun Y, Dowd SE, Shi H, Paré PW (2008) Soil bacteria confer plant salt tolerance by tissue-specific regulation of the sodium transporter HKT1. Mol Plant Microbe Interact 21:737–744

Zwanenburg B, Pospíšil T (2013) Structure and activity of strigolactones: new plant hormones with a rich future. Mol Plant 6:38–62

8 Emerging Insights on Rhizobacterial Functions

P. Jishma and E. K. Radhakrishnan

Abstract

Plant-rhizobacterial interaction is one of the complex bio-communications in the environment and is highly significant since plants are the primary producers on earth. The rhizosphere region is known to be a multifaceted environment remarkable for the various types of processes mediated by a wide array of biologically active molecules of both plant and microbial origin. Due to these, the design of the rhizosphere architecture can be determined by many factors, and a deeper understanding on the same will enable to modulate the functions related to plant growth and development. The signaling molecules produced by rhizobacteria can induce many beneficial changes in plant system including the enhancement of the nutrient uptake by plants, growth hormone production, stress tolerance, and protection from many pathogens. In addition to this, plant growth-promoting rhizobacteria (PGPR) can remove the heavy metals and detoxify the pesticides present in the contaminated soil. Hence, the exploration of PGPR can be a step toward conserving the greener environment, and for this, deeper insight into the signaling and communications that happen belowground is important.

Keywords

Rhizobacteria · Rhizosphere · Plant growth · Nutrients · Disease management

P. Jishma · E. K. Radhakrishnan (✉)
School of Biosciences, Mahatma Gandhi University, Kottayam, Kerala, India
e-mail: radhakrishnanek@mgu.ac.in

V. Kumar et al. (eds.), *Microbiome in Plant Health and Disease*,
https://doi.org/10.1007/978-981-13-8495-0_8

8.1 Introduction

8.1.1 Rhizosphere and Importance of Rhizobacteria

Rhizosphere is the nutrient pool in soil where the plant root is consistently in touch with plant-associated microorganisms. The active functioning of this region results in beneficial impact on plant growth (Vejan et al. 2016). Rhizosphere is also considered as a naturally constructed microbial treasure spot in the ecosystem, where the most actively functioning microorganisms can be observed. The soil environment, plant system, and microbiome interact with each other to influence the physiological, chemical, and biological functions. A vast variety of microbes inhabit at the rhizosphere, in which bacteria are the well studied. They exhibit symbiotic or non-symbiotic relation with plants, as determined by the chemico- biological communications involved (Gouda et al. 2018). The plant root system acts like a chemical factory from where the phenolic compounds are liberated to the surroundings to modulate the belowground interactions. The discharge of mucilage liberates a huge amount of active molecules into the zone of rhizosphere, which include the cell wall polymers of plants such as cellulose and pectin. These compounds along with sugars and organic acids can have a role as chemical attractants for a large number of diverse species of microbial communities. The composition of exudates depends upon the physiological status, species, and stages of plant growth (Kang et al. 2010). Here, the specific molecules discharged from the root are considered to play a role in regulating the microbial community structure. Several reports on the rhizomicrobiome and rhizodeposits of different plant species indicate the exudate molecules to have a critical role in designing the plant-microbe communication (Barriuso et al. 2008; Lareen et al. 2016).

Soil microbial communities perform key roles in cycling of carbon (C) and other nutrients to maintain plant growth (Bulgarelli et al. 2013). The deposited carbon compounds like organic acids and sugars at rhizosphere does not always promote the recruitment of beneficial microorganisms itself, as the same can be utilized as growth substrates by pathogens in the soil. However, plants can also be hypothesized to have an identification system to differentiate advantageous microbes from harmful ones. At present, many beneficial rhizomicrobiomes are at the end of risk due to fluctuating climate, soil degradation, and poor land management practices (Amundson et al. 2015).

8.1.2 Rhizobacteria: The Second Genome of Plants

The total number of microorganisms colonized in plants can attain cell population which can be much larger than the number of plant cells. Similarly, the number of rhizobacterial genes functioning collectively to provide beneficial effect to plant can be higher than that of plant genes itself. The whole plant-associated microbial communities are collectively known as the plant microbiome and it functionally act as the second genome of plants. In this perspective, plants can be observed as super living systems that rely in part on their microbiome for specific functions and characteristics (Mendes et al. 2013). Since the 2000s, research focused towards in-depth documentation of the great quantity and variety of the rhizomicrobiome by using advanced techniques like metagenomics. Reports from sequencing studies have shown the rhizosphere to be a niche of hotspot of biological activities, with roots of plants supporting a vast variety of microbial taxa (Bulgarelli et al. 2013). Since the microbiome strongly influences plant functions and its genome, it can be considered as a plant second genome (Turner et al. 2013).

The diversity of rhizospheric bacteria can be modulated by plant root exudates, and these bacteria can effectively influence the plant growth by producing regulatory molecules. Thus, rhizomicrobial system is regarded as a well-designed, exterior and efficient surrounding for plants (Philippot et al. 2013; Spence et al. 2014) with a function as plants second genome (Berendsen et al. 2012). Metagenomic approaches have also made it possible to study the plant-associated microorganisms at their environmental status. Plants harbor their own microbiome, and the region of rhizosphere contains diverse microbial density, and it forms a functionally significant part of microbiome (Mendes et al. 2011).

8.2 PGPR for Promotion of Plant Growth and Health Protection

Plant growth-promoting rhizobacteria (PGPR) are the various communities of bacteria inhabited in the rhizosphere region (Ahmad et al. 2008), and their colonization on the roots can be considered to augment plant growth. The presence of root exudates in the rhizosphere, which can act as a nutrient source for the growth of microorganisms, enhances the microbial diversity of the rhizosphere than its surrounding soil (Igiehon and Babalola 2018). Different soil types can differ in its physicochemical nature like structure and texture, organic matter content, pH, and nutrient composition. These characteristics of soil can have determining effect on rhizomicrobiome by generating and encouraging surroundings that support certain species of microorganisms. Hence, the regulation of composition of root exudates can affect the selection of microorganisms by plants (Berendsen et al. 2012).

Rhizosphere-associated plant growth-promoting bacteria (PGPB) are observed in all agro-ecosystems (Saraf et al. 2014) and are metabolically active as they enhance the plant growth through different mechanisms. The vital action of PGPR on plants enhances growth, enables tolerance to abiotic stress, supports nutrient availability and absorption, modulates growth regulators of plants, and mediates the degradation of environmental pollutants (Choudhary et al. 2011; Liu et al. 2016). However, the action of different species of PGPR on plants varies with the type and species of the host (Glick 2012; Shailendra Singh 2015; Vejan et al. 2016).

PGPR augment the plant growth indirectly through the inhibition of phytopathogens by producing antagonistic compounds. Antibiotics, biocidal volatiles, lytic enzymes, and detoxification enzymes and siderophores are general biochemicals synthesized and released by PGPR. Siderophores chelate iron available from the soil and make this available to the plant, antibiotics discourage pathogenic microbial colonization and biofilm formation, lytic enzymes lyse various organic compounds including chitin in the cell wall of pathogenic fungi, and detoxification enzymes neutralize the action of toxins released by pathogens and inhibit the damages caused to plants. Volatiles like hydrogen cyanide released from PGPR suppress the functions of fungal pathogens. This potential of PGPR to effectively compete with various pathogens for nutrients or specific niches on the root, and further induction of induced systemic resistance (ISR), is highly remarkable (Compant et al. 2005).

8.3 Rhizomicrobiome Recruitment

Plant root-rhizobacterial interaction and the subsequent functioning are beneficial to each other. Also, there exists an important selective colonization mechanism in the rhizosphere designed primarily by the plant itself. Selective recruitment of beneficial rhizobacteria by the plant is chemically mediated through the production of compounds which either attract the beneficial bacteria or repel antagonistic organisms (Cheng and Cheng 2015; Swamy et al. 2016). In plant-rhizobacterial interaction, both the partners produce signal molecules which are deposited to the rhizosphere. However, root exudates are the key determinants of rhizosphere microbial population (Badri et al. 2008). The carbon source limits the growth of microbes in the soil and is communicated to plants as “rhizosphere effect,” which thereby releases carbon-containing compounds known as rhizodeposits. These include a wide array of molecules that originate from mucilages, volatiles, sloughed-off root cells and tissues, and soluble lysates and are released from damaged and intact cells (Dennis et al. 2010). These compounds modify the physicochemical properties of the rhizospheric area, thereby recruiting selected bacteria with the promises to enhance plant growth and health, tolerance to stress, and defense responses to guard the plant from microbial infections and pest attack (Dennis et al. 2010). The potential of soil bacteria to meet these plant requirements enables them to get recruited selectively to the specific rhizosphere from the soil reserve.

8.4 Chemotaxis: Communication Between Plant Roots and Rhizosphere Microbiome

The exudates released from the roots modulate the complex biomolecular communications between the rhizomicrobiome and the plant root cells. These compounds build a communication network with rhizomicrobiome and plant roots through many physicochemical or biological interactions in the soil (Huang et al. 2014). High energy is required for the secretion and release of rhizodeposits from plant cells to the surrounding soil. Primarily, in the process of chemotaxis, carbon-containing root exudates with low molecular weight derived from photosynthates are released into the rhizosphere, and a nutrient gradient is developed in the soil.

This gradient chemotactically attract various species of motile bacteria (Scharf et al. 2016). Also, these rhizodeposits act as potent chemical messengers to make possible the chemotaxis process of rhizobacteria and mediate biological communications through various complex molecular networks (Xie et al. 2012; Haichar et al. 2014).

The attraction of rhizobacteria toward plant roots is mediated by exudates and is commonly known as chemotaxis. This is the first step in establishment and subsequent colonization of rhizobacteria on roots (Begonia and Kremer 1994). Sensing the specific ligands through methyl-accepting chemotaxis proteins (MCP) initiated the studies on chemotaxis-mediated reaction to root exudates. This is most significant for colonization of rhizobacteria on plant roots and has beneficial impact on functions on PGPR. Systematic recognition of chemoattractants in multiple root deposits and sensing by chemoreceptors in PGPR important to augment their selective recruitment and successful colonization (Feng et al. 2018). The whole genome sequencing of various species of bacteria associated with plant roots indicate the existence of multiple chemotaxis functional systems and several numbers of chemoreceptors. The ability of rhizobacteria to get attracted towards roots due to the exuded compounds and its ability to grow quickly are significant characters that facilitate a bacterial species to be aggressive in the rhizosphere.

For the successful proliferation and establishment of bacteria in the rhizosphere, they must have the potential to utilize root exudates and compete to survive with other microorganisms or surroundings and colonize on root or rhizosphere effectively. Motility is an important property for the effective movement of bacteria towards the root and its colonization (Dennis et al. 2010). As previously mentioned, chemotaxis is an energy-requiring activity and so colonization property can be minimized if either flagella or ATP production is distracted (Dekkers et al. 1998). Bacterial growth rate is another important trait for successful colonization, which is potentially dependent on their ability to obtain compounds which are significant for growth and maintenance. Also, genes in rhizobacteria involved in the nutrient uptake from soil are connected with its growth rate (Dennis et al. 2010). Certain factors like environmental conditions, stages of growth, and species of plant may affect the composition of root exudates, which lead to the difference in bacterial recruitment in the rhizosphere and growth stimulation by specific members of microbial communities (Lakshmanan et al. 2014). Studies also indicate that plants of different species actively recruit specific communities of microbes to the rhizosphere, including those which support plant growth even under stressed conditions possibly through the modulation of root exudate composition (Badri et al. 2009).

8.5 Biofilm Formation and Quorum Sensing

Various species of beneficial soil bacteria including rhizobacteria produce microcolonies or biofilm on plant roots. In biofilms, the bacterial cells are embedded in an extracellular polymeric compound matrix bonded to a surface (Branda et al. 2005). The processes of auto-aggregation and development of biofilm by bacteria are significant to both its survival and colonization on the host plant (Fig. 8.1). In the

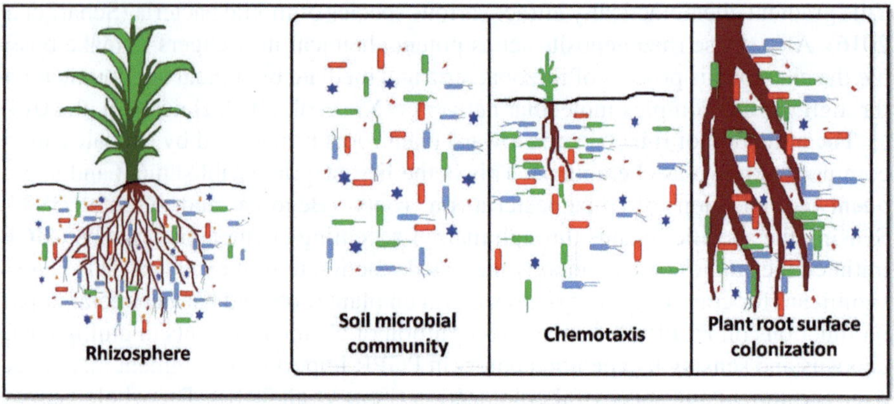

Fig. 8.1 Rhizobacterial chemotaxis and colonization on plant roots

interaction between plants and rhizobacteria, chemotaxis and adhesion of bacteria, colonization on plant, and cell-to-cell interaction are affected with genetic and environmental conditions. While attached on the surface of plant roots, rhizobacterial cells produce various extracellular polymeric substances (EPS) along with different exopolysaccharides, DNA, and proteins (Bogino et al. 2013). Biofilm cells are extremely reactive to numerous functions of their environment, and they alter their metabolic functions in response to nutrients and waste product gradients. They have specific cell-to-cell interaction and contact system with adjacent cells. The success of exploration and application of PGPR in agriculture is recognized to be dependent on their efficient colonization on plant roots (Bolwerk et al. 2003) and their subsequent ability to form microcolonies or biofilms. Hence, this can important role in effective existence for a successful plant-microbe interaction (Elsas et al. 2007).

In a beneficial plant-microbe interaction, PGPR colonization on plant roots develop into biofilm which provides protection against microbial pathogens especially soilborne. *Pseudomonas aeruginosa*, *Pseudomonas fluorescens*, *Paenibacillus polymyxa*, *Serratia* spp., and *Bacillus subtilis* develop biofilm, and the multicellular means of growth is likely to predominate in nature as a protecting shield against hostile environmental conditions. They mainly use quorum-sensing mechanisms to organize and control gene expression according to the local population density against specific microbial pathogens (Schuhegger et al. 2006; Moons et al. 2011). Moreover, the signaling system organizes and alters interaction between plant and rhizobacteria with regulated liberation of antibiotics and toxins.

In a bacterial community, the cells interact with each other to coordinate more functions rather than to live isolated from each other. This beneficial interaction is the basic mechanism of their survival and ability to adapt with the fluctuations in the environment (Fray 2002). Bacteria have constructed a complex network of signaling and communication to regulate the expression of specific genes dependent on cell density and is commonly called "quorum sensing" (QS) (Camara et al. 2002). Acylated homoserine lactone (AHL) is generally produced by Gram-negative bacteria as autoinducers. AHL are produced by a *LuxI*-type enzyme (signal synthase), which is encoded by the first gene of the lux operon (Nazzaro et al. 2013). Bacterial cells produce a basic level of AHL by using AHL synthase at low population density. AHLs are deposited in the soil when the cell density is enhanced, and after attaining a critical threshold concentration, the AHL molecule attaches to its associated receptor and, in turn, stimulates or suppresses the target gene expression (Steidle et al. 2001; Barriuso et al. 2008).

The bacterial population in the soil release specific metabolites into the environment to detect the nature of ecological niche like diffusion space and density and distribution of its population. This recognition mechanism favors the bacterial population to acclimatize with the conditions in the habitat by regulating their expression of genes and also their existence in the habitat (Hartmann et al. 2014). In addition to bacterial population, many plant species have been found to synthesize AHL-mimic substances or involved in activities stimulating QS of bacteria associated with plants (Gao et al. 2003). Flavonoids produced by leguminous plants enhance AHL synthesis gene expression in Rhizobia (Perez-Montano et al. 2011). Also, phytohormones such as indole acetic acid and cytokinin produced by *Gypsophila* were found to have the potential to stimulate and influence QS, type III secretion system, and development of the gall by *Pantoea agglomerans pv. gypsophilae* (Chalupowicz et al. 2009).

8.6 Rhizomicrobiome Functions

The recruited potential microbiome in the rhizosphere successfully establishes and beneficially acts on the plant system to support its growth and disease resistance. Rhizobacteria play a role in the enhancement of essential micronutrients in grains, improving the seed weight and yield, boosting the biomass of crop plants and leafy vegetables, reducing the application of chemical fertilizers in agricultural fields, modifying the soil nature, and inhibiting the growth and functions of various fungal and bacterial pathogens in plants by inducing systemic resistance and regulating the gene expression, which have important roles in signaling and pathogenesis-related pathways. Rhizobacteria can also degrade toxic organic pollutants and heavy metals in the polluted soil. These multiple functions of phyto-attracted rhizomicrobiome can be explored for various approaches like mineralization or solubilization of nutrients and supply to plants; production of phytohormones to support plant growth; designing of soil structure by reconstructing soil microbiome, nutrients, and other physiological characters; and production of hydrolytic enzymes to inhibit the survival and functioning of soilborne pathogens (Mathivanan et al. 2014). By modulating certain chemical properties such as content of organic matter, pH, and redox state, rhizobacteria can affect the metal bioavailability directly, leading to the leaching of contaminants from the soil. pH and organic matter of soil are the controlling factors of solubility of heavy metals which determine the bioavailability of heavy metals in the soil (Jing et al. 2007). Siderophores are the metal-chelating agents produced by rhizobacteria, which have significant function in the acquirement of heavy metals. These rhizobacterial siderophores play an important role in providing sufficient quantity of iron to the plants grown in the metal-polluted soil, which is iron deficient in nature. Such siderophores can chelate iron and regulate the availability of iron in the rhizosphere (Schalk et al. 2011).

8.6.1 PGPR-Mediated Crop Production Enhancement

Plant growth-promoting rhizobacteria have been applied in agricultural fields as a sustainable approach. Biological strengthening of crops for essential micronutrients is a feasible solution for the production of good quality crops. Sharma et al. (2014) have reported the potential of PGPR to increase zinc and iron deposition in the edible portion of rice (*Oryza sativa* L.) grains or endosperm. PGPR-mediated enhanced content of iron and zinc in the plants was also exhibited to associate with increased catalase and carbonic anhydrase activity. The activity of PGPR enhances the micronutrient deposition of the host plants directly or indirectly, and their suitable application has the potential to be explored for developing sustainable approach for biological strengthening of cereal grains (Sharma et al. 2014).

A previous study of Mathivanan et al. (2014) have reported the application of *Rhizobium* sp., *Pseudomonas* sp., and *Bacillus* sp. to enhance the number of seeds per plant, seed weight, and yield. Almost forty-five percent yield increase was observed with the combined inoculation of PGPR. Also, many studies have reported the application of PGPR to result in enhanced growth and yield of various types of plants including legumes.

Rhizosphere microbiome with plant probiotic potential including fertilizer mobilization has also been explored. In a previous study, rhizospheric *Pseudomonas fluorescens* was evaluated for its growth-promoting potential on commonly used leafy vegetable *Amaranthus tricolor* (L.) under field conditions for 1 month. *P. fluorescens*-treated *A. tricolor* exhibited increased growth traits such as leaf and root number, shoot length, and fresh weight. Also, the enhanced contents of nitrogen, phosphorus, and potassium improved soil fertility as a result of the treatment. Most remarkably, the inoculation of *P. fluorescens* alone and also with 50% of recommended NPK has revealed comparable growth of *A. tricolor* as that of full dose of NPK. Also, *P. fluorescens* combined with 50% NPK treatment enhanced the content of available nitrogen and phosphorus in the soil. This indicated the ability of selected rhizobacteria to enrich soil fertility and enhance crop productivity (Jimtha John et al. 2017).

Application of *Rhizobium* spp. and *Azospirillum* spp. has also shown to increase the plant tolerance to salinity conditions (Hamaoui et al. 2001). Also, *Azospirillum* spp. have the ability to provide drought tolerance in different species of plants. Treatment of *Azospirillum lipoferum* in wheat plants minimized the harmful effect of salinity in the soil. The application of PGPR can have tremendous promises in desert and semiarid areas, where drought stress reduces the growth of plants and crop production (Kramer and Boyer 1995; Bacilio et al. 2004). Greenhouse studies demonstrated the application of *Azospirillum brasilense* to maize seedlings to result in the alleviation of harmful effects of drought stress. The precise effects provided by this treatment were the enhancement in water content, decline in the decrease of water potential, enhancement in foliar area and total plant biomass, and augmentation of deposition of the osmoprotectant proline. In another experiment, treatment with *A. brasilense* in wheat seedlings has produced 12% of increase in yield in nonirrigated soil (Casanovas et al. 2002).

8.6.2 Enhanced Metabolite Production of Economically Important Crops

Plant growth hormone production, nitrogen fixation, ACC deaminase production, phosphate solubilization, and metabolite production with antimicrobial activity by rhizobacteria determine its function in plant growth and health. Hence, plant probiotic potential of rhizobacteria from ecologically diverse areas can have the promises to beneficially modulate biomass and active molecular composition of plants. In a previous study, *Proteus* spp. isolated from rhizosphere soil of *Pseudarthria viscida* and *Glycosmis arborea* were evaluated for their ability to enhance tuber size and active component diosgenin in *Dioscorea nipponica* during 1 year under field study. The tubers developed from plant treated with *Proteus* sp. exhibited remarkable increase in its size, number of roots, and diosgenin content than tubers in the control plants. This is due to the potential of used *Proteus* spp. to produce IAA, ammonia, siderophore and ACC deaminase. This study indicated the potential application of rhizobacteria as biofertilizer for the enhanced biomass and secondary metabolite production in an eco-friendly and cost-effective manner (Jimtha et al. 2017).

In the previous investigation of Singh et al. (2016), metabolites of *Streptomyces* sp. and *Trichoderma harzianum* were demonstrated to remarkably enhance the biomass in plants when compared to the control. In addition to the application of rhizobacterial isolates, secondary metabolites produced by candidate organisms can also enhance the production of plant secondary metabolites. Withanolide A content, deposition of lignin, total flavonoid, and phenolic contents of *Withania somnifera* have been described to be extremely induced with the application of *T. harzianum* metabolites. In addition to this, metabolites produced by *Trichoderma* and *Streptomyces* were found to have the application to support the augmentation of *in planta* composition and antioxidant molecules than the application with live rhizobacterial cells. The detection of new potential molecular elicitors from the rhizobacterial metabolites will enable its use as biofertilizers for the commercial cultivation of *W. somnifera* (Singh et al. 2016).

As the plant growth-promoting rhizobacteria have the ability to colonize plant roots, they enhance growth and development of plants by different mechanisms. The previous study reported beneficial activity of *Streptomyces* sp. and *Bacillus* sp. on enhanced growth and the production of capsaicin from *Capsicum annuum* L. This is a good practice for the acceleration of product yield with enhanced metabolite content for the plant cultivation, in an an eco-friendly approach (Datta et al. 2015). Also, preinoculation with *Bacillus subtilis* in tomato plants has demonstrated to significantly improve the fruit quality with high lycopene and texture after

15 days of harvest. This clearly indicated the ability of PGPR to enhance the nutritional content and shelf life of the fruits (Loganathan et al. 2014).

8.6.3 Rhizobacteria as an Inducer of Plant Defense Mechanism

Some of the rhizobacteria have the potential to suppress phytopathogens which otherwise cause disease to plant resulting in loss of crop yield. Various species of PGPR applied in agricultural field provide high protection to crop plants from various diseases. Induced resistance is an augmented state of defense response in plants developed by an external stimulation either by pathogens or beneficial microorganisms. Induced systemic resistance (ISR) and systemic acquired resistance (SAR) are the major types of induced resistance in plants, which are stimulated by prior infection that enhance resistance to subsequent exposure by a pathogen. PGPR with biocontrol property inhibit or reduce the disease severity by antagonistic interaction with soilborne pathogens and also resist root and foliar pathogens by inducing systemic resistance. ISR mediated by rhizobacteria exhibit similarity to that of SAR induced by pathogens, in which both types of induced resistance make the uninfected tissues of plant more resistant to a vast array of phytopathogens. Most of the rhizobacteria produce salicylic acid (SA) at the root tissues and induce SA-dependent SAR pathway, whereas other rhizobacteria induce diverse signaling pathway independent of SA. A previous study reported that in *Arabidopsis thaliana*, SA-independent ISR pathway with the involvement of jasmonic acid (JA) and ethylene signaling takes place. Rhizospheric *Pseudomonas* spp. stimulate systemic resistance after pathogen challenge in carnation, cucumber, radish, tobacco, and Arabidopsis, resulting in increased resistance. Synergistic induction of ISR and SAR can drastically enhance defensive potential against a broad spectrum of pathogens than ISR or SAR alone. In addition to *Pseudomonas* spp., *Bacillus* like *B. amyloliquefaciens*, *B. subtilis*, *B. pasteurii*, *B. cereus*, *B. pumilus*, *B. mycoides*, and *B. sphaericus* also stimulate defense response and reduce the incidence or severity of various diseases in different plants by modulating the plant immune response in diverse ways (Choudhary et al. 2007).

Ogata-Gutierrez et al. (2017) have studied the antiphytopathogenic effect of rhizospheric *Pseudomonas* spp. against *Alternaria alternata*, *Fusarium solani*, and *Curvularia lunata* in tomato plants. Biopriming of tomato seeds with the *Pseudomonas* spp. significantly boosted germination of seed, seedling emergence, plant growth and development and diminished disease incidence intensity caused by *A. alternata*. Also, an enhancement in the transcript expression level of three genes *AOS*, *ERF-2*, and *PR-P2* corresponding to jasmonate-, ethylene-, and pathogenesis-related pathways, respectively, was also observed in the presence of pathogen and rhizobacteria in tomato plants.

In the report by Jimtha et al. (2016), the preinoculation of *Bacillus* sp. to ginger rhizome resulted in significant protection to ginger rhizome from the challenged *Pythium myriotylum* which cause rhizome rot. In addition to this, rhizospheric *Bacillus* strain also exhibited synergistic antifungal activity with commercially procured biocontrol agent *Trichoderma* sp. The authors reported one of the reason for the antifungal effect of used *Bacillus* sp. to be due to the production of an antifungal compound pyrrolo[1,2-a]pyrazine-1,4-dione, hexahydro-3-(phenylmethyl) (Jimtha et al. 2016).

The *Pseudomonas* sp. isolated from rhizosphere also have antagonistic activity against *Clavibacter michiganensis* subsp. *michiganensis* which cause canker disease in tomato and *Pseudomonas syringae.* Here, inoculation with rhizospheric *Pseudomonas* sp. showed the potential to promote the growth and reduce severity of canker by stimulating systemic resistance in tomato seedlings as revealed by the expression analysis of marker genes like PR1a and ACO corresponding to salicylic acid and ethylene, respectively. This indicated that SA is involved as a signal molecule in the pathway for resistance induced by *Pseudomonas* sp. (Takishita et al. 2018).

8.6.4 The Function of Rhizobacteria in Disease-Suppressive Soil

In-depth understanding on the mechanisms of disease suppressiveness can lead to development of methods to improve the health of plants in unfavorable areas. The rhizobacteria of the genus *Pseudomonas*, the most dominating bacteria in the soil, colonize the plant root aggressively and exhibit high degree of disease suppression (Schroth and Hancock 1982). Numerous microbial species actively function for the disease suppressiveness of soils, but the complex communication network of the microbial interactions and belowground mechanisms remain indefinable for most disease-suppressive soils. Novel technologies like next-generation sequencing and other "omics" technologies have provided new insights into the microbial diversity of disease-suppressive soils and the detection of microbial consortia and its properties involved in disease suppressiveness.

Disease-suppressive soils are the suitable environment to describe microbiome-associated disease resistance in plants against soilborne pathogens. In this environment, the pathogens cannot survive or establish and sometimes can establish but cause no harmful impact or can establish and cause disease which is less severe and cause no disease even though the pathogen may exist in the soil. Pathogens can survive and rapidly multiply in non-suppressive soils, because of favorable abiotic and biotic conditions to the growth of the pathogen. Two types of disease suppressiveness were identified, general and specific suppressiveness. The activity of total microbiome in the soil has been attributed to the general suppressiveness of the soil, in which a competition between microbial community exist and these types of soil can be boosted by the application of organic matter (Weller et al. 2002). Specific suppressiveness is attributed to the remarkable functions of specific species of microbial community which interfere with the growth cycle of the soilborne pathogen, and this exhibit high degree of protection against phytopathogens than general suppressive soil (Tomihama et al. 2016; Mazzola and Freilich 2017).

One of the most important root diseases in wheat (*Triticum aestivum*) caused by *Gaeumannomyces graminis* var. *tritici* can be controlled by specific soil suppression and can be considered as a model system for biocontrol research. As effective fungicide and resistance cultivars are not available, this disease is normally managed by different cultural and biological approaches. The suppressive soil can be modulated by monocropping up to 5–7 years and application of organic matters into the soil. Also, this suppressiveness is due to the predominant presence of *Pseudomonas fluorescens* which have the ability to produce antimicrobial metabolites such as antibiotics, phenazine carboxylic acid, and 2,4-diacetylphloroglucinol.

Another study on the suppression of *Fusarium* wilt disease caused by *Fusarium oxysporum* in specific suppressive soil also exists. This study report the augmented growth of nonpathogenic *Fusarium* sp. to be influenced by soil pH. Here the organism elicit systemic resistance in the host and compete with the pathogen for nutrients and infection sites. The high population of *Pseudomonas fluorescens* in the suppressive soil can induce the resistance and produce siderophores to limit the availability of iron to other organisms in the soil. This synergistic action of both *Pseudomonas* sp. and *Fusarium* sp. make the soil with suppressive effect (Mousa and Raizada 2016).

8.7 Rhizoremediation: Degradation of Pesticides and Other Toxic Chemicals in the Soil by Rhizobacteria

Many synthetic organic compounds which include pesticides, polychlorinated biphenyls (PCBs), industrial solvents, petroleum products, dioxins and furans, explosives, and brominated flame retardants are considered as environmental contaminants. These synthetic organic compounds persist for long years once released into the ecosystem due to their specific chemical structures which are resistant to biological degradation processes. Because of this reason, such compounds are more toxic and lead to biomagnification through the food chain and cause serious health issues and harmful impact to all living forms especially humans. Thus, rhizoremediation is a vital solution for the removal of toxic chemicals from the soil. Rhizoremediation is a form of phytoremediation, in which plants and associated rhizospheric microorganisms are involved in the degradation of toxic chemicals. This process can happen either naturally or by bioaugmentation of toxic chemical-degrading microorganisms to the soil. *Pseudomonas* spp., *Burkholderia* spp., *Azospirillum* spp., *Enterobacter* spp., etc. are important rhizobacteria reported to be involved in the degradation of toxic chemicals (McGuinness and Dowling 2009). Such rhizobacteria produce certain specific enzymes capable of degrading the synthetic organic compounds.

With the advances in biotechnology and genetic engineering, different bacterial species can be genetically modified to synthesize enzymes, which can degrade the environmental pollutants. Genetic engineering of rhizobacteria is possible by natural gene transfer or recombinant DNA technology through which deficient bacteria can attain the ability to degrade contaminants. Application of genetically constructed rhizobacteria with phytoremediation potential can provide significant and promising step for the degradation and removal of toxic chemicals from the polluted site (Brazil et al. 1995; McGuinness and Dowling 2009).

Use of biodegrading microbial community in the soil for the removal of toxic pollutants provides an efficient, promising, and cost-effective approach. Application of such rhizobacteria can rupture the organic compounds completely into inorganic constituents. The microbial transformation of pollutants can be driven for their energy requirements, or a specific necessity. Unique nature, beneficial traits, population density and diversity, and potential catalytic property offer promises to explore rhizobacteria for remediation to remove pollutants from the soil (Paul et al. 2005). This process mainly involve three phases. In the first phase, the toxic parent compounds are moved through oxidation, reduction, or hydrolysis and form water-soluble and less toxic molecules than the parent compounds. In the second phase of degradation, pesticide metabolite is conjugated with sugar or amino acid and formed more water-soluble compound with reduced toxicity. Finally, the third phase converts the second phase metabolites into nontoxic secondary conjugates. Soil bacteria and fungi are the major candidates involved in the degradation processes because of their ability to produce intracellular or extracellular enzymes such as hydrolytic enzymes, oxygenases, and peroxidases (Van Eerd et al. 2003).

8.8 Nanotechnological Advances with Rhizobacteria

Nanotechnology is an emerging multidisciplinary research area, which is also established to have applications in the agricultural field. Nanoparticles are created by the controlled size manipulation and shape at the nanometer scale less than 100 nm with new features and higher property than the bulk materials (Medina-Pérez et al. 2019). The development of nanocomposites and nanoencapsules suggests their advantages to release small amounts of active components like fertilizers, herbicides, fungicides, or growth promoters in a stable form throughout the crop growth, avoiding overdoses and reducing input and waste (Nuruzzaman et al. 2016). Because of the increased demand for agricultural productivity with reduced input of cost and energy, the nanotechnology can be considered to be the most beneficial and novel approach in the agricultural sector. Remarkably, nanotechnology in terms of nanoparticles has promised potential applications such as nanofertilizer, nanopesticide, nanoherbicide, nanosensor, and also as specific delivery systems for the targeted and controlled release of agrochemicals (Campos et al. 2014; Grillo et al. 2016). It also offers several benefits in agriculture such as detection of pathogens, delivery of nanopesticides to the specific target sites, and enhanced absorption of nutrients in plants. The applications of nanotechnology have great potential to meet future agricultural challenges such as food security (Tripathi et al. 2018).

Sometimes nanoparticle treatment also has a negative impact on the soil. Most of the previous reports have also focused on analyzing the direct and indirect impact of accumulated nanoparticles on the structure of microbial community in the soil (Mishra et al. 2017). This limitation can be overcome by introducing biologically fabricated nanoparticles into the environment which is comparatively nontoxic and synthesized by cost-effective method. Numerous biological candidates such as bacteria, fungi, algae, and plant extracts have been explored for nanoparticle biosynthesis (Das et al. 2014). The advantage of the biological method for nanoparticle synthesis is that the complete process of synthesis is rapid and stable and uses non-toxic biomolecules with low cost, and most significantly, the synthesized nanoparticles are more stable. The shape and size of the nanoparticles during biosynthesis can be modulated by altering the pH and temperature of the reaction mix (Janardhanan et al. 2013; Hussain et al. 2016).

Additionally, numerous metal nanoparticles (Au, Ag, Fe, Pt, Ti, Zn, Mg, etc.) have been effectively synthesized using the biological method. Remarkably, biofabricated nanoparticles exhibited enhanced activity than those synthesized through physical and chemical methods (Kharissova et al. 2013). In an agricultural point of view, biofabricated nanoparticles have promising most potential and eco-friendly applications in the agricultural field especially for the promotion of plant growth, plant disease control, and management and tolerance to many biotic and abiotic stresses (Mishra et al. 2017). Raliya et al. (2015) have described exciting impact of biofabricated TiO_2 nanoparticles generated by *Aspergillus flavus* on growth of *Vigna radiata* and rhizomicrobiome. Similarly, Mishra et al. (2014) demonstrated significant antiphytopathogenic activity of biofabricated silver nanoparticles (AgNPs) against *Bipolaris sorokiniana*, the causative agent of spot blotch in wheat (*T. aestivum*). In addition to this, various previous reports have demonstrated the antimicrobial activity of AgNPs, synthesized biologically against different species of phytopathogens, which shows their significant promises in agriculture (Gopinath and Velusamy 2013; Paulkumar et al. 2014). Hence, exploration of biogenic nanoparticles into the field is a green approach for sustainable agricultural practices along with promises of PGPR.

8.9 Conclusion

The excess chemical input into the agricultural field in the form of pesticides and fertilizers leads to harmful consequences to the ecosystem. To retain the green environment, currently many new approaches are experimented. The most beneficial and influential method is the application of plant growth-promoting rhizobacteria with new technological advances without spoiling the health and yield of crops. The plant-rhizobacterial interaction is a network in which the microorganisms and plant cells communicate with each other through signaling molecules produced by both, and soil acts as a communication medium. The characteristics of this interaction are constructed by the type and species of plants and rhizobacteria involved. Through this complex linkage, the rhizomicrobiome augments the growth and development of plants and protects them from various pathogens causing disease in crops. This beneficial interaction can strongly hold by using new technological approach like nanotechnology. The application of nanoencapsulated PGPR in the agricultural field can provide controlled release of bacterial cells and nutrients, enhance absorption of nutrients, augment the germination of seeds, improve the delivery of agrochemicals into the targeted sites, and provide detection and inhibition of phytopathogens. Hence, the in-depth understanding of rhizobacterial functions and their mechanism will have significant effects for sustainable agriculture.

References

Ahmad F, Ahmad I, Khan MS (2008) Screening of free-living rhizospheric bacteria for their multiple plant growth promoting activities. Microbiol Res 163:173–181

Amundson R, Berhe AA, Hopmans JW, Olson C, Sztein AE, Sparks DL (2015) Soil and human security in the 21st century. Sci 348:1261071

Bacilio M, Rodriguez H, Moreno M, Hernandez JP, Bashan Y (2004) Mitigation of salt stress in wheat seedlings by a gfp-tagged *Azospirillum lipoferum*. Biol Fert Soils 40:40–188

Badri DV, Loyola-Vargas VM, Broeckling CD, De-la-Pena C, Jasinski M, Santelia D, Martinoia E, Sumner LW, Banta LM, Stermitz F, Vivanco JM (2008) Altered profile of secondary metabolites in the root exudates of *Arabidopsis* ATP-binding cassette transporter mutants. Plant Physiol 146:762–771

Badri DV, Weir TL, van der Lelie D, Vivanco JM (2009) Rhizosphere chemical dialogues: plant-microbe interactions. Curr Opin Biotechnol 20:642–650

Barriuso J, Ramos Solano B, Fray RG, Camara M, Hartmann A, Gutierrez Manero FJ (2008) Transgenic tomato plants alter quorum sensing in plant growth-promoting rhizobacteria. Plant Biotechnol J 6:442–452

Begonia MFT, Kremer RJ (1994) Chemotaxis of deleterious rhizobacteria to velvet leaf (*Abutilon theophrasti* Medik.) seeds and seedlings. FEMS Microbiol Ecol 15:227–236

Berendsen RL, Pieterse CM, Bakker PA (2012) The rhizosphere microbiome and plant health. Trends Plant Sci 17:478–486

Bogino P, Oliva M, Sorroche F, Giordano W (2013) The role of bacterial biofilms and surface components in plant-bacterial associations. Int J Mol Sci 14:15838–15859

Bolwerk A, Lagopodi AL, Wijfjes AH, Lamers GE, Chin AWTF, Lugtenberg BJ, Bloemberg GV (2003) Interactions in the tomato rhizosphere of two *Pseudomonas* biocontrol strains with the phytopathogenic fungus *Fusarium oxysporum* f. sp. *radicis-lycopersici*. Mol Plant-Microbe Interact 16:983–993

Branda SS, Vik S, Friedman L, Kolter R (2005) Biofilms: the matrix revisited. Trends Microbiol 13:20–26

Brazil GM, Kenefick L, Callanan M, Haro A, de Lorenzo V, Dowling DN, O'Gara F (1995) Construction of a rhizosphere pseudomonad with potential to degrade polychlorinated biphenyls and detection of bph gene expression in the rhizosphere. Appl Environ Microbiol 61:1946–1952

Bulgarelli D, Schlaeppi K, Spaepen S, van Themaat EVL, Schulze-Lefert P (2013) Structure and functions of the bacterial microbiota of plants. Ann Rev Plant Biol 64:807–838

Camara M, Williams P, Hardman A (2002) Controlling infection by tuning in and turning down the volume of bacterial small-talk. Lancet Infect Dis 2:667–676

Campos E, Oliveira J, Fraceto L (2014) Applications of controlled release Systems for Fungicides, herbicides, Acaricides, nutrients, and plant growth hormones: a review. Adv Sci Eng Med 6:373–387

Casanovas EM, Barassi C, Sueldo RJ (2002) *Azospirillum* inoculation mitigates water stress effects in maize seedlings. Cereal Res Comm 30(3):343–350

Chalupowicz L, Barash I, Panijel M, Sessa G, Manulis-Sasson S (2009) Regulatory interactions between quorum-sensing, auxin, cytokinin, and the Hrp regulon in relation to gall formation and epiphytic fitness of *Pantoea agglomerans* pv. *gypsophilae*. Mol Plant-Microbe Interact 22:849–856

Cheng F, Cheng Z (2015) Research progress on the use of plant allelopathy in agriculture and the physiological and ecological mechanisms of allelopathy. Front Plant Sci 6:1020

Choudhary DK, Prakash A, Johri BN (2007) Induced systemic resistance (ISR) in plants: mechanism of action. Indian J Microbiol 47:289–297

Choudhary DK, Sharma KP, Gaur RK (2011) Biotechnological perspectives of microbes in agro-ecosystems. Biotechnol Lett 33:1905–1910

Compant S, Duffy B, Nowak J, Clement C, Barka EA (2005) Use of plant growth-promoting bacteria for biocontrol of plant diseases: principles, mechanisms of action, and future prospects. Appl Environ Microbiol 71:4951–4959

Das VL, Thomas R, Varghese RT, Soniya EV, Mathew J, Radhakrishnan EK (2014) Extracellular synthesis of silver nanoparticles by the *Bacillus* strain CS 11 isolated from industrialized area. 3 Biotech 4:121–126

Datta M, Paul D, Sinha SN, Sengupta C (2015) Plant growth promoting Rhizobacteria improve the production and enhancement of alkaloid content in Chilli. Front Environ Microbiol 1:24–26

Dekkers LC, Phoelich CC, van der Fits L, Lugtenberg BJ (1998) A site-specific recombinase is required for competitive root colonization by *Pseudomonas fluorescens* WCS365. Proc Natl Acad Sci U S A 95:7051–7056

Dennis PG, Miller AJ, Hirsch PR (2010) Are root exudates more important than other sources of rhizodeposits in structuring rhizosphere bacterial communities? FEMS Microbiol Ecol 72:313–327

Elsas JDV, Jansson JK, Trevors JT (2007) Modern soil microbiology. CRC Press/Taylor & Francis, Boca Raton

Feng H, Zhang N, Du W, Zhang H, Liu Y, Fu R, Shao J, Zhang G, Shen Q, Zhang R (2018) Identification of chemotaxis compounds in root exudates and their sensing chemoreceptors in plant-growth-promoting Rhizobacteria *Bacillus amyloliquefaciens* SQR9. Mol Plant-Microbe Interact 31:995–1005

Fray RG (2002) Altering plant-microbe interaction through artificially manipulating bacterial quorum sensing. Ann Bot 89:245–253

Gao M, Teplitski M, Robinson JB, Bauer WD (2003) Production of substances by *Medicago truncatula* that affect bacterial quorum sensing. Mol Plant-Microbe Interact 16:827–834

Glick BR (2012) Plant growth-promoting Bacteria: mechanisms and applications. Scientifica 2012:1–15

Gopinath V, Velusamy P (2013) Extracellular biosynthesis of silver nanoparticles using *Bacillus* sp. GP-23 and evaluation of their antifungal activity towards *Fusarium oxysporum*. Spectrochim Acta A Mol Biomol Spectrosc 106:170–174

Gouda S, Kerry RG, Das G, Paramithiotis S, Shin H-S, Patra JK (2018) Revitalization of plant growth promoting rhizobacteria for sustainable development in agriculture. Microbiol Res 206:131–140

Grillo R, Chirakkuzhyil Abhilash P, Fraceto L (2016) Nanotechnology applied to bio-encapsulation of pesticides. J Nanosci Nanotechnol 16:1231–1234

Haichar FZ, Santaella C, Heulin T, Achouak W (2014) Root exudates mediated interactions belowground. Soil Biol Biochem 77:69–80

Hamaoui B, Abbadi JM, Burdman S, Rashid A, Sarig S, Okon Y (2001) Effects of inoculation with *Azospirillum brasilense* on chickpeas (*Cicer arietinum*) and faba beans (Vicia faba) under different growth conditions. Agronomie 21:553–560

Hartmann A, Rothballer M, Hense BA, Schroder P (2014) Bacterial quorum sensing compounds are important modulators of microbe-plant interactions. Front Plant Sci 5:131

Huang X-F, Chaparro JM, Reardon KF, Zhang R, Shen Q, Vivanco JM (2014) Rhizosphere interactions: root exudates, microbes, and microbial communities. Botany 92:267–275

Hussain I, Singh NB, Singh A, Singh H, Singh SC (2016) Green synthesis of nanoparticles and its potential application. Biotechnol Lett 38:545–560

Igiehon NO, Babalola OO (2018) Rhizosphere microbiome modulators: contributions of nitrogen fixing Bacteria towards sustainable agriculture. Int J Environ Res Public Health 15:574

Janardhanan A, Roshmi T, Varghese RT, Soniya EV, Mathew J, Radhakrishnan EK (2013) Biosynthesis of silver nanoparticles by a *Bacillus* sp. of marine origin, Mater Sci-Poland:31

Jimtha John C, Jishma P, Karthika NR, Nidheesh KS, Ray JG, Mathew J, Radhakrishnan EK (2017) *Pseudomonas fluorescens* R68 assisted enhancement in growth and fertilizer utilization of *Amaranthus tricolor* (L.). 3 Biotech 7:256

Jimtha JC, Jishma P, Arathy GB, Anisha C, Radhakrishnan EK (2016) Identification of plant growth promoting Rhizosphere *Bacillus* sp. WG4 antagonistic to *Pythium myriotylum* and its enhanced antifungal effect in association with *Trichoderma*. J Soil Sci Plant Nutr 16:578–590

Jimtha JC, Mathew J, Radhakrishnan EK (2017) Bioengineering of *Dioscorea nipponica* with rhizospheric Proteus spp. for enhanced tuber size and diosgenin content. 3 Biotech 7:261

Jing YD, He ZL, Yang XE (2007) Role of soil rhizobacteria in phytoremediation of heavy metal contaminated soils. J Zhejiang Univ Sci B 8:192–207

Kang BG, Kim WT, Yun HS, Chang SC (2010) Use of plant growth-promoting rhizobacteria to control stress responses of plant roots. Plant Biotechnol Rep 4:179–183

Kharissova OV, Dias HV, Kharisov BI, Perez BO, Perez VM (2013) The greener synthesis of nanoparticles. Trends Biotechnol 31:240–248

Kramer PJ, Boyer JS (1995) Water relations of plants and soils. Academic Press, San Diego

Lakshmanan V, Selvaraj G, Bais HP (2014) Functional soil microbiome: belowground solutions to an aboveground problem. Plant Physiol 166:689–700

Lareen A, Burton F, Schäfer P (2016) Plant root-microbe communication in shaping root microbiomes. Plant Mol Biol 90:575–587

Liu W, Wang Q, Hou J, Tu C, Luo Y, Christie P (2016) Whole genome analysis of halotolerant and alkalotolerant plant growth-promoting rhizobacterium *Klebsiella* sp. D5A. Sci Rep 6:26710

Loganathan M, Garg R, Venkataravanappa V, Saha S, Rai AB (2014) Plant growth promoting rhizobacteria (PGPR) induces resistance against Fusarium wilt and improves lycopene content and texture in tomato. Afr J Microbiol Res 8(11):1105–1111

Mathivanan S, Chidambaram A, Sundramoorthy P, Baskaran L, Kalaikandhan R (2014) Effect of combined inoculations of plant growth promoting Rhizobacteria (PGPR) on the growth and yield of groundnut (*Arachis hypogaea* L.). Int J Curr Microbiol Appl Sci 3:1010–1020

Mazzola M, Freilich S (2017) Prospects for biological Soilborne disease control: application of indigenous versus synthetic microbiomes. Phytopathology 107:256–263

McGuinness M, Dowling D (2009) Plant-associated bacterial degradation of toxic organic compounds in soil. Int J Environ Res Public Health 6:2226–2247

Medina-Pérez G, Fernández-Luqueño F, Vázquez Núñez E, López-Valdez F, Prieto-Mendez J, Madariaga-Navarrete A, Miranda-Arámbula M (2019) Remediating polluted soils using nanotechnologies: environmental benefits and risks. Pol J Environ Stud 28:1–17

Mendes R, Kruijt M, de Bruijn I, Dekkers E, van der Voort M, Schneider JH, Piceno YM, DeSantis TZ, Andersen GL, Bakker PA, Raaijmakers JM (2011) Deciphering the rhizosphere microbiome for disease-suppressive bacteria. Science 332:1097–1100

Mendes R, Garbeva P, Raaijmakers JM (2013) The rhizosphere microbiome: significance of plant beneficial, plant pathogenic, and human pathogenic microorganisms. FEMS Microbiol Rev 37:634–663

Mishra S, Singh BR, Singh A, Keswani C, Naqvi AH, Singh HB (2014) Biofabricated silver nanoparticles act as a strong fungicide against *Bipolaris sorokiniana* causing spot blotch disease in wheat. PLoS One 9:e97881

Mishra S, Keswani C, Abhilash PC, Fraceto LF, Singh HB (2017) Integrated approach of Agrinanotechnology: challenges and future trends. Front Plant Sci 8:471

Moons P, Van Houdt R, Vivijs B, Michiels CW, Aertsen A (2011) Integrated regulation of acetoin fermentation by quorum sensing and pH in Serratia plymuthica RVH1. Appl Environ Microbiol 77:4704–4704

Mousa WK, Raizada MN (2016) Natural disease control in cereal grains. Reference Module in Food Science, Elsevier

Nazzaro F, Fratianni F, Coppola R (2013) Quorum sensing and phytochemicals. Int J Mol Sci 14:12607–12619

Nuruzzaman M, Rahman MM, Liu Y, Naidu R (2016) Nanoencapsulation, Nano-guard for pesticides: a new window for safe application. J Agri Food Chem 64:1447–1483
Ogata-Gutierrez K, Chumpitaz-Segovia C, Lirio-Paredes J, Finetti-Sialer MM, Zuniga-Davila D (2017) Characterization and potential of plant growth promoting rhizobacteria isolated from native Andean crops. World J Microbiol Biotechnol 33:203
Paul D, Pandey G, Pandey J, Jain RK (2005) Accessing microbial diversity for bioremediation and environmental restoration. Trends Biotechnol 23:135–142
Paulkumar K, Gnanajobitha G, Vanaja M, Rajeshkumar S, Malarkodi C, Pandian K, Annadurai G (2014) *Piper nigrum* leaf and stem assisted green synthesis of silver nanoparticles and evaluation of its antibacterial activity against agricultural plant pathogens. Sci World J 2014:829894
Perez-Montano F, Guasch-Vidal B, Gonzalez-Barroso S, Lopez-Baena FJ, Cubo T, Ollero FJ, Gil-Serrano AM, Rodriguez-Carvajal MA, Bellogin RA, Espuny MR (2011) Nodulation-gene-inducing flavonoids increase overall production of autoinducers and expression of N-acyl homoserine lactone synthesis genes in rhizobia. Res Microbiol 162:715–723
Philippot L, Raaijmakers JM, Lemanceau P, van der Putten WH (2013) Going back to the roots: the microbial ecology of the rhizosphere. Nat Rev Microbiol 11:789–799
Raliya R, Biswas P, Tarafdar JC (2015) TiO2 nanoparticle biosynthesis and its physiological effect on mung bean (*Vigna radiata* L.). Biotechnol Rep (Amst) 5:22–26
Saraf M, Pandya U, Thakkar A (2014) Role of allelochemicals in plant growth promoting rhizobacteria for biocontrol of phytopathogens. Microbiol Res 169:18–29
Schalk IJ, Hannauer M, Braud A (2011) New roles for bacterial siderophores in metal transport and tolerance. Environ Microbiol 13:2844–2854
Scharf BE, Hynes MF, Alexandre GM (2016) Chemotaxis signaling systems in model beneficial plant–bacteria associations. Plant Mol Biol 90:549–559
Schroth MN, Hancock JG (1982) Disease-suppressive soil and root-colonizing Bacteria. Sci 216:1376–1381
Schuhegger R, Ihring A, Gantner S, Bahnweg G, Knappe C, Vogg G, Hutzler P, Schmid M, Van Breusegem F, Eberl L, Hartmann A, Langebartels C (2006) Induction of systemic resistance in tomato by N-acyl-L-homoserine lactone-producing rhizosphere bacteria. Plant Cell Environ 29:909–918
Shailendra Singh GG (2015) Plant growth promoting Rhizobacteria (PGPR): current and future prospects for development of sustainable agriculture. J Microb Biochem Technol 07
Sharma A, Shankhdhar D, Sharma A, Shankhdhar SC (2014) Micronutrient enhancement and localization in Rice grains under influence of plant growth promoting Rhizobacteria. J Crop Improv 28:502–517
Singh A, Gupta R, Srivastava M, Gupta MM, Pandey R (2016) Microbial secondary metabolites ameliorate growth, in planta contents and lignification in *Withania somnifera* (L.) Dunal. Physiol Mol Biol Plants 22:253–260
Spence C, Alff E, Johnson C, Ramos C, Donofrio N, Sundaresan V, Bais H (2014) Natural rice rhizospheric microbes suppress rice blast infections. BMC Plant Biol 14:130
Steidle A, Sigl K, Schuhegger R, Ihring A, Schmid M, Gantner S, Stoffels M, Riedel K, Givskov M, Hartmann A, Langebartels C, Eberl L (2001) Visualization of N-acylhomoserine lactone-mediated cell-cell communication between bacteria colonizing the tomato rhizosphere. Appl Environ Microbiol 67:5761–5770
Swamy MK, Akhtar MS, Sinniah UR (2016) Root exudates and their molecular interactions with Rhizospheric microbes. Interaction among Rhizospheric microbes, soil, and plant roots: Infl uence on micronutrient uptake and bioavailability. K.R. Hakeem and M. S. Akhtar, Springer International Publishing, Switzerland 2: 59–77
Takishita Y, Charron JB, Smith DL (2018) Biocontrol Rhizobacterium *Pseudomonas* sp. 23S induces systemic resistance in tomato (*Solanum lycopersicum* L.) against bacterial canker *Clavibacter michiganensis* subsp. *michiganensis*. Front Microbiol 9:2119
Tomihama T, Nishi Y, Mori K, Shirao T, Iida T, Uzuhashi S, Ohkuma M, Ikeda S (2016) Rice bran amendment suppresses potato common scab by increasing antagonistic bacterial community levels in the rhizosphere. Phytopathology 106:719–728

Tripathi M, Kumar S, Kumar A, Tripathi P, Kumar S (2018) Agro-nanotechnology: a future Technology for Sustainable Agriculture. Int J Curr Microbiol Appl Sci 7:196–200
Turner TR, James EK, Poole PS (2013) The plant microbiome. Genome Biol 14
Van Eerd LL, Hoagland RE, Zablotowicz RM, Hall JC (2003) Pesticide metabolism in plants and microorganisms. Weed Sci 51:472–495
Vejan P, Abdullah R, Khadiran T, Ismail S, Nasrulhaq Boyce A (2016) Role of plant growth promoting Rhizobacteria in agricultural sustainability—a review. Molecules 21:573
Weller DM, Raaijmakers JM, Gardener BB, Thomashow LS (2002) Microbial populations responsible for specific soil suppressiveness to plant pathogens. Annu Rev Phytopathol 40:309–348
Xie F, Williams A, Edwards A, Downie JA (2012) A plant arabinogalactan-like glycoprotein promotes a novel type of polar surface attachment by *Rhizobium leguminosarum*. Mol Plant-Microbe Interact 25:250–258

Microbiome in Plant Health and Disease: Challenges and Opportunities

9

Ranjini Ramesh

Abstract

In this chapter, the role played by the "plant microbiome" has been discussed in detail, with reference to their diversity, mechanisms of growth promotion in plants, and methods of their defense against predators and disease. The effects of plants on the diversity of the microbiome, and "co-adaptation" between the two have also been elaborated. The mechanisms of biocontrol detailed are "antagonism," "signal interference," "predation," "parasitism," "induced systemic resistance," and the role played by ferric ions. Direct plant growth promotion methods discussed are "rhizoremediation," "phytostimulation," and "stress control," among others. The diversity of the rhizosphere microbiome has also been detailed, with reference to *Azotobacter*, *Azospirillum*, mycorrhizae, and blue-green algae.

9.1 Introduction

"Sustainable agriculture" is the set of practices that comprise environment-friendly techniques and methods of farming, from the time of precultivation to the preparation of soil, cultivation, growing period of crops, harvest, and postharvest crop processing, storage, and distribution to the end users (UNEP 1996a, b, c).

In order to follow the above methods of farming, maintaining an optimum fertility level of soil, increasing soil fertility and plant growth, and, finally, keeping the crop/plant free of disease from the seed to grain stages are absolutely essential (Schlaeppi and Bulgarelli 2015). One of the primary requirements for this is the preservation and enhancement of natural microbial populations of the soil (bacteria

R. Ramesh (✉)
St Joseph's College, Autonomous, Bangalore, India

V. Kumar et al. (eds.), *Microbiome in Plant Health and Disease*,
https://doi.org/10.1007/978-981-13-8495-0_9

and fungi). By interacting with plants, they enhance the uptake of water, nutrients, and minerals from the soil by plants and make their absorption better to enhance growth parameters and keep the plants free of pathogenic microorganisms that kill them, or retard their growth and harvest products. These bacteria and fungi form the "plant microbiome," of which the "rhizosphere" is a significant part. Here, microbial consortia in the vicinity of the root system help in the abovementioned functions. They have mutualistic, commensalistic, or, sometimes, even parasitic associations with their host plants, to achieve the same.

There are other areas of interactions of plants and microorganisms, such as leaf surface ("phyllosphere"), inside leaf cells ("leaf endosphere"), inside root cells ("root endosphere"), and reproductive organs such as flowers and fruits, but the rhizosphere is the only area where the microbes are in direct contact with both the plant and the soil. Thus, the maximum number of beneficial interactions (with respect to the plants) occurs in the rhizosphere (Vorholt 2012).

9.2 What Is Sustainable Agriculture and Why Is It Necessary?

"Sustainable agriculture," as mentioned above, is the sum of all the methods and practices that define a particular way of farming using the most naturally available resources of land, water, and microbial consortia, to develop safe and nutritious food for the present population while also conserving these resources for the future generations (Fig. 9.1).

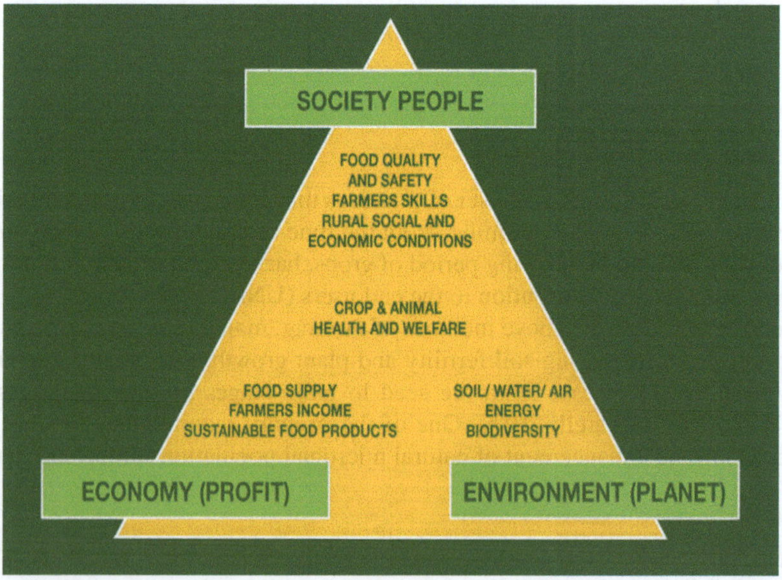

Fig. 9.1 The concept of sustainable agriculture

By 2020, there would be almost 1.5 billion people on Earth, over and above the present 8 billion, who need to be fed. This would be very similar to the challenge faced about 60–65 years back, for which "Green Revolution" practices were first implemented. The high-technology mechanized agricultural practices implemented then comprised the use of agrochemicals, high amounts of water, and high-yielding varieties of seeds/plants that increased the food production by two to three times; however, these practices have destroyed global agro-systems with respect to air, water, and soil quality. They have made the farmer a "prisoner of debt" and caught them in a repeating cycle of debt and loss of both income and land fertility. Today, decades of "chemical agriculture" have reduced the land and crop yields, especially in Asia and Africa, by nearly 15%–20% (UNEP 1996a, b, c).

9.3 The Plant Microbiome

The microbial consortia, that is, the sum total of all the microbial species that are hosted by the particular plant species, are mainly around the root system (Schlaeppi and Bulgarelli 2015).

Plants have several relationships within their "microbiome," particularly the "rhizosphere microbes," which could be "mutualism," "commensalism," or "parasitism." Microorganisms present in the "rhizosphere," "phyllosphere," "leaf endosphere," or "root endosphere" provide multiple benefits for their hosts (Vorholt2012), such as increased nutrient absorption from soil, resistance to factors of abiotic stress, hormone production, protection from pathogens, etc. (Schlaeppi and Bulgarelli 2015).

These microbes are known by the term: "PGPM" (*plant growth-promoting microorganisms*). Recent research in the plant microbiota field clearly supports the idea that PGPM represents only specific species of microbiota. However, the data available on PGPM is limited to studies on individual species in laboratory conditions (Bulgarelli et al. 2013); there is also less understanding as to how whole microbial communities can enhance plant growth. Research has proved the presence of the "core microbiome," which performs functions that are highly specific to each host (Vorholt 2012) such as water and nutrient absorption, fixation of nitrogen, uptake of phosphorus, pathogen antagonism, etc. These microbial consortia are important in phenomena such as "phytoremediation" (Yergeau et al. 2014). The benefit of added phosphorus and nitrogen is contributed by the "mycorrhizal fungi" and "rhizobium bacteria, " respectively (Oldroyd et al. 2011).

PGPM organisms that have been researched are *Azospirillum* spp., *Bacillus* spp., *Pseudomonas* spp., *Rhizobium* spp., *Serratia* spp., *Stenotrophomonas* spp., and various species of *Streptomyces*. The fungal species are *Ampelomyces* spp., *Coniothyrium* spp., and *Trichoderma* spp. (Franken 2012).

The addition of "microbial inoculants" to the soil, based on the characteristics of the microbiome, is one of the probable agricultural practices for the future; this will become a part of "sustainable agriculture." In order to understand this concept, it is essential to study the microbial part, in addition to the host interactions. This is

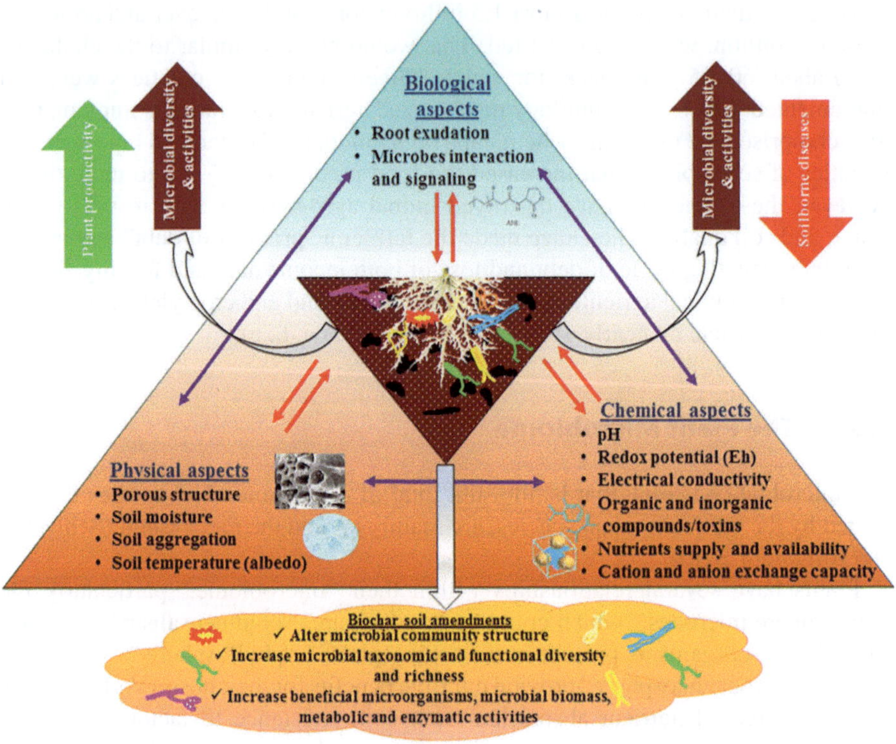

Fig. 9.2 Plant microbiome in the rhizosphere

because, in the future, the microbiome would emerge as a primary plant trait for regulation and optimization of growth parameters in plants (Bulgarelli et al. 2013). Thus, microbiomes of rhizosphere and phyllosphere would be an integral part of future programs in plant breeding. This would make cultivars of the next generation have an increased capacity to interact with microbes in the soil or with the added bacterial/fungal inoculants. This would make possible the breeding of new crop varieties that have improved responsiveness to the beneficial traits of microbes such as "augmented plant nutrient use efficiency" or "enhanced plant immunity." For controlling pathogens, the combined effectiveness of using beneficial plant microbiomes ("agricultural probiotics") and the plant's own immune function ("resistance genes") can give long-term protection from diseases for a long time (Dangl et al. 2013) – Fig. 9.2. shows the role of different microbiomes in the rhizospheric system.

9.4 The Rhizosphere Microbiome

It is the part of the soil colonized by the plant roots, which is much greater in microbial numbers and diversity than the remainder of the soil. It is the main area where the plant and natural population of soil microorganisms interact. Bacterial species mainly found in rhizosphere may be *Bacillus* spp., *Pseudomonas* spp., *Serratia* spp., *Stenotrophomonas* spp., *Streptomyces, Nitrosomonas* spp. and *Nitrobacter* spp., blue-green algae like *Azotobacter* spp., *Azospirillum* spp., *Rhizobium* spp. and *Nostoc* spp., in addition to mycorrhizal fungi like *Glomus* spp., *Acaulospora* spp., *Entrophospora* spp., *Gigaspora* spp., *Sclerocystis* spp., *Scutellospora* spp., *Ampelomyces* spp., *Coniothyrium* spp. and *Trichoderma* spp. (Franken 2012) and, finally, actinomycetes (Schlaeppi and Bulgarelli 2015).

9.4.1 Characteristic Features of the Plant Microbiome

The soil is known to have an enormous quantity of microbial diversity (Torsvik et al. 1990), and the latest studies have only reiterated this (Roesch et al. 2007). The diversity of genes and functions of the soil microbiome is only now being slowly understood and appreciated (Morales and Holben 2011). In a particular soil type, the native plants put selection pressure on this microbial biodiversity, restructuring them [reviewed in (Berg and Smalla 2009)]. Moreover, plants that are sensitive to microbial activity may show an increased or decreased performance, depending on the associated microbial species. These cross-effects are significant in modern agricultural systems. Some of these interactions are discussed below.

9.4.1.1 Root Exudation

In the plant rhizosphere, "root exudates" are an important aspect of interaction between plants and microbes (Badri and Vivanco 2009). The composition of the root exudate differs in various species of plants and even between cultivars of a species (Micallef et al. 2009). Because of this, the diversity of soil microbes is usually quite high (Salles et al. 2004).

Root exudates are generally made up of the following compounds: sugars, amino acids, flavonoids, proteins, and fatty acids (Badri and Vivanco 2009). These substances serve as "growth substrates" or "growth signals" for some of the microbial species and as "antimicrobials" or "growth deterrents" for other microbes (Bais et al. 2006).

An important obstacle in advancing concepts related to interactions between root exudates and soil microbes has been the ability to study root exudation in situ. However, recent developments are advancing in this area.

9.4.1.2 Extent of Plant-Driven Change to the Soil Microbiome

The effects of host plants have been detected in the "bulk soil microbiome" (Bremer et al. 2009), indicating that it could be possible to utilize plants to shape soil microbial communities more broadly than just in the rhizosphere. This potential is significant in agricultural systems, as host plants are switched during crop rotation. Microbial colonizers of the newly forming rhizosphere are drawn from the bulk soil community (Jones et al. 2004). The availability of beneficial colonizers in a newly forming rhizosphere may depend on the selective effects of the previous crop.

9.4.1.3 Plant-Microbiome Co-adaptation

The effects of host plants on the soil microbiome become more pronounced over time, and it is probable that their microbial partners undergo adaptations to the host. At least in cases of "rigid mutualistic symbiosis," there is evidence that plants and rhizosphere microbes have evolved simultaneously (Lambers et al. 2009). Global exchange of agricultural plant species provides an interesting aspect of studying such plant-microbiome co-adaptation. At the location of origin for a particular crop, long-term associations between the host plant and its microbial partners are possible. In contrast, movement of crop species to new parts of the world brings together soil microbiomes and host plants that may have no shared evolutionary history. The foreign host plant species may secrete a novel combination of exudates into the soil, some of which may have antimicrobial properties or behave as inefficient substrates for the local microbial community, thereby altering it irrevocably. Sudden and bulk replacement of host plants will change the selective pressures acting on the rhizosphere microbiome. Competitive advantage among microbial communities may switch, leading to a period of rearrangement; this has been observed in studies of invasive plants in their new habitats (Broz et al. 2007) through experimental host switching (Broeckling et al. 2008) and using plants with genetic defects related to root exudation (Badri et al. 2009).

9.4.1.4 Plant-Soil Feedback Mechanisms and Diffuse Mutualisms

Ecological studies of plant-soil feedback mechanisms through open pathways could be used in sustainable agriculture. There are two types of feedback mechanisms: "positive plant-soil feedback" and "'negative plant-soil feedback."

Recent research has showed that there is accumulation of pathogenic microbes after repeated cultivation of the same crop. This has become the basis of the need for "crop rotation" in agricultural systems (Hwang et al. 2009). This is an example of negative plant-soil feedback. On the other hand, studies have found that growth of one plant species enhances subsequent performance of the same species, which is an example of positive plant-soil feedback (Grunsven et al. 2009).

By decoding relevant plant traits and identifying suitable microorganisms (Mendes et al. 2011) or microbiome characteristics that are responsible for positive plant-soil feedbacks, it is possible to replicate these processes in agricultural systems.

9.5 Additional Characteristics of the Microbiome

To enrich soil with beneficial microbes or their secretions ("biofertilizers" or "soil conditioners"), it may be possible to cultivate such plant species that can shape the soil microbiome in a beneficial manner, in other words, microbes that promote plant health, enhance soil fertility, and mitigate soil pollution. Here, processes like "phytoremediation," with sub-processes like "phytostimulation," "phytoaccumulation," "rhizoremediation" (which may be achieved by rhizofiltration), "stress controllers," "mycorestoration," and "mycoremediation" are applicable. Microbial richness and,

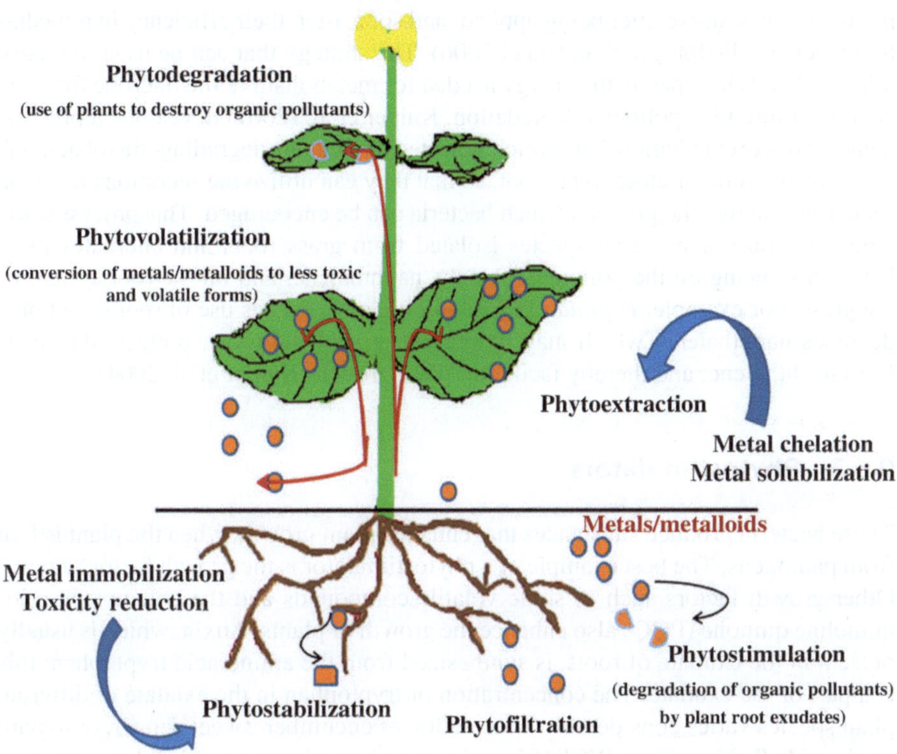

Fig. 9.3 The process of phytoremediation

paradoxically, its evenness of distribution in the rhizosphere are two important features that are particularly important for enhanced plant growth. Decreasing diversity leads to increasing evenness and thus more of similar functions being utilized in varying environments (Loreau et al. 2001). Many rare species of the microbiome cannot effectively carry out important functions. Thus, similarity (relative abundance) of members of the microbiome is very important (Van Elsas et al. 2008) – Fig. 9.3 shows the process of phytoremediation.

9.6 Direct Plant Growth Promotion by the Rhizosphere Microbiome

9.6.1 Rhizoremediators

These are microorganisms that remediate or treat polluted soil and degrade various classes of pollutants. The problem is that these bacteria and/or fungi, although they are effective under laboratory conditions, do not adapt so well to the conditions found in soil. In soil, their metabolism is mainly dependent on degrading a specific pollutant. They starve after being applied, and soon after, their efficiency in remediation is lessened (Bottiglieri and Keel 2006). The strategy that can be used to rectify this problem is to separate the energy needed for metabolism of the microbe from the energy required for pollutant degradation. Kuiper et al. (2001) developed a process, named "rhizoremediation," in which selected "pollutant-degrading rhizobacteria" reside on the root, or close to the root, so that they can utilize the secretions from the root, as nutrients. The growth of such bacteria can be encouraged. This process starts with a mixture of bacterial species isolated from grass roots and alternates them between growing on the pollutant, namely, naphthalene, and on the root system of the grass. For example, *P. putida* PCL1444 efficiently makes use of root secretions; degrades naphthalene, which may be present around the roots; protects the seeds from naphthalene; and thereby facilitates plant growth (Kuiper et al. 2004).

9.6.2 Phytostimulators

These bacteria produce substances that enhance plant growth, when the plant is free from pathogens. The best example of a phytostimulator is the growth factor "auxin." Other growth factors such as some volatile compounds and the cofactor pyrroloquinoline quinone (PQQ) also enhance the growth of plants. Auxin, which is usually present in the exudate of roots, is synthesized from the amino acid tryptophan; this is a part of the exudate. The concentration of tryptophan in the exudate of different plant species varies considerably. Inoculation of cucumber, sweet pepper, or tomato seeds with *P. fluorescens* WCS365 strain (which produces auxin) did not increase their root or shoot weights, but it did cause a significant increase in the weight of radish roots. Radish produced much more tryptophan in its exudate (per seedling) than cucumber, sweet pepper, or tomato (Kamilova et al. 2006). Thus, this is an example of "bacterial stimulation."

Azotobacter paspali, a nitrogen-fixing bacterium, has been isolated from a grass species in a subtropical climate. It is seen to improve the growth of a number of dicotyledon and monocotyledon species. Research has shown that plant growth is mainly due to growth factors such as IAA, gibberellins, and cytokinins and not so much due to nitrogen fixation (Okon et al. 1998).

Some rhizobacteria (such as) *B. subtilis*, *B. amyloliquefaciens*, and *Enterobacter cloacae* enhance the growth of plants by secreting certain volatile compounds (Ryu et al. 2003). The highest growth is seen with the release of 2,3-butanediol acetoin. Mutant strains of *B. amyloliquefaciens* IN937a and *B. subtilis* GB03 that cannot synthesize 2,3-butanediol acetoin are found to be inactive in enhancing plant growth. However, Zhang et al. (2008) found that *B. subtilis* GB03 increases the rate of photosynthesis and chlorophyll content of *A. thaliana* through the endogenous signaling of *glucose* and *abscisic acid*. Thus, it can be deduced that this bacterium regulates energy uptake in the plant.

In another example, the cofactor PQQ is known as a "plant growth promoter" (Choi et al. 2008). Synthetic PQQ is seen to enhance growth in tomato and cucumber. Studies show that PQQ plays the role of an antioxidant in these plants. It is, though, an indirect effect, as PQQ is a cofactor of many enzymes, which are involved in providing antifungal activity and induction of systemic resistance in the plant.

9.6.3 Stress Controllers

Bacterial strains that reduce the production of the enzyme, 1-aminocyclopropane-1-carboxylate (ACC) deaminase, enhance plant growth and development by reducing the level of ethylene, which are examples of "stress controllers." This is done by taking up the precursor of ethylene, ACC, and converting it into 2-oxobutanoate and ammonia. Various types of stress such as the presence of phytopathogenic bacteria, polyaromatic hydrocarbons, heavy metals like Ca^{2+} and Ni^{2+}, salt, and drought are relieved by strains that produce ACC deaminase (Glick et al. 2007).

Taking into consideration the above characteristics, one strategy is to shape the microbiome for optimal provision of very specific services, such as by offering a competitive advantage to particular microbes in order to enhance the rate of certain enzymatic transformations. From this perspective, the goal is to develop plants that are able to select specific beneficial microbes from among a broader community, which may include many members without any beneficial function to the plant.

9.7 Biological Control of Soilborne Plant Diseases by Rhizosphere Microbiome

Diseases cause annual crop loss of more than 200 billion dollars (Agrios 2005). The methods used to control diseases conventionally are development of resistance in plants and addition of chemicals. Resistance cannot be achieved against all diseases; moreover, the breeding of resistant plants successfully takes several years. In addition, genetically engineered resistant traits in plants are not easily accepted, and it is a sensitive issue in much of the world; it also defeats the very purpose of sustainable and environment-friendly agriculture. Today, with wide exposure to media, using agrochemicals is not accepted by producers or end users. It has also been banned by several governments.

The use of microbes to prevent or remedy diseases is known as "biological control" or "biocontrol." It is an environment-friendly approach to disease control. One microbe may be a natural enemy of another species, called the "pathogen." It would produce secondary metabolites to retard or stop the growth of the antagonistic species. The metabolites produce their effect only locally, viz., the part of the plant where it should particularly act. Agrochemicals do not reach the plant at all; instead, they are sprayed all over the soil. Biological molecules are biodegradable compared to agrochemicals that are designed to resist degradation by microbes. "Biocontrol" is not only controlling disease in the vegetative stage but also preventing disease during fruit and/or grain storage (called "postharvest control"). Studies conducted on controlling pathogens by "rhizobacteria" generally focus on pathogenic microorganisms. However, many rhizobacteria are active against weeds (Flores-Fargas and O'Hara 2006) and insects (Pechy-Tarr et al. 2008).

Soils where pathogens can cause diseases are known as "conducive soils." Natural control of plant diseases by various bacterial species occurs in many soil types. Certain soils, called "suppressive soils," have bacterial populations that protect plants against most fungal diseases. Adding a small quantity of suppressive soil to the conducive soil converts the latter into a suppressive one in due course. It is a complex process, involving the control microorganism, pathogen, plant, indigenous microorganisms, nematodes, protozoa, and the substrate, namely soil, stone wool, or vermiculite.

To be effective, the "control microbe" should be active in varying pH, temperatures, and ionic concentrations. This is not very easy to fulfill. Because of all these requirements, the efficacy of many commercial biocontrol agents (Copping 2004) has not been satisfactory. However, as we understand these mechanisms and the selection for active strains increases, biocontrol products will improve; thus, biocontrol has a good potential in the future (Compant et al. 2005; Haas and Defago 2005).

9.8 Mechanisms of Biocontrol

In this study, the disease researched was "tomato foot and root rot" (TFRR), caused by the fungal pathogen *Forl.* This has been used as a model system for studying mechanisms of biocontrol utilized by various microbes (Haas and Defago 2005). The following mechanisms of biocontrol have been distinguished from this study; however, they may not be conclusive in nature.

9.8.1 Antagonism

Bacterial species that produce "antibiotics" kill pathogens by a process known as "antagonism." This type of control mechanism is "biocontrol negative." The antibiotic has to be produced and secreted at the right micro-niche of the root surface of the bacteria, for it to be effective (Pliego et al. 2008). Nutrients secreted in the roots transfer the antibiotic along the entire root system (Chin-A-Woeng et al. 2003). The biocontrol agent should also be able to escape (in sufficient number) from microbial predators present in the rhizosphere, called the "protozoan grazers" (Jousset et al. 2006). Antibiotics that are known as "antagonistic gram-negative biocontrol bacteria" include compounds like hydrogen cyanide (HCN) (Haas and Keel, 2003); phenazines (Mavrodi et al. 2006), which are mainly phenazine-1-carboxylic acid and phenazine-1-carboxamide; 2,4-diacetyl phloroglucinol (Phl) (Thomashow and Weller 1996); pyoluteorin (Nowak-Thompson et al. 1999); pyrrolnitrin (Kirner et al. 1998); zwittermicin A (Emmert et al. 2004); and kanosamine' (Milner et al. 1996), produced by *Bacillus cereus*.

Studies have shown that D-gluconic acid (Kaur et al. 2006) and 2-hexyl-5-propyl resorcinol (Cazorla et al. 2006) are also secreted by antagonistic bacteria. Volatile compounds like 2,3-butanediol, a blend of volatile compounds produced by *Bacillus* spp. (Ryu et al. 2003) or fungi (Strobel 2006) may also offer protection. Lipopeptide "biosurfactants," produced by *B. subtilis* (Ongena et al. 2007) and pseudomonads (De Bruijn et al. 2007), are used for protection. Rhamnolipid and phenazine act synergistically against soilborne diseases, caused by *Pythium* spp. (Perneel et al. 2008).

9.8.2 Signal Interference

Many bacteria exhibit pathogenicity or virulence only at a high cell density. This can be detected when quorum-sensing molecules like "homoserine lactones" (AHLs) accumulate in the medium (Bassler 1999). AHLs are required for production of cell wall-degrading enzymes of the pathogen *Erwinia carotovora.* Signal interference is a biocontrol mechanism, which is based on the degradation of AHL (Lin et al. 2003) by AHL lactonases, produced by *B. thuringiensis* strains. They hydrolyze the lactone ring; this can also be achieved by AHL acylases that break the amide link. Studies show that AHL acylases play a role in the formation of "bacterial biofilms" (Shepherd and Lindow 2008). Thus, when biofilms are not formed, it makes it easier to control the pathogen.

9.8.3 Predation and Parasitism

"Predation" and "parasitism" are the main biocontrol mechanisms used by fungi like *Trichoderma* species, based on enzymatic destruction of pathogen cell wall (Harman et al. 2004). This mechanism has not been detected so far in bacteria. Even the fungal predator *Collimonas fungivorans* uses other methods to control TFRR (Kamilova et al. 2007).

9.8.4 Induced Systemic Resistance

Association of certain bacteria with plant roots can make the plants resistant to some pathogenic bacteria, fungi, and viruses; this phenomenon is known as "induced systemic resistance" (ISR); this differs from "systemic acquired resistance" (SAR). ISR was discovered by studies, where they found that resistance could be induced by the rhizobacterium *Pseudomonas* sp. strain WCS417r against the disease fusarium wilt of carnation (Van Peer et al. 1991); this was also seen in the selected rhizobacteria against the fungus *Colletotrichum orbiculare* in cucumber (Wei et al. 1991). ISR is mainly dependent on signaling of *jasmonic acid* and *ethylene* produced by the plant (Van Loon 2007).

Many bacterial metabolites induce ISR, viz., LPS, flagella, salicylic acid, and siderophores (Van Loon 2007). Cyclic lipopeptides (Ongena et al. 2007), the antifungal factor Phl (Iavicoli et al. 2003), signal molecule AHL (Shuhegge et al. 2006), and volatile blends produced by *B. subtilis* GB03 also induce ISR, in addition to individual volatile compounds such as acetoin and 2,3-butanediol (Ryu et al. 2003).

Compared to other methods of biocontrol, extensive colonization of the root system is not needed for inducing ISR (Kamilova et al. 2005). This is proved by the fact that certain strains of *B. cereus*, which are poor colonizers, are good biocontrol agents (Gilbert et al. 1994). Some antifungal metabolites like AFMs can also induce ISR. Research has speculated that several strains of *Bacillus* that are biocontrol agents do it through ISR rather than through antibiosis.

It has been reported by Rudrappa et al. (2008) that infection of the leaves of *A. thaliana* seedlings with the pathogen, *P. syringae* pv. *tomato* Pst DC3000 results in increased secretion of L-malic acid; this signals and helps in colonization of the plant by *B. subtilis* FB17, which is a biocontrol agent protecting the plant by ISR. De Weert et al. (2002) have reported that the bacterial species *P. fluorescens* WCS365 also functions through ISR (Kamilova et al. 2005) and shows strong chemotaxis toward citric acid.

9.8.5 Competition for Ferric Ions

If antibiosis happens on a medium with low concentration of ferric ions and the test strain inhibits growth of the fungus in the absence of added ferric ions, the bacterial strain produces a "siderophore," i.e., a molecule that chelates with ferric ions. After binding with ferric ions, the siderophore-ferric ion complex is further bound to "iron-limitation-dependent" receptors on the surface of the bacterial cell. The ferric ion is subsequently released and active in the cytoplasm as ferrous ion. Bacterial species that produce high concentrations of siderophores in the rhizosphere can inhibit the growth of fungal pathogens when the ferric ion concentration is low, e.g., in acidic soil (Schippers et al. 1987).

9.8.6 Competition for Nutrients and Niches

Competition of biocontrol bacteria with pathogens for nutrients and niches in the rhizosphere is a possible mechanism of biocontrol, but experimental proof is unavailable. Kamilova et al. (2005) say that if such a mechanism exists, these strains could be selected for biocontrol. In this experiment, they have applied a mixture of rhizosphere strains on surface-sterilized seeds; this was germinated in a gnotobiotic system (Simons et al. 1996). After 1 week, the root tip that contained the most competitive root colonizers was removed from the seedling and the bacteria in the removed root tip were allowed to multiply. This was further applied to fresh seeds; this is known as a "new enrichment cycle." After three cycles, the isolated bacteria were as good as, or even better, in colonization of the root tip than the control agent *P. fluorescens* WCS365. They also grew efficiently on the root exudate. Most of the isolates including *Pseudomonas* strains PCL1751 and PCL1760 control TFRR possibly by using the abovementioned mechanisms for biocontrol (Validov 2007).

Kamilova et al. (2005) have observed that the most competitive root tip-colonizing strains did not control TFRR. It has, thus, been concluded that efficient colonization of roots is not a criterion for biocontrol. An explanation comes from the work of Pliego et al. (2008) who have isolated two similar root colonizers, of which only one shows control of "white root rot" disease in avocado plants. The two strains colonize different areas on the root. The exact "mini-niche" on the root has to be colonized, in order to provide protection against the pathogen. A study on biocontrol of TFRR by the CNN strain *P. putida* PCL1760 in stone wool shows more cells of *P. putida* after 3 weeks of colonization (Validov 2007) than all other bacterial strains. This shows the high extent of protection provided by this strain.

9.8.7 Interference with Activity, Survival, Germination, and Sporulation of the Pathogen

Fusaric acid secreted by *Forl* hyphae acts as a chemical attractant for cells of *P. fluorescens* WCS365 (De Weert et al. 2003). During this experiment, the bacteria colonize *Forl* hyphae extensively to form micro-colonies (Bolwerk et al. 2003). This is likely to make the fungus less virulent.

It has been shown that *P. fluorescens* WCS365 also colonizes *Forl* hyphae when incubated in root exudate. It is seen that the poorer the growth medium, the more extensive the colonization (Kamilova et al. 2008). This observation supports the earlier suggestion (Kamilova et al. 2007) that bacteria colonize fungal hyphae and later utilize it as a food source. When incubated in root exudate, the microconidia of *Forl* germinate; these are used to spread the pathogen through air. The presence of *P. fluorescens* WCS365 inhibits the germination of spores, probably because nutrients are deprived. It also causes reduction of spore formation and, thus, inhibits spread of the pathogen. In conclusion, it is seen that the biocontrol agent *P. fluorescens* WCS365 inhibits the activity, survival, and germination of the pathogen; colonizes its hyphae; and prevents formation of new spores. Although all these mechanisms may not be present for all biocontrol agents, when plants are grown in sterile stone wool, these effects contribute significantly to reducing TFRR after *P. fluorescens* WCS365 is introduced (Validov et al. 2009).

9.9 Other Mechanisms of Biocontrol

To enhance the efficacy of disease control, two strains are inoculated into the seeds; each strain uses a different mechanism of biocontrol. However, such experiments do not result in better disease control. This is probably because the cell numbers of each species on the root are less than the threshold level required to cause control, due to competition or other reasons.

Bacteria that are native to the soil compete with biocontrol strains for root colonization and produce different factors or secrete antagonistic compounds that could reduce the beneficial effect of the biocontrol strain. In this study, sterile stone wool is used as the substrate. Because it is free from living microbes, it has the disadvantage that incoming pathogens destroy plants; the advantage is that such a system can be buffered with biocontrol bacteria. In one case, when stone wool was inoculated with *P. putida* PCL1760, it remained the dominant microbe for almost 3 weeks and had high affinity for the substrate (Validov 2007). A similar effect was seen in the saline desert soil in Uzbekistan, which is poor in organic matter and indigenous microflora. This soil is rich in plant pathogens as well as potential human pathogens. Under these circumstances, seeds that are inoculated with biocontrol strains, which adapt to stress conditions strongly, are found to decrease plant diseases and protect field-workers from exposure to pathogens (Egamberdiyeva et al. 2008).

9.10 The Biofertilizer Technology: An Application of the Rhizosphere Microbiome to Enhance Plant Health and Growth

A “biofertilizer” is a nitrogen-rich (or alternatively, phosphorus- or potassium-rich) metabolic product of a plant, animal, or microorganism, secreted into the environment. Biological nitrogen fixation utilizing symbiotic and nonsymbiotic microbes holds high potential for safe food production in the future. Every year, almost 139 billion tonnes of nitrogen is fixed in this way. There are varying categories of biofertilizers, such as “whole microbial species” (added as viable cells or spores), “biomanure,” “compost,” and “vermicompost.”

“Biofertilizer technology” is the process of domesticating wild microbial populations for commercial production of biofertilizers, or as in the case of biomanure and compost, using the metabolic processes of soil bacteria and fungi to degrade biological waste matter and enrich the soil. Vermicompost uses the natural ecological life cycle and adaptations of earthworm species to increase the bio value of compost.

Biofertilizers enhance the “physical soil structure” and “soil texture” (ratio of sand, slit, and clay particles in soil that affect features such as porosity, pore space, and water retention) of the soil in addition to chemical properties such as “cation exchange capacity,” “buffering capacity,” and “water holding capacity.”

Inoculation of crop plants with nitrogen fixers is accepted in countries like the USA, Germany, Brazil, Israel, Egypt, China, and India. They are slow compared to chemical fertilizers but sustainable over a longer period of time and cost-effective. They also provide a wider spectrum of nutrients, releasing them slowly compared to chemical fertilizers. However, organic fertilizers like manure can introduce pathogenic spores and cells, as also seeds of undesirable plants into the soil.

To achieve environmental, economic, and social sustainability in agriculture systems, all categories of biofertilizers and organic fertilizers (biomanure and compost) should be commercialized and produced in quantities equivalent to that of synthetic fertilizers.

9.11 Mycorestoration: Use of Fungi of the Rhizosphere Microbiome in Enhancement of Plant Health and Growth

It is the use of "mycorrhizal interactions of various fungal species with plant root systems" to increase the mobilization of phosphorus and potassium in soil and increase water availability, drought resistance, and protection from parasites (Tallapragada et al. 2011). *Glomus* spp. is one of the classical examples of mycorrhizal fungi in use today.

9.12 Biofertilizers of the Rhizosphere Microbiome

***Rhizobium* spp.**

The root nodules of legumes are miniature factories in producing soil nitrogen; they are engineered by the plant roots and the symbiotic bacterial species residing in them, namely, *Rhizobium* spp. They are mainly found in seven classes of legumes, namely, alfalfa (*Rhizobium meliloti*), clover (*R. trifolii*), pea (*R. leguminosarum*), bean (*R. phaseoli*), lupine (*R. lupini*), and soybean (*R. japonicum*).

Rhizobium spp. can fix free nitrogen from the atmosphere into the soil. It enters through the root hairs or the point of origin of secondary roots, colonizing the outer cortex. At this point, they become polymorphic, being known as "bacteroides." The host plant synthesizes "leghemoglobin," a compound that surrounds the bacteroides. This then helps them to metabolize molecular nitrogen, which they fix from the atmosphere; this nitrogen is also utilized by the plant for various reactions and growth. The bacterium, in turn, gets nutrition from the host.

Rhizobium biofertilizers can add almost 50–200 kilograms of nitrogen per hectare per year to the soil. This increases the yield of the crop by 25%–30% and the nitrogen content of soil by 40–80 kilograms; it covers the nitrogen requirement of subsequent harvests too (Zahran 1999). In India, several firms, private and government-owned, are involved in the commercial production of *Rhizobium* biofertilizers. A good-quality inoculum has about 10^8–10^9 cells per gram of the carrier material and a shelf life longer than 4 months. Plant seeds are coated with this inoculum; it is the most usual method of introduction into the soil, known as "seed pelleting." In the case of acidic and saline soils, the seeds are coated with gypsum or lime to protect the bacterial cells from effects of the acids and alkalis. *Rhizobium* species has the "nif" genes, which helps them fix atmospheric nitrogen through the above processes.

***Azotobacter* spp.**

In the case of nonlegume species like cotton, sorghum, pearl millet, maize, and various other cereals, *Azotobacter* spp. has been applied for many decades as a suitable biofertilizer. It is a dependable source of nitrogen by nonsymbiotic nitrogen fixation, which is also suitable for pot-based and field-based studies. The Tamil Nadu Agricultural University has recommended its use in rice cultivation. However, natural populations of *Azotobacter* spp. are usually low in soil, compared to other bacteria. In addition, it requires a substantial amount of energy to fix atmospheric nitrogen; the major energy source comes from the organic content of soil. Indian soils have very low organic content (0.1%–0.2%), and hence, this bacterial species is not of any practical use in Indian soil conditions. The most efficient strains of *Azotobacter* require about 1,000 kilograms of organic matter to fix 30 kilograms of nitrogen per hectare. Because of this, scientists have become reluctant to recommend its use. Its main advantage is that it secretes antibiotics into the soil, preventing plant infections, in addition to secretion of growth regulators such as IAA, IBA, NAA, and GA_1 to GA_3 (Steenhoudt and Vanderleyden 2000; Bashan and de-Bashan 2018).

***Azospirillum* spp.**

Azospirillum is found closely associated with root systems of most plants. It is associated with cereals like sorghum, maize, barley, oat, wheat, and some minor millets including fodder grass. They can either colonize the root surface or penetrate the roots and live in symbiotic association with the host. In a day, it fixes approximately 0.5–0.8 kilograms of nitrogen per hectare of soil. It secretes antibiotics, which behave like "biological pesticides," in addition to growth regulators (Steenhoudt and Vanderleyden 2000; Bashan and de-Bashan 2018).

Blue-Green Algae

The "blue-green algae" (BGA) are capable of nitrogen fixation and phosphate solubilization. They are a good substitute for synthetic nitrogen fertilizers, in the case of wet and semi-wet crop cultivation. This is especially true for rice fields, where it has replaced nearly 70% of chemical fertilizer application.

It is also capable of dissolving inorganic phosphorus, giving a double benefit to the crops. The secretion of organic acids by BGA cells increases availability of phosphorus in soil, as these organic acids can solubilize the inorganic calcium phosphate into soluble forms such as orthophosphates. *Anabaena* spp., *Nostoc* spp., *Aulosira* spp., and *Toypothrix* spp. are able to solubilize the extracellular phosphate, up to 2.27 mg P_2O_2 per 50 milliliters of soil water, every 20 days (Steenhoudt and Vanderleyden 2000; Bashan and de-Bashan 2018).

In addition, BGA cells also release various growth regulators like auxins (IAA, IBA, and NAA) and gibberellins (GA_1 to GA_3) that promote plant growth (Venkatraman and Neelkantan 1967). The growth rate of rice seedlings that were treated with algal filtrate of *Aulosira fertilissima* resembled that of seedlings treated with gibberellic acid (Singh and Trehan 1973).

BGA can also take up sodium from salt-affected soils. Organic acids secreted by BGA cells make $CaCO_3$ soluble, and the released calcium replaces sodium in the soil complex. Algalization of sodic and saline soil reduces the following: pH, electrical conductivity, and percentage of exchangeable sodium (Subhashini and Kaushik 1981).

BGA is a soil binder and improves the moisture-holding capacity. It improves soil aggregation by the release of complex polysaccharides such as glucose, galactose, xylose, arabinose, and rhamnose. An increase in organic matter by nearly 69%,

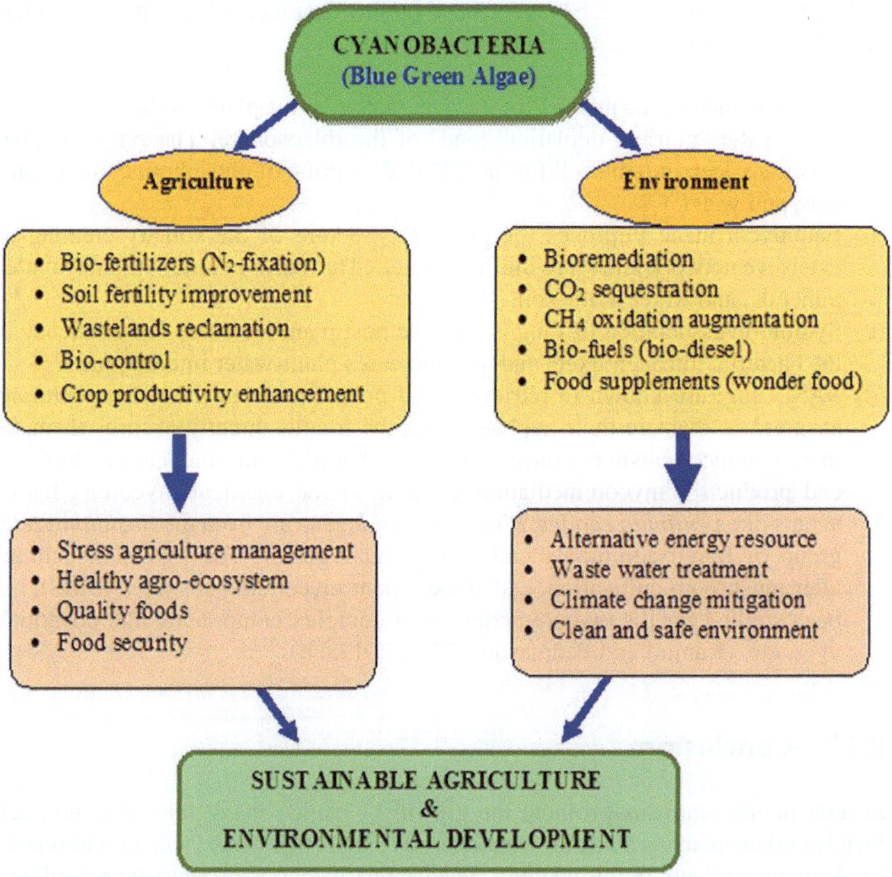

Fig. 9.4 Role of blue-green algae in sustainable agriculture

water holding capacity by 35%, and exchangeable calcium by 58% has been reported (Subhashini and Kaushik 1981) – Fig. 9.4. shows the role of blue-green algae (BGA) in sustainable agriculture.

Mycorrhizae

"Mycorrhizae" is a mutualistic association of fungi with the root systems of higher plants. Fungal species from *Basidiomycetes* and *Zygomycetes* form a mantle around the roots. The mycelium may be exogenous ("ectomycorrhizae") or endogenous ("endomycorrhizae"). In endomycorrhizae, the hyphae enter the root hairs, forming either highly branched systems ("arbuscles") and/or bladder-like structures ("vesicles"); it is thus known as "vesicular-arbuscular mycorrhizae" (VAM). The genera forming VAM associations are usually *Glomus* spp., *Acaulospora* spp., *Entrophospora* spp., *Gigaspora* spp., *Sclerocystis* spp., and *Scutellospora* spp. (Steenhoudt and Vanderleyden 2000; Bashan and de-Bashan 2018).

Mainly, the ways in which "mycorrhizal biofertilizer technology" increases plant growth and soil fertility are as follows:

(a) They enhance the acquisition of nutrient matter by the plant, as they grow much beyond the "nutrient depletion zone" of the rhizosphere. The fungal hyphae penetrate deep into the soil for increased absorption of phosphorus, micronutrients, and water.
(b) Ectomycorrhizae improves the physical structure of the soil by creating an extensive network known as the "hartig net." This helps to retain organic matter, minerals, and water in the soil.
(c) Symbiotic association of fungus with the host plant increases the efficiency of the bacterial nitrogen fixers and also increases plant-water interactions.
(d) VAM fungi are known to retrieve metal pollutants from the rhizosphere and accumulate them in their hyphae. Here, the fungus directly absorbs them for their own metabolism or converts them into harmless intermediates and/or final end products ("mycoremediation"). Some of the classical mycoremediation agents like *Lentinula edodes*, *Pleurotus florida*, etc. are from the *Basidiomycetes* group of fungi, commonly called the "mushrooms." The "spent mycelium" after mushroom cultivation, called the "spent mycelium substrate" (SMS), has been studied for the mycoremediation of complex compounds like phenolics, dyes, etc. (Ranjini and Padmavathi 2012, 2013a, b).

9.13 Conclusion

Several microorganisms promote the growth of plants; many microbial products that stimulate plant growth are marketed as "biofertilizers," "soil conditioners," "soil enhancers," etc. In this chapter, we have restricted ourselves to the role of bacterial and fungal genera present in the vicinity of the root system of the host, known as "rhizosphere." This is an important and significant portion of the "plant microbiome," which is the sum of all the microbial consortia associated with the host plant. Such bacteria are known as PGPR ("plant growth-promoting rhizobacteria"), while the fungal species are known as PGPF ("plant growth-promoting fungi"). The effects of these rhizobacteria and/or fungi are usually beneficial for the plants and may be direct or indirect. It also deals with bacterial and fungal interactions with the plant, in the microbiome, and the mechanisms of growth promotion, soil fertility enhancement, and disease control in the host. Finally, specific groups of bacteria and fungi have been discussed, with respect to their abovementioned roles, with special emphasis on nitrogen fixation, phosphate solubilization, etc. The mechanisms by which microbes can act beneficially on plant growth include (a) biofertilization, (b) stimulation of root growth, (c) rhizoremediation, and (d) plant stress control. Mechanisms of biological control by which rhizobacteria can promote plant growth indirectly are antibiosis, induction of systemic resistance, and competition for nutrients and niches. The above processes need to be encouraged in view of the fact that most of the global soils have been destroyed and depleted by the addition of chemical fertilizers, genetic modifications of seeds and plantlets, and pesticides to control diseases. The sustained usage of PGPR and PGPF organisms in agriculture can lead to "sustainable agriculture" in the global scenario and a meaningful role of the "plant microbiome" in this way of farming.

References

Agrios GN (2005) Plant pathology, 4th edn. Academic. 635 pp, Amsterdam

Badri DV, Vivanco JM (2009) Regulation and function of root exudates. Plant Cell Environ 32:666–681

Badri DV, Quintana N, El Kassis EG, Kim HK, Choi YH, Sugiyama A, Verpoorte R, Martinoia E, Manter DK, Vivanco JM (2009) An ABC transporter mutation alters root exudation of phytochemicals that provoke an overhaul of natural soil microbiota. Plant Phys 151:2006–2017

Bais HP, Weir TL, Perry LG, Gilroy S, Vivanco JM (2006) The role of root exudates in rhizosphere interactions with plants and other organisms. Annu Rev Plant Biol 57:233–266

Bashan Y, de-Bashan LE (2018) How the plant growth-promoting bacterium *Azospirillum* promotes plant growth-a critical assessment. Adv Agron 108:77–136

Bassler BL (1999) How bacteria talk to each other: regulation of gene expression by quorum sensing. Curr Opin Microbiol2: 582–587

Berg G, Smalla K (2009) Plant species and soil type cooperatively shape the structure and function of microbial communities in the rhizosphere. FEMS Microbiol Ecol 68:1–13

Bolwerk A, Lagopodi AL, Wijfjes AHM, Lamers GEM, Chin-A-Woeng TFC, Lugtenberg BJ, Bloemberg GV (2003) Interactions in the tomato rhizosphere of two *Pseudomonas* bio control strains with the phytopathogenic fungus *Fusarium oxysporum f.* sp. radicis-lycopersici. Mol Plant-Microbe Interact 16:983–993

Bottiglieri M, Keel C (2006) Characterization of PhlG, a hydrolase that specifically degrades the antifungal compound 2,4-diacetylphloroglucinol in the biocontrol agent *Pseudomonas fluorescens CHA0*. Appl Environ Microbiol 72:418–427

Bremer C, Braker G, Matthies D, Beierkuhnlein C, Conrad R (2009) Plant presence and species combination, but not diversity, influence denitrifier activity and the composition of K-type denitrifier communities in grassland soil. FEMS Microbiol Ecol 70:377–387

Broeckling CD, Broz AK, Bergelson J, Manter DK, Vivanco JM (2008) Root exudates regulate soil fungal community composition and diversity. Appl Environ Microbiol 74:738–744

Broz AK, Manter DK, Vivanco JM (2007) Soil fungal abundance and diversity: another victim of the invasive plant *Centaurea maculosa*. ISME J 1:763–765

Bulgarelli D, Schlaeppi K, Spaepen S, Ver Loren van Themaat E, Schulze-Lefert P (2013) Structure and functions of the bacterial microbiota of plants. Ann Rev Plant Biol 64: 807-838

Cazorla FM, Duckett SB, Bergstrom ET, Noreen S, Odijk R, Lugtenberg BJ, Thomas-Oates JE, Bloemberg GV (2006) Biocontrol of avocado Dematophora root rot by the antagonistic *Pseudomonas fluorescens PCL1606* correlates with the production of 2-hexyl 5-propyl resorcinol. Mol Plant-Microbe Interact 19:418–428

Chin-A-Woeng TF, Bloemberg GV, Lugtenberg BJ (2003) Phenazines and their role in biocontrol by Pseudomonas bacteria. New Phytol 57:503–523

Choi O, Kim J, Kim JG, Jeong Y, Moon JS, Park CS, Hwang I (2008) Pyrroloquinoline quinine is a plant growth promotion factor produced by *Pseudomonas fluorescens* B16. Plant Physiol 146:657–668

Compant S, Duffy B, Nowak J, Clement C, Barka EA (2005) Use of plant growth-promoting bacteria for bio control of plant diseases: principles, mechanisms of action, and future prospects. Appl Environ Microbiol 71:4951–4959

Copping LG (2004) The Manual of Biocontrol Agents. Alton, Hampshire, UK: Br. Crop Prot Counc Publ 702 pp. 3rd ed.

Dangl JL, Horvath DM, Staskawicz BJ (2013) Pivoting the plant immune system from dissection to deployment. Science 341:746–751

De Bruijn I, De Kock MJD, Yang M, De Waard P, Van Beek TA, Raaijmakers JM (2007) Genome-based discovery, structure prediction and functional analysis of cyclic lipopeptide antibiotics in *Pseudomonas species*. Mol Microbiol 63:417–428

De Weert S, Vermeiren H, Mulders IHM, Kuiper I, Hendrickx N, Bloemberg GV, Vanderleyden J, De Mot R, Lugtenberg BJ (2002) Flagella-driven chemotaxis towards exudate components is an important trait for tomato root colonization by *Pseudomonas fluorescens*. Mol Plant-Microbe Interact 15:1173–1180

De Weert S, Kuiper I, Lagendijk EL, Lamers GEM, Lugtenberg BJJ (2003) Role of chemotaxis toward fusaric acid in colonization of hyphae of *Fusarium oxysporum f.* sp. radicis-lycopersici by *Pseudomonas fluorescens WCS365*. Mol Plant-Microbe Interact 16:1185–1191

Egamberdiyeva D, Kamilova F, Validov S, Gafurova L, Kucharova Z, Lugtenberg B (2008) High incidence of plant growth-stimulating bacteria associated with the rhizosphere of wheat grown in salinated soil in Uzbekistan. Environ Microbiol 10:1–9

Emmert EA, Klimowicz AK, Thomas MG, Handelsman J (2004) Genetics of zwittermicin A production by *Bacillus cereus*. Appl Environ Microbiol 70:104–113

Flores-Fargas RD, O'Hara GW (2006) Isolation and characterization of rhizosphere bacteria with potential for biological control of weeds in vineyards. J Appl Microbiol 100:946–954

Franken P (2012) The plant strengthening root endophyte *Piriformospora indica*: potential application and the biology behind. Appl Microbiol Biotechnol 96:1455–1464

Gilbert GS, Handelsman J, Parke JL (1994) Root camouflage by disease control. Phytopathology 84:222–225

Glick BR, Cheng Z, Czarny J, Duan J (2007) Promotion of plant growth by ACC deaminase-producing soil bacteria. Eur J Plant Pathol 119:329–339

Grunsven RHA, Putten WH, Bezemer TM, Veenendaal EM (2009) Plant-soil feedback of native and range expanding plant species is insensitive to temperature. Oecologia 162:1059–1069

Haas D, Defago G (2005) Biological control of soil-borne pathogens by fluorescent pseudomonads. Nat Rev Microbiol 3:307–319

Haas D, Keel C (2003) Regulation of antibiotic production in root-colonizing *Pseudomonas* spp. and relevance for biological control of plant disease. Annu Rev Phytopathol 41:117–153

Harman GE, Howel CH, Viterbo A, Chet I, Lorito M (2004) *Trichoderma species*–opportunistic, avirulent plant symbionts. Nat Rev Microbiol 2:43–56

Hwang SF, Ahmed HU, Gossen BD, Kutcher HR, Brandt SA, Strelkov SE, Chang KF, Turnbull GD (2009) Effect of crop rotation on the soil pathogen population dynamics and canola seedling establishment. Plant Pathol J 8:106–112

Iavicoli A, Boutet E, Buchala A, Metraux JP (2003) Induced systemic resistance in *Arabidopsis thaliana* in response to root inoculation with *Pseudomonas fluorescens* CHA0. Mol Plant-Microbe Interact 16:851–858

Jones DL, Hodge A, Kuzyakov Y (2004) Plant and mycorrhizal regulation of rhizo deposition. New Phytol 163:459–480

Jousset A, Lara E, Wall LG, Valverde C (2006) Secondary metabolites help biocontrol strain *Pseudomonas fluorescens CHA0* to escape protozoan grazing. Appl Environ Microbiol 72:7083–7090

Kamilova F, Validov S, Azarova T, Mulders I, Lugtenberg B (2005) Enrichment for enhanced competitive plant root tip colonizers selects for a new class of biocontrol bacteria. Environ Microbiol 7:1809–1817

Kamilova F, Kravchenko LV, Shaposhnikov AI, Makarova N, Lugtenberg BJJ (2006) Effects of the tomato pathogen *Fusarium oxysporum f.* sp. *radicis-lycopersici* and of the bio control bacterium *Pseudomonas fluorescens* WCS365 on the composition of organic acids and sugars in tomato root exudate. Mol Plant-Microbe Interact 19:1121–1126

Kamilova F, Leveau JHJ, Lugtenberg B (2007) *Collimonas fungivorans*, an unpredicted *in vitro* but efficient in vivo bio control agent for the suppression of tomato foot and root rot. Environ Microbiol 9:1597–1603

Kamilova F, Lamers G, Lugtenberg B (2008) Biocontrol strain *Pseudomonas fluorescens* WCS365 inhibits germination of *Fusarium oxysporum* spores in tomato root exudate as well as subsequent formation of new spores. Environ Microbiol 10:2455–2461

Kaur R, Macleod J, Foley W, Nayudu M (2006) Gluconic acid, an antifungal agent produced by *Pseudomonas* species in biological control of take-all. Phytochemistry 67:595–604

Kirner S, Hammer PE, Hill DS, Altmann A, Fischer I, Weislo LJ, Lanahan M, van Pée KH, Ligon JM (1998) Functions encoded by pyrrolnitrin biosynthetic genes from *Pseudomonas fluorescens*. J Bacteriol 180:1939–1943

Kuiper I, Bloemberg GV, Lugtenberg BJJ (2001) Selection of a plant-bacterium pair as a novel tool for rhizostimulation of polycyclic aromatic hydrocarbon-degrading bacteria. Mol Plant-Microbe Interact 14:1197–1205

Kuiper I, Lagendijk EL, Bloemberg GV, Lugtenberg BJJ (2004) Rhizoremediation: a beneficial plant microbe interaction. Mol Plant-Microbe Interact 17:6–15

Lambers H, Mougel C, Jaillard B, Hinsinger P (2009) Plant-microbe-soil interactions in the rhizosphere: an evolutionary perspective. Plant Soil 321:83–115

Lin YH, Xu JL, Hu JY, Wang LH, Ong SL, Leadbetter JR, Zhang LH (2003) Acyl-homoserine lactone acylase from *Ralstonia strain XJ12B* represents a novel and potent class of quorum-quenching enzymes. Mol Microbiol 47:849–860

Loreau M, Naeem S, Inchausti P, Bengtsson J, Grime JP, Hector A, Hooper DU, Huston MA, Raffaelli D, Schmid B, Tilman D, Wardle DA (2001) Biodiversity and ecosystem functioning: current knowledge and future challenges. Science 294:804–808

Mavrodi DV, Blankenfeldt W, Thomashow LS (2006) Phenazine compounds in fluorescent *Pseudomonas* spp.: biosynthesis and regulation. Annu Rev Phytopathol 44:417–445

Mendes R, Kruijt M, De Bruijn I, Dekkers E, Van der Voor TM, Schneider JH, Piceno YM, De Santis TZ, Andersen GL, Bakker PA, Raaijmakers JM (2011) Deciphering the rhizosphere microbiome for disease-suppressive bacteria. Science 332:1097–1100

Micallef SA, Shiaris MP, Colon-Carmona A (2009) Influence of *Arabidopsis thaliana* accessions on rhizo bacterial communities and natural variation in root exudates. J Exp Bot 60:1729–1742

Milner J, Silo-Suh L, Lee JC, He H, Clardy J, Handelsman J (1996) Production of kanosamine by *Bacillus cereus UW85*. Appl Environ Microbiol 62:3061–3065

Morales SE, Holben WE (2011) Linking bacterial identities and ecosystem processes: can "omic" analyses be more than the sum of their parts? FEMS Microbiol Ecol 75:2–16

Nowak-Thompson B, Chaney N, Wing JS, Gould SJ, Loper JE (1999) Characterization of the pyoluteorin biosynthetic gene cluster of *Pseudomonas fluorescens* Pf-5. J Bacteriol 181:2166–2174

Okon Y, Bloemberg GV, Lugtenberg BJJ (1998) Biotechnology of bio fertilization and phytostimulation. In: Altman A (ed) Agricultural Biotechnology. Marcel Dekker, New York, pp 327–349

Oldroyd GE, Murray JD, Poole PS, Downie JA (2011) The rules of engagement in the legume-rhizobial symbiosis. Ann Rev Genet 45:119–144

Ongena M, Jourdan E, Adam A, Paquot M, Brans A, Joris B, Arpigny JL, Thonart P (2007) Surfactin and fengycin lipopeptides of *Bacillus subtilis* as elicitors of induced systemic resistance in plants. Environ Microbiol 9: 1084–1090

Pechy-Tarr M, Bruck DJ, Maurhofer M, Fisher E, Vogne C, Henkels MD, Donahue KM, Grunder J, Loper JE, Keel C (2008) Molecular analysis of a novel gene cluster encoding an insect toxin in plant-associated strains of *Pseudomonas fluorescens*. Environ Microbiol 10:2368–2386

Perneel M, D'Hondt L, De Maeyer K, Adiobo A, Rabaey K, Hofte M (2008) Phenazines and bio surfactants interact in the biological control of soil-borne diseases caused by *Pythium* spp. Environ Microbiol 10:778–788

Pliego C, De Weert S, Lamers G, De Vicente A, Bloemberg G, Cazorla FM, Ramos C (2008) Two similar enhanced root-colonizing Pseudomonas strains differ largely in their colonization strategies of avocado roots and Rosellinia neatrix hyphae. Environ Microbiol 10:3295–3304

Ranjini R, Padmavathi T (2012) Differential phenol tolerance and degradation by *Lentinula edodes* cultures in differing levels of carbon and nitrogen. Asiatic J Biotech Res 3 10(1):1457–1471

Ranjini R, Padmavathi T (2013a) A preliminary assessment of phenol tolerance and degradation by spent mycelium substrate (SMS) of novel edible mushroom *Hypsizygus ulmarius*. J Science Ind Res 72:767–771

Ranjini R, Padmavathi T (2013b) Phenol tolerance and degradation profile of novel edible mushroom *Hypsizygus ulmarius* in ligninolytic and non-ligninolytic media. Int J Pharma Biol Sci 3(4(B):987–994

Roesch LF, Fulthorpe RR, Riva A, Casella G, Hadwin AKM, Kent AD, Daroub SH, Camargo FA, Farmerie WG, Triplett EW (2007) Pyrosequencing enumerates and contrasts soil microbial diversity. ISME J 1:283–290

Rudrappa T, Czymmek KJ, Par'e PW, Bais HP (2008) Root-secreted malic acid recruits beneficial soil bacteria. Plant Physiol 148:1547–1556

Ryu CM, Farag MA, Hu CH, Reddy MS, Wie HX, Paré PW, Kloepper JW (2003) Bacterial volatiles promote growth of Arabidopsis. Proc Natl Acad Sci U S A 100:4927–4932

Salles JF, Van Veen JA, Van Elsas JD (2004) Multivariate analyses of *Burkholderia* species in soil: effect of crop and land use history. Appl Environ Microbiol 70:4012–4020

Schippers B, Bakker AW, Bakker PAHM (1987) Interactions of deleterious and beneficial rhizosphere microorganisms and the effect of cropping practices. Annu Rev Phytopathol 25:339–358

Schlaeppi K, Bulgarelli D (2015) The plant microbiome at work. The American Phytopathological Society, MPMI 28(3):212–217

Shepherd RW, Lindow SE (2008) Two dissimilar N-acyl-Homoserine lactone acylases of Pseudomonas syringae influence colony and biofilm morphology. Appl Environ Microbiol 75(1):45–53

Shuhegge R, Ihring A, Gantner S, Bahnweg G, Knaooe C, Vogg G, Hutzler P, Schmid M, Van Breusegem F, Eberl L, Hartmann A, Langebartels C (2006) Induction of systemic resistance in tomato by N-acyl-L-homoserine lactone-producing rhizosphere bacteria. Plant Cell Environ 29:909–918

Simons M, van der Bij Arjan J, Brand I, de Weger LA, Wijffelman CA, Lugtenberg BJJ (1996) Gnotobiotic system for studying rhizosphere colonization by plant growthpromoting Pseudomonas bacteria. Mol Plant Microbe Int 9(7):600–607

Singh VP, Trehan T (1973) Effect of extracellular products of *Aulosira fertilissima* on the growth of rice seedlings. Plant Soil 38:457–464

Steenhoudt O, Vanderleyden J (2000) *Azospirillum*, a free-living nitrogen-fixing bacterium closely associated with grasses: genetic, biochemical and ecological aspects. FEMS Microbiol Rev 24(4):487–506

Strobel G (2006) Harnessing endophytes for industrial microbiology. Curr Opin Microbiol 9(3):240–244

Subhashini D, Kaushik BD (1981) Amelioration of sodic soils with blue green algae. Aust J Soil Res 19:361–367

Tallapragada P, Rajiv R, Anjanappa V, Sardesai S, Selvaraj S, Khan S (2011) Comparing the potential of spent mycelium substrate of *Pleurotus florida* with biofertilizers to enhance growth of *Capsicum annuum*. Asian J Plant Sci Res 1(4):76–86

Thomashow LS, Weller DM (1996) Current concepts in the use of introduced bacteria for biological disease control: mechanisms and antifungal metabolites. In: Stacey G, Keen NT (eds) Plant-microbe interact, vol 1. Chapman & Hall, New York, pp 187–235

Torsvik V, Goksoyr J, Daae F (1990) High diversity in DNA of soil bacteria. Appl Environ Microbiol 56:782–787

UNEP (1996a) Farm the City. Our Planet (Jac Smit)

UNEP (1996b) Food for all: the world food summit. Our Planet (Jacques Diouf)

UNEP (1996c) Greening the fields. Our Planet (Hans Johnson)

Validov S (2007) Biocontrol of tomato foot and root rot by Pseudomonas bacteria in stonewool. PhD thesis. Leiden Univ. http://hdl.handle.net/1887/12480

Validov SZ, Kamilova F, Lugtenberg BJJ (2009) *Pseudomonas putida* strain PCL1760 controls tomato foot and root rot in stone wool under industrial conditions in a certified greenhouse. Biol Control 48:6–11

Van Elsas JD, Costa R, Jansson J, Sjoling S, Bailey M, Nalin R et al (2008) The metagenomics of disease-suppressive soils – experiences from the METACONTROL project. Trends Biotechnol 26:591–601

Van Loon LC (2007) Plant responses to plant growth-promoting bacteria. Eur J Plant Pathol 119:243–254

Van Peer R, Niemann GJ, Schippers B (1991) Induced resistance and phytoalexin accumulation in biological control of *Fusarium* wilt of carnation by *Pseudomonas* sp. strain WCS417r. Phytopathology 81:728–734

Venkatraman GS, Neelkantan S (1967) Effect of the cellular constituents of the nitrogen fixing blue-green algae *Cylindrosporum muscicola* on the root growth of rice seedlings. J Gen Appl Microbiol 13:53–61

Vorholt J (2012) Microbial life in the phyllosphere. Nature Rev Microbiol 10:828–840

Wei G, Kloepper JW, Tuzun S (1991) Induction of systemic resistance of cucumber to *Colletotrichum orbiculare* by select strains of plant growth-promoting rhizobacteria. Phytopathology 81:1508–1512

Yergeau E, Sanschagrin S, Maynard C, St-Arnaud M, Greer CW (2014) Microbial expression profiles in the rhizosphere of willows depend on soil contamination. ISME J 8:344–358

Zahran HH (1999) Rhizobium-legume symbiosis and nitrogen fixation under severe conditions and in an arid climate. Microbiol Mol Biol Rev 63(4):968–989

Zhang H, Xie X, Kim M-S, Kornyeyev DA, Holaday S, Par'e PW (2008) Soil bacteria augment *Arabidopsis* photosynthesis by decreasing glucose sensing and abscisic acid levels in plantae. Plant J 56:264–273

[illegible]

Waksman SA, Woodruff HB [illegible] [illegible]

[illegible]

[illegible]

[illegible]

Influence of the Rhizospheric Microbiome in Plant Health Management

10

Vankayalapati Vijaya Kumar

Abstract

The microbiome, a community of microorganisms that inhabit a particular environment, plays a vital role in maintaining the health of plants, humans, and other living beings. In plants, distinct microbiomes are associated with various niches—above ground (in the phyllosphere), in the internal tissues (in the endosphere), and below ground (in the rhizosphere)—of the same plant. The rhizospheric microbiome contains various microbes such as bacteria, fungi, actinomycetes, algae, protozoans, and nematodes. These microbes promote plant growth by nutrient acquisition, suppression of pathogens, and alleviation of abiotic stress.

This chapter reviews the function of the rhizospheric microbiome in plant health management and in sustainable agriculture.

Keywords

Microbiome · Plant health · Plant pathogens

10.1 Introduction

Plants are autotrophic organisms, which supply food, clothing, shelter, and other needs of living beings, including humans and animals. The microorganisms (or the combined genetic material of the microorganisms) present in a particular environment are described as the microbiome. The term "microbiome" was coined in 2001 by the Nobel Laureate Joshua Lederberg, who introduced the concept of the human microbiome. According to Lederberg the microbiome signifies "the ecological community of commensal, symbiotic, and pathogenic microorganisms that literally

V. V. Kumar (✉)
Trident Sugars Limited, Zaheerabad, Sangareddy, Telangana, India

V. Kumar et al. (eds.), *Microbiome in Plant Health and Disease*,
https://doi.org/10.1007/978-981-13-8495-0_10

share our body space and have been all but ignored as determinants of health and disease" (Lederberg and McCray 2001).

Plants also contain a diverse variety of microorganisms in various parts. The purposeful gene pool of prokaryotes, eukaryotes, and viruses accompanying the numerous niches of plants constitute the plant microbiome. These niches include whole plants or particular organs such as plant leaves, the root zone, the shoot zone, flowers, seeds, and the rhizosphere, which is the region of interaction between the soil and the roots (Rout and Southworth 2013). On the basis of the plant–microbiome interaction, the microorganisms are classified into three types: epiphytic, endophytic, and rhizospheric.

10.2 Epiphytic Microorganisms

Epiphytic microbes are the normal populations that can proliferate and survive on the plant surface. They grow in the base of the trichomes, in the substomatal chambers, in the hydathodes, and in the hollows and dents along the neighboring epithelial cell junctions. The leaf surface area is called the phyllosphere, with a wax-coated cuticle at the leaf–atmosphere interface. The phyllosphere is considered an unfavorable environment for supporting microorganisms, because of its exposure to ultraviolet light, temperature variations ranging from 5–10 °C during the night to 45–55 °C during the day, and low water and nutrient availability. In spite of this, typical microflora are established on the leaves as a result of contact with dust deposited on the leaves through air currents, aided by waxes, the cuticle, and appendages. Phyllospheric microorganisms survive on the leaves as a result of secretion of simple sugars such as glucose, fructose, and sucrose (Ajar et al. 2017; Hirano and Upper 2000; Lindow and Brandl 2003).

Filho et al. (2010) suggested that there was a reduction in bacterial spots (*Xanthomonas vesicatoria*) and early blight (*Alternaria solani*) in tomato plants after application of the biocontrol bacteria *Bacillus pumilus* and *Paenibacillus macerans*. Four days after prespraying of the tomato plants with benzalkonium chloride, epiphytic bacteria, and phosphate saline buffer, the plants were inoculated with the pathogens *Xanthomonas vesicatoria* and *A. solani*. Leaflet samples were collected and tested for pathogenic bacteria daily for 1 week, and it was found that the number of pathogenic bacteria was reduced by 70%. The bioassays also showed protection of the tomato plants by the epiphytic bacteria. Mercier and Lindow (2000) measured glucose and other sugars in bean leaves before and after inoculation with the bacteria *Pseudomonas fluorescens* strain A506. In uncolonized bean plants, averages of nearly 2.5 mg of whole sugar and 1.4 mg of glucose were observed per gram of leaves. After inoculation of bean plants with *P. fluorescens* strain A506 and incubation in humid conditions, the sugars were rapidly reduced to about 0.25 mg/g of leaf after 20 hours and the population of bacteria reached and was maintained at a size of about 1.73×10^7 colony-forming units per gram.

The following is a partial list of epiphytic microorganisms identified from different plants: *Pseudomonas* sp., *Stenotrophomonas* sp., and *Achromobacter* from

Hedera helix (Schreiber et al. 2005); *Pseudomonas*, *Stenotrophomonas*, *Bacillus*, and *Arthrobacter* from strawberry plants (Krimm et al. 2005); and *Citrobacter* sp., *Enterobacter*, *Bacillus* sp., *Pantoea* sp., *Raoultella* sp., *Serratia* sp., *Pseudomonas* sp., *Stenotrophomonas* sp., *Lysinibacillus* sp., and *Acinetobacter* sp. from various ethnomedicinal plant species in North India (Nongkhlaw and Joshi 2014).

10.3 Endophytic Microorganisms (Endophytes)

Endophytes are the associated microbes (such as fungi, bacteria, and viruses) that reside in the endosphere of plants throughout their life cycle or during a part of their life cycle without causing any obvious injury to the host plants (Kandel et al. 2017). Hardoim et al. (2015) defined endophytes as microbes—comprising fungi, bacteria, archaea, and protists—that inhabit the interior system of a plant irrespective of the outcome of this association. The presence of endophytic fungi in host plants reduces insect attacks. For example, protection of elm trees by the endophyte *Phomopsis oblonga*, which safeguarded the trees from the beetle *Physocnemum brevilineum*, was reported by Webber in 1981. This was due to a reduction in dispersion of the Dutch elm disease–causing fungus *Ceratocystis ulmi* by regulation of *P. brevilineum*, which is a carrier of this fungus. The report suggested that toxic compounds produced by *P. oblonga* were responsible for repelling the insects (Azevedo et al. 2000).

The entry of endophytic microorganisms into host plants occurs naturally during the growth of the plants or through wounds. They spread from parent to offspring or among individuals.

10.4 Rhizospheric Microorganisms

The German scientist Hiltner first described the rhizosphere in 1904 as the soil influenced by roots. It is the soil zone proximately adjoining the roots that maintains a high level of microbial activity. The soil zone manipulated by the plant roots and parts of the root tissues—influencing the soil's physical, chemical, and biological properties—is also included in the rhizosphere. The rhizosphere is subdivided into the following regions:

1. The endorhizosphere: root tissues with cortical layers and an endodermis.
2. The rhizoplane: root surfaces with adhering soil particles and microorganisms. This region is composed of the cortex, epidermis, and mucilage layers.
3. The ectorhizosphere: the outermost region, extending from the rhizoplane to the bulk soil (nonrhizospheric soil).

In plants associated with mycorrhizae, a zone in the rhizosphere is known as the mycorrhizosphere. In some plants, a densely adhered layer known as the rhizosheath is recognized (Prashar et al. 2014; Lindermann 1988).

This chapter reviews the role of the rhizospheric microbiome (hereinafter referred to as rhizospheric microorganisms/rhizospheric microbes) in plant health management through their biochemical activities in the rhizosphere.

10.5 Rhizodeposition

The microorganisms in the rhizosphere struggle to obtain sufficient nutrients, and sometimes they are starved of those nutrients. They compete with each other for the nutrients released by plant roots (rhizodeposits/root exudates). The process of release of carbonic compounds by plant roots is described as rhizodeposition, and the compounds released are known as rhizodeposits/root exudates. The rhizodeposits include a varied range of materials such as exudates, mucilaginous compounds, organic volatiles, and soluble lysate liberated from plant cells and tissues. The organic compounds released by plant roots consist of amino acids, sugar, fatty acid, organic acids, phytohormones and vitamins, sterols, enzymes, flavanones, purines/nucleotides, etc. Some of the compounds (such as glucose) are associated with activation of microorganisms or are metabolized by microorganisms, and others (such as flavonoids) are involved in the signaling and chemotaxis by which specific groups of organisms are activated. At the start of the twentieth century, it was estimated that root exudates produced 0.6–27% of the dry weight of the plants (Nguyen 2003; Dennis et al. 2010). The three main components of plant-resultant carbons that are allocated via plant roots in the rhizosphere include:

1. The root mass: either active or inactive
2. Rhizodeposits: plant-originated materials contained in the rhizospheric zone or the nearby soil, which are readily consumed and altered by rhizospheric microbes and mingled with the organic content of the soil
3. Carbon dioxide: liberated through respiration of the roots and the root microbiome (mycorrhizae and nodules) or microbial respiration using root-derived substrates (Cheng and Gershenson 2007)

The allocation of photosynthates to roots is estimated to range from 20–30% for wheat and barley (cereals) to 30–50% for pasture plants.

10.6 The Rhizospheric Microbiome

The rhizospheric microbiome is most widely studied and has been of particular interest to researchers in recent times, as it plays a vital role in maintaining the soil structure, texture, and fertility, thus improving plant health. Rhizospheric microbes include bacteria, fungi (including mycorrhizal fungi), oomycetes, protozoa, nematodes, algae, microarthropods, and archaea living in the rhizosphere. Many of these organisms do not have any adverse impact on the health of the plants. The impact of rhizospheric microorganisms on plants is either positive or negative, depending on their association (mutualistic or pathogenic). They influence plant health and growth by plant growth hormone secretion, supply of nutrients through decomposition and nutrient recycling (biofertilization), disease suppression, and plant immune system induction (Lakshmanan et al. 2014).

10.6.1 The Rhizosphere Effect

The rhizosphere effect is defined as improvement of soil microbial growth resulting from soil chemical and physical modifications and input of root secretions and organic debris of roots inside the rhizospheric zone. The rhizosphere contains more microorganisms than the bulk soil, as influenced by the availability of nutrients. In comparison of microbial populations in the rhizospheric soil (R) and the bulk soil (S), the magnitude of the rhizosphere effect can be calculated by employing the R/S ratio, resulting in the following observations regarding these populations: bacteria > fungi > actinomycetes > protozoa. Greater algal populations are observed in the bulk soil (Dotaniya and Meena 2015).

In soil, plants are associated with a variety of microbes that are free living or that live in intimate associations with the roots. These microbes can be saprophytic, parasitic, mutualistic, or pathogenic. Mycorrhizal fungi represent important mutualistic associations in the rhizosphere of the majority of plant species. Among mycorrhizae, arbuscular mycorrhizal fungi (AMFs) are well-known biotrophs, which are able to live and multiply in live plant roots and have a wide host range. With regard to bacteria, *Rhizobium*–legume symbiosis has been widely studied. A single plant root can harbor a diversity of fungi, bacteria, and archaea (Vandenkoornhuyse et al. 2007).

10.7 Mechanisms of the Rhizospheric Microbiome in Plant Health Management

Rhizospheric microorganisms that are helpful in plant growth promotion are known as plant growth–promoting microorganisms (PGPMs). Bacteria promoting plant growth are known as plant growth–promoting rhizobacteria (PGPRs). Fungi promoting plant growth are known as plant growth–promoting fungi (PGPFs). PGPMs are involved in plant growth promotion by various mechanisms. The direct mechanisms include biofertilization (biological nitrogen fixation; phosphate, potash, and zinc solubilization), production of phytohormones (such as auxins, gibberellins (GAs), and cytokinins (CKs)), production of siderophores for iron sequestration, and disease suppression. The indirect mechanisms include rhizosphere competition, induced systemic resistance (ISR), and production of stress-related phytohormones or plant progression regulators (such as cadaverine (Cad), abscisic acid (ABA), jasmonic acid (JA), and the ethylene (ET) catabolism enzyme 1-aminocyclopropane-1 -carboxylate deaminase (ACC)) (Cassán et al. 2013).

10.7.1 Biofertilization

About 100 years ago, Hellriegal and Wilfarth established that atmospheric nitrogen fixation takes place in legumes. In the early 1880s, the French agriculturalist Boussingault showed that leguminous plants are better than cereals for providing nitrogen to host plants. In 1888, the root-nodulating bacterium *Rhizobium* was isolated from leguminous root nodules by the Dutch scientist Beijerinck. Since that time, many biofertilizer organisms have been isolated. Strictly speaking, biofertilizers are not fertilizers, which give nutrients directly to plants. Instead, biofertilizers (also called as bioinoculants and microbial inoculants) contain living organisms of bacterial, fungal, or algal origin. These bacterial, fungal, or algal cultures are grown in a nutrient medium and packed in a carrier material. They help plants by supplying various nutrients through various biochemical processes such as N_2 fixation, P solubilization, K mobilization, Zn solubilization, phosphate and micronutrient mobilization. The term "biofertilizer"—or, more correctly, "bioinoculant" or "microbial inoculant"—usually describes a formulation containing living or latent cells of competent strains of N_2-fixing, P-solubilizing, or cellulolytic microbes used for seed or soil application with the aim of augmenting microbial numbers and gearing up the processes of microbes that boost nutrient availability and can be simply and effortlessly taken up by plants. Biofertilizers are composed of either:

1. Bacteria for nitrogen fixation (*Rhizobium*, *Azospirillum*, *Azotobacter*, *Glucanacetobacter*, *Frankia*, etc.), phosphate solubilization (*Bacillus megaterium*, *Pseudomonas*, *Arthrobacter*, *Paenibacillus*, *Serratia*, *Rhodococcus*, *Xanthomonas*, etc.), potash mobilization (*Bacillus mucilaginosus*, *Frateuria aurantia*, etc.), and zinc solubilization (*Bacillus* sp., *Pseudomonas* sp., *Xanthomonas* sp., *Enterobacter* sp., *Mycobacterium* sp., *Stenotrophomonas*); or
2. Fungi for phosphate solubilization (*Penicillium*, *Piriformospora indica*), phosphate solubilization, and micronutrient mobilization—arbuscular mycorrhizal fungi (Boraste et al. 2009; Kumar 2018)

Dey et al. (2017) reported that different *Azotobacter* isolates improved the growth of chili seedlings and also shown biocontrol ability against the pathogen *Rhizoctonia solani*.

10.7.2 Arbuscular Mycorrhizal Fungi

Mycorrhizal associations were discovered 100 years ago. The term "mycorrhiza" was coined by A.B. Frank and literally means "fungus root." A mycorrhiza is a symbiotic association between soil fungi and plant roots, in which the fungi are obligate symbionts. They grow and multiply in live plant roots. More than 85% of land plant families have AMF associations. An AMF enters the plant roots and forms arbuscules (highly dichotomously branched structures that aid nutrient exchange between the plant and the fungus) in the cortical cells and vesicles in the intercellular spaces (round or oval-shaped, deeply stained bodies that store phosphorus in the polyphosphate form). Polyphosphate is liberated into the plant roots by enzymatic mechanisms under phosphate-deficient conditions. When vesicles become old, they serve as reproductive structures. Some endomycorrhizal species form spores inside the roots, called intraradical spores. The AMF acts as a secondary root system for the plant and grows beyond the root zone because its hyphae are much thinner than the roots and can explore the soil for various nutrients and water that would otherwise be inaccessible to the plant. The presence of an AMF increases the effective surface area of the plant roots by 10 times and increases the nutrient-absorbing capacity by 60 times. The AMF assists in uptake of phosphorus and micronutrients (zinc, calcium, copper, magnesium, manganese, etc.) by the plant. Mycorrhizal plants can better withstand abiotic stresses (heavy metals, salinity, alkalinity, etc.) and biotic stresses (pathogens and pests). Mycorrhizae increase water uptake by plants. AMFs secrete glomalin, a proteinaceous substance that assists in the aggregation of soil particles. These soil aggregates increase the soil porosity and enhance aeration of the roots (Habte 2000; Kumar 2018; Menge 1985).

In an experiment conducted in 2006 and 2007, Farzaneh et al. (2011) reported uptake of magnesium (Mg), potassium (K), copper (Cu), phosphorus (P), and iron (Fe) in chickpea plants treated with Symbivit, a commercial mycorrhizal inoculum. A moderate level of mycorrhizal colonization (18–55% of the roots) was observed. Rani et al. (2011) reported improvements in germination (83.5%), the maximum shoot population (86,369 millable canes per hectare), cane yield (93.60 t/ha), sugar yield (11.89 t/ha), available phosphorus (51.94 kg/ha), and available potash (318 kg/ha) at 75% of the recommended dose of phosphorus and 12.5 kg/ha mycorrhizal application, in comparison with control treatment (germination 75%, 75,272 millable canes per hectare, cane yield 79.0 t/ha, sugar yield 9.89 t/ha, available phosphorus 45.54 kg/ha, and available potash 271 kg/ha).

10.7.3 Production of Plant Growth Hormones (Phytohormones)

A phytohormone is a carbon-based constituent, biosynthesized in well-defined plant organs, which is translocated to other plant parts, where the phytohormone activates precise morphological, physiological, and biochemical responses. Five classes of phytohormones—known as the "classical five"—are recognized: cytokinins, auxins, abscisic acid, gibberellins, and ethylene (Sharma and Kaur 2017).

Auxins were first described by Frits W. Went, a Dutch botanist, in 1926. He described a procedure for detecting them quantitatively by the *Avena* coleoptile curvature test. In 1934, Kögl, Haagen-Smit, and Erxleben isolated indole-3-acetic acid (IAA), which was identical to an auxin, an active substance from urine. In 1935, K.V. Thimann isolated IAA from a culture of the fungus *Rhizopus suinus*. Cytokinins were first discovered in autoclaved DNA samples in 1955 by F. Skoog, who demonstrated that they were active in promoting cell division in tobacco callus tissue. Gibberellins were discovered in maize and rice crops by E. Kurosawa in 1926, through the "bakanae effect" (pathological longitudinal growth). In

1934, Ethylene was recognized as a ripening hormone. Abscisic acid causes abscission in fruits and dormancy of buds, and was discovered around 1960.

Patel and Patel (2014) reported the production of IAA from isolates collected from the roots of desert plants. 16S ribosomal RNA (rRNA) gene sequencing revealed that the isolates showed 99% similarity to *Pseudomonas stutzeri* and *Bacillus* sp. Maximum IAA production was observed in a tryptone yeast extract broth medium with 200 μg/ml of tryptophan added. *Azospirillum brasilense* Spl3t SR2 produced IAA and indole lactic acid (ILA) in a nitrogen-free medium supplemented with tryptophan at 100 μg/ml of liquid medium. Increased IAA production from 1 μg/ml to 100 μg/ml was recorded with increased concentrations of tryptophan. Inoculation with liquid *Azospirillum* culture increased lateral root numbers, and all of the roots were covered densely with thick root hairs (Tien et al. 1979).

Azotobacter is a free-living bacteria, which fixes atmospheric nitrogen. *Azotobacter* sp. have been shown to produce IAA-like and GA3-like substances in bioassays. The highest detected concentration of IAA was 11 μg/ml. *Azotobacter* also solubilizes phosphate, secretes fungicidal substances, and produces siderophores (Sivashakthi et al. 2017).

GA3 production has been observed in a liquid medium (Czapek–Dox broth), and solid-state fermentation has been observed on *Jatropha* seed cake with use of *Fusarium moniliforme*. GA3 production started on the sixth day and reached the highest concentration (5.8 gm/l) on the eighth day in Czapek–Dox broth. Similarly, 105 mg/g of GA3 was formed on the fourth day on *Jatropha* seed cake, and this concentration remained constant (Rangaswamy 2012). A similar observation of GA3 production using *F. moniliforme* in solid-state fermentation on commercial wheat bran was reported by Panchal and Desai (2016). GA3 production of 154 μg/g was recorded after 168 h of incubation in a commercial wheat bran mineral salt acid bed. When soluble starch was added to the wheat bran, GA3 production was increased to 1160 μg/g after 168 h of incubation. Brown and Burlingham (1968) reported production of GA3 (0.03 μg/ml GA3 equivalent) and IAA by *Azotobacter chroococcum* strain A6 grown in a nitrogen-free mineral medium for 14 days. Inoculation of seedling roots with *Azotobacter* culture increased the stem length and leaf size until the formation of five true leaves.

10.7.4 Siderophore Production

Siderophores, or biochelators (from Greek, meaning "iron carriers"), are compounds of comparatively low molecular weight and are chelating agents for ferric ion, which is excreted by many fungal and bacterial species that grow under little or no iron stress. These biochelators obtain iron from their surroundings and make it accessible to microbial cells. We know that iron is a very important element. It is the fourth most abundant element in the earth's crust and is involved in many important enzymatic reactions (as a cofactor) and nonenzymatic reactions. Fe forms insoluble ferric (Fe^{3+}) complexes in neutral and alkaline pH conditions. Because of this, the Fe is not available to bacteria, fungi, oomycetes, and plants. Aerobic microorganisms require at least 1 μM of iron for proper growth. When siderophores come into contact with ferric Fe, they chelate the iron and transport it into the cell, where Fe^{3+} is reduced to its bioavailable Fe^{2+} form. Siderophores are classified as hydroxamate, catecholate, or mixed hydroxycarboxylic ligand groups. In acid soils, hydroxamate siderophores are produced mainly by fungi and *Streptomyces*, whereas in neutral to alkaline soils, both hydroxamate and catecholate siderophores are produced (Dimpka 2016; Gupta and Gopal 2008; Neilands 1995).

Different strategies have been adopted by plants and microbes for iron nutrition. Iron is an essential element for their metabolism and is available in low quantities in the rhizospheric region through chelation, acidification, and reduction. Acidification in the rhizosphere is increased by release of CO_2 during respiration of plants and microbes. As a result of this, the concentration of carbonic acid (H_2CO_3) increases, significantly acidifying the soil during its dissociation. Carbonic acid dissociates more in alkaline and neutral pH conditions. The involvement of respiration in soil acidification is particularly important in calcareous soils, which are known for their low iron availability. An active strategy for iron uptake (strategy I) has been developed by nongraminaceous plants and includes (1) secretion of protons, (2) reduction of Fe^{3+} to its more available Fe^{2+} form by plasmalemma-bound reductases, and (3) absorption of Fe^{2+} by iron transport through plasmalemma. Carboxylates—for example, citrate, oxalate, malate, and many others generally found in root exudates—contribute to the decrease in pH in the rhizosphere, when their exudation is coupled with proton efflux (Robin et al. 2008). Strategy II, adopted by graminaceous plants, involves secretion of phytosiderophores (PSs) for iron acquisition. In the apical root zone, phytosiderophores (mugineic acids) are secreted, and increased secretion is observed under conditions of Fe deficiency. Phytosiderophore secretion follows a diurnal pulse release; it lasts for 4–6 h approximately after the onset of light. Phytosiderophores are degraded by soil microbes, which use phytosiderophores as their only carbon resource (Schenkeveld et al. 2014). Takagi et al. (2008) tested phytosiderophores belonging to the mugineic family—mugineic acid (MA) and 2′-deoxymugineic acid (DMA)—along with ethylenediaminetetraacetic acid (EDTA), diethylenetriaminepentaacetic acid (DTPA), and deferriferrioxamine B (FOB) for their ability to extract iron from different soils. At an alkaline pH, MA and DMA showed the highest efficiency for Fe extraction. Mugineic acids are prone to microbial decomposition, while proving highly efficient in extracting Fe from calcareous, high-pH soils.

Siderophores have also been detected in aquatic environments. Jenifer et al. (2015) isolated 125 bacterial isolates from the Adayar River (Chennai, Tamil Nadu, India). Out of these, only 12 isolates produced siderophores detected by a chrome azurol sulfonate (CAS) assay. The isolates were recognized as *Escherichia coli* (six isolates, strains CH1–CH6) and *Pseudomonas aeruginosa* (six isolates, strains CH7–CH12). Siderophores produced by *E. coli* are of the hydroxamate type and those produced by *P. aeruginosa* are of the catecholate type. Gupta and Gopal (2008) reported siderophore production by different bacteria such as *P. fluorescens*, *Pseudomonas* sp., *Brevibacillus brevis*, *Enterobacter* sp., *Azospirillum brasilense*, and *Enterobacter* sp. on CAS agar plates. The highest production was observed with

P. fluorescens (with a 26-mm halo around the bacterial colony on CAS agar plates) and the lowest production was observed with *B. brevis* (with a 6-mm halo on CAS agar plates). *Trichoderma* also produces biochelators that effectively chelate iron and stop other fungal growth, making *Trichoderma* an effective biocontrol agent.

10.7.5 Disease Suppression

Because of their sedentary nature, plants cannot escape continuous attacks by pathogenic microorganisms and insect pests. It has been calculated that 20% of the productivity of food crops worldwide is lost due to diseases. The approaches and tactics used to regulate these diseases are pesticide use, breeding of disease-resistant crops, and crop rotation. The evolution of pesticide-tolerant pathogens, the banning of chemical pesticides, and public awareness and concern about genetically modified crops have created an urgent need to research and develop biological control agents for disease suppression (Doornbos et al. 2012). DeBach (1964) defined biological control as "the action of parasites, predators, or pathogens in maintaining another organism's population density at a longer average than would occur in their absence" (Sharma et al. 2013). Baker (1987) defined biological control as "the decrease of inoculum or the disease producing activity of a pathogen accomplished through one or more organisms, including the host plant but excluding man."

Plants tolerate biotic stress through direct mechanisms such as antibiosis, and they tolerate parasitism and competition for nutrients, trace elements, and microsites through indirect mechanisms such as induced systematic resistance.

10.7.6 Antibiosis

There are many bacteria and fungi in the rhizosphere, producing antibiotic or related compounds that control pathogens. Fewer aerobic bacteria are found in the rhizosphere, because of the prevailing low oxygen levels due to root respiration. The general microbial genera in the rhizosphere are *Azotobacter*, *Alcaligenes*, *Arthrobacter*, *Flavobacterium*, *Micrococcus*, *Pseudomonas*, *Mycobacterium*, *Agrobacterium*, *Cellulomonas*, etc. The dominating genera among them are *Pseudomonas* and *Agrobacterium*. Rhizospheric microbes produce a wide diversity of antibiotics or antimicrobial agents. These antibiotics are microbial secondary metabolite products in their stationary phase and are not involved in the growth and development of the microbes. Penicillin, streptomycin, cephalosporin, erythromycin, tetracycline, and polymyxin are antibiotics produced by various fungi and bacteria—namely, *Penicillium chrysogenum*, *Streptomyces griseus*, *Cephalosporium acremonium*, *Streptomyces erythreus*, *Streptomyces rimosus*, and *Bacillus polymyxa*, respectively (Geetanjali and Jain 2016). Fluorescent pseudomonads have been shown to produce secondary metabolites that are antifungal, such as hydrogen cyanide (HCN), 2,4-diacetylphloroglucinol (2,4-DAPG), pyoluteorin (PLT), pyrrolnitrin (PRN), siderophores, lytic enzymes (proteases), and phenazines

(Ahmadzadeh et al. 2006). Pandey and Malviya (2014) isolated different bacterial cultures from the rhizosphere and nonrhizospheric soils of two medicinal plants: *Aloe barbadensis* and *Ocimum tenuiflorum.* The cultures *Proteus vulgaris*, *Streptococcus equisimilis*, *Streptococcus epidermis*, *Streptococcus faecalis*, *Lactobacillus fermentum*, *Bacillus subtilis*, *Bacillus cereus*, and *Neisseria mucosa* were identified using Gram staining and Bergey's manual. *L. fermentum*, *B. cereus*, and *N. mucosa* secreted secondary metabolites—β-lactam antibiotics—which showed antagonistic effects against the pathogens *E. coli*, *Staphylococcus aureus*, *Salmonella typhi*, *Pseudomonas aeruginosa*, and *B. subtilis*.

Showkat et al. (2012) screened 136 rhizosphere samples from the Kashmir Valley (Jammu and Kashmir, India) and identified 52 isolates as *P. fluorescens*; seven of them showed significant antifungal activity against *Fusarium oxysporum* and *Aspergillus* sp. The Bandi6 and Bandi11 isolates, collected from the Bandipur region, showed the highest activity against *Fusarium* sp., with inhibition zones measuring 29 mm and 28 mm, respectively. The BG6 isolate, collected from the Budgam region, showed the largest inhibition zone (20 mm) against *Aspergillus*. All 52 isolates demonstrated production of siderophores.

B. subtilis IFS-01, identified from the rhizosphere of cereals by screening of 25 isolates, was shown to strongly inhibit the bacteria *Listeria monocytogenes* and *S. aureus,* whereas slight inhibition was noticed in *Erwinia carotovora*, *Pseudomonas syringae*, *Xanthomonas campestris*, *E. coli*, and *Salmonella arizonae*. *Salmonella typhimurium*, *Geotrichum candidum*, and *Rhizopus stolonifer* showed resistance to inhibition (Foldes et al. 2000).

Among fungi, *Trichoderma* is generally used as a biological control agent against many pathogenic fungi. It is also used as a biopesticide, biofertilizer, and soil amendment. It is a beneficial soil fungus, which is a saprophytic, opportunistic, and avirulent plant symbiont. It acts as a parasitic and antagonistic fungus against plant pathogenic fungi, protecting plants from phytopathogenic diseases (Vinalea et al. 2008). Biological control activity of *Trichoderma lignorum* (*viride*) against *R. solani* was demonstrated by Weindling in 1932. The same species of *Trichoderma* has since been shown to exert mycoparasitic activity against *Phytophthora*, *Pythium*, *Rhizopus*, and *Sclerotium rolfsii*.

10.7.7 Parasitism

Parasitism is a symbiotic relationship in which two organisms coexist for a prolonged period of time. In this relationship, generally the smaller organism (the parasite) benefits and the other, larger organism (the host) is harmed. Mycoparasitism is an interaction between an antagonistic fungus and a pathogen. Mycoparasitism of *T. lignorum* (*viride*) on *R. solani*, by coiling and killing, was observed by Weindling (1932).

Cell wall lytic enzyme production is a common mechanism involved in mycoparasitism. The following sequential steps are involved in mycoparasitism: (1) chemotropism and target recognition, (2) attachment to the target and coiling around,

(3) penetration of the cell wall, and (4) host cell digestion. In the environment, *Trichoderma* strains locate other fungi, grow toward the target fungi, and consecutively start producing cell wall–degrading enzymes, which are hydrolytic in nature. *Trichoderma* attaches to the target host and coils its hyphae around the target host fungus, forms appressoria on the surface of the host, penetrates the host cell wall, and finally causes collapse of the target host hyphae (Steyaert et al. 2003). A single pathogenic fungus can be attacked by many mycoparasitic fungi. For example, a powdery mildew pathogen is attacked by a number of mycoparasitic fungi such as *Acrodontium crateriforme*, *Acremonium alternatum*, *Gliocladium virens*, *Ampelomyces quisqualis*, and *Cladosporium oxysporum* (Heyadri and Pessarakli 2010; Naher et al. 2014).

10.7.8 Induced Systemic Resistance

Induced systemic resistance is an indirect mechanism of biocontrol. In some plants, systemic resistance is induced by rhizobacteria through phytohormones such as ethylene and jasmonic acid; in other plants it is induced via the salicylic acid (SA) pathway (Mendes et al. 2013) The PGPR *B. cereus* AR156 induces systemic resistance against a wide range of pathogens, including *P. syringae* pv. tomato DC3000. A study was conducted by Niu et al. (2010) to analyze the *B. cereus* AR156 strain, which induces systemic resistance against DC3000 in *Arabidopsis* ecotype Col-0 plants. Biomass incrementation and reductions in pathogen density and disease severity in the leaves of AR156-treated plants were observed. In the AR156-treated leaves, genes related to defense—PR1, PR2, PR5, and PDF 1.2—were expressed simultaneously, suggesting concurrent activation of jasmonic acid–dependent, salicylic acid–dependent, and ethylene-dependent signaling pathways by AR156.

Trichoderma asperellum was shown to induce systemic resistance against *P. syringae* pv. *lachrymans* in cucumber plants, activating two defense genes encoding hydroperoxide lyase and phenylalanine, leading to phytoalexin accumulation. Similarly, chitinase defense gene expression was induced in oil palm plants inoculated with both *Trichoderma harzianum* and *Ganoderma boninense*, but not in oil palm plants treated with *G. boninense* alone (Naher et al. 2014).

10.8 Conclusion

Plants are very important for the survival of all other organisms on earth. Continuous attacks by various pathogens and pests cause considerable losses in agricultural productivity, to the tune of about 20% globally. The development of chemical fertilizers and pesticides controlled diseases and pests initially, but, as time has passed, the resistance developed by pathogens and pests has increasingly posed major threats to agricultural productivity. The chemicals used in fertilizers and pesticides ultimately reach water bodies through runoff water and cause eutrophication resulting in harm to aquatic flora and fauna. To avoid the ill effects caused by chemical fertilizers and pesticides, scientists worldwide are working to isolate and identify microorganisms from natural sources such as soil (especially rhizospheric soil) and water. Many identified bacteria (*Azospirillum*, *Rhizobium*, *Azotobacter*, *Bacillus megaterium*, *Frateuria aurantia*, *Glucanacetobacter*, *Nitrosomonas*, *Nitrobacter*, etc.) and fungi (*Trichoderma* sp., *Gliocladium*, *Acremonium*, *Ampelomyces*. etc.), when applied to seeds or in the soil, have been shown to enhance the growth of plants through nutrient recycling (nitrogen fixation, phosphate solubilization, etc.) and release into the soil of various metabolites (such as hormones), siderophores for chelating iron and other micronutrients, antibiotic substances for biocontrol of pathogens and pests, etc.

In India, the government included biofertilizer organisms in its Fertilizer Control Order 1985, standards for relevant commercial products have been published, and biopesticides are regulated by the Insecticides Act 1968. These regulations prescribe quality standards as well as listing of lab facilities for establishing production units, and have made it mandatory to obtain prescribed licenses for production and sale of biofertilizers and biopesticides.

However, because of lack of awareness of the availability and use of biofertilizers and biopesticides, farmers continue to use chemical fertilizers and pesticides that cause degradations in soil fertility and in the yields and quality of produce. The advantages of biofertilizers and biopesticides are as follows:

1. They are low-cost products.
2. They are eco-friendly products, causing no harm to the environment, other organisms, or the persons using them.
3. They are biodegradable.
4. They can be easily stored and applied.
5. Soil fertility is improved by their use.

Governments and producer companies should increase awareness among farmers about the functions and usage of biofertilizers and biopesticides to improve soil fertility, thus maintaining the health of plants and improving soil fertility and productivity in an eco-friendly manner.

References

Ahmadzadeh M, Afsharmanesh H, Javan-Nikkhah M, Sharifi-Tehrani A (2006) Identification of some molecular traits in fluorescent pseudomonads with antifungal activity. Iran J Biotechnol 4:245–253

Ajar NY, Priyanka V, Divjot K, Kusam LR, Vinod K, Bhanumati S, Vinay SC, Sugitha TCK, Anil KS, Harcharan SD (2017) Plant microbiomes and its beneficial multifunctional plant growth promoting attributes. Int J Environ Sci Nat Res 3:555601. https://doi.org/10.19080/IJESNR.2017.03.555601.08

Azevedo JL, Maccheroni W Jr, Pereira JO, de Araújo WL (2000) Endophytic microorganism a review on insect control and recent advances on tropical plants. Electron J Biotechnol 1:40–65

Baker KF (1987) Evolving concepts of biological control of plant pathogens. Annu Rev Phytopathol 25:67–85

Boraste A, Vamsi KK, Jhadav A, Khairnar Y, Gupta N, Trivedi S, Patil P, Gupta G, Gupta M, Mujapara AK, Joshi B (2009) Biofertilizers: a novel tool for agriculture. Int J Microbiol Res 1:23–31

Brown M, Burlingham SK (1968) Production of plant growth substances by *Azotobacter chroococcum*. J Gen Microbiol 53:135–144

Cassán F, Vanderleyden J, Spaepen S (2013) Physiological and agronomical aspects of phytohormone production by model plant-growth-promoting rhizobacteria (PGPR) belonging to the genus *Azospirillum*. J Plant Growth Regul 33:440–459. https://doi.org/10.1007/s00344-013-9362-4

Cheng W, Gershenson A (2007) Carbon fluxes in the rhizosphere. In: Cardon ZG, Whitbeck JL (eds) The rhizosphere—an ecological perspective. Academic, San Diego, pp 31–56

DeBach P (1964) The scope of biological control. In: Paul DeBach, Evert I Schlinger. Biological control of insect pests and weeds. Reinhold Publishing Corporation, New York, pp 3–20

Dennis PG, Miller AJ, Hirsch PR (2010) Are root exudates more important than other sources of rhizodeposits in structuring rhizosphere bacterial communities? FEMS Microbiol Ecol 72:313–327

Dey R, Sarkar K, Dutta S, Murmu S, Mandal N (2017) Role of *Azotobacter* sp. isolates as a plant growth promoting agent and their antagonistic potentiality against soil borne pathogen (*Rhizoctonia solani*) under in vitro condition. Int J Curr Microbiol App Sci 6:2830–2836

Doornbos RF, van Loon CL, Bakker PAHM (2012) Impact of root exudates and plant defense signaling on bacterial communities in the rhizosphere: a review. Agron Sustain Dev 32:227–243

Dimpka C (2016) Microbial siderophores: production, detection and application in agriculture and environment. Endocytobiosis Cell Res 27:7–16

Dotaniya ML, Meena VD (2015) Rhizosphere effect on nutrient availability in soil and its uptake by plants: a review. Proc Natl Acad Sci, India, Sect A Biol Sci 85:1–12

Farzaneh M, Vierheilig H, Lössl A, Kaul HP (2011) Arbuscular mycorrhiza enhances nutrient uptake in chickpea. Plant Soil Environ 57:465–470

Filho RL, Romeiro RS, Alves E (2010) Bacterial spot and early blight biocontrol by epiphytic bacteria in tomato plants. Pesq Agropec Bras 45:1381–1387

Foldes T, Banhegyi I, Herpai Z, Varga L, Szigeti J (2000) Isolation of *Bacillus* strains from the rhizosphere of cereals and in vitro screening for antagonism against phytopathogenic, food-borne pathogenic and spoilage micro-organisms. J Appl Microbiol 89:840–846

Geetanjali, Jain P (2016) Antibiotic production by rhizospheric soil microflora—a review. Int J Pharm Sci Res 7:4304–4314

Gupta A, Gopal M (2008) Siderophore production by plant growth promoting rhizobacteria. Indian J Agric Res 42:153–156

Habte (2000) Mycorrhizal fungi and plant nutrition. In: Silva JA, Uchida R (eds) Plant nutrient management in Hawaii's soils, approaches for tropical and subtropical agriculture. College of Tropical Agriculture and Human Resources, University of Hawaii at Manoa, Manoa, pp 127–131

Heyadri A, Pessarakli M (2010) A review on biological control of fungal plant pathogens using microbial antagonists. J Biol Sci 10:273–290

Hardoim PR, van Overbeek LS, Berg G, Pirttilä AM, Compant S, Campisano A, Döring M, Sessitsch A (2015) The hidden world within plants: ecological and evolutionary considerations for defining functioning of microbial endophytes. Microbiol Mol Biol Rev 79:293–320

Hirano SS, Upper CD (2000) Bacteria in the leaf ecosystem with emphasis on *Pseudomonas syringae*—a pathogen, ice nucleus, and epiphyte. Microbiol Mol Biol Rev 64:624–653

Jenifer CA, Sharmili A, Anbumalarmathi J, Umamaheswari K, Shyamala K (2015) Studies on siderophore production by microbial isolates obtained from aquatic environment. Eur J Expt Biol 5:41–45

Lakshmanan V, Selvaraj G, Bais HP (2014) Functional soil microbiome: below ground solutions to an aboveground problem. Plant Physiol 66:689–700

Lederberg J, McCray AT (2001) 'Ome sweet' omics—a genealogical treasury of words. Scientist 15:8

Lindermann RG (1988) Mycorrhizal interactions with the rhizosphere microflora: the mycorrhizal effect. Phytopathology 78:366–371

Lindow SE, Brandl MT (2003) Microbiology of the phyllosphere. Appl Environ Microbiol 69:1875–1883

Kandel SL, Joubert PM, Doty SL (2017) Bacterial endophyte colonization and distribution within plants. Microorganisms 5:77. https://doi.org/10.3390/microorganisms5040077

Krimm U, Abanda-Nkpwatt D, Schwab W, Schreiber L (2005) Epiphytic microorganisms on strawberry plants (*Fragaria ananassa* cv. Elsanta): identification of bacterial isolates and analysis of their interaction with leaf surfaces. FEMS Microb Ecol 53:483–492. https://doi.org/10.1016/j.femsec.2005.02.004

Kumar VV (2018) Biofertilizers and biopesticides in sustainable agriculture. In: Meena VS (ed) Role of rhizospheric microbes in soil. Springer, Singapore, pp 377–398

Mendes R, Garbeva P, Raaijmakers JM (2013) The rhizosphere microbiome: significance of plant beneficial, plant pathogenic, and human pathogenic microorganisms. FEMS Microbiol Rev 37:634–663

Menge JA (1985) Mycorrhiza agriculture technologies. In: Innovative biological technologies for lesser developed countries—workshop proceedings. Office of Technology Assessment, Washington, DC, pp 185–203

Mercier J, Lindow SE (2000) Role of leaf surface sugars in colonization of plants by bacterial epiphytes. Appl Environ Microbiol 66:379–284

Naher L, Yusuf UK, Ismail A, Hossain A (2014) *Trichoderma* spp.: a biocontrol agent for sustainable management of plant diseases. Pak J Bot 46:1489–1493

Neilands JB (1995) Siderophores: structure and function of microbial iron transport compounds. J Biol Chem 270:26273–26726

Nguyen C (2003) Rhizodeposition of organic C by plant: mechanisms and controls. In: Lichtfouse E, Navarrete M, Debaeke P, Véronique S, Alberola C (eds) Sustainable agriculture. Springer, Dordrecht, pp 97–123

Niu D-D, Liu HX, Jiang CH, Wang YP, Wang QY, Jin HL, Guo JH (2010) The plant growth–promoting rhizobacterium *Bacillus cereus* AR156 induces systemic resistance in *Arabidopsis thaliana* by simultaneously activating salicylate- and jasmonate/ethylene-dependent signalling pathways. Mol Plant-Microbe Interact 24:533–542

Nongkhlaw FMW, Joshi SR (2014) Distribution pattern analysis of epiphytic bacteria on ethnomedicinal plant surfaces: a micrographical and molecular approach. J Microsc Ultrastruct 2:34–40. https://doi.org/10.1016/j.jmau.2014.02.003

Panchal RR, Desai PV (2016) Study of gibberellic acid production by solid state fermentation using *Fusarium moniliforme* Sheldon. Int J Appl Sci Biotechnol 4:402–407

Pandey A, Malviya T (2014) Production of antibiotics isolated from soil bacteria from rhizospheric and non-rhizospheric region of medicinal plants. Ind J App Res 4:25–32

Patel MV, Patel RK (2014) Indole-3-acetic acid (IAA) production by endophytic bacteria isolated from saline dessert, the little Runn of Kutch. CIB Tech J Microbiol 3:17–28

Prashar P, Kapoor N, Sachdeva S (2014) Rhizosphere: its structure, bacterial diversity and significance. Rev Environ Sci Biotechnol 13:63–77. https://doi.org/10.1007/s11157-013-9317-z

Rangaswamy V (2012) Improved production of gibberellic acid by *Fusarium moniliforme*. J Microbiol Res 2:51–55

Rani TU, Naidu NV, Ramalaxmi CS, Kumar VV, Sreelatha T (2011) Bioefficacy of mycorrhizae on yield and quality of sugarcane. J Soils Crops 21:1–8

Robin A, Vansuyt G, Hinsinger P, Meyer JM, Briat JF, Lemanceau P (2008) Iron dynamics in the rhizosphere: consequences for plant health and nutrition. Adv Agron 99:183–225

Rout ME, Southworth D (2013) The root microbiome influences scales from molecules to ecosystems: the unseen majority. Am J Bot 100:1689–1691

Schenkeveld Y, Schindlegger Y, Oburger E, Puschenreiter M, Hann S, Kraemer SM (2014) Geochemical processes constraining iron uptake in strategy II Fe acquisition. Environ Sci Technol 48:12662–12670

Schreiber L, Krimm U, Knoll D, Sayed M, Auling G, Kroppenstedt RM (2005) Plant–microbe interactions: identification of epiphytic bacteria and their ability to alter leaf surface permeability. New Phytol 166:589–594

Sharma A, Diwevidi VD, Singh S, Pawar KK, Jerman M, Singh M, Singh S, Srivastawa D (2013) Biological control and its important in agriculture. Int J Biotech Bioeng Res 4:175–180

Sharma S, Kaur M (2017) Plant hormones synthesized by microorganisms and their role in biofertilizer—a review article. Int J Adv Res 5(12):1753–1762

Showkat S, Murtaza I, Laila O, Ali A (2012) Biological control of *Fusarium oxysporum* and *Aspergillus* sp. by *Pseudomonas fluorescens* isolated from wheat rhizosphere soil of Kashmir. IOSR J Pharm Biol Sci 1:24–32

Sivasakthi S, Saranraj P, Sivasakthivelan P (2017) Biological nitrogen fixation by *Azotobacter* sp.—a review. Indo Asian J Multidiscip Res 3:1274–1284

Steyaert JM, Ridgway HJ, Elad Y, Stewart A (2003) Genetic basis of mycoparasitism: a mechanism of biological control by species of *Trichoderma*. New Zealand J Crop Hort Sci 31:281–291

Takagi S, Kamei S, Yu MH (2008) Efficiency of iron extraction from soil by mugineic acid family phytosiderophores. J Plant Nutr 11:643–651

Tien TM, Gaskins MH, Hubbell DH (1979) Plant growth substances produced by *Azospirillum brasilense* and their effect on the growth of pearl millet (*Pennisetum americanum* L.). Appl Environ Microbiol 37:1016–1024

Vandenkoornhuyse P, Mahe S, Ineson P, Staddon P, Ostlet N, Cliquet J-B, Francez AA-J, Fitter AH, Young JPW (2007) Active root-inhabiting microbes identified by rapid incorporation of plant-derived carbon into RNA. Proc Natl Acad Sci 104:16970–16975

Vinalea F, Sivasithamparamb K, Ghisalbertic EL, Marraa R, Wooa SL, Loritoa M (2008) *Trichoderma* plant pathogen interactions. Soil Biol Biochem 40:1–10

Role of Microbes in Plant Health, Disease Management, and Abiotic Stress Management

11

Zabta Khan Shinwari, Faouzia Tanveer, and Irum Iqrar

Abstract

Plant yield, productivity, and food quality are highly influenced by several abiotic and biotic stresses. Agricultural stresses and associated food security issues require the optimization of reliability, efficient use of resources, and mitigation of the environmental impacts of food production. Agricultural sustainability challenges are directly linked to social, environmental, and economic factors. Biotic factor of stresses, which derive from dealings with other microorganisms as well as macroorganisms, chiefly includes damage or infection by various pests or pathogens. The abiotic factors of stresses include severe temperatures, droughts, stagnation, environmental contaminants, and salinity. Plants undergo various physiological, molecular, and biochemical changes under these environmental stresses that impact overall plant development and growth. Different strategies and mechanisms may be used to control these stresses in plants, e.g., chemical pesticides, which, however, are inefficient and detrimental to the environment. Alternate or unusual answers target to develop ecofriendly approaches by employing biological or live agents that mitigate abiotic stresses and improve disease resistance by reinforcement of natural defense system of plants. In this chapter, we have focused on the role of microbes in plant health and disease and abiotic stress management by summarizing current knowledge of the field, covering all aspects of stress agriculture, and further discussing important mechanisms used by microbes in mitigating these stresses.

Keywords

Disease management · Abiotic stress · Microbes · Plant

Z. K. Shinwari (✉) · F. Tanveer · I. Iqrar
Department of Biotechnology, Quaid-i-Azam University, Islamabad, Pakistan

V. Kumar et al. (eds.), *Microbiome in Plant Health and Disease*,
https://doi.org/10.1007/978-981-13-8495-0_11

11.1 Introduction

Exposure of plants to both abiotic and biotic stressors leads to significant deficits in worldwide production of agricultural produce (Shinwari et al. 1998). Agronomists are facing issues such as climate change, reduced arable lands, exhausting natural resources, improper soil nutrition for plants, less responsiveness of crops to agrochemicals, and environmental degradation (Shinwari et al. 1998a; Sahu et al. 2018). The use of chemical pesticides and fertilizers degrades soil richness and causes environmental contamination. Therefore it is essential to find safe and ecologically viable solution for sustainable agricultural production (Kumar and Verma 2017). Recent ideas disclose a compact, often synergetic, association between microbes and host plants. Microbes can arouse the process of germination and growth of plants, counteract common diseases and stimulate resistance to stress and overall physical vigor and robustness, improve nutrient utilization efficiency, as well as prevent diseases and provide tolerance against abiotic stresses (Berg et al. 2017; Busby et al. 2017).

Microbial diversity has been recognized as a main feature in the disease prevention and can be executed as a biological marker in plant defense stratagems. Now, we know that there are several plant growth-enhancing bacteria as well as mycorrhizal fungi which enhance plant development and growth under biotic and abiotic stresses through various mechanisms. Generally, plant growth-promoting microbes (PGPM) accomplish the great feat of plant stress management through production of plant hormones, improved nutrient access, siderophore production, bioactive secondary metabolites, enhanced antioxidant system, etc. Moreover, in dealing with biotic stress, induced systemic resistance (ISR) and acquired systemic resistance (ASR) come into play (Narusaka et al. 1999). Plant-crop interactions can replace conventional cultivation practices in modern agriculture and thus offer a sustainable solution (Nakashima et al. 2000; Kumar and Verma 2017).

When plants are wide-open to external stresses such as water scarcity, high temperature, toxicity of heavy metals, low temperatures, soil salinity, insects, pests, or pathogens, useful features rendered by microbes may persuade tolerance or resistance and activate defense mechanisms, allowing host plant to attain adaptation or adjustment to such unwanted stressors (Gómez-Merino and Trejo-Téllez 2018). Plant growth-promoting rhizobacteria (PGPR) effectively mitigate the influence of many abiotic stresses (temperature, water stress, metal toxicity, salinity, and cold stress) on host plants. Dual symbiotic systems (endophytic, rhizospheric bacteria, and symbiotic fungi) and symbiotic fungi (*Arbuscular mycorrhizal* fungi) also tend to mitigate the abiotic stresses in plants (Milošević et al. 2012).

Microbes have the potential to provide manifold attributes of the system, embracing indispensable purposes as follows: (1) seed germination, growth, and development through hormone production; (2) nutrient supply like nitrogen fixation, mobilizing phosphorus, and minerals availability like iron; (3) resistance against biotic stresses (defense of pathogens and parasites); (4) resistance against abiotic stresses; and (5) production of bioactive secondary metabolites (Berg et al. 2017). In abovementioned examples of plant growth promotion, microorganisms are

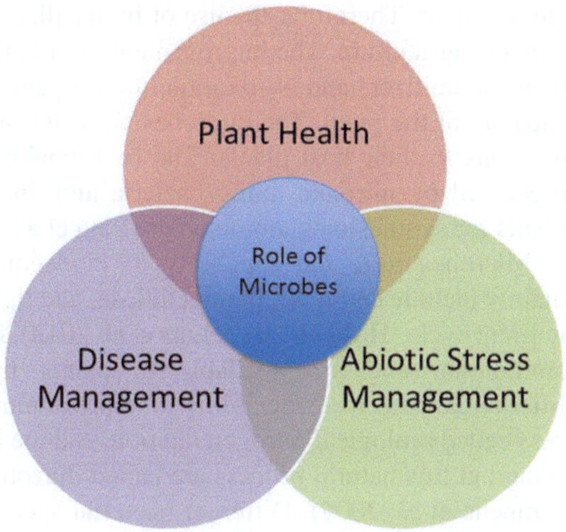

Fig. 11.1 Diagrammatic representation of plant health and disease and abiotic stress management by microbes

important, and this could be one of the reasons why these key microorganisms are transmitted vertically (Bragina et al. 2012; Truyens et al. 2015). Figure 11.1 shows the general picture about microbial role in plant growth management.

11.2 Microbial Role in Plant Health

It is well-known that the microbiome of plants is one of the chief determining factors of plant growth, productivity, and health (Berg et al. 2017). For nutrition and health of the plant, microbial activity is important as it allows the nutrient uptake and shows antagonism against wide range of phytopathogens (Matilla and Krell 2018). These microbes perform a key function in the acquisition and assimilation of micro- and macronutrients, soil texture improvement, and secretion and modulation of extracellular biomolecules such as secondary metabolites, phytohormones, antimicrobial compounds, and innumerable signaling compounds, which leads to improved plant development and growth. Healthy plants are associated with their microorganisms through metabolic cooperation and the exchange of hormones, signals, and nutrients (Berg et al. 2017). The microbes favor the growth of the plants through the solubilization of phosphate, acetic indole acid (IAA), cytokinins, gibberellins, ACC deaminase, production of siderophores, and supply of essential vitamins (Jha et al. 2011).

11.2.1 Fixation of Atmospheric N_2 and P Solubilization

The lack of essential minerals leads to the degradation of the soil. To avoid this problem, a large amount of chemical fertilizers are used which are expensive, deplete the natural resources during their production, and pose risks for man and the environment. Therefore, the use of biofertilizers is now considered to achieve sustainable agriculture. The improvement of mineral contents through the use of phosphate solubilizers and N_2-fixation microorganisms helps to improve the absorption capacity of the plant. The microbes allow the supply of micronutrients and macronutrients to their host plants. The root exudates of the plants are consumed and processed by nitrogen-fixing bacteria and, in turn, provide available nitrogen to plants for synthesizing amino acids (Lata et al. 2018).

Microbes have the capability to biodecompose organic components, which mainly include cellulose, hemicellulose, and lignin which expedite and simplify the nutrient cycle (He et al. 2012; Lata et al. 2018). Plants use N_2 in the form of NO_3 and ammonium and are available to plants by absorption through the roots. Microorganisms have the ability to improve nitrogen availability to plants by a process called biological nitrogen fixation, and the microorganisms responsible for carrying out this natural process are called diazotrophs or biological N_2-fixing agents (Franche et al. 2009). Different bacterial species exercise a valuable influence on plant development and growth such as *Azospirillum*, *Acinetobacter*, *Arthrobacter*, *Serratia*, *Bacillus*, *Enterobacter*, *Burkholderia*, *Erwinia*, *Pseudomonas*, *Rhizobium*, and *Flavobacterium* (Tilak et al. 2005; Egamberdiyeva 2005).

After nitrogen, phosphorus is the second most important nutrient needed for normal plant development and growth. The microbes intimately associated with plant roots excrete some organic acids which solubilize the insoluble forms of phosphate and convert them to bioavailable inorganic form of P (Waghunde et al. 2017). Phosphobacterin, an important commercial biofertilizer produced from *Bacillus megaterium* var. *phosphaticum*, was frequently utilized by Eastern European countries and India (Khan et al. 2007). *Colletotrichum tofieldiae* (an asymptomatic fungus) from *Arabidopsis* colonizes shoot and root, and it promotes growth and development only under phosphate-deficient conditions (Hiruma et al. 2016).

11.2.2 Phytohormone Production

Phytohormones are essential and needed for optimal plant development and growth and also exert beneficial effects on plants during stressful conditions (Davis 2004). Phytostimulators secrete phytohormones named IAA, ethylene, and cytokinins. Indole acetic acid (IAA), auxins, stimulates long-lasting and immediate responses in developing plantlets (Narusaka et al. 2001, 2003; Shi et al. 2009; Kapoor et al. 2012). It controls important physiological processes of plants. IAA stimulates cell enlargement and division, differentiates tissues, and affects plant response to gravity and light (Teale et al. 2005; Hamayun et al. 2010). Tryptophan is the precursor of IAA biosynthesis (Khamna et al. 2010; Kerkari et al. 2012). The prospective bacterial strains of *Escherichia*, *Micrococcus*, *Pseudomonas*, *Bacillus*, and *Staphylococcus* genera isolated from wild herbaceous flora were evaluated for increased production of endogenous IAA and their influence on the growth of wheat variety Inqalab-91. Inoculation of potential microbes enhanced wheat var. Inqalab-91's shoot and root length and also fresh weight of shoot (Ali et al. 2009).

11.2.3 ACC Deaminase

Under various abiotic and biotic stress conditions, plant produces ethylene as a stress signal. 1-Aminocyclopropane-1-carboxylate (ACC) deaminase, which is an ethylene precursor, acts as a catalyzer during the catalysis of ACC into ammonia and α-ketobutyrate (Glick et al. 2007; Li et al. 2011b). Rhizobacteria-producing acetylsalicylic acid deaminases can improve ethylene gas-induced abiotic stress due to saline conditions, toxic chemicals, floods, heavy metals, drought, and plant pathogens (Glick et al. 2007; Hardoim et al. 2008). *Bacillus* spp. having plant growth–promoting (PGP) activities like phytase, siderophore, cyanogens, lytic enzymes, IAA, and ACC deaminase solubilized organic and inorganic source of phosphates also hindered the growth of various plant pathogens such as *Fusarium solani*, *Fusarium oxysporum*, *Rhizoctonia solani*, and *Sclerotinia sclerotiotum* (Kumar et al. 2012).

11.2.4 Crop Weed Competition

Weeds possess several varied adaptations and growth habits and thus have the ability to grow in various environmental niches where other plants cannot grow or flourish (Ramesh et al. 2017). The weeds generally absorb more mineral nutrients than the cultivated plants and pile up a large quantity of elements in their tissue, i.e., *Echinochloa crus-galli* absorbs more nitrogen than rice crops (Talbert and Burgos 2007). The extract of polygonal leaves infiltrate comprises the flavonoids, which are harmful for germination of plumule and the radical of weeds such as the spiny amaranth (Saddique et al. 2018). The development and growth of *Digitaria sanguinalis* and *Amaranthus* sp. were reduced by inhibitors produced by the decomposition of *Sorghum halepense* rhizomes (Lajter et al. 2015).

11.3 Role of Microbes in Disease Management

The quantity and quality of foods, fibers, and foods obtained from crops of agronomic importance can be maintained by controlling their pathogenic diseases. The most common plant pathogeneses are caused by bacterial species, fungal species, oomycetes, viruses, nematodes, and higher parasitic plants. New knowledge reveals an impressive microbial diversity among all plants and antagonist microorganisms for plant pathogens (Berg et al. 2017). The most viable and environment-friendly strategy encompasses the use of biological control agents to decrease agricultural chemical input and their residues in the environment (Haggag and Abdel-Latif 2007; Sharma et al. 2015).

Extracellular products play a considerable role in the rhizospheric zone, while plant root–associated bacteria play a considerable role against phytopathogens (Lugtenberg and Kamilova 2009). It is well-documented that microbe-based biocontrol agents are able to control plant diseases, promote plant growth, and manage various types of stresses (Dodd and Pérez-Alfocea 2012; Egamberdieva et al. 2013). Common means and processes employed by biological control agents to suppress plant pathogenesis are generally categorized as indirect and direct antagonism, such as antifungal metabolites production, proteolytic enzymes that biodegrade the plant cell walls, resistance induced to the host, and struggle for niches and nutrients (Li et al. 2016). Bacteria or fungi may also persuade systemic resistance against pathogens in plants after colonization and active entry into the plant host, causing alterations in plant morphology and physiology or by stimulating the biosynthesis of bioactive components (Bailey et al. 2006; Melnick et al. 2008).

11.3.1 Direct Antagonism

Direct mechanisms for enhanced plant growth are accredited to the suppression of harmful microflora by introducing plant-friendly bacteria. In other words, direct antagonism results from high selectivity of antagonistic agent or physical contact with the pathogens (Pundir and Jain 2015). Antagonism against phytopathogens is due to extracellular enzymes (β-1,3-glucanase, proteases, and chitinase), antibiotics, siderophores, and hydrogen cyanide produced by microbes (Bhatia et al. 2005; Dutta and Khurana 2015). These microbes may also act as rivals of plant pathogens for nutrient acquisition and root colonization (Haas and Défago 2005). Microbes can show biological control activities through well-described mechanisms involving antibiosis, competition, and hyperparasitism. Fungal strains and different bacteria gathered from varied ecological niches, such as soils, sediments, plants, and animals, have been isolated for numerous metabolites, and it has been shown that these metabolites have potent bioactivity (Laatsch 2010).

(a) *Hyperparasitism*
Hyperparasitism is an ecological approach offered by microbes to defend the host plant and is considered as a direct form of animosity (Pal and Gardener 2006). In hyperparasitism, the disease-causing agents are attacked directly by the microorganisms that kill them or their disease-producing propagules (Waghunde et al. 2016).

(b) *Competition*
From a microbial point of view, the soils and surfaces of living plants are usually the environments with limited nutrients. Therefore, to colonize the plant surface, a microorganism must efficiently face a competition for accessible nutrients (Pal and Gardener 2006). Microbes compete with the pathogen niches and nutrients which might delay or reduce root colonization by the pathogen and increase competition for the mineral, for example, sequestration of iron by siderophores is an efficient system of gathering siderophores (Santoro et al. 2015).

(c) *Antibiosis*
Antibiosis results from the production of a secondary toxic metabolite by the microbes for another microorganism (disease-causing agent) and is a significant

feature for the disease suppression (Pundir and Jain 2015). Antibiotics cause direct influence on plants and may lead to systemic resistance induction (Bakker et al. 2003) and are considered as conventional phenomena for many biocontrol agents activities of, e.g., *Streptomyces*, *Bacillus* spp., *Pseudomonas*, and *Trichoderma* spp.

11.3.2 Indirect Antagonism

Indirect antagonism occurs as a result of activities that are not involved in detection of a pathogen by the biocontrol agent (Pundir and Jain 2015). For example, encouragement of host plant defense pathways by a nonpathogenic biocontrol agent is an indirect form of antagonism following mechanisms of induction and competition of host resistance (Pal and Gardener 2006). Mechanisms of antagonism, i.e., induced systematic resistance (ISR) and systematic acquired resistance (SAR), translate host plant mechanism of chemical activation or physical defense by inducers and pathogenic monitoring (Singh and Pathak 2015). The systematic acquired resistance pathway is commonly persuaded by pathogenic attack linked to the buildup of pathogen-related proteins (PRPs) and mediated by salicylic acid. PRPs comprise enzyme diversity, some of which might play a direct role to lyse disease-causing agents (microbial pathogens) such as beta-1,3-glucanase and chitinase, strengthen the cell wall to resist infection, or persuade confined cell fatality (Waghunde et al. 2017).

The ISR is persuaded by a certain nonpathogenic activity and is caused by ethylene or jasmonic acid with no link to accumulation of PRP (Tripathi et al. 2008). The substances involved in the ISR are in part same as those involved in antagonistic activity by the microbes such as production of volatile organic components (VOCs), antibiotics, siderophores, and N-acyl-homoserine lactones (Prasad et al. 2015). The induced systemic resistance is induced by the bacterial genera (*Pseudomonas*, *Bacillus*, and *Serratia*) in various plant pathogens and signaling mechanisms intricate with defense priming (Pieterse et al. 2014; Waghunde et al. 2017). Although it has also been stated that different bacteria induce an ISR mediated by salicylic acid, phytohormones (especially ethylene and jasmonic acid) play a dynamic role in ISR initiation (Pieterse et al. 2012).

11.3.3 Consortium of Microbes in Plant Disease Management

Plant development and growth promoters cohabit with different microbial strains in rhizosphere or soil in dissimilar amalgamations (Vacheron et al. 2013). Mixing of biological control agents of various microbial species that carryout activities promoting plant growth can be a better practice as compared to the employment of single microorganism for the management of plant disease and to achieve the desired agricultural results. Furthermore, the employment of microorganisms in a consortium can enhance the effectiveness, dependability, and uniformity of microbes in different soil conditions and environments (Stockwell et al. 2011). *Bacillus*,

Rhizobium, *Glomus*, *Pseudomonas*, and *Trichoderma* have been used to create microbial consortia (Prasad et al. 2015). Some of the examples of microbial role in plant development, health, and disease management are given in Table 11.1.

Table 11.1 Microbial role in plant health and disease management using various mechanisms of action

Stress	Microorganism	Plant	Mechanism	References
Nutrient deficiency	*Azotobacter chroococcum* and *Piriformospora indica*	Wheat	Improved mineral and nutrient uptake (exclusively zinc)	Abadi and Sepehri (2015)
Fusarium wilt disease	*Alcaligenes faecalis* S18 and *Bacillus cereus* S42	Tomato	Production of volatiles	Abdallah et al. (2016) and Waghunde et al. (2017)
Fusarium oxysporum and *Aspergillus niger*	*Serratia marcescens MOSEL-w2, Enterobacter cloacae MOSEL-w7, Paenibacillus sp. MOSEL-w13*	Canola	Production of ACC deaminase IAA, hydrogen cyanide, ammonia, and siderophores	Afzal et al. (2015)
	Streptomyces alboniger, Pseudomonas taiwanensis, and Pseudomonas geniculata			Afzal et al. (2017)
Rice blast disease	*Stenotrophomonas maltophilia*	Rice seedlings	Production of volatile and diffusible antibiotics	Etesami and Alikhani (2016)
Root-rot phytopathogens, i.e., *Fusarium flocciferum*, *Epicoccum nigrum*, *Phoma herbarum*, *P. notoginseng*, and *Scytalidium lignicola*	*Trichoderma koningiopsis* YIM PH30002	Host plant	Mycoparasitism and production of volatile organic compound (VOCs)	Chen et al. (2016)
Gummy stem blight	*Pseudomonas aeruginosa* 231-1	Watermelon plants	Hyperparasitism	Waghunde et al. (2016)
Collar rot pathogen *Sclerotium rolfsii*	*Trichoderma* (THU0816), fluorescent *Pseudomonas* (PHU094), and *Rhizobium* (RL091) strains (microbial consortium)	Chickpea	Physiological defense responses activation	Singh et al. (2013)

11.4 Role in Abiotic Stress Management

The population of world is estimated to reach up to 9.7 billion people by the year 2050, thus increasing the amount of food required for human consumption by 70% with rapidly depleting natural resources (Masood et al. 2005; Cole et al. 2018). Climate change has greatly threatened food security by imposing additional external pressures which directly impact the agricultural output. On the global scale, it also impacts the people whose livelihood depends on agriculture and that in fact includes majority of the world's poor. Thus, it has become even more pertinent to improve the resource use, develop adaptive capability of agronomists, and counter the ecological impacts of food production to ensure food and health security in the middle of climate change. It is now more challenging for crop plants to deal with changing climatic conditions using their fundamental biological mechanisms (Lipper et al. 2014; Meena et al. 2017), which require finding alternative solutions. Figure 11.2 shows the factors affecting plant productivity and different functions of microbes in stress management.

Abiotic stress is the stress condition encountered by plants from certain environmental factors. It comprises the nonliving element of ecosystem affecting the living part of the system. The basic abiotic factors that influence plant growth in an agricultural ecosystem include temperature, water, salts, essential nutrients, and pH. Many of these factors are linked to each other and at large are influenced by climate change, e.g., increased rainfall leads to flooding, whereas less or no rainfall results in drought, imposing negative impact on crops. In order to counter drought effects, farmers turn toward irrigation, which may add more salts in the soil causing salinity (Enebe and Babalola 2018). Naturally occurring microorganisms from varied environments show enormous metabolic potential to deal with abiotic pressures. As part of the natural ecosystem, microbial interactions with plants can regulate local as well as systemic responses in their host to ever-changing environmental conditions. These complex cellular mechanisms underlying plant-microbe interactions are increasingly studied at physiological, biochemical, and molecular levels to better understand their symbiotic relationship (Meena et al. 2017). Plant growth-enhancing bacterial strains isolated from stress-tolerant wild plants have served as successful inoculants for the agriculturally important crops (Coleman-Derr and Tringe 2014).

Abiotic stresses cause changes in the production of growth as well as stress-related phytohormones hence affecting the normal functioning of plants. Growth-promoting microbes employ numerous means and methods for plant development, growth enhancement, and tolerance of abiotic stresses. Plant growth-enhancing bacteria directly influence plant growth and development by production of phytohormones, e.g., gibberellins, ethylene, auxins (indole-3-acetic acid [IAA]), and cytokinins under stress conditions (Fahad et al. 2015). Under abiotic stresses, general mechanism of action by PGPB also involves production of siderophore, hydrogen cyanide, phosphatase, and nitrogenase (involved in phosphate solubilization and nitrogen fixation, respectively). ACC deaminase-synthesizing plant growth-promoting bacteria (PGPB) elicit plant tolerance to abiotic stresses by regulating

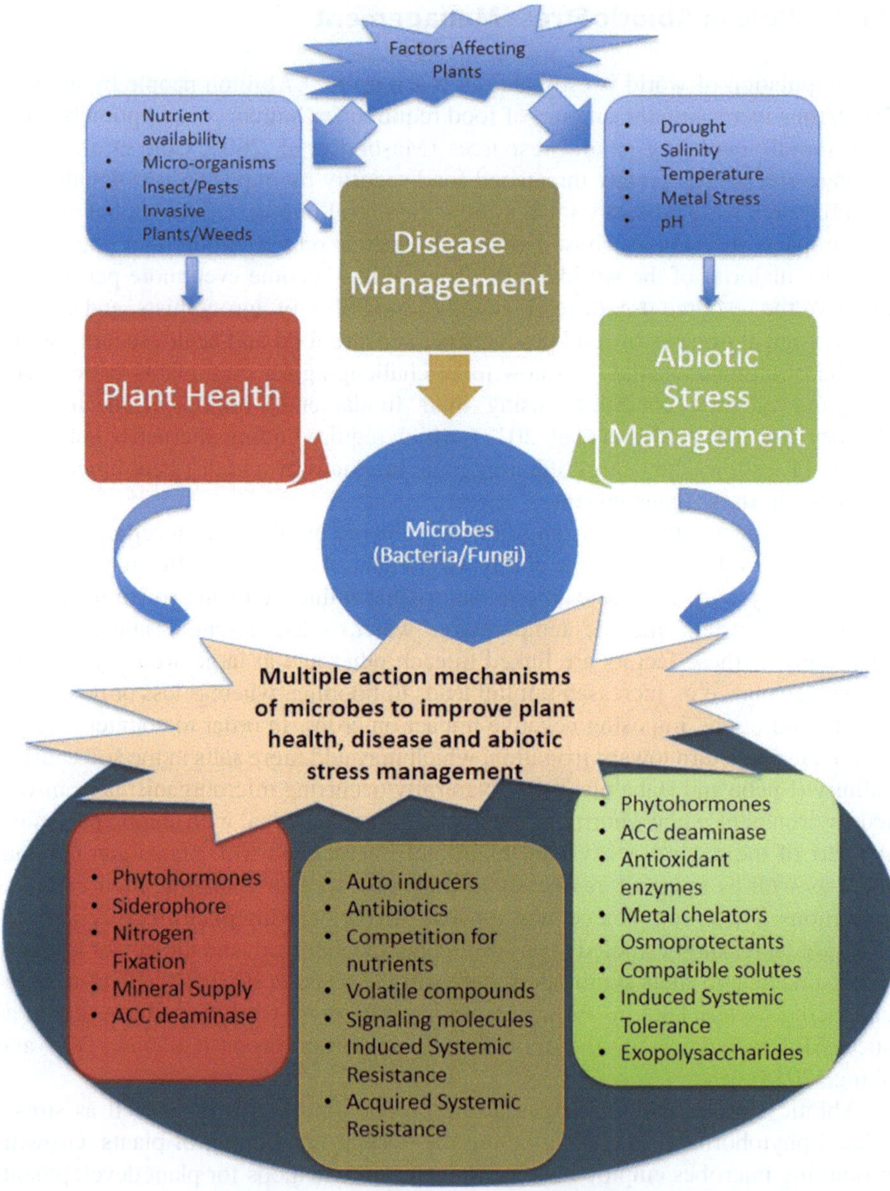

Fig. 11.2 Diagrammatic representation of factors affecting plant productivity and different roles of microbes in stress management with multiple action mechanisms (mechanisms of action for each defined role are represented by the same color as the stresses indicated above)

the concentration of ACC in plant tissues. A number of studies report the function of ACC deaminase-producing bacterial strains in managing salinity, heavy metal, drought, and flooding stress (Fahad et al. 2015; Saikia et al. 2018). Some latest examples of plant-friendly microbes inducing abiotic stress tolerance with their corresponding mechanisms are given in Table 11.2.

Table 11.2 Role of microbes in abiotic stress mitigation using various mechanisms of action

Stress	Microorganism	Plant	Mechanism	References
Heat stress	*Pseudomonas putida*	Wheat	Phytohormone, HCN, ammonia, siderophore and P-solubilization, and accumulation of metabolites like proline, sugars, starch, amino acids, and proteins	Ali et al. (2011)
Salinity stress	*Curtobacterium albidum*	*Oryza sativa* L. *Arabidopsis thaliana*	Modulation of osmolytes and antioxidative enzymes, and induction of systemic tolerance	Vimal et al. (2018)
	Enterobacter sp.	Alfafa	Production of 2-keto-4-methylthiobutyric acid (KMBA)	de Zélicourt et al. (2018)
Drought stress	*Bacillus* spp., *Enterobacter* spp., *Moraxella* spp., and *Pseudomonas* spp.	Wheat	Auxin production	Raheem et al. (2018)
	Consortium (*Ochrobactrum pseudogrignonense*, *Pseudomonas* sp., and *Bacillus subtilis*)	*Vigna mungo* L. and *Pisum sativum* L.	ACCD production, reactive oxygen scavenging enzymes, and osmolytes	Saikia et al. (2018)
Cadmium and Iron toxicity	*Enterobacter* sp.	*Hibiscus cannabinus*	Metal immobilization (siderophore production, IAA)	Chen et al. (2017)
Lead toxicity	*Pseudomonas gessardii*, *Pseudomonas fluorescens*	Sunflower	Lead uptake (increase in ascorbate peroxidase, catalase, superoxide dismutase, glutathione reductase, and proline contents)	Saleem et al. (2018)
Arsenic toxicity	*Achromobacter* sp.	Rice	ACCD (Arsenic uptake)	Corsini et al. (2018)

11.4.1 Drought Stress

Drought greatly affects the plant yield and productivity. It reduces the metabolic and physiological functions of plants. It reduces plant development, progress, nodulation, yield, and chlorophyll content (Enebe and Babalola 2018) as well as limits nutrient availability and transport during water-deficient conditions (Vurukonda et al. 2016). It also leads to an increase in reactive oxygen species (ROS) and oxidative stress, which happens because of an imbalance created between the rate of electron transport and reducing power activity for metabolic consumption (Beck et al. 2007; Kasim et al. 2013). ROS further induce alterations in membrane structure and function, enzyme stability, and lipid peroxidation (Tiwari et al. 2016). PGPB help moderate antioxidant machinery of plant by regulating concentration of antioxidant enzymes, thus conferring plant tolerance to abiotic pressures (Ghosh et al. 2018). Plant growth-enhancing rhizobacteria alleviate the drought condition by causing biochemical and physiological changes in plants by a progression known as rhizobacterial-induced drought endurance and resilience (RIDER). This process comprises regulation of phytohormones and antioxidants, secretion of exopolysaccharides (EPS), and matching organic solutes, e.g., amino acids, sugars, and polyamines, and/or manufacturing of volatile organic compounds, dehydrins, and heat shock protein (Kaushal and Wani 2016a). These mechanisms help plants survive drought stress by maintaining plant growth, membrane integrity, and enzyme stability as well as effectively regulating the water potential and nutrient uptake by increased root surface area (Kumar and Verma 2018; Vacheron et al. 2013).

11.4.2 Salinity Stress

Salinity is an important abiotic factor affecting the world's agricultural lands (Masood et al. 2005). Excessive accumulation of sodium chloride and other salts induces water-deficient conditions due to uncontainable stomata closure causing osmotic stress to plant roots. It results in ionic imbalance which causes reduced shoot and leaf growth, premature leaf death, and necrosis (Enebe and Babalola 2018; Munns and Tester 2008; Julkowska and Testerink 2015). Decreased water uptake and increased concentration of salts such as sodium, potassium, magnesium, calcium, and chloride within the cell increase ion toxicity. Under high salt stress, the process of nodulation is also negatively affected as the activity of nitrogenase enzyme involved in nitrogen fixation is reduced (Kumar and Verma 2018; Suzuki et al. 2016). PGPB including both endophytes and rhizobacteria have been found effective in alleviating salinity stress. Direct mechanisms include phytohormones production, nutrient uptake, siderophore production, and nitrogen fixation. Some mechanisms of action are similar to those found in drought stress as osmotic balance is important in both conditions, e.g., buildup of osmolytes such as trehalose, glycine betaine, and proline, production of volatile organic compounds, and EPS production. These mechanisms help promote plant growth by maintaining the ion homeostasis. Phosphate solubilization is also an important trait as high salt concentration restricts the phosphorus uptake in plants which is essential for plant growth. PGPR augment plant lenience to salinity stress through induced systemic tolerance (IST) (Kaushal and Wani 2016b; Kumar and Verma 2018). Furthermore, it has been reported that using either plant growth-promoting bacteria producing ACC deaminase enzyme or transgenic plants which express the corresponding *acdS* gene, growth development, production of seeds, and improvement of quality of *Camelina sativa* on a marginal land not appropriate for cultivation due to high salinity can be facilitated (Heydarian et al. 2016).

11.4.3 Heat and Cold Stress

Ever-changing climatic conditions have increased the intensity of heat and cold stress. The temperature stress causes changes in membrane, water potential, and photosynthetic activity in plants. Microbes adapted to cold or hot environment can better mitigate adverse effects of temperature stress. Microbes have specific enzymatic machinery that helps regulate their metabolism to adapt to changing temperature and thus are able to maintain their membrane integrity and enzyme stability. Heat and cold shock proteins are overexpressed under these environments. Molecular chaperons provide defense against heat stress (Alam et al. 2017; Kumar and Verma 2018). Protein denaturation during extreme temperature conditions can be dealt with trehalose found in microbes which forms a gel-like web to protect plants from dehydration caused by heat stress. It also plays its part in salinity and drought stress (Shameer and Prasad 2018). In high altitude agroecosystem, cold-adapted microbes have an immense potential to help plants cope with the challenging climatic conditions. A study reported psychrophilic and psychrotolerant bacteria from a cold desert of the Himalaya, India, that showed plant growth-stimulating traits, including *Pseudomonas*, *Bacillus*, *Arthrobacter*, *Exiguobacterium*, *Sanguibacter*, *Sporosarcina*, *Staphylococcus*, *Providencia*, *Bosea*, *Psychrobacter*, *Burkholderia*, *Janthinobacterium*, *Aeromicrobium*, *Brevundimonas*, *Citricoccus*, *Jeotgalicoccus*, *Methylobacterium*, *Pantoea*, *Aeromonas*, *Plantibacter*, *Sphingobacterium*, *Variovorax*, *Rhodococcus*, *Janibacter*, and *Kocuria* (Yadav et al. 2015). Similarly, heat-tolerant plant-associated bacteria isolated from wheat showed diverse plant development and growth-encouraging traits at higher temperature and included bacterial genera such as *Arthrobacter*, *Alcaligenes*, *Bacillus*, *Methylobacterium*, *Delftia*, and several pseudomonads (Verma et al. 2016).

11.4.4 Contaminants Stress

Organic and inorganic contaminants are constantly being added into our environment by human activities including industrial discharge and agricultural practices, e.g., adding excessive fertilizers and pesticides to soil. These contaminants are causing significant risks to environment and human health. Phytoremediation based on combined action of plants and associated microbes is a promising remediation technology. Microbe-assisted phytoremediation has been recognized as an effective strategy to clean up heavy metal-polluted soils and biodegradation of organic pollutants (Feng et al. 2017). PGPB counter heavy metal stress using mechanisms including metal mobilization, immobilization, volatilization, bioaccumulation, enzymatic detoxification, and EPS complexation in addition to phytohormone production, phosphate solubilization, siderophore, ACC deaminase, and nitrogen fixation (Glick 2010; Ma et al. 2016).

Microbes can affect metal solubility as well as availability in soil. Metal pollutants cannot be degraded, so they must be either extracted or stabilized in the soil. The mechanism of nutrient mobilization and uptake through metal-chelating siderophores and enzymes involved in phosphate solubilization also facilitate heavy metal uptake in stress conditions (Ullah et al. 2015). Chelating compounds such as siderophores produced by growth-promoting microbes may lower soil pH and increase metal solubility by complex formation. Organic acids, e.g., gluconic acid, citric acid, and oxalic acid, produced by these microbes, can increase metal mobilization to ensure its uptake and accumulation in plant shoots, a process called as phytoextraction. Bioavailability of metals can also be enhanced by redox processes, e.g., reduction of Fe (III) to Fe (II) and Mn (IV) to Mn (III), correspondingly rendering them less toxic. Biosurfactants and phytochelatins also play a role in increasing bioavailability by complexing heavy metals (Abou-Shanab et al. 2019; Gadd 2000, 2010; Ullah et al. 2015; Yong et al. 2014; Złoch et al. 2016). Plant development and growth-enhancing bacteria may reduce metal availability in a process called phytostablization which is particularly important in highly metal-contaminated soils. It involves changes in metal speciation, adsorption of metals on their cell wall, or exclusion through precipitation. A blend of different phytotechnologies with a cost-effective and sustainable use can provide huge benefits in restoration of metal-contaminated lands, a strategy termed as phytomanagement (Burges et al. 2018; Kong and Glick 2017).

11.4.5 Alkalinity Stress

Alkalinity imposes its own inhibitory challenges upon crop plants in alkaline soils and affects plants at biological and physiological level. Other than sodium chloride stress, there are salts such as sodium carbonate and sodium hydrogen carbonate which are detrimental to crops at high concentration. High pH in alkaline soils reduces the bioavailability of essential macro- and micronutrients such as phosphorus, manganese, zinc, copper, and iron causing nutrient deficiency and osmotic imbalance (Chen et al. 2011). Application of bioinoculants provides an attractive alternative to ameliorate high pH stress. PGPB can increase nodule formation in plants by boosting nitrogenase activity for effective nitrogen fixation (Abd-Alla et al. 2014).

11.5 Conclusion

Rise in temperature of atmosphere and constant fluctuations in rainfall pattern due to climate change are severely affecting food production worldwide. Abiotic aspects such as salinity, drought, soil pollutants, and poor nutrient availability and uptake affect crop growth and development. Similarly, plant pathogens such as bacteria, viruses, and fungi cause reduced crop biomass and yield. The most common soil management methods used to curb these stresses do not follow any target-oriented approach. Usually, nutrient-deficient soils are managed by the use of excessive amount of fertilizers, whereas plant pathogens are controlled by application of pesticides. These methods may increase crop yield but are also a major cause of environmental pollution and ecosystem disintegration, and they also pose serious threats to human health (Majeed et al. 2018). Plant microbe symbiosis can be exploited for general plant health promotion and fitness as well as conferring stress tolerance to crops in agricultural lands. Through detailed study of microbial mechanisms, more understanding can be gained to develop appropriate and well-targeted management methods using microbial inoculants in stress agriculture. It can also deliver a novel approach to alleviate the increasing effects of change in climate.

References

Abadi VAJM, Sepehri M (2015) Effect of *Piriformospora indica* and *Azotobacter chroococcum* on mitigation of zinc deficiency stress in wheat (*Triticum aestivum* L.). Symbiosis 69:9–19

Abd-Alla MH, El-Enany AWE, Nafady NA, Khalaf DM, Morsy FM (2014) Synergistic interaction of *Rhizobium leguminosarumbv. viciae* and arbuscular mycorrhizal fungi as a plant growth promoting biofertilizers for faba bean (*Vicia faba* L.) in alkaline soil. Microbiol Res 169(1):49–58

Abdallah RA, Mokni-Tlili S, Nefzi A et al (2016) Biocontrol of Fusarium wilt and growth promotion of tomato plants using endophytic bacteria isolated from *Nicotiana glauca* organs. Biol Control 97:80–88

Abou-Shanab RA, El-Sheekh MM, Sadowsky MJ (2019) Role of rhizobacteria in phytoremediation of metal-impacted sites. In: Emerging and eco-friendly approaches for waste management. Springer, Singapore, pp 299–328

Afzal I, Iqrar I, Shinwari ZK, Yasmin A (2017) Plant growth-promoting potential of endophytic bacteria isolated from roots of wild *Dodonaea viscosa* L. Plant Growth Regul 81(3):399–408

Afzal I, Shinwari ZK, Iqrar I (2015) Selective isolation and characterization of agriculturally beneficial endophytic bacteria from wild hemp using canola. Pak J Bot 47(5):1999–2008

Alam MA, Seetharam K, Zaidi PH, Dinesh A, Vinayan MT, Nath UK (2017) Dissecting heat stress tolerance in tropical maize (*Zea mays* L.). Field Crops Res 204:110–119

Ali B, Sabril AN, Ljung K, Hasnain S (2009) Auxin production by plant associated bacteria: impact on endogenous IAA content and growth of *Triticum aestivum* L. Lett Appl Microbiol 48:542–547

Ali SZ, Sandhya V, Grover M, Linga VR, Bandi V (2011) Effect of inoculation with a thermotolerant plant growth promoting *Pseudomonas putida* strain AKMP7 on growth of wheat (*Triticum* spp.) under heat stress. J Plant Interact 6(4):239–246

Bailey BA, Bae H, Strem MD, Roberts DP, Thomas SE, Crozier J, Samuels GJ, Choi IY, Holmes KA (2006) Fungal and plant gene expression during the colonization of cacao seedlings by endophytic isolates of four *Trichoderma* species. Planta 224:1449–1464

Bakker PAHM, Ran LX, Pieterse CMJ, van Loon LC (2003) Understanding the involvement of rhizobacteria mediated induction of systemic resistance in biocontrol of plant diseases. Can J Plant Pathol 25:5–9

Beck EH, Fettig S, Knake C, Hartig K, Bhattarai T (2007) Specific and unspecific responses of plants to cold and drought stress. J Biosci 32(3):501–510

Berg G, Köberl M, Rybakova D, Müller H, Grosch R, Smalla K (2017) Plant microbial diversity is suggested as the key to future biocontrol and health trends. FEMS Microbiol Ecol 93(5)

Bhatia S, Dubey RC, Maheshwari DK (2005) Enhancement of plant growth and suppression of collar rot of sunflower caused by *Sclerotium rolfsii* through fluorescent Pseudomonas. Indian Phytopathol 58:17–24

Bragina A, Berg C, Cardinale M (2012) Sphagnum mosses harbour highly specific bacterial diversity during their whole lifecycle. ISME J 6:802–813

Burges A, Alkorta I, Epelde L, Garbisu C (2018) From phytoremediation of soil contaminants to phytomanagement of ecosystem services in metal contaminated sites. Int J Phytoremed 20(4):384–397

Busby PE, Soman C, Wagner MR, Friesen ML, Kremer J, Bennett A, Morsy M, Eisen JA, Leach JE, Dangl JL (2017) Research priorities for harnessing plant microbiomes in sustainable agriculture. PLoS Biol 15(3):e2001793

Chen JL, Liu K, Miao CP, Sun SZ, Chen YW, Xu LH, Zhao LX (2016) Salt tolerance of endophytic *Trichoderma koningiopsis* YIM PH30002 and its volatile organic compounds (VOCs) allelopathic activity against phytopathogens associated with *Panax notoginseng*. Ann Microbiol 66(3):981–990

Chen Y, Yang W, Chao Y, Wang S, Tang YT, Qiu RL (2017) Metal-tolerant Enterobacter sp. strain EG16 enhanced phytoremediation using *Hibiscus cannabinus* via siderophore-mediated plant growth promotion under metal contamination. Plant Soil 413(1–2):203–216

Chen L, Yin H, Xu J, Liu X (2011) Enhanced antioxidative responses of a salt-resistant wheat cultivar facilitate its adaptation to salt stress. Afr J Biotechnol 10(74):16887–16896

Cole MB, Augustin MA, Robertson MJ, Manners JM (2018) The science of food security. npj Sci Food 2(1):14

Coleman-Derr D, Tringe SG (2014) Building the crops of tomorrow: advantages of symbiont-based approaches to improving abiotic stress tolerance. Front Microbiol 5:283

Corsini A, Colombo M, Gardana C, Zecchin S, Simonetti P, Cavalca L (2018) Characterization of As (III) oxidizing *Achromobacter* sp. strain N2: effects on arsenic toxicity and translocation in rice. Ann Microbiol 68(5):295–304

Davis PJ (2004) Plant hormones: biosynthesis, signal transduction, action! Kluwer Academic, Dordrecht

de Zélicourt A, Synek L, Saad MM, Alzubaidy H, Jalal R, Xie Y, Andres-Barrao C, Rolli E, Guerard F, Mariappan KG, Daur I, Colcombet J, Benhamed M, Depaepe T, Straeten DVD, Hirt H (2018) Ethylene induced plant stress tolerance by *Enterobacter* sp. SA187 is mediated by 2-keto-4-methylthiobutyric acid production. PLoS Genet 14(3):e1007273

Dodd IC, Pérez-Alfocea F (2012) Microbial alleviation of crop salinity. J Exp Bot 63:3415–3428

Dutta S, Khurana SP (2015) Plant growth-promoting rhizobacteria for alleviating abiotic stresses in medicinal plants. In: Plant-growth-promoting rhizobacteria (PGPR) and medicinal plants. Springer, Cham, pp 167–200

Egamberdieva D, Jabborova D, Wirth S (2013) Alleviation of salt stress in legumes by coinoculation with *Pseudomonas* and *Rhizobium*. In: Plant microbe symbiosis: fundamentals and advances. Springer, New Delhi, pp 291–303

Egamberdiyeva D (2005) Plant growth promoting rhizobacteria isolated from a calsisol in semi-arid region of Uzbekistan: biochemical characterization and effectiveness. J Plant Nutr Soil Sci 168:94–99

Enebe MC, Babalola OO (2018) The influence of plant growth-promoting rhizobacteria in plant tolerance to abiotic stress: a survival strategy. Appl Microbiol Biotechnol 102(18):7821–7835

Etesami H, Alikhani HA (2016) Suppression of the fungal pathogen *Magnaporthe grisea* by *Stenotrophomonas maltophilia*, a seed-borne rice (*Oryza Sativa* L.) endophytic bacterium. Arch Agron Soil Sci 62(9):1271–1284

Fahad S, Hussain S, Bano A, Saud S, Hassan S, Shan D, Khan FA, Khan F, Chen Y, Wu C, Tabassum MA (2015) Potential role of phytohormones and plant growth-promoting rhizobacteria in abiotic stresses: consequences for changing environment. Environ Sci Pollut Res 22(7):4907–4921

Feng NX, Yu J, Zhao HM, Cheng YT, Mo CH, Cai QY, Li YW, Li H, Wong MH (2017) Efficient phytoremediation of organic contaminants in soils using plant–endophyte partnerships. Sci Total Environ 583:352–368

Franche C, Lindström K, Elmerich C (2009) Nitrogen-fixing bacteria associated with leguminous and non-leguminous plants. Plant and Soil 321:35–59

Gadd GM (2000) Bioremedial potential of microbial mechanisms of metal mobilization and immobilization. Curr Opin Biotechnol 11(3):271–279

Gadd GM (2010) Metals, minerals and microbes: geomicrobiology and bioremediation. Microbiology 156(3):609–643

Ghosh D, Sen S, Mohapatra S (2018) Drought-mitigating *Pseudomonas putida* GAP-P45 modulates proline turnover and oxidative status in *Arabidopsis thaliana* under water stress. Ann Microbiol 68:1–16

Glick BR (2010) Using soil bacteria to facilitate phytoremediation. Biotechnol Adv 28(3):367–374

Glick BR, Todorovic B, Czarny J, Cheng Z, Duan J, McConkey B (2007) Promotion of plant growth by bacterial ACC deaminase. Crit Rev Plant Sci 26(5-6):227–242

Gómez-Merino FC, Trejo-Téllez LI (2018) The role of beneficial elements in triggering adaptive responses to environmental stressors and improving plant performance. In: Biotic and abiotic stress tolerance in plants. Springer, pp 137–172

Haas D, Défago G (2005) Biological control of soil-borne pathogens by fluorescent pseudomonads. Nat Rev Microbiol 3(4):307

Haggag WM, Abdel-latif HAM (2007) Biotechnological aspects of microorganisms used in plant biological control. World J Agric Sci 3:771–776

Hamayun M, Sohn EY, Khan SA, Shinwari ZK, Khan AL, Lee IJ (2010) Silicon alleviates the adverse effects of salinity and drought stress on growth and endogenous plant growth hormones of Soybean (*Glycine Max* L.). Pak J Bot 42(3):1713–1722

Hardoim PR, Van Overbeek LS, Elsas JD (2008) Properties of bacterial endophytes and their proposed role in plant growth. Trends Microbiol 16:463–471

He X, Han G, Lin Y, Tian X, Xiang C, Tian Q, Wang F, He Z (2012) Diversity and decomposition potential of endophytes in leaves of a *Cinnamomum camphora* plantation in China. Ecol Res 27:273–284

Heydarian Z, Yu M, Gruber M, Glick BR, Zhou R, Hegedus DD (2016) Inoculation of soil with plant growth promoting bacteria producing 1-aminocyclopropane-1-carboxylate deaminase or expression of the corresponding acdS gene in transgenic plants increases salinity tolerance in *Camelina sativa*. Front Microbiol 7:1966

Hiruma K, Gerlach N, Sacristán S (2016) Root endophyte *Colletotrichum tofieldiae* confers plant fitness benefits that are phosphate status dependent. Cell 165:464–474

Jha Y, Subramanian RB, Patel S (2011) Combination of endophytic and rhizospheric plant growth promoting rhizobacteria in *Oryza sativa* shows higher accumulation of osmoprotectant against saline stress. Acta Physiol Plant 33:797–802

Julkowska MM, Testerink C (2015) Tuning plant signaling and growth to survive salt. Trends Plant Sci 20:586–594

Kapoor R, Kumar RA, Kumar A, Patil S, Pratush A, Kaur M (2012) Indole acetic acid production by fluorescent *Pseudomonas* isolated from the rhizospheric soils of Malus and Pyrus. Recent Res Sci Technol 4(1):06–09

Kasim WA, Osman ME, Omar MN, El-Daim IAA, Bejai S, Meijer J (2013) Control of drought stress in wheat using plant-growth-promoting bacteria. J Plant Growth Regul 32(1):122–130

Kaushal M, Wani SP (2016a) Plant-growth-promoting rhizobacteria: drought stress alleviators to ameliorate crop production in drylands. Ann Microbiol 66(1):35–42

Kaushal M, Wani SP (2016b) Rhizobacterial plant interactions: strategies ensuring plant growth promotion under drought and salinity stress. Agric Ecosyst Environ 231:68–78

Kerkari S, Raikeri L, Tiwari A, Mayilraj S, Dastager S (2012) Biofilm associated indole acetic acid producing bacteria and their impact in the proliferation of biofilm mats in solar salterns. Biologia 67(3):454–460

Khamna S, Yokota A, Peberdy JF, Lumyong S (2010) Indole-3-acetic acid production by *Streptomyces* sp. isolated from some Thai medicinal plant rhizosphere soils. Eurasia J Biosci 4:23–32

Khan M, Zaidi SA, Wani PA (2007) Role of phosphate-solubilizing microorganisms in sustainable agriculture – a review. Agron Sustain Dev 27:29–24

Kong Z, Glick BR (2017) The role of plant growth-promoting bacteria in metal phytoremediation. Adv Microb Physiol 71:97–132

Kumar A, Verma JP (2017) Does plant-Microbe interaction confer stress tolerance in plants: a review? Microbiol Res 207:41–52

Kumar P, Dubey RC, Maheshwari DK (2012) *Bacillus* strains isolated from rhizosphere showed plant growth promoting and antagonistic activity against phytopathogens. Microbiol Res 167:493–499

Kumar A, Verma JP (2018) Does plant—microbe interaction confer stress tolerance in plants: a review? Microbiol Res 207:41–52

Laatsch H, AntiBase (2010) A data base for rapid structural determination of microbial natural products, and annual updates, chemical concepts, Weinheim, Germany; see http://wwwuser.gwdg.de/~ucoc/laatsch/AntiBase.htm

Lajter I, Kusz N, Hohmann J, Vasas A (2015) Isolation, characterisation and chemotaxonomic significance of secondary metabolites from *Polygonum persicaria* L. Planta Med 81(16):118

Lata R, Chowdhury S, Gond SK, White JF Jr (2018) Induction of abiotic stress tolerance in plants by endophytic microbes. Lett Appl Microbiol 66(4):268–276

Li N, Alfiky A, Vaughan MM, Kang S (2016) Stop and smell the fungi: fungal volatile metabolites are overlooked signals involved in fungal interaction with plants. Fungal Biol Rev 30(3):134–144

Li R, Zhang H, Liu W, Zheng X (2011a) Biocontrol of postharvest gray and blue mold decay of apples with *Rhodotorula mucilaginosa* and possible mechanisms of action. Int J Food Microbiol 146(2):151–156

Li Z, Chang S, Lin L, Li Y, An Q (2011b) A colorimetric assay of 1-aminocyclopropane-1-carboxylate (ACC) based on ninhydrin reaction for rapid screening of bacteria containing ACC deaminase. Lett Appl Microbiol 53:178–185

Lipper L, Thornton P, Campbell BM, Baedeker T, Braimoh A, Bwalya M, Caron P, Cattaneo A, Garrity D, Henry K, Hottle R (2014) Climate-smart agriculture for food security. Nat Clim Change 4(12):1068

Lugtenberg B, Kamilova F (2009) Plant growth promoting rhizobacteria. Annu Rev Microbiol 63:541–556

Ma Y, Rajkumar M, Zhang C, Freitas H (2016) Beneficial role of bacterial endophytes in heavy metal phytoremediation. J Environ Manage 174:14–25

Majeed A, Muhammad Z, Ahmad H (2018) Plant growth promoting bacteria: role in soil improvement, abiotic and biotic stress management of crops. Plant Cell Rep:1–11

Masood S, Seiji Y, Shinwari ZK, Anwar R (2005) Mapping quantitative trait loci (QTLs) for salt tolerance in rice (*Oryza sativa*) using RFLPs. Pak J Bot 36(4):825–834

Matilla MA, Krell T (2018) Plant growth promotion and biocontrol mediated by plant-associated bacteria. In: Plant microbiome: stress response. Springer, pp 45–80

Meena KK, Sorty AM, Bitla UM, Choudhary K, Gupta P, Pareek A, Singh DP, Prabha R, Sahu PK, Gupta VK, Singh HB, Krishanani KK, Minhas PS (2017) Abiotic stress responses and microbe-mediated mitigation in plants: the omics strategies. Front Plant Sci 8:172

Milošević NA, Marinković JB, Tintor BB (2012) Mitigating abiotic stress in crop plants by microorganisms. Matica Srpska Proc Natl Sci 123:17–26

Melnick RL, Zidack NK, Bailey BA, Maximova SN, Guiltinan M, Backman PA (2008) Bacterial endophytes: *Bacillus* spp. from annual crops as potential biological control agents of black pod rot of cacao. Biol Control 46:46–56

Munns R, Tester M (2008) Mechanisms of salinity tolerance. Annu Rev Plant Biol 59:651–681

Nakashima K, Shinwari ZK, Miura S, Sakuma Y, Seki M, Yamaguchi-Shinozaki K, Shinozaki K (2000) Structural organization, expression and promoter activity of an Arabidopsis gene family encoding DRE/CRT binding proteins involved in dehydration and high salinity responsive gene expression. Plant Mol Biol 42(4):657–665

Narusaka Y, Nakashima K, Shinwari ZK, Shinozaki K, Shinozaki K (2001) Promoter analysis of DREB1 genes encoding transcription factors involved in cold responsive gene expression in Arabidopsis. Plant Cell Physiol 42:94

Narusaka Y, Nakashima K, Shinwari ZK, Sakuma Y, Furihata T, Abe H, Narusaka M, Shinozaki K, Shinozaki KY (2003) Interaction between two cis-acting elements, ABRE and DRE, in ABA-dependent expression of Arabidopsis rd29A gene in response to dehydration and high salinity stresses. Plant J 34(2):137–149

Narusaka Y, Shinwari ZK, Nakashima K, Yamaguchi-Shinozaki K, Shinozaki K (1999) The roles of the two cis-acting elements, DRE and ABRE in the dehydration, high salt and low temperature responsive expression of the Rd29a gene in Arabidopsis thaliana. Plant Cell Physiology 40:91

Pal KK, Gardener BM (2006) Biological control of plant pathogens. The Plant Health Instructor 2:1117–1142

Pieterse CMJ, Van der Does D, Zamioudis C, Leon-Reyes A, Van Wees SCM (2012) Hormonal modulation of plant immunity. Annu Rev Cell Dev Biol 28

Pieterse CMJ, Zamioudis C, Berendsen RL, Weller DM, Van Wees SCM, Bakker PAHM (2014) Induced systemic resistance by beneficial microbes. Annu Rev Phytopathol 52:347–375

Prasad R, Kumar M, Varma A (2015) Role of PGPR in soil fertility and plant health. In: Plant-growth promoting rhizobacteria (PGPR) and medicinal plants. Springer, Cham, pp 247–260

Pundir RK, Jain P (2015) Mechanism of prevention and control of medicinal plant-associated diseases. In: Plant growth promoting rhizobacteria (PGPR) and medicinal plants. Springer, Cham, pp 231–246

Raheem A, Shaposhnikov A, Belimov AA, Dodd IC, Ali B (2018) Auxin production by rhizobacteria was associated with improved yield of wheat (*Triticum aestivum* L.) under drought stress. Arch Agron Soil Sci 64(4):574–587

Ramesh K, Rao AN, Chauhan BS (2017) Role of crop competition in managing weeds in rice, wheat, and maize in India: a review. Crop Prot 95:14–21

Saddique M, Kamran M, Shahbaz M (2018) Differential responses of plants to biotic stress and the role of metabolites. In: Plant metabolites and regulation under environmental stress, pp 69–87

Saikia J, Sarma RK, Dhandia R, Yadav A, Bharali R, Gupta VK, Saikia R (2018) Alleviation of drought stress in pulse crops with ACC deaminase producing rhizobacteria isolated from acidic soil of Northeast India. Sci Rep 8(1):3560

Saleem M, Asghar HN, Zahir ZA, Shahid M (2018) Impact of lead tolerant plant growth promoting rhizobacteria on growth, physiology, antioxidant activities, yield and lead content in sunflower in lead contaminated soil. Chemosphere 195:606–614

Santoro MV, Cappellari L, Giordano W, Banchio E (2015) Systemic induction of secondary metabolite biosynthesis in medicinal aromatic plants mediated by rhizobacteria. In: Plant-growth-promoting rhizobacteria (PGPR) and medicinal plants. Springer, Cham, pp 263–285

Shameer S, Prasad TNVKV (2018) Plant growth promoting rhizobacteria for sustainable agricultural practices with special reference to biotic and abiotic stresses. Plant Growth Regul 84(3):603–615

Sharma S, Singh V, Kumar V, Devi S, Shukla KP, Tiwari A, Singh J, Bisht S (2015) Plant growth-promoting rhizobacteria (PGPR): emergence and future facets in medicinal plants. In: Plant-growth-promoting rhizobacteria (PGPR) and medicinal plants. Springer, Cham, pp 109–131

Shi Y, Lou K, Li C (2009) Promotion of plant growth by phytohormone-producing endophytic microbes of sugar beet. Biol Fertil Soils, 45:645–653

Shinwari ZK, Nakashima K, Miura S, Kasuga M, Seki M, Yamaguchi-Shinozaki K, Shinozaki K (1998) An Arabidopsis gene family encoding DRE binding protein involved in low temperature – responsive gene expression. Biochem Biophysical Res Commun 250:161–170

Shinwari ZK, Nakashima K, Miura S, Kasuga M, Shinozaki KY, Shinozaki K (1998a) Identification of a gene family encoding dehydration-responsive-element (DRE) binding proteins in *Arabidopsis thaliana* and analysis of its promoter. Plant Cell Physiol 39:106–111

Singh A, Sarma BK, Upadhyay RS, Singh HB (2013) Compatible rhizosphere microbes mediated alleviation of biotic stress in chickpea through enhanced antioxidant and phenylpropanoid activities. Microbiol Res 168(1):33–40

Singh SK, Pathak R (2015) Ecological manipulations of rhizobacteria for curbing medicinal plant diseases. In: Plant-growth-promoting rhizobacteria (PGPR) and medicinal plants. Springer, Cham, pp 217–230

Stockwell VO, Johnson KB, Sugar D, Loper JE (2011) Mechanistically compatible mixtures of bacterial antagonists improve biological control of fire blight of pear. Phytopathology 101:113–123

Suzuki N, Bassil E, Hamilton JS, Inupakutika MA, Zandalinas SI, Tripathy D, Luo Y, Dion E, Fukui G, Kumazaki A, Nakano R (2016) ABA is required for plant acclimation to a combination of salt and heat stress. PLoS One 11(1):e0147625

Talbert RE, Burgos NR (2007) History and management of herbicide-resistant barnyardgrass (*Echinochloa crusgalli*) in Arkansas rice. Weed Technol 21(2):324–331

Teale WD, Paponov IA, Ditengou F, Palme K (2005) Auxin and the developing root of *Arabidopsis thaliana*. Physiol Plant 123:130–138

Tilak KVBR, Ranganayaki N, Pal KK, De R, Saxena AK, Nautiyal CS, Mittal S, Tripathi AK, Johri BN (2005) Diversity of plant growth and soil health supporting bacteria. Curr Sci 89:136–150

Tiwari S, Lata C, Chauhan PS, Nautiyal CS (2016) *Pseudomonas putida* attunes morphophysiological, biochemical and molecular responses in *Cicer arietinum* L. during drought stress and recovery. Plant Physiol Biochem 99:108–117

Tripathi S, Kamal S, Sheramati I (2008) Mycorrhizal fungi and other root endophytes as biocontrol agents against root pathogens. Mycorrhiza 3:281–306

Truyens S, Weyens N, Cuypers A (2015) Bacterial seed endophytes: genera, vertical transmission and interaction with plants. Environ Microbiol Rep 7:40–50

Ullah A, Heng S, Munis MFH, Fahad S, Yang X (2015) Phytoremediation of heavy metals assisted by plant growth promoting (PGP) bacteria: a review. Environ Exp Bot 117:28–40

Vacheron J, Desbrosses G, Bouffaud ML, Touraine B, Moënne-Loccoz Y, Muller D, Legendre L, Wisniewski-Dyé F, Prigent-Combaret C (2013) Plant growth promoting rhizobacteria and root system functioning. Front Plant Sci 4:356

Verma P, Yadav AN, Khannam KS, Mishra S, Kumar S, Saxena AK, Suman A (2016) Appraisal of diversity and functional attributes of thermotolerant wheat associated bacteria from the peninsular zone of India. Saudi J Biol Sci

Vimal SR, Patel VK, Singh JS (2018) Plant growth promoting *Curtobacterium albidum* strain SRV4: an agriculturally important microbe to alleviate salinity stress in paddy plants. Ecol Indic. In press

Vurukonda SSKP, Vardharajula S, Shrivastava M, SkZ A (2016) Enhancement of drought stress tolerance in crops by plant growth promoting rhizobacteria. Microbiol Res 184:13–24

Waghunde RR, Shelake RM, Sabalpara AN (2016) Trichoderma: a significant fungus for agriculture and environment. Afr J Agric Res 11(22):1952–1965

Waghunde RR, Shelake RM, Shinde MS, Hayashi H (2017) Endophyte microbes: a weapon for plant health management. In: Microorganisms for green revolution. Springer, Singapore, pp 303–325

Yadav AN, Sachan SG, Verma P, Saxena AK (2015) Prospecting cold deserts of north western Himalayas for microbial diversity and plant growth promoting attributes. J Biosci Bioeng, 119(6):683–693

Yong X, Chen Y, Liu W, Xu L, Zhou J, Wang S, Chen P, Ouyang P, Zheng T (2014) Enhanced cadmium resistance and accumulation in *Pseudomonas putida* KT 2440 expressing the phytochelatin synthase gene of *S chizosaccharomyces* pombe. Lett Appl Microbiol 58(3):255–261

Złoch M, Thiem D, Gadzała-Kopciuch R, Hrynkiewicz K (2016) Synthesis of siderophores by plant-associated metallotolerant bacteria under exposure to $Cd2^{+}$. Chemosphere 156:312–325

Plant-Microbiome Interactions in Agroecosystem: An Application

12

Maqshoof Ahmad, Sajid Mahmood Nadeem, and Zahir Ahmad Zahir

Abstract

Global food security is the major challenge for agricultural scientists, but it should not be on the cost of depletion of nonrenewable resources such as soil. Due to the decrease in agricultural land, the use of synthetic chemical fertilizers to increase crop productivity has placed extra strain on fragile agroecosystem, thereby deteriorating its health. Plant-associated microbial communities interact with plants positively or negatively. These interactions are affected by the quality of root exudates and physicochemical properties of soil. Beneficial soil microbes have a number of plant development and growth-endorsing characteristics including biological nitrogen fixation, phytohormone production, nutrient mobilization and solubilization, biocontrol activity, production of hydrolytic enzymes, and stress tolerance induction. These traits of beneficial microbes can be harnessed with better soil health, improved plant growth and productivity, and improved stress tolerance of crop plants. Improvement in beneficial microbial populations through rhizosphere engineering or use of microbial inoculants and/or their metabolites can be helpful to modify the soil microbiome, leading to increased productivity of agroecosystem. Present review highlights the significance of soil microbiome with special reference to plant health. The symbiotic plant microbial communications and the most prominent plant growth-promoting mechanisms used by soil microbes are discussed. The potential applications of plant-microbe interactions for improving crop productivity under natural as well

M. Ahmad
Department of Soil Science, University College of Agriculture and Environmental Sciences, The Islamia University of Bahawalpur, Bahawalpur, Pakistan

S. M. Nadeem
University of Agriculture Faisalabad, Burewala-Vehari, Pakistan

Z. A. Zahir (✉)
Institute of Soil and Environmental Sciences, University of Agriculture, Faisalabad, Pakistan

V. Kumar et al. (eds.), *Microbiome in Plant Health and Disease*,
https://doi.org/10.1007/978-981-13-8495-0_12

as stressful situations to maintain the sustainability of agroecosystem have been explained with examples, followed by their future prospects.

Keywords
Agroecosystem · Microbes · Plant · Crop production · Nutrients · Stress

12.1 Plant-Microbe Interactions: Introduction

Structural community of microbes and their diversity in the rhizospheric regions of plants are essential for plant development, growth, and health. Owing to critical role in plant health, the scientists termed the microbial community associated with plant as second genome of plants (Berendsen et al. 2012) that is much larger than that of plant. Microbes vary in their number and diversity which constitute in order of tens of thousands of species in fertile agricultural soils. In general, soil microbial communities include algae, protozoa, nematodes, fungi, bacteria, and microarthropods (Lynch 1990; Raaijmakers 2001). Most of these microbes have neutral effects on plants, but they are considered as important players of the food web as they utilize most of the carbon released by plant roots as rhizodeposits. The remaining less than 10% of the total rhizosphere microbes exert beneficial or harmful effects on plants. The pathogenic microorganisms in soil include pathogenic fungi and bacteria, oomycetes, and nematodes, while the beneficial microbial community may consist of free-living, associative symbiotic and symbiotic plant growth-endorsing bacterial genera, endophytic AM fungi, and algae. Recent research in plant-microbe interactions shows that host-specific microbial species are associated with dissimilar species of plant growing in the same environment (Berendsen et al. 2012). The number and diversity of beneficial and deleterious microorganisms depend upon the amount and characteristic of exudates from roots (Somers et al. 2004) as these root exudates along with soil physicochemical properties shape the rhizosphere microbial community structure and thus overall health of the plant.

Among beneficial soil-plant-microbe interactions, symbiotic plant-microbe interactions are most important which involve dynamic changes in the genome of interacting partners, through establishment of metabolic and signaling network. In plant-microbe interactions, two symbiotic associations, i.e., root nodule (RN) symbiosis and arbuscular mycorrhizal fungi (AMF) association, have been extensively studied during previous two decades (Kawaguchi and Minamisawa 2010). A third type of microorganisms called endophytes has also been recognized in this regard during recent years. The endophytes reside within tissues of plant without triggering any symptoms of disease and are called as "endosymbionts." They provide novel bioactive metabolites including phenolic acids, alkaloids, terpenoids, tannins, saponins, steroids, and quinones (Gouda et al. 2016). All these associations are significant for better plant development and growth.

Soil microbes have different plant growth-promoting mechanisms through which they are indirectly or directly implicated in improving plant development and growth. Some mechanisms are very usual and conventional among the culturable microorganisms; however, other microbes are strain-specific. Under natural agroecosystems, vegetation cover, plant species, temperature, and soil moisture regimes, environmental and soil physicochemical conditions induce large fluctuations in microbial population. The fluctuations in growth conditions cause the induction or suppression of plant development and growth-fostering characters of microorganisms. The most common plant development and growth-endorsing features include fixation of biological nitrogen, phytohormone manufacture, solubilization of nutrients, biocontrol activity, excretion of hydrolytic enzymes, and stress tolerance induction. The application of beneficial soil microbes for increasing soil health and crop production is vital for agroecological systems due to their environment-friendly nature, cost-effectiveness, and minimization of the dependence on nonrenewable resources (Sathya et al. 2017).

Plant-microbe synergism in the rhizospheric region is modulated by edaphic features. Such synergism might be valuable, advantageous, or detrimental for one or both of the partners. These interactions can be harnessed with better soil health, improved plant productivity and growth, and induced stress tolerance in crop plants. Improvement in beneficial microbial populations through rhizosphere engineering or use of microbial inoculants and/or their metabolites can be helpful to modify the soil microbiome (Velmourougane et al. 2017), leading to increased crop productivity and agroecosystem sustainability. The use of these beneficial microbes can not only enable plants to maintain their growth and productivity under various kinds of environments but also improve soil health that can be beneficial in maintaining agroecosystem sustainability.

12.2 Soil Microbiome and Plant Health

Soil microbial communities constitute the diverse populations which carry out key functions in ecosystem vital for human, plant, and animal health. Pathogenic microbes can have severe negative impact on plant growth; however, beneficial plant-soil-microbe interactions are vital for sustainable agriculture. Unfortunately, most of the beneficial functions carried out by soil microbes are threatened by climate change, land degradation, and poor management practices (Amundson et al. 2015). The manipulation of soil microbiome is critical to restore ecosystem function (Calderon et al. 2017) for agriculture sustainability. A comprehensive study of soil microbiome interactions under different conditions can create an opportunity to manage ecosystem services and soil microbial metabolism. In rhizosphere, soil microbes interact directly with plant roots and have significant influence on plant health. Rhizosphere is the thin zone of soil around roots that is manipulated and persuaded by root exudates and may harbor up to 10^{11} microbial cells/g of roots (Egamberdieva et al. 2008) or rhizosphere soil. Disease-suppressive soils have more distinct evidence of impact of soil microbiomes on plant health, where beneficial soil microbes are involved directly in the pathogenic microorganism's suppression.

In general, all soils naturally have some ability to suppress pathogenic microorganisms depending upon the number and diversity of beneficial microbes present in the soil. This phenomenon is termed as general disease suppression. The general disease suppression in a soil can be enhanced through stimulation of beneficial microbial community using organic amendments (Hoitink and Boehm 1999). However, soils can also have the ability to suppress specific kinds of pathogens termed "specific suppression" (Raaijmakers et al. 2009) that is attributed to the production of metabolites by beneficial microorganisms which are toxic to certain kind of pathogens while not to the others. In addition to inhibition of pathogens, beneficial microbes can also modulate and boost the defensive mechanism of plants' aboveground parts (Zamioudis and Pieterse 2012) that is known as induced systemic resistance (ISR). The ISR response is associated with priming to accelerate defense-related gene expression (Van der Ent et al. 2009). Although specific microorganisms protect plants against pathogens through direct or indirect mechanisms, the effectiveness of these microbes is mainly manipulated and induced by rest of the community microbes. To be effective against pathogens, these microbes should be there in appropriately good population (Raaijmakers et al. 1995). Most of these microorganisms live as commensals since they neither harm nor directly help the plant; however, they effectively compete with pathogens, thus suppressing them.

Root microbiome is shaped by plant species as plants excrete up to 40% of their photosynthates in the rhizospheric zone (Bais et al. 2006) which directly influences the microbial growth. Rhizosphere soil has much more microbial number than bulk soil (Costa et al. 2006); however, in general, there is less diversity of microbes in rhizosphere than bulk soil that might be attributed to the presence of specific kinds of metabolites in host root exudates which ultimately favor the growth of certain kinds of microbes while suppressing the others. There can be suppression effect of these metabolites on certain microbial species that favors the growth of other microbes. For example, Wang et al. (2018a, b) compared the rhizospheric microorganism's population of four *Ferula* species at different soil depths. They reported that rhizosphere bacteria vary with depth of soil Ferula therapeutic value. The specific rhizosphere bacterial population increased with the medicinal value of *Ferula* species, while soil depth showed negative effect on bacterial abundance. Microbial communities of diverse species of plants growing in the same ground are different (Garbeva et al. 2008; Berg and Smalla 2009), while the same species of plants can induce same communities of microbes in diverse soils (Miethling et al. 2000) even within plant species, there is also genotypic variation in inducing the rhizosphere microbial community (Micallef et al. 2009), suggesting that microbial community structure is shaped by root exudates. As plants can induce the microbial community, it can be concluded that rhizosphere microbial community is host specific that contributes substantially to plant health through suppression of pathogens, provision of growth hormones, and solubilization of nutrients along with performing other important functions.

12.3 Symbiotic Plant-Microbe Interactions

Symbiosis is the biological association between two organisms that involves dynamic changes in the genome of both partners, through establishment of metabolic and signaling network. In plant-microbe interactions, two symbiotic associations, i.e., root nodule (RN) symbiosis and arbuscular mycorrhizal (AM) symbiosis, have been extensively studied during previous two decades (Kawaguchi and Minamisawa 2010). A third type of microorganisms called endophytes has also been recognized in this regard during recent years. The endophytes reside within the tissues of plant without instigating any disease and are called as "endosymbionts." They provide novel bioactive metabolites including phenolic acids, alkaloids, terpenoids, tannins, saponins, steroids, and quinones (Gouda et al. 2016). All these associations are significant for better development and growth of plants. A list of microbes showing plant growth promotion has been presented in Table 12.1.

12.4 Rhizobial Associations

The root nodule symbiosis involves the development of specialized structures called as root nodules formed through communication between plants and atmospheric nitrogen-fixing bacteria. The "rhizobia" are motile, Gram-negative, rods, do not form spores, and generally belong to the order *Rhizobiales* of class *Alphaproteobacteria*, but several bacteria occur in the order *Burkholderiales* of the class *Betaproteobacteria*. These mutual N_2-fixing bacterial genera include mostly *Allorhizobium*, *Rhizobium*, *Mesorhizobium*, *Bradyrhizobium*, *Azorhizobium*, and *Sinorhizobium*. Most plant species from legume family have the capability to fix N_2 through RN symbiosis (Andrews and Andrews 2017) that gives them competitive advantage under low-nitrogen conditions (Andrews et al. 2013). The atmospheric nitrogen fixed by leguminous plants shares a major proportion of available nitrogen pool in agricultural ecosystems (Andrews et al. 2011).

For the initiation of nodulation process, the host plant produces a number of organic components, mostly flavonoids, which in turn encourage the biosynthesis of protein *NodD* by rhizobial species (Wang et al. 2012; Downie 2014). The amount and type of organic compounds produced by host plants depend upon the legume species. The protein *NodD* induces the transcription of other important genes implicated in the nodulation process and production of Nod factors (*nodABC* genes). The Nod factors such as lipopolysaccharides, lipochito-oligosaccharides, and exopolysaccharides are produced by the rhizobia as signal molecules for plants to initiate the process of nodulation (Jones et al. 2007; Oldroyd and Downie 2008). The basic structure of Nod factors released by different rhizobia is the same but differs in length (Wang et al. 2012) and is modified by species-specific proteins. The Nod factor receptors in legume host perceive the signal and respond accordingly (Wang et al. 2012; Downie 2014). The rhizobia enter the host roots through root hair infection (Sprent et al. 2013) and the host root cell wall material grows and infection thread is formed. In general, rhizobia are attached to the tip of root infection thread

Table 12.1 Effective strains of bacteria (associative and endophytic) and fungi that form association with plant and promote plant growth

Species	Crop	References
Associative bacteria		
Azospirillum spp.	Sorghum (*Sorghum bicolor)*	Pereira et al. (1988)
Azospirillum brasilense		Dobbelaere et al. (2001)
Azospirillum brasilense strain *Sp7*		Molla et al. (2001)
Azospirillum spp.	Grass	Moreira et al. (2008)
Azospirillum brasilense	Wheat (*Triticum aestivum*)	Dobbelaere et al. (2001)
Bacillus subtilis SU47, *Arthrobacter* sp.		Upadhyay et al. (2012)
Azospirillum brasilense strain *Sp7*	Banana (*Musa acuminata*)	Mia et al. (2007)
Bacillus sphaericus strain UPMB10		Mia et al. (2007)
Herbaspirillum spp.	Sugarcane (*Saccharum officinarum*)	Weber et al. (1999)
Gluconacetobacter diazotrophicus		Suman et al. (2005)
Nitrospirillum amazonense		Schwab et al. (2018)
Bacillus vietnamensis MG43		Govindarajan et al. (2008)
Bacillus japonicum SEMIA 5079 and Azospirillum brasilense Ab-V5	Soybean (*Glycine max*)	Hungria et al. (2013)
Bradyrhizobium elkanii		Kuykendall et al. (1992)
Rhizobium faba	Faba bean (*Vicia faba*)	Tian et al. (2008)
Rhizobium leguminosarum	Pea (*Pisum sativum*)	Frank (1889)
Rhizobium alamii	*Medicago ruthenica*	Berge et al. (2009)
Rhizobium endophyticum	Common bean (*Phaseolus vulgaris*)	Lopez-Lopez et al. (2010)
Mesorhizobium opportunistum	Chickpea (*Cicer arietinum*)	Nandasena et al. (2009)
Azospirillum lipoferum	Rice (*Oryza sativa*)	Ladha et al. (1982)
Gluconacetobacter diazotrophicus		Muthukumarasamy et al. (2005)
Azospirillum sp. *B510*		Bao et al. (2013)
Halobacillus spp.		Rima et al. (2018)
Gluconacetobacter diazotrophicus	Maize (*Zea mays*)	Tian et al. (2009)
Bacillus spp.		Calvo et al. (2017)
Agrobacterium tumefaciens	Cucumber (*Cucumis sativus*)	My et al. (2015)
Azotobacter chroococcum 76A	Tomato (*Solanum lycopersicum*)	Van Oosten et al. (2018)
Pseudomonas putida	Safflower (*Carthamus tinctorius*)	Nosheen et al. (2018)
Rhizobium sp.		Saghafi et al.(2018)
Endophytic bacteria		
Rhizobium leguminosarum bv. Trifolii	Rice (*Oryza sativa*)	Yanni et al. (1997)
Serratia marcescens		Gyaneshwar et al. (2001)
Bacillus pumilus		Bacilico-Jimenz et al. (2003)
Trichoderma spp.		Doni et al. (2014)

(continued)

Table 12.1 (continued)

Species	Crop	References
Bacillus sphaericus	Banana (*Musa acuminata*)	Mia et al. (2007)
Bacillus sp.	Rose (*Rosa damascena trigintipetala*)	El-Deeb et al. (2012)
Paenibacillus polymyxa	Wheat (*Triticum aestivum*)	Beck et al. (2003)
Klebsiella pneumoniae 342		Fouts et al. (2008)
Enterobacter sp.		Tian et al. (2017)
Achromobacter sp. and *Acinetobacter* sp.		Patel and Archana (2017)
Azospirillum sp.		Singh et al. (2017)
Burkholderia phytofirmans PsJN	Tomato (*Solanum lycopersicum*)	Weilharter et al. (2011)
Burkholderia phytofirmans PsJN	Maize (*Zea mays*)	Weilharter et al. (2011)
Klebsiella pneumoniae 342		Fouts et al. (2008)
Klebsiella, *Enterobacter*, and *Pantoea* sp.		Rodrigues and Forzani (2016)
Serratia proteamaculans 568	Soybean (*Glycine max*)	Taghavi et al. (2009)
Gluconacetobacter diazotrophicus	Sugarcane (*Saccharum officinarum*)	Rouws et al. (2010)
Burkholderia phytofirmans	Onion (*Allium cepa*)	Zuniga et al. (2013)
Bacillus, *Pantoea* and *Serratia* genus	Pistachio trees (*Pistacia vera*)	Etminani and Harighi (2018)
Bacillus sp.	*Wedelia trilobata*	Dai et al. (2016)
Enterobacter sp. *Cronobacter* sp.	*Withania coagulans*	Ullah et al. (2018)
Bacillus sp.	*Ammodendron bifolium*	Zhu and She (2018)
Bacillus Pseudomonas sp.	Jerusalem artichoke (*Helianthus tuberosus* L.)	Khamwan et al. (2018)
Fungi		
Glomus versiforme	Chickpea (*Cicer arietinum*)	Alloush et al. (2000)
Glomus intraradices	Pepper (*Capsicum annuum)*	Martin and Stutz (2004) and Beltrano et al. (2013)
Dive versiformis	White clover (*Trifolium repens)*	Lu and Wu (2017)
Glomus intraradices	Rangpur lime (*Citrus limonia*)	Nogueira and Cordoso (2006)
Glomus caledonium	Cucumber (*Cucumis sativus*)	Ortas (2010)
Glomus mosseae, Glomus intraradices, or Glomus versiforme		Wang et al. (2008)
Rhizophagus irregularis	Wheat (*Triticum aestivum*)	Perez-de-Luque et al. (2017)
Glomus mosseae	Garlic (*Allium sativum)*	Sari et al. (2002)
Glomus intraradices and *Glomus mosseae.*	Maize (*Zea mays*)	Lone et al. (2015)

(continued)

Table 12.1 (continued)

Species	Crop	References
Glomus intraradices and *Glomus mosseae*	Potato (*Solanum tuberosum*)	Lone et al. (2015)
Glomus sp.	Onion (*Allium cepa*)	Shuab et al. (2014)
Glomus mosseae, *Glomus versiforme*, and *Paraglomus occultum*	Peach (*Prunus persica* L. Batsch)	Wu et al. (2010)
Rhizophagus irregularis	Tomato (*Solanum lycopersicum* L.)	Khalloufi et al. (2017)

and moves inside where they multiply and differentiate into pleomorphic forms known as bacteroids, the N_2-fixing form. The whole nodulation process includes initiation of nodule, infection of rhizobia, organogenesis, fixation of atmospheric nitrogen, senescence, and feedback regulation (Oldroyd and Downie 2008; Kouchi et al. 2010). All this process, from the release of chemical signals to the start of N_2 fixation, takes about 6–15 days depending upon crop species.

Generally, the legume species are highly restricted in nature with respect to their plant host symbionts (Liu et al. 2012), in some cases; however, in grain legumes, rhizobial strains from distinctive genera in *Alphaproteobacteria* and *Betaproteobacteria* can nodulate the same legume host (Guimaraes et al. 2012). It is well documented that lateral gene transfer of specific symbiosis genes within rhizobial genera is crucial to allow leguminous plants to form symbiotic association with rhizobial genera under specific soil conditions that sustain symbiosis genes' specificity between rhizobia and legume species (Andrews and Andrews 2017). The nodulation and N_2-fixation process consumes high amount of metabolic energy from the host plant, thus legumes strictly control the number of nodules and nitrogen fixation. The RN symbiosis is not the only process that benefits the legume crop during growth under field conditions. Diverse microbes are associated with legumes as endophytes and epiphytes under natural environments which help in plant development and growth enhancement under dissimilar fertility level and soil physicochemical conditions.

12.5 Plant-Fungi Associations

The AM symbiosis has been recognized as the most common and widely spread ecological synergism between microbes and plants. The endophytic AM fungi are a heterogeneous fungal group of the phylum *Glomeromycota* which make symbiotic relationship with more than 90% of all higher plant families (Bonfante and Genre 2010). The AM fungi synergism is the base of all plant root endosymbioses that originated roughly about 400 million years ago, in the early period of Devonian (Parniske 2008). The AM fungi are a heterogeneous group of diverse fungal taxa, which are associated with the plant roots of over 90% species. They can colonize a wide range of environments including croplands, grasslands, tropical forests, and alpine and boreal zones. These fungi play a significant role in cycling of nutrients and help plants in the absorption of these nutrients, including nitrogen and phosphorous, using their extra radical hyphae and arbuscules (Parniske 2008; Selosse and Roy 2009). The arbuscules are branched structures which are enveloped in the periarbuscular membrane. The phosphate in plants is absorbed through mycorrhizae-induced phosphorous transporter gene such as MtPT4; these genes are upregulated in arbuscules of plant root cells. Some of these transporter genes are essential for establishment of AM fungi symbiosis and also acquisition of phosphate from the surroundings (Javot et al. 2007).

In case of AM fungi symbiosis, both the partners (plant and fungi) get benefited from the association as the AM fungi improve host plants' growth through manipulating water absorption, mineral uptake, and inducing resistance against diseases while the host plants' presence is compulsory for growth and reproduction of the fungi (Smith and Read 2008). In natural ecosystems, the mycorrhizal fungi help plants to survive better by improving the overall plant growth and fitness. It has been a well-known fact that mycorrhizal fungal genera significantly improve the uptake of nutrients, induce abiotic and biotic stresses mitigation in host plants, and increase plant biomass as compared to artificially induced nonsymbiotic conditions; the AM host plant can survive without AM fungal partner (Smith and Read 2008). Contrariwise, the AM fungal symbionts are obligate biotrophs which cannot grow without host plant, showing that these fungi strictly depend on host plants for their growth and reproduction. The AM fungi are important in ecosystems establishment and play a critical role in early stages of the life cycle of host plant (Knappova et al. 2016). In addition to helping in phosphorus acquisition, the mycorrhizal fungi also aid in the uptake and transfer of considerable amount of nitrogenous compound to host plant via fungal hyphae.

12.6 Endophytes

Endophytes, called as endosymbionts, are a group of endosymbiotic microorganisms colonizing plant tissues. The bacterial endophytes were first reported by Darnel in 1904 in plants, which can also provide a number of novel bioactive compounds including phenolic acids, alkaloids, terpenoids, tannins, saponins, steroids, and quinones (Gouda et al. 2016). A huge number of bacterial and fungal genera colonizing the intercellular and/or intracellular locations of plants have been identified (Singh and Dubey 2015). They complete all or part of their life cycle within tissues of host plant without producing any obvious symptom of disease. The endophytes improve the plant growth and nutrient concentration and have the capability to persuade stress tolerance against various types of biotic and abiotic stresses in addition to fixation of N_2, as in case of rhizobia (Beneduzi et al. 2013).

With almost every plant species, the endosymbionts are associated and have integral role in life of plant. The endosymbiosis is considered crucial for plant's survival. It is documented that from per gram of fresh shoot and root weight, about 10^5 cfu of endophytic bacteria can be isolated, and they are so diverse in nature that around 70–80% of them are still waiting for their identification despite advancement in the sciences. Among the important functions of endosymbionts are defense from plant pathogens, communication with other associated microbes, involvement in stimulating the plant defense processes against abiotic and biotic ecological stresses, and volatile compound production. Bacterial endophytes are also reported to produce allelopathic compounds, and these compounds act as natural biocontrol for diverse pests (González and Lopez 2013) in addition to fixation of N_2, as in case of rhizobia. The blend of all these growth-enhancing properties augments immunity level of plant against pests and diseases (Hayat et al. 2010). In addition to symbiotic fungi and rhizobia, roots of plants are also inhabited by a diverse variety of bacterial species from other genera of bacteria, such as *Azotobacter*, *Paenibacillus*, *Pseudomonas*, *Bacillus*, *Burkholderia*, *Rhizobium*, and many more, which consecutively function together and mutually promote plant development and growth (Maheshwari 2013) as endosymbionts. The endophytic fungi have been classified into nonclavicipitaceous and clavicipitaceous endophytes and belong to the *Ascomycota* or *Basidiomycota* group (Jalgaonwala et al. 2011). These endophytic fungi have the ability to produce a number of bioactive compounds including antibiotics and can be a good bioresource to develop biopesticides. Among the soil-inhabiting microorganisms, nonsymbiotic endophytic bacteria are less studied for their potential roles and plant growth-promoting aspects (Rosenblueth and Martinez-Romero 2006). The endosymbiont inhabitants in plant species differs with developmental stage of host plant, host crop species, and environmental conditions (Dudeja and Giri 2014).

12.7 Plant Growth-Promoting Mechanisms of Soil Microbes

Soil is heterogeneous in nature and has diversity of microorganisms. Soil-plant-microbe interactions are important for ecosystem sustainability. About 5% of the total microorganisms in soil have beneficial impact on plant growth. These beneficial microbes have different plant growth-promoting mechanisms through which they are indirectly or directly intricate in improving plant development and growth (Nadeem et al. 2013). Some mechanisms are very usual and customary among the microbes which are cultured in labs, while others are strain-specific. Under natural agroecosystems, vegetation cover, plant species, temperature and soil moisture regimes, and environmental and soil physicochemical conditions induce large fluctuations in microbial population. The fluctuations in growth conditions cause the induction or suppression of plant growth-enhancing phenomenon of microorganisms. The most conventional plant growth-enhancing characteristics include fixation of atmospheric nitrogen, production of phytohormones, solubilization of nutrients, biocontrol activity, making hydrolytic enzymes, and stress tolerance induction. The schematic view of plant growth-enhancing mechanism by soil microbes is presented in Fig. 12.1.

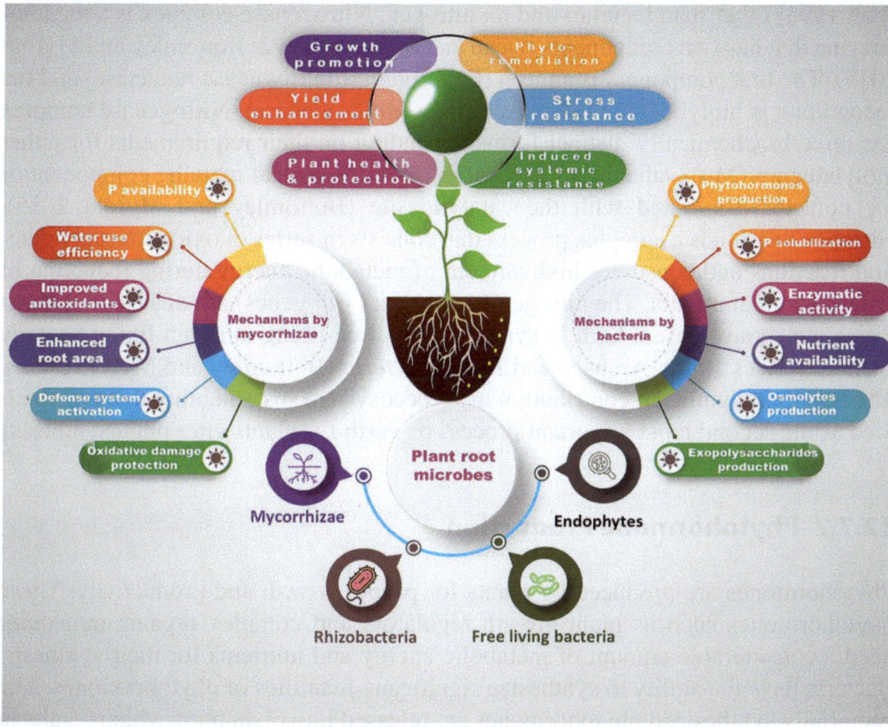

Fig. 12.1 Common plant growth-enhancing mechanism by soil microbes

12.7.1 Biological Nitrogen Fixation

The atmospheric nitrogen shares the major proportion of its total contents on earth that is not directly available to plants. It needs to be reduced artificially to NH_3 (ammonia gas) by Haber-Bosch procedure (Rubio and Ludden 2008) or through natural means such as thunderstorms and biological nitrogen fixation (BNF). During BNF, atmospheric N_2 is reduced to NH_3 by N_2-fixing microorganisms through nitrogenase activity (Kim and Rees 1994). This biologically fixed N_2 accounts for around 66% of the total fixed N_2 through all means worldwide. Biologically, N_2 is fixed by nitrogen-fixing bacteria called as rhizobia. The bacteria involved in BNF are categorized into free-living, associate, and symbiotic bacteria. Although free-living N_2 fixers such as *Gluconacetobacter*, *Azospirillum*, and *Azotobacter* spp. abundantly exist in natural ecosystem (Bashan and Levanony 1990), the contribution of these bacteria is negligible when compared with total BNF. The symbiotic nitrogen-fixing bacteria called as "rhizobia" contribute the highest proportion of BNF (Zahran 2001).

In addition to rhizobia, other plant growth-promoting rhizobacteria (PGPR) such as diazotrophs also have the nitrogenase complex which fix N_2 in nonleguminous plants. These diazotrophs form nonobligate interactions with their host plants (Glick et al. 1998) other than legumes and fix nitrogen. Nitrogenase complex is a metalloenzyme that has two components (Dean and Jacobson 1992; Bottomley and Myrold 2015). The first component is an iron (Fe) protein (dinitrogenase reductase) and the second one is molybdenum (Mo)-Fe protein (dinitrogenase). Nitrogenase complex has three biochemically distinct forms depending on their requirements for either molybdenum (Mo), vanadium (V), or iron (Fe) as a critical metallic component of the cofactor associated with the catalytic site (Bottomley and Myrold 2015). Nitrogen fixation is a complex process that consists of series of oxidation and reduction reactions and consumes high amount of metabolic energy during reduction of dinitrogen to ammonia. The nitrogen-fixing genes (*nif* genes) are of several distinct forms which are present both in symbiotic and free-living nitrogen-fixing bacteria (Kim and Rees 1994), Archaea and *Proteobacteria* (Bottomley and Myrold 2015). The BNF has significant contribution in agroecosystem sustainability that is considered as the second most important process on earth for plants after photosynthesis.

12.7.2 Phytohormone Production

Phytohormones are produced by plants for proper growth and productivity. These phytohormones such as plant growth regulators and complex organic molecules need a considerable amount of metabolic energy and nutrients for their synthesis. Bacteria have the ability to synthesize significant quantities of phytohormones. The bacterially synthesized phytohormones are released into plant body which results in significant positive effects on plant growth and development. It is well documented in literature that bacteria can produce up to 60 times higher amount of plant growth regulators as compared to plants (Camerini et al. 2008).

The important phytohormones which are produced by soil microbes include auxins, gibberellins, abscisic acid, ethylene, and cytokinins. These phytohormones help in plant growth during cell division, cell enlargement, seed germination, root formation, and stem elongation (Taiz and Zeiger 2000; Khalid et al. 2006; Kang et al. 2010). These microbially produced phytohormones meet the plant's hormonal requirements and save much needed plant's metabolic energy, thus improve crop growth and productivity (Zahir et al. 2010; Jamil et al. 2018).

Auxins are effective under stress but some plants are unable to produce enough auxins to cope with adverse conditions, resulting in failure to alleviate stress conditions. Under such conditions, exogenous application of auxins or inoculation with microbes having ability to produce auxins can help for resumption of normal metabolic functions (Ahmad et al. 2013c). Jamil et al. (2018) evaluated the exogenous application of L-tryptophan in combination with *Pseudomonas fluorescens* under drought conditions that resulted in significant increase in physiological parameters and yield.

Abscisic acid (ABA) is also a stress hormone (Zhang et al. 2006) and plays a critical role in photoperiodic induction of flowering (Wilmowicz et al. 2008). Gibberellins (GA) are involved in leaf expansion and stem elongation of plants. Exogenously applied GA promotes parthenocarpy in fruits, bolting of the plants, breaks tuber dormancy, and increases the number of buds and fruit size. A number of soil microorganisms have been reported to produce GA which can have positive or negative effects on plant growth and nodulation. They have the ability to induce nodule organogenesis but can inhibit nodulation at infection stage (McAdam et al. 2018). Cytokinin has been reported to be involved in plant cell division, root development, root hair formation, and chloroplast development, shoot growth, and leaf senescence. It also controls cell division in plants (Arkhipova et al. 2007; Oldroyd 2007) and regulates nodulation and nitrogen fixation (Kisiala et al. 2013). Ethylene is a stress hormone produced in plants that regulates plant physiological processes and induces stress tolerance in plants (Arshad and Frankenberger 2002). The higher concentration of ethylene under stress negatively affects plant growth (Zahir et al. 2008). Bacterial strains have been reported to regulate ethylene production in plants through 1-aminocyclopropane-1-carboxylic acid (ACC) deaminase activity (Ahmad et al. 2011; Khan et al. 2013; Singh et al. 2015).

Literature reports the production of phytohormones such as auxins, ABA, cytokinins, and gibberellins by *Rhizobium*, *Sinorhizobium*, *Bradyrhizobium*, *Azospirillum*, *Bacillus*, *Paenibacillus*, and *Pseudomonas* (Bottini et al. 2004; Boiero et al. 2007; Afzal et al. 2010; Zahir et al. 2010; Gamalero and Glick 2011; Ahmad et al. 2011; Mumtaz et al. 2017) which improve plant growth and productivity under natural conditions (Ahmad et al. 2015, 2016; Mumtaz et al. 2018).

12.7.3 Nutrient Solubilization

Nutrient cycling is the major function of soil microorganisms. Crop residues when incorporated into the soil are attacked by microorganisms for carbon, energy, and nutrient source. The mineral nutrients from decomposed crop residues enter in to the soil while carbon is released as CO_2 into the atmosphere. Moreover, microbes also solubilize chemically fixed nutrients such as potassium (K), phosphorus (P), iron (Fe), and zinc (Zn). Microorganisms release extracellular enzymes such as phosphatases which solubilize the inorganic phosphate in soil. Microorganisms also produce organic acids which lower the soil pH in microclimate, thus causing the solubilization of nutrients such as P, K, Fe, and Zn (Jennings 1994; Ahmad et al. 2016). It has been well documented that bacteria produce gluconic acid and its derivatives which solubilize the Zn and inorganic phosphate in soil (Gadd and Sayer 2000; Saravanan et al. 2007). Soil microbes also secrete low-molecular-mass iron-chelating compounds, siderophores, which can solubilize iron thus making it bioavailable (Machuca et al. 2007).

Scientists around the globe are working to identify the microbial strains responsible for the solubilization of insoluble nutrients in soil. For example, *Bacillus aryabhattai* and *Bacillus* sp. (AZ6) have been recognized as potential candidates for Zn solubilization from insoluble Zn resources (Ramesh et al. 2014; Hussain et al. 2015). Similarly, Mumtaz et al. (2017) screened 70 isolates and reported that 4 isolates can be the potential strains for solubilization of insoluble Zn in soil. They reported these strains as *Bacillus aryabhattai* S10, *Bacillus* sp. ZM20, *B.*

aryabhattai ZM31, and *B. subtilis* ZM63 after identification through 16S rDNA sequencing. In addition to *Bacillus*, the strains from other genera can also solubilize inorganic Zn compounds. For example, *Acinetobacter* sp. (AGM3), *Gluconacetobacter diazotrophicus*, and *Exiguobacterium aurantiacum* have been reported to solubilize inorganic Zn and Fe compounds, thus increasing Zn and Fe contents in grain crops (Ramesh et al. 2014; Gandhi and Muralidharan 2016; Shaikh and Saraf 2017). Secretion of chelating ligands, amino acids, organic acids, phytohormones, and vitamins by microbial strains can be the possible mechanisms for solubilization of inorganic compounds in addition to oxidoreductive systems and proton extrusion (Wakatsuki 1995; Saravanan et al. 2007).

Phosphate solubilization in soil depends upon the organic and inorganic nature of phosphate complexes that can be catalyzed by organic acid secretion and proton discharge. The P availability in soil depends upon pH and amount and nature of soil minerals. Under acidic conditions, P makes complexes with Fe and aluminum (Al), while at alkaline pH, it makes complexes with calcium (Goldstein 2000). The phosphate-solubilizing bacteria (PSB) solubilize Fe/Al-phosphate complexes by releasing proton, altering the negative charge at exchange sites, and thus facilitating the release of phosphate ions from complexes. The decreased adsorption of phosphates facilitates the release of primary and secondary orthophosphates (Henri et al. 2008). Moreover, the PSB can secrete carboxylic acid that releases carboxyl ions, thus replacing P in precipitated complexes through ligand exchange. Under alkaline conditions, PSB solubilize calcium phosphate complexes through secreting the organic acids thus acidifying the surrounding environment. The calcareous soils have high buffering capacity that can reduce the efficiency of PSB in releasing P (Stephen and Jisha 2009). From organic complexes, PSB release P through secretion of extracellular enzymes such as phosphatases (Dodor and Tabatabai 2003).

12.7.4 Biocontrol Activity

Soil microorganisms are effectively involved in the suppression of soil pathogens. The production of siderophores, antibiotics, hydrolytic enzymes, and competition for nutrients along with triggering the ISR in plants are important biocontrol mechanisms which soil microbes employ to improve plant productivity (Lugtenberg and Kamilova 2009; Kotan et al. 2009; Glick 2012). Moreover, degradation of fungal cell wall through hydrolytic enzymes is also used by soil microorganisms (Ramyasmruthi et al. 2012). The cell wall-degrading enzymes are important weapons to control phytopathogenic fungi in soil (Picard et al. 2000). The well-known fungal cell wall-degrading enzymes include chitinase, lyase, and cellulase (Nadeem et al. 2013). These enzymes are important for suppression of diseases. For example, inoculation with *Pseudomonas* sp. containing chitinase can inhibit the growth of *Rhizoctonia solani* through degradation of cell wall (Nielsen et al. 2000).

The antibiotics produced by soil microbes are mostly effective to control fungal pathogens (Glick 2012). A number of important antibiotics and antifungal metabolites are produced by soil microorganisms. These include 2,4-diacetyphlaroglucinol, phenazines, tensin, pyoluteorin, viscosinamide, pyrrolnitrin, and hydrogen cyanide (Haas and Keel 2003; Mazurier et al. 2009; Bhattacharyya and Jha 2012; Glick 2012). Hydrogen cyanide (HCN) is a volatile antibiotic produced by bacteria that works synergistically with other antibiotics to improve their efficiency for the suppression of plant diseases. It has been observed that inoculation with *Pseudomonas* strain having ability to produce HCN can suppress black rot of tobacco (Voisard et al. 1981).

Siderophores are Fe-chelating compounds which bind the available Fe in soil thus making it unavailable for use by soilborne pathogens. Siderophore-producing bacteria have been recognized as useful tools for biocontrol, as plant Fe requirement is much lower than most of the microorganisms (O'Sullivan and O'Gara 1992). Moreover, many plant species have the potential uptake Fe complexed with siderophores (Wang et al. 1993) that is not available to pathogen. These siderophore-producing bacteria reduce the availability of iron to fungi (Sayyed et al. 2008), thus suppressing its growth (Arora et al. 2001). It has been observed by Matthijs et al. (2007) that inoculation with *Pseudomonas* strain having ability to produce siderophores suppressed the disease caused by *Pythium* sp. by decreasing iron availability for fungal growth. It has been well documented that fungi are unable to use Fe complexed with siderophores (Solano et al. 2009). It might be due to high affinity of siderophores for Fe that limits its availability for fungal growth (Glick 2012).

Soil microbes can induce resistance in plants against pathogens, leading to induced systemic resistance (ISR). The ISR is phenotypically similar to the systemic acquired resistance (SAR) that is plant's internal mechanism to respond to infection by pathogens (Pieterse et al. 2009). Siderophore-producing bacteria are also involved in ISR, thus enhancing plant's defensive mechanism against pathogenic microorganisms. Literature reports the effectiveness of siderophores producing PGPR to enhance ISR against fungal and viral diseases (Saravanakumar et al. 2007). In addition to siderophores, jasmonate and ethylene signaling also stimulate the host plant's defense mechanisms against pathogens (Verhagen et al. 2004). Other ISR compounds released by soil microbes include lactones, homoserine, cyclic lipopeptides, 2,4-diacetylphloroglucinol, lipopolysaccharides (LPS), and some other volatile compounds (Lugtenberg and Kamilova 2009). Some antibiotics produced by soil microbes are also directly involved in ISR that can enhance the efficiency of antibiotics, thus increasing resistance against pathogens (Jha et al. 2011). From above discussion, it can be concluded that in addition to other plant growth-promoting mechanisms, soil microorganisms can also be effective in protecting plants against pathogens by suppressing their growth.

12.7.5 Enzymatic Activity

Production of enzymes by soil microorganisms is an important aspect that has been extensively explored by scientist during recent years. A number of bacterial strains have been documented which produce certain enzymes such as ACC-deaminase, catalase, cellulase, phosphatase, and chitinase. These enzymes can help plants to withstand different kinds of biotic and abiotic stresses. For example, ethylene is one of the phytohormones that has specific role in plant senescence and maturity. It is required for plant metabolism during normal growth and development, but in lesser amount (Khalid et al. 2006). It is also produced under stress (Saleem et al. 2007) that causes the change in normal metabolic processes of plants leading to its rescue from stressed conditions. Higher level of ethylene is produced under stresses that decreases the root and shoot growth of plants. For example, Ahmad et al. (2011) reported a decrease in root and shoot length and increase in stem diameter of mung bean under salinity stress, and they linked it to increase in ethylene concentrations. The ACC-deaminase has been reported in a number of bacterial strains belonging to genera *Pseudomonas*, *Bacillus*, and *Enterobacter* (Nadeem et al. 2010b; Ahmad et al. 2011). The improvement in plant growth due to inoculation with ACC-deaminase-containing bacteria under stressful environments has been well documented (Mayak et al. 2004; Zahir et al. 2010; Ahmad et al. 2012; Glick 2012). ACC is the immediate precursor of ethylene and cleaves it into α-ketobutyrate and NH_3 (Glick et al. 1998).

Chitinase is also an important enzyme that is produced by soil microbes and has the potential to suppress plant diseases (Glick et al. 2007). Similarly, another microbial enzyme cellulase can help in the penetration of rhizobia into root hairs during nodulation, thus increasing the nitrogen fixation in legumes (Sindhu and Dadarwal 2001). Phosphatases are also produced by phosphate-solubilizing soil microbes which help in the release of inorganic P from complexes, thus increasing P availability to plants (Dodor and Tabatabai 2003). The hydrolytic enzymes, such as chitinases, proteases, lipases, and glucanases, are also produced by soil microbes which are effective in biocontrol of pathogens. These enzymes are involved in fungal cell wall dissolution, thus suppressing their growth. The effectiveness of biocontrol mechanism of soil bacteria has been well documented against different pathogens (Kim et al. 2008; Glick 2012).

12.7.6 Stress Tolerance Induction

Soil microbes have adapted to a wide range of environments, thus can tolerate higher degree of environmental stresses. For example, *Rhizobium* can tolerate up to 64 dS m^{-1} salinity in solution culture (Forawi 1994) that enables these bacteria to develop successful symbiosis with legume crops, thus increasing nodulation under salt-stressed conditions (Ahmad et al. 2011). It has been reported that rhizobia are more tolerant to environmental stresses than their host plants (Elsheikh 1992). The PGPR also have remarkable tolerance against different stresses such as drought and salinity (Sandhya et al. 2009; Ahmad et al. 2011). Fungi can also be used as good tool to induce stress tolerance. For example, endophytic fungi including *Rhizodermea veluwensis*, *Phialocephala fortinii*, and *Rhizoscyphus* sp. enhanced the heavy metal stress tolerance in *Clethra barbinervis* by increasing the K uptake and decreasing the heavy metal concentrations in plant parts, thus enabling it to grow naturally at mine sites (Yamaji et al. 2016). The exact mechanisms of higher levels of stress tolerance of soil microbes have not yet been explored (Spaepen et al. 2009); however, these might be the production of exopolysaccharides (EPS) by soil bacteria that protect them against stresses, thus enhancing their survival (Upadhyay et al. 2011). Literature also reports the accumulation of poly-β-hydroxybutyrate, proline, and ectoine in bacterial body as protective measures for their survival under stress conditions (Bernard et al. 1993; Arora et al. 2006). The ACC-deaminase activity of soil bacteria and fungi has also been well documented as a mechanism for stress tolerance induction in crop plants (Glick et al. 2007; Nadeem et al. 2010a, b; Ahmad et al. 2011; Aban et al. 2017; Saravanakumar et al. 2018). The use of these stress-tolerant strains can be effective to improve soil fertility and crop growth (Egamberdieva and Kucharova 2009; Ahmad et al. 2013). It is evident from the above literature that beneficial soil microbes can withstand variable soil and environmental conditions that enable them to live better in competitive environment. These mechanisms make beneficial soil microbes a useful tool to maintain soil fertility and increase crop productivity and agroecosystem sustainability.

12.8 Application of Plant-Microbe Interactions in Agroecosystem

Plant-microbe interactions in the rhizosphere are modulated by edaphic factors. Rhizosphere is the zone of maximum biological activity, and plant roots and soil microbes communicate with each other. These interactions might be beneficial or harmful for one or both of the partners.

The beneficial plant-microbe interactions can be harnessed with better soil health, improved plant growth and productivity, and induced stress tolerance in crop plants. Improvement in beneficial microbial populations through rhizosphere engineering or use of microbial inoculants and/or their metabolites can be helpful to modify the soil microbiome (Velmourougane et al. 2017), leading to increased crop productivity and agroecosystem sustainability. Crop improvement through inoculation with beneficial soil microbes under normal and stressful environments has been widely studied and reviewed by many scientists (Glick et al. 2007; Singh et al. 2013; Nadeem et al. 2011, 2013, 2014, 2015; Ahmad et al. 2016). However, application of modern techniques to improve performance of soil microbes can be a key to agriculture sustainability by improving crop productivity, balanced nutrition, soil fertility, and plant stress tolerance (Gouda et al. 2018). Some selected examples of plant growth promotion by mycorrhizae and PGPR are presented in Table 12.2.

12.9 Plant Growth Promotion under Normal Conditions

Soil microbiome can effectively be exploited for improving the productivity of agroecosystems. Previous section reports the important plant growth-promoting mechanisms which directly or indirectly improve crop yield and soil health, thus overall productivity of the system. Manipulation of rhizosphere microbiome

Table 12.2 Effectiveness of microbes for promoting plant growth

Growth condition	Crop	Response	References
(a) the impact of bacterial inoculation on crop growth under normal and stress conditions			
Normal (field trial)	Chickpea (*Cicer arietinum*)	Dual inoculation of bacteria enhanced nodule number, nodule fresh weight, shoot N content, and yield	Valverde et al. (2006)
Normal (pot and field condition)		Inoculation improved nodulation and yield of chickpea	Elkoca et al. (2008)
Drought stress (pot trial)	Pea (*Pisum sativum*)	Rhizobacteria containing ACC-deaminase enhanced the growth of pea plant by reducing the negative impact of drought	Zahir et al. (2008)
Normal (field trial)	Rice (*Oryza sativa*)	Significant increase in growth due to improving number of tillers and shoot length	Bao et al. (2013)
Normal (field trial)		Inoculation with phosphate-solubilizing bacteria enhanced the growth and yield parameters, and significant increase in yield parameters was observed	Chamani et al. (2015)
Normal (pot and field condition)		Significant increase in number of tiller and seed yield	Isawa et al. (2010)
Normal (pot and field condition)	Maize (*Zea mays*)	Significant increase in growth was observed in pot and field trials	Ferreira et al. (2013)
Normal (pot trial)		Endophytic and rhizobacteria associated with sugarcane enhanced the maize growth and indole acetic acid content	Rodrigues and Forzani (2016)
Normal (pot trial)		Nitrogen-fixing *Gluconacetobacter diazotrophicus* colonized the inbred grain corn lines and sweet corn varieties. A positive correlation was observed between plant sucrose content and colonization efficiency	Tian et al. (2009)
Nutrient deficiency (pot trial)		Multistrain bacterial consortium significantly improved the growth of maize by enhancing the availability of P and K	Abou-el-Seoud and Abdel-Megeed (2012)
Drought stress (pot study)		Inoculation with *Azospirillum* helped maize seedling tolerate drought stress to a higher level as compared to uninoculated plants	Garcia et al. (2017)

(continued)

Table 12.2 (continued)

Growth condition	Crop	Response	References
Salinity stress (pot trial)	Wheat (*Triticum aestivum*)	Inoculation reduced sodium uptake and improved plant growth, sugar, and proline content	Upadhyay et al. (2012)
Normal (field trial)		Endophyte inoculation significantly increase the root length, root fresh weight, and root dry weight	Singh et al. (2017)
Normal (field trial)		Inoculation of wheat with bacteria in the biofertilizer enhanced the growth and productivity	Hussain (2016)
Drought stress (pot trial)		Under drought stress, endophytic *Burkholderia phytofirmans* PsJN improved the growth of wheat by maintaining ion balance	Naveed et al. (2014)
Salt and drought stress (hydroponic study)		*Arthrobacter protophormiae* (SA3) and *Dietzia natronolimnaea* (STR1) improved salt tolerance, while *Bacillus subtilis* (LDR2) provide protection against drought tolerance	Barnawal et al. (2017)
Normal (field trial)	Sugarcane (*Saccharum officinarum*)	Inoculation enhanced germination, growth, and sugarcane juice content	Beneduzi et al. (2013)
Normal (field trial)	Soybean (*Glycine max*)	Rhizobium inoculation enhanced the soybean yield compared to uninoculated	Hungria et al. (2013)
Abiotic stress (lab study)	Carrot (*Daucus carota)*	Inoculated bacteria showed biocontrol potential and significantly enhanced and promoted root formation on carrot slices	Etminani and Harighi (2018)
Normal (lab study in glass vial)	Cucumber (*Cucumis sativus*)	Nitrogen-fixing bacteria showed nitrogen-fixing ability and caused positive effect on plant growth	My et al. (2015)
Salinity stress (lab study)		Improved the growth of cucumber by reducing the impact of salinity. Inoculated plant showed better growth compared to inoculated plants	Nadeem et al. (2016)
Normal (lab study)	*Wedelia trilobata*	Endophytic *Bacillus* significantly enhanced the growth of inoculated plant. Effect of endophyte was different in case of invasive and native clonal plants	Dai et al. (2016)

(continued)

Table 12.2 (continued)

Growth condition	Crop	Response	References
Normal (pot study)	Potato *Solanum tuberosum*	The rhizobacterial strains showed variable response and caused significant positive impact on potato growth	Dawwam et al. (2013)
Nutrient stress (pot study)	*Okra(Abelmoschus esculentus)*	Inoculation enhanced the root and shoot growth of okra compared to no inoculation	Prajapati et al. (2013)
Normal (pot study)	Century plant *Agave americana* L.	Significant increase in plant growth and sugar content was observed due to phytohormone production and nutrient-solubilizing ability of bacteria	Torre-Ruiz et al. (2016)
Metal stress (pot study)	Mustard greens (*Brassica juncea*)	Inoculation enhanced the phytoremediation efficiency of plant and improved growth compared to uninoculated one	Qiu et al. (2014)
Normal and metal stress (pot study)	Pearl millet *Pennisetum glaucum*	Mitigate the negative impact of temperature and salinity stress and improve growth by the production of phytohormones and phosphorus availability	Misra et al. (2012)
Salinity stress (pot trial)	Barley (*Hordeum vulgare* L) and pearl millet (*Pennisetum glaucum*)	Inoculation improved the phytoremediation activity of the plant. Less electrolyte leakage and more membrane stability was observed in inoculated plants	Jodeh et al. (2015)
Salt and drought stress	Tomato (*Solanum lycopersicum*)	*Azotobacter* strains showed high tolerance to salt and drought stresses and alleviated the negative effects exerted by stress on tomato plants	Viscardi et al. (2016)
Salt stress (pot study)	Camelina (*Camelina sativa*)	Improved salinity tolerance of inoculated plant was due to several mechanisms. Salinity tolerance and presence of ACC-deaminase enzyme is responsible for reducing stress-induced ethylene	Heydarian et al. (2018)
Salinity stress (pot trial)	Oat seedlings (*Avena sativa*)	Inoculation modulated the expression profile of *rbcL* and *WRKY1* genes and enhanced plant's stress tolerance against salinity	Sapre et al. (2018)

(continued)

Table 12.2 (continued)

Growth condition	Crop	Response	References
Salt stress (pot study)	Citrus (*Citrus macrophylla*)	Both rhizobacterial strains reduce the negative impact of stress, and lower contents of abscisic acid (ABA) and salicylic acid (SA) were observed in inoculated plants under salt stress	Vives-Peris et al. (2018)
(b) the impact of mycorrhizal inoculation on crop growth under normal and stress conditions			
Normal (pot trial)	Onion (*Allium cepa*)	Mycorrhizae enhanced chlorophyll content as well as fresh and dry weight of onion	Shuab et al. (2014)
Normal (pot trial)	White clover (*Trifolium repens*)	Significant increase in nodule number, root length, volume and number of lateral roots, and chlorophyll content	Lu and Wu (2017)
Normal (pot trial)	Lettuce (*Lactuca sativa*)	Rhizophagus intraradices enhanced the Zn uptake of lettuce grown at two P levels; however, *Funneliformis mosseae* did not affect Zn content	Konieczny and Kowalaska (2016)
Abiotic stress (lab studies)	Common milkweed (*Asclepias syriaca*)	Mycorrhizae influenced plant resistance phenotype and a key factor for determining the outcome of plant herbivore	Vannette and Hunter (2013)
Drought stress (pot study)	Lime (*Citrus aurantifolia*)	Improved growth through its significant positive impact on chlorophyll contents and photosynthesis activity of the plant	Shahsavar et al. (2016)
Normal (pot study)	Tomato (*Solanum lycopersicum*)	Enhanced the plant resistance against bacterial wilt and improved its growth	Tahat et al. (2012)
Drought stress (lab study)		Mycorrhizal inoculation positively affects the tomato tolerance to water stress. A group of fungal genes play a key role in the water-transport process	Chitarra et al. (2016)
Normal (pot study)	Tomato (*Solanum lycopersicum*) and pepper (*Capsicum annuum*)	Caused significant impact on plant biomass, P accumulation, and improved fruit yield	Padmavathi et al. (2015)
Normal (pot study)	Sorghum (*Sorghum bicolor*) and chili pepper (*Capsicum annuum*)	Mycorrhizae caused significant differences in the growth of the host plant which shows preference of host plant for fungus.	Lee and Eom (2015)

(continued)

Table 12.2 (continued)

Growth condition	Crop	Response	References
Normal (pot study)	*Melberry (Morus alba)*	Improved growth through its significant positive impact on chlorophyll contents and photosynthesis and stomatal conductance	Shi et al. (2016)
Normal (pot study)	*Chinese Wedelia* (*Wedelia chinensis*)	Among seven indigenous AM fungi, *Glomus fasciculatum* improved plant nutrition and improved plant growth	Nisha and Rajeshkumar (2010)
Salinity stress (pot study)	Hangbaiju (*Chrysanthemum morifolium)*	Inoculation enhanced root length, shoot and root dry weight, and root N content. Nitrogen uptake could be the mechanism responsible for salinity tolerance	Wang et al. (2018a, b)
Drought stress (pot study)	Soybean (*Glycine max*)	Mitigated the impact of water stress. Arbuscule formation was higher in the unimproved than improved genotypes	Salloum et al. (2017)
Drought stress (pot study)	Maize (*Zea mays*)	Mycorrhizae together with rhizobacteria enhanced the vegetative and reproductive traits, root colonization, the grain yield of maize, content of P and N	Ghorchiani et al. (2018)
Salt stress (lab study)	*Acacia gerrardii*	Mycorrhizae alone and in combination with bacteria promoted plant growth by enhancing N, P, K, Mg, and Ca contents and phosphatase activities and reducing Na and Cl concentration	Hashem et al. (2016)

changes the soil microbial diversity and population that improves plant performance through change in water dynamics and enzyme activities in soil (Ahmadi et al. 2018). Rhizosphere engineering through augmentation can help to enhance root colonization that increases the availability of nutrients, reduces the use of chemical fertilizer, and conserves organic systems (Ahemad and Kibret 2014). It has been observed that combined use of rhizobacteria, endophytic bacteria, and arbuscular mycorrhizal fungi (AMF) significantly enhanced crop productivity by less use of chemical fertilizers (Pérez et al. 2007).

The PGPR can enhance crop productivity and nutrient availability through fixing atmospheric N_2, solubilizing inorganic P, production of Fe(III)-specific chelating siderophores, and phytohormones such as cytokinins, auxins, and gibberellins (Fravel 2005). Siderophore-producing bacterial strains *Stenotrophomon asche-latiphaga* and *Myristica yunnanensis* significantly improved plants' zinc and phosphorous contents of canola and maize plants (Ghavami et al. 2016). Results showed a significant increase in root and shoot Zn contents, thus improving crop growth and productivity. They suggested these strains as potential bioinoculant for improving plant productivity that can reduce the use of chemical fertilizers. This can also be a possible option to correct the nutrient deficiency in canola and maize crops, leading to agroecosystem stability (Ghavami et al. 2016).

Solubilization of nutrients is an important mechanism used by soil microbes to improve growth, yield, and quality of crop plants. In alkaline calcareous soils, the decreased efficiency of fertilizers especially Zn and phosphorus is an issue. Especially formation of insoluble zincate complex upon Zn fertilization is considered a serious threat to soil-plant nutrition. The issue can be resolved by inoculation with Zn-solubilizing bacteria. For example, Mumtaz et al. (2018) evaluated four Zn-solubilizing PGPR strains *Bacillus* spp. (ZM20), *B. subtilis* (ZM63), and *B.* aryabhattai (ZM31 and S10) for their effectiveness to improve growth, yield, and quality of maize grains. It was observed that Zn-solubilizing *Bacillus* strains significantly improved the plant growth, yield, and nutrient concentration in maize grains. Use of phosphate-solubilizing bacteria can also be helpful to improve crop productivity and fertilizer efficiency in alkaline calcareous soils. Recently, Ahmad et al. (2018) evaluated the phosphate-solubilizing *Bacillus* strains to improve cotton growth under alkaline conditions. They reported that bacterial strains varied in their growth-promoting traits and they differed in P-solubilization efficiency. Efficient root colonization of these strains in cotton under salt-affected soils helped plants to uptake more phosphorus thus improving cotton growth.

The combined use of PGPR with other soil microbes and/or organic and inorganic sources of nutrients can be effective to improve crop productivity by sustaining soil fertility. Ahmad et al. (2015) evaluated *Pseudomonas fluorescens* in combination with different sources of organic manure and chemical fertilizer for enhancing the productivity of cucumber (*Cucumis sativus* L.). They reported significant improvement in growth, fruit quality, and yield of cucumber by combined application of *P. fluorescens*, organic manure, biogas slurry, and chemical fertilizer. So, the combined use of organic sources and *P. fluorescens* can be used to enhance cucumber productivity that can also sustain soil fertility for future. Similarly, the effectiveness of ACC-deaminase-based biofertilizer consisting of *Rhizobium* and *Pseudomonas* strains was evaluated in combination with P-enriched compost under field conditions to improve the productivity of chickpea on marginal soils in Bahawalpur. The combined use of ACC-deaminase-containing biofertilizer and P-enriched compost effectively improved chickpea productivity on marginal soils under field conditions and can be used as effective strategy to cope with scenario of limited water availability and sustaining agroecological systems (Ahmad et al. 2017).

Mycorrhizal associations use different growth-promoting characteristics (Smith and Read 2008) such as improvement in rhizobial activities for N_2 fixation (Krapp 2015), improvement in photosynthetic rates (Hashem et al. 2015), enhancing phosphatase activity in soil (Liu et al. 2015), producing bioactive substances (Goicoechea et al. 1997), detoxification of heavy metals (Zong et al. 2015), reducing the effect of stresses through osmotic adjustments (Xun et al. 2015), and increase in resistance to abiotic (Hashem et al. 2015) and biotic (Yuan et al. 2016) stresses. Fungi enhanced the nitrogen status of plants when applied in combination with PGPR, rhizobia, or both (Barnawal et al. 2014; Armada et al. 2015; Barrett et al. 2015). The AM fungi enhance surface area of plant roots through symbiotic associations (Kaiser et al. 2015) and help in the exchange of nutrients between soil and plant roots (Buscot 2015), thus enhancing nutrient uptake and plant growth. Rice is the crop with high water requirement and is severely affected by water scarcity and climate change. The AM fungi can establish strong symbiotic associations with roots of rice crop. Rice has been studied as a model for molecular determinants regarding establishment and functioning of AM symbiosis to provide insights into potential breeding target for improving the crop interaction with AMF. There are strong evidences which show the beneficial effects of AM fungi on performance of rice crop under field conditions (Mbodj et al. 2018).

Moreover, multistrain biofertilizers can be more efficient than single-strain inoculants due to their multifarious traits. For example, Zahir et al. (2018) evaluated the effectiveness of multistrain biofertilizer to enhance growth, nodulation, and productivity of ten genotypes of mung bean under field conditions. They also evaluated the effect of biofertilizer on total bacterial DNA in soil and reported increase in nodulation, growth, and yield of mung bean as compared to uninoculated control. The genotypes also varied in their productive potential and responded differently to biofertilizer under field conditions.

12.10 Role of Soil Microbes under Stress

Climate change and anthropogenic activities breed a number of environmental stresses which can seriously affect the productivity of agroecosystems (Vimal et al. 2017). These stresses are classified as abiotic (salinity, drought, flooding, temperature, wounding, and heavy metal stresses) and biotic (insect and pathogenic stresses). These stresses can significantly reduce the productivity of cropping systems. Soil microbes can be successfully used to reduce the effect of these stresses on crop productivity. Mechanisms used by these microbes for reducing the impact of environmental stresses are summarized in above sections. A number of reports are available regarding field application of these mechanisms for inducing stress tolerance in crop plants (Glick et al. 2007; Ahmad et al. 2012; Nadeem et al. 2014), thus enabling plants to maintain normal metabolic processes. This section summarizes some of the selected studies regarding use of soil microbes under abiotic and biotic stresses.

Soil microbes can reduce the effect of stresses on plant growth by releasing 1-aminocyclopropane-1-carboxylic acid (ACC) deaminase which consumes the ACC; the immediate precursor of ethylene thus suppresses the stress-induced production of ethylene (Abou-Shanab et al. 2006; Glick et al. 1998). Other well-known plant growth-promoting mechanisms used by soil microbes under stress include alteration of root morphology, increase in water uptake, antibiotics production, and induction of plant defense mechanisms (Kidd et al. 2017). The PGPR also influence mobility and phytoavailability of trace metals in soil (Sessitsch et al. 2013).

12.10.1 Abiotic Stress

Abiotic stresses such as drought, salinity, and heavy metals are among the most common problems of our agroecological systems. Many efforts are being made to cope with these stresses. Plants also regulate internal metabolic processes to adapt to the stressed environments, however, by compromising their yield. Higher levels of ethylene production, imbalance of ionic ratios, nutritional imbalance, and production of toxic reactive oxygen species are among the major changes which occur in plants under different kinds of stresses. Beneficial soil microbes can effectively be used to regulate the metabolic processes of plants under stress, thus maintaining their yields (Nadeem et al. 2014). Use of these microbes can be durable, cost-effective, and environment-friendly that not only enhances crop yield but also improves soil health.

Environmental stresses can also affect the growth of beneficial soil microbes; however, they have adapted to a wide range of environments through their particular characteristics such as production of exopolysaccharides and ACC-deaminase activity. For example, *Rhizobium* can tolerate up to 64 dS m^{-1} salinity in solution culture (Forawi 1994) that enables these bacteria to develop successful symbiosis with legume crops, thus increasing nodulation under salt-stressed conditions (Ahmad et al. 2011). The stress-tolerant soil microbes can also be effective to induce tolerance in crop plants against abiotic stresses such as salinity, drought, high temperature, and heavy metal toxicity (Grover et al. 2011). The exopolysaccharides released by soil microbes can also protect plants from desiccation by forming protective layer around soil aggregates (Tisdall and Oades 1982). Exopolysaccharides also increase root colonization of microbes (Santaella et al. 2008) by improving soil aggregation (Sandhya et al. 2009) and improving water and nutrients availability to plants (Tisdall and Oades 1982). The inoculation with ACC-deaminase containing bacteria has the potential to reduce negative effect of ethylene on pepper and tomato plants, thus improving growth of these crops (Mayak et al. 2004). It has been reported that the combined use of *Rhizobium* with PGPR-containing ACC-deaminase minimized the negative effect of ethylene on mung bean, thus improving nodulation under salinity stress (Ahmad et al. 2011).

The use of bacteria with ACC-deaminase activity is helpful in improving crop productivity under stresses, and the efficiency of microbial inoculants can be improved by using them in combination with exogenous plant growth regulators. For example, Jamil et al. (2018) evaluated the effectiveness of using *Pseudomonas fluorescens* strain containing ACC-deaminase in combination with L-tryptophan to reduce the effect of drought stress on wheat crop. They reported that using L tryptophan at 25 ppm along with *P. fluorescens* is more effective than their separate application. They suggested that the approach could be effective to improve productivity of wheat under water scarcity. In another study, Nadeem et al. (2017) reported that *P. fluorescens* in combination with compost and biochar improved the water stress tolerance in cucumber. They conducted a greenhouse experiment for evaluating the effectiveness of integrated use of biochar, compost, and *P. fluorescens* to alleviate the effect of water-deficit stress. They used three levels of water, i.e., field capacity (D0), 75% field capacity (D1), and 50% field capacity (D2), and concluded that integrated use of these sources was an effective strategy to alleviate the deleterious effects of water stress on cucumber growth. They, however, proposed field studies to further investigate the biotechnology for its long-term impact on agroecosystem sustainability.

Soil microbes present in rhizosphere of hyperaccumulating plants are distinct with higher genetic diversity and have high level of resistance to metal stress, which can effectively be used to improve crop performance and phytoremediation of heavy metal-contaminated soils (Thijs et al. 2017; Benizri and Kidd 2018). Recently, Ghasemi et al. (2018) evaluated the effect of bacterial inoculation on plant health, growth, and Ni phytoextraction ability of three Ni-hyper accumulator species, *Odontarrhena inflate*, *O. bracteata*, and *O. serpyllifolia* using five rhizobacterial strains isolated from *O. serpyllifolia*. They reported that bacterial strains effectively enhanced the Ni removal by stimulating plant growth and/or increasing shoot Ni concentration. However, the efficacy of these strains varied with soil type, plant species, and bacterial strain. Antioxidative enzymes and malondialdehyde (MDA) and H_2O_2 concentration was also lower in inoculated plants, indicating protective effect of these strains on plants. In another study, Alvarez-Lopez et al. (2017) evaluated the effect of combined use of composted sewage sludge and bacterial inoculation on the growth and heavy metals (Cd and Zn) accumulation ability of *Salix caprea* and *Nicotiana tabacum* in contaminated mine tailings. Bacterial inoculation improved biomass of tobacco in compost-amended soil, while it did not work so efficiently in unamended soil.

The AM fungi have been reported to improve the carbon and nitrogen cycling in alpine grasslands (Li et al. 2015). The hyphal networks of AM fungi help plants in uptake of water and nutrients in stressed environments and restrict the availability of heavy metals to plant roots (Miransari 2011). The fungal associations can be helpful in the restoration of degraded lands and forests. The use of fungi in combination with organic sources or bacteria has been reported to be helpful in the restoration of soil fertility and organic matter contents in degraded soils (Rashid et al. 2016).

12.10.2 Biotic Stress

The use of soil microbes can also be effective to control pests and diseases of field crops. For example, Prabhukarthikeyan et al. (2014) evaluated the combined use of PGPR and endophytic bacteria (*Bacillus* strains) and reported that the combination was effective in controlling the fusarium wilt and fruit borer in tomatoes in the absence of pesticide. In another report, Bandi and Sivasubramanian (2012) reported the ability of *Pseudomonas fluorescens* to induce systemic resistance against thrips (*Thrips tabaci* L.). He regarded *Pseudomonas fluorescens* as effective biocontrol agent against pests. Soil microbes have the ability to produce allelopathic substances, which are effective against various pests of crop plants (Sessitsch et al. 2004). Different metabolites synthesized by soil microbes suppress growth and prevalence of plant pathogens that indicates their potential to be used as effective biopesticides. It has been reported that beneficial soil microbes can suppress growth of pathogenic fungi, bacteria and viruses, weeds, nematodes, and insect pests through production of antibiotics and hydrolyzing enzymes or ISR (Gao et al. 2015).

The use of soil microbes as biocontrol agent is regarded as an environment-friendly approach as these microbes are very specific to their host pathogens (Kachhawa 2017). The use of soil microbes could decrease agrochemical use, helping to foster environmental sustainability by reducing the harmful effects of toxic chemical compounds. The use of plant growth mechanisms of beneficial soil microbes is economical and ecofriendly approach to protect plants against stress conditions. These plant-microbe interactions are vital for sustainable agriculture because this approach depends upon biological processes and can replace conventional agricultural practices (Kumar and Verma 2018).

The above discussion shows the effectiveness of soil microbes for enhancing crop productivity under normal as well as stressed environments. It is evident from the literature that soil microbes use a number of direct and indirect mechanisms for improving crop productivity. The use of these beneficial microbes can not only enable plants to maintain their growth and productivity under various kinds of environments but also improve soil health that can be beneficial in maintaining agroecosystem sustainability.

12.11 Conclusions and Future Prospects

Above discussed literature indicates that soil microbiome has strong implications on plant growth. Soil-plant-microbe interactions can be harnessed with good crop productivity and ecosystem sustainability. Soil microbes interact with plant roots positively or negatively and thus have significant effects on plant growth and productivity and soil health. Symbiotic plant microbe interactions have been well documented which can significantly improve plant growth. These interactions are affected by quality of root exudates and physicochemical properties of soil.

Beneficial soil microbes have a number of plant growth-promoting mechanisms including biological nitrogen fixation, phytohormone production, and nutrient solubilization. These traits of beneficial microbes can be harnessed for better soil health, improved plant growth and productivity, and improved stress tolerance of crop plants. Improvement in beneficial microbial populations through rhizosphere engineering or use of microbial inoculants and/or their metabolites can be helpful in modifying the soil microbiome, leading to increased productivity of agroecosystem. The soil microbes are equally effective to enhance plant growth under normal as well as stress conditions. Soil microbes also protect plants from biotic and abiotic stresses through ACC deaminase activity, exopolysaccharides production, and production of hydrolytic enzymes and volatile compounds.

Future research should focus on understanding the mechanisms involved in bacterial-induced growth promotion. Research should also be conducted to investigate why the same isolate with specific PGP traits performs differently and could not induce the same plant response under distinct soil conditions. Strategic improvement in plant-microbe interactions through bioinformatics, molecular genetics, and modeling tools should be carried out for improving crop productivity and agroecosystem sustainability.

References

Aban JL, Barcelo RC, Oda EE, Reyes GA, Balangcod TD, Gutierrez RM, Hipol RM (2017) Auxin production, phosphate solubilisation and ACC deaminase activity of root symbiotic fungi (RSF) from *Drynaria quercifolia* L. Bull Env Pharmacol Life Sci 6(5):26–31

Abou-el-Seoud II, Abdel-Megeed A (2012) Impact of rock materials and biofertilizations on P and K availability for maize (*Zea maize*) under calcareous soil conditions. Saudi J Biol Sci 1:55–63

Abou-Shanab R, Angle J, Chaney R (2006) Bacterial inoculants affecting nickel uptake by Alyssum murale from low, moderate and high Ni soils. Soil Biol Biochem 38:2882–2889

Afzal A, Bano A, Fatima M (2010) Higher soybean yield by inoculation with N-fixing and P- solubilizing bacteria. Agron Sustain Dev 30:487–495

Ahemad M, Kibret M (2014) Mechanisms and applications of plant growth promoting rhizobacteria: current perspective. J King Saud Univ Sci 26(1):1–20

Ahmad M, Zahir ZA, Asghar HN, Asghar M (2011) Inducing salt tolerance in mung bean through co-inoculation with *Rhizobium* and PGPR containing ACC-deaminase. Can J Microbiol 57(7):578–589

Ahmad M, Zahir ZA, Asghar HN, Arshad M (2012) The combined application of rhizobial strains and plant growth promoting rhizobacteria improves growth and productivity of mung bean (*Vigna radiata* L.) under salt-stressed conditions. Ann Microbiol 62:1321–1330

Ahmad M, Zahir ZA, Khalid M, Nazli F, Arshad M (2013) Efficacy of *Rhizobium* and *Pseudomonas* strains to improve physiology, ionic balance and quality of mungbean under salt-affected conditions on farmer's fields. Plant Physiol Biochem 63:170–176

Ahmad M, Zahir ZA, Nazli F, Akram F, Arshad M, Khalid M (2013c) Effectiveness of halo-tolerant, auxin producing *Pseudomonas* and *Rhizobium* strains to improve osmotic stress tolerance in mung bean (*Vigna radiata* L.). Braz J Microbiol 44(4):1341–1348

Ahmad M, Zahir ZA, Zeshan MSH, Nasim M, Nadeem SM, Nazli F, Jamil M (2015) Improving the productivity of cucumber through combined application of organic fertilizers and *Pseudomonas fluorescens*. Pak J Agri Sci 52(4):1011–1016

Ahmad M, Nadeem SM, Naveed M, Zahir ZA (2016) Potassium solubilizing bacteria and their application in agriculture. In: Meena VS, Maurya BR, Verma JP, Meena RS (eds) Potassium solubilizing microorganisms for sustainable agriculture. Springer, New Delhi, pp 293–313

Ahmad M, Zahir ZA, Jamil M, Nazli F, Iqbal Z (2017) Field application of ACC-deaminase biotechnology for improving chickpea productivity in Bahawalpur. Soil Environ 36(2):197–206

Ahmad M, Ahmad I, Hilger TH, Nadeem SM, Akhtar MF, Jamil M, Hussain A, Zahir ZA (2018) Preliminary study on phosphate solubilizing *Bacillus subtilis* strain Q3 and *Paenibacillus* sp. strain Q6 for improving cotton growth under alkaline conditions. PeerJ. https://doi.org/10.7717/peerj.5122

Ahmadi K, Razavi BS, Maharjan M, Kuzyakov Y, Kostka SJ, Carminati A, Zarebanadkouki M (2018) Effects of rhizosphere wettability on microbial biomass, enzyme activities and localization. Rhizosphere 7:35–42

Alloush GAZ, Zeto SK, Clark RB (2000) Phosphorus source, organic matter and arbuscular mycorrhiza effects on growth and mineral acquisition of chickpea grown in acidic soil. J Plant Nutr 23:1351–1369

Alvarez-Lopez V, Prieto-Fernandez A, Roiloa S, RodriguezGarrido B, Herzig R, Puschenreiter M, Kidd PS (2017) Evaluating phytoextraction efficiency of two high biomass crops after soil amendment and inoculation with rhizobacterial strains. Environ Sci Pollut Res 24:7591–7606

Amundson R, Berhe AA, Hopmans JW, Olson C, Sztein AE, Sparks DL (2015) Soil and human security in the 21st century. Science 348:1261071-1–1261071-6

Andrews M, Andrews ME (2017) Specificity in legume-rhizobia symbioses. Int J Mol Sci 18(4):1–39

Andrews M, James EK, Sprent JI, Boddey RM, Gross E, dos Reis FB Jr (2011) Nitrogen fixation in legumes and actinorhizal plants in natural ecosystems: values obtained using ^{15}N natural abundance. Plant Ecol Divers 4:131–140

Andrews M, Raven JA, Lea PJ (2013) Do plants need nitrate? The mechanisms by which nitrogen form affects plants. Ann Appl Biol 163:174–199

Arkhipova TN, Prinsen EA, Veselov SU, Martinenko EV, Melentiev LV, Kudoyarova GR (2007) Cytokinin producing bacteria enhance plant growth in drying soil. Plant Soil 292:305–315

Armada E, Azcon R, Lopez-Castillo OM, Calvo-Polanco M, JM R'ı-L (2015) Autochthonous arbuscular mycorrhizal fungi and *Bacillus thuringiensis* from a degraded Mediterranean area can be used to improve physiological traits and performance of a plant of agronomic interest under drought conditions. Plant Physiol Biochem 90:64–74

Arora NK, Kang SC, Maheshwari DK (2001) Isolation of siderophore-producing strains of *Rhizobium meliloti* and their biocontrol potential against *Macrophomina phaseolina* that causes charcoal rot of groundnut. Curr Sci 8:673–677

Arora NK, Singhal V, Maheshwari DK (2006) Salinity-induced accumulation of poly-b- hydroxybutyrate in rhizobia indicating its role in cell protection. World J Microbiol Biotechnol 22:603–606

Arshad M, Frankenberger WT Jr (2002) Ethylene: agricultural sources and applications. Ann Bot 90(3):424

Bacilico-Jimenz M, Aguiler S, Ventura-zapta E, Perez-Campos E, Bouquelet S, Zenteno E (2003) Chemical characterization of root exudates from rice (*Oryza sativa*) and their effects on the chemotactic response of endophytic bacteria. Plant Soil 249:271–277

Bais HP, Weir TL, Perry LG, Gilroy S, Vivanco JM (2006) The role of root exudates in rhizosphere interactions with plants and other organisms. Annu Rev Plant Biol 57:233–266

Bandi S, Sivasubramanian P (2012) Management of *Thrips tabaci* Lindeman in onion using *Pseudomonas fluorescens* Migula through induced resistance. J Biopest 5:1–3

Bao Z, Sasaki K, Okubo T, Ikeda S, Anda M, Hanzawa E, Kaori K, Tadashi S, Hisayuki M, Minamisawa K (2013) Impact of *Azospirillum* sp. B510 inoculation on rice-associated bacterial communities in a paddy field. Microbes Environ 28:487–490

Barnawal D, Bharti N, Maji D, Chanotiya CS, Kalra A (2014) ACC deaminase-containing *Arthrobacter protophormiae* induces NaCl stress tolerance through reduced ACC oxidase activity and ethylene production resulting in improved nodulation and mycorrhization in *Pisumsativum*. J Plant Physiol 171:884–894

Barnawal D, Bharti N, Pandey SS, Pandey A, Chanotiya CS, Kalra A (2017) Plant growth-promoting rhizobacteria enhance wheat salt and drought stress tolerance by altering endogenous phytohormone levels and TaCTR1/TaDREB2 expression. Physiol Planta 161:502–514

Barrett LG, Bever JD, Bissett A, Thrall PH (2015) Partner diversity and identity impacts on plant productivity in acacia-rhizobial interactions. J Ecol 103:130–142

Bashan Y, Levanony H (1990) Current status of *Azospirillum* inoculation technology: *Azospirillum* as a challenge for agriculture. Can J Microbiol 36:591–608

Beck HC, Hansen AM, Lauritsen FR (2003) Novel pyrazine metabolites found in polymyxin biosynthesis by *Paenibacillus polymyxa*. FEMS Microbiol Lett 220:67–73

Beltrano J, Ruscitti M, Arango MC, Ronco M (2013) Effects of arbuscular mycorrhiza inoculation on plant growth, biological and physiological parameters and mineral nutrition in pepper grown under different salinity and p levels. J Soil Sci Plant Nutr 13

Beneduzi A, Moreira F, Costa PB, Vargas LK, Lisboa BB, Favreto R, Baldani JI, Passaglia LMP (2013) Diversity and plant growth promoting evaluation abilities of bacteria isolated from sugarcane cultivated in the south of Brazil. Appl Soil Ecol 63:94–104

Benizri E, Kidd PS (2018) The role of the rhizosphere and microbes associated with hyperaccumulator plants in metal accumulation. In: Van der Ent A, Echevarria G, Baker A, Morel JL (eds) Agromining: farming for metals extracting unconventional resources using plants. Springer, Berlin, pp 157–188

Berendsen RL, Pieterse CMJ, Bakker PAHM (2012) The rhizosphere microbiome and plant health. Trends Plant Sci 17(8):478–486

Berg G, Smalla K (2009) Plant species and soil type cooperatively shape the structure and function of microbial communities in the rhizosphere. FEMS Microbiol Ecol 68:1–13

Berge O, Lodhi A, Brandelet G, Santaella C, Roncato MA, Christen R, Heulin T, Achouak W (2009) *Rhizobium alamii* sp. nov., an exopolysaccharide-producing species isolated from legume and non-legume rhizospheres. Int J Systematic Evolu Microbiol 59:367–372

Bernard T, Jebbar M, Rassouli Y, Himdi KS, Hamelin J, Blanco C (1993) Ectoine accumulation and osmotic regulation in *Brevibacterium linens*. J Gen Microbiol 139:129–138

Bhattacharyya PN, Jha DK (2012) Plant growth-promoting rhizobacteria (PGPR): emergence in agriculture. World J Microbiol Biotechnol 28:1327–1350

Boiero L, Perrig D, Masciarelli O, Penna C, Cassan F, Luna V (2007) Phytohormone production by three strains of *Bradyrhizobium japonicum* and possible physiological and technological implications. Appl Microbiol Biotechnol 74:874–880

Bonfante P, Genre A (2010) Mechanisms underlying beneficial plant-fungus interactions in mycorrhizal symbiosis. Nat Commun 1:48. https://doi.org/10.1038/ncomms1046

Bottini R, Cassan F, Piccoli P (2004) Gibberellin production by bacteria and its involvement in plant growth promotion and yield increase. Appl Microbiol Biotechnol 65:497–503

Bottomley PJ, Myrold DD (2015) Biological N inputs. In: Paul EA (ed) Soil microbiology, ecology and biochemistry, 4th edn. Academic Press, pp 447–470

Buscot F (2015) Implication of evolution and diversity in arbuscular and ectomycorrhizal symbioses. J Plant Physiol 172:55–61

Calderon K, Spor A, Breuil MC, Bru D, Bizouard F, Violle C, Barnard RL, Philippot L (2017) Effectiveness of ecological rescue for altered soil microbial communities and functions. ISME J 11:1–12

Calvo P, Watts DB, Kloepper JW, Torbert HA (2017) Effect of microbial-based inoculants on nutrient concentrations and early root morphology of corn (*Zea mays*). J Plant Nutr Soil Sci 180:56–70

Camerini S, Senatore B, Lonardo E, Imperlini E, Bianco C, Moschetti G, Rotino GL, Campion B, Defez R (2008) Introduction of a novel pathway for IAA biosynthesis to rhizobia alters vetch root nodule development. Arch Microbiol 190:67–77

Chamani HE, Yasari E, Pirdashti H (2015) Response of yield and yield components of rice (*Oryza sativa* L. cv. Shiroodi) to different phosphate solubilizing microorganisms and mineral phosphorous. Int J Biosci 6:70–75

Chitarra W, Pagliarani C, Maserti B, Lumini E, Siciliano I, Cascone P, Schubert A, Gambino G, Balestrini R, Guerrieri E (2016) Insights on the impact of arbuscular mycorrhizal symbiosis on tomato tolerance to water stress. Plant Physiol 1:00307

Costa R, Gotz M, Mrotzek N, Lottmann J, Berg G, Smalla K (2006) Effects of site and plant species on rhizosphere community structure as revealed by molecular analysis of microbial guilds. FEMS Microbiol Ecol 56:236–249

Dai ZC, Fu W, Wan LY, Cai HH, Wang N, Qi SS, Du DL (2016) Different growth promoting effects of endophytic bacteria on invasive and native clonal plants. Front Plant Sci 7:706

Dawwam GE, Elbeltagy A, Emara HM, Abbas IH, Hassan MM (2013) Beneficial effect of plant growth promoting bacteria isolated from the roots of potato plant. Annal Agric Sci 58:195–201

Dean DR, Jacobson MR (1992) Biochemical genetics of nitrogenase. In: Stacey G, Burris RH, Evans HJ (eds) Biological nitrogen fixation. Chapman and Hall, New York, pp 763–834

Dobbelaere S, Croonenborghs A, Thys A, Ptacek D, Vanderleyden J, Dutto P, Labandera-Gonzalez C, Caballero-Mellado J, Aguirre JF, Kapulnik Y, Brener S (2001) Responses of agronomically important crops to inoculation with *Azospirillum*. Func Plant Biol 28:871–879

Dodor DE, Tabatabai MA (2003) Effect of cropping systems on phosphatases in soils. J Plant Nutr Soil Sci 166:7–13

Doni F, Isahak A, Zain CR, Yusoff WM (2014) Physiological and growth response of rice plants (*Oryza sativa* L.) to *Trichoderma* spp. inoculants. AMB Express 4:45

Downie JA (2014) Legume nodulation. Curr Biol 24:184–190

Dudeja SS, Giri R (2014) Beneficial properties, colonization, establishment and molecular diversity of endophytic bacteria in legume and non-legume. Afr J Microbiol Res 8:1562–1572

Egamberdieva D, Kucharova Z (2009) Selection for root colonising bacteria stimulating wheat growth in saline soils. Biol Fertil Soils 45:563–571

Egamberdieva D, Kamilova F, Validov S, Gafurova L, Kucharova Z, Lugtenberg B (2008) High incidence of plant growth-stimulating bacteria associated with the rhizosphere of wheat grown on salinated soil in Uzbekistan. Environ Microbiol 10:1–9

El-Deeb B, Bazaid S, Gherbawy Y, Elhariry H (2012) Characterization of endophytic bacteria associated with rose plant (*Rosa damascena trigintipetala*) during flowering stage and their plant growth promoting traits. J Plant Interact 3:248–253

Elkoca E, Kantar F, Sahin F (2008) Influence of nitrogen fixing and phosphorus solubilizing bacteria on the nodulation, plant growth, and yield of chickpea. J Plant Nut 31:157–171

Elsheikh EAE (1992) Effect of salinity on growth, nodulation and nitrogen yield of inoculated and nitrogen fertilized chickpea (Cicer arietinum L.). Arch Biotechnol 1:17–28

Etminani F, Harighi B (2018) Isolation and identification of endophytic bacteria with plant growth promoting activity and biocontrol potential from wild pistachio trees. Plant Pathol J 34:208–217

Ferreira AS, Pires RR, Rabelo PG, Oliveira RC, Luz JMQ, Brito CH (2013) Implications of *Azospirillum brasilense* inoculation and nutrient addition on maize in soils of the Brazilian Cerrado under greenhouse and field conditions. Appl Soil Ecol 72:103–108

Forawi HAS (1994) Effects of salinity on nodulation and nitrogen fixation of fenugreek (*Trigonella foenumgraecum*). MSc (Agri) Thesis, Faculty of Agriculture, University of Khartoum, Sudan

Fouts DE, Tyler HL, DeBoy RT, Daugherty S, Ren Q, Badger JH, Durkin AS, Huot H, Shrivastava S, Kothari S, Dodson RJ, Mohamoud Y, Khouri H, Roesch LF, Krogfelt KA, Struve C, Triplett EW, Methe BA (2008) Complete genome sequence of the N2-fixing broad host range endophyte *Klebsiella pneumoniae* 342 and virulence predictions verified in mice. PLoS Genet 4:e1000141

Frank B (1889) Über die Pilzsymbiose der Leguminosen. Berichte der Deutschen Botanischen Gesellschaft 7:332–346

Fravel D (2005) Commercialization and implementation of biocontrol 1. Annu Rev Phytopathol 43:337–335

Gadd GM, Sayer GM (2000) Fungal transformations of metals and metalloids. In: Lovley DR (ed) Environmental microbe-metal interactions. American Society of Microbiology, Washington, DC, pp 237–256

Gamalero E, Glick BR (2011) Mechanisms used by plant growth-promoting bacteria. In: Maheshwari DK (ed) Bacteria in agrobiology: plant nutrient management. Springer, Berlin/Heidelberg, pp 17–46

Gandhi A, Muralidharan G (2016) Assessment of zinc solubilizing potentiality of *Acinetobacter* sp. isolated from rice rhizosphere. Eur J Soil Biol 76:1–8

Gao Y, Liu Q, Zang P, Li X, Ji Q, He Z et al (2015) An endophytic bacterium isolated from Panax ginseng CA Meyer enhances growth, reduces morbidity, and stimulates ginsenoside biosynthesis. Phytochem Lett 11:132–138

Garbeva P, van Elsas JD, van Veen JA (2008) Rhizosphere microbial community and its response to plant species and soil history. Plant Soil 302:19–32

Garcia JE, Maroniche G, Creus C, Suarez-Rodríguez R, Ramirez-Trujillo JA, Groppa MD (2017) In vitro PGPR properties and osmotic tolerance of different *Azospirillum* native strains and their effects on growth of maize under drought stress. Microbiol Res 202:21–29

Ghasemi Z, Ghaderian SM, Rodriguez-Garrido B, Prieto-Fernandez A, Kidd PS (2018) Plant species-specificity and effects of bioinoculants and fertilization on plant performance for nickel phytomining. Plant Soil 425:265–285

Ghavami N, Alikhani HA, Pourbabaee AA, Besharati H (2016) Study the effects of siderophore-producing bacteria on zinc and phosphorous nutrition of canola and maize plants. Comm Soil Sci Plant Anal 47:1517–1527

Ghorchiani M, Etesami H, Alikhani HA (2018) Improvement of growth and yield of maize under water stress by co-inoculating an arbuscular mycorrhizal fungus and a plant growth promoting rhizobacterium together with phosphate fertilizers. Agric, Ecosys Environ 258:59–70

Glick BR (2012) Plant growth promoting bacteria: mechanisms and applications. Scientifica 2012:1–15. https://doi.org/10.6064/2012/963401

Glick BR, Penrose DM, Li J (1998) A model for the lowering of plant ethylene concentrations by plant growth-promoting bacteria. J Theor Biol 190:63–68

Glick BR, Cheng Z, Czarny J, Cheng Z, Duan J (2007) Promotion of plant growth by ACC deaminase- producing soil bacteria. Eur J Plant Pathol 119:329–339

Goicoechea N, Antol'ın MC, S'anchez-D'ıaz M (1997) Gas exchange is related to the hormonal balance in mycorrhizal or nitrogen-fixing alfalfa subjected to drought. Physiol Plant 100:989–997

Goldstein AH (2000) Bioprocessing of rock phosphate ore: essential technical considerations for the development of a successful commercial technology. In: Proceedings of the 4th international fertilizer association technical conference. IFA, Paris. (Vol. 220)

Gonzâlez MBR, Lopez JG, (2013) Beneficial plant-microbial interactions: ecology and applications. CRC Press, Taylor and Francis Group, LLC

Gouda S, Das G, Sen SK, Shin H, Patra JK (2016) Endophytes: a treasure house of bioactive compounds of medicinal importance. Front Microbiol 7:1538. https://doi.org/10.3389/fmicb.2016.01538

Gouda S, Kerry RG, Das G, Paramithiotis S, Shin H, Patra JK (2018) Revitalization of plant growth promoting rhizobacteria for sustainable development in agriculture. Microbiol Res 206:131–140

Govindarajan M, Balandreau J, Kwon SW, Weon HY, Lakshminarasimhan C (2008) Effects of the inoculation of *Burkholderia vietnamensis* and related endophytic diazotrophic bacteria on grain yield of rice. Microb Ecol 55:21–37

Grover M, Ali Sk Z, Sandhya V, Abdul Rasul, Venkateswarlu B (2011) Role of microorganisms in adaptation of agriculture crops to abiotic stresses. World J Microbiol Biotechnol 27:1231–1240

Guimaraes AA, Jaramillo PMD, Nóbrega RSA, Florentino LA, Silva KB, de Moreira FMS (2012) Genetic and symbiotic diversity of nitrogen-fixing bacteria isolated from agricultural soils in the Western Amazon by using cowpea as the trap plant. Appl Environ Microbiol 78:6726–6733

Gyaneshwar P, James EK, Mathan N, Reddy PM, Reinhold-Hurek B, Ladha JK (2001) Endophytic colonization of rice by a diazotrophic strain of *Serratia marcescens*. J Bacteriol 183:2634–2645

Haas D, Keel C (2003) Regulation of antibiotic production in root-colonizing *Pseudomonas* spp. and relevance for biological control of plant disease. Annu Rev Phytopathol 41:117–153

Hashem A, Abd Allah EF, Alqarawi AA, Aldubise A, Egamberdieva D (2015) Arbuscular mycorrhizal fungi enhances salinity tolerance of *Panicum turgidum* Forssk by altering photosynthetic and antioxidant pathways. J Plant Interact 10:230–242

Hashem A, Abd_Allah EF, Alqarawi AA, Al-Huqail AA, Wirth S, Egamberdieva D (2016) The interaction between arbuscular mycorrhizal fungi and endophytic bacteria enhances plant growth of Acacia gerrardii under salt stress. Front Microbiol 7:1089

Hayat R, Ali S, Amara U, Khalid R, Ahmed I (2010) Soil beneficial bacteria and their role in plant growth promotion: a review. Ann Microbiol 60:579–598

Henri F, Laurette NN, Annette D, John Q, Wolfgang M, François-Xavier E, Dieudonne N (2008) Solubilization of inorganic phosphates and plant growth promotion by strains of *Pseudomonas fluorescens* isolated from acidic soils of Cameroon. Afr J Microbiol Res 2:171–178

Heydarian Z, Gruber M, Glick BR, Hegedus DD (2018) Gene expression patterns in roots of camelina sativa with enhanced salinity tolerance arising from inoculation of soil with plant growth

promoting bacteria producing 1-aminocyclopropane-1-carboxylate deaminase or expression the corresponding *acds* gene. Front Microbiol 9
Hoitink H, Boehm M (1999) Biocontrol within the context of soil microbial communities: a substrate-dependent phenomenon. Annu Rev Phytopathol 37:427–446
Hungria M, Nogueira MA, Araujo RS (2013) Co-inoculation of soybeans and common beans with rhizobia and azospirilla: strategies to improve sustainability. Biol Fertil Soils 49:791–801
Hussain A, Arshad M, Zahir ZA, Asghar M (2015) Prospects of zinc solubilizing bacteria for enhancing growth of maize. Pak J Agri Sci 52:915–922
Hussain M, Asgher Z, Tahir M, Ijaz M, Shahid M, Ali H, Sattar A (2016) Bacteria in combination with fertilizers improve growth, productivity and net returns of wheat (*Triticum aestivum* L.). Pak J Agri Sci 53:633–664
Isawa T, Yasuda M, Awazaki H, Minamisawa K, Shinozaki S, Nakashita H (2010) *Azospirillum* sp. strain B510 enhances rice growth and yield. Microbes Environ 25:58–61
Jalgaonwala RE, Mohite BV, Mahajan RT (2011) Natural products from plant associated endophytic fungi. J Microbiol Biotechnol Res 1:21–32
Jamil M, Ahamd M, Anwar F, Zahir ZA, Kharal MA, Nazli F (2018) Inducing drought tolerance in wheat through combined use of L-tryptophan and *Pseudomonas fluorescens*. Pak J Agri Sci 55:331–337
Javot H, Pumplin N, Harrison MJ (2007) Phosphate in the arbuscular mycorrhizal symbiosis: transport properties and regulatory roles. Plant Cell Environ 30:310–322
Jennings DH (1994) Translocation in fungal mycelia. In: Wessels JGH, Meinhardt F (eds) The Mycota: growth, differentiation and sexuality. Springer, Berlin, pp 163–173
Jha CK, Aeron A, Patel BV, Maheshwari DK, Saraf M (2011) Enterobacter: role in plant growth promotion. In: Maheshwari DK (ed) Bacteria in agrobiology: plant growth responses. Springer, Heidelberg, pp 159–182
Jodeh S, Alkowni R, Hamed R, Samhan S (2015) The study of electrolyte leakage from barley (*Hordeum vulgare* L) and pearl millet using plant growth promotion (PGPR) and reverse osmosis. J Food Nutr Res 3:422–429
Jones KM, Kobayashi H, Davies BW, Taga ME, Walker GC (2007) How rhizobial symbionts invade plants: the *Sinorhizobium-Medicago* model. Nat Rev Microbiol 5:619–633
Kachhawa D (2017) Microorganisms as a biopesticides. J Entomol Zool Studies 5:468–473
Kaiser C, Kilburn MR, Clode PL, Fuchslueger L, Koranda M, Cliff JB, Solaiman ZM, Murphy DV (2015) Exploring the transfer of recent plant photosynthates to soil microbes: mycorrhizal pathway vs direct root exudation. New Phytol 205:1537–1551
Kang BG, Kim WT, Yun HS, Chang SC (2010) Use of plant growth-promoting rhizobacteria to control stress responses of plant roots. Plant Biotechnol Rep 4:179–183
Kawaguchi M, Minamisawa K (2010) Plant-microbe communications for symbiosis. Plant Cell Physiol 51:1377–1380
Khalid A, Akhtar MJ, Mahmood MH, Arshad M (2006) Effect of substrate-dependent microbial ethylene production on plant growth. Microbiology 75:231–236
Khalloufi M, Martinez-Andujar C, Lachaal M, Karray-Bouraoui N, Perez-Alfocea F, Albacete A (2017) The interaction between foliar GA3 application and arbuscular mycorrhizal fungi inoculation improves growth in salinized tomato (*Solanum lycopersicum* L.) plants by modifying the hormonal balance. J Plant Physiol 214:134–144
Khamwan S, Boonlue S, Jogloy S, Mongkolthanaruk W (2018) Characterization of endophytic bacteria and their response to plant growth promotion in *Helianthus tuberosus* L. Biocataly Agri Biotechnol 13:153–159
Khan MY, Asghar HN, Jamshaid MU, Akhtar MJ, Zahir ZA (2013) Effect of microbial inoculation on wheat growth and phyto-stabilization of chromium contaminated soil. Pak J Bot 45:27–34
Kidd PS, Álvarez-López V, Becerra-Castro C, Cabello-Conejo M, Prieto-Fernández Á (2017) Chapter three-potential role of plant-associated bacteria in plant metal uptake and implications in phytotechnologies. Adv Bot Res 83:87–126
Kim J, Rees DC (1994) Nitrogenase and biological nitrogen fixation. Biochemistry 33:389–397

Kim YC, Jung H, Kim KY, Park SK (2008) An effective biocontrol bioformulation against *Phytophthora* blight of pepper using growth mixtures of combined chitinolytic bacteria under different field conditions. Eur J Plant Pathol 120:373–382

Kisiala A, Laffont C, Emery RJN, Frugier F (2013) Bioactive cytokinins are selectively secreted by *Sinorhizobium meliloti* nodulating and nonnodulating strains. Mol Plant-Microbe Interact 26:1225–1231

Knappova J, Pankova H, Munzbergova Z (2016) Roles of arbuscular mycorrhizal fungi and soil abiotic conditions in the establishment of a dry grassland community. PLoS One 11(7):1–24

Konieczny A, Kowalaska I (2016) The role of arbuscular mycorrhiza in zinc uptake by lettuce grown at two phosphorus levels in the substrate. Agri Food Sci 25:124–137

Kotan R, Sahin F, Demirci E, Eken C (2009) Biological control of the potato dry rot caused by fusarium species using PGPR strains. Biol Control 50:194–198

Kouchi H, Imaizumi-Anraku H, Hayashi M, Hakoyama T, Nakagawa T, Umehara Y, Suganuma N, Kawaguchi M (2010) How many peas in a pod? Legume genes responsible for mutualistic symbioses underground. Plant Cell Physiol 51:1381–1397

Krapp A (2015) Plant nitrogen assimilation and its regulation: a complex puzzle with missing pieces. Curr Opin Plant Biol 25:115–122

Kumar A, Verma JP (2018) Does plant-microbe interaction confer stress tolerance in plants: a review? Microbiol Res 207:41–52

Kuykendall LD, Saxena B, Devine TE, Udell SE (1992) Genetic diversity in *Bradyrhizobium japonicum* Jordan 1982 and a proposal for *Bradyrhizobium elkanii* sp. nov. Can J Microbiol 38:501–505

Ladha JK, Barraquio WL, Watanabe I (1982) Immunological techniques to identify Azospirillum associated with wetland rice. Can J Microbiol 28:478–485

Lee EH, Eom AH (2015) Growth characteristics of *Rhizophagus clarus* strains and their effects on the growth of host plants. Mycobiology 43:444–449

Li XL, Zhang JL, Gai JP, Cai XBL, Christie P, Li X (2015) Contribution of arbuscular mycorrhizal fungi of sedges to soil aggregation along an altitudinal alpine grassland gradient on the Tibetan plateau. Environ Microbiol 17:2841–2857

Liu WYY, Ridgway HJ, James TK, Premaratne M, Andrews M (2012) Characterisation of rhizobia nodulating *Galega officinalis* (goat's rue) and *Hedysarum coronarium* (sulla) NZ. Plant Prot 65:192–196

Liu H, Yuan M, Tan SY, Yang XP, Lan Z, Jiang QY, Ye ZH, Jing YX (2015) Enhancement of arbuscular mycorrhizal fungus (*Glomus versiforme*) on the growth and cd uptake by cd-hyperaccumulator *Solanumnigrum*. Appl Soil Ecol 89:44–49

Lone R, Shuab R, Sharma V, Kumar V, Mir R, Koul KK (2015) Effect of arbuscular mycorrhizal fungi on growth and development of potato (*Solanum tuberosum*) plant. Asian J Crop Sci 7:233–243

Lopez-Lopez A, Rogel MA, Ormeño-Orrillo E, Martínez-Romero J, Martínez-Romero E (2010) *Phaseolus vulgaris* seed-borne endophytic community with novel bacterial species such as rhizobium endophyticum sp. nov. Syst Appl Microbiol 33(6):322–327

Lu L, Wu Q (2017) Mycorrhizas promote plant growth, root morphology and chlorophyll production in white clover. Biotechnology 16:34–39

Lugtenberg B, Kamilova F (2009) Plant-growth-promoting rhizobacteria. Annu Rev Microbiol 63:541–556

Lynch J (1990) The rhizosphere. Wiley, London, p 458

Machuca A, Pereira G, Aguiar A, Milagres AM (2007) Metal-chelating compounds produced by ectomycorrhizal fungi collected from pine plantations. Lett Appl Microbiol 44:7–12

Maheshwari DK (2013) Bacteria in agrobiology: disease management. Springer Science & Business Media, Springer-Verlag Berlin Heidelberg

Martin CA, Stutz JC (2004) Interactive effects of temperature and arbuscular mycorrhizal fungi on growth, P uptake and root respiration of *Capsicum annuum* L. Mycorrhiza 14:241–244

Matthijs S, ATehrani KA, Laus G, Jackson RW, Cooper RM, Cornelis P (2007) Thioquinolobactin, a *Pseudomonas* siderophore with antifungal and anti-Pythium activity. Environ Microbiol 9:425–434

Mayak S, Tirosh T, Glick BR (2004) Plant growth-promoting bacteria confer resistance in tomato plants to salt stress. Plant Physiol Biochem 42:565–572

Mazurier S, Corberand T, Lemanceau P, Raaijmakers JM (2009) Phenazine antibiotics produced by fluorescent pseudomonads contribute to natural soil suppressiveness to *Fusarium* wilt. ISME J 3:977–991

Mbodj D, Effa-Effa B, Kane A, Manneh B, Gantet P, Laplaze L, Diedhiou AG, Grondin A (2018) Arbuscular mycorrhizal symbiosis in rice: establishment, environmental control and impact on plant growth and resistance to abiotic stresses. Rhizosphere 8:12–26

McAdam EL, Reid JB, Foo E (2018) Gibberellins promote nodule organogenesis but inhibit the infection stages of nodulation. J Exp Bot 69:2117–2130

Mia MA, Shamsuddin ZH, Zakaria W, Marziah M (2007) Associative nitrogen fixation by *Azospirillum* and *Bacillus spp*. in bananas. Infomusa 16:11–15

Micallef SA, Shiaris MP, Colón-Carmona A (2009) Influence of *Arabidopsis thaliana* accessions on rhizobacterial communities and natural variation in root exudates. J Exp Bot 60:1729–1742

Miethling R, Wieland G, Backhaus H, Tebbe CC (2000) Variation of microbial rhizosphere communities in response to crop species, soil origin, and inoculation with *Sinorhizobium meliloti* L33. Microb Ecol 40:43–56

Miransari M (2011) Hyperaccumulators, arbuscular mycorrhizal fungi and stress of heavy metals. Biotechnol Adv 29:645–653

Misra N, Gupta G, Jha PN (2012) Assessment of mineral phosphate-solubilizing properties and molecular characterization of zinc-tolerant bacteria. J Basic Microbiol 52:549–558

Molla AH, Shamsuddin ZH, Halimi MS, Morziah M, Puteh AB (2001) Potential for enhancement of root growth and nodulation of soybean co-inoculated with *Azospirillum* and *Bradyrhizobium* in laboratory systems. Soil Biol Biochem 33:457–463

Moreira FMS, Lange A, Klauberg-Filho O, Siqueira JO, Nobrega RSA, Lima AS (2008) Associative diazotrophic bacteria in grass roots and soils from heavy metal contaminated sites. Anais da Academia Brasileira de Ciencias 80:749–761

Mumtaz MZ, Ahmad M, Jamil M, Hussain T (2017) Zinc solubilizing *Bacillus* spp. potential candidates for biofortification in maize. Microbiol Res 202:51–60

Mumtaz MZ, Ahmad M, Jamil M, Asad SA, Hafeez F (2018) *Bacillus* strains as potential alternate for zinc biofortification of maize grains. Int J AgriBiol 20:1779–1786

Muthukumarasamy R, Cleenwerck I, Revathi G, Vadivelu M, Janssens D, Hoste B, Gum KU, Park KD, Son CY, Sa T, Caballero-Mellado J (2005) Natural association of *Gluconacetobacter diazotrophicus* and diazotrophic *Acetobacter peroxydans* with wetland rice. Systematic Appl Microbiol 28:277–286

My PT, Manucharova NA, Stepanov AL, Pozdnyakov LA, Selitskaya OV, Emtsev VT (2015) *Agrobacterium tumefaciens* as associative nitrogen fixing bacteria. Moscow Univ Soil Sci Bull 70:133–138

Nadeem SM, Zahir ZA, Naveed M, Ashraf M (2010a) Microbial ACC-deaminase: prospects and applications for inducing salt tolerance in plants. Crit Rev Plant Sci 29:360–393

Nadeem SM, Zahir ZA, Naveed M, Asghar HN, Arshad M (2010b) Rhizobacteria capable of producing ACC-deaminase may mitigate the salt stress in wheat. Soil Sci Soc Am J 74:533–542

Nadeem SM, Ahmad M, Zahir ZA, Ashraf M (2011) Microbial ACC-deaminase biotechnology: perspectives and applications in stress agriculture. In: Maheshwari DK (ed) Bacteria in agrobiology: stress management. Springer, Berlin, pp 141–185

Nadeem SM, Naveed M, Zahir ZA, Asghar HN (2013) Plant-microbe interactions for sustainable agriculture: fundamentals and recent advances. In: Arora NK (ed) Plant microbe symbiosis: fundamentals and advances. Springer, New Delhi, pp 51–103

Nadeem SM, Ahmad M, Zahir ZA, Javaid A, Ashraf M (2014) The role of mycorrhizae and plant growth promoting rhizobacteria (PGPR) in improving crop productivity under stressful environments. Biotechnol Adv 32:429–448

Nadeem SM, Naveed M, Ahmad M, Zahir ZA (2015) Rhizosphere bacteria for crop production and improvement of stress tolerance: mechanisms of action, applications, and future prospects. In: Arora NK (ed) Plant microbes symbiosis: applied facets. Springer, New Delhi, pp 1–36. https://doi.org/10.1007/978-81-322-2068-8_1

Nadeem SM, Ahmad M, Naveed M, Imran M, Zahir ZA, Crowley DE (2016) Relationship between in vitro characterization and comparative efficacy of plant growth-promoting rhizobacteria for improving cucumber salt tolerance. Arch Microbiol 198:379–387

Nadeem SM, Imran M, Naveed M, Khan MY, Ahmad M, Zahir ZA, Crowley DE (2017) Synergistic use of biochar, compost and plant growth-promoting rhizobacteria for enhancing cucumber growth under water deficit conditions. J Sci Food Agri. https://doi.org/10.1002/jsfa.8393

Nandasena KG, O'Hara GW, Tiwari RP, Willems A, Howieson JG (2009) *Mesorhizobium australicum* sp. nov. and *Mesorhizobium opportunistum* sp. nov., isolated from *Biserrula pelecinus* L. in Australia. Int J Systematic Evolu Microbiol 59:2140–2147

Naveed M, Hussain MB, Zahir ZA, Mitter B, Sessitsch A (2014) Drought stress amelioration in wheat through inoculation with Burkholderia phytofirmans strain PsJN. Plant Growth Regul 73:121–131

Nielsen TH, Thrane C, Christophersen C, Anthoni U, Sorensen J (2000) Structure, production characteristics and fungal antagonism of tensin - a new antifungal cyclic lipopeptide from Pseudomonas fluorescens strain 96.578. J Appl Microbiol 89:992–1001

Nisha MC, Rajeshkumar S (2010) Effect of arbuscular mycorrhizal fungi on growth and nutrition of *Wedilia chinensis* (Osbeck) Merril. Indian J Sci Technol 6:676–678

Nogueira MA, Cardoso EJBN (2006) Plant growth and phosphorus uptake in mycorrhizal Rangpur lime seedlings under different levels of phosphorus. Pesq Agropec Bras Brasilia 41:93–99

Nosheen A, Yasmin H, Naz R, Bano A, Keyani R, Hussain I (2018) *Pseudomonas putida* improved soil enzyme activity and growth of kasumbha under low input of mineral fertilizers. Soil Sci Plant Nutr 12:1–6

O'Sullivan DJ, O'Gara F (1992) Traits of fluorescent *Pseudomonas* spp. involved in suppression of plant root pathogens. Microbiol Rev 56:662–676

Oldroyd GED (2007) Nodules and hormones. Science 315(5808):52–53

Oldroyd GE, Downie JA (2008) Coordinating nodule morphogenesis with rhizobial infection in legumes. Annu Rev Plant Biol 59:519–546

Ortas I (2010) Effect of mycorrhiza application on plant growth and nutrient uptake in cucumber production under field conditions. Spanish J Agric Res 8:116–122

Padmavathi T, Dikshit R, Seshagiri S (2015) Effect of *Rhizophagus* spp. and plant growth promoting *Acinetobacter junii* on *Solanum lycopersicum* and *Capsicum annuum*. Brazilian J Bot 38:273–280

Parniske M (2008) Arbuscular mycorrhiza: the mother of plant root endosymbiosis. Nat Rev Microbiol 6:763–775

Patel JK, Archana G (2017) Diverse culturable diazotrophic endophytic bacteria from Poaceae plants show cross-colonization and plant growth promotion in wheat. Plant Soil 417:99–116

Pereira JA, Cavalcante VA, Baldani JI, Dobereiner J (1988) Field inoculation of sorghum and rice with *Azospirillum spp.* and *Herbaspirillum seropedicae*. In: Nitrogen fixation with non-legumes. Springer, Dordrecht, pp 219–224

Pérez E, Sulbarán M, Ball MM, Yarzabál LA (2007) Isolation and characterization of mineral phosphate-solubilizing bacteria naturally colonizing a limonitic crust in the southeastern Venezuelan region. Soil Biol Biochem 39:2905–2914

Perez-de-Luque A, Tille S, Johnson I, Pascual-Pardo D, Ton J, Cameron DD (2017) The interactive effects of arbuscular mycorrhiza and plant growth-promoting rhizobacteria synergistically enhance host plant defences against pathogens. Sci Reports 7:16409. https://doi.org/10.1038/s41598-017-16697-4

Picard C, Di Cello F, Ventura M, Fani R, Guckert A (2000) Frequency and biodiversity of 2,4-diacetylphloroglucinol-producing bacteria isolated from the maize rhizosphere at different stages of plant growth. Appl Environ Microbiol 66:948–955

Pieterse CMJ, Leon-Reyes A, van der Ent S, van Wees SCM (2009) Networking by small-molecule hormones in plant immunity. Nat Chem Biol l5:308–316

Prabhukarthikeyan R, Saravanakumar D, Raguchander T (2014) Combination of endophytic *Bacillus* and *Beauveria* for the management of *Fusarium* wilt and fruit borer in tomato. Pest Manag Sci 70:1742–1750

Prajapati K, Sharma MC, Modi HA (2013) Growth promoting effect of potassium solubilizing microorganisms on *Abelmoschus esculentus*. Intl J Agric Sci 3:181–188

Qiu Z, Tan H, Zhou S, Cao L (2014) Enhanced phytoremediation of toxic metals by inoculating endophytic Enterobacter sp. CBSB1 expressing bifunctional glutathione synthase. J Hazard Mater 267:17–20

Raaijmakers JM (2001) Rhizosphere and rhizosphere competence. In: Maloy OC, Murray TD (eds) Encyclopedia of plant pathology. Wiley, New York, pp 859–860

Raaijmakers JM, Leeman M, van Oorschot M, van der Sluis I, SchippersB, Bakker P (1995) Dose-response relationships in biological-control of fusarium-wilt of radish by *Pseudomonas* spp. Phytopathology 85: 1075–1081

Raaijmakers JM, Paulitz TC, Steinberg C, Alabouvette C, Moënne-Loccoz Y (2009) The rhizosphere: a playground and battlefield for soilborne pathogens and beneficial microorganisms. Plant Soil 321:341–361

Ramesh A, Sharma SK, Sharma MP, Yadav N, Joshi OP (2014) Inoculation of zinc solubilizing *Bacillus aryabhattai* strains for improved growth, mobilization and biofortification of zinc in soybean and wheat cultivated in vertisols of Central India. Appl Soil Ecol 73:87–96

Ramyasmruthi S, Pallavi O, Pallavi S, Tilak K, Srividya S (2012) Chitinolytic and secondary metabolite producing *Pseudomonas fluorescens* isolated from Solanaceae rhizosphere effective against broad spectrum fungal phytopathogens. Asian J Plant Sci Res 2:16–24

Rashid MA, Mujawar LH, Shahzad T, Almeelbi T, Ismail IMI, Oves M (2016) Bacteria and fungi can contribute to nutrients bioavailability and aggregate formation in degraded soils. Microbiol Res 183:26–41

Rima FS, Biswas S, Sarker PK, Islam MR, Seraj ZI (2018) Bacteria endemic to saline coastal belt and their ability to mitigate the effects of salt stress on rice growth and yields. Ann Microbiol 68:525–535

Rodrigues AA, Forzani MV (2016) Isolation and selection of plant growth-promoting bacteria associated with sugarcane. Pesq Agropec Trop Goiania 46:149–158

Rosenblueth M, Martinez-Romero E (2006) Bacterial endophytes and their interactions with hosts. Mol Plant-Microbe Interact 19:827–837

Rouws LFM, Meneses CHSG, Guedes HV, Vidal MS, Baldani JI, Schwab S (2010) Monitoring the colonization of sugarcane and rice plants by the endophytic diazotrophic bacterium *Gluconacetobacter diazotrophicus* marked with gfp and gusA reporter genes. Lett Appl Microbiol 51:325–330

Rubio LM, Ludden PW (2008) Biosynthesis of the iron-molybdenum cofactor of nitrogenase. Annu Rev Microbiol 62:93–111

Saghafi D, Ghorbanpour M, Lajayer BA (2018) Efficiency of rhizobium strains as plant growth promoting rhizobacteria on morpho-physiological properties of *Brassica napus* L. under salinity stress. J Soil Sci Plant Nutr 18:253–268

Saleem M, Arshad M, Hussain S, Bhatti AS (2007) Perspective of plant growth promoting rhizobacteria (PGPR) containing ACC deaminase in stress agriculture. J Indian. Microbiol Biotechnol 34:635–648

Salloum MS, Menduni MF, Luna CM (2017) A differential capacity of arbuscular mycorrhizal fungal colonization under well-watered conditions and its relationship with drought stress mitigation in unimproved vs. improved soybean genotypes. Bot 96:135–144

Sandhya V, Ali SKZ, Grover M, Reddy G, Venkateswarlu B (2009) Alleviation of drought stress effects in sunflower seedlings by the exopolysaccharides producing *Pseudomonas putida* strain GAP-P45. BiolFertil Soils 46:17–26

Santaella C, Schue M, Berge O, Heulin T, Achouak W (2008) The exopolysaccharide of *Rhizobium* sp. YAS34 is not necessary for biofilm formation on *Arabidopsis thaliana* and *Brassica napus* roots but contributes to root colonization. Environ Microbiol 10:2150–2163

Sapre S, Gontia-Mishra I, Tiwari S (2018) *Klebsiella* sp. confers enhanced tolerance to salinity and plant growth promotion in oat seedlings (*Avena sativa*). Microbiol Res 206:25–32

Saravanakumar D, Vijayakumar C, Kumar N, Samiyappan R (2007) PGPR-induced defense responses in the tea plant against blister blight disease. Crop Prot 26:556–565

Saravanakumar K, Ali DM, Kathiresan K, Wang M (2018) An evidence of fungal derived 1-amin ocyclopropane-1-carboxylate deaminase promoting the growth of mangroves. Beni-Suef Univ J Basic Appl Sci. https://doi.org/10.1016/j.bjbas.2018.03.013

Saravanan VS, Madhaiyan M, Thangaraju M (2007) Solubilization of zinc compounds by the diazotrophic, plant growth promoting bacterium *Gluconacetobacter diazotrophicus*. Chemosphere 66:1794–1798

Sari N, Ortas I, Yetisir H (2002) Effect of mycorrhizae inoculation on plant growth, yield, and phosphorus uptake in garlic under field conditions. Commun Soil Sci Plant Anal 33:2189–2201

Sathya A, Vijayabharathi R, Gopalakrishnan S (2017) Plant growth-promoting actinobacteria: a new strategy for enhancing sustainable production and protection of grain legumes. Biotechnology 7:102. https://doi.org/10.1007/s13205-017-0736-3

Sayyed RZ, Patil AS, Gangurde NS, Bhamare HM, Joshi SA, Fulpagare UG (2008) Siderophore producing *A. faecalis*: a potent biofungicide for the control of ground phytopathogens. J Res Biotechnol 4:411–413

Schwab S, Terra LA, Baldani JI (2018) Genomic characterization of *Nitrospirillum amazonense* strain CBAmC, a nitrogen-fixing bacterium isolated from surface-sterilized sugarcane stems. Mole Gene Geno 293:997–1016

Selosse MA, Roy M (2009) Green plants that feed on fungi: facts and questions about mixotrophy. Trends Plant Sci 14:64–70

Sessitsch A, Reiter B, Berg G (2004) Endophytic bacterial communities of field-grown potato plants and their plant-growth-promoting and antagonistic abilities. Can J Microbiol 50:239–249

Sessitsch A, Kuffner M, Kidd P, Vangronsveld J, Wenzel WW, Fallmann K, Puschenreiter M (2013) The role of plant associated bacteria in the mobilization and phytoextraction of trace elements in contaminated soils. Soil Biol Biochem 60:182–194

Shahsavar AR, Refahi A, Zarei M, Aslmoshtagh E (2016) Analysis of the effects of *Glomus etunicatum* fungi and *Pseudomonas fluorescence* bacteria symbiosis on some morphological and physiological characteristics of Mexican lime (*Citrus aurantifolia* L.) under drought stress conditions. Adv Hort Sci 30:39–45

Shaikh S, Saraf M (2017) Biofortification of *Triticum aestivum* through the inoculation of zinc solubilizing plant growth promoting rhizobacteria in field experiment. Biocatalysis Agricul Biotechnol 9:120–126

Shi SM, Chen K, Gao Y, Liu B, Yang XH, Huang XZ, Liu GX, Zhu LQ, He XH (2016) Arbuscular mycorrhizal fungus species dependency governs better plant physiological characteristics and leaf quality of mulberry (*Morus alba* L.) seedlings. Front Microbiol 7:1–11

Shuab R, Lone R, Naidu J, Sharma V, Imtiyaz S, Koul KK (2014) Benefits of inoculation of arbuscular mycorrhizal fungi on growth and development of onion (*Allium cepa*) plant. Am Eur J Agric Environ Sci 14:527–535

Sindhu SS, Dadarwal KR (2001) Chitinolytic and cellulolytic *Pseudomonas* sp. antagonistic to fungal pathogens enhances nodulation by *Mesorhizobium* sp. *Cicer*in chickpea. Microbiol Res 156:353–358

Singh R, Dubey AK (2015) Endophytic actinomycetes as emerging source for therapeutic compounds. Indo Global J Pharm Sci 5:106–116

Singh D, Singh H, Prabha R (2013) Plant-microbe interactions in agro-ecological perspectives. Volume 2: microbial interactions and agro-ecological impacts. Springer, Singapore

Singh RP, Shelke GM, Kumar A, Jha PN (2015) Biochemistry and genetics of ACC deaminase: a weapon to stress ethylene produced in plants. Front Microbiol 6:1–14

Singh D, Geat N, Rajawat MVS, Prasanna R, Saxena AK, Kaushik R (2017) Isolation and characterization of plant growth promoting endophytic diazotrophic bacteria from wheat genotypes and their influence on plant growth promotion. Int J Curr Microbiol App Sci 6:1533–1540

Smith SE, Read DJ (2008) Mycorrhizal symbiosis, 3rd edn. Academic Press, Boston. 800p

Solano BR, Maicas JB, Gutierrez Manero J (2009) Biotechnology of the rhizosphere. In: Kirakosyan A, Kaufman PB (eds) Recent advances in plant biotechnology. Springer, Dordrecht/ Heidelberg/ New York, pp 137–162

Somers E, Vanderleyden J, Srinivasan M (2004) Rhizosphere bacterial signaling: a love parade beneath our feet. Crit Rev Microbiol 30:205–240

Spaepen S, Vanderleyden J, Okon Y (2009) Plant growth-promoting actions of rhizobacteria. Adv Bot Res 51:283–320

Sprent JI, Ardley JK, James EK (2013) From north to south: a latitudinal look at legume nodulation processes. S Afr J Bot 89:31–41

Stephen J, Jisha MS (2009) Buffering reduces phosphate solubilizing ability of selected strains of bacteria. World J Agric Sci 5:135–137

Suman A, Gaur A, Shrivastava AK, Yadav RL (2005) Improving sugarcane growth and nutrient uptake by inoculating Gluconacetobacter diazotrophicus. Plant Growth Regul 47:155–162

Tahat MM, Sijam R, Othman R (2012) The potential of endomycorrhizal fungi in controlling tomato bacterial wilt *Ralstonia solanacearum* under glasshouse conditions. Afr J Biotechnol 11:13085–13094

Taghavi S, Garafola C, Monchy S, Newman L, Hoffman A, Weyens N, Barac T, Vangronsveld J, vander Lelie D (2009) Genome survey and characterization of endophytic bacteria exhibiting a beneficial effect on growth and development of poplar trees. Appl Environ Microbiol 75:748–757

Taiz L, Zeiger E (2000) Plant physiology, 2nd edn. Benjamin Cumings Publishing Company, San Francisco

Thijs S, Langill T, Vangronsveld J (2017) Chapter two- the bacterial and fungal microbiota of hyperaccumulator plants: small organisms, large influence. Adv Bot Res 83:43–86

Tian CF, Wang ET, Wu LJ, Han TX, Chen WF, Gu CT, Gu JG, Chen WX (2008) Rhizobium fabae sp. nov., a bacterium that nodulates *Vicia faba*. Int J Syst Evol Microbiol 58:2871–2875

Tian G, Pauls P, Dong Z, Reid LM, Tian L (2009) Colonization of the nitrogen-fixing bacterium *Gluconacetobacter diazotrophicus* in a large number of Canadian corn plants. Can J Plant Sci 89:1009–1019

Tian B, Zhang C, Ye Y, Wen J, Wu Y, Wang H, Li H, Cai S, Cai W, Cheng Z et al (2017) Beneficial traits of bacterial endophytes belonging to the core communities of the tomato root microbiome. Agric Ecosyst Environ 247:149–156

Tisdall JM, Oades JM (1982) Organic matter and water stable aggregates in soils. J Soil Sci 33:141–163

Torre-Ruiz NDL, Ruiz-Valdiviezo VM, Rincon-Molina CI, Rodriguez-Mendiola M, Arias-Castroa C, Gutierrez-Miceli FA, Palomeque-Dominguez H, Rincon-Rosales R (2016) Effect of plant growth-promoting bacteria on the growth and fructan production of *Agave americana* L. Brazilian J Microbiol 47:587–596

Ullah A, Mushtaq H, Ali U, Hakim, Ali E, Mubeen S (2018) Screening, isolation, biochemical and plant growth promoting characterization of endophytic bacteria. Microbiol Curr Res 2:62–68

Upadhyay SK, Singh JS, Singh DP (2011) Exopolysaccharide-producing plant growth- promoting rhizobacteria under salinity condition. Pedosphere 21:214–222

Upadhyay SK, Singh JS, Saxena AK, Singh DP (2012) Impact of PGPR inoculation on growth and antioxidant status of wheat under saline conditions. Plant Biol 14:605–611

Valverde A, Burgos A, Fiscella T, Rivas R, Velázquez E, Rodriguez-Barrueco C, Cervantes E, Chamber M, Igual J-M (2006) Differential effects of coinoculations with *Pseudomonas jessenii* PS06 (a phosphate-solubilizing bacterium) and *Mesorhizobium ciceri* C-2/2 strains on the growth and seed yield of chickpea under greenhouse and field conditions. Plant Soil 287:43–50

Van der Ent S, Van Wees SC, Pieterse CM (2009) Jasmonate signaling in plant interactions with resistance-inducing beneficial microbes. Phytochemistry 70:1581–1588

Van Oosten MJ, Di Stasio E, Cirillo V, Silletti S, Ventorino V, Pepe O, Raimondi G, Maggio A (2018) Root inoculation with *Azotobacter chroococcum* 76A enhances tomato plants adaptation to salt stress under low N conditions. BMC Plant Biol 18:205

Vannette RL, Hunter MD (2013) Hunter mycorrhizal abundance affects the expression of plant resistance traits and herbivore performance. J Ecol 101:1019–1029

Velmourougane K, Saxena G, Prasanna R (2017) Plant-microbe interactions in the rhizosphere: mechanisms and their ecological benefits. In: Singh D, Singh H, Prabha R (eds) Plant-microbe interactions in agro-ecological perspectives, volume 2: microbial interactions and agro-ecological impacts. Springer, Singapore, pp 193–219

Verhagen BWM, Glazebrook J, Zhu T, Chang HS, van Loon LC, Pieterse CMJ (2004) The transcriptome of rhizobacteria-induced systemic resistance in *Arabidopsis*. Mol Plant-Microbe Interactions 17:895–908

Vimal SR, Singh JS, Arora NK, Singh S (2017) Soil-plant-microbe interactions in stressed agriculture management: a review. Pedosphere 27:177–192

Viscardi S, Ventorino V, Duran P, Maggio A, De Pascale S, Mora ML, Pepe O (2016) Assessment of plant growth promoting activities and abiotic stress tolerance of *Azotobacter chroococcum* strains for a potential use in sustainable agriculture. J Soil Sci Plant Nutr 16:848–863

Vives-Peris V, Gomez-Cadenas A, Perez-Clemente RM (2018) Salt stress alleviation in citrus plants by plant growth-promoting rhizobacteria *Pseudomonas putida* and *Novosphingobium* sp. Plant Cell Rep 30:1–3

Voisard C, Keel C, Haas D, Defago G (1981) Cyanide production in *Pseudomonas fluorescens* helps suppress black root rot of tobacco under gnotobiotic conditions. EMBO J 8:351–358

Wakatsuki T (1995) Metal oxidoreduction by microbial cells. J Ind Microbiol Biotechnol 14:169–177

Wang Y, Brown HN, Crowley DE, Szaniszlo PJ (1993) Evidence for direct utilization of a siderophore, ferrioxamine B, in axenically grown cucumber. Plant Cell Environ 16:579–585

Wang C, Li X, Zhou J, Wang G, Dong Y (2008) Effects of arbuscular mycorrhizal fungi on growth and yield of cucumber plants. Commun Soil Sci Plant Anal 39

Wang D, Yang S, Tang F, Zhu H (2012) Symbiosis specificity in the legume: rhizobial mutualism. Cell Microbiol 14:334–342

Wang X, Wang Z, Jiang P, He Y, Mu Y, Lv X, Zhuang L (2018a) Bacterial diversity and community structure in the rhizosphere of four *Ferula* species. Sci Reports 8:5345. https://doi.org/10.1038/s41598-018-22802-y

Wang Y, Wang M, Li Y, Wu A, Huang J (2018b) Effects of arbuscular mycorrhizal fungi on growth and nitrogen uptake of *Chrysanthemum morifolium* under salt stress. PLoS One 13(4):e0196408

Weber OB, Baldani VLD, Teixeira KRS, Kirchhof G, Baldani JI, Dobereiner J (1999) Isolation and characterization of diazotrophic bacteria from banana and pineapple plants. Plant Soil 210:103–113

Weilharter A, Mitter B, Shin MV, Chain PS, Nowak J, Sessitsch A (2011) Complete genome sequence of the plant growth-promoting endophyte *Burkholderia phytofirmans* strain PsJN. J Bacteriol 193:3383–3384

Wilmowicz E, Kesy J, Kopcewicz J (2008) Ethylene and ABA interactions in the regulation of flower induction in *Pharbitis nil*. J Plant Physiol 165:1917–1928

Wu QS, Li GH, Zou YN (2010) Roles of arbuscular mycorrhizal fungi on growth and nutrient acquisition of peach (*Prunus persica* L. Batsch) seedlings. J Anim Plant Sci 21:746–750

Xun FF, Xie BM, Liu SS, Guo CH (2015) Effect of plant growth promoting bacteria (PGPR) and arbuscular mycorrhizal fungi (AMF) inoculation on oats in saline-alkali soil contaminated by petroleum to enhance phytoremediation. Environ Sci Pollut R 22:598–608

Yamaji K, Watanabe Y, Masuya H, Shigeto A, Yui H, Haruma T (2016) Root fungal endophytes enhance heavy-metal stress tolerance of *Clethra barbinervis* growing naturally at mining sites via growth enhancement, promotion of nutrient uptake and decrease of heavy-metal concentration. PLoS One 11(7):1–24

Yanni YG, Rizk RY, Corich V, Squartini A, Ninke K, Philip-Hollingsworth S, Orgambide G, De Bruijn F, Stoltzfus J, Buckley D, Schmidt TM (1997) Natural endophytic association between

Rhizobium leguminosarum bv. *trifolii* and rice roots and assessment of its potential to promote rice growth. Plant Soil 194:99–114

Yuan SF, Li MY, Fang ZY, Liu Y, Shi W, Pan B, Wu K, Shi JX, Shen B, Shen QR (2016) Biological control of tobacco bacterial wilt using *Trichoderma harzianum* amended bio-organic fertilizer and the arbuscular mycorrhizal fungi *Glomus mosseae*. Biol Control 92:164–171

Zahir ZA, Munir A, Asghar HN, Shaharoona B, Arshad M (2008) Effectiveness of rhizobacteria containing ACC deaminase for growth promotion of peas (*Pisum sativum*) under drought conditions. J Microbiol Biotechnol 18:958–963

Zahir ZA, Shah MK, Naveed M, Akhter MJ (2010) Substrate-dependent auxin production by *Rhizobium phaseoli* improves the growth and yield of *Vigna radiata* L. under salt stress conditions. J Microbiol Biotechnol 20:1288–1294

Zahir ZA, Ahmad M, Hilger TH, Dar A, Malik SR, Abbas G, Rasche F (2018) Field evaluation of multistrain biofertilizer for improving the productivity of different mung bean genotypes. Soil Environ 37(1):45–52

Zahran HH (2001) Rhizobia from wild legumes: diversity, taxonomy, ecology, nitrogen fixation and biotechnology. J Biotechnol 91:143–153

Zamioudis C, Pieterse CMJ (2012) Modulation of host immunity by beneficial microbes. Mol Plant-Microbe Interact 25:139–150

Zhang J, Jia W, Yang J, Ismail AM (2006) Role of ABA in integrating plant responses to drought and salt stresses. Field Crop Res 97:111–119

Zhu Y, She X (2018) Evaluation of the plant-growth-promoting abilities of endophytic bacteria from the psammophyte *Ammodendron bifolium*. Can J Microbiol 64:253–264

Zong K, Huang J, Nara K, Chen YH, Shen ZG, Lian CL (2015) Inoculation of ectomycorrhizal fungi contributes to the survival of tree seedlings in a copper mine tailing. J For Res 20:493–500

Zuniga A, Poupin MJ, Donoso R, Ledger T, Guiliani N, Gutierrez RA, Gonzalez B (2013) Quorum sensing and indole-3-acetic acid degradation play a role in colonization and plant growth promotion of Arabidopsis thaliana by *Burkholderia phytofirmans* PsJN. Mol Plant-Microbe Interact 26:546–553

Biodiversity and Biotechnological Applications of Microorganisms Associated with Tropical Plants

13

João Lúcio Azevedo and Maria Carolina Quecine

Abstract

According to the World Trade Organization, the largest agrarian producers in the world are in the tropics. Moreover, the appreciated global resources of biological diversity also occur in tropical areas, such as the Amazon, which is considered one of the chief rain forests and plays a significant role in protection and discovery of novel microbial, plant, and animal species. However, data from tropical regions are scarce. Little is known about the great biodiversity of microorganisms and the application of these resources to improve tropical agriculture. Tropical agriculture presents more problems of greater complexity than agriculture in temperate climates. One approach for solving some agricultural problems in the tropics is sustainable use of microorganisms. Efficient microbes have shown potential in the agricultural field for use in plant development and growth promotion, mostly by delivering valuable compounds to their host plants. Microbial plant growth promoters may improve crop development and growth in numerous biological ways, such as production of secondary metabolites (for example, plant development hormones, including auxins), production of siderophores, phosphate solubilization, and nitrogen fixation. Microorganisms have also received considerable attention because of their potential for use as agents for biological control of various plant fungal pathogens and bacterial diseases. There are numerous processes associated with plant–pathogen hostility. The competition for space and nutrients in the host plant, in the rhizospheric region, and in the soil can be fierce, as can the production of antimicrobial complexes that directly affect plant pathogens. These mainly include lytic enzymes, antibiotics, volatile organic compounds, and other compounds. This review discusses studies on the microorganism diversity that is present in well-established tropical

J. L. Azevedo (✉) · M. C. Quecine
Departamento de Genética, Escola Superior de Agricultura "Luiz de Queiroz", Universidade de São Paulo, Piracicaba, São Paulo, Brazil
e-mail: jlazevedo@usp.br

V. Kumar et al. (eds.), *Microbiome in Plant Health and Disease*,
https://doi.org/10.1007/978-981-13-8495-0_13

crops, with examples of successful use of microorganisms in solving tropical problems, with the aim being to achieve more sustainable agriculture.

Keywords
Biodiversity · Biocontrol · Cacao · Disease · Microbe

13.1 Introduction

Microorganisms are able to live in all environments—including iceboxes and thermal and hypersaline waters—and they can also colonize human beings, other animals, and plants. In their association with plants, microorganisms play crucial roles and have negative, neutral, or beneficial effects. Damaging effects are observed in interactions between pathogens and their hosts. Beneficial microorganisms have been less investigated but play crucial roles in plant fitness and can be classified as endophytic or rhizospheric microorganisms.

Endophytes play important roles within their plant hosts, protecting them from insect pests and pathogenic microorganisms. Some endophytic and rhizospheric microorganisms may also promote growth through synthesis of particular compounds (such as insoluble mineral phosphates), production of plant hormones (such as indole acetic acid (IAA)), symbiotic nitrogen fixation, and manufacture of siderophores, enzymes, antibiotics, and other antagonists that act against plant pathogens. They also cause solubilization and mineralization of nutrients. The term "endophyte" is used to refer to the microbiota residing partly or during the its whole life inside the plant tissues and does not cause apparent or visible indications of any disease. Many kinds of such microorganisms were defined by the end of the last century as microbes that can be segregated from surface-sterilized plant tissues or obtained from plant inner tissues and do not cause harm to their host plants (Wilson 1995; Halmann et al. 1997). Not too long ago, Mendes and Azevedo (2007) considered both culturable and nonculturable microorganisms as endophytes but suggested dividing them into two kinds, with category I being those that do not acquire outside structures and category II being those that do acquire outside structures, such as mycorrhizal fungi and N_2-fixing bacterial species that form nodules.

The plant microbiome also includes rhizospheric microorganisms, which inhabit the external parts of plant roots in close contact with the surrounding soil. Initial isolation and study of endophytic microbiota has usually involved host plants from temperate regions. As shown by Azevedo and Araujo (2007), at the beginning of the present century, scant information was available about endophytic microbiota isolated from tropical areas. However, several authors did address the abundance of endophytes isolated from tropical regions, and, more recently, many reports describing isolation of fungi and bacteria from tropical host plants have been published (Azevedo and Araujo 2007; Lacava and Azevedo 2014; Batista et al. 2017). It is important to mention that most of the published papers involving tropical endophytic microorganisms have addressed the importance of their use as beneficial microorganisms mainly in terms of their role in reducing environmental pollution. Use of chemicals in agriculture increases damage to microorganisms. Chemical fertilizers, herbicides, insecticides, fungicides, and other chemical products have been frequently used in tropical agriculture, damaging the environment and increasing costs (Sansawal 2017). Use of biofertilizers and biological control could mitigate the application of these synthetic products. Bacteria and fungi may perform as plant development and growth promoters through direct or indirect mechanisms—for instance, through phytohormone production and acquisition of nitrogen, phosphorus, and iron. Additionally, growth can be indirectly promoted by a reduction in insect damage and inhibition of plant pathogenic microorganisms. Agricultural pests and plant diseases may also be controlled by members of the endophytic microbiota, reducing the use of chemicals in agriculture. The literature in this specific area is increasing. This review therefore presents research concerning the microbiota associated with typical and/or well-adapted tropical plants, mainly in Brazil. It describes the use of these endophytic and rhizospheric microorganisms in controlling pests and plant pathogenic diseases and in promoting plant growth to mitigate the application of synthetic products and to help reduce costs and ecological damage.

13.2 Cacao Endophytes and Control of "Witches' Broom" Disease

Cacao (*Theobroma cacao*) is a native Brazilian crop used to produce chocolate. Currently, the main producers are African countries, which provide 72% of the global production, with the Ivory Coast and Ghana being the biggest producers. The Americas contribute 16% of global cacao production, followed by Asia and Oceania. Brazil used to be one of the major world producers until 1989, with the state of Bahia being the main producer. However, since the appearance of "witches' broom" disease, caused by the fungus *Moniliophthora perniciosa* (initially named *Crinipellis perniciosa*), the productivity has decreased over the years; thus, Brazil contributed only 4% of the world's total production in 2016. *M. perniciosa* is a basidiomycete fungus; the disease is hard to contain and has caused several farmers to abandon cultivation of cacao. The idea to use endophytic microorganisms for biocontrol of the cacao pathogen was suggested by Arnold et al. in 2003 (Arnold et al. 2003). Two years later, in Brazil, Rubini et al. (2005) isolated endophytes from cacao to evaluate their potential for biocontrol of *M. perniciosa*. The endophytic fungal population was isolated from branches of healthy resistant, susceptible, and symptomatic cacao plants. After DNA isolation and characterization of the endophytic fungi, 23 different genera were obtained (Table 13.1).

A total of 265 endophytes were assessed versus *M. perniciosa* in vitro, and 43 isolates derived from resistant, susceptible, and affected branches were capable of inhibiting the development of the pathogenic fungus to some degree. Screening of these endophytes was conducted through inoculation of cacao seedlings with the isolated endophytes followed by inoculation with the pathogenic fungus. Among all

Table 13.1 Endophytic fungi isolated from branches of healthy resistant, susceptible, and symptomatic cacao (*Theobroma cacao*) plants

Phylum	Class	Genus
Ascomycota	Incertae sedis	*Acremonium*
		Pseudofusarium
	Dothideomycetes	*Botryosphaeria*
		Cladosporium
		Lasiodiplodia
		Rhizopycnis
	Eurotiomycetes	*Blastomyces*
	Sordariomycetes	*Geotrichum*
		Cordyceps
		Colletotrichum
		Fusarium
		Gliocladium
		Gibberella
		Diaporthe
		Monilochoetes
		Pestalotiopsis
		Nectria
		Trichoderma
		Phomopsis
		Verticillium
		Xylaria
Basidiomycota	Agaricomycetes	*Pleurotus*
Mucoromycota	Mucoromycetes	*Syncephalastrum*

Adapted from Rubini et al. (2005)

of the endophytes, only 14 were able to reduce the symptoms of witches' broom disease. The inoculations were also tested under greenhouse conditions, and one isolate from the species *Gliocladium catenulatum* gave the best results, reducing the symptoms of witches' broom disease by approximately 71% (Rubini et al. 2005).

Cabral et al. (2009) suggested that some yeasts isolated from the Amazon and Atlantic forests could be used to control witches' broom disease. Lana et al. (2011) studied the physiological and genetic variation of *M. perniciosa* isolated from diseased and healthy branches of *T. cacao*, using the random amplification of polymorphic DNA (RAPD) technique, and tested its virulence in plants in addition to the production of enzymes; they concluded that it was not possible to distinguish pathogenic and endophytic isolates by means of enzyme production. This was the first study showing detection of endophytic *M. perniciosa*, opening up the possibility of using nonpathogenic isolates to reduce disease. A total of 69 bacterial endospore isolates were tested against *M. perniciosa*, and some of them reduced the effects of the disease caused by *M. perniciosa* in cacao (Melnick et al. 2011).

Using a *Bacillus subtilis* strain, Falcaeo et al. (2014) inoculated germinating cacao seeds and evaluated seedling growth 30 days after the inoculation. They found antagonistic effects of the bacteria against *M. perniciosa* and other pathogenic fungi, showing that this *B. subtilis* strain promoted the development of the aboveground parts of cacao seedlings and possessed antimicrobial characteristics, producing an antifungal compound. Growth promotion was also observed in cacao with use of *Enterobacter cloacae* and *B. subtilis* endophytes derived from vigorous plants. It was concluded that these results could have important implications for lessening the occurrence of pathogenic fungi in *T. cacao* (Leite et al. 2013). *Trichoderma* endophytes and mycorrhizae were also used together in organic and nonorganic production areas in Peru. Higher yields were found in at least one area, indicating synergy between these factors (Tuesta-Pinedo et al. 2016). Finally, in Brazil, Almeida et al. (2018) isolated fungal *Trichoderma* species associated with cacao trees. *T. lentiforme* and *T. parareesei* were the most abundant, and it was concluded that they could, after further studies, be important for biological control of cacao diseases.

13.3 Revealing the Cultivable Endophytic Community Associated with Banana

The banana plant (*Musa* spp.) is another of the most significant fruits in the global market. India is the main producer, followed by China, other countries located in Asia, and all tropical areas in the Americas (Actualitix 2013). Several studies of endophytic isolation from the genus *Musa* have already been published, with such studies involving isolation of endophytes from distinct regions and countries, including fungi (Pereira et al. 1999; Photita et al. 2001; Cao et al. 2002; Ting et al. 2008; Zakaria and Rahman 2011; Thangaleva and Gopi 2015) and bacteria (Martinez et al. 2003; Ting et al. 2008; Thomas and Soly 2009; Ngamau et al. 2012; Sousa et al. 2013, 2017; Sekhar and Thomas 2015; Karthik et al. 2017; Su et al. 2017). These reports, in addition to discussing the isolation of distinct genera and species, also indicated that some diseases, such as those caused by *Fusarium*, can be controlled by endophytes (Kavino and Manoranjitham 2018).

To date, most of the studies related to endophytes as control agents were likely conducted using plants cultivated with use of fertilizers and other agrochemicals to increase production and reduce pests. It is therefore possible that some endophytes were absent because of inhibition by chemicals. A recent study showed that endophytes isolated from banana leaves cultivated under organic management included some genera and species that had not been previously isolated (Souza Junior et al. 2018). These endophytic fungi and bacteria are now being tested for their potential to control diseases and promote plant growth. The use of endophytes isolated from organic cultures may provide a way to discover new beneficial uses of endophytic microorganisms that could not be previously isolated for growth promotion and control of pests and diseases, including some diazotrophic bacteria from the genus *Herbaspirillum* and fungal species with possible actions for controlling diseases by producing antimicrobial compounds. Tables 13.2 and 13.3 list the endophytic fungi and bacteria isolated from *Musa* spp.

Table 13.2 Endophytic fungi isolated from *Musa* spp. cultivated under conventional and organic management

Genus	Management type	References
Alternaria	Traditional	Pereira et al. (1999)
		Cao et al. (2002)
Aspergillus	Traditional/organic	Pereira et al. (1999);
		Cao et al. (2002)
		Souza Junior et al. (2018)
Cephalosporium	Traditional	Cao et al. (2002)
Cladosporium	Traditional	Photita et al. (2001)
		Cao et al. (2002)
Colletotrichum	Traditional	Pereira et al. (1999)
		Photita et al. (2001)
Cordana	Traditional	Pereira et al. (1999) Photita et al. (2001)
Curvularia	Traditional	Pereira et al. (1999)
		Photita et al. (2001)
Dactylaria	Traditional	Photita et al. (2001)
Deightoniella	Traditional	Photita et al. (2001)
		Cao et al. (2002)
Drechslera	Traditional	Pereira et al. (1999)
Epicoccum	Traditional	Pereira et al. (1999)
Fusarium	Traditional	Pereira et al. (1999)
		Photita et al. (2001)
		Zakaria and Rahman (2011)
Gloeosporium	Traditional	Cao et al. (2002)
Glomerella	Traditional	Pereira et al. (1999)
Guignardia	Traditional	Photita et al. (2001)
Humicola	Traditional	Pereira et al. (1999)
Myxosporium	Traditional	Cao et al. (2002)
Nigrospora	Traditional	Pereira et al. (1999)
		Photita et al. (2001)
Penicillium	Traditional	Cao et al. (2002)
Periconia	Traditional	Pereira et al. (1999)
Phomopsis	Traditional	Pereira et al. (1999)
Phyllosticta	Traditional	Pereira et al. (1999)
Pyriculariopsis	Traditional	Photita et al. (2001)
Sarcinella	Traditional	Pereira et al. (1999)
		Cao et al. (2002)
Spicaria	Traditional	Cao et al. (2002)
Trichoderma	Traditional	Pereira et al. (1999)
		Thangavelu and Gopi (2015)
Uncinula	Traditional	Cao et al. (2002)
Xylaria	Traditional	Pereira et al. (1999)
Acrocalymma	Organic	Souza Junior et al. (2018)
Byssochlamys	Organic	Souza Junior et al. (2018)

(continued)

Table 13.2 (continued)

Genus	Management type	References
Hypocreales	Organic	Souza Junior et al. (2018)
Myrothecium	Organic	Souza Junior et al. (2018)
Nigrospora	Organic	Souza Junior et al. (2018)
Peniophora	Organic	Souza Junior et al. (2018)
Periconia	Organic	Souza Junior et al. (2018)
Peroneutypa	Organic	Souza Junior et al. (2018)
Saccaricola	Organic	Souza Junior et al. (2018)

13.4 Controlling Citrus Variegated Chlorosis Using the Host Microbial Community

The bacterium *Xylella fastidiosa* was described for the first time in 1884 as a plant pathogenic species attacking grapes in California (USA), causing a disease later recognized as Pierce's disease (Pierce 1892). Some similar diseases caused by *X. fastidiosa* were described in several other plants (mainly in North and South America (Hopkins 1989)) and, more recently, in fruit trees in other parts of the world (Azevedo et al. 2016). Another problematic disease of citrus in Brazil was reported in 1987, known as citrus variegated chlorosis (CVC), and it has turned out to be very important, affecting sweet oranges in the country (Rosseti et al. 1990).

Brazil is the world's largest producer of citrus fruits, accounting for 25% of global production. In the final years of the last century, CVC disease was discovered in approximately 90% of the plantations in the country (Lambais et al. 2000). At that time, the economic losses were increasing, reaching millions of US dollars per year. This motivated an increasing number of studies and publications aiming to identify ways to reduce damage caused by *X. fastidiosa* in citrus. A program was launched by the São Paulo Research Foundation (FAPESP) in Brazil, aiming to study several aspects of the causative agent of CVC. After approximately 3 years of research, *X. fastidiosa* became the first plant pathogenic species of bacteria to be completely genome sequenced (Simpson et al. 2000). Other new features of this species were described—for instance, the existence of plasmids and viruses, recombination methods, and similarities and differences among the strains of the species. However, despite the increased knowledge of the genome and other characteristics of *X. fastidiosa*, the genetic means of the plant host characteristics were not explicated (Almeida and Nunney 2015). The disease continued to cause economic damage, and it was necessary for FAPESP to launch another program—titled "Functional Control of CVC"—to find other ways to diminish the damage caused by the pathogen. Research showed that in citrus orchards attacked by *X. fastidiosa*, some plants did not develop symptoms of the disease. However, the studies showed that these plants were not genetically resistant mutants, and research was therefore carried out to study the endophytic bacterial community by comparing endophytes found in attacked plants and those in plants that did not show disease symptoms.

Table 13.3 Endophytic bacteria isolated from *Musa* spp. cultivated under conventional and organic management

Genus	Management	References
Acinetobacter	Traditional	Thomas and Soly (2009)
		Su et al. (2017)
Agrobacterium	Traditional	Souza et al. (2013)
Aneurinibacillus	Traditional	Souza et al. (2013)
Arthrobacter	Traditional	Sekhar and Thomas (2015)
Bacillus	Traditional/natural	Thomas and Soly (2009)
		Ngamau et al. 2012
		Souza et al. (2013)
		Sekhar and Thomas (2015)
		Su et al. (2017)
		Souza Junior et al. (2018)
Brevibacterium	Traditional	Sekhar and Thomas (2015)
Brevundimonas	Traditional	Sekhar and Thomas (2015)
Citrobacter	Traditional	Martinez et al. (2003)
		Su et al. (2017)
Corynebacterium	Traditional	Sekhar and Thomas (2015)
Curtobacterium	Traditional	Sekhar and Thomas (2015)
Enterobacter	Traditional	Martinez et al. (2003)
		Ngamau et al. (2012)
		Souza et al. (2013)
		Sekhar and Thomas (2015)
		Su et al. (2017)
Evingella	Traditional	Ngamau et al. (2012)
Klebsiella	Traditional	Martinez et al. (2003)
		Souza et al. (2013)
		Sekhar and Thomas (2015)
Kokuria	Traditional	Sekhar and Thomas (2015)
Kytococcus	Traditional	Sekhar and Thomas (2015)
Lysinibacillus	Traditional	Souza et al. (2013)
Microbacterium	Traditional	Su et al. (2017)
Micrococcus	Traditional	Thomas and Soly (2009)
		Souza et al. (2013)
		Sekhar and Thomas (2015)
Naumannella	Traditional	Sekhar and Thomas (2015)
Paenibacillus	Traditional	Thomas and Soly (2009) Souza et al. (2013)
Pantoea	Traditional/natural	Souza Junior et al. (2018)
Pseudacidovorax	Traditional	Souza Junior et al. (2018)
Pseudomonas	Traditional	Ngamau et al. (2012)
		Sekhar and Thomas (2015)
		Su et al. (2017)
Rahnella	Traditional	Ngamau et al. (2012)

(continued)

Table 13.3 (continued)

Genus	Management	References
Rhizobium	Traditional	Martinez et al. (2003)
		Souza et al. (2013)
Rhodococcus	Traditional	Souza et al. (2013)
Rothia	Traditional	Sekhar and Thomas (2015)
Serratia	Traditional/natural	Ting et al. (2008)
		Ngamau et al. (2012)
		Sekhar and Thomas (2015)
		Souza Junior et al. (2018)
Sphingomonas	Traditional	Sekhar and Thomas (2015)
Staphylococcus	Traditional	Thomas and Soly (2009)
		Sekhar and Thomas (2015)
Streptomyces	Traditional	Su et al. (2017)
Yersinia	Traditional	Ngamau et al. (2012)
Arsenicicoccus	Natural (organic)	Souza Junior et al. (2018)
Brevibacterium	Natural	Souza Junior et al. (2018)
Herbaspirillum	Natural	Souza Junior et al. (2018)
Lactococcus	Natural	Souza Junior et al. (2018)
Neisseria	Natural	Souza Junior et al. (2018)
Pseudorhodoferax	Natural	Souza Junior et al. (2018)
Sphingobacterium	Natural	Souza Junior et al. (2018)
Stenotrophomonas	Natural	Souza Junior et al. (2018)
Streptococcus	Natural	Souza Junior et al. (2018)
Variovorax	Natural	Souza Junior et al. (2018)

Endophytic bacteria from distinct citrus root stocks were isolated by Araujo et al. (2001, 2002). Bacteria from symptomatic citrus plants included more species of the genus *Methylobacterium* than healthy plants. *Methylobacterium extorquens* was found only in affected plants. Lacava et al. (2004, 2006a, b) showed that *Methylobacterium mesophylicum* was present in healthy hosts. In addition, Lacava et al. (2004) demonstrated that *M. mesophylicum* reduced the reproduction of *X. fastidiosa*, while *M. extorquens* had no influence on the growth of the pathogen. Using the plant *Catharanthus roseus* as a standard replica, Lacava et al. (2006a) showed that the occurrence of the pathogenic bacterium *X. fastidiosa* was lessened by inoculation with the endophytic bacterium *M. mesophylicum*. This suggested that these endophytic bacteria may compete for colonization and nutrient opportunities within the citrus plant host. Therefore, development of CVC symptoms was reduced by the endophytic bacterial population. In addition, the environmental conditions also affected the host physiology (Lacava et al. 2004; Dourado et al. 2015). Another endophytic bacterium, *Curtobacterium flaccumfaciens*, isolated from diseased and healthy citrus, was also found to act as a biocontrol agent (Araujo et al. 2002; Lacava et al. 2004). This suggested that the growth of *X. fastidiosa* may be reduced by the presence of *C. flaccumfaciens* and that this bacterial endophyte could be used to control CVC in citrus (Lacava et al. 2007; Azevedo et al. 2016).

On the basis of previous studies, another proposed strategy is the use of paratransgenesis to control CVC disease. Paratransgenesis is the genetic modification or amendment of synergetic microflora that are transmitted by an insect. This inclusive tactic to prevent disease is known as mutual or synergetic control and is a variant of symbiotic treatment or therapy (Ahmed 2003; Beard et al. 1998, 2001; Rio et al. 2004). In citrus woods (groves) in Brazil, *Dilobopterus costalimai* Young, *Oncometopia facialis* (Signoret), and *Acrogonia citrina* Marucci & Cavichioli are the sharpshooters that are customarily found, whereas in citrus nurseries and young woods (groves), *Bucephalogonia xanthophis* (Berg) is frequently found (Redak et al. 2004). In this way, as mentioned in previous results, *Methylobacterium* spp., segregated as bacterial endophytes from citrus plants, were genetically modified. *C. roseus* (model plant) seedlings were inoculated with a globular fluorescent protein–marked strain of *M. mesophilicum*, and the movement, colonization, and functions of the bacterial strain were later observed in the xylem vessels of the model plants. It was also demonstrated that *M. mesophilicum* inhabits the same place as *X. fastidiosa* subsp. *pauca* inside the host plants, but it might also be passed on by *B. xanthophis*. Understanding of the ecological niche of *M. mesophilicum* is a prerequisite for understanding and examination of the possible application of synergetic regulation to interrupt transmission of *X. fastidiosa* subsp. *pauca*. It is a bacterial pathogen that causes CVC disease through insects that act as vectors (Gai et al. 2009).

It is known that inside plant vessels, *X. fastidiosa* cells establish in the form of biofilms formed by numerous cells, which are surrounded by an extracellular matrix (Tyson et al. 1985). This extracellular matrix is composed of a polymer known as an exopolysaccharide (EPS), and this EPS is named fastidian gum. In (1999), Nankai and coworkers explained the means of xanthan gum production. This gum is also an EPS, and the enzyme β 1,4-D-endoglucanase, produced by *Bacillus* sp., is involved in depolymerization of the gum. Later on, in 2005, Lima and colleagues were able to clone the endoglucanase A gene of the citrus bacterial endophyte *B. pumilus*, and this cloned gene was capable of degrading xanthan gum. Consequently, on the basis of associated (symbiotic) regulation, one more citrus bacterial endophyte, *M. extorquens* strain AR1.6/2, was also modified genetically. Later on, its ability to inoculate and colonize a prototype plant and its dealings with *X. fastidiosa* were assessed. Further, this bacterial strain was improved to produce a new strain, AREgIA, that expressed an endoglucanase enzyme. With use of fluorescence microscopy, it was revealed that a green fluorescent protein (GFP)–marked bacterial strain, ARGFP, was capable of colonizing the xylem vessels of *C. roseus* seedlings. Further, by use of scanning electron microscopy (SEM), it was observed that *X. fastidiosa* and AREglA might coinhabit the xylem vessels of the model plant *C. roseus*. Interestingly, *M. extorquens* was detected in the xylem, along with the plant pathogen *X. fastidiosa*, and seemed to lessen biofilm development (Ferreira-Filho et al. 2012).

Now, approximately 20 years after the spread of CVC caused by *X. fastidiosa*, the situation has changed. The percentage of citrus trees showing disease symptoms decreased from 52.6% in 2009 to 37.6% in 2012 and only 3.0% in 2016 (FUNDECITRUS 2017). The bacterium and endophytes are transported by approximately 12 insect species. Simple ways to control the disease were introduced, using healthy plants derived from disease-free seedings produced in protected greenhouses that prevented contact with insects that could transmit the disease. Control of infected insects was also performed in areas near orchards, including eradication of affected parts of citrus trees. Therefore, with use of simple and traditional techniques, the problem has now been partially solved. However, other plant pathogens, such as those causing citrus canker and greening, have become major problems for citrus production in Brazil. For management of CVC, detection, identification, and other methods are fundamental. In addition, other alternatives, such as use of endophytic microorganisms that inhabit the same niche as vascular pathogens, are favored for biocontrol (Eljournaidi et al. 2016).

13.5 Bioprospecting for Microorganisms Associated with Guarana—A Typical Tropical Crop—And Control of Anthracnose

Paullinia cupana Mart. var. *sorbilis*—commonly known as "guarana"—is a native plant of the Amazon. The *Paullinia* genus includes around 200 species, which are mostly limited to the Amazon region, but there are some exceptions, meaning that this genus is also found in subtropical and tropical American regions (Schimpl et al. 2013). Guarana seeds are used as a common and important component in commercial products and have stimulant properties, which are due to their high caffeine concentration. Guarana plants are cultivated commercially in Brazil, and the seeds are used in popular carbonated drinks and marketed natural goods (Silva et al. 2016).

To improve the knowledge concerning the microbial community associated with this tropical crop, efforts have been made to assess cultivable and noncultivable guarana microorganisms. For instan ce, 96 bacteria were isolated from the guarana rhizosphere. These bacterial species were subjected to various biochemical tests and identified by a partial or total 16S rRNA sequencing technique. Firmicutes and Proteobacteria were the prevailing and major rhizospheric phyla that were identified, and *Bacillus* and *Burkholderia* were the most dominant genera. Out of 13 bacterial strains, only four exhibited plant growth–promoting traits; interestingly, the majority of them belonged to the *Burkholderia* genus.

From the guarana rhizosphere, two multitrait plant growth–promoting strains (the *Bacillus* strain RZ2MS9 and the *Burkholderia ambifaria* strain RZ2MS16) dramatically enhanced soybean and corn plant growth in greenhouse environments. Increases in the corn root dry weight of 136.9% and 247.8% were obtained with RZ2MS16 and RZ2MS9 inoculation, respectively, in comparison with uninoculated controls, after 60 days of growth. The dry shoot weights of the treated soybean and corn plants were also significantly higher than those of the uninoculated plants, representing an increase of more than 47% for both bacterial strains and plants (Batista et al. 2018).

It is known that guarana production has been drastically diminished by a *Colletotrichum* sp. fungal pathogen, which is a causal agent of anthracnose. This disease manifests as round, orange-colored necrotic lesions on leaf surfaces (Silva et al. 2004). Aiming to identify ways to control this pathogen, Bonatelli et al. (2016) selected 15 fungal species from guarana leaf lesions infested with anthracnose, which belonged to five genera: *Fusarium*, *Leptosphaeria*, *Microdochium*, *Pestalotiopsis*, and *Phomopsis*. Four *Fusarium* isolates, one *Pestalotiopsis* isolate, and one *Microdochium* isolate constantly hindered the growth of anthracnose fungi under in vitro conditions. With the exception of the *Microdochium* isolate, the isolates were also capable of inhibiting the pathogen growth in in vivo assays, which were performed on separated guarana plant leaves. Some of the mechanisms resulting in pathogen growth inhibition were identified. The *Fusarium* isolates produced chitinase enzymes, whereas the *Pestalotiopsis* and *Fusarium* isolates manufactured antagonistic volatile organic compounds (VOCs).

In similar work, Bogas et al. (2015) assessed bacterial communities associated with guarana leaves and verified increased bacterial diversity in comparison with asymptomatic plants. Comprehensive examination of bacterial endophyte organization by use of culture-dependent and 16S rRNA clone libraries disclosed the occurrence of Actinobacteria, Acidobacteria, Firmicutes, Bacteroidetes, and Proteobacteria phyla. The Firmicutes phyla covered the bulk of the isolates in asymptomatic plants. The authors suggested that anthracnose can restructure bacterial endophyte communities through a preference for particular strains in the phyllospheric zone of *P. cupana*. The knowledge of such communications is very significant for strategy development and improvement of biological control of *Colletotrichum*.

In a complementary study, the community of bacteria associated with anthracnose-symptomatic leaves of guarana were assessed using a culture-independent technique based on partial massive sequencing of 16S rRNA (Bonatelli 2012). The cultured bacterial strains were evaluated for inhibition of *Colletotrichum* sp. growth, as well as enzyme and siderophore production. On the basis of the culture-independent technique, Proteobacteria was the most abundant phylum. Many sequences were assigned as unclassified, suggesting a completely new community associated with guarana. The presence of anthracnose disease benefited the cultured *Acinetobacter* and *Pseudomonas* strains, which were significantly more abundant in symptomatic leaves. In vitro, 11.38% of the strains inhibited *Colletotrichum* sp. growth, and most of them were obtained from symptomatic leaves. Asymptomatic leaves hosted several amylase and polygalacturonase producers that could be related to antagonistic effects or mechanisms for survival on the leaves (Bonatelli 2012).

Silva et al. (2016) isolated endophytic bacteria from guarana plant seeds collected in the Amazon region and in the northeast state of Bahia. In these regions, this pathogen is not problematic for guarana farms. From these seeds, 102 bacterial isolates were assessed in in vitro conditions against a plant pathogenic *Colletotrichum* sp. These bacterial isolates were also examined for production of enzymes such as esterase, amylase, protease, cellulase, lipase, and pectinase. Almost 15% of the bacterial isolates demonstrated greater inhibitory activity against the fungal pathogen.

Production of at least one enzyme by the bacterial isolates was confirmed by the process of partial 16S rRNA sequencing. The *Bacillus* genus was most commonly detected, followed by other bacterial strains such as *Microbacterium, Paenibacillus, Stenotrophomonas*, and *Ochrobactrum*. The majority of the bacterial isolates exhibited biocontrol activities against the fungal pathogen. Therefore, a suitable biological control mechanism employing potential endophytes of guarana could be used in the future to prevent the spread of fungal pathogens.

13.6 Improving Sugarcane Fitness by Use of Bacteria

In Brazil, sugarcane is one of the most significant crops. Its importance is increasing because of the gradual substitution of fossil fuels with renewable and cleaner energy sources such as ethanol. Worldwide, sugarcane cultivation occupies a total area of over 25 million hectares (Meyer and Clowes 2013). It is understood that by the year 2030, sugarcane and its by-products will become the world's second largest energy resource (after petrochemicals and their by-products), covering 20% of the global energy environment. Consequently, biotechnological approaches that can mediate sugarcane–bacterium interactions must be evaluated. Studies in tropical countries have indicated that bacterial–host plant synergism may increase the fitness and growth of accompanying crop plants.

Brazil is the leading sugarcane producer worldwide. In growing sugarcane, Brazil uses only 90–120 kg of nitrogen per hectare, and its total area of sugarcane cultivation is more than 10 million hectares (Joris 2015). This amount of nitrogen application and total land area of cultivation results in a sugarcane yield of around 70.6 $t.ha^{-1}$ (FAO 2016). The main difference in sugarcane cultivation between India and Brazil is the application of valuable and advantageous microorganisms in Brazil. These beneficial microbes, when applied to plants, convert atmospheric nitrogen into ammonium ions, which are easily taken up by plants, by a process known as biological nitrogen fixation (BNF). Therefore, the dependence on chemical fertilizers is reduced to a greater extent (Mehnaz 2013). These beneficial bacteria also promote sugarcane growth by processes such as solubilization of phosphate (Singh et al. 2007) and production of phytohormones (especially IAA (Videira et al. 2012)); moreover, they produce low molecular weight proteinaceous compounds, which sequester iron from the environment (Tailor and Joshi 2012). *Beijerinckia* was the first plant growth–promoting bacterium (PGPB) isolated from sugarcane to be widely examined and researched (Döbereiner and Ruschel 1958; Döbereiner 1959, 1961). Others include *Gluconacetobacter* (Cavalcante and Döbereiner 1988), *Herbaspirillum* (Baldani et al. 1992), *Burkholderia* (Boddey et al. 2003; Caballero-Mellado et al. 2004; Reis et al. 2004), and *Azospirillum* (Tejera et al. 2005). Figure 13.1 shows the microbial diversity and overall benefits of microbes found on plants.

Some PGPBs are able to inhabit numerous diverse agriculturally important crops. This phenomenon is called cross-colonization (Zakria et al. 2008; Quecine et al. 2012). One of these beneficial bacteria, *Pantoea agglomerans*

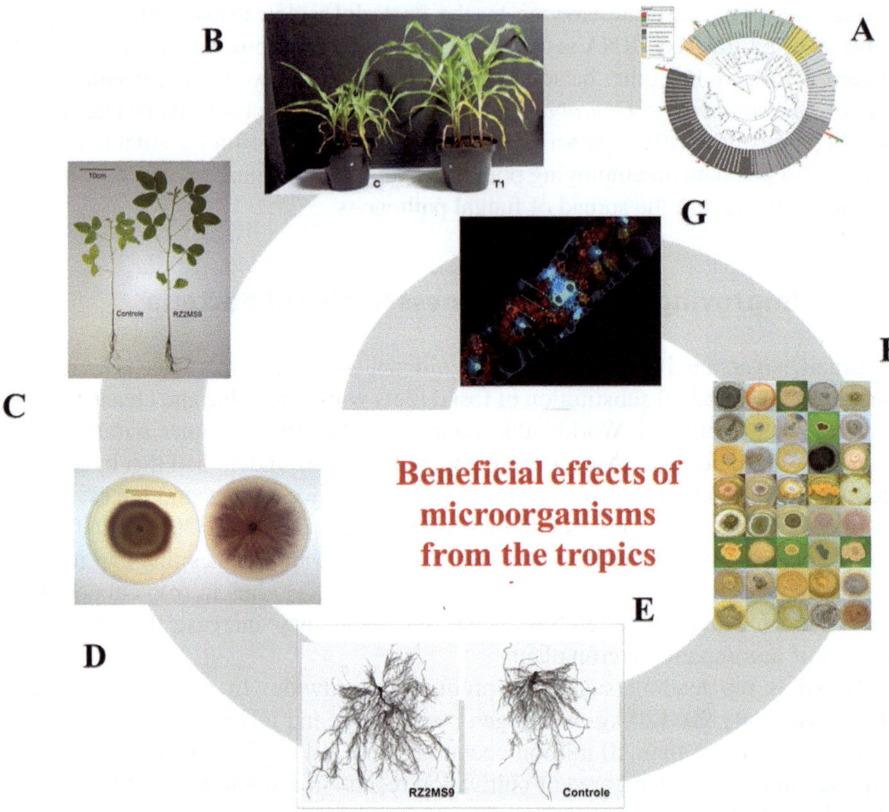

Fig. 13.1 An overview of studies of microorganisms in the tropics reveals their high diversity (**a** and **f**), reflecting the great potential of these microorganisms as plant growth promoters (**b**, **c**, **e**) and agents of disease control (**d**) inside the plant host (**g**)

strain 33.1—which was previously isolated from *Eucalyptus grandis* and is a known eucalyptus growth promoter—was evaluated regarding the biotechnological aspects of the sugarcane–bacterial association. Strain 33.1 was able to promote sugarcane growth and to induce production of resistance proteins in inoculated plants. The production of plant hormones and phosphatase by strain 33.1 was associated with sugarcane growth–promoting bacterial mechanisms. With the aim of elucidating the behavior of strain 33.1 during its interaction with sugarcane, this endophyte, harboring an integrative pNKGFP plasmid, 33.1::pNKGFP, was added to a liquid medium and then to a substrate in which sugarcane seedlings were growing. The highest 33.1::pNKGFP density was observed in the plant rhizosphere. Moreover, the addition of strain 33.1 and 33.1::pNKGFP to the substrate had no effects on the sugarcane-associated bacterial density (Quecine et al. 2012).

The bacterial endophyte *P. agglomerans* (33.1) was also modified genetically using the plasmid pJTT, resulting in expression of the *cry1Ac7* gene (Quecine et al. 2014). Bioassays of *Diatraea saccharalis* control by 33.1:pJTT were carried out in a simulated medium. With confirmation of the part regulation of larvae by 33.1:pJTT, a new methodology was developed. 33.1:pJTT was applied to sugarcane stalks containing *D. saccharalis* larvae. The results showed that 33.1:pJTT was capable of increasing the larval fatality of *D. saccharalis* fed on an artificial diet and sugarcane stalks. The larval development of *D. saccharalis* was also impaired, and the larval weights were significantly reduced by 33.1:pJTT. A desirable alternative is to conduct bioassays that better mimic the natural environment. 33.1:pJTT was reisolated from sugarcane stalks and *D. saccharalis* larvae. Sugarcane seedling reinfection by 33.1:pJTT was also confirmed. All of these results showed the potential of *P. agglomerans* (33.1) to express the Cry protein, which is essential to control of the *D. saccharalis* sugarcane borer.

With an aim to improve the understanding of microorganism communities associated with sugarcane, Souza et al. (2016) explained the bacterial community structure connected with several host tissues. The researchers identified 23,811 bacterial operational taxonomic units (OTUs) inhabiting the endophytic and exophytic compartments of the roots, shoots, and leaves. The researchers also observed that different bacterial and fungal communities preferentially colonized different tissues in the sugarcane plants. In general, Rhodospirillaceae and Chitinophagaceae were the predominant taxa in the rhizosphere, endophytic roots, and young shoots. Families such as Sinobacteraceae, Cytophagaceae, and Hyphomicrobiaceae were more plentiful in endophytes from the root compartment of the plant. In young shoots, Acidobacteriaceae representatives were more abundant. The plant stalks were found to be colonized by yeast groups demonstrating more than 12% of the total comparative abundance. Members of the families Pseudomonadaceae, Enterobacteriaceae, and Moraxellaceae were discovered to be preferential colonizers of endophytic leaves and stalks. The majority of the bacterial communities characterized earlier, whose diversity was unexplored, proved the significance of the microbiome findings. The efficient ones could be utilized as potential biotechnological tools.

Armanhi et al. (2018) used corn as a model plant for artificial inoculation. They recovered 399 exclusive cultured bacterial species, which represented 15.9% of the rhizospheric zone microbiota and 61.6–65.3% of the microbiome of the endophytic zone of sugarcane stalks. The researchers demonstrated that when corn plants were inoculated, the members of the artificial microbial community efficiently colonized the plant parts. This artificial inoculation displaced the innate microbiome and dominated about 53.9% of the rhizospheric microbial communities. Subsequently, the plants inoculated with beneficial microbes increased their biomass by 3.4 times in comparison with the uninoculated control plants.

13.7 Mangroves as a Reservoir of New Beneficial Microorganisms

The ecosystem of mangrove plants is a littoral tropical biome, situated in the zone of transition between the sea and the land, which is distinguished by cyclic flooding. This periodic flooding bestows specific and unique ecological circumstances upon this region. In such bionetworks, the flora are controlled by a specific plant group species. Such species provide exclusive environmental conditions that shelter miscellaneous microbial groups, including endophytic microbes. Because of their close alliance with host plants, endophytic microbes can be found and investigated for production of biotechnologically important products such as phytohormones, proteins, enzymes, antibiotics, VOCs, and other beneficial complexes.

Castro et al. (2014) isolated endophytic microbes from two *Rhizophora mangle* and *Avicennia nitida* mangrove species. These two mangrove species are found in the Bertioga and Cananéia regions of Brazil. From these two species, *Bacillus* was the major obtainable bacterial genus, although other commonly found endophytic bacterial genera such as *Enterobacter*, *Curtobacterium*, and *Pantoea* were also reported. After identification of these bacterial isolates, the endophytic bacterial communities were assessed for the production of enzymes. Protease enzyme activity was detected in 75% of the bacterial species, while endoglucanase enzyme activity was observed in 62% of them. This research enriched our knowledge of various types of dominant microbes, which exist as endophytes and are involved in the production of significant enzymes, in the ecosystem of mangrove plants.

Among the isolated bacterial strains, a huge number of these strains—115 in total—were assessed for their capability to fix atmospheric nitrogen and solubilize insoluble phosphorus. These bacterial strains, which had both of the aforementioned properties, were further tested for phytohormone (auxin) production. Out of these 115, only two isolates were selected on the basis of higher IAA production. These bacterial strains were used for inoculation of trees for reforestation purposes, and the potential of these bacterial isolates was assessed under field conditions. One of the bacterial isolates was *Pseudomonas fluorescens* (strain MCR1.10) which exhibited a lesser phosphorus solubilization property, whereas this property was quite strong in another isolate, *Enterobacter* strain MCR1.48. The researchers applied these two isolates for reforestation of *Acacia polyphylla* trees. The results indicated that application of the isolate MCR1.48 endophyte increased *Acacia polyphylla* dry shoot biomass, demonstrating that this isolate could efficiently promote *Acacia* tree development, fitness, and growth; therefore, it could be applied to enhance growth and development of seedlings of this *Acacia* plant (Castro et al. 2018).

13.8 Concluding Remarks

Microbiomes make up a huge percentage and a huge amount of the biodiversity on this planet, and they play significant roles in the function and structure of farming systems all over the world. Thus, on the basis of efforts to achieve sustainable management in agriculture, microorganisms are being targeted because of their close interactions with plants. An increasing number of scientists are attempting to isolate new microorganisms and to explicate the means of plant growth and development, biocontrol, and biological remediation facilitated by those organisms. Information on the accompanying microbiome is indispensable for promotion of plant growth and development because it can be used to obtain potential bioinoculants.

Bioinoculants not only are cost effective and environmentally friendly but also enhance production of crop plants in an eco-friendly and natural fashion, which is a step toward organic farming.

References

Actualitix (2013) https://en.actualitix.com/country/wld/banana-producing-countries.php. Accessed June 2019

Ahmed FE (2003) Genetically modified probiotics in foods. Trends Biotechnol 21:491–497

Almeida RPP, Nunney L (2015) How do plant diseases caused by *Xylella fastidiosa* emerge. Plant Dis 99:1457–1467

Almeida KA, Armesto C, Monteiro FP, Souza JT (2018) Diversity of *Trichoderma* species isolated from dead branches and sapwood of *Theobroma cacao* trees. Trop Plant Pathol 43:90–94

Araujo WL, Saridakis HO, Barroso PAV, Aguilar-Vildoso CI, Azevedo JL (2001) Variability and interactions between endophytic bacteria and fungi isolated from leaf tissues of citrus rootstocks. Can J Microbiol 47:229–236

Araujo WL, Marcon J, Maccheroni W Jr, Van Elsas JD, Azevedo JL (2002) Diversity of endophytic bacterial population and their interaction with *Xylella fastidiosa* in citrus plants. Appl Environ Microbiol 68:4096–4914

Armanhi JSL, Souzal RSC, Damasceno NB, Araújo LM, Imperial J, Arruda P (2018) A community-based culture collection for targeting novel plant growth–promoting bacteria from the sugarcane microbiome. Front Plant Sci 8:2191. https://doi.org/10.3389/fpls.2017.02191

Arnold AE, Mejia LC, Kyllo D, Rojas EI, Maynard Z, Robbins N, Herre EA (2003) Fungal endophytes limit pathogen in a tropical tree. Proc Natl Acad Sci 100:15649–15654

Azevedo JL, Araujo WL (2007) Diversity and applications of endophytic fungi isolated from tropical plants. In: Ganguli BN, Desmuckh SK (eds) Fungal multifaceted microbes. Anamaya Publishers, New Delhi, pp 207–231

Azevedo JL, Araujo WL, Lacava PT (2016) The diversity of citrus endophytic bacteria and their interactions with *Xylella fastidiosa* and host plants. Genet Mol Biol 39:476–491

Baldani VLD, Baldani JI, Olivares FL, Döbereiner J (1992) Identification and ecology of *Herbaspirillum seropedicae* and closely related *Pseudomonas rubrisubalbicans*. Symbiosis 13:65–73

Batista BD, Almeida JR, Bezerra TE, Azevedo JL, Quecine MC (2017) Describing the unexplored microrganisms associated with guarana: a typical tropical plant. In: de Azevedo JL, Quecine MC (eds) Diversity and benefits of microorganisms from the tropics. Springer, Cham, pp 293–312

Batista BD, Lacava PT, Ferrari A, Teixeira-Silva NS, Bonatelli ML, Tsui S, Mondin M, Kitajima EW, Pereira JO, Azevedo JL, Quecine MC (2018) Screening of tropically derived, multi-trait plant growth-promoting rhizobacteria and evaluation of corn and soybean colonization ability. Microbiol Res 206:33–42

Beard CB, Durvasula RV, Richards FF (1998) Bacterial symbiosis in arthropods and the control of disease transmission. Emerg Infect Dis 4:581–591

Beard CB, Dotson EM, Pennington PM, Eichler S, Cordon-Rosales C, Durvasula RV (2001) Bacterial symbiosis and paratransgenic control of vector-borne Chagas disease. Int J Parasitol 31:621–627

Boddey RM, Urquiaga S, Alves BJ, Reis V (2003) Endophytic nitrogen fixation in sugarcane: present knowledge and future applications. Plant Soil 252:139–149

Bogas AC, Ferreira AJ, Araujo WL, Astolfi-Filho S, Kitajima EW, Lacava PT, Azevedo JL (2015) Endophytic bacterial diversity in the phyllosphere of Amazon *Paullinia cupana* associated with asymptomatic and symptomatic anthracnose. Springerplus 4:258

Bonatelli M (2012) Bactérias endofíticas e epifiticas cultivadas e não cultivadas do guaranazeiro e o controle da antracnose. Dissertation, Escola superior de Agricultura "Luiz de Queiroz", Universidade de São Paulo
Bonatelli ML, Tsui S, Marcon J, Batista BD, Kitajima EW, Pereira JO, Azevedo JL, Quecine MC (2016) Antagonistic activity of fungi from anthracnose lesions on *Paullinia cupana* against *Colletotrichum* sp. J Plant Pathol 98:197–205
Caballero-Mellado J, Martínez-Aguilar L, Paredes-Valdez G, Estrada-De Los Santos P (2004) *Burkholderia unamae* sp. nov., an N_2-fixing rhizospheric and endophytic species. Int J Syst Evol Microbiol 54:1165–1172
Cabral AS, Carvalho PM, Pinotti T, Hagler AN, Mendonça-Hagler LG, Macrea A (2009) Killer yeasts inhibit the growth of the phytopathogen *M. perniciosa*, the causal agent of witches' broom disease. Braz J Microbiol 40:108–110
Cao LX, You JL, Zhou SN (2002) Endophytic fungi from *Musa acuminata* leaves and roots in South China. World J Microbiol Biotechnol 18:69–171
Castro RA, Quecine MC, Lacava PT, Batista BD, Luvizzoto DM, Marcon J, Melo IS, Azevedo JL (2014) Isolation and enzyme bioprospection of endophytic bacteria associated with plants of Brazilian mangrove ecosystem. Springerplus 3:382
Castro RA, Dourado MN, Almeida JR, Lacava PT, Nave A, Melo IS, Azevedo JL, Quecine MC (2018) Mangrove endophyte promotes reforestation tree (*Acacia polyphylla*) growth. Braz J Microbiol 49:59–66
Cavalcante VA, Döbereiner J (1988) A new acid-tolerant nitrogen-fixing bacterium associated with sugarcane. Plant Soil 108:23–31
Döbereiner J (1959) Sobre a ocorrência de *Beijerinckia* em alguns solos do Brasil. Rev Bras Biol 19:151–160
Döbereiner J (1961) Nitrogen fixing bacteria of the genus *Beijerinckia* Drex. in the rhizosphere of sugarcane. Plant Soil 15:211–216
Döbereiner J, Ruschel AP (1958) Uma nova espécie de *Beijerinckia*. Revista de Biologia 1:261–272
Dourado MN, Santos DS, Nunes RL, Oliveira LRB, Oliveira MB, Araujo WL (2015) Differential gene expression in *Xylella fastidiosa* 9a5c during co-cultivation with the endophytic bacterium *Methylobacterium mesophilicum* SR1.6/6. J Basic Microbiol 55:1357–1366
Eljounaidi K, Lee SK, Bae H (2016) Bacterial endophytes as potential biocontrol agents of vascular wilt diseases: review and future prospects. Biol Control 103:62–68
Falcaeo LL, Silva-Werneck JO, Vilarinho BR (2014) Antimicrobial and plant growth–promoting properties of the cacao endophyte *Bacillus subtilis* ALB629. J Appl Microbiol 116:1584–1592
FAO [Food and Agriculture Organization of the United Nations] (2016) FAOSTAT beta: food and agriculture data. http://faostat.fao.org/beta/en/#home. Accessed 16 Oct 2016
Ferreira-Filho AS, Quecine MC, Bogas AC, Rossetto PB, Lima AOS, Lacava PT, Azevedo JL, Araújo WL (2012) Endophytic *Methylobacterium extorquens* expresses a heterologous β-1,4-endoglucanase A (EglA) in *Catharanthus roseus* seedlings, a model host plant for *Xylella fastidiosa*. World J Microbiol Biotechnol 28:1475–1481
FUNDECITRUS (2017) http://www.fundecitrus.com.br/levantamentos
Gai CS, Lacava PT, Quecine MC, Auriac MC, Lopes JRS, Araújo WL, Miller TA, Azevedo JL (2009) Transmission of *Methylobacterium mesophilicum* by *Bucephalogonia xanthophis* for paratransgenic control strategy of citrus variegated chlorosis. J Microbiol 47:448–454
Hallmann J, Quadt-Hallmann A, Mahhaffee WF, Kçoepper JW (1997) Bacterial endophytes in agriculture crops. Can J Microbiol 43:895–914
Hopkins DL (1989) *Xylella fastidiosa*: xylem limited bacterial pathogen of plants. Annu Rev Phytopathol 27:271–290
Joris HAW (2015) Nitrogênio na produção de cana-de-açúcar: aspectos agronômicos e ambientais. Dissertation, Instituto Agronômico de Campinas
Karthik M, Pushpakanth R, Krishnamoorthy R, Senthikumar M (2017) Endophytic bacteria associated with banana cultivars and their inoculation effect in plant growth. J Hortic Sci Biotechnol 92:568–576

Kavino M, Manoranjithan SK (2018) In vitro bacterization of banana (*Musa* spp.) with native endophytic and rhizospheric bacterial isolates novel ways to combat *Fusarium* wilt. Eur J Plant Pathol 151:371–387

Lacava PT, Azevedo JL (2014) Biological control of insect-pests and diseases by endophytes. In: Verma VC, Gange AC (eds) Advances in endophytic research. Springer, New Delhi, pp 231–243

Lacava PT, Araujo WL, Marcon J, Maccheroni W Jr, Azevedo JL (2004) Interaction between endophytic bacteria from citrus plants and the phytopathogenic bacterium *Xylella fastidiosa* causal agent of citrus variegated chlorosis. Lett Appl Microbiol 39:55–59

Lacava PT, Li WB, Araujo WL, Azevedo JL, Hartung JS (2006a) Rapid specific and quantitative assay for the detection of the endophytic bacterium *Methylobacterium mesophilicum* in plants. J Microbiol Methods 65:535–541

Lacava PT, Andreote FD, Araujo WL, Azevedo JL (2006b) Caracterização da comunidade bacteriana endofitica de citrus por isolamento, PCR especifico e DGGE. Pesq Agropec Bras 41:637–642

Lacava PT, Li WB, Araujo WL, Azevedo JL, Hartung JS (2007) The endophytic *Curtobacterium flacumfaciens* reduces symptoms caused by *Xylella fastidiosa* in *Catharanthus roseus*. J Microbiol 45:385–393

Lambais MR, Goldman MHS, Camargo LEA, Goldman GH (2000) A genomic approach to the understanding of *Xylella fastidiosa* pathogenicity. Curr Opin Microbiol 3:459–462

Lana TG, Azevedo JL, Pomella AWV, Monteiro RTR, Silva CB, Araujo WL (2011) Endophytic and pathogenic isolates of the cacao fungal pathogenic *Moniliophthora perniciosa* (tricholomataceae) are indistinguishable based on genetic and physiological analysis. Genet Mol Res 10:326–334

Leite HAC, Silva AB, Gomes FP, Gramacho KP, Faria JC, Siuza JT, Loguercio LL (2013) *Bacillus subtilis* and *Enterobacter cloacae* endophytes from healthy *Theobroma cacao* L. trees can systematically colonize seedlings and promote growth. Appl Microbiol Technol 57:2639–2651

Lima AOS, Quecine MC, Fungaro MHP, Andreote F, Arau'jo WL, Silva-Filho MC, Pizzirani-Kleiner AA, Azevedo JL (2005) Molecular characterization of a novel beta-1,4 endoglucanase from endophytic *Bacillus pumilus* strain. Appl Microbiol Biotechnol 68:57–65

Martinez L, Caballero-Melado J, Orozco J, Martinez-Romero E (2003) Diazotrophic bacteria associated with banana (*Musa* spp.). Plant Soil 257:35–47

Mehnaz S (2013) Microbes—friends and foes of sugarcane. J Basic Microbiol 53:954–971

Melnick RL, Suarez C, Bailey BA (2011) Isolation of endopytic endospore-forming bacteria from *Theobroma cacao* as potential biological control agents of cacao disease. Biol Control 57:236–245

Mendes R, Azevedo JL (2007) Valor bitecnologico de fungos endofiticos isolados de plantas de interesse econômico. In: Maia LC, Malosso E, Iano-Melo AM (eds) Micologia: avanços no conhecimento. Sociedade Brasileira de Microbiologia, Recife, pp 129–140

Meyer J, Clowes M (2013) Sugarcane and its environment. In: Meyer J, Rein P, Turner P, Mathias K (eds) Good management practices for the cane sugar industry. Bartens

Nankai H, Hashimoto W, Miki H, Kawai S, Kousaku M (1999) Microbial system for polysaccharide depolymerization: enzymatic route for xanthan depolymerization by *Bacillus* sp. strain GL1. Appl Environ Microbiol 65:2520–2526

Ngamau CN, Matiru VN, Tani A, Muthuri CW (2012) Isolation and identification of endophytic bacteria of banana (*Musa* spp.) in Kenya and their potential as biofertilizers for sustainable banana production. Afr J Microbiol Res 6:6414–6422

Pereira JO, Vieira MLC, Azevedo JL (1999) Endophytic fungi from *Musa acuminata* and their reintroduction in axenic plants. World J Microbiol Biotechnol 15:37–40

Photita W, Lumyong S, Lumyong P, Hyde KD (2001) Endophytic fungi from wild banana (*Musa acuminata*) at Doi Suthep Pul National Park, Thailand. Mycol Res 105:1508–1513

Pierce NB (1892) The California vine disease. USDA Division of Vegetable Pathology bulletin number 2. US Department of Agriculture, Washington, DC. 222p

Quecine MC, Araujo WL, Rossetto PB, Ferreira A, Tsui S, Lacava PT, Mondin M, Azevedo JL, Pizzirani-Kleiner AA (2012) Sugarcane growth promotion by the endophytic bacterium *Pantoea agglomerans* 33.1. Appl Environ Microbiol 78:7511–7518

Quecine MC, Araújo WL, Tsui S, Parra JRP, Azevedo JL, Pizzirani-Kleiner AA (2014) Control of *Diatraea saccharalis* by the endophytic *Pantoea agglomerans* 33.1 expressing cry1Ac7. Arch Microbiol 196:227–234

Redak RA, Purcell AH, Lopes JRS, Blua MJ, Mizell RF III, Andersen PC (2004) The biology of xylem fluid feeding insect vectors of *Xylella fastidiosa* and their relation to disease epidemiology. Annu Rev Entomol 49:243–270

Reis VM, Estrada-De Los Santos P, Tenorio-Salgado S, Vogel J, Stoffels M, Guyon S et al (2004) *Burkholderia tropica* sp. nov., a novel nitrogen-fixing, plant-associated bacterium. Int J Syst Evol Microbiol 54:2155–2162

Rio RVM, Hu Y, Aksoy S (2004) Strategies of the hometeam: symbioses exploited for vector-borne disease control. Trends Microbiol 12:325–336

Rossetti V, Garnier M, Bove JM, Beretta MJG, Teixeira ARR, Quaggio JA, de Negri JD (1990) Presence de bacteria dans le xyleme de oranges atteintes de chlorose variegee une novelle maladie des agrumes au Brasil. C R Acad Sci 310:345–349

Rubini MR, Silva-Ribeiro RT, Pomella AWV, Maki CS, Araujo WL, Santos DR, Azevedo JL (2005) Diversity of endophytic fungal community of cacao (*Theobroma cacao* L.) and biological control of *Crinipellis perniciosa*, causal agent of witches' broom disease. Int J Biol Sci 1:24–33

Sansawal R (2017) Role of endophytic in agriculture. Chem Sci Rev Lett 6:2397–2409

Schimpl FC, Silva JF, Gonçalves JFC, Mazzafera P (2013) Guarana: revisiting a highly caffeinated plant from the Amazon. J Ethnopharmacol 150:14–31

Sekhar AC, Thomas P (2015) Isolation and identification of shoot-tip associated endophytic bacteria from banana Grand Naime and testing for antagonistic activity *Oxysporum* f. sp. *cubense*. Am J Plant Sci 6:943–954

Silva KS, Reboucas TNH, Lemos OL, Bomfim MP, Bomfim AA, Esquivel GL, Barreto APP, Sao Jose AR, Dias NO, Tavares GM (2004) Patogenicidade causada pelo fungo *Colletotrichum gloesporioides* (Penz) em diferentes especies frutiferas. Rev Bras Frutic 28:131–133

Silva MCS, Polonio JC, Quecine MC, Almeida TT, Bogas AC, Pamphile JA, Pereira JO, Astolfi-Filho S, Azevedo JL (2016) Endophytic cultivable bacterial community obtained from the *Paullinia cupana* seed in Amazonas and Bahia regions and its antagonistic effects against *Colletotrichum gloeosporioides*. Microb Pathog 98:16–22

Simpson AJG et al (2000) The genome sequence of the plant pathogen *Xylella fastidiosa*. Nature 406:151–159

Singh KP, Suman A, Singh PN, Lal M (2007) Yield and soil nutrient balance of a sugarcane plant–ratoon system with conventional and organic nutrient management in sub-tropical India. Nutr Cycl Agroecosyst 79:209–219

Souza Junior CA, Marcon J, Andrade PAM, Silva JÁ, Faraldo MIF, Verdi MCQ, Melo Filho AA, Azevedo JL (2018) Endophytic bacterial and fungi associated to banana leaves (*Musa* spp.) cultivated under organic management. J Agric Sci 10:460–467

Souza AS, Xavier AA, Costa MR, Cardoso MAS, Pereira MCT, Nietshe S (2013) Endophytic bacterial diversity in banana Prata Ana (*Musa* spp.) roots. Genet Mol Biol 36:252–264

Souza RSC, Okura VK, Armanhi JSL, Jorrín B, Silva MJ, González-Guerrero M, Araújo LM et al (2016) Unlocking the bacterial and fungal communities assemblages of sugarcane microbiome. Sci Rep 6:28774

Souza GLOD, Silva DF, Nietshe S, Xavier AA, Pereira MCT (2017) Endophytic bacteria used as bioinoculants in micropropagated banana seedlings. Rev Bras Frutic 39:2

Su L, Shen Z, Ruan Y, Tao C, Chao Y, Li R, Shen Q (2017) Isolation of antagonistic endophytes from banana roots against *Meloydogine javanica* and their effects on soil nematode community. Front Microbiol 8:2070

Tailor AJ, Joshi B (2012) Characterization and optimization of siderophore production from *Pseudomonas fluorescens* strain isolated from sugarcane rhizosphere. J Environ Res Dev 6:688–694

Tangavelu R, Gopi M (2015) Combined application of native *Trichoderma* isolates possessing multiple functions for the control of *Fusarium* wilt disease in banana Grand Naine. Biocontrol Sci Tech 25:1147–1164

Tejera N, Lluch C, Martinez-Toledo MV, Gonzalez-Lopez J (2005) Isolation and characterization of *Azotobacter* and *Azospirillum* strains from the sugarcane rhizosphere. Plant Soil 270:223–232

Thomas T, Soly TA (2009) Endophytic bacteria associated with growing shoottips of banana (*Musa* spp.) cv. Grand Naine and the affinity of endophytes to the host. Microb Ecol 58:952

Ting ASY, Meon S, Kadir J, Radu S, Singh G (2008) Endophytic microorganisms as potential growth promoters of banana. BioControl 53:541–553

Tuesta-Pinedo AL, Trigozo-Barta E, Torres JJC, Arevalo-Gardini E, Arevalo-Hernandez CO, Cernandez LBZ, Leon-Ttacca B (2016) Optimization of organic and inorganic fertilization cocoa (*Theobroma cacao* L.) with the inclusion of *Trichoderma* endophyte and arbuscular mycorrhizae. Tecnol Marcha 30:67–78

Tyson GE, Stojanovic BJ, Kuklinski RF, DiVittorio TJ, Sullivan ML (1985) Scanning electron microscopy of Pierce's disease bacterium in petiolar xylem of grape leaves. Phytopathology 75:264–269

Videira SS, de Oliveira DM, de Morais RF, Borges WL, Baldani VLD, Baldani JI (2012) Genetic diversity and plant growth promoting traits of diazotrophic bacteria isolated from two *Pennisetum purpureum* Schum. genotypes grown in the field. Plant Soil 356:51–66

Wilson D (1995) Endophyte: the evolution of the term and clarification of its use and definition. Oikos 73:274–276

Zakaria L, Rahman NHA (2011) Endophytic *Fusarium* spp. from wild banana (*Musa acuminata*) roots. Afr J Microbiol Res 5:3600–3602

Zakria KU, Ogawa T, Yamamoto A, Saeki Y, Akao S (2008) Influence of inoculation technique on the endophytic colonization of rice by *Pantoea* sp. isolated from sweet potato and by *Enterobacter* sp. isolated from sugarcane Muhammad. Soil Sci Plant Nutr 54:224–236

Rhizobia for Biological Control of Plant Diseases

14

Camila Gazolla Volpiano, Bruno Brito Lisboa,
Camille Eichelberger Granada,
Jackson Freitas Brilhante São José,
Andreia Mara Rotta de Oliveira, Anelise Beneduzi,
Yelena Perevalova, Luciane Maria Pereira Passaglia,
and Luciano Kayser Vargas

Abstract

Rhizobia are a diverse group of nodule-forming bacteria known for inhabiting the soil and establishing functional symbiotic associations with legume plants. Rhizobial inoculants are widely employed in agricultural practices to reduce nitrogen fertilizer inputs on legume crops due to rhizobial ability to fix atmospheric nitrogen. Here we argue that rhizobia should also be considered an alternative method to agricultural pesticides use in plant disease management. Several rhizobial strains have been reported leading to disease resistance, while also promoting plant yield and biomass increases. The biocontrol properties of rhizobia could be associated with lytic enzymes and antimicrobial secondary metabolite production, especially when regarding diseases affecting root systems of plants. Aside from the action of antifungal molecules, suppression of plant diseases could be related to rhizobial plant growth promotion and/or symbiotic efficiency. Moreover, rhizobia have been found to induce systemic resistance to immunize plants, which is a valuable process, considering foliar and viral diseases. This review will focus on rhizobial mechanisms and efficacy to biocontrol diseases caused by different classes of pathogens affecting leguminous and even non-leguminous plants.

C. G. Volpiano · Y. Perevalova · L. M. P. Passaglia
Department of Genetics, Universidade Federal do Rio Grande do Sul – UFRGS,
Porto Alegre, Rio Grande do Sul, Brazil

B. B. Lisboa · J. F. B. S. José · A. M. R. de Oliveira · A. Beneduzi ·
L. K. Vargas (✉)
Departament of Agricultural Research and Diagnosis, Secretariat of Agriculture,
Livestock and Irrigation of Rio Grande do Sul, Porto Alegre, Rio Grande do Sul, Brazil
e-mail: luciano-kayser@seapi.rs.gov.br

C. E. Granada
Universidade do Vale do Taquari – UNIVATES, Lajeado, Rio Grande do Sul, Brazil

V. Kumar et al. (eds.), *Microbiome in Plant Health and Disease*,
https://doi.org/10.1007/978-981-13-8495-0_14

Keywords
Biocontrol · Rhizobium · Legume · Plant disease

14.1 Introduction

In order to sustain the increasing global demand for food, feed and fibre, from the eighteenth century onwards, farmers have been gradually replacing organic agriculture practices with synthetic fertilizers and pesticides. However, the use of agrochemicals has been connected with the contamination of the environment, as well as food products (Aguilar et al. 2017; Blankson et al. 2016; Yadav et al. 2015), poisoning of farmers (Piccoli et al. 2016; Sankoh et al. 2016), development of pesticide resistance (Bass et al. 2015) and elimination of non-target organisms and positive plant-microbe associations (Fox et al. 2007; Franco-Andreu et al. 2016; Rivera-Becerril et al. 2017).

The growing concern over this environmental impact and more strict legislation about agrochemicals stimulates a transition to a sustainable agriculture, which undoubtedly requires the development of alternatives to agrochemicals. One such possibility is the utilization of "biological control agents" or "biocontrol agents" to curtail the population of a pest and/or its effects on crops. An important source of biocontrol agents is the portion of soil in the strict vicinity of plant roots known as the rhizosphere. In the rhizosphere, organisms such as bacteria, fungi, algae and nematodes could affect the physical, chemical and biological properties of soil and plants (Brevik et al. 2015). Some beneficial bacteria could result in changes in the rhizosphere and plants, leading to the improvement of plant development, growth and productivity, as well as inhibiting the development of plant diseases (Glick 2012). Among these beneficial microorganisms, rhizobium is a group comprised of bacteria able to establish symbiotic relationship with leguminous plants, playing a significant role in the maintenance of soil fertility (Herridge et al. 2008). Rhizobia provide fixed nitrogen to leguminous plants as a result of the biological nitrogen fixation (BFN) process, reducing or eliminating synthetic nitrogen fertilizer inputs on crops (Mercante et al. 2017; Zilli et al. 2006). In addition, some rhizobial strains have been described as biocontrol agents, representing an opportunity to also reduce pesticide inputs in agricultural systems. Thus, this review brings together the available information about rhizobia efficacy and mechanisms of control of several diseases caused by different plant pathogens.

14.2 The Well-Known Story of Rhizobia-Legume Symbiosis and Nitrogen Fixation

Rhizobium is a group of nodule-forming bacteria able to establish symbiosis with plants from the Fabaceae (or Leguminosae) family (Lindström and Martinez-Romero 2005). Inside the nodule structure, rhizobia provide fixed nitrogen to the plant in exchange for access to plant-derived carbon sources. In addition, rhizobial strains could also be found in the soil or inhabiting as plant endophytes and/or epiphytes. In some non-legumes rhizobia were also reported promoting plant growth (Antoun et al. 1998; Mitra et al. 2016). Currently, rhizobia consist of several species distributed within at least 14 genera. Reports include a range of α-proteobacteria comprising *Bradyrhizobium*, *Ensifer*, *Mesorhizobium* and *Rhizobium* and a few β-proteobacteria such as *Paraburkholderia* and *Cupriavidus* (Table 14.1).

The symbiotic relationship initiates with an exchange of molecular signals between both partners, leading to mutual recognition and development of symbiotic structures. The symbiotic legume plant secretes flavonoids that elicit the expression of rhizobial nodulation (nod) genes, which are responsible for synthesizing nod factors (NFs). NFs are lipochitooligosaccharides responsible for triggering the nodule developmental process, which could be summarized in the following steps: (i) root hair curling, (ii) formation of the infection thread, (iii) infestation of root cells and continuous cell proliferation and iv) formation of the root nodule (Cerri et al. 2016; Desbrosses and Stougaard 2011). Inside the nodules, rhizobia transform atmospheric nitrogen (N_2) into usable, fixed nitrogen (NH_3), in a complex process known as BNF. This process is carried out by nitrogenases, a family of enzymes that catalyse the breaking of the triple covalent bond of N_2 molecules and the addition of three hydrogen atoms to each nitrogen atom. For further information on root nodule development and BFN, see De Bruijn (2015), Masson-Boivin et al. (2009) and Poole et al. (2018).

As a result of BNF, rhizobia are widely used as a microbial inoculant to enhance legume production in different production systems and are considered the best option among microbial technology in agricultural practices (Vargas et al. 2017). The rhizobial inoculation practice could bring a huge positive financial impact for farmers. In Brazil, rhizobial inoculation of soybean (*G. max*) crops has been reported to supply up to 300 kg of nitrogen per hectare, resulting in savings of approximately US$ 7 billion per year (Hungria et al. 2006; Hungria et al. 2013).

14.3 The Little-Known Story of Rhizobia Protecting Plants from Diseases

Rhizobia inoculants are commercialized worldwide, leading to significant contributions to the productivity of agricultural systems. Although the basic reasoning behind rhizobial application on crops is to increase nitrogen availability, rhizobial strains also have been found to induce resistance to several diseases in leguminous and even non-leguminous plants (Fig. 14.1).

14.4 Rhizobia Effects on Plant Diseases Caused by Fungi

Plants are susceptible to numerous fungal diseases, such as seed and seedling blights, root rots and wilts. Plant-pathogenic fungi are well known to prevent germination, kill seedlings and reduce plant growth. In attempts to reduce fungicide use in the management of fungal diseases, rhizobia have been evaluated as biocontrol

Table 14.1 Genera with known rhizobial bacteria and examples of some rhizobial species and their hosts

Genus	Species	Nodulation host	References
Azorhizobium	*A. caulinodans*	*Sesbania rostrata*	Souza Moreira et al. (2006) and Dreyfus et al. (1988)
	A. doebereinerae	*Sesbania virgata*	
Agrobacterium	*A. deltaense*	*Sesbania cannabina*	Yan et al. (2017a, b)
	A. salinitolerans		
Bradyrhizobium	*B. japonicum*	*Glycine max*	Jordan 1982, Li et al. (2015) and Stajković et al. (2010)
		Vigna angularis	
	B. guangdongense	*Arachis hypogaea*	
		Lablab purpureus	
Paraburkholderia (Burkholderia)	*P. mimosarum*	*Mimosa pigra*	Bournaud et al. (2017), Chen et al. (2006), Dall'Agnol et al. (2016) and Vandamme et al. (2002)
	P. nodosa	*Phaseolus vulgaris*	
	P. piptadeniae	*Piptadenia gonoacantha*	
	P. tuberum	*Aspalathus carnosa*	
Cupriavidus	*C. necator*	*Mimosa caesalpiniaefolia*	Silva et al. (2012)
		Leucaena leucocephala	
		P. vulgaris	
		Vigna unguiculata	
Devosia	*D. neptuniae*	*Neptunia natans*	Bautista et al. (2010) and Rivas et al. (2003)
	D. yakushimensis	*Pueraria lobata*	
Ensifer (Sinorhizobium)	*E. meliloti*	*Medicago truncatula*	Jones et al. (2008), Li et al. (2011) and Rome et al. (1996)
	E. sojae	*G. max*	
		V. unguiculata	
	E. medicae	*Medicago polymorpha*	
Mesorhizobium	*M. loti*	*Lotus japonicus*	Lu et al. (2009), Yokota et al. (2009) and Zhou et al. (2010)
	M. robiniae	*Robinia pseudoacacia*	
	M. shangrilense	*Caragana microphylla*	
		P. vulgaris	
Methylobacterium	*M. nodulans*	*Crotalaria glaucoides*	Jourand et al. (2004)
		Crotalaria perrottetii	
		Crotalaria podocarpa	

(continued)

Table 14.1 (continued)

Genus	Species	Nodulation host	References
Microvirga	*M. lupini*	*Lupinus texensis*	Ardley et al. (2012) and Radl et al. (2014)
	M. lotononidis	*Listia angolensis*	
	M. zambiensis		
	M. vignae	*V. unguiculata*	
Ochrobactrum	*O. lupini*	*Lupinus albus*	Trujillo et al. (2005) and Zurdo-Piñeiro et al. (2007)
	O. cytisi	*P. vulgaris*	
Pararhizobium	*P. giardinii*	*P. vulgaris*	Amarger et al. (1997) and Wang et al. (2011)
	P. herbae	*Albizia julibrissin*	
Phyllobacterium	*P. trifolii*	*Trifolium repens*	Jiao et al. (2015) and Valverde et al. 2005)
		L. albus	
	P. sophorae	*Sophora flavescens*	
Rhizobium	*R. tropici*	*P. vulgaris*	Baraúna et al. (2016), Dall'Agnol et al. (2014), Martínez-Romero et al. (1991) and Taha et al. (2018)
	R. altiplani	*Mimosa pudica*	
	R. laguerreae	*Lens culinaris*	
	R. paranaense	*P. vulgaris*	

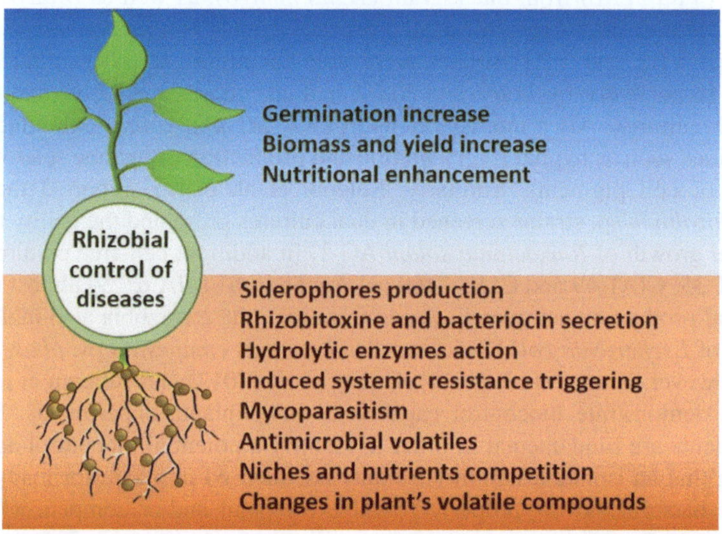

Fig. 14.1 Mechanisms employed by rhizobia leading to plant disease resistance

agents, especially against legume root rots, considering they occupy the same ecological niche as the disease-causing fungi.

Rhizobia were reported to parasitize, deform and inhibit hyphae and reproductive structures of fungi. Tu (1978) evaluated the effect of *B. japonicum* strain on *Phytophthora megasperma* root rot in the soybean. In pot tests with different fungi concentrations, rhizobial population increases corresponded to higher nodulation and plant biomass numbers and also decreases in disease severity. Employing electron microscopy, the rhizobia were found colonizing the surface and the inside of hyphal tips (Tu 1978). Similarly, Ganesan et al. (2007) evaluated six *Rhizobium* sp. (RI-1–RI-6) isolates from the peanut (*A. hypogaea*) for inhibition of *Sclerotium rolfsii* mycelial growth in dual cultures. The most effective rhizobial strains, RI-2 and RI-3, inhibited 60.5% and 62.5% of the mycelium diameter, respectively. These rhizobial strains also reduced up to 11% of *S. rolfsii* growth through volatile compound production. Moreover, peanut plants infected with *S. rolfsii* and treated with RI-2 and RI-3 showed significant increases in shoot length and root length compared to infected controls.

The secretion of hydrolytic enzymes could also be employed by rhizobia in order to antagonize fungal pathogens. Kumar et al. (2011) isolated five bacterial strains (TR1–TR5) from root nodules of fenugreek (*Trigonella foenumgraecum*). Isolates TR1, TR2 and TR4 inhibited the growth of *Fusarium oxysporum*, causing loss of structural integrity of the mycelium, hyphal perforation, lysis, fragmentation and degradation. TR1 and TR4 reported chitinase production, while TR2 reported β-1,3-glucanase activity. Likewise, Dubey et al. (2012) evaluated six *Bradyrhizobium* sp. isolates (VR1–VR6) from black gram (*Vigna mungo*), as well as *Bradyrhizobium* sp. NAIMCC-B-00262 for antifungal properties against *Macrophomina phaseolina*. Rhizobial VR2 and VR1 isolates were able to inhibit *M. phaseolina* mycelial growth, respectively, by 71.5% and 50.5% in dual cultures and 37.6% and 49.2% in cell-free cultures. After interaction with VR2, *M. phaseolina* exhibited several deformities such as hyphal fragmentation and production of hyaline sclerotium due to loss of cell pigments. Similarly, Kelemu et al. (1995) reported that all 15 *Bradyrhizobium* sp. strains screened in dual cultures presented the ability to inhibit mycelial growth of *Rhizoctonia solani* AG-1. In addition, cell-free culture filtrates of CIAT 35, CIAT 49 and CIAT 2469 (=LAB 504 =SEMIA 6129) almost inhibited sclerotial production completely. Aside from this, these rhizobia also inhibited the growth of *Escherichia coli* DH5α and *Xanthomonas campestris* pv. *phaseoli* CIAT 555. However, Kumar et al. (2011), Dubey et al. (2012) and Kelemu et al. (1995) did not demonstrate biocontrol capabilities in in planta experiments. In planta experiments are fundamental in order to identify a potential biocontrol agent considering that an isolate tested in vitro may not succeed in planta for many reasons (i.e. the bacteria did not properly colonize the plant and/or compete with native microbiota).

The production of antimicrobial secondary metabolites has also been reported as a mechanism employed by rhizobial strains to achieve biological control of fungal pathogens. Chakraborty and Purkayastha (1984) reported that a *Bradyrhizobium* strain significantly reduced charcoal rot disease caused by *M. phaseolina* in different soybean cultivars. For example, at 28 incubation days, the root rot indexes of 0.75 and 0.25 correspondingly decrease to 0.50 and 0.10 for the cultivars Soymax and UPSM-19, respectively, while infected controls maintained these indexes. Whole culture extracts of the rhizobial strain yielded a compound identified as rhizobitoxine through chromatographic, ultraviolet and infra-red spectrophotometric analyses. Moreover, a dosage response curve with rhizobitoxine showed the antifungal properties of this compound. There are some other reports describing *Bradyrhizobium* strains that are able to produce rhizobitoxine (Fuhrmann 1990; Owens et al. 1972; Yuhashi et al. 2000).

Some rhizobia also have been shown to produce small, high-affinity iron-chelating compounds known as siderophores (Bhagat et al. 2014; Datta and Chakrabartty 2014; Roy and Chakrabartty 2000), which could increase bacteria competition ability under iron-deficient conditions, consequentially limiting the availability of iron for pathogenic fungi (Crowley 2006; Martínez-Viveros et al. 2010). Siderophore-producing bacteria can also promote plant growth by directly improving iron acquisition by plants (Grobelak and Hiller 2017). Arora et al. (2001) tested 12 rhizobial isolates from *Mucuna pruriens* and found that between all isolates, only the siderophore-producing RMP3 and RMP5 were able to inhibit *M. phaseolina* in vitro growth, with inhibition rates reaching up to 77%. In addition, in peanut plants infected with *M. phaseolina*, treatments with these isolates were able to enhance plant seed germination from 58% (infected control) up to 88.6% and seedling biomass from 3.24 g per plant (infected control) up to 11.78 g per plant. Similarly, Deshwal et al. (2003) evaluated ten strains of peanut-nodulating *Bradyrhizobium* and found three strains (AHR-2, AHR-5 and AHR-6) able to produce siderophores, as well as performing antagonistic action against *M. phaseolina*.

Besides rhizobial ability to stimulate plant growth via nitrogen fixation, rhizobia can also secrete plant growth inducer molecules analogous to plant hormones, such as indole acetic acid (IAA; Ghosh et al. 2015). IAA synthesis is considered a common feature in soil-beneficial bacteria and part of their plant colonization strategy. In a biocontrol context, the phytostimulation action of IAA produced by bacteria could be helpful. Moreover, rhizobia could also directly affect the growth of plant pathogens by IAA production. Volpiano et al. (2018) screened 78 rhizobial strains from the *SEMIA Culture Collection* for antagonism towards *S. rolfsii*. Thirty-three antagonistic strains were detected, 16 of which were able to inhibit more than 84% fungus mycelial growth. Antagonistic strains produced up to 36.5 μg mL^{-1} of IAA. Volpiano et al. (2018) found a direct relationship between in vitro bacterial IAA production and *S. rolfsii* mycelial growth inhibition ($r = 0.447$, $p = 0.011$). The effect of exogenous IAA on the growth of *S. rolfsii* was also studied, and *S. rolfsii* growth was reported decreased at the tested IAA concentrations of 250 (43.8 μg mL^{-1}) and 500 μM (87.6 μg mL^{-1}). Volpiano et al. (2018) selected ten *Rhizobium* spp. antagonistic strains for exploratory in planta tests under pot and field conditions. Common bean plants grown on pots with *S. rolfsii*-infested soil and inoculated with the strains SEMIA 4032, 4077, 4088, 4080 and 4085 presented no disease symptoms. The most efficient strains detected under field conditions, SEMIA 439 and 4088, were reported to decrease disease incidence by 18.3% and 14.5% of the *S. rolfsii*-infested control.

Mixtures of biocontrol agents with different plant colonization patterns are hypothesized to be useful for the biocontrol of various plant pathogens via different disease suppression mechanisms. Yuttavanichakul et al. (2012) assessed 765 peanut-nodulating rhizobial isolates including commercial *Bradyrhizobium* sp. TAL 173 and 350 soil-isolated plant growth-promoting rhizobacteria (PGPR) strains for the ability to inhibit *Aspergillus niger* growth in plate assays. No rhizobia and only 11 PGPR isolates could inhibit *A. niger* growth. However, the effects on the control of crown and root rot disease caused by the antagonist PGPR *Bacillus* spp. strains A20 and A45 were increased with co-inoculation with the rhizobia TAL 173. Individual inoculants of A20 and A45 were able to reduce the disease severity score from 2.83 (uninoculated, infected control) to 1.72. Treatments composed of A20 + TAL173, A45 + TAL173 and A20 + A45 + TAL173 presented 1.16, 0.61 and 0.33 disease severity scores, respectively. However, *Bradyrhizobium* sp. TAL 173 does not have in vitro antagonistic activity against *A. niger*.

Singh et al. (2010) evaluated the nodulating *Rhizobium leguminosarum* strain Rl and arbuscular mycorrhizal fungi (AMF) as biocontrol agents against *F. oxysporum* f. sp. *ciceris* (Foc) in the chickpea (*Cicer arietinum*). R1 was able to maintain infected plants at the same plant height and with the same shoot and root dry weight of uninoculated, uninfected plants. Infected plants co-inoculated with rhizobia and AMF presented plant height and shoot and root dry weight superior from uninoculated, uninfected plants. Akhtar et al. (2010) examined the effects of the inoculation of rhizobial strain *Rhizobium* sp. AQ07, *Bacillus pumilus* (MTCC No. 1640) and *Pseudomonas alcaligenes* (MTCC No. 493) on wilt disease caused by *F. oxysporum* in lentils (*L. culinaris*). The wilting index was reduced from 4 (infected control) to (i) 3 with *B. pumilus* or *P. alcaligenes* individual inoculation, (ii) 1 with *B. pumilus* and *P. alcaligenes* co-inoculation, (iii) 2 with *Rhizobium* inoculation and (iv) 1 with *Rhizobium* inoculated with *B. pumilus* and *P. alcaligenes* (individual or combined). Samavat et al. (2011) reported that treatments with two *P. fluorescens* isolates (UTPF68 and UTPF109) applied individually or in combination with the culture filtrates of five rhizobia isolates (RH3–RH7) were able to reduce the disease caused by *R. solani* (AG-4) on the common bean. RH4 + UTPF109 treatment gave the lowest severity of damping off. Moreover, treatments with both *P. fluorescens* isolates, individually or in combined treatments (especially RH4+UTPF109 and RH6+UTPF68), significantly improved shoot and root weights. These bacteria produce different amounts of siderophores, hydrocyanic acid, indole acetic acid, exopolysaccharides and chitinases.

Rhizobial strains are usually employed in co-inoculation with non-rhizobial antagonistic microorganisms, exclusively attempting nitrogen fixation. However, the biocontrol effects of rhizobia on plant pathogens must not be neglected. An example of this is the detection of *Bradyrhizobium* sp. SEMIA 6144 as a biocontrol agent against *S. rolfsii*. Initially, *Bacillus* sp. CHEP5, a strain antagonistic towards *S. rolfsii* growth, isolated from leaves of peanut, was reported by Tonelli et al. (2011) as able to induce systemic resistance (ISR) in the peanut. Figueredo et al. (2017) evaluated the effect of co-inoculation of CHEP5 with SEMIA 6144, hypothesizing compounding a defensive response (via CHEP5) with nitrogen supply (via SEMIA 6144) in peanut plants. The authors were surprised in finding that plants inoculated with SEMIA 6144 individual cultures presented *S. rolfsii*-promoted disease incidence reduced by 52%, reaching a value similar to CHEP5-inoculated plants. Disease incidences from SEMIA 6144 and CHEP5 co-inoculated plants were not statistically different from SEMIA 6144 or CHEP5 individual inocula treatments.

Interestingly, plants inoculated with a SEMIA 6144 derivative mutant unable to produce Nod factors showed higher disease incidence and lower shoot dry weight than plants inoculated with the wild-type strain, indicating a role for these molecules in the biocontrolling phenotype. In planta experiments were not able to detect "indirect antagonism" mechanisms, such as plant growth promotion, ISR and competition for ecological plant niches and for nutrients (Elbadry et al. 2006; Knudsen et al. 1997; Pang et al. 2009) in dual culture screens.

Notably, rhizobial strains have been reported with biocontrol properties against fungal pathogens even in non-legumes. For example, Chandra et al. (2007) reported that the strong hydrocyanic acid (HCN) producer *Mesorhizobium loti* MP6, isolated from root nodules of *M. pudica*, inhibited up to 75% of *Sclerotinia sclerotiorum* growth. *Brassica campestris* seeds inoculated with rhizobia MP6 presented 70% germination rates, while those grown on *S. sclerotiorum*-infested soil showed only 42% germination rates. Moreover, the incidence of *S. sclerotiorum*-promoted disease declined 99% with MP6 treatment. Similarly, Omar and Abd-Alla (1998) evaluated 21 bacterial strains in order to identify biocontrol agents against *R. solani*, *M. phaseolina* and *Fusarium solani* in okra (*Abelmoschus esculentus*), sunflower (*Helianthus* spp.) and soybean. All the tested strains significantly suppressed the in vitro growth of the three soil-borne root-infecting fungi, with the rhizobial strains *Bradyrhizobium* sp. TAL 377 (= SEMIA 5028 and USDA 138), *Bradyrhizobium* sp. WPBS 3211 D and *Rhizobium* sp. TAL 182 (= SEMIA 4021) being the most effective. TAL 182 treatments presented 100% of relative efficiency (difference in healthy seedlings relative to pathogen control) in okra, while TAL 377 presented 116% and 86% of relative efficiency in soybean and sunflower, respectively.

Despite the promising results obtained so far on rhizobia leading plants to resistance of fungal diseases, studies lack an evaluation of the efficacy variability induced by different field situations. The only study we have found was of Jensen et al. (2002) which conducted field experiments in 1997, 1998 and 1999 at Staples and Verndale (Minnesota, USA) in order to compare the effect of fungicides and other treatments including the inoculation with *R. tropici* UMR 1899 (= CIAT 899, SEMIA 4077, ATCC 49672), on the control of root rot caused by *F. solani* f. sp. *phaseoli* (Fsp11), *F. oxysporum* (Fo11) and *R. solani* AG-4 in the common bean (*P. vulgaris*). In greenhouse studies, rhizobial treatment promoted a 50% reduction in root rot disease severity, aside from increasing shoot and root dry weight. In the first field experiment (Staples 1997), the disease severity and percentage of emergence in plants inoculated with rhizobia were similar to the infected control. However, rhizobial treatment promoted an increase of 765 to 1.463 kg per hectare on yield, a result that was equal or better compared to the treatments with the fungicide Captan 400 with *Bacillus subtilis* GBO3 or *Trichoderma harzianum* T-22 and the fungicide Vitavax 200 when combined with GBO3, which were able to significantly decrease disease severity. In the second field test (Verndale 1998), rhizobial inoculation increased shoot dry weight and promoted a reduction from 6.7 to 4.6 on the root rot disease severity scale. However, the effect on yield was not statistically significantly different from the infected control. In the third field test (Staples 1999), only a coformulation of *B. subtilis* MBI600 with *R. tropici* significantly reduced disease severity and enhanced yield.

As previously mentioned, several modes of action have been identified and are believed to play a role in rhizobial protective effect against plant diseases. However, further studies must be carried out to clearly demonstrate the impact of these modes of action on rhizobial efficacy for biological control. In addition, studies must evaluate the rhizobial effects on plants diseased by a diversified population of pathogenic organisms in order to verify the stability and durability of their efficacy. The importance of identifying types of biological control agents with lower risk of efficacy loss is highlighted by Bardin et al. (2015).

14.5 Rhizobia Effects on Plant Diseases Caused by Bacteria

Similar to rhizobia, plant-pathogenic bacteria establish compatible interactions with plants to obtain nutrients from the host upon colonization. Indeed, rhizobia and pathogenic bacteria have adopted similar strategies to colonize, invade and establish a chronic infection in the plant host (Soto et al. 2006, 2011). The presence or absence of a single gene, i.e. nifH, essential for BNF could be the difference between an efficient rhizobial strain and an opportunistic bacterial strain (Westhoek et al. 2017).

The research on rhizobia effects on diseases caused by plant-pathogenic bacteria is limited; however, rhizobia treatments were already reported to demonstrate biocontrol properties against plant-pathogenic bacteria. Osdaghi et al. (2011) evaluated a *Rhizobium leguminosarum* bv. *phaseoli* strain treatment on the common bacterial blight (CBB) caused by *Xanthomonas axonopodis* pv. *phaseoli* in the common bean. In the tests, one common bean CBB-susceptible cultivar, one susceptible line (cv. Khomein and line Ks21479) and two CBB-tolerant lines (Ks51103 and BF13607) were employed. In greenhouse and field conditions, *R. leguminosarum* bv. *phaseoli* were reported to significantly lower disease severity in the cultivar and 3 lines. Moreover, the effect of *R. leguminosarum* bv. *phaseoli* and urea fertilizer was equal in improving plant dry shoot and root weights, the number of pods per plant and the number of seeds per pod. As mentioned before, rhizobial strains also have been reported to ISR in legumes plants (Elbadry et al. 2006). Considering that ISR is not targeted towards any specific pathogens, it would be a suitable mechanism for rhizobia to protect plants from pathogenic bacteria, especially regarding foliar bacterial diseases. However, the direct action of antimicrobial molecules produced by rhizobia must be evaluated. For example, Mourad et al. (2009) tested rhizobial strains isolated from root nodules of *Medicago ciliaris* and *M. polymorpha* for antimicrobial activity against *Pseudomonas savastanoi*, the agent responsible for olive knot disease. *Rhizobium* sp. ORN 24 and ORN83 produced inhibition halos of 10 and 25 mm in dual cultures with *P. savastanoi*. ORN24 was also found to produce a heat-resistant bacteriocin-like substance.

These studies provide a good preliminary base to justify the study of the effects of rhizobia on bacteria-induced diseases.

14.6 Rhizobia Effects on Plant Diseases Caused by Nematodes

Parasitic nematodes and rhizobia have a common ability to establish interactions with plants, leading to the formation of complex, novel structures on roots. In contrast to rhizobial-induced nodules, root-knot nematodes induce the formation of highly polyploid expanded cells. In many plants, the cortical and pericyclic cells around these giant cells expand and divide, resulting in the formation of galls (Bird and Koltai 2000; Williamson and Hussey 1996). Moreover, nematodes could affect rhizobia-legume symbiosis. Evaluating different plant and nematodes genotypes, Wood et al. (2018) demonstrated that there is a genetic conflict for *M. truncatula* plants between attracting *E. meliloti* Em1022, a highly effective nitrogen fixer, and repelling *Meloidogyne hapla* nematodes. There was a genetic correlation between the number of nodules and the number of galls. Nematode-infected plants formed fewer nodules and had less nodule biomass than uninfected plants. On the other hand, soil nematodes also could mediate the interaction between plant and rhizobia in a positive way. Horiuchi et al. (2005) reported that *Caenorhabditis elegans* are able to act as a vector to transfer *E. meliloti* cells to the roots of *M. truncatula* in response to attractive plant-released volatiles.

Rhizobia strains were already reported acting as biocontrol agents of diseases caused by parasitic nematodes through direct and/or indirect mechanisms. Reitz et al. (2000) demonstrated that lipopolysaccharides secreted by *Rhizobium etli* strain G12 triggered ISR to infection in potato roots against the potato cyst nematode *Globodera pallida*. Siddiqui et al. (2007) evaluated the biocontrol properties of *Meloidogyne javanica* in lentils inoculated with a commercial culture of *Rhizobium* (Lentil strain), *Pseudomonas putida*, *P. alcaligenes*, *Paenibacillus polymyxa* and/or *B. pumilus*. Considering individual inocula, the most effective treatment was *Rhizobium* inoculation, which promoted the reduction from 72 (infested control) to 40 galls per root system and 14960 (infected control) to 7520 nematodes per kg of soil, aside from causing a greater increase in plant growth in absence of *M. javanica*. Considering co-inoculations, the most effective treatment, *Rhizobium* plus *P. putida*, promoted the reduction in galling to 28 galls per root system and nematode population to 5260 nematodes per kg of soil. Ashoub and Amara (2010) demonstrated the ability of a broad bean (*Vicia faba*) *Rhizobium* isolate to achieve 100% *Meloidogyne incognita* juvenile mortality in vitro at 72 h. Moreover, *Rhizobium* inoculation was able to reduce about 96% of galls in eggplant (*Solanum melongena*) roots infected with *M. incognita*.

We consider that the potential qualities of rhizobia as a nematode antagonist present a promising line of research.

14.7 Rhizobia Effects on Plant-Insect Herbivores Interactions

The effect of rhizobia on plant-insect herbivores interactions is largely unexplored. However, rhizobial treatments were already reported to affect (either negatively or positively) insect herbivore interactions with leguminous plants. Kempel et al. (2009) compared the performance of the herbivorous insects *Spodoptera littoralis* and *Myzus persicae* in consuming white clover (*T. repens*) in the presence and absence of a *R. leguminosarum* bv. *trifolii* strain. Two independent greenhouse experiments were conducted: the first with nodulating and non-nodulating acyanogenic white clover strains and the second with acyanogenic and cyanogenic nodulating white clover strains. In the first experiment, the presence of rhizobia in nodulating plants resulted in plant biomass increase. However, rhizobia had a positive effect on *S. littoralis* larval growth and number of *M. persicae* offspring, possibly due to the improved plant food quality. In the second experiment, plant biomass was also increased with rhizobial treatment, but no positive effect on herbivore performance was found on treatment with cyanogenic plant strains.

Kempel et al. (2009) hypothesized that the nitrogen provided in the rhizobial treatment could lead to insect herbivore control as a result of additional production of defensive substances that contain nitrogen such as the cyanogenic compounds. In fact, cyanogenesis in plants is known to highly demand nitrogen, i.e. 15% of leaf nitrogen could be allocated to cyanogenesis in *Eucalyptus cladocalyx* (Gleadow et al. 1998; Miller and Woodrow 2008). In corroboration, Thamer et al. (2011) reported that a root nodule *Rhizobium* spp. isolate improved the resistance of cyanogenic lima bean (*Phaseolus lunatus*) against the insect herbivore *Epilachna varivestis* Mulsant (Mexican bean beetle), as well as promoted plant growth.

In addition to cyanogenic compounds, some legume plants are also able to produce volatile organic compounds (VOCs), which are comprised mainly of fatty acid derivatives, terpenoids, phenyl propanoids and benzenoids (Ballhorn et al. 2011; Winter and Rostás 2010). In lima beans, VOCs were reported repelling herbivorous beetle in response to feeding damage (Heil 2004). Ballhorn et al. (2013) reported that the composition of jasmonic acid (JA)-induced volatiles was different in lima bean plants inoculated with a *Bradyrhizobium* spp. strain. After induction with JA, Mexican bean beetles significantly avoided inoculated plants compared to non-inoculated plants.

14.8 Rhizobia Effects on Parasitic Plants

The protection of legume plants from pathogenic plants is perhaps one of the most unexpected effects of rhizobial inoculation. To our knowledge, up to date, all reports of such cases regard the parasitic broomrape weeds (*Orobanche* spp.), which represent a limiting factor for production of food legumes in Mediterranean areas in Asia and Southern Europe (Abu-Irmaileh 1998).

Orobanche infestation can lead to pea (*Pisum sativum*) yield losses of up to 80% (Rubiales et al. 2003, 2005). Mabrouk et al. (2007b) evaluated the performance of

Orobanche crenata in peas with inoculations of four *Rhizobium* isolates. P.MleTem.92 and P.OM1.92 isolates produced no or few nodules in peas. In contrast, plants treated with P.1236 and P.SOM isolates were reported displaying approximately 30 to 65 nodules per plant, with a twofold increase in shoot dry mass and total nitrogen content in comparison with uninoculated plants. Inoculation with the isolates P.MleTem.92 and P.OM1.92 was not able to influence the germination rate of *O. crenata* seeds germinated in Petri dishes in co-culture with pea plants. However, the weed germination was reduced by a factor of 2.5 and 5 in the presence of P.SOM and P.1236 isolates, respectively. Moreover, the germinated *O. crenata* seeds stopped developing close to the pea roots with P.SOM and P.1236 treatments. In pot experiments, P.OM1.92 and P.MleTem.92 did not reduce pea susceptibility to *O. crenata*. In contrast, tubercles structures that are formed after the parasite penetrates the roots were rarely formed on P-SOM and P1236 treatments. Moreover, the reduction in infection was reported to be associated with an enhanced activity of peroxidase and phenylalanine ammonia lyase, enzymes that are responsible for plant defence reaction. The biocontrol findings with P.SOM and P.1236 were corroborated in an additional report (Mabrouk et al. 2007a). Afterwards, higher concentrations of phenolic compounds and lignin were reported in pea roots inoculated with P.SOM (Mabrouk et al. 2010). Similarly, Bouraoui et al. (2012) evaluated the performance of *Orobanche foetida* in the broad bean with inoculations of ten *Rhizobium* isolates. In hydroponic co-cultures, *O. foetida* germination was significantly decreased by 75% after inoculation with the isolate Mat. Moreover, the Mat isolate promoted an 89% reduction of tubercle number in inoculated broad beans compared to the control plants. In pot experiments, the number of emerged parasites was significantly decreased with all *Rhizobium* isolates inoculation. These provocative findings with *O. foetida* indicate the necessity of further study in other parasitic plant species.

14.9 Rhizobia Effects on Viral Diseases of Plants

Viruses are the smallest of the plant infectious agents. Different from the other crop pathogens presented here, the management of viral diseases must be accomplished through the induction of the plant natural defence since a direct control through chemical application is not yet available. Various reports have been employing non-rhizobial bacteria to promote ISR and achieve protection against viral viruses in non-legume plants. *Pseudomonas fluorescens* strains have been already employed against tomato spotted wilt virus (TSWV) in tomato (*Solanum lycopersicum*) (Kandan et al. 2005), tobacco necrosis virus (TNV) in tobacco (*Nicotiana glutinosa* and *Nicotiana tabacum*) (Maurhofer et al. 1994) and banana bunchy top virus (BBTV) in banana (*Musa* spp.) (Kavino et al. 2008). Treatment with the *Bacillus amyloliquefaciens* strain EXTN-1 was reported to significantly reduce the number of pepper mild mottle virus (PMMoV) symptomatic tobacco plants (Ahn et al. 2002). Even treatments with the plant growth-promoting fungus *Penicillium simplicissimum* GP17-2 were reported as able to promote the ISR against cucumber mosaic virus (CMV) in *Arabidopsis thaliana* and tobacco (Elsharkawy et al. 2013).

The research concerning rhizobia and plants viruses has so far dealt mainly with how some viral diseases affect the nodulation process, BNF and consequently the nitrogen content in a few plants (Abd El-Ghaffar et al. 2011; Chowdhury et al. 1987; Huang 2001; Orellana and Fan 1978; Orellana et al. 1978, 1980; Tu et al. 1970). However, we were able to find a few reports that have employed rhizobial treatments against plant viral diseases. Elbadry et al. (2006) verified the occurrence of ISR against bean yellow mosaic potyvirus (BYMV) in broad bean inoculated with *R. leguminosarum* bv. *viciae* FBG05 and *P. fluorescens* FB11. Plants showed a significant reduction in disease incidence from 91.33% (infected control) to 43% and 27.7% when inoculated with rhizobia and *Pseudomonas* FB11 strains, respectively. Moreover, serological examination for the BYMV concentration in challenged plants was evaluated with DAS-ELISA method, where rhizobia and FB11 treatments showed ELISA values of 0.75 and 0.60, respectively, while the challenged control showed an ELISA value of 1.74. Singh and Srivastava (1983) hypothesized that the increase in nitrogen nutrition, promoted via the *Rhizobium phaseoli* strain Dangeard inoculation, could affect the replication and symptomatic expression of common bean mosaic virus (CBMV) in mung beans (*Vigna radiata*). In pot tests with mung beans grown with differential synthetic nitrogen addition, the CBMV activity coincided with the amount of nitrogen supplied in both uninoculated and inoculated plants. However, rhizobial treatments were able to decrease the mean number of CBMV lesions: from 5.16 to 4.33 at 50 days after inoculation (DAI) with no synthetic nitrogen added and from 23.00 to 18.00 at 40 DAI with 784 mg/L of synthetic nitrogen added. Further investigation is needed about the mechanisms involved in the rhizobial effects on viral propagation on plants.

14.10 Conclusions

Rhizobia are perhaps the most extensively and practically investigated bacteria in agricultural practices due to their ability to form effective symbiosis with leguminous plants. In addition, inoculation with biological control rhizobia represents an efficient, safe and economic alternative to chemical control in plant disease management. However, despite considerable successes achieved thus far, we consider the use of rhizobia for biocontrol of plant diseases is still one of the most underexplored and promising niches in rhizobial research. Undoubtedly, further research is required to reveal further characteristics of rhizobia which could be practically valuable in achieving the maximum benefits from such an organism. Methods to directly identify biocontrol agents active against target pathogens should be employed to detect such agents among the rhizobia already stored in different culture collection centres around the world to broaden the spectrum of viable alternatives to ecologically damaging pesticides. Perhaps, the greatest challenge facing rhizobial application as biocontrol agents will be farmers sticking to the use of familiar pesticides. However, as more data is gathered, and the application of rhizobia as biocontrol agents, as well as an economical alternative to nitrogen supplements, these bacteria could eventually be presented as an irresistible alternative to the agricultural industry.

References

Abd El-Ghaffar MH, Abo-El Maaty SA, Mahmoud SY (2011) Identification and detectability of broad bean stain virus in broad bean seeds and effects on nodulation. Arch Phytopathol Plant Protect 44:390–403. https://doi.org/10.1080/03235400903092925

Abu-Irmaileh B (1998) Present status of *Orobanche* control in the Near East. In: Current problems in *Orobanche* researches. Proceedings of the 4th International *Orobanche* Workshop September 23–26, 1998 Albena, Bulgaria, pp 425–430

Aguilar JAP, Andreu V, Campo J, Picó Y, Masiá A (2017) Pesticide occurrence in the waters of Júcar River, Spain from different farming landscapes. Sci Total Environ 607:752–760. https://doi.org/10.1016/j.scitotenv.2017.06.176

Ahn IP, Park K, Kim CH (2002) Rhizobacteria-induced resistance perturbs viral disease progress and triggers defense-related gene expression. Mol Cells 13:302–308

Akhtar MS, Shakeel U, Siddiqui ZA (2010) Biocontrol of *Fusarium* wilt by *Bacillus pumilus, Pseudomonas alcaligenes* and *Rhizobium* sp. on lentil. Turk J Biol 34:1–7

Amarger N, Macheret V, Laguerre G (1997) *Rhizobium gallicum* sp. nov. and *Rhizobium giardinii* sp. nov., from *Phaseolus vulgaris* nodules. Int J Syst Evol Microbiol 47:996–1006

Antoun H, Beauchamp CJ, Goussard N, Chabot R, Lalande R (1998) Potential of *Rhizobium* and *Bradyrhizobium* species as plant growth promoting rhizobacteria on non-legumes: effect on radishes (*Raphanus sativus* L.). In: Molecular microbial ecology of the soil. Springer, Dordrecht, pp 57–67

Ardley JK, Parker MA, De Meyer SE, Trengove RD, O'Hara GW, Reeve WG, Yates RJ, Dilworth MJ, Willems A, Howieson JG (2012) *Microvirga lupini* sp. nov., *Microvirga lotononidis* sp. nov. and *Microvirga zambiensis* sp. nov. are alphaproteobacterial root-nodule bacteria that specifically nodulate and fix nitrogen with geographically and taxonomically separate legume hosts. Int J Syst Evol Microbiol 62:2579–2588. https://doi.org/10.1099/ijs.0.035097-0

Arora N, Kang S, Maheshwari D (2001) Isolation of siderophore-producing strains of *Rhizobium meliloti* and their biocontrol potential against *Macrophomina phaseolina* that causes charcoal rot of groundnut. Curr Sci 81:673–677

Ashoub A, Amara M (2010) Biocontrol activity of some bacterial genera against root-knot nematode, *Meloidogyne incognita*. J Am Sci 6:321–328

Ballhorn DJ, Reisdorff C, Pfanz H (2011) Quantitative effects of enhanced CO_2 on jasmonic acid induced plant volatiles of lima bean (*Phaseolus lunatus* L.). J Appl Bot Food Qual 84:65–71

Ballhorn DJ, Kautz S, Schädler M (2013) Induced plant defense via volatile production is dependent on rhizobial symbiosis. Oecologia 172:833–846. https://doi.org/10.1007/s00442-012-2539-x

Baraúna AC, Rouws LFM, Simoes-Araujo JL, dos Reis Junior FB, Iannetta PPM, Maluk M, Goi SR, Reis VM, James EK, Zilli JE (2016) *Rhizobium altiplani* sp. nov., isolated from effective nodules on *Mimosa pudica* growing in untypically alkaline soil in central Brazil. Int J Syst Evol Microbiol 66:4118–4124. https://doi.org/10.1099/ijsem.0.001322

Bardin M, Ajouz S, Comby M, Lopez-Ferber M, Graillot B, Siegwart M, Nicot PC (2015) Is the efficacy of biological control against plant diseases likely to be more durable than that of chemical pesticides? Front Plant Sci 6. https://doi.org/10.3389/fpls.2015.00566

Bass C, Denholm I, Williamson MS, Nauen R (2015) The global status of insect resistance to neonicotinoid insecticides. Pestic Biochem Physiol 121:78–87

Bautista VV, Monsalud RG, Yokota A (2010) *Devosia yakushimensis* sp. nov., isolated from root nodules of *Pueraria lobata* (Willd.) Ohwi. Int J Syst Evol Microbiol 60:627–632. https://doi.org/10.1099/ijs.0.011254-0

Bhagat D, Sharma P, Sirari A, Kumawat K (2014) Screening of *Mesorhizobium* spp. for control of *Fusarium wilt* in chickpea *in vitro* conditions. Int J Curr Microbiol Appl Sci 3:923–930

Bird DM, Koltai H (2000) Plant parasitic nematodes: habitats, hormones, and horizontally-acquired genes. J Plant Growth Regul 19:183–194

Blankson G, Osei-Fosu P, Adeendze E, Ashie D (2016) Contamination levels of organophosphorus and synthetic pyrethroid pesticides in vegetables marketed in Accra, Ghana. Food Control 68:174–180. https://doi.org/10.1016/j.foodcont.2016.03.045

Bouraoui M, Abbes Z, Abdi N, Hmissi I, Sifi B (2012) Evaluation of efficient *Rhizobium* isolates as biological control agents of *Orobanche foetida* Poir. parasitizing *Vicia faba* L. minor in Tunisia. Bulg J Agric Sci 18:557–564

Bournaud C, Moulin L, Cnockaert M, de Faria S, Prin Y, Severac D, Vandamme P (2017) *Paraburkholderia piptadeniae* sp. nov. and *Paraburkholderia ribeironis* sp. nov., two root-nodulating symbiotic species of *Piptadenia gonoacantha* in Brazil. Int J Syst Evol Microbiol 67:432–440. https://doi.org/10.1099/ijsem.0.001648

Brevik E, Cerdà A, Mataix-Solera J, Pereg L, Quinton J, Six J, Van Oost K (2015) The interdisciplinary nature of SOIL. Soil 1:117–129. https://doi.org/10.5194/soil-1-117-2015

Cerri MR, Frances L, Kelner A, Fournier J, Middleton PH, Auriac M-C, Mysore KS, Wen J, Erard M, Barker DG, Oldroyd GE, de Carvalho-Niebel F (2016) The symbiosis-related ERN transcription factors act in concert to coordinate rhizobial host root infection. Plant Physiol. https://doi.org/10.1104/pp.16.00230

Chakraborty U, Purkayastha R (1984) Role of rhizobitoxine in protecting soybean roots from *Macrophomina phaseolina* infection. Can J Microbiol 30:285–289

Chandra S, Choure K, Dubey RC, Maheshwari DK (2007) Rhizosphere competent *Mesorhizobiumloti* MP6 induces root hair curling, inhibits *Sclerotinia sclerotiorum* and enhances growth of Indian mustard (*Brassica campestris*). Braz J Microbiol 38:124–130. https://doi.org/10.1590/S1517-83822007000100026

Chen WM, James EK, Coenye T, Chou J-H, Barrios E, de Faria SM, Elliott GN, Sheu S-Y, Sprent JI, Vandamme P (2006) *Burkholderia mimosarum* sp. nov., isolated from root nodules of *Mimosa* spp. from Taiwan and South America. Int J Syst Evol Microbiol 56:1847–1185. https://doi.org/10.1099/ijs.0.64325-0

Chowdhury J, Srivastava R, Singh R (1987) Effect of common bean mosaic virus infection on nitrogenase activity in root nodules of mung bean. J Plant Dis Protect 94:126–129

Crowley DE (2006) Microbial siderophores in the plant rhizosphere. In: Iron nutrition in plants and rhizospheric microorganisms. Springer, Dordrecht, pp 169–198

Dall'Agnol RF, Ribeiro RA, Delamuta JRN, Ormeño-Orrillo E, Rogel MA, Andrade DS, Martínez-Romero E, Hungria M (2014) *Rhizobium paranaense* sp. nov., an effective N_2-fixing symbiont of common bean (*Phaseolus vulgaris* L.) with broad geographical distribution in Brazil. Int J Syst Evol Microbiol 64:3222–3229. https://doi.org/10.1099/ijs.0.064543-0

Dall'Agnol RF, Plotegher F, Souza RC, Mendes IC, dos Reis Junior FB, Béna G, Moulin L, Hungria M (2016) *Paraburkholderia nodosa* is the main N_2-fixing species trapped by promiscuous common bean (*Phaseolus vulgaris* L.) in the Brazilian 'Cerradão'. FEMS Microbiol Ecol 92. https://doi.org/10.1093/femsec/fiw108

Datta B, Chakrabartty PK (2014) Siderophore biosynthesis genes of *Rhizobium* sp. isolated from *Cicer arietinum* L. 3 Biotech 4:391–401. https://doi.org/10.1007/s13205-013-0164-y

De Bruijn FJ (2015) Biological nitrogen fixation. In: Principles of plant-microbe interactions. Springer, Leiden, pp 215–224

Desbrosses GJ, Stougaard J (2011) Root nodulation: a paradigm for how plant-microbe symbiosis influences host developmental pathways. Cell Host Microbe 10:348–358. https://doi.org/10.1016/j.chom.2011.09.005

Deshwal V, Dubey R, Maheshwari D (2003) Isolation of plant growth-promoting strains of *Bradyrhizobium* (*Arachis*) sp. with biocontrol potential against *Macrophomina phaseolina* causing charcoal rot of peanut. Curr Sci:443–448. https://doi.org/10.1016/j.chom.2011.09.005

Dreyfus B, Garcia J-L, Gillis M (1988) Characterization of *Azorhizobium caulinodans* gen. nov., sp. nov., a stem-nodulating nitrogen-fixing bacterium isolated from *Sesbania rostrate*. Int J Syst Evol Microbiol 38:89–98

Dubey P, Gupta G, Dubey R (2012) Culture filtrates of plant growth promoting *Bradyrhizobium* sp. (Vigna) strains VR1 and VR2 inhibit growth and sclerotia germination of *Macrophomina phaseolina in vitro*. New York Sci J 5:1–9

Elbadry M, Taha R, Eldougdoug KA, Gamal-Eldin H (2006) Induction of systemic resistance in faba bean (*Vicia faba* L.) to bean yellow mosaic potyvirus (BYMV) via seed bacterization with plant growth promoting rhizobacteria. J Plant Dis Protect 113:247–251

Elsharkawy M, Shimizu M, Takahashi H, Hyakumachi M (2013) Induction of systemic resistance against Cucumber mosaic virus by *Penicillium simplicissimum* GP17-2 in *Arabidopsis* and tobacco. Plant Pathol 61:964–976. https://doi.org/10.5423/PPJ.SI.07.2012.01

Figueredo MS, Tonelli ML, Ibáñez F, Morla F, Cerioni G, del Carmen Tordable M, Fabra A (2017) Induced systemic resistance and symbiotic performance of peanut plants challenged with fungal pathogens and co-inoculated with the biocontrol agent *Bacillus* sp. CHEP5 and *Bradyrhizobium* sp. SEMIA6144. Microbiol Res 197:65–73. https://doi.org/10.1016/j.micres.2017.01.002

Fox JE, Gulledge J, Engelhaupt E, Burow ME, McLachlan JA (2007) Pesticides reduce symbiotic efficiency of nitrogen-fixing rhizobia and host plants. Proc Natl Acad Sci U S A 104:10282–10287. https://doi.org/10.1073/pnas.0611710104

Franco-Andreu L, Gómez I, Parrado J, García C, Hernández T, Tejada M (2016) Behavior of two pesticides in a soil subjected to severe drought. Effects on soil biology. Appl Soil Ecol 105:17–24. https://doi.org/10.1016/j.apsoil.2016.04.001

Fuhrmann J (1990) Symbiotic effectiveness of indigenous soybean bradyrhizobia as related to serological, morphological, rhizobitoxine, and hydrogenase phenotypes. Appl Environ Microbiol 56:224–229

Ganesan S, Kuppusamy RG, Sekar R (2007) Integrated management of stem rot disease (*Sclerotium rolfsii*) of groundnut (*Arachis hypogaea* L.) using *Rhizobium* and *Trichoderma harzianum* (ITCC-4572). Turk J Agric For 31:103–108

Ghosh PK, Kumar De T, Maiti TK (2015) Production and metabolism of indole acetic acid in root nodules and symbiont (*Rhizobium undicola*) isolated from root nodule of aquatic me-dicinal legume *Neptunia oleracea* Lour. J Bot 2015:1–11. https://doi.org/10.1155/2015/575067

Gleadow R, Foley W, Woodrow I (1998) Enhanced CO_2 alters the relationship between photosynthesis and defence in cyanogenic *Eucalyptus cladocalyx* F. Muell. Plant Cell Environ 21:12–22

Glick BR (2012) Plant growth-promoting bacteria: mechanisms and applications. Scientifica. https://doi.org/10.6064/2012/963401

Grobelak A, Hiller J (2017) Bacterial siderophores promote plant growth: screening of catechol and hydroxamate siderophores. Int J Phytoremediation 19:825–833. https://doi.org/10.1080/15226514.2017.1290581

Heil M (2004) Direct defense or ecological costs: responses of herbivorous beetles to volatiles released by wild lima bean (*Phaseolus lunatus*). J Chem Ecol 30:1289–1295

Herridge DF, Peoples MB, Boddey RM (2008) Global inputs of biological nitrogen fixation in agricultural systems. Plant Soil 311. https://doi.org/10.1007/s11104-008-9668-3

Horiuchi J, Prithiviraj B, Bais HP, Kimball BA, Vivanco JM (2005) Soil nematodes mediate positive interactions between legume plants and rhizobium bacteria. Planta 222:848–857

Huang JS (2001) Rhizobium-legume symbiosis and the effects of diseases on nodulation and nitrogen fixation. In: Plant pathogenesis and resistance. Springer, Dordrecht, pp 175–236

Hungria M, Franchini JC, Campo RJ, Crispino CC, Moraes JZ, Sibaldelli RNR, Mendes IC, Arihara J (2006) Nitrogen nutrition of soybean in Brazil: contributions of biological N_2 fixation and N fertilizer to grain yield. Can J Plant Sci 86:927–939

Hungria M, Nogueira MA, Araujo RS (2013) Co-inoculation of soybeans and common beans with rhizobia and azospirilla: strategies to improve sustainability. Biol Fertil Soils 49:791–801. https://doi.org/10.1007/s00374-012-0771-5

Jensen CE, Percich J, Graham P (2002) Integrated management strategies of bean root rot with *Bacillus subtilis* and *Rhizobium* in Minnesota. Field Crops Res 74:107–115

Jiao YS, Yan H, Jil HJ, Liu YH, Sui XH, Zhang XX, Wang ET, Chen WX, Chen WF (2015) *Phyllobacterium sophorae* sp. nov., a symbiotic bacterium isolated from root nodules of *Sophora flavescens*. Int J Syst Evol Microbiol 65:399–406. https://doi.org/10.1099/ijs.0.067017-0

Jones KM, Sharopova N, Lohar DP, Zhang JQ, VandenBosch KA, Walker GC (2008) Differential response of the plant *Medicago truncatula* to its symbiont *Sinorhizobium meliloti* or an exopolysaccharide-deficient mutant. Proc Natl Acad Sci U S A 105:704–709. https://doi.org/10.1073/pnas.0709338105

Jordan D (1982) Transfer of *Rhizobium japonicum* Buchanan 1980 to *Bradyrhizobium* gen. nov., a genus of slow-growing, root nodule bacteria from leguminous plants. Int J Syst Evol Microbiol 32:136–139

Jourand P, Giraud E, Béna G, Sy A, Willems A, Gillis M, Dreyfus B, de Lajudie P (2004) *Methylobacterium nodulans* sp. nov., for a group of aerobic, facultatively methylotrophic, legume root-nodule-forming and nitrogen-fixing bacteria. Int J Syst Evol Microbiol 54:2269–2273. https://doi.org/10.1099/ijs.0.02902-0

Kandan A, Ramiah M, Vasanthi V, Radjacommare R, Nandakumar R, Ramanathan A, Samiyappan R (2005) Use of *Pseudomonas fluorescens*-based formulations for management of tomato spotted wilt virus (TSWV) and enhanced yield in tomato. Biocontrol Sci Technol 15:553–569

Kavino M, Harish S, Kumar N, Saravanakumar D, Samiyappan R (2008) Induction of systemic resistance in banana (*Musa* spp.) against Banana bunchy top virus (BBTV) by combining chitin with root-colonizing *Pseudomonas fluorescens* strain CHA0. Eur J Plant Pathol 120:353–362. https://doi.org/10.1007/s10658-007-9223-8

Kelemu S, Thomas RJ, Moreno CX, Ocampo GI (1995) Strains of *Bradyrhizobium* from tropical forage legumes inhibit *Rhizoctonia solani* AG-1 *in vitro*. Australas Plant Pathol 24:168–172

Kempel A, Brandl R, Schädler M (2009) Symbiotic soil microorganisms as players in aboveground plant–herbivore interactions–the role of rhizobia. Oikos 118:634–640. https://doi.org/10.1111/j.1600-0706.2009.17418.x

Knudsen IMG, Hockenhull J, Funck Jensen D, Gerhardson B, Hökeberg M, Tahvonen R, Teperi E, Sundheim L (1997) Selection of biological control agents for controlling soil and seed-borne diseases in the field. Eur J Plant Pathol 103:775–784

Kumar H, Dubey R, Maheshwari D (2011) Effect of plant growth promoting rhizobia on seed germination, growth promotion and suppression of *Fusarium* wilt of fenugreek (*Trigonella foenum-graecum* L.). Crop Prot 30:1396–1403. https://doi.org/10.1016/j.cropro.2011.05.001

Li QQ, Wang ET, Chang YL, Zhang YZ, Zhang YM, Sui XH, Chen WF, Chen WX (2011) *Ensifer sojae* sp. nov., isolated from root nodules of *Glycine max* grown in saline-alkaline soils. Int J Syst Evol Microbiol 61:1981–1988. https://doi.org/10.1099/ijs.0.025049-0

Li YH, Wang R, Zhang XX, Young JPW, Wang ET, Sui XH, Chen WX (2015) *Bradyrhizobium guangdongense* sp. nov. and *Bradyrhizobium guangxiense* sp. nov., isolated from effective nodules of peanut. Int J Syst Evol Microbiol 65:4655–4661. https://doi.org/10.1099/ijsem.0.000629

Lindström K, Martinez-Romero M (2005) International Committee on Systematics of Prokaryotes; Subcommittee on the taxonomy of *Agrobacterium* and *Rhizobium*. Int J Syst Evol Microbiol 55:1383–1383

Lu YL, Chen WF, Wang ET, Han LL, Zhang XX, Chen WX, Han SZ (2009) *Mesorhizobium shangrilense* sp. nov., isolated from root nodules of *Caragana* species. Int J Syst Evol Microbiol 59:3012–3018. https://doi.org/10.1099/ijs.0.007393-0

Mabrouk Y, Zourgui L, Sifi B, Belhadj O (2007a) The potential of *Rhizobium* strains for biological control of *Orobanche crenata*. Biologia 62:139–143

Mabrouk Y, Zourgui L, Sifi B, Delavault P, Simier P, Belhadj O (2007b) Some compatible *Rhizobium leguminosarum* strains in peas decrease infections when parasitised by *Orobanche crenata*. Weed Res 47:44–53. https://doi.org/10.1111/j.1365-3180.2007.00548.x

Mabrouk Y, Mejri S, Hemissi I, Simier P, Delavault P, Saidi M, Belhadj O (2010) Bioprotection mechanisms of pea plant by *Rhizobium leguminosarum* against *Orobanche crenata*. Afr J Microbiol Res 4:2570–2575

Martínez-Romero E, Segovia L, Mercante FM, Franco AA, Graham P, Pardo MA (1991) *Rhizobium tropici*, a novel species nodulating *Phaseolus vulgaris* L. beans and *Leucaena* sp. Trees. Int J Syst Evol Microbiol 41:417–426

Martínez-Viveros O, Jorquera M, Crowley D, Gajardo G, Mora M (2010) Mechanisms and practical considerations involved in plant growth promotion by rhizobacteria. J Soil Sci Plant Nut 10:293–319. https://doi.org/10.4067/S0718-95162010000100006

Masson-Boivin C, Giraud E, Perret X, Batut J (2009) Establishing nitrogen-fixing symbiosis with legumes: how many rhizobium recipes? Trends Microbiol 17:458–466. https://doi.org/10.1016/j.tim.2009.07.004

Maurhofer M, Hase C, Meuwly P, Metraux J-P, Defago G (1994) Induction of systemic resistance of tobacco to tobacco necrosis virus by the root-colonizing *Pseudomonas fluorescens* strain CHA0: Influence of the gacA gene and of pyoverdine production. Phytopathology 84:139–147

Mercante FM, Otsubo AA, Brito OR (2017) New native rhizobia strains for inoculation of common bean in the Brazilian savanna. Rev Bras Ciênc Solo 41. https://doi.org/10.1590/18069657rbcs20150120

Miller RE, Woodrow IE (2008) Resource availability and the abundance of an N-based defense in Australian tropical rain forests. Ecology 89:1503–1509

Mitra S, Mukherjee A, Wiley-Kalil A, Das S, Owen H, Reddy PM, Ané J-M, James EK, Gyaneshwar P (2016) A rhamnose deficient lipopolysaccharide mutant of *Rhizobium* sp. IRBG74 is defective in root colonization and beneficial interactions with its flooding-tolerant hosts *Sesbania cannabina* and wetland rice. J Exp Bot 67:5869–5884. https://doi.org/10.1093/jxb/erw354

Mourad K, Fadhila K, Chahinez M, Meriem R, Philippe DL, Abdelkader B (2009) Antimicrobial activities of *Rhizobium* sp. strains against *Pseudomonas savastanoi*, the agent responsible for the olive knot disease in Algeria. Grasas y Aceites 60:139–146

Omar S, Abd-Alla M (1998) Biocontrol of fungal root rot diseases of crop plants by the use of rhizobia and bradyrhizobia. Folia Microbiolo 43:431–437

Orellana R, Fan F (1978) Nodule infection by bean yellow mosaic virus in *Phaseolus vulgaris*. Appl Environ Microbiol 36:814–818

Orellana R, Fan F, Sloger C (1978) Tobacco ringspot virus and *Rhizobium* interactions in soybean: impairment of leghemoglobin accumulation and nitrogen fixation. Phytopathology 68:577–582

Orellana R, Weber D, Cregan P (1980) N_2-fixing competence of *Rhizobium japonicum* strains in soybean infected with tobacco ringspot virus. Physiol Mol Plant Pathol 17:381–388

Osdaghi E, Shams-Bakhsh M, Alizadeh A, Mr L, Hatami Maleki H (2011) Induction of resistance in common bean by *Rhizobium leguminosarum* bv. *phaseoli* and decrease of common bacterial blight. Phytopathol Mediterr 50:45–54. https://doi.org/10.14601/Phytopathol_Mediterr-8524

Owens LD, Thompson JF, Pitcher R, Williams T (1972) Structure of rhizobitoxine, an antimetabolic enol-ether amino-acid from *Rhizobium japonicum*. J Chem Soc Chem Commun:714–714

Pang Y, Liu X, Ma Y, Chernin L, Berg G, Gao K (2009) Induction of systemic resistance, root colonisation and biocontrol activities of the rhizospheric strain of *Serratia plymuthica* are dependent on N-acyl homoserine lactones. Eur J Plant Pathol 124:261–268. https://doi.org/10.1007/s10658-008-9411-1

Piccoli C, Cremonese C, Koifman RJ, Koifman S, Freire C (2016) Pesticide exposure and thyroid function in an agricultural population in Brazil. Environ Res 151:389–398. https://doi.org/10.1016/j.envres.2016.08.011

Poole P, Ramachandran V, Terpolilli J (2018) Rhizobia: from saprophytes to endosymbionts. Nat Rev Microbiol 16:29–303. https://doi.org/10.1038/nrmicro.2017.171

Radl V, Simões-Araújo JL, Leite J, Passos SR, Martins LMV, Xavier GR, Rumjanek NG, Baldani JI, Zilli JE (2014) *Microvirga vignae* sp. nov., a root nodule symbiotic bacterium isolated from cowpea grown in semi-arid Brazil. Int J Syst Evol Microbiol 64:725–730. https://doi.org/10.1099/ijs.0.053082-0

Reitz M, Rudolph K, Schröder I, Hoffmann-Hergarten S, Hallmann J, Sikora R (2000) Lipopolysaccharides of *Rhizobium etli* strain G12 act in potato roots as an inducing agent of systemic resistance to infection by the cyst nematode *Globodera pallida*. Appl Environ Microbiol 66:3515–3518

Rivas R, Willems A, Subba-Rao NS, Mateos PF, Dazzo FB, Kroppenstedt RM, Martínez-Molina E, Gillis M, Velázquez E (2003) Description of *Devosia neptuniae* sp. nov. that nodulates and fixes nitrogen in symbiosis with *Neptunia natans*, an aquatic legume from India. Syst Appl Microbiol 26:47–53

Rivera-Becerril F, van Tuinen D, Chatagnier O, Rouard N, Béguet J, Kuszala C, Soulas G, Gianinazzi-Pearson V, Martin-Laurent F (2017) Impact of a pesticide cocktail (fenhexamid, folpel, deltamethrin) on the abundance of *Glomeromycota* in two agricultural soils. Sci Total Environ 577:84–93. https://doi.org/10.1016/j.scitotenv.2016.10.098

Rome S, Fernandez MP, Brunel B, Normand P, Cleyet-Marel JC (1996) *Sinorhizobium medicae* sp. nov., isolated from annual *Medicago* spp. Int J Syst Evol Microbiol 46:972–980

Roy N, Chakrabartty PK (2000) Effect of aluminum on the production of siderophore by *Rhizobium* sp.(*Cicer arietinum*). Curr Microbiol 41:5–10

Rubiales D, Pérez-de-Luque A, Cubero J, Sillero J (2003) Crenate broomrape (*Orobanche crenata*) infection in field pea cultivars. Crop Prot 22:865–872

Rubiales D, Moreno M, Sillero J (2005) Search for resistance to crenate broomrape (*Orobanche crenata* Forsk.) in pea germplasm. Genet Resour Crop Evol 52:853–861

Samavat S, Besharati H, Behboudi K (2011) Interactions of rhizobia cultural filtrates with *Pseudomonas fluorescens* on bean damping-off control. J Agric Sci Technol 13:965–976

Sankoh AI, Whittle R, Semple KT, Jones KC, Sweetman AJ (2016) An assessment of the impacts of pesticide use on the environment and health of rice farmers in Sierra Leone. Environ Int 94:458–466. https://doi.org/10.1016/j.envint.2016.05.034

Siddiqui ZA, Baghel G, Akhtar M (2007) Biocontrol of *Meloidogyne javanica* by *Rhizobium* and plant growth-promoting rhizobacteria on lentil. World J Microbiol Biotechnol 23:435–441. https://doi.org/10.1007/s11274-006-9244-z

Silva K, Florentino LA, Silva KB, De Brandt E, Vandamme P, de Souza Moreira FM (2012) *Cupriavidus necator* isolates are able to fix nitrogen in symbiosis with different legume species. Syst Appl Microbiol 35:175–182. https://doi.org/10.1016/j.syapm.2011.10.005

Singh R, Srivastava R (1983) Effect of nitrogen nutrition and *Rhizobium* on multiplication and symptom expression of common bean mosaic virus in mung bean. J Plant Dis Prot 207:212

Singh PK, Singh M, Vyas D (2010) Biocontrol of *Fusarium* wilt of chickpea using arbuscular mycorrhizal fungi and *Rhizobium leguminosorum* Biovar. Caryologia 63:349–353. https://doi.org/10.1080/00087114.2010.10589745

Soto MJ, Sanjuan J, Olivares J (2006) Rhizobia and plant-pathogenic bacteria: common infection weapons. Microbiology 152:3167–3174. https://doi.org/10.1099/mic.0.29112-0

Soto M, Nogales J, Pérez-Mendoza D, Gallegos M-T, Olivares J, Sanjuán J (2011) Pathogenic and mutualistic plant-bacteria interactions: ever increasing similarities. Open Life Sci 6:911–917. https://doi.org/10.2478/s11535-011-0069-x

Souza Moreira FM et al (2006) *Azorhizobium doebereinerae* sp. nov. microsymbiont of *Sesbania virgata* (Caz.) Pers. Syst Appl Microbiol 29:197–206. https://doi.org/10.1016/j.syapm.2005.09.004

Stajković O, Kuzmanović D, Miličić B (2010) Nodulation and N_2 fixation effectiveness of *Bradyrhizobium* strains in symbiosis with Adzuki Bean, *Vigna angularis*. Braz Arch Biol Technol 53:293–299. https://doi.org/10.1590/S1516-89132010000200007

Taha K, Berraho EB, El Attar I, Dekkiche S, Aurag J, Béna G (2018) *Rhizobium laguerreae* is the main nitrogen-fixing symbiont of cultivated lentil (*Lens culinaris*) in Morocco. Syst Appl Microbiol 41:113–121. https://doi.org/10.1016/j.syapm.2017.09.008

Thamer S, Schädler M, Bonte D, Ballhorn DJ (2011) Dual benefit from a belowground symbiosis: nitrogen fixing rhizobia promote growth and defense against a specialist herbivore in a cyanogenic plant. Plant Soil 341:209–219. https://doi.org/10.1007/s11104-010-0635-4

Tonelli ML, Furlan A, Taurian T, Castro S, Fabra A (2011) Peanut priming induced by biocontrol agents. Physiol Mol Plant Pathol 75:100–105. https://doi.org/10.1016/j.pmpp.2010.11.001

Trujillo ME, Willems A, Abril A, Planchuelo A-M, Rivas R, Ludeña D, Mateos PF, Martínez-Molina E, Velázquez E (2005) Nodulation of *Lupinus albus* by strains of *Ochrobactrum lupini* sp. nov. Appl Environ Microbiol 71:1318–1327. https://doi.org/10.1128/AEM.71.3.1318-1327.2005

Tu JC (1978) Protection of soybean from severe *Phytophthora* root rot by *Rhizobium*. Physiol Mol Plant Pathol 12:233–240

Tu J, Ford R, Grau C (1970) Some factors affecting the nodulation and nodule efficiency in soybeans infected by soybean mosaic virus. Phytopathology 60:1653–1656

Valverde A, Velázquez E, Fernández-Santos F, Vizcaíno N, Rivas R, Mateos PF, Martínez-Molina E, Igual JM, Willems A (2005) *Phyllobacterium trifolii* sp. nov., nodulating *Trifolium* and *Lupinus* in Spanish soils. Int J Syst Evol Microbiol 55:1985–1989. https://doi.org/10.1099/ijs.0.63551-0

Vandamme P, Goris J, Chen W-M, De Vos P, Willems A (2002) *Burkholderia tuberum* sp. nov. and *Burkholderia phymatum* sp. nov., nodulate the roots of tropical. Syst Appl Microbiol 25:507–512. https://doi.org/10.1078/07232020260517634

Vargas LK, Volpiano CG, Lisboa BB, Giongo A, Beneduzi A, Passaglia LMP (2017) Potential of rhizobia as plant growth-promoting rhizobacteria. In: Microbes for legume improvement. Springer, Vienna, pp 153–174

Volpiano CG, Lisboa BB, São José JFB, de Oliveira AMR, Beneduzi A, Passaglia LMP, Vargas LK (2018) *Rhizobium* strains in the biological control of the phytopathogenic fungi *Sclerotium (Athelia) rolfsii* on the common bean. Plant Soil. https://doi.org/10.1007/s11104-018-3799-y

Wang ET, Chen WF, Sui XH, Zhang XX, Liu HC, Chen WX (2011) *Rhizobium herbae* sp. nov. and *Rhizobium giardinii*-related bacteria, minor microsymbionts of various wild legumes in China. Int J Syst Evol Microbiol 61:1912–1920. https://doi.org/10.1099/ijs.0.024943-0

Westhoek A, Field E, Rehling F, Mulley G, Webb I, Poole PS, Turnbull LA (2017) Policing the legume-*Rhizobium* symbiosis: a critical test of partner choice. Sci Rep 7:1419. https://doi.org/10.1038/s41598-017-01634-2

Williamson VM, Hussey RS (1996) Nematode pathogenesis and resistance in plants. Plant Cell 8:1735

Winter TR, Rostás M (2010) Nitrogen deficiency affects bottom-up cascade without disrupting indirect plant defense. J Chem Ecol 36:642–651. https://doi.org/10.1007/s10886-010-9797-z

Wood CW, Pilkington BL, Vaidya P, Biel C, Stinchcombe JR (2018) Genetic conflict with a parasitic nematode disrupts the legume–rhizobia mutualism. Evol Lett 2:233–245. https://doi.org/10.1002/evl3.51

Yadav IC, Devi NL, Syed JH, Cheng Z, Li J, Zhang G, Jones KC (2015) Current status of persistent organic pesticides residues in air, water, and soil, and their possible effect on neighboring countries: A comprehensive review of India. Sci Total Environ 511:123–137. https://doi.org/10.1016/j.scitotenv.2014.12.041

Yan J, Li Y, Han XZ, Chen WF, Zou WX, Xie Z, Li M (2017a) *Agrobacterium deltaense* sp. nov., an endophytic bacteria isolated from nodule of *Sesbania cannabina*. Arch Microbiol 199:1003–1009. https://doi.org/10.1007/s00203-017-1367-0

Yan J, Li Y, Yan H, Chen WF, Zhang X, Wang ET, Han XZ, Xie ZH (2017b) *Agrobacterium salinitolerans* sp. nov., a saline alkaline tolerant bacterium isolated from root nodule of *Sesbania cannabina*. Int J Syst Evol Microbiol 67:1906–1911. https://doi.org/10.1099/ijsem.0.001885

Yokota K, Fukai E, Madsen LH, Jurkiewicz A, Rueda P, Radutoiu S, Held M, Hossain MS, Szczyglowski K, Morieri G, Oldroyd GED, Downie JA, Nielsen MW, Rusek AM, Sato S, Tabata S, James EK, Oyaizu H, Sandal N, Stougaard J (2009) Rearrangement of actin cytoskeleton mediates invasion of *Lotus japonicus* roots by *Mesorhizobium loti*. Plant Cell 21:267–284. https://doi.org/10.1105/tpc.108.063693

Yuhashi KI, Ichikawa N, Ezura H, Akao S, Minakawa Y, Nukui N, Yasuta T, Minamisawa K (2000) Rhizobitoxine production by *Bradyrhizobium elkanii* enhances nodulation and competitiveness on *Macroptilium atropurpureum*. Appl Environ Microbiol 66:2658–2663

Yuttavanichakul W, Lawongsa P, Wongkaew S, Teaumroong N, Boonkerd N, Nomura N, Tittabutr P (2012) Improvement of peanut rhizobial inoculant by incorporation of plant growth promoting rhizobacteria (PGPR) as biocontrol against the seed borne fungus, *Aspergillus niger*. Biol Control 63:87–97. https://doi.org/10.1016/j.biocontrol.2012.06.008

Zhou PF, Chen WM, Wei GH (2010) *Mesorhizobium robiniae* sp. nov., isolated from root nodules of *Robinia pseudoacacia*. Int J Syst Evol Microbiol 60:2552–2556. https://doi.org/10.1099/ijs.0.019356-0

Zilli JÉ, Valicheski RR, Rumjanek NG, Simões-Araújo JL, Freire Filho FR, Neves MCP (2006) Eficiência simbiótica de estirpes de *Bradyrhizobium* isoladas de solo do Cerrado em caupi. PAB 41:811–818

Zurdo-Piñeiro JL, Rivas R, Trujillo ME, Vizcaíno N, Carrasco JA, Chamber M, Palomares A, Mateos PF, Martínez-Molina E, Velázquez E (2007) *Ochrobactrum cytisi* sp. nov., isolated from nodules of *Cytisus scoparius* in Spain. Int J Syst Evol Microbiol 57:784–788. https://doi.org/10.1099/ijs.0.64613-0

15 Bioactive Compounds Produced by Biocontrol Agents Driving Plant Health

Miguel O. P. Navarro, Amanda C. M. Piva, Ane S. Simionato, Flávia R. Spago, Fluvio Modolon, Janaina Emiliano, Anabela Marisa Azul, Andreas Lazaros Chryssafidis, and Galdino Andrade

Abstract

In nature, biocontrol of particular organisms by certain microbial agents depends essentially on competition for space and resources. Microbial metabolism provides a large number of bioactive compounds that can be used in control of plant diseases, mainly produced by insects, nematodes, viruses, fungi, and bacteria. Bioprospecting for microbial bioactive compounds from biocontrol agents is one of the alternatives currently being studied for plant protection, especially in species of agronomic importance. Here, we review several biological compounds and how, in general, they were discovered and have been used to improve plant health.

Keywords

Bioactive compounds · Biocontrol · Bacteria · Plant health

M. O. P. Navarro · A. C. M. Piva · A. S. Simionato · F. Modolon · J. Emiliano · G. Andrade (✉)
Microbial Ecology Laboratory, Department of Microbiology, State University of Londrina, Londrina, Paraná, Brazil
e-mail: andradeg@uel.br

F. R. Spago
Microbial Ecology Laboratory, Department of Fishing Engineering, Federal Institute of Espírito Santo, Campus Piúma, Piúma, Brazil

A. M. Azul
Centre for Functional Ecology (CFE), Department of Life Sciences, University of Coimbra, Coimbra, Portugal

A. L. Chryssafidis
Agroveterinary Sciences Center, State University of Santa Catarina, Lages, Santa Catarina, Brazil

V. Kumar et al. (eds.), *Microbiome in Plant Health and Disease*,
https://doi.org/10.1007/978-981-13-8495-0_15

15.1 Introduction

Plant diseases due to infections caused by fungi, bacteria, and viruses are of great importance in agriculture. Phytopathogen microorganisms cause different types of diseases that limit agricultural productivity, with direct decreases in quantity and quality, causing significant economic losses. However, the pesticide inputs used against phytopathogens also influence nontarget saprophytic microbes and beneficial microbiomes. Use of chemical products is the main strategy for controlling phytopathogens to guarantee high yields and the quality of the products. Ecological concerns about using chemicals to control pests, and their growing costs, have encouraged the search for alternatives that are cost effective as well as being less harmful or not harmful to the environment and to the health of living beings.

Thus, the boundary is the cost, and researchers are looking for low-cost options with a low environmental impact. One of the most interesting alternatives is biological control, or biocontrol, defined as use of resident or introduced living organisms, or parts of them, to suppress population density or the impact of a specific organism, making it less abundant or less harmful (Eilenberg et al. 2001). Direct action occurs by several mechanisms such as production of antimicrobial compounds, insecticides, or nematicides; production of hydrolytic enzymes; competition and/or immobilization of micronutrients such as Fe by siderophore production; competition for space; inactivation of germination factors; degradation of pathogenic factors; or parasitism of the pathogen (Govindasamy et al. 2011).

Bioprospecting for microbial bioactive compounds (MBCs) produced by biocontrol agents is very important to drive plant health. In this chapter, we review several microbial biological control products and how, in general, they were discovered and have been used to improve plant health.

15.2 Brief History of Bioactive Compounds Researched for Plant Protection

The main problems related to the extensive use of chemical fungicides in agriculture include selection of resistant phytopathogens, contamination of soil and the environment, toxicity for beneficial microbial communities present in the soil, negative effects on the ozone layer, and hazards to human and animal health. Therefore, is important to include toxicity assessments in protocols for selection and characterization of new MBCs before applying them in the field.

The general data requirements to support the registration of an MBC at the US Environmental Protection Agency include the biochemical product characterization and mammalian toxicology, non-target-organism testing studies, identification of allergenicity potential, and product performance in terms of environmental and public health (Leahy et al. 2014). The stages in developing a new MBC are divided into several phases, the first stage being the discovery phase (which includes bioprospecting, production, characterization, and bioactivity in vitro) and the phase of efficacy and toxicology testing.

The discovery phase involves basic research for bioprospecting to select strains, production, and purification and bioactivity evaluation in vitro. This phase is also the first boundary to select the best candidates to produce a new MBC. Actually, from the approximately 23,000 MBCs known, only about 150 compounds have direct uses in human medicine, veterinary medicine, and/or agriculture (Bérdy 2012).

Bioprospecting is known as a system to locate, evaluate, and systematically explore the microbial diversity that exists in the environment, with exploration of genetic and biochemical resources for commercial purposes as the main goal. Bioprospecting strategies employ the biodiversity of culturable and nonculturable microorganisms to identify genes, enzymes, metabolites, and/or microorganisms for biotechnological applications (Simionato et al. 2017a, b).

Microbial bioprospecting is divided into two categories depending on whether the microorganism is culturable or nonculturable. In the case of a culturable microorganism, the process involves its cultivation and storage in a collection base with a comprehensive description of its respective habitat(s) for the purpose of documenting the microbial biodiversity on Earth. Selection of microorganisms from specific environmental conditions contributes to facilitating their optimization for biotechnological applications on the basis of documented information about physiology, metabolism, and microbial ecology through use of large-scale phenotyping techniques (Simionato et al. 2017a, b).

Bioprospecting for nonculturable microorganisms cannot use current techniques. Uncultivated bacteria are the most abundant species on the planet; a common problem of uncultivated bacteria is that they do not grow in laboratory conditions. The reason is that the chief growth factors and nutrients that are present in the natural environment are produced by other organisms. Researchers have opted mainly to use metagenomic, coculture helping, and in situ culture for microbial "domestication." The development of those innovative culture methods has allowed access to new soil and marine microorganisms for further biotechnological applications in industry or in research (Kaeberlein et al. 2002).

Subsequent to bioprospecting, large-scale production, analytical methods, characterization, the identity, and the biological properties of the MBC must be evaluated to guarantee the safety and economic sustainability of the product. During bioprospecting, production or purification processes are important to monitor the activity of the MBC. Thus, it becomes necessary to conduct antimicrobial susceptibility tests against target microorganisms, to carry out bioassays, and to determine the resistance against the product. The major susceptibility and resistance tests are detailed in Table 15.1. In the same way, data from small-scale laboratory experiments and representative field trials are necessary to ensure that the agent or product is effective for plant protection.

Furthermore, toxicity and ecotoxicology testing, crop residue analyses, and testing of the environmental impact on nontarget microorganisms are necessary for MBC registration. All tests are described in Table 15.1. Therefore, it is imperative that the processes of production, characterization, toxicity, regulation, and use of products based on the MBC are well detailed before marketing of the product.

Table 15.1 Requirements of biological control agents

Tests	Objectives	Methods	References
Antagonistic test	Instances of antibiosis to determine the ability of the MBC to suppress pathogen growth	Direct antagonism, agar diffusion well, bioautography, agar disk diffusion, broth dilution	Balouiri et al. (2016)
Production	Microorganism or metabolite production by a specific fermentation process	Fermentation process	Simionato et al. (2017a, b)
Identification	Structural identification, taxonomic description, and species affiliation of the MBC at the strain level	Chromatography, spectroscopy (mass spectrometry, nuclear magnetic resonance, x-ray, infrared), sequencing, computational methods	Marx (2016) and Simionato et al. (2017a, b)
Biological properties	Provision of information on biology, biogeography, and ecology to determine the effects and risks of the MBC	Cytotoxicity, activity, omics analysis, diverse array technology, microarrays, microscopy methods	OECD (2016)
Mammalian health	Identification of aspects of the MBC needed to ensure animal and human protection	Acute toxicity, higher-tier studies, sensitization, genotoxicity	WHO (2017)
Residues	Description of quantities of the microbiological agent or its toxins remaining in food products	Analytical methods (e.g., liquid chromatography–mass spectrometry)	FAO/WHO (2017) and OECD (2016)
Environment and ecotoxicology	Evaluation of the risks of using the MBC in ecological compartments other than those where it naturally occurs	Specificity, pathogenicity, effects on nontarget organisms	FAO/WHO (2017)

MBC microbial bioactive compound

15.3 Microbial Bioactive Compounds Produced by Gram-Negative Bacteria

15.3.1 *Pseudomonas* Species

Pseudomonas sp. are Gram-negative bacteria and facultative aerobic microorganisms; under appropriate nutritional and environmental conditions, they can grow in the absence and in the presence of oxygen, but they grow faster in aerobic conditions. *Pseudomonas* sp. are easy to grow in vitro, and one can change the phenotype of this bacterium by using the tools of molecular biology (Chin-A-Woeng et al. 2002). These Gram-negative bacteria are saprophytic with little pathogenicity potential. Because of their ability to adapt to the most diverse environmental conditions, the genus *Pseudomonas* is present in virtually all ecosystems, from water and soil to animals and plants (Madigan et al. 2010). The versatility of the genus *Pseudomonas* is associated with a vast number of genotypic and phenotypic adaptation mechanisms. Such cellular and molecular mechanisms include production of a wide diversity of intra- and extracellular metabolites, some of which are antibiotics, growth promoters, or resistance inducers. Sections 15.3.1.1, 15.3.1.2, 15.3.1.3, 15.3.1.4, 15.3.1.5, 15.3.1.6 discuss the main groups of metabolites of agronomic interest.

15.3.1.1 Phenazines

Phenazines are a group of aromatic heterocyclic compounds containing nitrogen and brightly colored pigments. Phenazines are also produced by *Burkholderia*, *Streptomyces*, *Brevibacterium*, *Mycobacterium*, and *Xanthomonas* (Pierson and Pierson 2010), and they are easily extracted from a microbial culture.

Phenazines are biosynthesized during secondary metabolism, and their biological function is intriguing; they have no functions attributed to cell growth, energy, or reserve. One hypothesis is that bacteria use phenazines as a subsistence ability strategy to compete for nutrients or to improve other survival conditions (Laursen and Nielsen 2004).

Although the mechanism of action of phenazines is not completely understood, it is known that they diffuse through the cell wall and membrane, and they act as a reducing agent in uncoupling of oxidative phosphorylation and generation of intracellular superoxide radicals and hydrogen peroxide, which are fatal to the cell (Chin-A-Woeng et al. 2002; Blankenfeldt and Parsons 2014). They also interfere with electron flow and functional enzymes related to cellular respiration (Yu et al. 2018). Small modifications in the structural base of phenazines may give rise to different pigments, from dark red (aeruginosin A) to bright blue (pyocyanin (PYO)), with different biological actions (Prince-Whelan et al. 2006).

The antifungal action of phenazines is very well known. Their antibacterial activity has been demonstrated only against Gram-positive bacteria in association with silver nanoparticles, which showed synergistic action (Cardozo et al. 2013). In addition, they have antiparasitic and antitumor properties. Tumor cells are more susceptible to respiratory interference and generation of reactive oxygen species (ROS) by phenazine compounds (Pierson and Pierson 2010). In plants, phenazines have also demonstrated induced systemic resistance (ISR) activity against many pathogens and may influence growth.

Environmental and growth conditions influence the type and number of phenazine compounds that are synthesized by bacterial strains (Shanmugaiah et al. 2010). A change in the pH of the culture medium, for example, alters the antimicrobial activity of phenazines. At a neutral pH, phenazine-1-carboxyamide (PCN) was shown to be 10 times more potent that phenazine-1-carboxylic acid (PCA), and PCA showed no significant activity against *Fusarium oxysporum* f. sp. *radicis-lycopersici*. However, when the pH was decreased below 6.0, PCA increased its activity, and it was more potent than PCN at pH 3 (Thomas et al. 1998b; Gheorghe et al. 2017). Depending on the functional group, there is a change in solubility, as well as changes in the pigment and electron transfer capacity, at different pH values (Cezairliyan et al. 2013; Yu et al. 2018), altering the biological activities (Dharni et al. 2012).

PCA, a bright lemon yellow pigment, was first isolated and identified from *Pseudomonas aureofaciens* in the 1930s by Dr. Kluyver, who published a report on it in 1956 (Kluyver, 1956). It was the first phenazine to be synthesized, mainly in *Pseudomonas* sp. PCA acts as a metabolic intermediate of biosynthesis of other types of phenazine. For example, through the action of a transamidase catalyzed by the PhzH portion, PCA is converted into PCN (Pierson and Pierson 2010); through the action of a methyltransferase and a flavin containing monooxygenase, it is converted into PYO; and the enzyme PhzS converts it into 1-hydroxyphenazine, among other reactions (Pierson et al. 2013). Thus, a single microorganism can produce many different kinds of phenazine.

The oxyreduction activity and superoxide accumulation of PCA give it wide antimicrobial activity against various phytopathogenic fungi, including *Phytophthora capsici* in plants (Lee et al. 2003); *Phellinus noxius*, which causes brown rot in rubber trees (Huang et al. 2016); *F. oxysporum* f. sp. *radicis-lycopersici* in tomatoes; and *Gaeumannomyces graminis* var. *tritici*, which causes take-all disease in wheat (Puopolo et al. 2013).

Lee and collaborators (2003) tested the antifungal activity of PCA against anthracnose (caused by *Colletotrichum orbiculare*) on leaves of cucumber pre- and post-treatment in vitro, which showed higher efficiency in preventive treatment (10 μg mL^{-1}) and curative treatment (500 μg mL^{-1}) than a commercial control. Simionato et al. (2017a, b) used scanning electron microscopy to observe that treatment of strawberries and grapes with 12.5 μg mL^{-1} of PCA inhibited mycelial growth of *Botrytis cinerea*, causing distortion and damage of hyphae, besides absence of exopolysaccharide formation (which is one of the main factors in the virulence of the fungus), probably decreasing pathogenic and necrotic activity in the fruit.

PCA has eminent potential in the development of new antimicrobials for agriculture, with higher efficiency against several phytopathogens and a low environmental impact. In China, PCA has received pesticide registration certification, under the name shenqinmycin, from the Ministry of Agriculture to control *Fusarium* sp. wilt in watermelon, *Phytophthora* sp. blight in pepper, and sheath blight (ShB) in rice (Yuan et al. 2008; Xu et al. 2015).

PYO (5-methyl-1-hydroxyphenazine), which is bright blue, was the first phenazine to be discovered, in 1859, and it has been the best explored phenazine for its well-known antibiotic action. PYO was initially identified in immunosuppressed patients with chronic infections caused by *Pseudomonas aeruginosa*, with secretions with a bluish pigment (Prince-Whelan et al. 2006), and it is known as one of the virulence factors of *Pseudomonas*, with toxic potential against eukaryotic and prokaryotic cells.

PYO modulates the redox cycle and, like the other compounds present in the phenazine group, it is reduced nonenzymatically by reduced nicotinamide adenine dinucleotide phosphate (NADPH). The products of this reduction react with oxygen and generate ROS, which can cause significant oxidative stress, affecting all homeostasis and cellular respiration (Ho Sui et al. 2012; Barakat et al. 2013).

In plants, ROS play an important role in defense against phytopathogens. Infection by living organisms increases production of ROS, which usually exist at low concentrations in plant cells, leading the plant to experience a hypersensitivity reaction, promoting the induction of resistance (Resende et al. 2003; Audenaert et al. 2002). It has been reported that the action of PYO can stimulate ISR in some plants (such as beans (Abeysinghe 1999) and tomato (Audenaert et al. 2002)) against *B. cinerea*, and it can stimulate ISR in rice against *Magnaporthe grisea*, also known as rice blast fungus (de Vleesschauwer et al. 2008).

PYO has wide activity against phytopathogen fungi such as *Colletotrichum falcatum*, which causes red rot in sugarcane leaves; *F. oxysporum*, which causes *Fusarium* wilt in several plants; *Sclerotium rolfsii*, which causes root rot (Rane et al. 2008); and *Macrophomina phaseolina*, which causes gray root rot in peanuts and soybean (Khare and Arora 2011).

PCN is a yellowish-green pigment with strong antagonistic action against fungal phytopathogens (Chin-A-Woeng et al. 2000; Peng et al. 2018). Girard et al. (2006) reported that PCN, produced by *Pseudomonas chlororaphis*, reduced insoluble mineral metals such as Fe (III) and manganese, which was advantageous for the microorganism because their availability is increased by dissolution of these minerals, which are found only in limited concentrations in the soil. A small amount of PCN can reduce a large amount of these minerals by *P. chlororaphis* culture, and the redox properties also indicate that PCN is recycled several times (Hernandez et al. 2004).

Shanmugaiah et al. (2010) observed antibacterial activity of PCN against *Xanthomonas oryzae* pv. *oryzae*, which cause bacterial leaf blight (BLB) in rice, achieving greater control than the commercial compound rifamycin. In addition to antibacterial activity, they observed antifungal action against *Rhizoctonia solani*, which causes sheath blight in rice, indicating the potential of this compound to control both sheath blight and bacterial leaf blight (Shanmugaiah et al. 2010). This compound is also efficient for control of *F. oxysporum* f. sp. *radicis-lycopersici*, the causal agent of tomato foot and root rot (Chin-A-Woeng et al. 1998), and has been shown to control *B. cinerea* in field trials in strawberry (Zhang et al. 2015).

1-Hydrophenazine (1-OHPZ), which is a yellow-brown color, exhibits antimicrobial activity against phytopathogens and was first reported by Saosoong et al. (2009), who isolated it from *P. aeruginosa* and observed its activity against *Xanthomonas campestris* pv. *vesicatoria*, the causal agent of mancha bacteriana (bacterial leaf spot) in diverse cultures. Others authors have observed antifungal activity against several phytopathogens such as *Curvularia andropogonis*, *Bipolaris australiensis*, *Alternaria alternata*, *Alternaria solani*, *Colletotrichum acutatum*, and *F. oxysporum* (Dharni et al. 2012).

15.3.1.2 Pyrroles

The pyrrole pyrrolnitrin (3-chloro-4-(2′-nitro-3′-chlorophenyl) pyrrole) is the most prominent compound. Pyrroles are produced during secondary metabolism in some bacteria, by the genera *Pseudomonas* and *Burkholderia*, and by derivatives of the aromatic amino acid tryptophan. Pyrrolnitrin was first isolated in the 1960s by Arima et al. (1964) from *Pseudomonas pyrrocinia* and later demonstrated its antimicrobial action (Kilani and Fillinger 2014). This yellow pigment compound blocks transport of electrons between succinate/nicotinamide adenine dinucleotide reductase (NADH) and coenzyme Q, an enzyme responsible for transport of electrons to complex III of the respiratory chain in the mitochondria, and it also acts to prevent oxidation of lipids, proteins, and DNA (Bentinger et al. 2007; Gomes 2012).

Pyrroles have been shown to have antifungal activities against some important plant pathogens such as *R. solani* (El-Banna and Winkelmann 1998), *Alternaria* sp., *Fusarium* sp., *Verticillium dahliae*, *Thielaviopsis basicola* (Howell and Stipanovic 1979), and *B. cinerea* (Hammer and Evensen 1993).

Because of its photoinstability, photo-oxidation of the pyrrole ring inactivating molecules occurs, and this compound has not been widely used in the field (Corran et al. 2008). Through chemical synthesis of analogous molecules, it has been possible to reverse this problem, resulting in the phenylpyrroles (Kilani and Fillinger 2014). Among the tested analogues, the compound that was highlighted was fludioxonil, which has already been registered and is present in several commercial products for phytopathogenic fungal treatments pre- and postharvest. This compound has the same in vitro activity as pyrrolnitrin of natural origin, but in the greenhouse and in the field it exhibits more efficient activity due to its photostability (Corran et al. 2008).

15.3.1.3 Siderophores

These are compounds of low molecular weight (around 400–2000 Da) with high affinity for iron, an important element for growth and metabolism of organisms (Fedrizzi 2006). A function of siderophores is to capture Fe (III) under low–free iron conditions (Neilands 1995). In addition to connecting with iron, they can also capture other metals such as copper, molybdenum, and aluminum (Fedrizzi 2006).

The genus *Pseudomonas* produces a diverse variety of siderophores. Pyochelin (Pch) and pyoverdin (Pvd) are the ones that have been most intensively studied with a focus on agriculture. These compounds are known to perform biological control by withdrawing the iron available for phytopathogens to grow, causing nutrient competition and population decline (Weller 2007; Scavino and Pedraza 2013).

Pyochelin [2-(2-oxyphenyl-2-thiazolin-4-yl)-3-methylthiazolidine-4-carboxylic acid] is the first siderophore to be produced and, when it is identified that the iron concentration is low, it is converted into pyoverdin, which has higher affinity for the metal (Dumas et al. 2013).

The first report on pyoverdin production was published in 1892, when Gessard and coworkers observed the presence of a yellowish-green fluorescent pigment. However, its action was established only in 1978 by Meyer and Hornsperger. This compound is widely known as one of the virulence factors of this genus for controlling biofilm formation, quorum sensing communication, and regulation of other virulence factors (Imperi et al. 2009), but it also acts beneficially in plants, promoting growth (Fedrizzi 2006).

More than 100 types of pyoverdin from *Pseudomonas* sp. have now been described, and they are composed of three parts: the chromophore, which is conserved in all pyoverdins; a side chain attached to the chromophore; and a peptide part that is specific to each type, being able to be linear or cyclic (Cézard et al. 2015), which prevents its degradation by proteolytic enzymes (Scavino and Pedraza 2013). When the chromophore center binds to iron, this compound undergoes a color change to dark brown (Schalk et al. 1999).

Some scientists claim that pyoverdin can stimulate ISR to some pathogens. Studies with *P. fluorescens* WCS358 deficient in the synthesis of pyoverdin demonstrated that there was no induction of resistance in beans, tomato, tobacco, or eucalyptus, indicating that this compound can act as a systemic resistance elicitor (Ran et al. 2005; van Loon et al. 2008).

15.3.1.4 Hydrogen Cyanide

Hydrogen cyanide (HCN) is an extremely volatile compound and, when in contact with air or water, it produces highly toxic cyanide anions. This compound can be produced during secondary metabolism in some bacteria, mainly Gram-negative ones such as *Pseudomonas* and *Burkholderia* (Fernando et al. 2005).

Some studies have reported that HCN is produced by *Pseudomonas* sp. (Devi and Kothamasi 2009). In addition, it may be able to stimulate ISR in plants (Devi and Kothamasi 2009). Its mechanism of action involves inhibition of the activity of cytochrome c oxidase, which is responsible for transport of electrons in cellular respiration, preventing production of adenosine triphosphate (ATP) (Blumer and Haas 2000; Spence et al. 2014).

In addition to antimicrobial activity, this compound shows antinematode activity. Nandi et al. (2015) observed that the presence of HCN together with pyrrolnitrin repelled and contributed to the death of the model nematode *Caenorhabditis elegans*. Kang et al. (2018) worked with a mutant strain of *P. chlororaphis* O6 without HCN production and observed a reduction in biocontrol of gall nematodes, proving that HCN production is correlated with biocontrol of nematodes.

15.3.1.5 Diacetylphloroglucinol

Diacetylphloroglucinol (2,4-DAPG) is a phenolic compound of natural origin produced during secondary metabolism in Gram-negative bacteria, mainly the genus *Pseudomonas* (Meyer et al. 2009), with a broad spectrum of action (Khan and Parmar 2013).

At low concentrations, this compound can act as a molecular signal for expression of plant protectors. Besides stimulating ISR (Iavicoli et al. 2003; Weller et al. 2012) it stimulates exudate production by plant roots (Combes-Meynet et al. 2011), becoming an important protector in disease-suppressive soils.

Antimicrobial activity has been observed against *G. graminis* var. *tritici* in tobacco, *T. basicola* in wheat (Keel et al. 1992), and *Pseudomonas syringae* pv. *tomato* in tomato (Weller et al. 2012). Antinematode activity has been observed against *Globodera rostochiensis* (Cronin et al. 1997), *Meloidogyne incognita* in tomato (Siddiqui and Shaukat 2003a), and *Meloidogyne javanica* (Siddiqui and Shaukat 2003b).

15.3.1.6 New Compounds

Because of the increased demand for new products with antimicrobial activity and natural origins, many studies are being conducted in search of such compounds. A compound produced during secondary metabolism in a *P. aeruginosa* isolate from an orchard with a high incidence of citrus canker demonstrated high antimicrobial potential (de Oliveira et al. 2016). This compound, which is dark green and highly stable, belongs to the organometallic family. Its structure has not yet been fully elucidated. A semipurified fraction named F4a (containing PCA, PCN, and indolinone compounds) showed activity against *Xanthomonas citri* pv. *citri* in oranges (de Oliveira et al. 2011), *X. axonopodis* in eucalyptus (Lopes et al. 2012), *X. arboricola* pv. *pruni* (Vasconcellos et al. 2014), and *Pectobacterium carotovorum* subsp. *carotovorum* in tomato (Munhoz et al. 2017). This isolated compound showed high activity at a low concentration (0.125 μg mL^{-1}) against *X. citri* pv. *citri* (de Oliveira et al. 2016). These results are promising in the search for alternatives for control of different phytopathogens.

15.3.2 *Burkholderia* Species

The *Burkholderia* genus is extremely versatile, being able to inhabit diverse environments. It also has diverse interactions with several hosts, among which are plant and animal species. Therefore, *Burkholderia* spp. are ubiquitous microorganisms. Some species are opportunistic pathogens in humans; however, a range of them are nonpathogenic environmental bacteria. This genus is known to produce molecules with antifungal and/or antibacterial activity, siderophores, phenazines, and auxins (Vial et al. 2007). The production of these bioactive metabolites is mainly regulated via quorum sensing LuxI/R family genes in *Burkholderia* (Choudhary et al. 2013). In addition, some species have nif genes for biological nitrogen fixation (BNF) (Minerdi et al. 2001).

There is great interest in research on *Burkholderia* metabolites with antimicrobial activity. These can be used in agriculture indirectly (with application of the various studied strains as biocontrol agents) and directly, leading to production and purification of these compounds to be applied alone in crop fields. These resources can be used for prevention and/or treatment of phytodiseases caused by both fungi and bacteria, and even by insect infestations.

Thus, in Sects. 15.3.2.1, 15.3.2.2, 15.3.2.3, 15.3.2.4, 15.3.2.5, 15.3.2.6, 15.3.2.7, 15.3.2.8, 15.3.2.9, 15.3.2.10, 15.3.2.11, 15.3.2.12, 15.3.2.13, 15.3.2.14, 15.3.2.15, 15.3.2.16 and 15.3.2.17 we focus on summarizing the known metabolites produced by *Burkholderia* spp. that mainly have antimicrobial activity and that can be used directly or indirectly for treatment of the most diverse diseases in crop fields.

15.3.2.1 Pyrrolnitrin and Its Analogues

Burkholderia pyrrocinia 2327 was the first bacterial strain used on an industrial scale for production of the antifungal pyrrolnitrin (Kwak and Shin 2015). Isolated from the soil in 1963 in Japan, it was used to create the product Pyroace® (Fujisawa

Pharmaceutical Co., Ltd., Japan) (van Pée and Ligon 2000), which is used against pathogenic fungal in humans. However, it has also been observed that this compound has strong activity against fungi of agronomic interest (Jung et al. 2018; Okada et al. 2005).

Pyrrolnitrin biosynthesis involves tryptophan as a precursor molecule (Floss et al. 1971). El-Banna and Winkelmann (1998) elucidated the mechanism of action of pyrrolnitrin in *Neurospora crassa*, which acts on the electron transport chain of the fungus. However, Okada et al. 2005 showed that a primary mechanism of this compound occurred because of interference with the osmotic signal transduction pathway. The authors also evaluated the antimicrobial effect against phytopathogenic fungi in vitro, where it was possible to observe a strong effect of pyrrolnitrin against these frequent causes of disease in plants.

Jung et al. (2018) prospected for *Burkholderia cepacia* strain JBK9 in soil samples in the Republic of Korea, which showed strong antagonism against phytopathogenic fungi. The bacterium was cultured and the supernatant was subjected to extraction with hexane. The hexane phase showed significant activity against *R. solani*, *P. capsici*, and *F. oxysporum*, with 58.11%, 31.27%, and 59.43% inhibitory effects on mycelial growth, respectively. The authors purified the bioactive compound and identified it as pyrrolnitrin.

However, this compound is stable for only about 30 days in soil after application (Howell and Stipanovic 1979) and is light sensitive (Gordee and Westhead 1972), which makes it necessary to increase the number of applications of the product. To improve its efficiency, Syngenta AG developed two molecules analogous to pyrrolnitrin: fenpiclonil and fludioxonil. Both are 3-cyano-4-phenylpyrrol analogues of pyrrolnitrin that show enhanced photostability and similar antifungal activity (Corran et al. 2011).

Other analogues derived from pyrrolnitrin have been identified. Sultan et al. (2008) isolated [3-chloro-4-(3-chloro-2-nitrophenyl)-5-methoxy-3-pyrrolin-2-one] and [4-chloro-3-(3-chloro-2-nitrophenyl)-5-methoxy-3-pyrrolin-2-one] from *B. cepacia* K87 secondary metabolism. These compounds showed marginal activities when compared with pyrrolnitrin, and the researchers suggested that both were biodegraded derivatives of pyrrolnitrin.

15.3.2.2 Phenazines

A few phenazines are also known to be synthetized by *Burkholderia* spp., such as iodinin (Bell and Turner 1973), 4,9-dihydroxyphenazine-1,6-dicarboxylic acid dimethyl ester (Cartwright et al. 1995), and the more recently identified phencomycin (Han et al. 2014).

Isolated from *B. cepacia* 5.5B, a purple pigment identified as 4,9-dihydroxyphenazine-1,6-dicarboxylic acid dimethyl ester is a phenazine that has shown high activity against *R. solani*, similar to the activity of pyrrolnitrin (Cartwright et al. 1995).

Han et al. (2014) isolated and identified phencomycin, for the first time, from the metabolism of *Burkholderia glumae* 411gr-6, which has activity against many phytopathogenic fungi such as *Alternaria brassicicola*, *Aspergillus oryzae*, *B. cinerea*,

Cladosporium cucumerinum, *Colletotrichum gloeosporioides*, *C. orbiculare*, *Cylindrocarpon destructans*, *Diaporthe citri*, *F. oxysporum*, *Magnaporthe oryzae*, *P. capsici*, *Rhizopus stolonifer*, and *Sclerotinia sclerotiorum*, besides having activity against plant pathogenic bacteria such as *P. syringae*, *Ralstonia solanacearum*, and *X. campestris*.

15.3.2.3 Siderophores

Pyochelin is well known to be produced by *Pseudomonas* (as mentioned in Sect. 15.3.1.3), and its antimicrobial activity and ISR capacity have been characterized. It is known that the *Burkholderia* genus also produces pyochelin (Meyer et al. 1995), besides other siderophores: ornibactin (Deng et al. 2017), malleobactin (Franke et al. 2015), cepaciachelin (Barelmann et al. 1996), azurechelin (Sokol et al. 1992), and cepabactin (Meyer et al. 1989). However, although all siderophores have the characteristic of being able to sequester Fe, which gives them the potential for antimicrobial activity, there are no records of trials showing antimicrobial activity of these other siderophores, with the exception of ornibactin. Deng et al. (2017) found that the antibacterial activity of the strain *Burkholderia contaminans* MS14 is closely related to production of ornibactin. This siderophore showed activity against *X. citri* pv. *malvacearum*, *P. carotovorum* WSCH1, *R. solanacearum*, *P. syringae* B301, *Erwinia amylovora*, *B. glumae* 291, *Escherichia coli*, and *Clavibacter michiganensis* subsp. *michiganensis* (Herrera 2017).

15.3.2.4 Xylocandins

Xylocandins (or cepacidins) are a complex of antifungal peptides isolated originally from *B. cepacia* ATCC 39277 (Bisacchi et al. 1987; Meyers et al. 1987) and further isolated from *B. cepacia* AF2001 (Lee et al. 1994; Lim et al. 1994).

In in vitro assays, Lee et al. (1994) demonstrated antifungal activity of cepacidin A against clinical isolates (filamentous fungi and yeasts) and against phytopathogenic fungi such as *Aspergillus niger*, *F. oxysporum*, and *R. stolonifer*, and they noted that the antifungal activity was strong, with minimal inhibitory concentration (MIC) values equal to or less than those seen with amphotericin B. Furthermore, in semi-greenhouse conditions, Lee et al. (2000) tested the biocontrol potential of AF2001, which showed excellent growth suppression of *Pythium ultimum* in cucumbers and cotton plants; moreover, minor activity against *R. solani* in cotton plants was verified.

15.3.2.5 Burkholdines

Bk-1229 and Bk-1097 are octapeptides composed of nonproteinogenic amino acids (β-hydroxytyrosine, β-hydroxyasparagine) and a fatty acyl amino acid isolated from a culture of *Burkholderia ambifaria* 2.2 N; both are known as burkholdines (Tawfik et al. 2010). These compounds have shown potent antifungal activity against a range of fungal phytopathogens such as *B. cinerea*, *A. solani*, *Phytophthora infestans*, and *Mycosphaerella fijiensis* (an agent of black sigatoka disease in banana plants) (Tawfik et al. 2010).

Furthermore, Lin et al. (2012) isolated three new burkholdines from 2.2 N—Bk-1119, Bk-1213, and Bk-1215—which also showed antifungal activity. However, Bk-1119 show more pronounced activity than the other two. The authors suggested that burkholdines are potential scaffolds for development of new antifungal compounds with selective targets (Lin et al. 2012; Tawfik et al. 2010).

15.3.2.6 Occidiofungins

Occidiofungins are cyclic glycosylated oligopeptides, synthesized by a nonribosomal peptide synthetase (NRPS), and are structurally similar to xylocandins. The production of these compounds was identified from *B. contaminans* MS14 (Lu et al. 2016) and *B. pyrrocinia* Lyc2 (Wang et al. 2016).

These compounds have high activity against animal pathogenic fungi and, mainly, vegetal pathogenic fungi such as *A. alternata*, *Aspergillus fumigatus*, *R. solani*, various species in the *Phytium* genus (Lu et al. 2009), *Cochliobolus heterostrophus*, *C. acutatum*, *G. graminis*, *Geotrichum candidum*, *Glomerella cingulata*, and *T. basicola* (Wang et al. 2016).

15.3.2.7 Cepalycin

Cepalycin I and cepalycin II were isolated from *B. cepacia* JN106, and both show antifungal and hemolytic activities. It has been suggested that both have deleterious interactions with cholesterol in fungal and erythrocyte membranes, contributing to their activities (Abe and Nakazawa 1994).

15.3.2.8 2-Hydroxymethyl-Chroman-4-One

This compound shows strong antifungal activity against members of the phycomycetes group—such as *P. ultimum*, *P. capsici*, and *S. sclerotiorum*—and significant antagonism against *B. cinerea*, *R. solani*, and *Alternaria panax* (Kang et al. 2004). This molecule was isolated from secondary metabolism in *Burkholderia* sp. MSSP (which was isolated from *Mimosa pudica* by Kang et al. (2004)) and was identified as 2-hydroxymethyl-chroman-4-one by the authors, utilizing spectrometric approaches. To date, there are no records of this compound being isolated from another microbial strain.

15.3.2.9 AFC-BC-11

An antimicrobial compound produced by *B. cepacia* BC11 shows strong activity against the following phytopathogen fungi: *R. solani*, *P. ultimum*, *Colletotrichum* sp., *Helminthosporium maydis*, *B. cinerea*, *Fusarium* sp., *R. stolonifer*, *Rhodotorula glutinis*, *S. rolfsii*, and *Scopulariopsis brevicaulis*. The molecular structure of the compound was partially elucidated, and the authors named it AFC-BC-11 (Kang et al. 1998). Furthermore, Kang et al. (1998) tested and proved the efficacy of this antifungal compound in control of cotton damping-off disease caused by *R. solani.*

15.3.2.10 Quinolone Derivates

Recently, two new 2-alkylquinolones were isolated from secondary metabolism in *Burkholderia* sp. MBAF1239, and their molecular structures were elucidated as

[(E)-2-(hept-2-en-1-yl) quinolin-4(1H)-one] and [(E)-2-(non-2-en-1-yl) quinolin-4(1H)-one]. These compounds show activity against *Rhizopus oryzae*, a phytopathogen that causes rice seedling blight (Li et al. 2018).

15.3.2.11 Quinoline Derivates

Moon et al. (1996) demonstrated the effects of two quinoline derivatives—HQM and NMQ—isolated from *B. cepacia* PCII. In vitro assays of HQM and NMQ showed strong activity against *P. capsici*, *R. solani*, and *F. oxysporum*. In bioassays where inoculation with PCII was conducted, suppression of the *Phytophthora* blight of red pepper disease (caused by *P. capsici*) was observed and inoculation with PCII also promoted plant growth. Both compounds had previously been isolated from *B. cepacia* RB425 (Yoshihisa et al. 1989).

15.3.2.12 Altericidins

Altericidins are a complex of oligopeptides isolated from *Burkholderia* metabolism. They have shown strong activity against *Alternaria kikuchiana* and potential to control black spot pear disease caused by this fungus (Kirinuki et al. 1977). The mechanism of their action is related to the cell wall and changes in membrane permeability, with selective activity against filamentous fungi (Kirinuki and Ichiba 1986).

15.3.2.13 Cepaciamides A and B

Cepaciamides are compounds with antifungal activity against *B. cinerea* and *Penicillium expansum*, the casual agents of storage rot disease in beetroot (Toshima et al. 1999). Cepaciamide A and cepaciamide B were isolated from *B. cepacia* D-202 by Toshima et al. (1999). Both compounds are 3-amino-2-piperidinone-containing lipids. However, the production of these substances from bacterial metabolism is very low. The authors described the total synthetic synthesis of both molecules and analogues.

15.3.2.14 Syrbactin

Syrbactin is a generic term used to describe the family of the antibiotics syringolin, glidobactin, and cepafungin, which have similar biosynthesis pathways and identical mechanisms of action (Krahn et al. 2011). While syringolin is synthetized by *P. syringae* pv. *syringae* (Amrein et al. 2004), glidobactin and cepafungin are generated during metabolism in *Burkholderia* (Shoji et al. 1990). However, glidobactin has also been identified in *Photorhabdus* (Bozhüyük et al. 2016) and *Polyangium brachysporum* metabolism (Oka et al. 1988).

These syrbactin compounds show strong and irreversible inhibition of proteasomes, and this mechanism is directly linked to their different biological effects (Krahn et al. 2011). It has been reported that syringolin is able to stimulate ISR in rice plants and hence to control fungal infection by *Pyricularia oryzae* (Wäspi et al. 1998). Glidobactin and cepafungin have shown potent antifungal activity, as well as anticancer properties (Oka et al. 1988; Terui et al. 1990). Cepafungin I, cepafungin III, and glidobactin A have shown strong activity against *A. fumigatus*, *Penicillium digitatum*, *Microsporum canis*, *Trichophyton rubrum*, *Trichophyton mentagrophytes*, *T. asteroides*, and *Candida* sp. (Shoji et al. 1990).

15.3.2.15 Phenylacetic Acid, Hydrocinnamic Acid, 4-Hydroxyphenylacetic Acid, and 4-Hydroxyphenylacetate Methyl Ester

Mao et al. (2006) isolated *Burkholderia* sp. strain MP-1 in the region of Naju, South Korea. It demonstrated strong antagonism against several phytopathogenic fungal species. Four different biomolecules from secondary bacterial metabolism were detected, characterized, and identified: phenylacetic acid (PA), hydrocinnamic acid (HA), 4-hydroxyphenylacetic acid (4HPA), and 4-hydroxyphenylacetate methyl ester (4HPME). However, it is possible that 4HPME originated from esterification of 4HPA, caused by treatment of the samples with methanol during the purification process. Nevertheless, all of these compounds showed inhibitory activity against the fungi *A. brassicicola*, *B. cinerea*, *C. gloeosporioides*, *Chaetomium globosum*, *Didymella bryoniae*, *Pestalotiopsis* sp., *F. oxysporum* f. sp. *cucumerinum*, *P. capsici*, *R. solani*, and *Stemphylium vesicarium*.

Among these metabolites, PA had already been isolated from *Pseudomonas* sp. (Kang 1999), *Bacillus licheniformis* (Kim et al. 2004), *Streptomyces humidus* (Hwang et al. 2001), *Enterobacter cloacae* S11:T:07 (Burkhead et al. 1998), and *G. cingulata* (Hirota et al. 1992). In addition to antifungal activity, PA has activity against the pinewood nematode *Bursaphelenchus xylophilus* (Kawazu et al. 1996).

15.3.2.16 Hydrogen Cyanide

HCN is a volatile secondary metabolite from several Gram-negative bacteria such as *Pseudomonas* (Laville et al. 1998) and *Burkholderia* (Gilchrist et al. 2013). HCN is a potent inhibitor of cytochrome c oxidase and many other metalloenzymes, giving it a strong nonselective antimicrobial effect (Blumer and Hass 2000).

15.3.2.17 Unknown Compounds

Many substances with antimicrobial activity produced by *Burkholderia* strains are still unknown. Dichloromethane and methanol extracts of *B. pyrrocinia* strain RV1R2 demonstrated insecticide activity against *Tenebrio molitor* (Silva et al. 2015). Additionally, two unidentified compounds isolated from the ethyl acetate extract of RV1R2 showed high activity against *R. solani* and *S. sclerotiorum* (Silva 2018).

Lassie et al. (2018) recently described antiyeast activity of *Burkholderia* sp. strain RV7S3 extracts. Also, RV7S3 demonstrated antifungal properties against *R. solani*, suppressing the root rot disease caused by this fungus in tobacco plants under greenhouse conditions (Nunes 2018).

15.4 Microbial Bioactive Compounds from Gram-Positive Bacteria

15.4.1 *Bacillus* Species

The main mechanisms by which biocontrol agents suppress pathogens are antibiosis, with production of substances with antimicrobial activity; competition; promotion of plant growth; and induction of acquired resistance (Xu et al. 2013).

Among the bacterial antagonists are Gram-positive bacteria that produce a wide variety of substances with antimicrobial activity. Many of these belong to the genus *Bacillus*—for example, *B. subtilis*, *B. amyloliquefaciens*, and *B. thuringiensis*. *Bacillus*, at present, is a genus that contains 377 species (http://www.bacterio.net/bacillus.html). Belonging to the family Bacillaceae, phylum Firmicutes, these bacteria form endospores—resistant structures that can withstand severe environmental variations (Marquez et al. 2011; Hoyles et al. 2012).

Some of the substances with antimicrobial activity produced by *Bacillus* are ribosomal antibiotics, which include subtilisin A, subtilin, sublancin, chitinase, and TasA. Other compounds are produced by facilitation of NRPS or polyketide synthases, such as chlorotetain, bacilysin, mycobacillin, difficidin, rhizocticins, and bacillaene; and cyclic lipopeptides (CLPs), including iturins, surfactins, and fengycins with potential biotechnological applications due to their tensoactive properties (Xu et al. 2013; Ye et al. 2012; Shafi et al. 2017).

In the lipopeptide family, iturins and fengycins exhibit strong antifungal activity against phytopathogens, being considered the main compounds with antagonistic activity (Ye et al. 2012).

15.4.1.1 Iturins

The group of iturins is subdivided into iturin, bacillomycin, and mycosubtilin, which cause cell leakage by forming pores in the cytoplasmic membrane. All of them are cyclic peptides with seven α-amino acids (A1–A7) and one unique β-amino fatty acid (βAA) (Gong et al. 2014; Ye et al. 2012; Shafi et al. 2017). Iturin A has seven isomers called iturin A2–A8, all of which are considered potent antibiotics, but there are just a few field applications of these molecules, possibly because of limitations in production and purification of such substances (Ye et al. 2012).

Iturin A isomers have shown high antifungal activity against *R. solani*, *Penicillium chrysogenum*, *Penicillium italicum*, *Penicillium vindicatum*, *Aspergillus ochraceus*, *Aspergillus versicolor*, *A. niger*, *F. oxysporum*, *C. gloeosporioides*, *V. dahliae*, *Alternaria mali*, and *P. oryzae*, interacting with the cytoplasmic membrane of the target cells and leading to increased permeability to K^+ (Yu et al. 2002, Gong et al. 2006; Kim et al. 2010).

Ye et al. (2012) purified iturin A2, a molecule with antifungal activity, extracted from the growth of *B. subtilis* B47 and tested against *Bipolaris maydis*. In in vitro tests, iturin A2 completely inhibited the fungus at a concentration of 300 mg kg^{-1}. In field experiments, iturin A2 was more effective when used for inoculation 1 day before application of the phytopathogen, also in a concentration of 300 mg kg^{-1}, indicating a better preventive effect than a curative effect; hence, for effective control, its application is recommended before the onset of disease or at the earliest signs of disease.

15.4.1.2 Bacillomycin

Bacillomycin D is a member of the iturin lipopeptide family and has a cyclic ring in its chemical structure. It is considered the most potent metabolite with antifungal action, inhibiting mycelial growth, productivity, and spore germination, with severe ultrastructural changes, damaging the cell wall and membrane. It has shown activity against the fungi *Absidia corymbifera*, *A. niger*, *Candida albicans*, *Kluyveromyces bulgaris*, and *Saccharomyces cerevisiae* (Chowdhury et al. 2015, Gong et al. 2014; Gu et al. 2017; Ye et al. 2012).

Bacillomycin D promotes changes in the morphology of the plasmatic membrane and cell wall of *Fusarium graminearum*, and such changes promote accumulation of ROS, causing fungal cell death (Gu et al. 2017).

Bacillus subtilis AU195 and *B. subtilis* B-FS06 produce bacillomycin D analogues with high inhibitory activity against *Aspergillus flavus*, which can be used as a grain and feed preservative, avoiding aflatoxin contamination. Bacillomycin D, produced by *B. amyloliquefaciens* SQR9 and *B. amyloliquefaciens* NJN-6, demonstrated antifungal activity against *F. oxysporum* but only a limited antagonistic in vitro effect against *R. solanacearum* (Moyne et al. 2001; Yuan et al. 2012; Zhang et al. 2008; Gong et al. 2014; Xu et al. 2013).

15.4.1.3 Mycosubtilin

Mycosubtilin exhibits antifungal activity against some crop pests such as *B. cinerea*, *F. oxysporum*, *Pythium aphanidermatum*, and *R. solani*, acting on the cytoplasmic membrane of target cells and forming ion-conducting pores that promote permeability to electrolytes (Yu et al. 2002; Leclère et al. 2005).

Mycosubtilin was tested for control of lettuce mildew caused by *Bremia lactucae*. Mycosubtilin at 100 mg L^{-1} protected the lettuces, producing about seven times more healthy seedlings than the control samples, with no signs of phytotoxicity in the treated plants. Synergistic efficacy was evaluated with mycosubtilin and surfactin, with both compounds being applied in doses of 50 mg L^{-1}, indicating better effects and allowing a reduction in the dose of mycosubtilin by half (Deravel et al. 2014).

15.4.1.4 Fengycins

Fengycins, which are also called plastathins, are composed of a hydroxylated fatty acid and ten amino acids, comprising fengicins A and B. Like iturins, they have strong antifungal activity, but they act more specifically against filamentous fungi (Yánez-Mendizábal et al. 2012; Gong et al. 2014)

The action of fengycins is less well known in comparison with other lipopeptides. The antifungal activity of fengycins is due to their ability to interact with lipid components of the fungal cytoplasmic membrane, such as ergosterol, and to alter its structure (packaging) and permeability in a dose-dependent manner, with an interesting application in biological control of fruit diseases (Touré et al. 2004, Ongena et al. 2005). Fengycins produced by *B. subtilis* CPA-8 have an antifungal effect and potential for biological control of peach rot caused by *Monilinia laxa* and *Monilinia fructicola* (Yánez-Mendizábal et al. 2012).

15.4.1.5 Surfactins

Surfactins are composed of a hydroxylated fatty acid and seven amino acids. Surfactins are not fungitoxic by themselves, but they express some antifungal activity in synergism with iturin A, which is essential for formation of a biofilm matrix and has surfactant properties, with the capacity to increase penetration of some substances (Deravel et al. 2014; Yánez-Mendizábal et al. 2012; Gong et al. 2014).

Surfactin and, to a lesser extent, fengycin may act as elicitors of host plant immunity and contribute to increased resistance to pathogenesis in bean and tomato plants (Raaijmakers et al. 2010). Similarly, low concentrations of surfactin have been shown to induce several plant defense events in tobacco cells (Jourdan et al. 2009; Chowdhury et al. 2015).

15.4.1.6 Lytic Enzymes

In addition to producing antibiotics against a variety of pathogenic diseases in plants, *Bacillus* species are capable of producing enzymes with very strong lytic activity, degrading fungal cell walls. Chitin, an insoluble polysaccharide, is an important element of fungal cell walls, formed by polysaccharide glycosidic bonds. *Bacillus* spp. produce chitinases, glucanases, and chitosanases, which hydrolyze the cell walls of fungal pathogens, presenting great potential for disease management in plants, since plant cells do not contain chitin (Shafi et al. 2017).

Gomaa (2012) demonstrated that the chitinase produced by *B. thuringiensis* was more effective against phytopathogenic fungi than that produced by *B. licheniformis*. In the presence of chitinase produced by *B. thuringiensis*, fungal growth of *A. flavus*, *A. niger*, *A. terreus*, *F. oxysporum*, *Fusarium* sp., *R. solanacearum*, and *Rhizopus* sp. ranged from 15.11% to 44.66%, although the chitinase produced by *B. licheniformis* was effective only against *A. flavus*, *A. niger*, and *A. terreus*, the growth of which ranged from 17.06% to 35.79%.

Varieties of *Bacillus* species produce many compounds that can be used against different plant pathogens. Their bioactive metabolites require a stabilizer that can improve their activities under field conditions. This strategy is more effective for control of phytopathogens than use of the bacterium itself in the field. Therefore, it is essential to understand the bacterial active compounds so that a stable and effective formulation can be developed (Shafi et al. 2017).

15.4.2 Actinobacteria

Ants, like humans, are farmers, and leaf-cutter ants (Attini) started farming 50 million years ago. They harvest leaves from the surface, drag them into their colony, and chew the pieces to a pulp, which they use to grow a mushroom (Basidiomycota: Agaricales: Lepiotaceae and Pterulaceae) in vast underground gardens that supply them with sugars. Fungus-growing ants (*Acromyrmex octospinosus*) protect the mushroom against a devastating mold (*Escovopsis* sp.) with antimicrobials produced by an Actinobacteria species (e.g., *Pseudonocardia* spp.), which lives in a patch on their skin. Ants use multidrug therapy to maintain their fungal cultivars. In the same way that ants have evolved along with Actinobacteria in a symbiotic association to keep their crops safe from pathogens, humans have used actinomycetes in recent decades to guarantee the quality of agricultural products.

The Actinobacteria phylum is a group of ubiquitous bacteria, which are very abundant in soils, in animals, in aquatic environments, and on practically any natural surface. They are Gram-positive bacteria and mostly strict aerobes, which grow well at a pH between 5 and 9, and at temperatures of 25–35 °C, with high frequencies of guanine and cytosine in their DNA, which is why they show significant morphological diversity (El-Tarabily and Sivasithamparam 2006). Actinobacteria develop filaments during their growth. These branch and produce mycelia, which can be vegetative or aerial mycelia, similar to those in fungi. At the end of such filaments, asexual reproductive spores are formed; when these spores reach adequate substrates, they form new colonies (Hwang et al. 2014). These structures are very resistant to adverse conditions, a feature that allows them to persist in environments that show conditions of abiotic stress.

These microorganisms, which are abundant in soil, are important saprophytes of plants because of their lytic enzymes. Thus, Actinobacteria play an important role in the decomposition process of organic material, being capable of degrading complex molecules and recalcitrant substances, such as cellulose, lignocellulose, xylan, and lignin (Sousa et al. 2008; Zhou et al. 2009).

Actinomycetes can produce enzymes with antimicrobial activity, suggesting they have great potential as biological controllers of plant pathogens, especially pathogenic fungi. Actinobacteria secrete enzymes, proteases, and hydrolases such as glucanases and chitinases, which degrade fungal cell walls, causing hyphal lysis and making the pathogen more susceptible to attack by other antifungal metabolites (El-Tarabily and Sivasithamparam 2006). There are many reports indicating the importance of such enzymes produced by actinomycetes in the suppression of plant diseases. A chitinolytic actinomycete, *Streptomyces vinaceusdrappus*, showed in vitro antifungal activity against sclerotia producing the pathogen *R. solani*, and a similar effect was observed in greenhouse experiments (Yandigeri et al. 2015). The antifungal potential of *Streptomyces griseorubens*, a chitinase producer microorganism, showed pronounced activity against the phytopathogenic fungus *F. oxysporum* f. sp. *lypopersici* (the causative agent of wilt disease in tomato), in both in vitro and field evaluations (Rashad et al. 2017). Kamil et al. (2018) isolated three actinomycetes strains—two belonging to *Streptomyces* and one to *Micromonospora* sp.—which showed the strongest in vitro inhibitory effects against *Lasiodiplodia theobromae*. Subsequently, a significant reduction was observed in the number of defoliated leaves and conidia counts of *L. theobromae* in mango seedlings treated with *S. samsunensis* (Kamil et al. 2018).

Besides enzymes, Actinobacteria produce metabolites that enhance solubilization, fixation, and availability of minerals and nutrients, improving plant health. Some actinomycetes are able to solubilize phosphates, which is a very important effect, since 95–99% of phosphorus is in an insoluble form and cannot be used by plants. Lack of phosphorus is one of the main limitations in plant growth in organic production (Otero Jiménez 2011). Actinomycetes convert insoluble phosphate into a soluble form by acidification, chelation, and exchange reactions. Within the Actinobacteria, the genus *Frankia* comprises nitrogen-fixing bacteria present in a free-living environment and in symbiosis with diverse angiosperms (actinorhizal plants), enabling them to grow well even in nitrogen-poor soils (Barka et al. 2016; Lewin et al. 2016).

Furthermore, actinomycetes produce siderophores, which can solubilize and chelate iron from the soil. Iron is another element required by plants. The iron-limiting conditions in soil stimulate actinomycetes to produce siderophores, which compete and thus inhibit the growth of pathogenic microorganisms (Caballero-Mellado 2006). Additionally, catechol and/or hydroxamate siderophores produced by Actinobacteria can promote plant growth and help the plant to assimilate iron, reducing pathogenic competitors (Tank et al. 2012; Sharma and Salwan 2018).

The most studied biocontrol mechanism is antibiosis mediated by production of secondary metabolites and their antagonistic interaction. Actinomycetes are well known for their ability to produce antimicrobial compounds, which promote inhibition of plant pathogens. They produce approximately 70% of all known antibiotics, and the majority of these compounds are produced by *Streptomyces*, isolated from both marine and terrestrial habitats (Manivasagan et al. 2014; Lasudee et al. 2018; Sharma and Salwan 2018). Several antibiotics produced by actinomycetes are currently used in biological control of plant pests and diseases (Barka et al. 2016).

In addition to the direct mechanism of biological control, actinomycetes are involved in ISR in different plant–pathogen systems. Salicylic acid, jasmonic acid, and ethylene are major players in regulation of signaling pathways involved in ISR, mainly through increased levels of pathogenesis-related proteins. For instance, inoculation of tomato plants with *Micromonospora* sp. stimulated ISR and reduced leaf infection caused by *B. cinerea* (Martínez-Hidalgo et al. 2015). Inoculation of *Arabidopsis* seeds with *Streptomyces* sp. leads to upregulation of the PR-1 or PDF1.2 transcripts and to increased protection against the necrotrophic *A. brassicicola* (Lewin et al. 2016). Likewise, nitrous oxide production by *Streptomyces* has been suggested to activate plant defenses, improving the plant's protection against pathogens (Vaishnav et al. 2018).

Volatile organic compounds (VOCs) produced by Actinobacteria have great potential in agriculture as biopesticides. A large number of VOCs have been characterized from actinomycetes, esters, alcohols, ketones, alkanes, alkenes, isoprenes, and terpenoids; these VOCs were reported to inhibit *Cladosporium cladosporioides*, *Fusarium* spp., *A. niger*, *Penicillium citrinum*, *R. solani*, *Pyricularia grisea*, *Bipolaris oryzae*, *S. sclerotiorum*, and *B. cinerea* in tomato and strawberry fruit (Sharma and Salwan 2018).

In conclusion, inhibition of pathogens mediated by antimicrobial compounds is generally the primary focus in efforts to suppress plant diseases through use of Actinobacteria. However, biocontrol of plant pathogens may involve a diversity of other mechanisms, including ISR, production of cell wall–degrading enzymes (glucanases and chitinases), inhibition of pathogen growth through production of VOCs, production of siderophores, competition for nutrient resources, and destructive parasitism, offering great versatility for controlling fungal, bacterial, and parasitic pathogens.

15.5 Microbial Bioactive Compounds Produced by Fungi

The use of beneficial microorganisms as plant inoculants for biofertilization, phytostimulation, and biocontrol has increased because of the needs to reduce use of chemical fertilizers and pesticides, and to maintain a healthy ecosystem through development of sustainable agriculture (Naznin et al. 2013). In agriculture, plant growth–promoting and biocontrol microorganisms have emerged as safe alternatives to chemical pesticides. Fungi and their metabolites may have great potential as agents for controlling various phytopathogens (Vurukonda et al. 2018).

Fungi play an important role in the ecosystem because of its nutritional versatility and their different forms of interaction with plants. They are important decomposers and recyclers of organic matter and interact with different plant tissues, affecting the host positively or negatively (Zeilinger et al. 2016). Fungi interact with plants by forming associations and by acting as pathogens or as plant growth promoters associated with the rhizosphere. Like bacteria, fungi produce a series of metabolic compounds that can positively affect plant growth, either through pathogen antagonism or by reducing plant stress. These metabolites activate plant protection mechanisms, induce systemic resistance to pathogens, protect plants from herbivory, stimulate phytohormone synthesis, and improve the efficiency of water and nutrient transport (Fig. 15.1).

Fungi associated with plants are a source of natural bioactive compounds that have been utilized for applications in agriculture, medicine, and the food industry. Although most studies have focused on isolating these metabolites for development

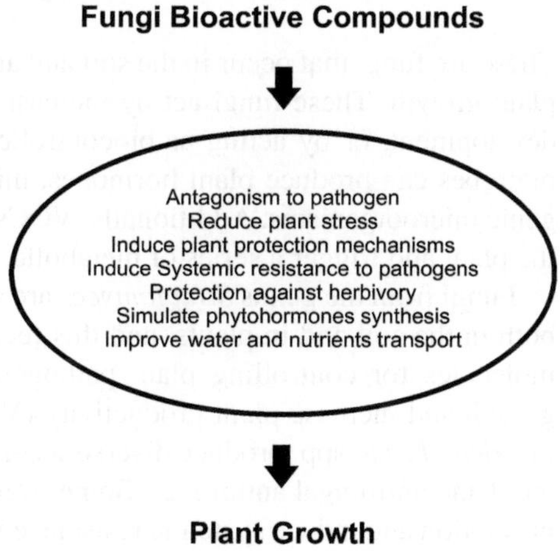

Fig. 15.1 Mechanisms of action of bioactive compounds produced by fungi: pathogen antagonism, reduction of plant stress, induction of plant protection mechanisms, induction of systemic resistance, protection against herbivory, stimulation of phytohormone synthesis, and improvement of water and nutrient transport

of new products with antimicrobial, anticancer, cytotoxic, or insecticidal activities, these substances also act as promoters of plant growth and resistance.

15.6 Plant-Associated Fungi that Produce Bioactive Compounds

15.6.1 Mycorrhizal Fungi

The term "mycorrhizal symbiosis" refers to coexistence of fungi with the roots of vascular plants. Root colonization by endomycorrhizal fungi causes changes in the quantity and quality of exudates produced by roots, affecting plants' health status, their competitiveness and success in the ecosystem, formation of soil aggregates, increased resistance to abiotic and biotic stresses, and activation of immune responses in plants (Jamiołkowska et al. 2017). These fungi are capable of producing bioactive compounds that influence their plant host. For instance, glomalin plays different functions, contributing to immobilization of pollutants in the soil–hyphal interface and reducing palatability for predators, thus protecting the plant (Souza et al. 2012).

Plants associated with mycorrhizal fungi can withstand drought-induced oxidative stress by increased production of antioxidant compounds that scavenge ROS and enhance the activities of antioxidant enzymes (Rapparini and Penuelas 2014). Mycorrhizal fungi can also produce phytohormones and volatile compounds that increase plant resistance to pathogenic fungi, bacteria, and nematodes. They cause physiological changes in the host, increasing concentrations of phosphorus, phenols, sulfur, amino acids, etc. (Saranya and Kumutha 2011).

Mycorrhizal fungi can effectively activate plant immune responses locally and systemically (Jamiołkowska et al. 2017).

15.6.2 Plant Growth–Promoting Fungi

These are fungi that occur in the soil and are able to colonize plant roots, promoting plant growth. These fungi act by increasing seed germination, biomass, and plant development, or by acting as biocontrollers. These activities occur because these microbes can produce plant hormones, mineralize substrates, and suppress pathogenic microorganisms. Additionally, VOCs produced by these fungi are detected by the plant and trigger a series of metabolic responses (Naznin et al. 2013).

Fungi from the genus *Streptomyces* are active producers of antibiotics and VOCs, both in the soil and in plants, and this feature is helpful for identifying active biomolecules for controlling plant pathogens. These fungi can also promote plant growth and increase plant productivity (Vurukonda et al. 2018). Numerous strains of *Trichoderma* spp. produce diverse secondary metabolites, which include antibacterial and antifungal antibiotics. Some strains play an important role in plant growth promotion and induce systemic resistance in plants (Singh et al. 2017). *Trichoderma* can take up ACC secreted by the plant root and convert it into α-ketobutyrate and ammonia; thus, this fungus can protect plants against stress caused by flooding, salination, drought, waterlogging, heavy metals, toxic organic compounds, and pathogens (Lugtenberg et al. 2013).

15.6.3 Endophytic Fungi

These are fungi that spend their entire life cycle, or part of it, colonizing the intra- and/or intercellular spaces of healthy tissues of host plants, without causing disease. They are able to promote plant growth and produce secondary metabolites; in return, they receive nutrients and housing. It is estimated that there are more than one million endophytic fungi colonizing all plant species; hence, they are important components of the plant microecosystem (Zhao et al. 2011).

Endophytic fungi are the most studied fungal group in relation to production of bioactive compounds. Most endophytes positively affect plant growth, providing nutrients and exhibiting antagonism to pathogens, as well as decreasing stress effects on plants. Beneficial effects have been obtained by using Beauveria bassiana, Piriformospora indica, F. oxysporum, Ophiostoma-like fungi, Phialocephala fortinii, Trichoderma harzianum, and other Trichoderma species (Samarina et al. 2017).

Endophytic fungi are capable of producing the original compounds of their hosts or similar compounds. These bioactive compounds have been shown to assist plants with resistance to stress caused by biotic and abiotic factors, increasing the immune response of the plants to pathogens (Rajamanikyam et al. 2017). Many endophytes can also protect their hosts by inducing a defense mechanism, producing antibiotics that inhibit the growth of pathogens or competing with the pathogens for space and nutrients (Alurappa et al. 2018).

Production of bioactive substances by endophytic fungi is directly related to the independent evolution of these microorganisms, which may have incorporated genetic information from higher plants, allowing these fungi to better adapt to the host and to carry out some primary functions, such as protection against pathogens, insects, and herbivores. They are chemical synthesizers within plants (Pimentel et al. 2011).

Jia et al. (2016) presented the interactions between endophytic plants and fungi according to three aspects:

1. Endophytic fungi producing hormones that promote the growth of the host plant; for example, the endophytic fungi *Phoma glomerata* LWL2 and *Penicillium* sp. LWL3 significantly promote shoot and allied growth attributes of Gas-deficient dwarf mutant Waito-C and Dongjinbeyo rice by producing gibberellins and indoleacetic acid (Waqas et al. 2012).
2. Endophytic fungi producing bioactive compounds that increase resistance of the host plant to stress conditions; for example, the endophytic fungus *Paecilomyces formosus* LWL1 in *japonica* rice cultivar Dongjin significantly improves plant

growth attributes such as plant height, fresh weight, dry weight, and chlorophyll content by producing secondary metabolites under heat stress (Waqas et al. 2015).

3. Endophytic fungi promoting accumulation of secondary metabolites originally produced by the host plant; for example, endophytic fungi residing in medicinal plants are capable of biosynthesizing pharmacologically active secondary metabolites that are similar to or identical to those produced by their host medicinal plant (Venieraki et al. 2017).

Table 15.2 lists some substances isolated from endophytic fungi and their roles in plant development.

Table 15.2 Substances produced by endophytic fungi and their importance for their host plants

Fungi	Host plants	Metabolites	Functions	References
Diaporthe helianthi	*Luehea divaricata*	2-(4 hydroxyphenyl) ethanol or tyrosol	Antioxidant activity Antifungal activity Antibacterial activity	Specian et al. (2012)
Botryosphaeria rhodina	*Bidens pilosa*	Complex of four depsidones (botryorhodines A–D) and the auxin indole carboxylic acid	Antifungal activity	Abdou et al. (2010)
Alternaria sp. *Aspergillus* sp. *Botryodiplodia theobromae* *Botrytis* sp. *Cladosporium cladosporioides* *Ectostroma* sp. *Fusarium* sp. *Metarhizium anisopliae* *Monochaetia* sp. *Mucor rouxianus* *Ozonium* sp. *Papulaspora* sp. *Periconia* sp. *Pestalotia bicilia* *Pestalotiopsis* sp. *Phyllosticta* sp. *Pithomyces* sp. *Taxomyces* sp. *Tubercularia* sp.	*Cardiospermum helicacabum* *Citrus medica* *Cupressus* sp. *Ginkgo biloba* *Hibiscus rosa-sinensis* *Podocarpus* sp. *Taxus* sp. *Terminalia arjuna* *Wollemia nobilis*	Paclitaxel (Taxol) [a tetracyclic diterpenoid]	Antifungal activity	Wagner and Flores (1994) Zhao et al. (2010)
315 endophytic fungi isolates	*Swietenia macrophylla*	Not identified	Antimicrobial activity	Ibrahim et al. (2014)

(continued)

Table 15.2 (continued)

Fungi	Host plants	Metabolites	Functions	References
Fusarium sp. *Penicillium* sp. *Guignardia mangiferae* *Xylaria* sp. *Penicillium paxilli* *Aspergillus aculeatus* *Phomopsis* sp. *Eutypella scoparia* *Botryosphaeria* sp.	*Garcinia* sp.	Not identified	Antibacterial activity Antifungal activity	Phongpaichit et al. (2006)
Alternaria sp. *Fusarium* sp. *Monilia* sp. *Penicillium* sp. *Phialocephala fortinii* *Trametes hirsuta*	*Diphylleia* sp. *Dysosma* sp. *Sabina* sp. *Sinopodophyllum* sp.	Podophyllotoxin Aryltetralin lignan	Stimulation of plant defense responses Antibacterial activity	Moraes et al. (2002) Zhao et al. (2010)
Entrophospora infrequens *Fusarium solani* *Neurospora* sp.	*Camptotheca acuminata* *Nothapodytes* sp. *Merrilliodendron megacarpum* *Ophiorrhiza mungos* *Osmia pumila* *Eravatamia heyneana* *Mostuea brunonia* *Apodytes dimidiata*	Camptothecin [a pentacyclic quinoline alkaloid]	Plant growth regulation during seed development, seed hydration, seed germination, and early seedling growth Antifungal activity Antiparasitic activity	Patil et al. (2015) Tao and Buta (1986) Zhao et al. (2010)
Eutypella spp. *Alternaria* *Aspergillus* *Chaetomium* *Colletotrichum* *Dothideomycetes* *Eutypa* *Flavodon* *Fusarium* *Talaromyces*	*Catharanthus roseus*	Vinblastine and vincristine [terpenoid indole alkaloids derived from coupling of vindoline and catharanthine monomers]	Stimulation of plant defense responses by triggering of metabolite synthesis	Kuriakose et al. (2016) Palem et al. (2015)
Shiraia sp.	*Huperzia serrata*	Huperzine A	Plant growth regulator	Wang et al. (2011)

(continued)

Table 15.2 (continued)

Fungi	Host plants	Metabolites	Functions	References
Fusarium oxysporum	*Dioscorea zingiberensis*	Diosgenin	Precursor of steroids Antiviral activity	Li et al. (2011)
Thielavia subthermophila	*Hypericum perforatum*	Hypericin Emodin	Chromophore	Jendželovská et al. (2016)
Unidentified	*Melia azedarach*	Toosendanin	Antiparasitic activity Insecticidal activity	Nicoletti and Fiorentino (2015)
Penicillium chermesinum	*Hertieri littoralis*	2-Chloro-3,4,7-trihydroxy-9-methoxy-1-methyl-6H-benzo[c] chromen-6-one	Antibacterial activity Antifungal activity Antioxidant activity	Darsih et al. (2017)
Nectria sp. *Fusarium* sp. *Rhizopycnis* sp. *Acremonium* sp. *Penicillium* sp.	*Dioscorea zingiberensis*	Not identified	Antibacterial activity	Xu et al. (2007)

The bioactive compounds produced by fungi derive from their pathways of biosynthesis and degradation, and can be acids, alkaloids, alcohols, cytochalasins, depsipeptides, esters, steroids, phenols, furandiones, glycosides, hydrocarbons, isocoumarins, isoprenoids, ketones, lactones, lignans, lipids, peptides, perylene derivatives, polyketides, proteins, quinones, shikimates, terpenoids, and xanthones.

This diversity of bioactive compounds with unique structural and biological specificity, produced by endophytic fungi, is still a source of antimicrobial, antiviral, anticancer, antioxidant, neuroprotective, and antifungal activities in drug discovery and development (Darsih et al. 2017).

15.7 Conclusion

The production of a certain antimicrobial compound for a particular microorganism depends essentially on space and resource competition. Microbial metabolism has inspired a large number of studies of bioactive compounds that can be used in biological control of plant diseases. The vast majority of natural antimicrobials come from microbial secondary metabolism, which provides benefits for the plants. These metabolites have the potential to be used in agroindustry, especially in relation to pathogen control and environmental sustainability.

There is a significant demand for production of new bioactive compounds that will replace the agrochemicals currently used in control of plant diseases. Many problems have already been reported, such as increasing microbial resistance to currently used chemicals, confirming the need for development of new biological agents, from natural sources, that can be used in control and treatment of plant diseases.

References

Abdou R, Scherlach K, Dahse HM, Sattler I, Hertweck C (2010) Botryorhodines A–D, anti-fungal and cytotoxic depsidones from *Btryosphaeria rhodina*, an endophyte of the medicinal plant *Bidens pilosa*. Phytochemistry 71:110–116

Abe M, Nakazawa T (1994) Characterization of hemolytic and antifungal substance, cepalycin, from *Pseudomonas cepacia*. Microbiol Immunol 38:1–9

Abeysinghe S (1999) Investigation of metabolites of the rhizobacterium *Pseudomonas aeruginosa* strain SA44 involved in plant protection. Dissertation, Vrije Universiteit Brussel

Alurappa R, Chowdappa S, Narayanaswamy R, Sinniah UR, Mohanty SK, Swamy MK (2018) Endophytic fungi and bioactive metabolites production: an update. In: Patra JK, Das G, Shin H (e) (eds) Microbial biotechnology: volume 2. Application in food and pharmacology. Singapore, Springer, pp 455–482

Amrein H, Makart S, Granado J, Shakya R, Schneider-Pokorny J, Dudler R (2004) Functional analysis of genes involved in the synthesis of syringolin A by *Pseudomonas syringae* pv. *syringae* B301D-R. Mol Plant-Microbe Interact 17:90–97. https://doi.org/10.1094/MPMI.2004.17.1.90

Arima K, Imanaka I, Kousaka M, Fukuta A, Tamura G (1964) Pyrrolnitrin, a new antibiotic substance, produced by *Pseudomonas*. Agric Biol Chem 28:575–576

Audenaert K, Pattery T, Cornelis P, Höfte M (2002) Induction of systemic resistance to *Botrytis cinerea* in tomato by *Pseudomonas aeruginosa* 7NSK2: role of salicylic acid, pyochelin, and pyocyanin. Mol Plant-Microbe Interact 15:1147–1156

Balouiri M, Sadiki M, Ibnsouda SK (2016) Methods for in vitro evaluating antimicrobial activity: a review. J Pharm Anal 6:71–79. https://doi.org/10.1016/j.jpha.2015.11.005

Barakat R, Goubet I, Manon S, Berges T, Rosenfeld E (2013) Unsuspected pyocyanin effect in yeast under anaerobiosis. Microbiology 3:1–14. https://doi.org/10.1002/mbo3.142

Barelmann I, Meyer JM, Taraz K, Budzikiewicz H (1996) Cepaciachelin, a new catecholate siderophore from *Burkholderia* (*Pseudomonas*) *cepacia*. Z Naturforsch C 51:627–630. https://doi.org/10.1515/znc-1996-9-1004

Barka EA, Vatsa P, Sanchez L, Gaveau-Vaillant N, Jacquard C, Klenk H-P, Clément C, Ouhdouch Y, van Wezel GP (2016) Taxonomy, physiology, and natural products of Actinobacteria. Microbiol Mol Biol Rev 80:1–43. doi:10.1128/MMBR.00019-15

Bell SC, Turner JM (1973) Iodinin biosynthesis by a pseudomonad. Biochem Soc Trans 1:751–753. https://doi.org/10.1042/bst0010751

Bentinger M, Brismar K, Dallner G (2007) The antioxidant role of coenzyme Q. Mitochondrion 7:41–50

Bérdy J (2012) Thoughts and facts about antibiotics: where we are now and where we are heading. J Antibiot (Tokyo) 65:441–441. https://doi.org/10.1038/ja.2012.54

Bisacchi GS, Hockstein DR, Koster WH, Parker W, Rathnum ML, Unger SE (1987) Xylocandin: a new complex of antifungal peptides. J Antibiot 40:1520–1529. https://doi.org/10.7164/antibiotics.40.1520

Blankenfeldt W, Parsons JF (2014) The structural biology of phenazine biosynthesis. Curr Opin Struct Biol 29:26–33. https://doi.org/10.1016/j.sbi.2014.08.013

Blumer C, Haas D (2000) Mechanism, regulation, and ecological role of bacterial cyanide biosynthesis. Arch Microbiol 173:170–177

Bozhüyük KA, Zhou Q, Engel Y, Heinrich A, Pérez A, Bonde HB (2016) Natural products from *Photorhabdus* and other entomopathogenic bacteria. In: The molecular biology of *Photorhabdus* bacteria. Springer, Cham, pp 55–79. https://doi.org/10.1007/82_2016_24

Burkhead KD, Slininger PJ, Schisler DA (1998) 1589701. Biological control bacterium enterobacter cloacae S11: T: 07 (NRRL B-21050) produces the antifungal compound phenylacetic acid in sabouraud maltose broth culture. Soil Biol Biochem 30:665–667

Caballero- Mellado J (2006) Microbiología agrícola e interacciones microbianas con plantas. Rev Latinoam Microbiol 48(2):154–155

Cardozo VF, Oliveira AG, Nishio EK, Perugini MRE, Andrade CGTJ, Silveira WD, Durán N, Andrade G, Kobayashi RKT, Nakazato G (2013) Antibacterial activity of extracellular compounds produced by a *Pseudomonas* strain against methicillin-resistant *Staphylococcus aureus* (MRSA) strains. Ann Clin Microbiol Antimicrob 12:1–8

Cartwright DK, Chilton WS, Benson DM (1995) Pyrrolnitrin and phenazine production by *Pseudomonas cepacia*, strain 5.5 B, a biocontrol agent of *Rhizoctonia solani*. Appl Microbiol Biotechnol 43:211–216. https://doi.org/10.1007/BF00172814

Cezairliyan B, Vinayavekhin N, Grenfell LD, Yuen GJ, Saghatelian A, Ausubel FM (2013) Identification of *Pseudomonas aeruginosa* phenazines that kill *Caenorhabditis elegans*. PLoS Pathog 9. https://doi.org/10.1371/journal.ppat.1003101

Cézard C, Farvacques N, Sonnet P (2015) Chemistry and biology of pyoverdines, *Pseudomonas* primary siderophores. Curr Med Chem 21:165–186

Chin-A-Woeng TFC, Bloemberg GV, Mulders IHM, Dekkers LC, Lugtenberg BJJ (2000) Root colonization by Phenazine-1-Carboxamide-producing bacterium PCL1391 is essential for biocontrol of tomato foot and root rot. Mol Plant-Microbe Interact 13(12):1340–1345

Chin-A-Woeng TF, Bloemberg GV, Mulders IH, Dekkers lC, Lugtenberg BJ (2002) Root colonization by phenazine-1-carboxamide-producing bacterium *Pseudomonas chlororaphis* PCL1391 is essential for biocontrol of tomato foot and root rot. Mol Plant-Microbe Interact 13:1340–1345

Chin-A-Woeng TF, Bloemberg GV, van der Bij AJ, van der Drift KMGM, Schripsema J, Kroon B, Scheffer RJ, Keel C, Bakker PAHM, Tichy H, Bruijin FJ, Thomas-Oates JE, Lugtenberg BJJ (1998) Biocontrol by phenazine-1-carboxamide-producing *Pseudomonas chlororaphis* PCL1391 of tomato root rot caused by *Fusarium oxysporum* f. sp. *radicis-lycopersici*. Mol Plant-Microbe Interact 11:1069–1077

Choudhary KS, Hudaiberdiev S, Gelencsér Z, Gonçalves Coutinho B, Venturi V, Pongor S (2013) The organization of the quorum sensing luxI/R family genes in *Burkholderia*. Int Mol Sci 14:13727–13747. https://doi.org/10.3390/ijms140713727

Chowdhury SP, Hartmann A, Gao XW, Borriss R (2015) Biocontrol mechanism by root associated *Bacillus amyloliquefaciens* FZB42—a review. Front Microbiol 6:780

Combes-Meynet E, Pothier JF, Moënne-Loccoz Y, Prigent-Combaret C (2011) The *Pseudomonas* secondary metabolite 2,4-diacetylphloroglucinol is a signal inducing rhizoplane expression of *Azospirillum* genes involved in plant-growth promotion. Mol Plant-Microbe Interact 24:271–284. https://doi.org/10.1094/MPMI-07-10-0148

Corran A, Knauf-Beiter G, Zeun R (2008) Fungicides acting on signal transduction. In: Krämer W, Schirmer U (eds) Modern crop protection compounds. Wiley-VCH Verlag GmbH, Weinheim, pp 561–580

Corran A, Knauf-Beiter G, Zeun R (2011) Fungicides acting on signal transduction. In: Krämer W, Schirmer U, Jeschke P, Witschel M (eds) Modern crop protection compounds. Wiley-VCH, Weinheim, pp 715–737. https://doi.org/10.1002/9783527644179.ch17

Cronin D, Moënne-Loccoz Y, Fenton A, Dunne C, Dowling DN, O'Gara F (1997) Role of 2,4-diacetylphloroglucinol in the interactions of the biocontrol pseudomonad strain F113 with the potato cyst nematode *Globodera rostochiensis*. Appl Environ Microbiol 63:1357–1361

Darsih C, Prachyawarakorn V, Mahidol C, Ruchirawat S, Kittakoop P (2017) A new polyketide from the endophytic fungus *Penicillium chermesinum*. Indones J Chem 17:360–364

de Oliveira AG, Spago FR, Simionato AS, Navarro MOP, da Silva CS, Barazetti AR, Cely MVT, Tischer CA, San Martin JAB, de Jesus Andrade CGT, Novello CR, Mello JCP, Andrade G

(2016) Bioactive organocopper compound from *Pseudomonas aeruginosa* inhibits the growth of *Xanthomonas citri* subsp. *citri*. Front Microbiol 7:1–12

de Oliveira AG, Murate LS, Souza PB, Spago FR, Lopes LP, Beranger JPO, JAB SM, Nogueira MA, JCP M, CGTJ A, Andrade G (2011) Evaluation of the antibiotic activity of extracellular compounds produced by the *Pseudomonas* strain against the *Xanthomonas citri* pv. *citri* 306 strain. Biol Control 56:125–131

de Vleesschauwer D, Djavaheri M, Bakker PAHM, Höfte M (2008) *Pseudomonas fluorescens* WCS374r-induced systemic resistance in rice against *Magnaporthe oryzae* is based on pseudobactin mediated priming for a salicylic acid–repressible multifaceted defense response. Plant Physiol 148:1996–2012

Deng P, Foxfire A, Xu J, Baird SM, Jia J, Delgado KH, Shin R, Smith L, Lu SE (2017) Siderophore product ornibactin is required for the bactericidal activity of *Burkholderia contaminans* MS14. Appl Environ Microbiol. https://doi.org/10.1128/AEM.00051-17

Deravel J, Lemière S, Coutte F, Krier F, Hese NV, Béchet M, Sourdeau N, Höfte M, Leprêtre A, Jacques P (2014) Mycosubtilin and surfactin are efficient, low ecotoxicity molecules for the biocontrol of lettuce downy mildew. Appl Microbiol Biotechnol 98:6255–6264

Devi KK, Kothamasi D (2009) *Pseudomonas fluorescens* CHA0 can kill subterranean termite *Odontotermes obesus* by inhibiting cytochrome c oxidase of the termite respiratory chain. FEMS Microbiol 300:195–200. https://doi.org/10.1111/j.1574-6968.2009.01782.x

Dharni S, Alam M, Kalani K, Khaliq A, Samad A, Srivastava SK, Patra DD (2012) Production, purification, and characterization of antifungal metabolite from *Pseudomonas aeruginosa* SD12, a new strain obtained from tannery waste polluted soil. Microb Biotechnol 22:674–683. https://doi.org/10.4014/jmb.1109.0906

Dumas Z, Ross-Gillespie A, Kummerli R (2013) Switching between apparently redundant iron-uptake mechanisms benefits bacteria in changeable environments. Proc Biol Sci 280:20131055. https://doi.org/10.1098/rspb.2013.1055

Eilenberg J, Hajek A, Lomer C (2001) Suggestions for unifying the terminology in biological control. BioControl 46:387–400. https://doi.org/10.1023/A:1014193329979

El-Banna N, Winkelmann G (1998) Pyrrolnitrin from *Burkholderia cepacia*: antibiotic activity against fungi and novel activities against streptomycetes. J Appl Microbiol 85:69–78

El-Tarabily KA, Sivasithamparam K (2006) Non streptomycete actinomycetes as biocontrol agents of soil borne fungal plant pathogens and as plant growth promoters. Soil Biol Biochem 38:1505–1520. https://doi.org/10.1016/j.soilbio.2005.12.017

FAO/WHO [Food and Agriculture Organization of the United Nations/World Health Organization] (2017) International code of conduct on pesticide management. http://www.fao.org/fileadmin/templates/agphome/documents/Pests_Pesticides/Code/Code_ENG_2017updated.pdf. Accessed July 11, 2018

Fedrizzi SMG (2006) Produção de metabolitos antimicrobianos e sideróforos de isolados provenientes de Terra Preta Antropogênica da Amazônia Ocidental. Dissertation, Universidade de São Paulo

Fernando WGD, Nakkeeran S, Zhang Y (2005) Biosynthesis of antibiotics by PGPR and its relation in biocontrol of plant diseases. In: Siddiqui ZA (ed) PGPR: biocontrol and biofertilization. Springer, Dordrecht, pp 67–109

Floss HG, Manni PE, Hamill RL, Mabe JA (1971) Further studies on the biosynthesis of pyrrolnitrin from tryptophan by *Pseudomonas*. Biochem Biophys Res Commun 45:781–787. https://doi.org/10.1016/0006-291x(71)90485-2

Franke J, Ishida K, Hertweck C (2015) Plasticity of the malleobactin pathway and its impact on siderophore action in human pathogenic bacteria. Chem Eur J 21:8010–8014. https://doi.org/10.1002/chem.201500757

Gheorghe I, Popa M, Marutescu L, Saviuc C, Lazar V, Chifiriuc MC (2017) Lessons from interregn communication for development of novel, ecofriendly pesticides. In: Grumezescu AM (ed) New pesticides and soil sensors. Academic, London, pp 1–46

Gilchrist FJ, Sims H, Alcock A, Jones AM, Bright-Thomas RJ, Smith D, Spanel P, Webb AK, Lenney W (2013) Is hydrogen cyanide a marker of *Burkholderia cepacia* complex? J Clin Microbiol. https://doi.org/10.1128/JCM.02157-13

Girard G, van Rij ET, Lugtenberg BJJ, Bloemberg GV (2006) Regulatory roles of psrA and rpoS in phenazine-1-carboxamide synthesis by *Pseudomonas chlororaphis* PCL1391. Microbiology 152:43–58

Gomaa EZ (2012) Chitinase production by *Bacillus thuringiensis* and *Bacillus licheniformis*: their potential in antifungal biocontrol. J Microbiol 50:103–111

Gomes F (2012) A síntese de coenzima Q e a estabilidade de DNA mitocondrial em *Saccharomyces cerevisiae*. Dissertation, Universidade de São Paulo

Gong M, Wang JD, Zhang J, Yang H, Lu XF, Pei Y, Cheng JQ (2006) Study of the antifungal ability of *Bacillus subtilis* strain PY-1 in vitro and identification of its antifungal substance (iturin A). Acta Biochim Biophys Sin 38:233–240

Gong Q, Zhang C, Lu F, Zhao H, Bie X, Lu Z (2014) Identification of bacillomycin D from *Bacillus subtilis* fmbJ and its inhibition effects against *Aspergillus flavus*. Food Control 36:8–14

Gordee RS, Westhead J (1972) Pyrrolnitrin. In: Gordee RS, Westhead J (eds) Analytical microbiology New York: Academic, pp 329–338. doi: https://doi.org/10.1016/B978-0-12-403502-7.50034-X

Govindasamy V, Senthilkumar M, Magheshwaran V et al (2011) *Bacillus* and *Paenibacillus* spp.: potential PGPR for sustainable agriculture. In: Maheshwari DK (ed) Plant growth and health promoting bacteria. Springer, Berlin, pp 333–336

Gu Q, Yang Y, Yuan Q, Shi G, Wu L, Lou Z, Huo R, Wu H, Borriss R, Gao X (2017) Bacillomycin D produced by *Bacillus amyloliquefaciens* is involved in the antagonistic interaction with the plant pathogenic fungus *Fusarium graminearum*. Appl Environ Microbiol. https://doi.org/10.1128/AEM.01075-17

Hammer PE, Evensen KB (1993) Post-harvest control of *Botrytis cinerea* on cut flowers with pyrrolnitrin. Plant Dis 77:283–286

Han JW, Kim JD, Lee JM, Ham JH, Lee D, Kim BS (2014) Structural elucidation and antimicrobial activity of new phencomycin derivatives isolated from *Burkholderia glumae* strain 411gr-6. J Antibiot 67:721. https://doi.org/10.1038/ja.2014.50

Hernandez ME, Kappler A, Newman DK (2004) Phenazines and other redox active antibiotics promote microbial mineral reduction. Appl Environ Microbiol 70:921–928

Herrera K (2017) Purifying an unknown *Burkholderia contaminans* MS14 bactericidal. Thesis, Texas A&M University. http://hdl.handle.net/1969.1/164551. Accessed

Hirota A, Horikawa T, Fujiwara A (1992) Isolation of phenylacetic acid from a phytopathogenic fungus, *Glomerella cingulata*. Biosci Biotech Biochem 57:492

Ho Sui SJ, Lo R, Fernandes AR, Caulfield MDG, Lerman JA, Xie L, Bourne PE, Baillie DL, Brinkman FSL (2012) Raloxifene attenuates *Pseudomonas aeruginosa* pyocyanin production and virulence. Int J Antimicrob Agents 40:246–251

Howell CR, Stipanovic RD (1979) Control of *Rhizoctonia solani* on cotton seedlings with *Pseudomonas fluorescens* and with an antibiotic produced by the bacterium. Phytopathology 69:480–482. https://doi.org/10.1094/Phyto-69-480

Hoyles L, Honda H, Logan NA, Halket G, La Ragione RM, McCartney AL (2012) Recognition of greater diversity of *Bacillus* species and related bacteria in human faeces. Res Microbiol 163:3–13

Huang H, Sun L, Bi K, Zhong G, Hu M (2016) The effect of phenazine-1-carboxylic acid on the morphological, physiological, and molecular characteristics of *Phellinus noxius*. Molecules 21:1–10. https://doi.org/10.3390/molecules21050613

Hwang BK, Lim SW, Kim BS, Lee JY, Moon SS (2001) Isolation and in vivo and in vitro antifungal activity of phenylacetic acid and sodium phenylacetate from *Streptomyces humidus*. Appl Environ Microbiol 67:3739–3745

Hwang K, Kim H, Charusanti P, Palsson BO, Lee SY (2014) Systems biology and biotechnology of *Streptomyces* species for the production of secondary metabolites. Biotechnol Adv 32:255–268. https://doi.org/10.1016/j.biotechadv.2013.10.008

Iavicoli A, Boutet E, Buchala A, Métraux JP (2003) Induced systemic resistance in *Arabidopsis thaliana* in response to root inoculation with *Pseudomonas fluorescens* CHA0. Mol Plant-Microbe Interact 16:851–858

Ibrahim D, Lee CC, Sheh-Hong L (2014) Antimicrobial activity of endophytic fungi isolated from *Swietenia macrophylla* leaves. Nat Prod Commun 9:247–250

Imperi F, Tiburzi F, Visca P (2009) Molecular basis of pyoverdine siderophore recycling in *Pseudomonas aeruginosa*. Proc Natl Acad Sci U S A 106:20440–20445. https://doi.org/10.1073/pnas.0908760106

Jamiołkowska A, Księżniak A, Hetman B, Kopacki M, Skwaryło-Bednarz B, Gałązka A, Thanoon AH (2017) Interactions of arbuscular mycorrhizal fungi with plants and soil microflora. Acta Sci Pol Hortorum Cultus 16:89–95

Jendželovská Z, Jendželovský R, Kuchárová B, Fedorocko P (2016) Hypericin in the light and in the dark: two sides of the same coin. Front Plant Sci:560–567. https://doi.org/10.3389/fpls.2016.00560

Jia M, Chen L, Xin H, Zheng C, Rahman K, Han T, Qin L (2016) A friendly relationship between endophytic fungi and medicinal plants: a systematic review. Front Microbiol 7:1–14

Jourdan E, Henry G, Duby F, Dommes J, Barthélemy JP, Thonart P (2009) Insights into the defense-related events occurring in plant cells following perception of surfactin type lipopeptide from *Bacillus subtilis*. Mol Plant-Microbe Interact 22:456–468

Jung BK, Hong SJ, Park GS, Kim MC, Shin JH (2018) Isolation of *Burkholderia cepacia* JBK9 with plant growth–promoting activity while producing pyrrolnitrin antagonistic to plant fungal diseases. Appl Biol Chem 61:173–180. https://doi.org/10.1007/s13765-018-0345-9

Kaeberlein T, Lewis K, Epstein SS (2002) Isolating "uncultivable" microorganisms in pure culture in a simulated natural environment. Science 296:1127–1129. https://doi.org/10.1126/science.296.5570.1055

Kamil FH, Saeed EE, El-Tarabily KA, AbuQamar SF (2018) Biological control of mango dieback disease caused by *Lasiodiplodia theobromae* using streptomycete and non streptomycete Actinobacteria in the United Arab Emirates. Front Microbiol 9:1–19. https://doi.org/10.3389/fmicb.2018.00829

Kang BR, Anderson AJ, Kim YC (2018) Hydrogen cyanide produced by *Pseudomonas chlororaphis* O6 exhibits nematicidal activity against *Meloidogyne hapla*. Plant Pathol J 34:35–43. https://doi.org/10.5423/PPJ.OA.06.2017.0115

Kang JG (1999) Production of the antifungal compound phenylacetic acid by antagonistic bacterium *Pseudomonas* sp. Agric Chem Biotechnol 42:197–201

Kang JG, Shin SY, Kim MJ, Bajpai V, Maheshwari DK, Kang SC (2004) Isolation and antifungal activities of 2-hydroxymethyl-chroman-4-one produced by *Burkholderia* sp. MSSP. J Antibiot 57:726–731. https://doi.org/10.7164/antibiotics.57.726

Kang Y, Carlson R, Tharpe W, Schell MA (1998) Characterization of genes involved in biosynthesis of a novel antibiotic from *Burkholderia cepacia* BC11 and their role in biological control of *Rhizoctonia solani*. Appl Environ Microbiol 64:3939–3947

Kare E, Arora NK (2011) Dual activity of pyocyanin from *Pseudomonas aeruginosa*—antibiotic against phytopathogen and signal molecule for biofilm development by rhizobia. Can J Microbiol 57:708–713. https://doi.org/10.1139/W11-055

Kawazu K, Zhang H, Kanzaki H (1996) Accumulation of benzoic acid in suspension cultured cells of *Pinus thunbergii* Parl. in response to phenylacetic acid administration. Biosci Biotechnol Biochem 60:1410–1412. https://doi.org/10.1271/bbb.60.1410

Keel C, Schnider U, Maurhofer M, Voisard C, Laville J, Burger U, Wirthner P, Haas D, Défago G (1992) Suppression of root diseases by *Pseudomonas fluorescens* CHA0: importance of the bacterial secondary metabolite 2,4-diacetylphloroglucinol. Mol Plant-Microbe Interact 5:4–13

Khan H, Parmar N (2013) Bioinoculants: understanding chickpea rhizobia in providing sustainable agriculture. In: Maheshwari DK, Saraf M, Aeron A (eds) Bacteria in agrobiology: crop productivity. Springer, London, pp 185–216

Kilani J, Fillinger S (2014) Phenylpyrroles: 30 years, two molecules and (nearly) no resistance. Front Microbiol 7:1–10. https://doi.org/10.3389/fmicb.2016.02014

Kim PI, Ryu J, Kim YH, Chi YT (2010) Production of biosurfactant bipopeptides iturin A, fengycin, and surfactin A from *Bacillus subtilis* CMB32 for control of *Colletotrichum gloeosporioides*. J Biotechnol 20:138–145

Kim Y, Cho JY, Kuk JH, Moon JH, Cho JI, Kim YC, Park KH (2004) Identification and antimicrobial activity of phenylacetic acid produced by *Bacillus licheniformis* isolated from fermented soybean, Chungkook Jang. Curr Microbiol 48:312–317. https://doi.org/10.1007/s00284-003-4193-3

Kirinuki T, Ichiba T (1986) Effects of altericidins on the leakage of cellular constituents from *Alternaria kikuchiana* and *Ustilago maydis*. J Pest Sci 11:453–458. https://doi.org/10.1584/jpestics.11.453

Kirinuki T, Iwamura K, Suzuki N, Fukami H, Ueno T (1977) Altericidins, a complex of polypeptide antibiotics, produced by *Pseudomonas* sp. and their effect for the control of black spot of pear caused by *Alternaria kikuchiana* Tanaka. Sci Rep Fac Agric Kobe Univ (Japan). https://doi.org/10.24546/00228538

Kluyver AJ (1956) *Pseudomonas aurefaciens* nov. spec. and its pigments. J Bacteriol 72:406–411

Krahn D, Ottmann C, Kaiser M (2011) The chemistry and biology of syringolins, glidobactins and cepafungins (syrbactins). Nat Prod Rep 28:1854–1867. https://doi.org/10.1039/c1np00048a

Kuriakose GC, Palem PPC, Jayabaskaran C (2016) Fungal vincristine from *Eutypella* spp CrP14 isolated from *Catharanthus roseus* induces apoptosis in human squamous carcinoma cell line -A431. BMC Complement Altern Med 16:302

Kwak Y, Shin JH (2015) Complete genome sequence of *Burkholderia pyrrocinia* 2327T, the first industrial bacterium which produced antifungal antibiotic pyrrolnitrin. J Biotechnol 211:3–4. https://doi.org/10.1016/j.jbiotec.2015.06.420

Lassie FS, Emiliano J, Simionato AS, Navarro MOP, Gionco B, Silva CS, Niekawa ETG, Silva FM, Dealis ML, Valoto MV, Andrade G (2018) Bioactive substances produced by *Burkholderia* sp.with antifungal action in *Candida* spp. Agronomy Sci Biotechnol 4:70–78

Lasudee K, Tokuyama S, Lumyong S, Pathomaree W (2018) Actinobacteria associated with arbuscular mycorrhizal *Funneliformis mosseae* spores, taxonomic characterization and their beneficial traits to plants: evidence obtained from mung bean (*Vigna radiata*) and Thai jasmine rice (*Oryza sativa*). Front Microbiol 9:1247. https://doi.org/10.3389/fmicb.2018.01247

Laursen JB, Nielsen J (2004) Phenazine natural products: biosynthesis, synthetic analogues, and biological activity. Chem Rev 104:1663–1685

Laville J, Blumer C, Von Schroetter C, Gaia V, Défago G, Keel C, Haas D (1998) Characterization of the hcnABC gene cluster encoding hydrogen cyanide synthase and anaerobic regulation by ANR in the strictly aerobic biocontrol agent *Pseudomonas fluorescens* CHA0. J Bacteriol 180:3187–3196

Leahy J, Mendelsohn M, Kough J, Jones R (2014) Biopesticide oversight and registration at the US Environmental Protection Agency. In: Gross AD, Coats JR, Duke SO, Seiber JN (eds) Biopesticides: state of the art and future opportunities. American Chemical Society, Washington, DC, pp 3–18. https://doi.org/10.1021/bk-2014-1172.ch001

Leclère V, Béchet M, Adam A, Guez JS, Wathelet B, Ongena M, Thonart P, Gancel F, CholletImbert M, Jacques P (2005) Mycosubtilin overproduction by *Bacillus subtilis* BBG 100 enhances the organism's antagonistic and biocontrol activities. Appl Environ Microbiol 8:4577–4584

Lee CH, Kempf HJ, Lim YH, Cho YH (2000) Biocontrol activity of *Pseudomonas cepacia* AF2001 and anthelmintic activity of its novel metabolite, cepacidine A. J Microbiol Biotechnol 10:568–571

Lee CH, Kim S, Hyun B, Suh JM, Yon C, Kim C, Lim Y, Kim C (1994) Cepacidine A, a novel antifungal antibiotic produced by *Pseudomonas cepacia*. J Antibiot 47:1402–1405. https://doi.org/10.7164/antibiotics.47.1402

Lee JY, Moon SS, Hwang BK (2003) Isolation and in vitro and in vivo activity against *Phytophthora capsici* and *Colletotrichum orbiculare* of phenazine-1-carboxylic acid from *Pseudomonas aeruginosa* strain GC-B26. Pest Manag Sci 59:872–882. https://doi.org/10.1002/ps.688

Lewin GR, Carlos C, Chevrette MG, Horn HA, McDonald BM, Stankey RJ, Fox BG, Currie CR (2016) Evolution and ecology of Actinobacteria and their bioenergy applications. Annu Rev Microbiol 70:235–254. https://doi.org/10.1146/annurev-micro102215-095748

Li D, Oku N, Hasada A, Shimizu M, Igarashi Y (2018) Two new 2-alkylquinolones, inhibitory to the fish skin ulcer pathogen *Tenacibaculum maritimum*, produced by a rhizobacterium of the genus *Burkholderia* sp. Beilstein J Org Chem 14:1446–1451. https://doi.org/10.3762/bjoc.14.122

Li P, Mao Z, Lou J, Li Y, Mou Y, Lu S, Peng Y, Zhou L (2011) Enhancement of diosgenin production in *Dioscorea zingiberensis* cell cultures by oligosaccharides from its endophytic fungus *Fusarium oxysporum* Dzf17. Molecules 16:10631–10644

Lim Y, Suh J-W, Kim S, Hyun B, Kim C, Lee C-h (1994) Cepacidine A, a novel antifungal antibiotic produced by Pseudomonas cepacia. II. Physicochemical properties and structure elucidation. J Antibiot 47(12):1406–1416

Lin Z, Falkinham JO III, Tawfik KA, Jeffs P, Bray B, Dubay G, Coz JE, Schmidt EW (2012) Burkholdines from *Burkholderia ambifaria*: antifungal agents and possible virulence factors. J Nat Prod 75:1518–1523. https://doi.org/10.1021/np300108u

Lopes LP, de Oliveira AG, Beranger JPO, Góis CG, Vasconcellos FCS, San Martin JAB, Andrade CGTJ, Mello JCP, Andrade G (2012) Activity of extracellular compounds of *Pseudomonas* sp. against *Xanthomonas axonopodis* in vitro and bacterial leaf blight in eucalyptus. Trop Plant Pathol 37:233–238

Lu S, Smith JL, Austin F, Gu G (2016) US Patent 14/806,121

Lu SE, Novak J, Austin FW, Gu G, Ellis D, Kirk M, Wilson-Stanford S, Tonelli M, Smith L (2009) Occidiofungin, a unique antifungal glycopeptide produced by a strain of *Burkholderia contaminans*. Biochemistry 48:8312–8321. https://doi.org/10.1021/bi900814c

Lugtenberg BJJ, Malfanova N, Kamilova F, Berg G (2013) Plant growth promotion by microbes. In: de Bruijn FJ (ed) Molecular microbial ecology of the rhizosphere. Wiley, Hoboken, pp 559–573

Madigan MT, Martinko JM, Dunlap PV, Clark DP (2010) Microbiologia industrial. In: Madigan MT, Martinko JM, Dunlap PV, Clark DP (eds) Microbiologia de Brock. Porto Alegre, Artmed, pp 734–760

Manivasagan P, Venkatesan J, Sivakumar K, Kim SK (2014) Pharmaceutically active secondary metabolites of marine Actinobacteria. Microbiol Res 169:262–278. https://doi.org/10.1016/J.MICRES.2013.07.014

Mao S, Lee SJ, Hwangbo H, Kim YW, Park KH, Cha GS, Park RD, Kim KY (2006) Isolation and characterization of antifungal substances from *Burkholderia* sp. culture broth. Curr Microbiol 53:358–364. https://doi.org/10.1007/s00284-005-0333-2

Marquez MC, Sanchez-Porro C, Ventosa A (2011) Hallophilic and haloakalophilic, aerobic endospore forming bacteria in soil. In: Logan NA, De Vos P (eds) Endospore forming soil bacteria. Springer, New York, pp 309–339

Martínez-Hidalgo P, García JM, Pozo MJ (2015) Induced systemic resistance against *Botrytis cinerea* by *Micromonospora* strains isolated from root nodules. Front Microbiol 6:1–11. https://doi.org/10.3389/fmicb.2015.00922

Marx V (2016) Microbiology: the road to strain-level identification. Nat Methods 13:401–404. https://doi.org/10.1038/nmeth.3837

Meyer JM, Hohnadel D, Hall F (1989) Cepabactin from *Pseudomonas cepacia*, a new type of siderophore. Microbiology 135:1479–1487. https://doi.org/10.1099/00221287-135-6-1479

Meyer JM, Hornsperger JM (1978) Role of pyoverdine, the iron binding fluorescent pigment of *Pseudomonas fluorescens*, in iron transport. J Gen Microbiol 107:329–331

Meyer JM, Van Van T, Stintzi A, Berge O, Winkelmann G (1995) Ornibactin production and transport properties in strains of *Burkholderia vietnamiensis* and *Burkholderia cepacia* (formerly *Pseudomonas cepacia*). Biometals 8:309–317. https://doi.org/10.1007/BF00141604

Meyer SLF, Halbrendt JM, Carta LK, Skantar AM, Liu T, Abdelnabby HME, Vinyard BT (2009) Toxicity of 2,4-diacetylphloroglucinol (DAPG) to plant parasitic and bacterial feeding nematodes. J Nematol 41:274–280

Meyers E, Bisacchi GS, Dean L, Liu WC, Minassian B, Slusarchyk DS, Sykes RB, Tanaka SK, Trejo W (1987) Xylocandin: a new complex of antifungal peptides. J Antibiot 40:1515–1519. https://doi.org/10.7164/antibiotics.40.1515

Minerdi D, Fani R, Gallo R, Boarino A, Bonfante P (2001) Nitrogen fixation genes in an endosymbiotic *Burkholderia* strain. Appl Environ Microbiol 67:725–732

Moon SS, Kang PM, Park KS, Kim CH (1996) Plant growth promoting and fungicidal 4-quinolinones from *Pseudomonas cepacia*. Phytochemistry 42:365–368. https://doi.org/10.1016/0031-9422(95)00897-7

Moraes RM, Lata H, Bedir E, Maqbool M, Cushman K (2002) The American mayapple and its potential for podophyllotoxin production. In: Janick J, Whipkey A (eds) Trends in new crops and new uses. ASHS, Alexandria, pp 527–532

Moyne AL, Shelby R, Cleveland TEE, Tuzun S (2001) Bacillomycin D: an iturin with antifungal activity against *Aspergillus favus*. J Appl Microbiol 90:622–629

Munhoz LD, Fonteque JP, Santos IMO, Navarro MOP, Simionato AS, Goya ET, Rezende MI, Balbi-Peña MI, de Oliveira AG, Andrade G (2017) Control of bacterial stem rot on tomato by extracellular bioactive compounds produced by *Pseudomonas aeruginosa* LV strain. Cogent Food Agric 1:1–2

Nandi M, Selin C, Brassinga AKC, Belmonte MF, Fernando WD, Loewen PC, De Kievit TR (2015) Pyrrolnitrin and hydrogen cyanide production by *Pseudomonas chlororaphis* strain PA23 exhibits nematicidal and repellent activity against *Caenorhabditis elegans*. PLoS One 10:e0123184

Naznin HA, Kimura M, Miyazawa M, Hyakumachi M (2013) Analysis of volatile organic compounds emitted by plant growth promoting fungus *Phoma* sp. GS8-3 for growth promotion effects on tobacco. Microbes Environ 28:42–49

Neilands JB (1995) Siderophores—structure and function of microbial iron transport compounds. J Biol Chem 270:26723–26726

Nicoletti R, Fiorentino A (2015) Plant bioactive metabolites and drugs produced by endophytic fungi of Spermatophyta. Agriculture 5:918–970

Nunes TF (2018) Avaliação de substâncias extracelulares de *Burkholderia pyrrocinia* (RV7S3) no controle de *Rhizoctonia solani* em *Nicotina tabacum*. Thesis, State University of Londrina

OECD (2016) Guidance document on storage stability of microbial pest control products. Series on pesticides no. 85. Environment Directorate. http://www.oecd.org/officialdocuments/publicdisplaydocumentpdf/?cote=env/jm/mono(2016)54&doclanguage=en

Oka M, Yaginuma K, Numata K, Konishi M, Oki T, Kawaguchi H (1988) Glidobactins a, b and c, new antitumor antibiotics. J Antibiot 41:1338–1350. https://doi.org/10.7164/antibiotics.41.1338

Okada A, Banno S, Ichiishi A, Kimura M, Yamaguchi I, Fujimura M (2005) Pyrrolnitrin interferes with osmotic signal transduction in *Neurospora crassa*. J Pestic Sci 30:378–383. https://doi.org/10.1584/jpestics.30.378

Ongena M, Jacques P, Touré Y, Destain J, Jabrane A, Thonart P (2005) Involvement of fengycin-type lipopeptides in the multifaceted biocontrol potential of *Bacillus subtilis*. Appl Microbiol Biotechnol 69:29–38

Otero Jiménez V (2011) Aislamiento, selección e identificación de actinomicetos, bacterias fotosintéticas no sulfurosas y bacterias ácido lácticas con potencial biofertilizante, a partir de suelos asociados al cultivo de plátano en la Costa Atlántica Colombiana. Thesis, Universidad Nacional de Colombia

Palem PPC, Kuriakose GC, Jayabaskaran C (2015) An endophytic fungus, *Talaromyces radicus*, isolated from *Catharanthus roseus*, produces vincristine and vinblastine, which induce apoptotic cell death. PLoS One 10(12):e0153111

Patil AS, Patil SR, Kale AS (2015) Influence of growth regulators on germination and growth of endangered medicinal plant *Nothapodytes nimmoniana* J. Graham under shade net conditions. Global J Res Med Plants Indigen Med 4:1–9

Peng H, Zhang P, Bilal M, Wang W, Hu H, Zhang X (2018) Enhanced biosynthesis of phenaine-1-carboxamide by engineered *Pseudomonas chlororaphis* HT66. Microb Cell Fact 17(1):117. https://doi.org/10.1186/s12934-018-0962-3

Phongpaichit S, Rungjindamai N, Rukachaisirikul V, Sakayaroj J (2006) Antimicrobial activity in cultures of endophytic fungi isolated from *Garcinia* species. FEMS Immunol Med Microbiol 48:367–372

Pierson EA, Wang D, Pierson LS III (2013) Roles and regulation of phenazines in the biological control strain *Pseudomonas chlorophis*. In: Chincholkar S, Thomashow L (eds) Microbial phenazines: biosynthesis, agriculture and health. Springer, London, pp 141–162

Pierson LS III, Pierson EA (2010) Metabolism and function of phenazines in bacteria: impacts on the behavior of bacteria in the environment and biotechnological processes. Appl Microbiol Biotechnol 86:1659–1670

Pimentel MR, Molina G, Dionísio AP, Maróstica Junior MR, Pastore GM (2011) The use of endophytes to obtain bioactive compounds and their application in biotransformation process. Biotechnol Res Int 2011:576286

Prince-Whelan A, Dietrich LE, Newman DK (2006) Rethinking 'secondary' metabolism: physiological roles for phenazine antibiotics. Nat Chem Biol 2:71–78

Puopolo G, Masi M, Raio A, Andolfi A, Zoina A, Cimmino A, Evidente A (2013) Insights on the susceptibility of plant pathogenic fungi to phenazine-1-carboxylic acid and its chemical derivatives. Nat Prod Res 27:956–966. https://doi.org/10.1080/14786419.2012.696257

Raaijmakers J, De Bruin I, Nybroe O, Ongena M (2010) Natural functions of cyclic lipopeptides from *Bacillus* and *Pseudomonas*: more than surfactants and antibiotics. FEMS Microbiol 34:1037–1062

Rajamanikyam M, Vadlapudi V, Amanchy R, Upadhyayula SM (2017) Endophytic fungi as novel resources of natural therapeutics. Braz Arch Biol Technol 60. https://doi.org/10.1590/1678-4324-2017160542

Ran LX, van Loon LC, Bakker PAHM (2005) No role for bacterially produced salicylic acid in rhizobacterial induction of systemic resistance in *Arabidopsis*. Phytopathology 95:3515–3518

Rane MR, Sarode PD, Chaudhari BL, Chincholkar SB (2008) Exploring antagonistic metabolites of established biocontrol agent of marine origin. Appl Biochem Biotechnol 151:665–675. https://doi.org/10.1007/s12010-008-8288-y

Rapparini F, Penuelas J (2014) Mycorrhizal fungi to alleviate drought stress on plant growth. In: Miransari M (ed) Use of microbes for the alleviation of soil stresses. Springer, New York, pp 21–42

Rashad YM, Al-Askar AA, Ghoneem KM, Saber WIA, Hafez EE (2017) Chitinolytic *Streptomyces griseorubens* E44G enhances the biocontrol efficacy against *Fusarium* wilt disease of tomato. Phytoparasitica 45:227–237. https://doi.org/10.1007/s12600-017-0580-3

Resende MLV, Salgado SML, Chaves ZM (2003) Espécies ativas de oxigênio na resposta de defesa de plantas a patógenos. Fitopatol Bras 28:123–130

Samarina LS, Malyarovskaya VI, Rogozhina EV, Malyukova LS (2017) Endophytes, as promotors of in vitro plant growth (review). Sel'skokhozyaistvennaya Biologiya 52(5):917–927

Saosoong K, Wongphathanakul W, Poasiri C, Ruangviriyachai C (2009) Isolation and analysis of antibacterial substance produced from *P. aeruginosa* TISTR 781. KKU Science J 37:163–172

Saranya K, Kumutha K (2011) Synergistic interactions of AM fungi with essential groups of microbes in plant growth promotion. Int J Curr Res 3(7):026–030

Scavino AF, Pedraza RO (2013) The role of siderophores in plant growth promoting bacteria. In: Maheshwari DK, Saraf M, Aeron A (eds) Bacteria in agrobiology: crop productivity. Springer, Berlin, pp 265–286

Schalk IJ, Kyslik P, Prome D, van Dorsselaer A, Poole K, Abdallah MA, Pattus F (1999) Copurification of the FpvA ferric pyoverdin receptor of *Pseudomonas aeruginosa* with its iron free ligand: implications for siderophore-mediated iron transport. Biochemist 38:9357–9365

Shafi J, Tian H, Ji M (2017) *Bacillus* species as versatile weapons for plant pathogens: a review. Biotechnol Biotechnol Equip 31:446–459

Shanmugaiah V, Mathivanan N, Varghese B (2010) Purification, crystal structure and antimicrobial activity of phenazine-1-carboxamide produced by a growth-promoting biocontrol bacterium, *Pseudomonas aeruginosa* MML2212. J Appl Microbiol 108:703–711

Sharma V, Salwan R (2018) Biocontrol potential and applications of Actinobacteria in agriculture. In: Singh B, Gupta V, Passari A (eds) Actinobacteria: diversity and biotechnological applications, new and future developments in microbial biotechnology and bioengineering, Ist edn. Elsevier, San Diego, pp 93–108

Shoji JI, Hinoo H, Kato T, Hattori T, Hirooka K, Tawara K, Shiratori O, Terui Y (1990) Isolation of cepafungins I, II and III from *Pseudomonas* species. J Antibiot 43:783–787. https://doi.org/10.7164/antibiotics.43.783

Siddiqui IA, Shaukat SS (2003a) Plant species, host age and host genotype effects on *Meloidogyne incognita* biocontrol by *Pseudomonas fluorescens* strain CHA0 and its genetically modified derivatives. Phytopathology 151:231–238

Siddiqui IA, Shaukat SS (2003b) Suppression of root-knot disease by *Pseudomonas fluorescens* CHA0 in tomato: importance of bacterial secondary metabolite, 2,4-diacetylpholoroglucinol. Soil Biol Biochem 35:1615–1623

Silva CS (2018) Avaliação da atividade fungicida de compostos produzidos por *Burkholderia pyrrocinia* cepa RV1R2 contra *Sclerotinia sclerotiorum* e *Rhizoctonia solani*. Thesis, State University of Londrina

Silva CS, Vivan ACP, Simões G, Simionato AS, Navarro MOP, Panagio LA, Almeida RSC, de Oliveira AG, Andrade G (2015) Metabólitos secundários com atividade inseticida sobre *Tenebrio molitor* Linnaeus (Coleoptera: Tenebrionidae) produzidos por isolado do complexo *Burkholderia cepacia*. Blucher Biochem Proc 1:353–353. https://doi.org/10.5151/biochem-vsimbbtec-22107

Simionato AS, Navarro MOP, Barazetti AR et al (2017a) Strategies for biological control and antagonisms. In: Singh DP, Singh HB, Prabha R (eds) Plant–microbe interactions in agro-ecological perspectives. Springer, Singapore, pp 221–244

Simionato AS, Navarro MOP, de Jesus MLA, Barazetti AR, da Silva CS, Simões GC, BalbiPeña MI, de Mello JCP, Panagio LA, de Almeida RSC, Andrade G, de Oliveira AG (2017b) The effect of phenazine-1-carboxylic acid on mycelial growth of *Botrytis cinerea* produced by *Pseudomonas aeruginosa* LV strain. Front Microbiol 8:1102. https://doi.org/10.3389/fmicb.2017.01102

Singh J, Rajput RS, Bisen K, Singh S, Singh HB (2017) Role of *Trichoderma* secondary metabolites in plant growth promotion and biological control. In: Singh HB, Sarma BK, Keswani C (eds) Advances in PGPR research. CABI, Oxfordshire, pp 411–426

Sokol PA, Lewis CJ, Dennis JJ (1992) Isolation of a novel siderophore from *Pseudomonas cepacia*. J Med Microbiol 36:184–189. https://doi.org/10.1099/00222615-36-3-184

Sousa CS, Soares ACF, Garrido MS (2008) Characterization of streptomycetes with potential to promote plant growth and biocontrol. Sci Agric 65:50–55. https://doi.org/10.1590/S0103-90162008000100007

Souza CS, Menezes RSC, Sampaio EVSB, Lima FS (2012) Glomalin: characteristics, production, limitations and contribution to soils. Semina: Ciências Agrárias 33(1):3033–3044

Spence CA, Raman V, Donofrio NM, Bais HP (2014) Global gene expression in rice blast pathogen *Magnaporthe oryzae* treated with a natural rice soil isolate. Planta 239:171–185. https://doi.org/10.1007/s00425-013-1974-1

Specian V, Sarragiotto MH, Pamphile JA, Clemente E (2012) Chemical characterization of bioactive compounds from the endophytic fungus Diaporthe helianthi isolated from *Luehea divaricata*. Braz J Microbiol 43(3):1174–1182

Sultan MZ, Park K, Lee SY, Park JK, Varughese T, Moon SS (2008) Novel oxidized derivatives of antifungal pyrrolnitrin from the bacterium *Burkholderia cepacia* K87. J Antibiot 61:420–425

Tank N, Rajendran N, Patel B, Saraf M (2012) Evaluation and biochemical characterization of a distinctive pyoverdin from a *Pseudomonas* isolated from chickpea rhizosphere. Braz J Microbiol 43:639–648. https://doi.org/10.1590/S1517-83822012000200028

Tao KLJ, Buta JG (1986) Differential effects of camptothecin and interactions with plant hormones on seed germination and seedling growth. Plant Growth Regul 4:219–226

Tawfik KA, Jeffs P, Bray B, Dubay G, Falkinham JO III, Mesbah M, Youssef D, Khalifa S, Schmidt EW (2010) Burkholdines 1097 and 1229, potent antifungal peptides from *Burkholderia ambifaria* 2.2 N. Org Lett 12:664–666. https://doi.org/10.1021/ol9029269

Terui Y, Nishikawa J, Hinoo H, Kato T, Shoji JI (1990) Structures of cepafungins I, II and III. J Antibiot 43:788–795. https://doi.org/10.7164/antibiotics.43.788

Toshima H, Maru K, Saito M, Ichihara A (1999) Study on fungitoxic 3-amino-2-piperidinone-containing lipids: total syntheses of cepaciamides A and B. Tetrahedron Lett 40:939–942. https://doi.org/10.1016/S0040-4039(98)02494-0

Touré Y, Ongena M, Jacques P, Guiro A, Thonart P (2004) Role of lipopeptides produced by *Bacillus subtilis* GA1 in the reduction of grey mould disease caused by *Botrytis cinerea* on apple. J Appl Microbiol 96:1151–1160

Vaishnav A, Sharma SK, Choudhary DK, Sharma KP, Ahmad E, Sharma MP, Ramesh A, Saxena AK (2018) Nitric oxide as a signalling molecule in plant–bacterial interactions. In: Egamberdieva D, Ahmad P (eds) Plant microbiome: stress response. Springer, Singapore, pp 183–199

van Loon LC, Bakker PA, van der Heijdt WH, Wendehenne D, Pugin A (2008) Early responses of tobacco suspension cells to rhizobacterial elicitors of induced systemic resistance. Mol Plant-Microbe Interact 21:1609–1621

van Pée KH, Ligon JM (2000) Biosynthesis of pyrrolnitrin and other phenylpyrrole derivatives by bacteria. Nat Prod Rep 17:157–164. https://doi.org/10.1039/A902138H

Vasconcellos FCS, de Oliveira AG, Lopes-Santos L, Beranger JPO, Cely MVT, Simionato AS, Pistori JF, Spago FR, de Mello JCP, San Martin JAB, Jesus Andrade CGT, Andrade G (2014) Evaluation of antibiotic activity produced by *Pseudomonas aeruginosa* LV strain against *Xanthomonas arboricola* pv. *pruni*. Agric Sci 5:71–76

Venieraki A, Dimou M, Katinakis P (2017) Endophytic fungi residing in medicinal plants have the ability to produce the same or similar pharmacologically active secondary metabolites as their hosts. Hellenic Plant Prot J 10:51–66

Vial L, Groleau MC, Dekimpe V, Deziel E (2007) *Burkholderia* diversity and versatility: an inventory of the extracellular products. J Microbiol Biotechnol 17:1407–1429

Vurukonda SSKP, Giovanardi D, Stefani E (2018) Plant growth promoting and biocontrol activity of *Streptomyces* spp. as endophytes. Int J Mol Sci 19:1–26

Wagner LJ, Flores HE (1994) Effect of taxol and related compounds on growth of plant pathogenic fungi. Biochem Cell Biol 84:1173–1178

Wang XQ, Liu AX, Guerrero A, Liu J, Yu XQ, Deng P, Ma L, Baird SM, Smith L, Li XD, Lu SE (2016) Occidiofungin is an important component responsible for the antifungal activity of *Burkholderia pyrrocinia* strain Lyc2. J Appl Microbiol 120:607–618. https://doi.org/10.1111/jam.13036

Wang Y, Zeng QG, Zhang ZB, Yan RM, Wang LY, Zhu D (2011) Isolation and characterization of endophytic huperzine A–producing fungi from *Huperzia serrata*. J Ind Microbiol Biotechnol 38:1267–1268

Waqas M, Khan AL, Kamran M, Hamayun M, Kang S, Kim Y, Lee I (2012) Endophytic fungi produce gibberellins and indoleacetic acid and promotes host-plant growth during stress. Molecules 17:10754–10773

Waqas M, Khan AL, Shahzad R, Ullah I, Khan AR, Lee J (2015) Mutualistic fungal endophytes produce phytohormones and organic acids that promote *japonica* rice plant growth under prolonged heat stress. J Zhejiang Univ-SCIENCE B (Biomedicine & Biotechnol) 16:1011–1018

Weller DM (2007) Phytopathology 97(2):250–256

Weller DM, Mavrodi DV, van Pelt JA, Pieterse CMJ, van Loon LC, Bakker PAHM (2012) Induced systemic resistance in *Arabidopsis thaliana* against Pseudomonas syringae pv. Tomato by 2,4-diacetylphloroglucinol-producing *Pseudomonas fluorescens*. Phytopathology 102:403–412

Wäspi U, Blanc D, Winkler T, Rüedi P, Dudler R (1998) Syringolin, a novel peptide elicitor from *Pseudomonas syringae* pv. *syringae* that induces resistance to *Pyricularia oryzae* in rice. Mol Plant-Microbe Interact 11:727–733. https://doi.org/10.1094/MPMI.1998.11.8.727

WHO [World Health Organization] (2017) WHO pesticide evaluation scheme (WHOPES), Guidelines for testing [online]. WHO, Geneva

Xu L, Zhou L, Zhao J, Li J, Li X, Wang J (2007) Fungal endophytes from *Dioscorea zingiberensis* rhizomes and their antibacterial activity. Appl Microbiol 46:68–72

Xu Z, Shao J, Li B, Shen Q, Zhang R (2013) Bacillomycin D in *Bacillus amyloliquefaciens* SQR9 contributes to antifungal activity and biofilm formation. Appl Environ Microbiol 79:755–781

Xu S, Pan X, Luo J, Wu J, Zhou Z, Liang X, He Y, Zhou M (2015) Effects of phenazine-1-carboxylic acid on the biology of the plant-pathogenic bacterium Xanthomonas oryzae pv. Oryzae. Pestic Biochem Physiol 117:39–46

Yandigeri MS, Malviya N, Solanki MK, Shrivastava P, Sivakumar G (2015) Chitinolytic *Streptomyces vinaceusdrappus* S5MW2 isolated from Chilika Lake, India enhances plant growth and biocontrol efficacy through chitin supplementation against *Rhizoctonia solani*. World J Microbiol Biotechnol 31:1217–1225. https://doi.org/10.1007/s11274-015-1870-x

Yánez-Mendizábal V, Zeriouh H, Viñas I, Torres R, Usall J, Vicente A, Pérez-García A, Teixidó N (2012) Biological control of peach brown rot (*Monilinia* spp.) by *Bacillus subtilis* CPA-8 is based on production of fengycin like lipopeptides. Eur J Plant Pathol 132:609–619

YE Y-f, LI Q-q, FU G, YUAN G-q, MIAO J-h, LIN W (2012) Identification of antifungal substance (Iturin A_2) produced by *Bacillus subtilis* B47 and its effect on southern corn leaf blight. J Integr Agric 11(1):90–99

Yoshihisa H, Zenji S, Fukushi H, Katsuhiro K, Haruhisa S, Takahito S (1989) Production of antibiotics by *Pseudomonas cepacia* as an agent for biological control of soil borne plant pathogens. Soil Biol Biochem 21:723–728. https://doi.org/10.1016/0038-0717(89)90070-9

Yu GY, Sinclair JB, Hartman GL, Bertagolli BL (2002) Production of iturin A by *Bacillus amyloliquefaciens* suppressing *Rhizoctonia solani*. Soil Biol Biochem 34:955–963

Yu JM, Wang D, Pierson LS III, Pierson EA (2018) Effect of producing different phenazines on bacterial fitness and biological control in *Pseudomonas chlororaphis* 30-84. Plant Pathol J 34:44–58. https://doi.org/10.5423/PPJ.FT.12.2017.0277

Yuan J, Li B, Zhang N, Waseem R, Shen Q, Huang Q (2012) Production of bacillomycin and macrolactin type antibiotics by *Bacillus amyloliquefaciens* NJN-6 for suppressing soil borne plant pathogens. J Agric Food Chem 60:2976–2981

Yuan L, Li Y, Wang Y, Zhang X, Xu Y (2008) Optimization of critical medium components using response surface methodology for phenazine-1-carboxylic acid production by *Pseudomonas* sp. M-18Q. J Biosci Bioeng 3:232–237. https://doi.org/10.1263/jbb.105.232

Zeilinger S, Gupta VK, Dahms TES, Silva RN, Singh HB, Upadhyay RS, Gomes EV, Tsui CK-M, Chandra Nayak S, van der Meer JR (2016) Friends or foes? Emerging insights from fungal interactions with plants. FEMS Microbiol Rev 40(2):182–207

Zhang T, Shi ZQ, Hu LB, Cheng LG, Wang F (2008) Antifungal compounds from *Bacillus subtilis* B-FS06 inhibiting the growth of *Aspergillus flavus*. World J Microbiol Biotechnol 24:783–788

Zhang Y, Wang C, Su P, Liao X, Borkovich KA (2015) Control effect and possible mechanism of the natural compound Phenazine-1-Carboxamide against Botrytis cinerea. PLoS One 10(10):e0140380

Zhao J, Shan T, Mou Y, Zhou L (2011) Plant derived bioactive compounds produced by endophytic fungi. Mini-Rev Med Chem 11:159–168

Zhao J, Zhou L, Wang J, Shan T, Zhong L, Liu X, Gao X (2010) Endophytic fungi for producing bioactive compounds originally from their host plants. In: Mendez-Vilas A (ed) Current research, technology and education topics in applied microbiology and microbial biotechnology, vol 1. Formatex Research Center, Badajoz, pp 567–576

Zhou Q, Li K, Jun X, Bo L (2009) Role and functions of beneficial microorganisms in sustainable aquaculture. Bioresour Technol 100:3780–3786. https://doi.org/10.1016/j.biortech.2008.12.037

The Continuous Story of Truffle-Plant Interaction

16

Ghoson Mosbah Daba, Waill Ahmed Elkhateeb, Ting-Chi Wen, and Paul William Thomas

Abstract

Truffles are symbiotic, ectomycorrhizal fungi that grow in the specific climates over a wide range of host plants. Truffles belong to genus *Tuber* and some species are famous for their high market value. In this chapter, the interaction between the genus *Tuber* and their host plants was deliberated.

Keywords

Truffle · *Tuber* · Life cycle · Host plant

16.1 Introduction

The term "truffle" is most often used to define the edible hypogeal fruiting bodies of fungi belonging to the *Tuber* genus (ascomycetes). Within the *Ascomycota* phylum, these fungi are further characterized by being dependent on a living plant partner in order to complete their lifecycle. Although the exact relationship of some species within the genus with their plant host has not yet been fully elucidated, we know that the ability to form structures with their host plants' root system known as

G. M. Daba · W. A. Elkhateeb
Chemistry of Natural and Microbial Products Department, Pharmaceutical Industries Researches Division, National Research Center, Giza, Egypt

T.-C. Wen
The Engineering Research Center of Southwest Bio–Pharmaceutical Resources, Ministry of Education, Guizhou University, Guiyang, Guizhou Province, China

P. W. Thomas (✉)
Mycorrhizal Systems Ltd, Lancashire, UK

University of Stirling, Stirling, UK
e-mail: paul@plantationsystems.com

V. Kumar et al. (eds.), *Microbiome in Plant Health and Disease*,
https://doi.org/10.1007/978-981-13-8495-0_16

mycorrhizae is widespread within the genus. Consisting of a dense hyphal mass surrounding a root tip (the mantel), the association is further characterized as ectomycorrhizal (ECM) with a network of highly branched hyphae extending from the mantel to occupy the space between root epidermal and cortical cells (the Hartig net). This structure allows a large surface area for the mutually beneficial exchange of resources between the plant host and the truffle fungus (Smith and Read 1997). Further, the formation of ECM induces changes in root morphology, which may appear as formation of lateral roots and/or in some plants root dichotomous branching in meristems (Peterson and Bonfante 1994). Different fungal strains affect plant growth in different ways and behave dissimilarly during the formation of ectomycorrhiza (Thoen et al. 1990). Moreover, there is variation among strains of the same truffle species in their capability of colonizing a single host (Giomaro et al. 2000).

Revered for their complex aroma, truffles are originally known from the northern hemisphere but occupy a range of different climatic zones, from the Mediterranean climate of northern Morocco to the temperate oceanic climate of the UK (Bonito et al. 2013). However, with the advent of cultivation efforts, fruiting bodies are also now harvested in a number of southern hemisphere countries (Zambonelli et al. 2015; Hall et al. 2017). In those areas where truffles occur naturally or are cultivated, they are often from a key component of local cultural identity, and through harvesting and associated activities, they may also be a valued contributor to the local economy (Samils et al. 2008). Further, in other regions, truffles may be locally scarce, with associated legislative protection, and are utilized as key species for local conservation initiatives.

Although around 40 species of truffles have been discovered, only few of them have a significant market value (Fig. 16.1), and these include the summer/autumn truffle (*Tuber aestivum syn. uncinatum*) and the black truffle (*Tuber melanosporum Vitt.*). Fresh *Tuber indicum* produced in China are more than 300 tonnes every year. Truffles occupy a high position in the list of the globally most expensive foods due to the gap between production (hundreds of tonnes) and market demands. The cost of one kilogram of white truffles can surge to $5000 (Patel et al. 2017). Recently, a trio of white truffles weighing 1.9 pounds was sold at a charity auction in Italy for $85,600 in November 2017 (New York Post 2017). The genome of *Tuber melanosporum* has recently been published, and this has given scientists unique opportunities to study more about the natural science of this precious edible fungus (Ursula Kües and Francis Martin 2011).

16.2 Why Truffles Are Expensive?

Truffles are famous for being among the world's most expensive food products. This high value is due to the inability of production to meet the market demand; this is true even for the widely cultivated species *T. melanosporum* and also for *T. magnatum*, a species that has so far eluded cultivation attempts. Cultivation of *T. melanosporum* began, in earnest, in Europe in the 1970s and today is so successful that over 90% of all black perigord truffles originating from France now come from

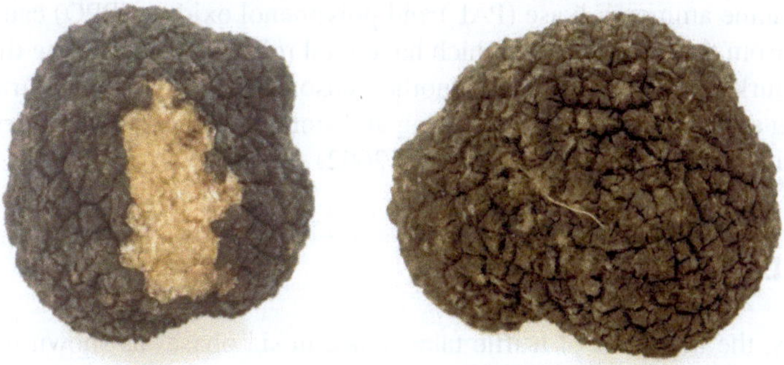

Tuber aestivum syn. uncinatum

Tuber melanosporum

Fig. 16.1 Main truffles of market value. (Photographs taken by Paul W. Thomas and Ting-Chi Wen)

purposely planted truffle orchards (Reyna and Garcia-Barreda 2014). Although cultivation occurs on a large scale in France, this undertaking has only been enough to stem the decline in natural truffle production rather than increase yields overall. Further, climate change is already impacting truffle production across Europe (Thomas and Büntgen 2019), and this is forecasted to continue. From the point at which cultivation begins in a new truffle orchard, fruiting bodies may take over 10 years to be produced even if the management is good. However, truffles remain a crop that requires a significant investment in time, money, and technical expertise, factors which create a barrier to entry for many would-be cultivators.

Aside from production, there are other factors that contribute to the high price point of truffles. This includes the lack of harvest mechanization, as cultivators are dependent on specially trained scent-detection dogs to locate their subterranean harvest (Splivallo et al. 2012). Further, the perishable nature of truffles limits their availability through distribution networks. Some metabolites in truffles such as phenylalanine ammonia-lyase (PAL) and polyphenol oxidase (PPO) cause melanin creation from the polyphenols, which have vital role in compromising the aesthetic appeal (Burke and Cairney 2002). Another reason for the high price of truffles is the damaging effect of industrial processing and storage methods such as drying, chilling, pickling, and canning (Murcia et al. 2002).

16.3 Life Cycle of Tubers

Generally, the life cycle of truffle takes place in six phases as shown in Fig. 16.2. Phases 1 and 2 start with a limited extraradical phase of vegetative growth in which the hyphae proliferate before coming into contact with roots of the host plant. Once this contact is initiated, phase 3 or the symbiotic phase begins, leading to phase 4 which is characterized by the development of a new organ (ectomycorrhiza). In the final phases 5 and 6, the mycelium is organized into the fruit body which is responsible of producing sexual fructifications to be dispersed in the environment. After that, vegetative mycelia are developed from these fructifications, creating a new extraradical phase and completing the truffle life cycle. Fruiting bodies are generally collected during phases 5 and 6 in the autumn/winter (the period between October and December). The mycelium may produce an additional sexual fructification during summer time (between June and August), which signifies the same growing phase but appears in a dissimilar season (Vita et al. 2015).

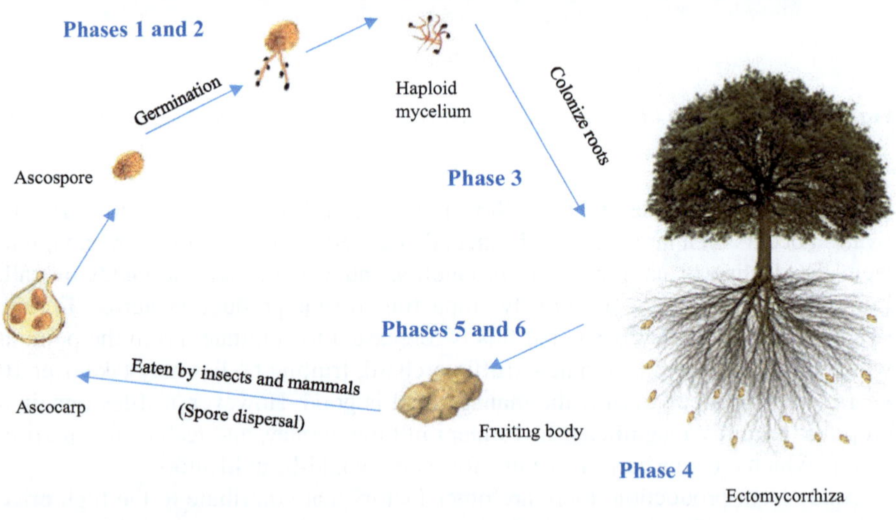

Fig. 16.2 Major phases in *Tuber* life cycle

16.4 Interaction Between *Tuber* and Their Host Plant

Basically, tubers grow near the roots of trees, especially oak trees; that's the reason for the nomenclature (mycorrhizae), which appeared in 1885 by Frank, to describe the symbiotic relation existing between the roots of trees and *Tuber*. This relation was experimentally proven in 1967 by Fassi and Fontana who inoculated *Pinus strobus* L. with the *Tuber maculatum* Vittad and hence induced production of ectomycorrhizas.

Many studies highlighted the critical need of *Tuber* to be associated with a host plant in order to produce mature fruiting bodies and complete their life cycle (Payen et al. 2014). Beyond this, a lot of what is known about the *Tuber* genus is confined to studies on just a handful of species.

Developing truffle ascocarps require carbon, and this is photosynthetically derived, being provided by a host plant, and then accumulating in ascocarps until they become totally mature (Le Tacon et al. 2013, Payen et al. 2014). It was thought that the ascocarps are originated using the host plant dead tissues or some organic matter from the soil (Callot 1999). Le Tacon et al. 2013 have proven the dependence of *Tuber* ascocarps on their hosts during their growth and development. *Tuber* ascocarps may remain as a carbon sink for many months even after the photosynthetic carbon assimilation in the host plant ceases. Studies using ^{13}C tracer described that even after the fall of host plant leaves, *Tuber* mycorrhizas obtain plant carbon and passage sugars through mycelium channels for the growth of fruiting bodies (Le Tacon et al. 2013). This resource distribution is not a one-way street as the truffle fungus also transports nutrients to the host plant and may contribute in the nitrogen-fixation process (Belfiori et al. 2012; Le Roux et al. 2016). Both *Tuber* and the plant can profit from the interaction and absorb major nutrients (e.g., nitrogen, phosphate).

Some truffle fungi may also benefit their plant host in other ways as some volatile organic compounds released by *Tuber* may negatively impact plant growth. For example, volatiles emitted by some truffles (*Tuber* spp.) inhibited the growth of *Arabidopsis thaliana,* causing bleaching of *A. thaliana* leaves, and/or inhibition of leaf and root growth (Splivallo et al. 2007a). This inhibition is often observed in the wild as well as truffle orchards and is visible as barren patches of ground, extending beyond the tree canopy in which many plant species will not grow. These barren areas are often termed "brûlés" (French for burn), and it indicates the ability of the truffle fungi to remove potential resource competitors to their plant host (Splivallo et al. 2011). Several studies propose that soil microarthropods play a vital role in fungi spreading, but much remains unknown about the interaction between truffle and soil microarthropods (Menta et al. 2014). Surveys suggest that some organisms, such as some *Collembola*, might find a promising environment inside the brûlé, while others may not (Menta et al. 2014; Pinto et al. 2017). Further, Aleksandar et al. (2013) proposed that bacteria may be pivotal in improving truffle nutrition, ascocarp degradation and facilitating relationships with other soil fungi.

16.5 Signaling Between Truffles and Plants

Generally, communication between microorganisms and plants in soil takes place by exchanging chemical volatile compounds or solutes signals throughout the rhizosphere. More than 200 of VOCs have been isolated from the fruiting bodies and mycelia of *Tuber* (Zeppa et al. 2004) and from ectomycorrhizal roots (Menotta et al. 2004). They are hydrocarbons characterized by a high vapor pressure that generally include alcohol, aldehyde, and/or ketone functional groups and often contain sulfur atoms (Splivallo et al. 2011). Altered proportions of such sulfurous compounds lead to the difference in aroma among various *Tuber* species (Wenke et al. 2010). Truffles live in symbiosis with several yeasts, which also participate in characteristic scent of those truffles (Buzzini et al. 2005). Moreover, some bacteria are loosely or tightly associated with mycorrhizal associations, complementing the roles of the external mycelium by transferring the mineral nutrients (Frey-Klett et al. 2007) and creating VOCs that contribute to truffle aroma in association with other *Tuber*-associated microbes (yeasts and other fungi) (Buzzini et al. 2005; Tarkka and Piechulla 2007).

So VOCs are communication tools involved in the signaling between truffle and plants (Splivallo et al. 2007b), and they have a potential role as mycorrhization signals (Menotta et al. 2004). Fruiting bodies' VOCs shortened primary roots of plants (Splivallo et al. 2007b). Compounds such as trans-2-octenal, 3-octenol, and 1-octen-3-ol cause decolorizing and root growth inhibition depending on the *Tuber* species. The mycorrhizal symbiosis is based on exchange of resources: *Tuber* provides some nutrients in return for organic carbon structures that it gets from the plant. Formation of fruiting bodies by *Tuber* critically depends on forming a symbiotic relationship with plant roots and establishing ectomycorrhizas (Splivallo et al. 2011).

Tubers release the auxin class hormone indole-3-acetic acid (IAA) and ethylene, which work together encouraging ectomycorrhiza creation. Studies revealed that at an early stage of interaction and before direct contact, truffles release both hormones at levels that induces morphological changes such as root shortening, lateral root creation, enlarged branching, and root hair elongation in the roots of the host plant *Cistus* and the nonhost *Arabidopsis* (Splivallo et al. 2009; Fu et al. 2015). IAA and ethylene are the main signals guiding root progress before contact with the mycelium. Other signals might also be involved at this early interaction stage or later on such as the instable intermediate of ethylene, α-Keto-γ-(methylthio) butyric acid (KMBA). It is suggested that KMBA produced by truffle could be degraded by the action of enzymes such as peroxidases (Chague´ et al. 2002) either after being uptaken inside roots or by other microorganisms existing in the rhizosphere.

16.6 Conclusion

Truffle is the edible hypogeal fruiting body of fungi belonging to the *Tuber* genus (ascomycetes). Truffles are famous for being among the world's most expensive food products. Generally, the life cycle of truffle takes place in six phases. *Tuber* mycorrhizas gain plant carbon and passage sugars through mycelium channels so as to develop fruiting bodies. The truffle fungus also transports nutrients to the host plant and may contribute in the nitrogen-fixation process. Some truffle fungi may also benefit their plant host in other ways as some volatile organic compounds released by *Tuber* may negatively impact plant growth. Formation of fruiting bodies by *Tuber* critically depends on forming symbiotic relationships with plant roots and establishing ectomycorrhizas. Communication between microorganisms and plants in soil takes place by exchanging chemical volatile compounds or solutes signals throughout the rhizosphere. In excess of 200 of VOCs have been isolated from the fruiting bodies and mycelia of *Tuber.* The interaction among the genus *Tuber* and their host plants need more full enquiry.

References

Aleksandar P, Stanković S, Saljnikov E, Krüger D, Buscot F, Tarkka M, Marjanović Ž (2013) Actinobacteria may influence white truffle (*Tuber magnatum*, Pico) nutrition, ascocarp degradation and interactions with other soil fungi [J]. Fungal Ecol 6(6):527–538

Belfiori B, Riccioni C, Tempesta S, Pasqualetti M, Paolocci F, Rubini A (2012) Comparison of ectomycorrhizal communities in natural and cultivated *Tuber melanosporum* truffle grounds. FEMS Microbiol Ecol 81:547–561

Bonito G, Smith ME, Nowak M, Healy RA, Guevara G, Cázares E, Kinoshita A, Nouhra ER, Domínguez LS, Tedersoo L, Murat C (2013) Historical biogeography and diversification of truffles in the Tuberaceae and their newly identified southern hemisphere sister lineage. PloS One 8(1):e52765

Burke R, Cairney J (2002) Laccases and other polyphenol oxidases in ecto-and ericoid mycorrhizal fungi. Mycorrhiza 12:105–116

Buzzini P, Gasparetti C, Turchetti B, Cramarossa MR, Vaughan-Martini A, Martini A, Pagnoni UM, Forti L (2005) Production of volatile organic compounds (VOCs) by yeasts isolated from the ascocarps of black (*Tuber melanosporum* Vitt.) and white (*Tuber magnatum* Pico) truffles. Arch Microbiol 184:187–193

Callot G (1999) La truffe, la terre, la vie, ed. INRA, París, 210 p

Chague´ V, Elad Y, Barakat R, Tudzynski P, Sharon A (2002) Ethylene biosynthesis in *Botrytis cinerea*. FEMS Microbiol Ecol 40:143–149

Fassi B, Fontana A (1967) Sintesi micorrizica tra Pinus strobus e Tuber maculatum – I. Micorrize e sviluppo dei semenzali nel secondo anno. Allionia 13:177–186

Frank AB (1885) Uber die auf Wurzelsymbiose beruhende Ernährung gewisser Bäume durch unterirdische Pilze. Ber Dtsch Bot Ges 3:128–145

Frey-Klett P, Garbaye J, Tarkka M (2007) The mycorrhiza helper bacteria revisited. New Phytol 176:22–36

Fu SF, Wei JY, Chen HW, Liu YY, Lu HY, Chou JY (2015) Indole-3-acetic acid: A widespread physiological code in interactions of fungi with other organisms. Plant Signal Behavior 10(8):e1048052

Giomaro G, Zambonelli A, Sisti D, Cecchini M, Evangelista V, Stocchi V (2000) Anatomical and morphological characterization of mycorrhizas of five strains of *Tuber borchii* Vittad. Mycorrhiza 10:107–114

Hall I, Fitzpatrick N, Miros P, Zambonelli A (2017) Counter-season cultivation of truffles in the southern hemisphere: an update. Ital J Mycol 46:21–36

Italian white truffles auctioned off for over $85K (2017) New York post, November 12

Le Roux C, Tournier E, Lies A, Sanguin H, Chevalier G, Duponnois R, Mousain D, Prin Y (2016) Bacteria of the genus *Rhodopseudomonas* (Bradyrhizobiaceae): obligate symbionts in mycelial cultures of the black truffles Tuber melanosporum and Tuber brumale. Springer Plus 5:1085–1094

Le Tacon F, Zeller B, Plain C, Hossann C, Bre´chet C, Robin C (2013) Carbon transfer from the host to *Tuber melanosporum* mycorrhizas and ascocarps followed using a 13C pulse-labeling technique. PLoS One 8(5):e64626

Menotta M, Gioacchini A, Amicucci A, Buffalini M, Sisti D, Stocchi V (2004) Headspace solid-phase microextraction with gas chromatography and mass spectrometry in the investigation of volatile organic compounds in an ectomycorrhiza synthesis system. Rapid Commun Mass Spectrom 18:206–210

Menta C, García-Montero LG, Pinto S, Conti FD, Baroni G, Maresi M (2014) Does the natural "microcosm" created by *Tuber aestivum*, affect soil microarthropods? A new hypothesis based on Collembola in truffle culture [culture J]. Appl Soil Ecol 84(84):31–37

Murcia MA, Martinez-Tome M, Jimenez AM, Vera AM, Honrubia M, Parras P (2002) Antioxidant activity of edible fungi (truffles and mushrooms): losses during industrial processing. J Food Prot 65:1614–1622

Patel S, Rauf A, Khan H, Khalid S, Mubarak MS (2017) Potential health benefits of natural products derived from truffles: a review. Trends Food Sci Technol 70:1–8

Payen T, Murat C, Bonito G (2014) Truffle phylogenomics: new insights into truffle evolution and truffle life cycle. Adv Bot Res 70:211–234

Peterson R, Bonfante P (1994) Comparative structure of vesicular arbuscular mycorrhizas and ectomycorrhizas. Plant Soil 159:79–88

Pinto S, Gatti F, Garcíamontero L, Menta C (2017) Does soil fauna like truffles just as humans do? One-year study of biodiversity in natural brûlés of *Tuber aestivum* Vittad. [J]. Sci Total Environ 584

Reyna, S., Garcia-Barreda, S., 2014. Black truffle cultivation: a global reality. For. Syst. 23 (2), 317–328.

Samils N, Olivera A, Danell E, Alexander SJ, Fischer C, Colinas C (2008) The socioeconomic impact of truffle cultivation in rural Spain. Econ Bot 62(3):331–337

Smith S, Read D (1997) Mycorrhizal symbiosis, Ed 2. Academic Press, London

Splivallo R, Bossi S, Maffei M, Bonfante P (2007a) Discrimination of truffle fruiting body versus mycelial aromas by stir bar sorptive extraction. Phytochem 68:2584e2598

Splivallo R, Novero M, Bertea CM, Bossi S, Bonfante P (2007b) Truffle volatiles inhibit growth and induce an oxidative burst in *Arabidopsis thaliana*. New Phytol 175:417–424

Splivallo R, Fischer U, Göbel C, Feussner I, Karlovsky P (2009) Truffles regulate plant root morphogenesis via the production of auxin and ethylene. Plant Physiol 150(4):2018–2029

Splivallo R, Ottonello S, Mello A, Karlovsky P (2011) Truffle volatiles: from chemical ecology to aroma biosynthesis. New Phytol 189:688–699

Splivallo R, Valdez N, Kirchhoff N, Ona MC, Schmidt JP, Feussner I, Karlovsky P (2012) Intraspecific genotypic variability determines concentrations of key truffle volatiles. New Phytol 194:823–835

Tarkka M, Piechulla B (2007) Aromatic weapons: truffles attack plants by the production of volatiles. New Phytol 175(3):381–383

Thoen D, Sougoufara B, Dommergues Y (1990) In vitro mycorrhization of Casuarina and Allocasuarina species by *Pisolithus* isolates. Can J Bot 86:2537–2542

Thomas, P., & Büntgen, U. (2019). A risk assessment of Europe's black truffle sector under predicted climate change. Science of The Total Environment, 655, 27-34.

Ursula K, Francis M (2011) On the road to understanding truffles in the underground. Fungal Genet Biol 48(6):555–560

Vita F, Taiti C, Pompeiano A, Bazihizina N, Lucarotti V, Mancuso S, Alpi A (2015) Volatile organic compounds in truffle (*Tuber magnatum* Pico): comparison of samples from different regions of Italy and from different seasons. Sci Rep 5:12629

Wenke K, Kai M, Piechulla B (2010) Belowground volatiles facilitate interactions between plant roots and soil organisms. Planta 231:499–506

Zambonelli A, Iotti M, Hall I (2015) Current status of truffle cultivation: recent results and future perspectives. Ital J Mycol 44(1):31–40

Zeppa S, Gioacchini AM, Guidi C, Guescini M, Pierleoni R, Zambonelli A, Stocchi V (2004) Determination of specific volatile organic compounds synthesised during *Tuber borchii* fruit body development by solid-phase microextraction and gas chromatography/mass spectrometry. Rapid Commun Mass Spectrom 18:199–205

Can Soil Microorganisms Reduce Broomrape (*Orobanche* spp.) Infestation in Cropping Systems?

17

Gholamreza Mohammadi

Abstract

Among the parasitic plants around the world, broomrape is proposed as the most serious threat for crop production and food security. It attacks important crops belonging to different families such as Solanaceae, Fabaceae, Asteraceae, etc. which have a substantial contribution to supply people's food in a global scale. Sometimes, destructive effects of the parasite lead to the complete loss of the crop yield. In recent years, broomrape infestation has notably extended in various parts of the world including Iran. Conventional methods to control broomrape are usually expensive and in most cases inefficient. Moreover, the chemical compounds used to suppress the parasite such as methyl bromide and chloropicrin can severely damage beneficial soil organisms and destroy atmospheric ozone layer. Recently, soil microorganisms have been proposed as effective and environmentally sound agents to control broomrape and reduce its damaging effects in agroecosystems. They can be divided into two main groups including pathogenic and nonpathogenic microorganisms which can affect the parasite directly and indirectly, respectively. Among the pathogenic microorganisms, *Fusarium* spp. are the most important candidates, and among the nonpathogenic ones, two famous symbionts, i.e., mycorrhiza and *Rhizobium* spp., are mostly proposed. However, there are no many reports on the role of soil microorganisms as biocontrol agents for broomrape. In this chapter, some important microorganisms having controlling effects on this parasitic weed and the mechanisms by which their effects can be achieved are discussed.

Keywords

Broomrape · *Orobanche* · Microbe · Biocontrol

G. Mohammadi (✉)
Department of Plant Production and Genetic Engineering, College of Agriculture and Natural Resources, Razi University, Kermanshah, Iran

V. Kumar et al. (eds.), *Microbiome in Plant Health and Disease*,
https://doi.org/10.1007/978-981-13-8495-0_17

17.1 Introduction

Broomrape is an important constraint to produce a wide range of crop plants in many parts of the world such as Mediterranean region, southeast Europe, North Africa, and the Middle East (Parker 2012; Parker and Riches 1993). It has several species which can parasite the root of the host plant. Because of the exclusive characteristics of this plant parasite including a direct and close connection to the host plant, the control of broomrape can be very problematic. To date, different approaches have been recommended to control this parasite. Some of them are application of fumigants such as methyl bromide to sterilize the soil, hand weeding, the use of resistant cultivars, soil solarization, etc. In most cases, these approaches are hazardous to the environment, cost ineffective, and mostly inefficient. Recently, soil microorganisms have been proposed as safe, effective, and environmentally sound tools to reduce broomrape infestation in cropping systems which consequently lead to increased crop yield and economic return. They can be divided into different groups including symbionts (e.g., mycorrhiza and *Rhizobium* spp.) which colonize root of host plants and non-symbionts and pathogenic microorganisms (e.g., *Fusarium* spp.) which directly attack the parasite.

These biocontrol agents can support crops against the parasite by different mechanisms such as nutrient supply (especially P and N) to crop, prevention of parasite seed germination via production of toxins, and reduction of the seed germination stimulants released by host plant roots. In general, their protection effects may be achieved by releasing diverse compounds that suppress the parasite or eliciting the immune system of the host plant which leads to a higher resistance against broomrape. In recent years, the use of these beneficial microorganisms as a promising strategy that can be included in an integrated broomrape management program has received notable attention.

17.2 Broomrape as a Serious Threat for Crop Production and Food Security Around the World

Nearly one percent of angiosperm plant species are able to parasite other plants (Kuijt 1969; Estabrook and Yoder 1998; Parker and Riches 1993). Among them, *Orobanche* species are known as obligate parasites that can survive only when they attach to the roots of their host plants. They attack different families of crops including Solanaceae, Fabaceae, Asteraceae, Brassicaceae, and Umbelliferae (Parker and Riches 1993; Westwood et al. 2010) which sometimes leads to the full destruction of their fields. About 16 million hectares of arable lands of the Middle East and Mediterranean basin have been infested by different species of broomrape (Sauerborn 1991). In Table 17.1, the most important broomrape species, their common host crops, and geographical areas infested by them are presented.

Broomrape species are holoparasites which do not have chlorophyll and, therefore, not able to synthesize their needed assimilates and are entirely dependent on their host plants (Nickrent and Musselman 2004).

Table 17.1 Most important broomrape species along with their common host crops and geographical infested areas

Broomrape species	Host crop	Infested area	Reference
Orobanche ramosa (syn. *O. aegyptiaca* and *Phelipanche aegyptiaca*)	Different crop families including legumes and many vegetables	Mediterranean region, middle east and Asia	Parker (2009), Musselman (1991)
Orobanche crenata	Grain and forage legumes	Mediterranean basin and middle east	Rubiales et al. (2009)
Orobanche minor	Clover	Eastern France Oregon, USA	Eizenberg et al. (2005)
Orobanche foetida	Faba bean, common vetch	Tunisia, Morocco	Kharrat et al. (1992), Rubiales et al. (2005)
Orobanche cumuna	Sunflower	Some European countries, China	Parker (2009), Fernandez Escobar et al. (2009), Melero et al. (2000), Shindrova (2006)
Orobanche cernua	Sunflower	Turkey	Kaya et al. (2004)

It has been reported that among the broomrape species, *O. ramosa* is the most damaging one as it can parasite about eleven different dicotyledon families with high economic importance. In Iran, different species of broomrape have been distributed throughout the country and severely damage many important crops, as observed in some cases where the field was completely destroyed. Heavy yield losses are also reported in other parts of the world. For instance, in Spain, *O. crenata* is a serious damaging weed in legume crops, as in 1996, the yield loss caused by this parasite in pea production systems in Seville province was about 80% (Garcia-Torres et al. 1998). Crops parasitized by *Orobanche* spp. may be severely damaged even before emergence of the parasite above the soil surface. Therefore, broomrape control is usually difficult because of delayed parasitic shoot appearance in the field, a direct and close connection to the host plant, and the ability to produce a large number of tiny (dust-like) seeds with a long-time viability which germinate only after the chemical signal is released by the host plant (Linke and Saxena 1991).

17.3 Host-Parasite Interactions in Rhizosphere

Broomrape has a high potential to produce seeds and form a rich seed bank in the soil. A single plant can produce more than 500,000 seeds which may be viable for several decades in the soil. This can lead to a high genetic diversity resulting in a notable adaptability to the environmental conditions and different control strategies. Some of these strategies consist of the use of resistant cultivars and herbicide application (Joel et al. 2007). However, broomrape seeds are very small with minimum nutrient reserves and therefore are not able to germinate in the absence of a suitable host plant because they must receive chemical signals from their hosts before they

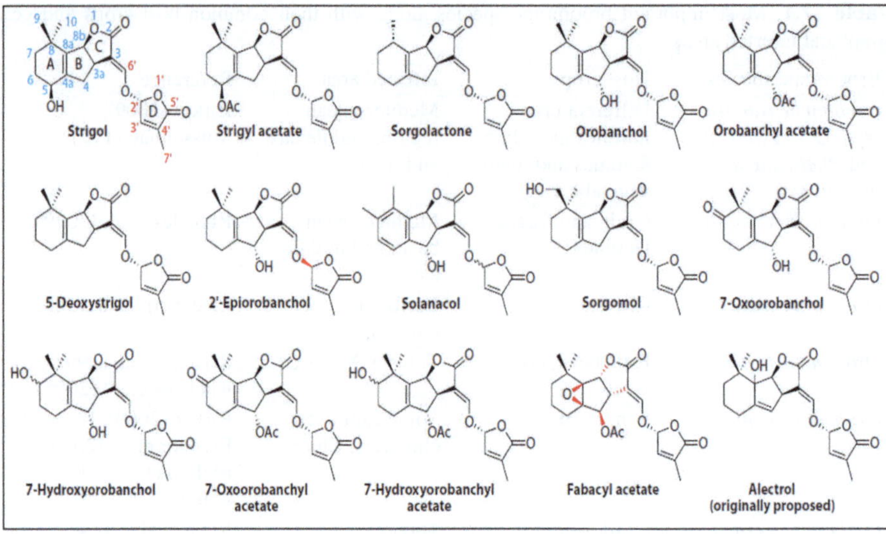

Fig. 17.1 Different forms of SLs. (Adapted from Xie et al. 2010)

can germinate. This is a very special detection mechanism in which strigolactones (SLs) play a key role. SLs are the compounds released by host plant roots and stimulate broomrape seed germination (Bouwmeester et al. 2003; Joel et al. 2007; Cardoso et al. 2011). By now, different forms of SLs have been known, some of which are mentioned in Fig. 17.1.

The parasite seeds are very sensitive to SLs as their germination can be stimulated at very low concentrations of these compounds (10^{-8} to 10^{-12} M) (Bouwmeester et al. 2003; Hirsch et al. 2003; Bouwmeester et al. 2007). Strigolactones are secondary metabolites which originate from the carotenoids (Matusova et al. 2005). They play different roles such as spore germination and hyphal branching of arbuscular mycorrhizal fungi (AM fungi) and increase AM fungi mitochondrial activity and respiration (Besserer et al. 2006, 2008). They can also regulate plant shoot branching (Gomez-Roldan et al. 2008; Umehara et al. 2008) by reducing the number of branches especially under stress conditions.

As mentioned above, SLs have enhancing effects on cell mitochondrial activity in AM fungi, and therefore, it can be expected that they also cause similar influence on broomrape seeds which can ultimately lead to their germination. As shown in Fig. 17.2, both SLs and abscisic acid have the same origin, and two carotenoid cleavage dioxygenases, namely, CCD7 and CCD8, are responsible for the biosynthesis of SLs from β-carotene (Lopez-Raez et al. 2010). However, SLs show a dual action in the plant rhizosphere as they can stimulate the germination of both AM fungi and parasitic plants (Akiyama et al. 2005; Harrison 2005; Paszkowski 2006; Bouwmeester et al. 2007) such as broomrape.

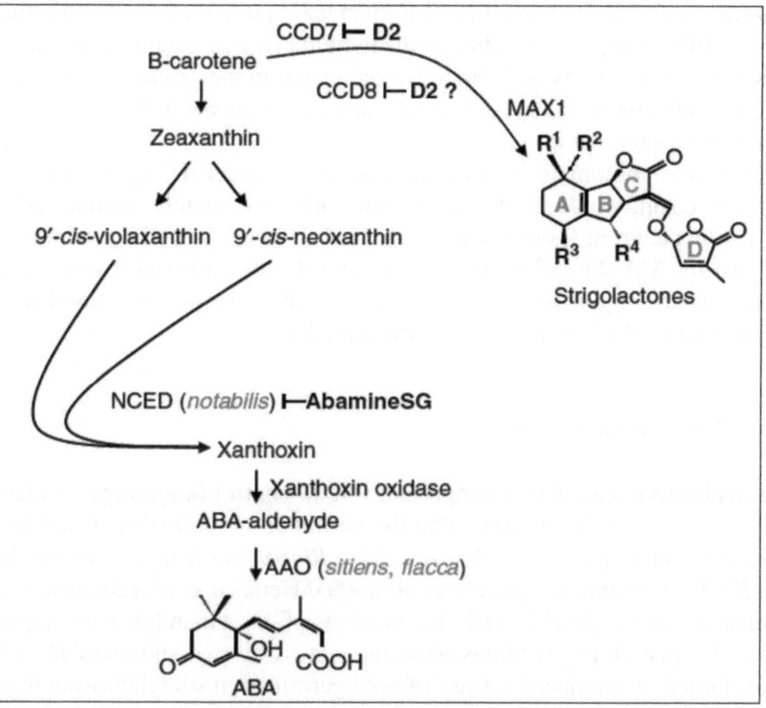

Fig. 17.2 Biosynthesis pathways of strigolactones and ABA from β-carotene. Abbreviations: *CCD7 and CCD8* carotenoid cleavage dioxygenase 7 and 8, respectively, *MAX1* corresponds to the cytochrome p450 shown to be involved in the biosynthesis of the branching inhibiting signal, *NCED* 9-*cis*-epoxycarotenoid dioxygenase, *ABA* abscisic acid, *AAO* aldehyde oxidase. (Adapted from Lopez-Raez et al. 2010)

17.4 Soil Microorganisms as Potential Promising Tools to Control Broomrape

In recent years, it has been demonstrated that soil microorganisms can act as biocontrol tools to reduce broomrape infestation in the field crops (Amsellem et al. 2001a, b; Boari and Vurro 2004; Sauerborn et al. 2007; Zermane et al. 2007; Hemissi et al. 2013; Iasur Kruh et al. 2017). This is a biological method which can be included in an integrated broomrape management program. Some efficient biocontrol agents belonging to different groups including beneficial (symbionts or free living) and pathogenic microorganisms are discussed in the following sections.

17.4.1 Mycorrhizae

Some workers have suggested that root infection by parasitic weeds can be reduced due to arbuscular mycorrhiza (AM) fungi colonization (Gworgwor and Weber 2003;

Lendzemo et al. 2005). According to Fernandez-Aparicio et al. (2010b), the germination of different species of *Orobanche* in field pea was notably reduced when pea plants were colonized by AM fungi. It was related to the production of exudates by the plant roots colonized by AM which showed suppressing effects on the parasite seed germination process.

Colonization of tomato plants by arbuscular mycorrhiza fungi notably decreased SLs produced and exuded by the plant roots, and consequently, a reduced *O. ramosa* germination occurred (Lopez-Raez et al. 2011a). In another study, sunflower plants colonized by AM showed a higher protection level against *O. cumana*. This was attributed to the suppressing effect on the parasite germination caused by the root exudates from colonized plants (Louarn et al. 2012).

17.4.2 Rhizobium

Soil rhizobacteria can play an important role to control broomrape species in crop fields. Some studies have shown that the germination of *Orobanche crenata* seeds was reduced in response to colonization by *Rhizobium leguminosarum* (Mabrouk et al. 2007c; Fernández-Aparicio et al. 2009). Hemissi et al. (2013) reported that germination of *O. foetida* and the number of its tubercles were significantly decreased when chickpea plants were inoculated by two strains of *Rhizobium*. It was explained by a reduced release of seed germination stimulant from the roots of the inoculated chickpea plants. According to Mabrouk et al. (2007c), pea plants inoculated by some strains of *R. leguminosarum* developed better and showed less vulnerability to *O. crenata*.

Ahonsi et al. (2003) found that some strains of *Bradyrhizobium* bacterium are able to increase legume resistance against parasitic weed attack. The reduction of pea plants infected by *O. crenata* was reported in the presence of *Rhizobium* which was attributed to a change in the rates of oxidative lipoxygenase (Lox) and phenylpropanoid/isoflavonoid pathways and releasing toxic substances such as phenolics and pisatin which can prevent seed germination of the parasite (Mabrouk et al. 2007a, b). Bouraoui et al. (2012) proposed two strains of *R. leguminosarum* as biocontrol agents to increase faba bean resistance against *O. foetida* as their inhibitory effects were confirmed in different experimental environments including field.

In chickpea, the infection level caused by *O. foetida* was lower in the presence of *Rhizobium* sp. strain Pch AZm. This was related to higher levels of defensive enzymes such as phenylalanine ammonia lyase (PAL) and peroxidase in the inoculated plants (Mabrouk et al. 2016). Moreover, an enhanced level of phenolics was observed in Rhizobial chickpea plant roots when they were subject to *O. foetida* (Mabrouk et al. 2016).

In a study, Fernandez-Aparicio et al. (2010a) found that non-Rhizobial and non-mycorrhizal mutants of pea and barrel medic (*Medicago truncatula*) showed more infection level caused by *O. crenata* when compared with Rhizobial and mycorrhizal mutants indicating that the symbiosis pathways for *Rhizobium* and mycorrhiza in the host plants may somewhat control broomrape invasion intensity (Fernandez-Aparicio et al. 2010a).

17.4.3 Pathogenic Microorganisms

In comparison with the symbionts, notable more studies have been conducted to investigate the controlling effects of pathogenic microorganisms on broomrape. These microorganisms have shown high potentials to suppress broomrape species in different cropping systems. It has been reported that there are nearly thirty genera of fungi which can attack different broomrape species (Boari and Vurro 2004). Previous studies indicated that the use of *Fusarium oxysporum* f. sp. orthoceras can effectively suppress *Orobanche* species in different crop fields (Bedi and Donchev 1991; Bedi 1994; Thomas et al. 1998). The suppressing effect of *F. lateritium* on two species of broomrape in tobacco was reported by Bozoukov and Kouzmnova (1994). Amsellem et al. (2001a, b) also found that treatment of tomato seeds or seedlings with some strains of *F. oxysporum* or *F. arthrosporioides* notably reduced the infestation level caused by two species of broomrape.

According to Shabana et al. (2003), there are six *Fusarium* species including *F. oxysporum*, *F. arthrosporioides*, *F. nygamai*, *F. oxysporum f.* sp. *orthoceras*, *F. semitectum* var. *majus*, and *F. solani* which can notably infect different species of broomrape and are potential candidates to biocontrol of this parasite. Cohen et al. (2002) found a decrease in infected tomato plants by *O. aegyptiaca* in the presence of *F. oxysporum* and *F. arthrosporioides*.

Despite the different *Fusarium* species which have showed controlling effects on broomrape, most of the studies in this context have been focused on two species, i.e., *F. oxysporum* and *F. solani* (Muller-Stöver et al. 2004; Dor and Hershenhorn 2009). In general, longevity and host specificity are two main reasons to consider *Fusarium* species as biocontrol agents for broomrape.

Al-Menoufi (1986) observed that broomrape tissues can severely be rotted due to application of *Alternaria, Gliocladium, Fusarium*, and *Sclerotinia*. Other workers reported the pathogenic effects of *Rhizoctonia solani*, *Verticillium microsporum*, and *Fusarium* species on various parts of broomrape species (Barloy and Pelhate 1962; Duafala et al. 1975, 1976; Talsakh Yan and Grigoryan 1978). Abdel-Kader and EL-Mougy (2009) showed that broomrape infestation in pea field was significantly decreased by using two species of *Trichoderma*. In other studies, Abdel-Kader et al. (1996, 1998) observed a notable reduction in broomrape density when the parasite tissues were colonized by *Trichoderma* spp. Abdel-Kader and EL-Mougy (2001) also reported that *Trichoderma* can attack broomrape species before and after emergence and notably decrease the infection level caused by this parasitic plant.

Thomas et al. (1999a) recognized a *Fusarium* strain which could infect *Orobanche crenata* Waller during the entire growing cycle of the parasite. Muller-Stover et al. (2002) found that both *Ulocladium botrytis* and *F. oxysporum* can cause necrosis signs on some broomrape species. In a study, broomrape germination was prevented due to toxins released by *Fusarium* spp. (Zonno and Vurro 2002). According to Mohammadi (2014), *F. solani* could suppress seed germination and germinated seed growth of *Orobanche* spp. Based on the results reported by Bedi and Donchev (1991), this suppression can be attributed to deterioration of broomrape seeds, germ tubes, and tubercles caused by *F. solani*. In a field study, pathogenic *Rhizoctonia* significantly decreased the germination of *O. ramosa* (Duafala et al. 1976).

Mazaheri and Vaziri (1991) showed that the number of tobacco plants infected by *Orobanche* spp. was substantially diminished in the presence of *F. oxysporum*. According to Abouzeid and El-Tarabily (2010), *Orobanche cernua* plants attacked by *Fusarium oxysporum* showed a dark color and infected plants rapidly lost. Hadj seyed hadi et al. (2005) reported similar results on *O. aegyptiaca* infected by *F. solani*. In another study, the parasitizing effect of two *Fusarium* species on *O. ramosa, O. aegyptiaca*, and *O. crenata* was identified by Amsellem et al. (2001a, b).

Some of the other pathogenic microorganisms which can attack and parasitize broomrape species consist of *Botrytis cinerea* found on *O. fasciculate* (Shaw 1973; Farr et al. 1989), *Thielaviopsis basicola* isolated from *O. ramosa* (Popova 1929), *Colletotrichum lagenarium* effective on *O. aegyptiaca* (Stojanovic and Boric 1981), and *Ulocladium atrum* observed on *O. minor* and *O. crenata* (Linke et al. 1992; Muller-Stöver and Kroschel 2005). The inhibiting effects of different pathogenic fungi species including *Fusarium, Sclerotinia, Rhizoctonia, Macrophomina*, and *Alternaria* isolated from the soil of tomato fields on broomrape germination and growth were reported by Karampur et al. (2004).

El-Kassas et al. (2005) isolated *Myrothecium verrucaria* from faba bean rhizosphere which could prevent *O. crenata* germination. This was resulted from macrocyclic trichothecene verrucarin A produced by this fungus. Seven macrocyclic trichothecenes including verrucarin A, B, M, and L acetate, roridin A, trichoverrol B, and isotrichoverrin B produced by *M. verrucaria* and another compound, namely, neosoloaniol monoacetate, isolated from *F. compactum* have been identified in which all of them showed preventing effects on germination of *O. ramosa*, and among them, roridin A was found as the strongest inhibitor metabolite without any toxic effect on animals (Andolfi et al. 2005).

Efficacy of microorganisms as biocontrol agents can be improved via using a suitable formulation which causes a successful storage, transport, and application. In addition, pathogenic microorganisms can show a notable high inhibitory effect on broomrape when they are applied together (Charudattan 2001). Dor et al. (2003) observed an enhanced controlling effect on *O. cumana* when *F. oxysporum* was used along with *F. solani* in sunflower field. Genetic manipulation techniques to develop the hypervirulent strains of host specific pathogens may be proposed as a promising approach to biocontrol broomrape in cropping systems (Gressel et al. 2004).

Boari and Vurro (2004) proposed soil microorganisms as useful biological tools to control broomrape because of their ability to infect parasite seeds in the early stages of its growth cycle, i.e., when the host plant has not seriously been damaged by broomrape. Moreover, soil microorganisms show a lower vulnerability to the environmental stresses compared to the above ground ones.

Other researchers reported harmful effects of different pathogenic microorganisms on broomrape young plants, flowering stalks, and flowers (Barloy and Pelhate 1962; Duafala et al. 1975, 1976; Talsakh Yan and Grigoryan 1978).

Sauerborn et al. (2007) suggested that in order to control broomrape, pathogenic microorganisms can be applied alone or as a component of an integrated management program. However, the use of pathogenic microorganisms to control broomrape should be done carefully due to their potential to infect non-target plants including field crop species. For example, Murasheva (1995) and Murasheva and Sizova (1995) suggested that application of *Fusarium oxysporum* to control broomrape may be risky due to its ability to attack some crops such as tomato, wheat, and sunflower. Moreover, in some cases, the application of these microorganisms to manage broomrape may not create a completely desirable result.

17.4.4 Other Microorganisms

Iasur Kruh et al. (2017) reported that a strain of *Pseudomonas* isolated from tomato rhizosphere reduced seed germination of *Phelipanche aegyptiaca* by 80% and could protect tomato against broomrape attack. This was attributed to the diverse compounds released by the bacteria which inhibit broomrape and improve the immune system of the host plant (Iasur Kruh et al. 2017). In another study, the germination of seeds and radicle elongation of broomrape were inhibited by *Azospirillum brasilense* (Zermane et al. 2007). They also found that the number of *O. foetida* and its biomass were notably reduced by *Pseudomonas fluorescens*.

Dadon et al. (2004) observed that seed germination and radicle growth of *P. aegyptiaca* were prevented when the host plant was colonized by a N-fixer microorganism, namely, *Azospirillum brasilense*. According to El-Kassas et al. (2005), the presence of *Myrothecium verrucaria* in faba bean rhizosphere could prevent the germination of *O. crenata*. This prevention was related to phenylpropanoid/isoflavonoid pathways induced by *M. verrucaria* (Mabrouk et al. 2007b, c, 2010).

It was revealed that some auxin-like substances are able to prevent the germination of the parasitic weeds such as *Striga* (Keyes et al. 2000). The auxin-producing ability by some rhizosphere dweller microorganisms including *Azotobacter* spp., *Pseudomonas putida*, *Azospirillum brasilense*, and *Klebsiella* spp. was reported by Frankenberger and Muhammed Arshad (1995). Therefore, these microorganisms can be proposed as potential biocontrol agents for *Orobanche* spp. In another study, ethylene-producing bacteria such as *Pseudomonas* spp. have been identified as efficient tools to control parasitic weeds (Ahonsi et al. 2003).

The controlling effect of *Pseudomonas fluorescens* (strain Bf7-9) on two species of broomrape has been shown by Zermane et al. (2007) who reported that this microorganism notably decreased the number of shoots and dry weights produced by parasitic weed species. Three species of *Pseudomonas* including *P. fluorescens*, *P. marginalis*, and *P. putida* have shown high suppressing potentials on *Orobanche* spp. In addition, they have a notable ability to colonize the roots of host plants (Seenivasan and Lakshmanan 2003). *P. fluorescens* has been introduced as a strong root colonizer (Chapon et al. 2002). It can effectively inhibit a wide spectrum of weed species without any negative influence on non-target plants (Kennedy et al. 2001). Decreasing effects of four rhizobacteria including *P. fluorescens*, *P.*

aeruginosa, *Bacillus subtilis*, and *B. atrophaeus* on radicle elongation of two broomrape species have been reported by Barghouthi and Salman (2010).

Necrotic signs on *O. cernua* infected by *Aspergillus alliaceus* were observed which led to the reduced number of the parasitic attachments, tubercles, and shoots in the sunflower field. Moreover, a lower number of *O. crenata* shoots was also recorded in the presence of *A. alliaceus* (Aybeke et al. 2014). In a study, the germinated seeds of *O. cumana* were notably decreased when the parasite was infected by *Streptomyces enissocaesilis*. Moreover, this microorganism caused a decrease in tubercle number formed by *O. cumana*, while it enhanced polyphenol oxidase activity, a defensive enzyme in sunflower roots, and improved beneficial microflora in rhizospheric environment of the host plant (Chen et al. 2016). The suppressing ability of *Pseudomonas* strain PhelS 10 isolated from tomato rhizosphere on *P. aegyptiaca* germination was suggested by Iasur Kruh et al. (2017) who observed a reduced number of the parasite shoots in the presence of the bacterium.

17.5 Mechanisms by Which Soil Phytomicrobiomes Can Dilute Broomrape Infestation in Cropping Systems

Soil phytomicrobiomes can affect broomrape emergence and growth both directly and indirectly. In the case of pathogenic microorganisms such as *Fusarium* spp., a direct effect can be defined. Pathogenic microorganisms attack seeds or germinated seedlings of the parasite which cause harmful consequences. This can be achieved by production of different inhibitor compounds such as verrucarin A (El-Kassas et al. 2005). According to Zonno and Vurro (2002), *Fusarium* fungus can produce some toxic compounds which inhibit *O. ramosa* seed germination. Muller-Stöver et al. (2002) also reported necrotic signs on *O. cumana* and *O. crenata* caused by *F. oxysporum*. Aybeke (2017) found that *F. oxysporum* can harm its host via production of reactive oxygen species, notable and irreversible genotoxic disorders on DNA, disruption of protein synthesis and metabolism, and induction of apoptosis in broomrape.

Thomas et al. (1999a, b) suggested that *F. oxysporum* is able to penetrate broomrape seeds and demolish their contents. In general, different parts of broomrape including juvenile plant, tubercles, and germ tubes of the parasite seeds can be attacked by pathogenic *F. oxysporum* (Abdel-Kader and El-Mougy 2009). In another study, different broomrape parts were seriously rotted in response to infection caused by some pathogenic fungi (Al-Menoufi 1986).

Indirect effect is usually achieved via establishment of a symbiotic relationship between host plant and soil microorganisms or changing the rhizosphere condition caused by them. Mycorrhiza and *Rhizobium* are the most important symbionts which can provide P and N to host plant in lieu of receiving the assimilates produced by their partners. These reciprocal relationships can significantly affect the level of invasion raised from the parasitic weeds.

Some workers have shown that soil nutrient status can notably regulate the amount of strigolactones produced by the host plant (Balzergue et al. 2011;

Yoneyama et al. 2012). In other words, in a stressful environment such as a nutrient-deficient soil, plants usually produce higher levels of SLs which help them to adjust the harmful stress effects (Umehara et al. 2008; Kohlen et al. 2011). Yoneyama et al. (2007) observed a higher SLs level released from red clover when it was grown in a soil with low phosphate. This plays a pivotal role to attract symbiont microorganisms such as AM as a P providing phytomicrobiome. However, this is a two-edged sword, because parasitic weeds such as broomrape are also able to receive these stimulating chemical signals, germinate, and attack the host plant (Fig. 17.3). Many studies have shown that plants inoculated by AM fungi produce lower levels of SLs and then are less prone to infection arising from broomrape (Fernandez-Aparicio et al. 2010b; Lopez-Raez et al. 2011a; Gworgwor and Weber 2003; Lendzemo et al. 2005). This can be explained by the ability of AM to provide P for the plant partner that encourages it to produce a lower level of SLs.

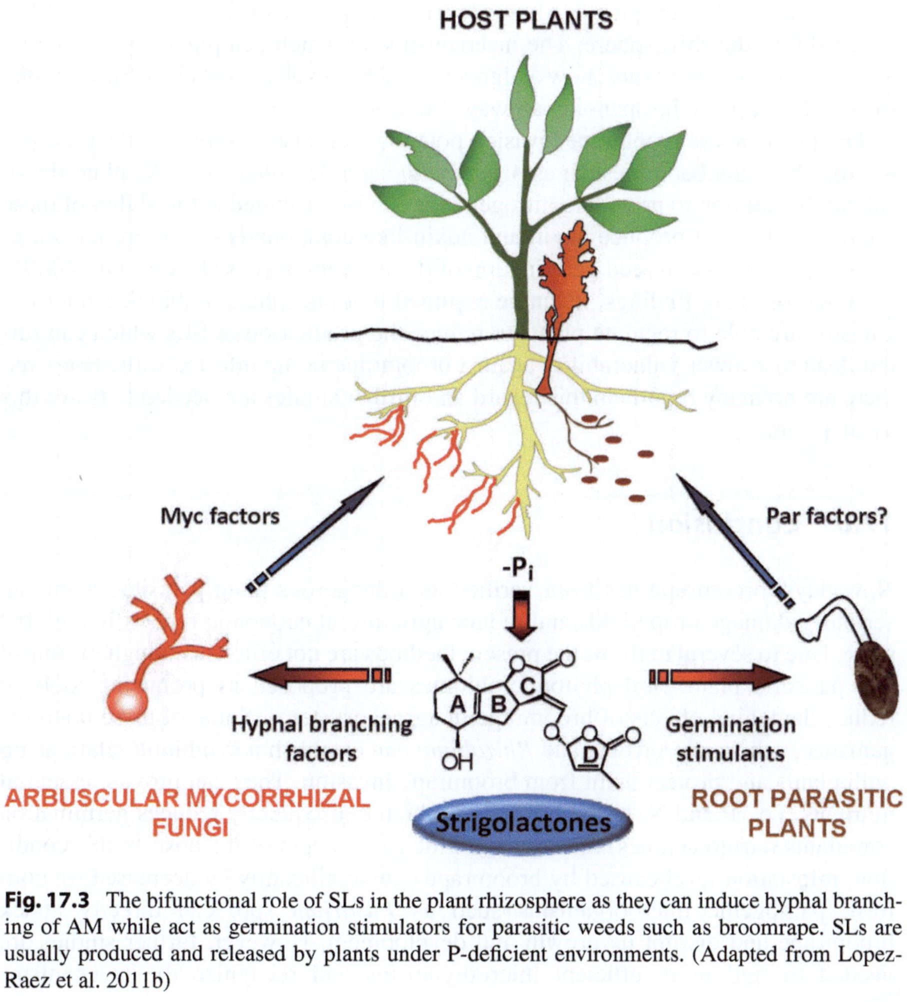

Fig. 17.3 The bifunctional role of SLs in the plant rhizosphere as they can induce hyphal branching of AM while act as germination stimulators for parasitic weeds such as broomrape. SLs are usually produced and released by plants under P-deficient environments. (Adapted from Lopez-Raez et al. 2011b)

Some surveys have demonstrated that phosphate accessibility is a key factor which regulates the SL level produced by plants (Umehara et al. 2010; Yoneyama et al. 2012). The plants grown under limited P condition show a dramatical increase in SL production (over tenfold) in order to develop an effective symbiotic relationship with AM fungi and consequently an enhanced access to soil P reserve (Gu et al. 2011).

A similar viewpoint has been proposed for N by Yoneyama et al. (2007, 2011) who reported that SL production can significantly be increased by plants grown in a nitrate-deficient environment. In addition to P, recent studies have revealed that plants colonized by AM have an increase access to N (Whiteside et al. 2012) leading to a lower SLs produced by them and subsequently a reduced susceptibility to broomrape. Reduced infection levels caused by broomrape were reported by some researchers when a symbiotic relationship was established between *Rhizobium* and plant species (Mabrouk et al. 2007c; Fernández-Aparicio et al. 2009; Hemissi et al. 2013). This may be explained by nitrogen providing ability of the bacterium to its plant partner and consequently a lower level of the stimulant chemical signals (SLs) released into the rhizosphere. The mechanisms by which pea plants inoculated by *Rhizobium leguminosarum* show a higher resistance level against *O. crenata* are due to an induced phenylpropanoid pathway (Mabrouk et al. 2010).

Of course, a less broomrape invasion potential was also observed in the presence of other N-fixing bacteria such as *Azospirillum* and *Azotobacter* in the plant rhizosphere. In addition to providing nitrogen, this can be attributed to the ability of these microorganisms to produce auxin and auxin-like compounds which are known as effective inhibitors on seed germination of the parasitic weeds (Keyes et al. 2000).

Based on these findings, it can be assumed that phosphate-mobilizing microorganisms are able to regulate plants to reduce the production of SLs which can further lead to a lower vulnerability against broomrape in the infested soils. However, there are no many reports in this regard and further studies are needed to prove this assumption.

17.6 Conclusion

Nowadays, broomrape has been verified as a dangerous plant parasite which can seriously damage crop yields and reduce agricultural economic returns in a global scale. Due to several reasons, the present methods are not efficient enough to control this parasitic plant. Soil phytomicrobiomes are proposed as promising tools to reduce damaging effects of broomrape in agroecosystems. Some of these microorganisms such as mycorrhiza and *Rhizobium* can establish a symbiotic relationship with plants and protect them from broomrape invasion. They can provide essential nutrients, i.e., P and N, to the host plant which consequently reduces germination stimulants (strigolactones) released from the root system of the host. In this condition, infestation level caused by broomrape can significantly be decreased. In contrast, pathogenic microorganisms such as *Fusarium* spp. can directly attack broomrape and disrupt its growth and development. However, further studies are needed to find more efficient microorganisms and recognize the mechanisms involved in their suppressing ability against broomrape. Moreover, their efficiency as broomrape biocontrol agents can be improved by using some approaches such as development of suitable formulations and application methods.

References

Abdel-Kader MM, El-Mougy NS (2001) Evaluation of different approaches of mycoherbicidal application for controlling *Orobanche crenata* in pea field. Egypt J Phytopathol 29:69–82

Abdel-Kader MM, El-Mougy NS (2009) Prospects of mycoherbicides for control of broomrapes (*Orobanche* spp.) in Egypt. J Plant Prot Res 49:63–75

Abdel-Kader MM, Diab MM, Ismail BR, Hassan E, Arafat KH (1996) *In vitro* test of different isolates of fungal genera for their pathogenicity against *Orobanche* spp. p. 907–911. In: Moreno MT, Cubero JI, Berner D, Joel D, Musselman LJ, Parker C (eds) Advances in Parasitic Plant Research. Proceeding of 6th international parasitic weed symposium. Cordoba, Spain, April 16–18

Abdel-Kader MM, Ismail BR, Diab MM, Hassan EA (1998) Preliminary evaluation of some soil-borne fungi parasitizing *Orobanche crenata* in greenhouse. pp 127–132. In: Proceedings of 6th Mediteranean symposium. EWRS, Montpellier, pp 13–15

Abouzeid MA, El-Tarabily KA (2010) *Fusarium* spp. suppress germination and parasitic establishment of bean and hemp broomrapes. Phytopathol Mediterr 49:51–64

Ahonsi MO, Berner DK, Emechebe AM, Lagoke ST, Sanginga N (2003) Potential of ethylene-producing *pseudomonads* in combination with effective N2-fixing bradyrhizobial strains as supplements to legume rotation for *Striga hermonthica* control. Biol Control 28:1–10

Akiyama K, Matsuzaki K, Hayashi H (2005) Plant sesquiterpenes induce hyphal branching in arbuscular mycorrhizal fungi. Nature 435:824–827

Al-Menoufi OA (1986) Studies on *Orobanche* spp. 2: Fungi associated with *Orobanche crenata* Forsk. Alexandria J Agric Res 31:297–310

Amsellem Z, Barghouthi S, Cohen B, Goldwasser I, Gressel J, Hornok L, Kerenyi Z, Kleifeld I, Klein O, Kroschel J, Sauerborn J, Müller-Stöver D, Thomas H, Vurro M, Zonno MC (2001a) Recent advances in the biocontrol of *Orobanche* (broomrape) species. BioControl 46:211–228

Amsellem Z, Kleifeld Y, Kereny Z, Hornok L, Goldwasser Y, Gressel J (2001b) Isolation, identification and activity of mycoherbicidal pathogens from juvenile broomrape plants. Biol Control 21:274–284

Andolfi A, Boari A, Evidente A, Vurro M (2005) Metabolites inhibiting germination of *Orobanche ramosa* seeds produced by *Myrothecium verrucaria* and *Fusarium compactum*. J Agric Food Chem 53:1598–1603

Aybeke M (2017) *Fusarium* infection causes genotoxic disorders and antioxidant-based damages in *Orobanche* spp. Microbiol Res 201:46–51

Aybeke M, ¸Sen B, Okten S (2014) *Aspergillus alliaceus*, a new potential biological control of the root parasitic weed *Orobanche*. J Basic Microbiol 54:S93–S101

Balzergue C, Puech-Page's V, Be'card G, Rochange SF (2011) The regulation of arbuscular mycorrhizal symbiosis by phosphate in pea involves early and systemic signaling events. J Exp Bot 62:1049–1060

Barghouthi S, Salman M (2010) Bacterial inhibition of *Orobanche aegyptiaca* and *Orobanche cernua* radical elongation. Biocontrol Sci Tech 20:423–435

Barloy J, Pelhate AH (1962) Preliminary observations on the parasites and diseases of hemp in Anjou. Ann Epiphytopatol 13:117–149. (Horticultural Abstracts 32: Abstr. No. 7160, 1963)

Bedi JS (1994) Further studies on control of sunflower broomrape with *Fusarium oxysporum* f. sp. *Orthoceras*- a potential mycoherbicide. In: Pieterse AH, Verkleij JAC, ter Borg SJ (eds) Proceedings of the 3rd international workshop on *Orobanche* and related *Striga* research. Royal Tropical Institute, Amsterdam, pp 539–544

Bedi JS, Donchev N (1991) Results on mycoherbicide control of sunflower broomrape (*Orobanche cumana* Wall.) under glasshouse and field conditions. In: Ransom JK, Musselman LJ, Worsham AD, Parker C (eds) Proceedings of 5th international symposium parasitic weeds. Nairobi, Kenya: CIMMYT, pp 76–82

Besserer A, Puech-Pages V, Kiefer P, Gomez-Roldan V, Jauneau A, Roy S, Portais JC, Roux C, Be'card G, Se'jalon-Delmas N (2006) Strigolactones stimulate arbuscular mycorrhizal fungi by activating mitochondria. PLoS Biol 4:e226

Besserer A, Be'card G, Jauneau A, Roux C, Se'jalon-Delmas N (2008) GR24, a synthetic analog of strigolactones, stimulates the mitosis and growth of the arbuscular mycorrhizal fungus *Gigaspora rosea* by boosting its energy metabolism. Plant Physiol 148:402–413

Boari A, Vurro M (2004) Evaluation of *Fusarium* spp. and other fungi as biological control agents of broomrape (*Orobanche ramosa*). Biol Control 30:212–219

Bouraoui M, Abbes Z, Abdi N et al (2012) Evaluation of efficient *Rhizobium* isolates as biological control agents of *Orobanche foetida* Poir. parasitizing Vicia faba L. minor in Tunisia. Bulg J Agric Sci 18:557–564

Bouwmeester HJ, Matusova R, Zhongkui S, Beale MH (2003) Secondary metabolite signalling in host–parasitic plant interactions. Curr Opin Plant Biol 6:358–364

Bouwmeester HJ, Roux C, Lo'pez-Ra'ez JA, Be'card G (2007) Rhizosphere communication of plants, parasitic plants and AM fungi. Trends Plant Sci 12:224–230

Bozoukov H, Kouzmnova I (1994) Biological Control of tobacco broomrape (*Orobanche* spp.) by mean of some fungi of the genus *Fusarium*. pp 534–538 In: Pieterse AH, Verkleij JAC, Ter-Borg SJ (eds) Biology and Management of *Orobanche*. Proceedings of 3th international workshop on *Orobanche*, and related *Striga* Research. Amsterdam, The Netherlands, Royal Tropical Institute

Cardoso C, Ruyter-Spira C, Bouwmeester HJ (2011) Strigolactones and root infestation by plant-parasitic *Striga*, *Orobanche* and *Phelipanche* spp. Plant Sci 180:414–420

Chapon A, Guillerm AY, Delalande L, Lebreton L, Sarniguet A (2002) Dominant colonisation of wheat roots by *Pseudomonas fluorescens* Pf29A and selection of the indigenous microflora in the presence of the take-all fungus. Eur J Plant Pathol 108:449–459

Charudattan R (2001) Biological control of weeds by means of plant pathogens: significance for integrated weed management in modern agroecology. BioControl 46:229–260

Chen J, Xue QH, McErlean CSP et al (2016) Biocontrol potential of the antagonistic microorganism *Streptomyces enissocaesilis* against *Orobanche cumana*. BioControl 61:781–791

Cohen B, Amsellem Z, Lev-Yadun S, Gressel J (2002) Infection of tubercles of the parasitic weed *Orobanche aegyptiaca* by mycoherbicidal *Fusarium* species. Ann Bot 90:567–578

Dadon T, Nun NB, Mayer AM (2004) A factor from *Azospirillum brasilense* inhibits germination and radicle growth of *Orobanche aegyptiaca*. Isr J Plant Sci 52:83–86

Dor E, Hershenhorn J (2009) Evaluation of the pathogenicity of microorganisms isolated from Egyptian broomrape (*Orobanche aegyptiaca*) in Israel. Weed Biol Manag 9:200–208

Dor E, Vurro M, Hershenhorn J (2003) The efficacy of a mixture of fungi to control Egyptian and sunflower broomrape. In: Cost Action 849: Parasitic plant management in sustainable agriculture. Meeting on biology and control of broomrape 30 Oct–2 Nov Athens, Greece

Duafala T, Gold AH, Sagen J, Wilhein S (1975) Apparent biological control of branched broomrape *Orobanche ramosa*. Proc Am Phytopathol Soc:113

Duafala T, Wilhein S, Gold AH, Sagen J (1976) *Rhizoctonia* disease of broomrape, a possible biological control. Proc Am Phytopathol Soc 3:272

Eizenberg H, Colquhoun JB, Mallory-Smith CA (2005) A predictive degree-days model for small broomrape (*Orobanche minor*) parasitism in red clover in Oregon. Weed Sci 53:37–40

El-Kassas R, Karam El-Din Z, Beale MH, Ward JL, Strange RN (2005) Bioassay-led isolation of *Myrothecium verrucaria* and verrucarin A as germination inhibitors of *Orobanche crenata*. Weed Res 45:212–219

Estabrook EM, Yoder JI (1998) Plant–plant communications: rhizosphere signalling between parasitic angiosperms and their hosts. Plant Physiol 116:1–7

Farr DF, Bills GF, Chamuris GP, Rossman AY (1989) Fungi on plants and plant products in the United States. The American Phytopathological Society (APS) Press, St Paul, MN

Fernandez Escobar J, Rodriguez Ojeda MI, Fernandez Martinez JM, Alonso LC (2009) Sunflower broomrape *(Orobanche cumana* Wallr.*)* in castillaleon, a traditionally non-infested area in northern Spain. Helia 32:57–64

Fernández-Aparicio M, Soto MJ, Rubiales D, Ocampo JA, García-Garrido JM (2009) The potential of *Rhizobium* mutants for biological controlof *Orobanche crenata*. Haustorium 54:3–4

Fernandez-Aparicio M, Rispail N, Prats E et al (2010a) Parasitic plant infection is partially controlled through symbiotic pathways. Weed Res 50:76–82

Fernandez-Aparicio M, García-Garrido JM, Ocampo JA, Rubiales D (2010b) Colonization of field pea roots by arbuscular mycorrhizal fungi reduces *Orobanche* and *Phelipanche* species seed germination. Weed Res 50:262–268

Frankenberger WT, Muhammed Arshad JR (1995) Phytohormones in soil microbial production and function. Marcel Deker, New York, p 503

Garcia-Torres L, Lopez-Granados F, Jurado-Exposito M, Diaz- Sanchez J (1998) The present state of *Orobanche* spp. infestation in Andalusia and the prospects for its management. In: Proceedings of 6th Mediterranean symposium. EWRS, Montpellier, pp 141–145

Gomez-Roldan V, Fermas S, Brewer PB et al (2008) Strigolactone inhibition of shoot branching. Nature 455:189–194

Gressel J, Hana A, Head G, Marasas W, Obilana B, Ochanda J, Souissi T, Tzotzos G (2004) Major heretofore intractable biotic constraints to African food security that may be amenable to novel biotechnological solutions. Crop Prot 23:661–689

Gu M, Chen A, Dai X, Liu W, Xu G (2011) How does phosphate status influence the development of the arbuscular mycorrhizal symbiosis? Plant Signal Behav 6:1300–1304

Gworgwor NA, Weber HC (2003) Arbuscular mycorrhizal fungi–parasite-host interaction for the control of Striga hermonthica (Del.) Benth. in sorghum. Mycorrhiza 13:277–281

Hadj Seyed Hadi M, Fazeli F, Taghi Darzi M, Shahmoradi R (2005) Biological control of *Orobanche aegyptiaca* in tomato. Paper presented at the the fourth world congress on Allelopathy, NSW Australia, 21–26 August

Harrison MJ (2005) Signaling in the arbuscular mycorrhizal symbiosis. Annu Rev Microbiol 59:19–42

Hemissi I, Mabrouk Y, Abdi N, Bouraoui M, Saidi M, Sifi B (2013) Growth promotion and protection against *Orobanche foetida* of chickpea (*Cicer aerietinum*) by two *Rhizobium* strains under greenhouse conditions. Afr J Biotechnol 12:1371–1377

Hirsch AM, Bauer WD, Bird DM, Cullimore J, Tyler B, Yoder JI (2003) Molecular signals and receptors: controlling rhizosphere interactions between plants and other organisms. Ecology 84:858–868

Iasur Kruh L, Lahav T, Abu-Nassar J, Achdari G, Salami R, Freilich S, Aly R (2017) Host-parasite-bacteria triangle: the microbiome of the parasitic weed *Phelipanche aegyptiaca* and tomato-*Solanum lycopersicum* (Mill.) as a host. Front Plant Sci 8:1–9

Joel DM, Hershenhorn J, Eizenberg H, Aly R, Ejeta G, Rich PJ, Ransom JK, Sauerborn J, Rubiales D (2007) Biology and management of weedy root parasites. Hort Rev 33:267–349

KaramPur F, Fasihi MT, Heydari A (2004) Identification of fungal pathogenic agents of broom rape in tomato fields. Research final report 29. in Persian

Kaya Y, Demerci M, Evci G (2004) Sunflower (*Helianthus annuus* L.) breeding in Turkey for broomrape (*Orobanche cernua* Loeffl.) and herbicide resistance. Helia 27:199–210

Kennedy AC, Johnson BN, Stubbs TL (2001) Host range of a deleterious rhizobacterium for biological control of downy brome. Weed Sci 49:792–797

Keyes WJ, Malley RO, Kim D, Lynn DG (2000) Signaling organogenesis in parasitic angiosperms: xenognosin generation, perception and response. J Plant Growth Regul 19:217–231

Kharrat M, Halila MH, Linke KH, Haddar T (1992) First report of *Orobanche foetida* Poiret on faba bean in Tunisia. FABIS Newsletter 30:46–47

Kohlen W, Charnikhova T, Liu Q, Bours R, Domagalska MA, Beguerie S, Verstappen F, Leyser O, Bouwmeester H, Ruyter-Spira C (2011) Strigolactones are transported through the xylem and play a key role in shoot architectural response to phosphate deficiency in nonarbuscular mycorrhizal host *Arabidopsis*. Plant Physiol 155:974–987

Kuijt J (1969) The biology of parasitic flowering plants. University of California, Berkeley

Lendzemo V, Kuyper TW, Kropff MJ, Vanast A (2005) Field inoculation with arbuscular mycorrhizal fungi reduces *Striga hermonthica* performance on cereal crops and has the potential to contribute to integrated *Striga* management. Field Crops Res 91:51–61

Linke KH, Saxena MC (1991) Study on viability and longevity of *Orobanche* seed under laboratory conditions. In: Wegmann K, Musselman LJ (eds) Proceedings of international workshop in *Orobanche* research. Tübingen, Germany: Eberhard-Karls-Universität, pp 110–114

Linke KH, Scheibel C, Saxena MC, Sauerborn J (1992) Fungi occurring on *Orobanche* spp. and their preliminary evaluation for *Orobanche* control. Trop Pest Manag 38:127–130

Lopez-Raez JA, Kohlen W, Charnikhova T, Mulder P, Undas AK, Sergeant MJ, Verstappen F, Bugg TDH, Thompson AJ, Ruyter-Spira C, Bouwmeester H (2010) Does abscisic acid affect strigolactone biosynthesis? New Phytol 187:343–354

Lopez-Raez JA, Charnikhova T, Fernandez I, Bouwmeester H, Pozo MJ (2011a) Arbuscular mycorrhizal symbiosis decreases strigolactone production in tomato. J Plant Physiol 168:294–297

Lopez-Raez JA, Pozo MJ, García-Garrido JM (2011b) Strigolactones: a cry for help in the rhizosphere. Botany 89:513–522

Louarn J, Carbonne F, Delavault P, Bécard G, Rochange S (2012) Reduced germination of *Orobanche cumana* seeds in the presence of arbuscular mycorrhizal fungi or their exudates. PLoS One. https://doi.org/10.1371/journal.pone.0049273

Mabrouk Y, Simier P, Arfaoui A, Sifi B, Delavault P, Zourgui L, Belhadj O (2007a) Induction of phenolic compounds in pea (*Pisum sativum* L.) inoculated by *Rhizobium leguminosarum* and infected with *Orobanche crenata*. J Phytopathol 155:728–734

Mabrouk Y, Simier P, Delavault P, Delgrange S, Sifi B, Zourgui L, Belhadj O (2007b) Molecular and biochemical mechanisms of defence induced in pea by *Rhizobium leguminosarum* against *Orobanche crenata*. Weed Res 47:452–460

Mabrouk Y, Zourgui L, Sifi B, Delavault P, Simier P, Belhadj O (2007c) Some compatible *Rhizobium leguminosarum* strains in peas decrease infections when parasitised by *Orobanche crenata*. Weed Res 47:44–53

Mabrouk Y, Mejri S, Hemissi I, Simier P, Delavault P, Saidi M, Belhadj O (2010) Bioprotection mechanisms of pea plant by *Rhizobium leguminosarum* against *Orobanche crenata*. Afr J Microbiol Res 23:2570–2575

Mabrouk Y, Mejri S, Hemissi I, Belhadj O (2016) Biochemical analysis of induced resistance in chickpea against broomrape (*Orobanche foetida*) by rhizobia inoculation. Phytopathol Mediterr 55:54–61

Matusova R et al (2005) The strigolactone germination stimulants of the plant-parasitic *Striga* and *Orobanche* spp are derived from the carotenoid pathway. Plant Physiol 139:920–934

Mazaheri A, Vaziri M (1991) Studies on the chemical control of broomrape (*Orobanche* spp) in tobacco fields, 10th edn. Plant Protection Congress of Iran Kerman 1–5 Sept 1991, p 201

Melero JM, Dominguez J, Fernandez Martinez JM (2000) Update on sunflower broomrape situation in Spain: racial status and sunflower breeding for resistance. Helia 23:45–56

Mohammadi A (2014) Biological control of *Orobanche Ramosa* by *Fusarium Solani*. Int J Adv Biol Biom Res 2:2751–2755

Müller-Stöver D, Kroschel J (2005) The potential of *Ulocladium botrytis* for biological control of *Orobanche* spp. Biol Control 33:301–306

Müller-Stöver D, Kroschel J, Thomas H, Sauerborn J (2002) Chlamydospores of *Fusarium oxysporum* Schlecht f. sp. *orthoceras* (Appel & Wollenw.) Bilai as inoculum for wheat-flour – Kaolin granules to be used for the biological control of *Orobanche cumana* Wallr. Eur J Plant Pathol 108:221–228

Müller-Stöver D, Thomas H, Sauerborn J, Kroschel J (2004) Two granular formulations of *Fusarium oxysporum* f. sp. Orthoceras to mitigate sunflower broomrape (*Orobanche cumana*). BioControl 49:595–602

Murasheva VN (1995) Biological peculiarities and identification of the broomrape fusariosis agent. Mikolog Fitopatol 29:53–85

Murasheva VN, Sizova TP (1995) Consequences of applying the causal agent of *Fusarium* wilt of broomrape to soil. Mikol Fitopatol 29:41–45

Musselman LJ (1991) *Orobanche ramosa* and *Orobanche aegyptiaca* in Flora Palaestina. In: Proceedings of international workshop on *Orobanche* research, Obermachtal Germany 19–22 August pp 1–5

Nickrent D, Musselman L (2004) Introduction to parasitic flowering plants. The Plant Health Instructor. https://doi.org/10.1094/PHI-I-2004-0330-01

Parker C (2009) Observations on the current status of *Orobanche* and *Striga* problems worldwide. Pest Manag Sci 65:453–459

Parker C (2012) Parasitic weeds: a world challenge. Weed Sci 60:269–276

Parker C, Riches CR (1993) Parasitic weeds of the world: biology and control. CAB International, Wallingford

Paszkowski U (2006) Mutualism and parasitism: the yin and yang of plant symbioses. Curr Opin Plant Biol 9:364–370

Popova AA (1929) Diseases of the Tobacco *Nicotiana rustica* L Morbi. Plantarum 18:45–53

Rubiales D, Sadiki M, Román D (2005) First report of *Orobanche foetida* on common vetch (*Vicia sativa*) in Morocco. Plant Dis 89:528

Rubiales D, Fernández-Aparicio M, Pérez-de-Luque A, Prats E, Castillejo MA, Sillero JC, Rispail N, Fondevilla S (2009) Breeding approaches for crenate broomrape (*Orobanche crenata* Forsk.) management in pea (*Pisum sativum* L.). Pest Manag Sci 65:553–559

Sauerborn J (1991) Parasitic flowering plants: ecology and management. Verlag Josef Margraf, Weikersheim

Sauerborn J, Müller-Stöver D, Hershenhorn J (2007) The role of biological control in managing parasitic weeds. Crop Prot 26:246–254

Seenivasan N, Lakshmanan PL (2003) Biocontrol potential of native isolates of *Pseudomonas fluorescens* against rice root nematode. J Ecobiol 15:69–72

Shabana YM, Müller-Stöver D, Sauerborn J (2003) Granular Pesta formulation of *Fusarium oxysporum* f. sp. orthoceras for biological control of sunflower broomrape: efficacy and shelf-life. Biol Control 26:189–201

Shaw CG (1973). Host fungus index for the Pacific Northwest. 1. Hosts Washington State Agricultural Experimental Station Bulletin No 766 pp 162 https://digitalcommons.usu.edu/aspen_bib/5290

Shindrova P (2006) Broomrape (*Orobanche cumana* Wallr.) in Bulgaria: distribution and race composition. Helia 29:111–120

Stojanovic D, Boric B (1981) Prilog proucavanju mikoflore parazitnih cvetnica iz rodova *Cuscuta I Orobanche*. [Contribution to the study of the mycoflora of parasitic phanerogams from the genera *Cuscuta* and *Orobanche*]. Zast Bilja 32:135–145

Talsakh Yan MG, Grigoryan SV (1978) Fungi found on broomrape in the Armenian SSR, USSR. Weeds Abstr 28:1063

Thomas H, Sauerborn J, Muller-Stover D, Ziegler A, Bedi JS, Kroschel J (1998) The potential of *Fusarium oxysporum* f. sp. *orthoceras* as a biological control agent for *Orobanche cumuna* in sunflower. Biol Control 13:41–48

Thomas H, Heller A, Sauerborn J, Muller-Stover D (1999a) *Fusarium oxysporum* f. sp. *orthoceras*, a potential mycoherbicide, parasitizes seeds of *Orobanche Cumana* (Sunflower Broomrape): a cytological study. Ann Bot 83:453–458

Thomas H, Sauerborn J, Muller-Stover D, Kroschel J (1999b) Fungi of *Orobanche aegyptica* in Nepal with potential as biological control agents. Biocontrol Sci Tech 9:379–381

Umehara M, Hanada A, Yoshida S, Akiyama K, Arite T, Takeda-Kamiya N, Magome H, Kamiya Y, Shirasu K, Yoneyama K et al (2008) Inhibition of shoot branching by new terpenoid plant hormones. Nature 455:195–200

Umehara M, Hanada A, Magome H, Takeda-Kamiya N, Yamaguchi S (2010) Contribution of strigolactones to the inhibition of tiller bud outgrowth under phosphate deficiency in rice. Plant Cell Physiol 51:1118–1126

Westwood JH, Yoder JI, Timko MP (2010) The evolution of parasitism in plants. Trends Plant Sci 15:227–235

Whiteside MD, Digman MA, Gratton E, Treseder KK (2012) Organic nitrogen uptake by arbuscular mycorrhizal fungi in a boreal forest. Soil Biol Biochem 55:7–13

Xie X, Yoneyama K, Yoneyama K (2010) The strigolactone story. Annu Rev Phytopathol 48:93–117

Yoneyama K, Yoneyama K, Takeuchi Y, Sekimoto H (2007) Phosphorus deficiency in red clover promotes exudation of orobanchol, the signal for mycorrhizal symbionts and germination stimulant for root parasites. Planta 225:1031–1038

Yoneyama K, Xie XN, Kisugi T, Nomura T, Sekimoto H, Yokota T, Yoneyama K (2011) Characterization of strigolactones exuded by *Asteraceae* plants. Plant Growth Regul 65:495–504

Yoneyama K, Xie X, Kim HI, Kisugi T, Nomura T, Sekimoto H, Yokota T, Yoneyama K (2012) How do nitrogen and phosphorus deficiencies affect strigolactone production and exudation? Planta 235:1197–1207

Zermane N, Souissi T, Kroschel J, Sikora R (2007) Biocontrol of broomrape (*Orobanche crenata* Forsk. and Orobanche foetida Poir.) by *Pseudomonas fluorescens* isolate Bf7-9 from the faba bean rhizosphere. Biocontrol Sci Tech 17:483–497

Zonno M, Vurro M (2002) Inhibition of germination of *Orobanche ramosa* seeds by *Fusarium* toxins. Phytoparasitica 30:519–524

Environmental Perspectives of Plant-Microbe Nexus for Soil and Water Remediation

18

Mahmoud Nasr

Abstract

Recently, the concerns about soil and water pollution have significantly enlarged due to the vast increase in urbanization, industrialization, population growth, and fossil fuel utilization. Exposure to high levels of pollution causes serious threats to ecological systems, natural environment, human health, and food chains. In this context, effective and promising treatment methods have been developed to avoid the deterioration of the soil and water systems. Environmental remediation has been introduced to overcome the drawbacks of conventional physical, chemical, and biological treatment processes. In this chapter, various remediation techniques including phytoremediation, bioremediation, phycoremediation, and mycoremediation are reviewed. Several forms of phytoremediation, e.g., phytodegradation, phytotransformation, phytoextraction, phytovolatilization, phytostabilization, phytofiltration, phytodesalination, and phytomining, that explore the involvement of plant-based technology for toxicants and pollutants removal are discussed. The activities of microbial species during the intrinsic remediation and bioaugmentation processes are demonstrated. The assemblage of rhizobacteria in the plant root system to detoxify contaminated soils and transform hazardous elements into harmless substances is also discussed. Environmental aspects related to microalgal cultures and fungal species for water remediation are demonstrated. The study objectives are reviewed in terms of previous investigations reported in the literature. Recommendations for future works in the field of environmental remediation are suggested.

Keywords

Plant · Microbes · Degradation · Environment · Remediation

M. Nasr (✉)
Sanitary Engineering Department, Faculty of Engineering, Alexandria University, Alexandria, Egypt

V. Kumar et al. (eds.), *Microbiome in Plant Health and Disease*,
https://doi.org/10.1007/978-981-13-8495-0_18

18.1 Introduction

Recently, exhaustive urbanization and industrialization, exponential population growth, and extensive utilization of fossil-based fuels (e.g., gas, coal, and oil) have resulted in rapid environmental deteriorations (Ali et al. 2013; Fawzy et al. 2018). Moreover, a wide range of natural and anthropogenic activities introduce hazardous elements, combustible substances, organic toxins, explosive and petroleum products, surfactants, heavy metals, and various inorganic and organic compounds to the soil and water environments (Agnello et al. 2016; Guarino et al. 2017; Luo et al. 2005). In addition, agricultural soils are negatively influenced by the disposal of cattle manure, fertilizers, crop residues, pesticides, sewage sludge, and agrochemical wastes (Singh et al. 2015). Hence, several associations such as Environmental Protection Agency, Food and Agriculture Organization of the United Nations, and World Health Organization have established allowable standards that can save the environment from the risks of elevated contaminant levels (FAO/WHO 2011). For example, the US permissible limits of some heavy metals in agricultural soils are (in mg/kg) as follows: 0.48 Cd, 200 Pb, 11 Cr, 1 Hg, 270 Cu, 1100 Zn, 72 Ni, and 0.11 As (Liu et al. 2018). Moreover, the weakly, moderately, and heavily polluted water bodies have nitrogen and phosphorus concentrations of 20 and 4 mg/L, 40 and 8 mg/L, and 85 and 15 mg/L, respectively (Rashidi et al. 2015).

Exposure to high levels of pollution causes adverse effects to the natural environment, living creatures, and food chains (Islam et al. 2015). For instance, human health risks from heavy metals include lung, brain, and kidney damages, carcinogenic infection, bone mineral loss, cardiovascular disease, nervous system disturbance, and intestine irritation (Fawzy et al. 2016). Moreover, contamination derives several concerns to food production, freshwater availability, and photosynthesis and respiration processes (Batista-García et al. 2017). Furthermore, the inhibition of crops' growth and germination, reduction of the microbial population, and alteration of enzymatic functions occur when the soil receives high concentrations of toxic elements (Ghasemi et al. 2018). Pollutants may seep into groundwater and endorse the transfer of pathogenic and undesirable microorganisms (Kuiper et al. 2004). Additionally, solids, nutrients, and organic pollutants that are released into water bodies cause dissolved oxygen depletion, clogging of fish gills, and eutrophication (Gupta et al. 2018). Accordingly, appropriate and promising treatment methods should be developed to protect the soil and water environmental systems.

Several physical, chemical, and biological processes have been effectively employed for the removal of pollutants from soil and aquatic ecosystems (Gonçalves et al. 2017). However, some drawbacks have been recognized to limit the wide application of physicochemical approaches. For example, the coagulation/flocculation technique involves high reagent utilization, and it generates an excessive amount of sludge comprising chemical compounds (Fawzy et al. 2016). Ultrafiltration and reverse osmosis systems demand high-energy consumption, chemical regeneration phases, and frequent backwashing (Rashidi et al. 2015). Ion exchange suffers from the incomplete removal of toxic substances and heavy metals, whereas adsorption retains the problem of exhausted adsorbent disposal (Fawzy et al. 2018).

Remediation techniques have been proposed to cope with the disadvantages of conventional treatment methods (Limmer and Burken 2016). Remediation is used to describe the removal of pollutants from the soil and water environments mainly via biological activities (Kulshreshtha et al. 2014). Remediation is used for the protection of human health and ecological systems, as well as for the recreation and restoration of contaminated lands (Srivastava et al. 2017). The most common in situ remediation techniques are phytoremediation (Jlassi et al. 2013), bioremediation (Faisal and Hasnain 2005), rhizoremediation (Fester et al. 2014), phycoremediation (Gupta et al. 2017), and mycoremediation (Singh et al. 2015). Phytoremediation is the use of plant-based technologies for contaminant removal (Sandhi et al. 2018), whereas bioremediation represents the elimination of pollutants via the action of microorganisms (Guarino et al. 2017). Liu et al. (2018) classified the phytoremediation mechanisms into phytoextraction and phytovolatilization for heavy metal elimination and phytoimmobilization and phytostabilization for the conversion of toxic metals into harmless forms. Other techniques such as phytodegradation (Al-Baldawi et al. 2015), phytofiltration (Islam et al. 2015), phytodesalination (Jlassi et al. 2013), and phytomining (Bani et al. 2015) have also been employed for the phytoremediation of soil and groundwater. Bioremediation can be classified into natural bioremediation (Curtis and Lammey 1998) and engineered remediation (or bioaugmentation) (Agnello et al. 2016). Rhizoremediation and rhizodegradation are the utilization of plant-microbe interaction for remediation purposes (Kuiper et al. 2004). The applications of algae and fungi for the remediation schemes have been recognized as phycoremediation (Ansari et al. 2018) and mycoremediation (Kulshreshtha et al. 2014), respectively. Although remediation is a cost-effective, simple, and environmentally friendly technique for the decontamination of soil and water, understanding the interaction between plants and microbes is still a focus of ongoing research.

Hence, this chapter briefly explores the application of plant- and microbial-based technologies for the remediation of soil and water environmental systems. The ability of microbial population associated with plant roots to eliminate pollutants from soil is also discussed. The objectives are highlighted regarding case studies reported in the literature. Recommendations for future investigations in the area of environmental remediation are also presented.

18.2 Soil and Water Pollution

Recently, the increase in organic and inorganic pollutants above the threshold levels has led to the deterioration of the soil and water environmental systems (Stout and Nüsslein 2010). Pollutants are originated from several domestic, commercial, agricultural, and industrial wastes (Rajasulochana and Preethy 2016). The chemical characteristics of wastes include organic matters such as phenols, pesticides, fats, oils, grease, carbohydrates, and proteins, as well as inorganic substances, viz., pH, nitrogen, phosphorus, chlorides, sulfur, oxygen, and alkalinity (Rashidi et al. 2015). Moreover, wastes are defined by physical properties such as total solids, temperature, color, and odor as well as biological constituents of pathogenic microorganisms, viruses, bacteria, protozoa, algae, and fungi (Gonçalves et al. 2017). Wastes also comprise heavy metals (e.g., Fe, Mn, Cu, Zn, Ni, Cd, Pb, As, Hg, and Cr) released into the environment from industrial sources, viz., electroplating, tanneries, battery manufacture, paints and pigments, plastic stabilizers, steel industries, and coal combustion (Rajkumar and Freitas 2008). The natural sources of heavy elements include volcanic eruptions, weathering of minerals, atmospheric deposition, and soil erosion (Prabhakaran et al. 2016). Chemical pollutants of nitrogen, phosphorus, and pesticides result from agricultural wastes, and they can reach groundwater through seepage (Wu et al. 2015). Furthermore, wastes contain contaminants of emerging concern such as pharmaceutical products, fire retardant chemicals, and cosmetic items.

18.3 Soil and Water Treatment

The purpose of treatment procedures is to reduce the concentrations of pollutants until reaching the acceptable limits, as identified by the regulatory agencies. The treatment technologies can be classified into three main groups, i.e., physical, chemical, and biological processes (Rashidi et al. 2015).

Physical treatment is used for solid-liquid separation via screening, sedimentation, floatation, and filtration. In the chemical treatment processes, specific reagents are added to water to react with the contaminants under consideration, converting pollutants into less harmful compounds (Fawzy et al. 2016). The chemical processes include coagulation/flocculation, adsorption, ion exchange, and chlorination.

Biological treatment methodologies include activated sludge process, trickling filters, rotating biological contactors, and aerated lagoons. Biological process involves aerobic, anaerobic, or facultative microorganisms in the form of suspended or attached fine particles (Nasr 2018). Biological treatment routes are highly effective for the reduction of suspended solids, nutrients, and soluble organic matters existing in wastewater, but they are inefficient to eliminate complicated and hazardous chemical pollutants. Moreover, aerobic treatment systems such as the activated sludge process require high amounts of energy for air supply and mixing in the aeration tanks (Nasr et al. 2014). Anaerobic digestion has been employed for the treatment of various waste types, along with the generation of methane as a value-added energy product. However, the toxic contaminants are insufficiently degraded as they can deactivate the anaerobic bacteria. Microalgae can also be used for nutrients removal from wastewater, and the produced biomass is employed for the generation of fertilizers, biofuel, and animal food (Gonçalves et al. 2017). In addition, some types of microorganisms have been observed to uptake and accumulate heavy metals in their cells (Rajasulochana and Preethy 2016).

18.4 Remediation

Recently, the remediation of polluted soil and water has been widely employed for in situ field applications. In this chapter, the remediation technologies are classified into phytoremediation, bioremediation, phycoremediation, and mycoremediation.

18.4.1 Phytoremediation

Phytoremediation is a term used to define the technologies that apply plants for the degradation, immobilization, and extraction of contaminants from soil and water environments (Viehweger 2014). Phytoremediation is considered as a green and ecofriendly approach for the treatment of various wastewater sources such as landfill leachate, domestic sewage, industrial effluent, and urban and agricultural runoff (Wu et al. 2015). Phytoremediation is employed to stabilize organic and inorganic contaminants remaining after secondary wastewater treatment (Jlassi et al. 2013). Phytoremediation can be used to substitute the conventional wastewater treatment systems especially in small populations and remote areas.

Some plants can endure moderate levels of toxic constituents by chelation, while hyperaccumulators have the tendency to sequester high levels of heavy metals in their cells and tissues (Muchate et al. 2018). Metallophytes are plants that can sustain and withstand heavy metal-containing soils. The accumulation of metal ions in plants is controlled by glutathione-phytochelatin-mediated resistance (Blum et al. 2007). Phytoremediation is influenced by several factors such as soil properties, texture, and structure, nutrients availability, pH degree, variation in temperature, and defense mechanisms against pathogens and herbivores (Islam et al. 2018).

The removal routes of mineral ions from the soil environment by plants undergo the following steps (Ali et al. 2013): (a) movement of ions from the soil solution into the root system, (b) some ions are absorbed into the root hairs that have a high surface area to volume ratio, (c) other ions translocate to the shoots through xylem vessels of the vascular system, (d) large quantities of ions are further deposited within plant vacuoles, and (e) some ions are sequestered into the vacuole, while other ions are degraded by enzymatic activities within the plant tissues.

Several types of macrophytes such as emergent, submerged, floating-leaved, and free-floating aquatic plants have been used for the phytoremediation process (Fresno et al. 2018). The commonly used emergent species are *Eleocharis*, *Iris*, *Juncus*, *Phragmites*, *Scirpus*, and *Typha*. The submerged plants include *Ceratophyllum demersum*, *Hydrilla verticillata*, *Myriophyllum verticillatum*, *Potamogeton crispus*, and *Vallisneria natans*. The floating-leaved plants include *Marsilea quadrifolia*, *Nymphaea tetragona*, *Nymphoides peltata*, and *Trapa bispinosa*, whereas the free-floating plants comprise *Eichhornia crassipes*, *Hydrocharis dubia*, *Lemna minor*, and *Salvinia natans*.

However, phytoremediation can be considered as a slow and partial process due to limitations in plant growth rate, time of contact between plant and pollutants, and the depth of plant roots as well as the variation of environmental and seasonal conditions. Moreover, the plant-metabolic capacity is sensitive toward toxic contaminants. The vegetation used in phytoremediation should be prohibited to become part of a food chain regarding animal feed or direct human consumption.

This chapter represents several forms of phytoremediation, viz., phytodegradation, phytoextraction, phytovolatilization, phytostabilization, phytofiltration, phytodesalination, and phytomining.

18.4.2 Phytodegradation (Phytotransformation)

In the phytodegradation process, some enzymes within the plant tissues such as dehalogenase and oxygenase tend to degrade organic pollutants (Al-Baldawi et al. 2015). Phytodegradation involves the secretion of a diversity of hydrolyzing and oxidizing enzymes that mineralize the environmentally persistent pollutants to CO_2, H_2O, NO_3^-, and simple inorganic substances. The metabolic activities of plants are associated with the accumulation of organic xenobiotics, resulting in the detoxification of polluted environments. However, phytodegradation is less effective for the removal of hazardous and nonbiodegradable substances.

Al-Baldawi et al. (2015) used *Scirpus grossus* for the phytodegradation of petroleum hydrocarbon in a constructed wetland subjected to diesel contamination (i.e., V_{diesel}/V_{water} increased from 0.0% as a control to 0.25%). After 72 days, petroleum hydrocarbon attained a maximum removal of 81.5% at V_{diesel}/V_{water} of 0.1%, and the hydrocarbon contents were 41.41 mg/kg in roots and 53.37 mg/kg in stems and leaves. The removal mechanisms included biodegradation by rhizobacteria, external adsorption onto plant tissues, and plant-microbe adaptation (Al-Baldawi et al. 2015).

Dolphen and Thiravetyan (2015) investigated the utilization of *Cyperus alternifolius* for the phytodegradation of ethanolamines with a concentration of 1400 mg/L for 12 days. The study (Dolphen and Thiravetyan 2015) elucidated that ethanolamines were mainly accumulated in the plant stems and that triethanolamine was degraded to diethanolamine and then to monoethanolamine followed by acetic acid formation. Monoethanolamine was totally eliminated after 12 days for soil treatment, 7 days for plant condition, and only 5 days for plant + soil interaction (Dolphen and Thiravetyan 2015).

18.4.3 Phytoextraction

Phytoextraction passes through three successive steps of phytoabsorption, phytoaccumulation, and phytosequestration (Ali et al. 2013). The initial stage of phytoextraction, i.e., phytoabsorption, is used to describe the absorption and uptake of toxic metals from soil or water by plant roots. Then, the pollutants are accumulated in the roots and translocated to above-ground parts and tissues such as shoots and leaves. This step is recognized as phytoaccumulation. Finally, the toxic substances are sequestrated in plant tissues through the phytosequestration process. The selected plants should adapt to severe environmental conditions, tolerate pathogens, toxic species, and pests, retain sufficient accumulation and translocation capabilities, and comprise widely branched root system and more above-ground biomass.

Luo et al. (2005) investigated the application of ethylenediaminetetraacetic acid (EDTA) and *S,S*-ethylenediaminedisuccinic acid (EDDS) as chelating agents to improve the phytoextraction of heavy metals from contaminated soil by corn and bean plants. It was depicted that the concentrations of Cu and Zn in shoots were greater in the plants treated with EDDS than in those subjected to EDTA, whereas EDTA was more effective than EDDS for the solubilization of Pb and Cd. The biodegradable chelating agents enhanced the solubilization of heavy metals in soil, in which the metals could readily transfer from the plant roots to the shoots (Luo et al. 2005).

Ghasemi et al. (2018) utilized three Ni-hyperaccumulators, viz., *O. serpyllifolia*, *Odontarrhena bracteata*, and *O. inflate*, for the phytoextraction of Ni from the serpentine soil. The study (Ghasemi et al. 2018) depicted that rhizobacterial inoculants improved Ni elimination and promoted plant growth and health.

18.4.4 Phytovolatilization

In phytovolatilization, pollutants are captured and translocated from the soil by plants, biologically converted into a volatile form, and then diffused into the atmosphere during transpiration (Fester et al. 2014). Phytovolatilization is highly effective for the vaporization of organic contaminants, volatile organic compounds, and certain heavy metals such as mercury and selenium. Direct phytovolatilization occurs for the volatilization of organics from stems or leaves, whereas indirect phytovolatilization undertakes plant root activities (Limmer and Burken 2016). However, this technique is not widely applied due to air quality degradation and the deposition of evaporated pollutants.

Sakakibara et al. (2010) investigated the utilization of *Pteris vittata* for the treatment of As-contaminated soils that accumulated arsenic compounds up to 6540 mg/kg-DW. The study (Sakakibara et al. 2010) depicted that the phytovolatilization process contributed to 90% of the total As removal from the soil.

Arnold et al. (2007) investigated the phytovolatilization of methyl *tert*-butyl ether ($C_5H_{12}O$) with a concentration of 200 mg/L by *Pinus* sp. in groundwater. The study (Arnold et al. 2007) found that the average reduction of $C_5H_{12}O$ was 96±2.9%, with an atmospheric half-life of about 4 days. The removal mechanism was mainly transpiration via diffusion into the atmosphere through tissues, followed by the phytodegradation and accumulation stages.

18.4.5 Phytostabilization

Phytostabilization is the control of the trace elements mobility in soils by covering the contaminated area with certain plants (Fresno et al. 2018). This process is also known as phytoimmobilization. The plant roots in the rhizosphere are able to inactivate heavy metals through several mechanisms such as soil stabilization, root sorption, rhizospheric reduction, and complexation/precipitation. Phytostabilization is also employed to maintain microbial diversity and enhance the physical and chemical properties of soil as well as to restore the functionality of polluted soils in the long term. Aided phytostabilization is the use of soil amendments combined with metal-tolerant plants to develop a healthy vegetation cover that can minimize the trace elements mobility (Touceda-González et al. 2017). However, this process is expensive due to the requirement of compost amendment and revegetation of affected soil.

Touceda-González et al. (2017) applied the aided phytostabilization technique to remediate Cu-rich mine tailings using municipal solid waste composting (as a soil amendment) and plants of *Populus nigra*, *Salix viminalis*, and *Salix caprea*. After 3 years, the extractable Cu concentration dropped from 33.0 to 0.4 mg/kg, and the available P improved from 2.1 to 133.3 mg/kg (Touceda-González et al. 2017). The soil enzymatic activities, shoot height, and nutrient concentrations in leaves and stems were also improved.

Fresno et al. (2018) investigated the remediation of metal-containing soil (As 2200 mg/kg and Cu 150 mg/kg) using an aided phytostabilization mechanism. The study (Fresno et al. 2018) used four composite amendments of (a) $FeSO_4$ + lime, (b) $FeSO_4$ + paper mill sludge, (c) $FeSO_4$ + olive mill waste compost, and (d) $FeSO_4$ + holm oak biochar, whereas the metal-excluding plant was *Lupinus albus*. After 48 days, the soluble As was reduced by 50–93%, and the extractable As and Cu were decreased by 50–89%; in addition, the nutrient content in plant tissues was improved (Fresno et al. 2018).

18.4.6 Phytofiltration

Phytofiltration is an effective phytoremediation technique used to describe the absorption (or adsorption) of pollutants from aqueous solutions onto several parts of the plant (Wu et al. 2015). For example, rhizofiltration defines the adsorption process by plant roots to minimize groundwater contamination. Moreover, the applications of plant shoots and seedlings for filtration are known as caulofiltration and blastofiltration, respectively.

Islam et al. (2015) examined the application of *Micrenthemum umbrosum* for the phytofiltration of carcinogenic pollutants (As and Cd) from contaminated water. The *Micrenthemum umbrosum* leaves could accumulate As and Cd of 1220 and 800 μg/g, respectively, from 1000 μg/L of both solutions, and the binding of As in plants followed the thiol formation mechanism.

Sandhi et al. (2018) found that *Warnstorfia fluitans* achieved As removal of 82% from arsenite- and arsenate-containing water with an As concentration of 74 μg/L during the initial hour. The plant living parts attained both adsorption and absorption mechanisms, whereas the dead portions achieved an adsorption process (Sandhi et al. 2018).

18.4.7 Phytodesalination

Phytodesalination refers to the elimination of salts from saline soil or groundwater using obligate halophytes via several mechanisms such as accumulation, exclusion, and excretion (Muchate et al. 2018). Most of the plant species can survive at an electrical conductivity up to 4 dS/m through physiological, morphological, and anatomical adaptations.

Jlassi et al. (2013) investigated the treatment of a salinized soil containing a moderate salt dose of 1.5 g-NaCl/kg using *Sulla carnosa*. The study (Jlassi et al. 2013) indicated that the plant productivity was 5.2 t-DW/ha compared to 4.7 t-DW/ha for the control. The phytodesalination capacity was found to be 320 kg-Na^+/ha.

Islam et al. (2018) used various halophytes for the phytodesalination of water having a salinity level of 0–7 dS/m. The study (Islam et al. 2018) depicted that the phytodesalination capacities (in kg-Na^+/ha) were 80 for *Ludwigia adscendens*, 130 for *Ipomoea aquatica*, and 105 for *Alternanthera philoxeroides*. The phytodesalination mechanism could be assigned to vacuolar sequestration as well as the existence of substomatal cavities and spongy mesophyll cells in leaf and xylem vessels (Islam et al. 2018).

Muchate et al. (2018) employed *Spinacia oleracea* for the desalination of a saline soil having an electrical conductivity of 0.3–12 dS/m. The study (Muchate et al. 2018) revealed that *Spinacia oleracea* could survive at an electrical conductivity up to 12 dS/m and that the protection mechanisms was described by antioxidant enzyme induction and osmotic modification.

18.4.8 Phytomining

Some metal-tolerant high biomass plants are cultivated to accumulate and concentrate specific metals from soil. Phytomining is a low-cost technology known to recover particular trace elements from contaminated or mineralized soils for commercial gain (Ali et al. 2013). Phytomining is constructively applied for soils that are characterized by insufficient fertility and productivity. It is also used for targeting low-grade ores that are expensively obtained by conventional mining methods.

Rosenkranz et al. (2018) found that the metal-accumulating plant species of *Nicotiana tabacum* and *Salix smithiana* were able to recover trace elements of Cu, Ni, and Zn from waste incineration bottom ash. The study (Rosenkranz et al. 2018) demonstrated that the injection of rhizobacterial strains improved the production of *Nicotiana tabacum* as well as the growth and nutritional status of *Salix smithiana*.

Bani et al. (2015) found that *Alyssum murale* could be used for Ni recovery from the ultramafic soil with an extraction yield of 105 kg/ha/year. The study (Bani et al. 2015) indicated that the net profit of Ni phytomining would be $1055/ha per year.

18.5 Bioremediation

Biological remediation, also termed as bioremediation, is the employment of the microorganisms' activities to eliminate metals and organic pollutants from the environment (García-Sánchez et al. 2018). Bioremediators are the microorganisms that implement the functions of bioremediation. Bioremediators can survive in harsh environments by evolving multiple stress responses and defense actions. The survival mechanisms of bioremediators include the formation of resistance genes, signalling pathways against heavy metals, sequestration by metallothioneins, biofilm development and aggregation, and generation of extracellular polymeric substances to bind the metal (Prabhakaran et al. 2016). Moreover, bioremediators contribute to soil heavy metal remediation through several mechanisms such as valence transformation, volatilization, biosorption, and extracellular chemical precipitation (Chandrangsu et al. 2017). Decomposition and degradation are used to describe the organics removal processes during bioremediation. However, bioremediation holds some limitations such as insufficient knowledge about the biodegradation mechanisms and microbial interactions as well as lack of experience in the process management, monitoring, and control.

Three major processes known as bioaugmentation, biostimulation, and fertilization have been used to mediate the bioremediation process (Prabhakaran et al. 2016). The term bioaugmentation expresses the enrichment (seeding) of indigenous microorganisms with a specific microbial culture, leading to enhance the biological degradation rate (Agnello et al. 2016). Biostimulation defines the addition of chemicals, oxygen, or macro- and micro-nutrients to the polluted environment to mediate the microbial degradation activities during bioremediation. Fertilization has also been reported to enhance the biodegradation of pollutants by improving the nitrogen, phosphorus, potassium, and carbon contents of the soil (Faisal and Hasnain 2005).

In this work, in situ bioremediation is classified into natural bioremediation and engineered bioremediation.

18.5.1 Natural Bioremediation

The indigenous and dominant microbial species in the soil and groundwater systems are able to acclimatize and partially or completely detoxify contaminants in the environment without human intervention (Guarino et al. 2017). This mechanism is known as natural remediation, intrinsic remediation, or passive attenuation. This process is achieved by the combination of physical routes (e.g., volatilization, diffusion, dispersion, and dilution); chemical attenuation schemes including abiotic reactions, uptake, and sorption; and biological degradation via aerobic and anaerobic activities (Liu et al. 2018). Some microorganisms use efflux pumps to remove metals such as cadmium, arsenate, and chromium that have entered the cell (Chandrangsu et al. 2017). The broadly known efflux systems are (a) ATPases pump, which pumps out metals using ATP to drive the reaction (known as active transport), and (b) chemiosmotic ion/proton pump, which uses the proton gradient to pump metals across the cell membrane (known as diffusion) (Nies 2003).

Curtis and Lammey (1998) indicated that the intrinsic remediation process was able to degrade 5.68 mg of hydrocarbons for a liter of groundwater. Several mechanisms such as sulfate decline, denitrification, ferrous iron decrease, and aerobic bioremediation occurred during the natural attenuation process (Curtis and Lammey 1998). However, the intrinsic remediation of environmental contaminants is a time-limiting process because indigenous microbes require a long-term adaptation (i.e., decades).

18.5.2 Bioaugmentation (Engineered Bioremediation)

The bioaugmentation process is employed to improve the biological degradation of contaminants by inoculating single strains or consortia of microorganisms having specific catalytic activities to the soil (Agnello et al. 2016).

Guarino et al. (2017) investigated the application of bioaugmentation for the bioremediation of soil subjected to petroleum hydrocarbons by adding 10^8 CFU of bacterial strains (e.g., *Sphingobium abikonense* and species of *Pseudomonas*) to 1 g of soil. The study (Guarino et al. 2017) indicated that the bioaugmentation process attained petroleum hydrocarbons reduction of 75–98% compared to 45–70% for natural attenuation.

Agnello et al. (2016) used *Pseudomonas aeruginosa* for the bioaugmentation of soil contaminated with petroleum hydrocarbons of 3800 mg/kg-DW. *Pseudomonas aeruginosa* has the advantages of producing metal-chelating siderophores that enhance metal bioavailability and biosurfactants (rhamnolipids). This tendency could facilitate the solubility and mobility of heavy elements. The study (Agnello et al. 2016) depicted that the bioaugmentation system attained petroleum hydrocarbons removal of 59% compared to 37% for natural attenuation and 47% for phytoremediation with alfalfa. The combination of bioaugmentation and phytoremediation achieved the highest hydrocarbons removal of 68% for 90 days, as well as, substantial reduction of heavy metals (i.e., Cu, Pb, and Zn) (Agnello et al. 2016).

García-Sánchez et al. (2018) studied the mycoaugmentation of soil contaminated with polycyclic aromatic hydrocarbons by using *Crucibulum leave* for 180 days. The investigation (García-Sánchez et al. 2018) indicated that the hydrocarbons content declined from 1132 to 696 μg/kg-DW for mycoaugmentation, and it reduced to 658, 617, and 475 μg/kg-DW for natural attenuation, phytoremediation using maize crops, and maize-*Crucibulum leave* integration, respectively.

18.6 Plant-Microbe Interaction

18.6.1 Rhizoremediation

Some soil microbial communities can adapt to environmental changes and build a firm biofilm around the plant roots via an ectophytic relationship (Blum et al. 2007). This plant-microbe interaction is known as rhizoremediation, and it has been recently used for improving the physical and chemical properties of soils. Rhizoremediation can also be employed to detoxify contaminated soils and transform hazardous elements into harmless constituents. In this process, microbes interact with the aquatic system in the soil environment to degrade the organic pollutants, whereas plants undertake mineral exchange and attain defense immunity against toxic constituents (Srivastava et al. 2017). Moreover, the plant roots can excrete nutrients that create a nutrient-rich environment for the stimulation of bacterial growth and survival. Subsequently, this action results in improving the removal efficiencies of pollutants (Wu et al. 2014).

Rhizoremediation is achieved by the combination of multiple mechanisms including biological degradation of organic contaminants by microorganisms, uptake of nutrients by plant tissues, and sorption of pollutants by plant roots. Viehweger (2014) addressed the mechanisms of heavy metal tolerance by plant-microbe adaptation including chelation, uptake/efflux, and transport. The elimination of heavy metals by rhizoremediation also include redox transformation and soil acidification (i.e., lowering of pH in the rhizosphere by bacteria), whereas the immobilization of heavy metals in soil includes volatilization, precipitation, sorption, and sequestration mechanisms. Faisal and Hasnain (2005) reported that bacteria could stimulate the plant growth through several mechanisms such as phosphate solubilization and atmospheric nitrogen fixation as well as the production of antifungal molecules, phytohormones, and siderophores.

Rhizoremediation is influenced by several factors such as (a) soil structure, texture, and physical properties, (b) availability of microbial community able to degrade the pollutant, (c) contaminant type and concentration, and (d) presence of sufficient oxygen (or an alternative electron acceptor) and nutrients.

Constructed wetlands (CWs) have been considered as an ideal example for rhizoremediation that involves plant-microbe interaction to reduce organics, assemble nutrients, accumulate heavy metals, and inhibit coliforms (Nasr and Ismail 2015; Wu et al. 2015). Based on the wetland hydrology, CWs can be categorized into subsurface flow (SSF) CWs and free-water surface (FWS) CWs. Referring to the flow direction, SSF CWs can be classified into vertical flow (VF) and horizontal flow (HF) CWs. Recently, conventional CWs have been modified into advanced configurations such as step-feed CWs, baffled-flow CWs, circular-flow corridor CWs, hybrid towery CWs, and artificial aerated CWs (Wu et al. 2014).

18.6.2 Rhizodegradation

Rhizodegradation represents the proliferation of rhizospheric microorganisms on the root system to breakdown organic pollutants in the soil, resulting in the enhancement of plant tolerance toward dissolved metals. The degradation of pollutants is improved by increasing the growth and metabolic activities of microorganisms at the root system, as compared to bulk soil. The part of plant roots remaining in the rhizoplane also releases specific enzymes that can breakdown organic contaminants in soils. Plant-associated bacteria can convert toxic trace metals into soluble elements that are readily captured by plant roots. For example, the reduction of selenate to elemental selenium is achieved by the microbial action, leading to the improvement of Se accumulation in plants.

Rajkumar and Freitas (2008) used plant growth-promoting rhizobacteria specified as *Pseudomonas* sp. (Ps29C) and *Bacillus megaterium* (Bm4C) to facilitate plant biomass production in Ni-contaminated soil. The bacterial strains protected plants against the inhibitory effects of nickel through the production of indole-3-acetic acid (IAA) and siderophores (Fe-chelating compounds) as well as the activity of phosphate solubilization in soil (Rajkumar and Freitas 2008).

18.7 Phycoremediation

Phycoremediation is a sustainable and environmentally friendly process that represents the application of microalgal cultures for the elimination and biotransformation of contaminants from wastewater. Microalgae can be used for the remediation of industrial effluents via the biosorption and bioaccumulation of heavy metals.

Gupta et al. (2017) represented the environmental sustainability of algae-based remediation for the elimination of nutrient and organic substances from wastewater and the sequestration of CO_2 from the air. The study (Gupta et al. 2017) also explored the economic viability of utilizing the algal biomass as an ecofriendly feedstock for biofuel production.

Gupta et al. (2018) indicated that the removal efficiencies of NH_4^+-N, PO_4^{3-}-P, BOD, and COD from domestic wastewater were 76.5%, 83.1%, 73.9%, and 42.8%, respectively, using *Scenedesmus* microalgae. The obtained biomass contained yields (dry cell weight) of 28.0% proteins, 12.4% carbohydrate, and 17.4% lipids (Gupta et al. 2018).

Ansari et al. (2018) found that the cost of oil production from *Scenedesmus obliquus* microalgae cultivated in municipal wastewater ranged from $0.883 to $2.088 per liter. The microalgae strain attained high pollutant removal efficiencies of 86.2% NH_4^+-N, ≈100.0% PO_4^{3-}-P, and 87.9% COD (Ansari et al. 2018).

18.8 Mycoremediation

Mycoremediation is used to define the involvement of fungal-based technology for the treatment of complex organic and chemical substances present in polluted soils and industrial effluents.

Kulshreshtha et al. (2014) reviewed the application of mushrooms as a basidiomycetous fungus for the mycoremediation of wastes. The removal mechanisms by mushrooms include bioconversion, biodegradation, and biosorption. The produced mushrooms are used as a protein-rich and highly nutritious food and can be utilized in industrial applications such as biopulping and biobleaching.

Batista-García et al. (2017) found that fungal species relevant to the zygomycota, ascomycota, and basidiomycota genera were able to degrade polycyclic aromatic hydrocarbons and phenols from industrial wastewater.

Singh et al. (2015) depicted that fungal strains of *Rhizomucor variabilis*, *Fusarium*, *Aspergillus nidulans*, *Emericella*, and *Aspergillus oryzae* could be used for the mycoremediation of As-contaminated soils. The mycoremediation mechanisms included bioaccumulation and biovolatilization with values of 0.023–0.259 g/kg and 0.23–6.40 mg/kg, respectively.

18.9 Recommendations

Environmental remediation is a promising, cost-effective, and environmentally friendly technology that can overcome the limitations of conventional treatment methods. However, several recommendations should be considered for the optimum application of plant-microbe remediation mechanisms:

(a) The heavy metal-loaded plants that are harvested after the remediation process should be properly and safely transported and disposed of, that is, to restrict the transfer of toxic elements into the food chain.
(b) The collected biomass should be adequately managed regarding various reuse and recycling routes, such as bio-ore production, pyrolysis, metal recovery, direct combustion, and composting applications.
(c) Researchers with different scientific disciplines, viz., microbiology, plant biology, soil chemistry, environmental engineering, and ecology, are encouraged to define the microbial activities, enzymatic functions, and interdisciplinary nature of rhizoremediation.
(d) Environmental remediation methodologies should be comprehensively investigated in terms of commercial and economic feasibility, risk assessment, value engineering, and life cycle assessment.
(e) Effective bioremediators and metal hyperaccumulators that can enhance the interactions between contaminants, soil, microbes, and plant roots should be discovered.
(f) Develop advanced spectroscopic and chromatographic techniques to explore the accumulation of heavy metals in plant tissues.

18.10 Conclusion

This study presented a brief description of the contributions of plant, microorganisms, and plant-microbe interaction to remediate the soil and water systems. Phytoremediation is the utilization of plant species to remediate contaminated soils, maintain microbial diversity, and restore the soil physical and chemical properties. Phytodegradation is employed to transform organic pollutants into CO_2, H_2O, NO_3^-, and simple inorganic matters. Phytoextraction is applied to eliminate heavy metals, whereas phytovolatilization is used to reduce volatile organic compounds and some heavy metals such as mercury and selenium. Phytoimmobilization and phytostabilization are used to control the trace elements mobility in soils by covering the contaminated zone with specific vegetation. Phytodesalination is used to eliminate salts from saline soil or groundwater, while phytomining is developed to recover particular trace elements from mineralized soils. The employment of the microorganisms' activities to reduce organic pollutants and metals from the environment is classified into natural bioremediation and bioaugmentation processes. Some soil microbial populations build a firm biofilm around the plant roots to improve the degradation efficiency of pollutants. This plant-microbe rhizoremediation provides several benefits including biological degradation of organic contaminants by microorganisms, uptake of nutrients by plant tissues, sorption of pollutants by plant roots, and enhancement of plant tolerance toward dissolved metals. Other sustainable and ecofriendly processes such as phycoremediation and mycoremediation are also explored.

References

Agnello AC, Bagard M, van Hullebusch ED, Esposito G, Huguenot D (2016) Comparative bioremediation of heavy metals and petroleum hydrocarbons co-contaminated soil by natural attenuation, phytoremediation, bioaugmentation and bioaugmentation-assisted phytoremediation. Sci Total Environ 563–564:693–703

Al-Baldawi I, Abdullah S, Anuar N, Suja F, Mushrifah I (2015) Phytodegradation of total petroleum hydrocarbon (TPH) in diesel-contaminated water using Scirpus grossus. Ecol Eng 74:463–473

Ali H, Khan E, Sajad M (2013) Phytoremediation of heavy metals--concepts and applications. Chemosphere 91(7):869–881

Ansari F, Gupta S, Nasr M, Rawat I, Bux F (2018) Evaluation of various cell drying and disruption techniques for sustainable metabolite extractions from microalgae grown in wastewater: a multivariate approach. J Clean Prod 182:634–643

Arnold C, Parfitt D, Kaltreider M (2007) Phytovolatilization of oxygenated gasoline-impacted groundwater at an underground storage tank site via conifers. Int J Phytoremediation 9(1):53–69

Bani A, Echevarria G, Sulçe S, Morel J (2015) Improving the agronomy of Alyssum murale for extensive phytomining: a five-year field study. Int J Phytoremediation 17(2):117–127

Batista-García RA, Kumar VV, Ariste A, Tovar-Herrera OE, Savary O, Peidro-Guzmán H, González-Abradelo D, Jackson SA, Dobson A, Sánchez-Carbente M, Folch-Mallol J, Leduc R, Cabana H (2017) Simple screening protocol for identification of potential mycoremediation tools for the elimination of polycyclic aromatic hydrocarbons and phenols from hyperalkalophile industrial effluents. J Environ Manage 198:1–11

Blum R, Beck A, Korte A, Stengel A, Letzel T, Lendzian K, Grill E (2007) Function of phytochelatin synthase in catabolism of glutathione-conjugates. Plant J 49(4):740–749

Chandrangsu P, Rensing C, Helmann J (2017) Metal homeostasis and resistance in bacteria. Nat Rev Microbiol 15(6):338–350

Curtis F, Lammey J (1998) Intrinsic remediation of a diesel fuel plume in Goose Bay, Labrador, Canada. Environ Pollut 103:203–210

Dolphen R, Thiravetyan P (2015) Phytodegradation of Ethanolamines by Cyperus alternifolius: effect of Molecular Size. Int J Phytoremediation 17(7):686–692

Faisal M, Hasnain S (2005) Chromate resistant Bacillus cereus augments sunflower growth by reducing toxicity of Cr(VI). J Plant Biol 48:187–194

FAO/WHO (2011) Joint FAO/WHO Food Standards Programme Codex Committee on Contaminants in Foods Fifth Session the Hague. the Netherlands, 21–25 March 2011: FAO/WHO.

Fawzy M, Nasr M, Abdel-Gaber A, Fadly S (2016) Biosorption of Cr(VI) from aqueous solution using agricultural wastes, with artificial intelligence approach. Sep Sci Technol 51(3):416–426

Fawzy M, Nasr M, Nagy H, Helmi S (2018) Artificial intelligence and regression analysis for Cd(II) ion biosorption from aqueous solution by Gossypium barbadense waste. Environ Sci Pollut Res 25(6):5875–5888

Fester T, Giebler J, Wick LY, Schlosser D, Kästner M (2014) Plant-microbe interactions as drivers of ecosystem functions relevant for the biodegradation of organic contaminants. Curr Opin Biotechnol 27:168–175

Fresno T, Moreno-Jiménez E, Zornoza P, Peñalosa J (2018) Aided phytostabilisation of As- and Cu-contaminated soils using white lupin and combined iron and organic amendments. J Environ Manage 205:142–150

García-Sánchez M, Košnář Z, Mercl F, Aranda E, Tlustoš P (2018) A comparative study to evaluate natural attenuation, mycoaugmentation, phytoremediation, and microbial-assisted phytoremediation strategies for the bioremediation of an aged PAH-polluted soil. Ecotoxicol Environ Saf 147:165–174

Ghasemi Z, Ghaderian S, Rodríguez-Garrido B, Prieto-Fernández Á (2018) Plant species-specificity and effects of bioinoculants and fertilization on plant performance for nickel phytomining. Plant Soil 425(1-2):265–285

Gonçalves A, Pires J, Simões M (2017) A review on the use of microalgal consortia for wastewater treatment. Algal Res 24:403–415

Guarino C, Spada V, Sciarrillo R (2017) Assessment of three approaches of bioremediation (natural attenuation, landfarming and bioagumentation – assisted landfarming) for a petroleum hydrocarbons contaminated soil. Chemosphere 170:10–16

Gupta S, Kumar N, Guldhe A, Ansari F, Rawat I, Nasr M, Bux F (2018) Wastewater to biofuels: comprehensive evaluation of various flocculants on biochemical composition and yield of microalgae. Ecol Eng 117:62–68

Gupta SK, Sriwastav A, Ansari FA, Nasr M, Nema AK (2017) Phycoremediation: an eco-friendly algal technology for bioremediation and bioenergy production. In: Bauddh K, Singh B, Korstad J (eds) Phytoremediation potential of bioenergy plants. Springer, Singapore

Islam M, Hosen M, Uddin M (2018) Phytodesalination of saline water using Ipomoea aquatica, Alternanthera philoxeroides and Ludwigia adscendens. Int J Environ Sci Technol. https://doi.org/10.1007/s13762-018-1705-z

Islam M, Saito T, Kurasaki M (2015) Phytofiltration of arsenic and cadmium by using an aquatic plant, Micranthemum umbrosum: phytotoxicity, uptake kinetics, and mechanism. Ecotoxicol Environ Saf 112:193–200

Jlassi A, Zorrig W, El Khouni A, Lakhdar A, Smaoui A, Abdelly C, Rabhi M (2013) Phytodesalination of a moderately-salt-affected soil by Sulla carnosa. Int J Phytoremediation 15(4):398–404

Kuiper I, Lagendijk E, Bloemberg G, Lugtenberg B (2004) Rhizoremediation: a beneficial plant-microbe interaction. Mol Plant Microbe Interact 17(1):6–15

Kulshreshtha S, Mathur N, Bhatnagar P (2014) Mushroom as a product and their role in mycoremediation. AMB Express 4(29). https://doi.org/10.1186/s13568-014-0029-8

Limmer M, Burken J (2016) Phytovolatilization of organic contaminants. Environ Sci Technol 50:6632–6643
Liu L, Li W, Song W, Guo M (2018) Remediation techniques for heavy metal-contaminated soils: principles and applicability. Sci Total Environ 633:206–219
Luo C, Shen Z, Li X (2005) Enhanced phytoextraction of Cu, Pb, Zn and Cd with EDTA and EDDS. Chemosphere 59(1):1–11
Muchate N, Rajurkar N, Suprasanna P, Nikam T (2018) Evaluation of Spinacia oleracea (L.) for phytodesalination and augmented production of bioactive metabolite, 20-hydroxyecdysone. Int J Phytoremediation 20(10):981–994
Nasr M (2018) Chapter 8: Modeling applications in environmental bioremediation studies. In: Phytobiont and ecosystem restitution, 1st edn. Springer, Singapore. https://doi.org/10.1007/978-981-13-1187-1_8
Nasr M, Elreedy A, Abdel-Kader A, Elbarki W, Moustafa M (2014) Environmental consideration of dairy wastewater treatment using hybrid sequencing batch reactor. Sustain Environ Res 24(6):449–456
Nasr M, Ismail S (2015) Performance evaluation of sedimentation followed by constructed wetlands for drainage water treatment. Sustain Environ Res 25(3):141–150
Nies D (2003) Efflux-mediated heavy metal resistance in prokaryotes. FEMS Microbiol Rev 781:1–27
Prabhakaran P, Ashraf M, Aqma W (2016) Microbial stress response to heavy metals in the environment. RSC Adv 6(111):109862–109877
Rajasulochana P, Preethy V (2016) Comparison on efficiency of various techniques in treatment of waste and sewage water – a comprehensive review. Resour Efficient Technol 2(4):175–184
Rajkumar M, Freitas H (2008) Effects of inoculation of plant-growth promoting bacteria on Ni uptake by Indian mustard. Bioresour Technol 99(9):3491–3498
Rashidi H, GhaffarianHoseini A, GhaffarianHoseini A, Nik Sulaiman N, Tookey J, Hashim N (2015) Application of wastewater treatment in sustainable design of green built environments: a review. Renew Sustain Energy Rev 49:845–856
Rosenkranz T, Kidd P, Puschenreiter M (2018) Effect of bacterial inoculants on phytomining of metals from waste incineration bottom ash. Waste Manage 73:351–359
Sakakibara M, Watanabe A, Inoue M, Sano S, Kaise T (2010) Phytoextraction and phytovolatilization of arsenic from as-contaminated soils by Pteris vittata. In: Proceedings of the annual international conference on soils, sediments, water and energy 12(26)
Sandhi A, Landberg T, Greger M (2018) Phytofiltration of arsenic by aquatic moss (Warnstorfia fluitans). Environ Pollut 237:1098–1105
Singh M, Srivastava P, Verma P, Kharwar R, Singh N, Tripathi R (2015) Soil fungi for mycoremediation of arsenic pollution in agriculture soils. J Appl Microbiol 119(5):1278–1290
Srivastava JK, Chandra H, Kalra SJS, Mishra P, Khan H, Yadav P (2017) Plant–microbe interaction in aquatic system and their role in the management of water quality: a review. Appl Water Sci 7(3):1079–1090
Stout L, Nüsslein K (2010) Biotechnological potential of aquatic plant-microbe interactions. Curr Opin Biotechnol 21(3):339–345
Touceda-González M, Álvarez-López V, Prieto-Fernández Á, Rodríguez-Garrido B, Trasar-Cepeda C, Mench M, Puschenreiter M, Quintela-Sabarís C, Macías-García F, Kidd PS (2017) Aided phytostabilisation reduces metal toxicity, improves soil fertility and enhances microbial activity in Cu-rich mine tailings. J Environ Manage 186.(Pt 2:301–313
Viehweger K (2014) How plants cope with heavy metals. Bot Stud 55:35
Wu H, Zhang J, Ngo HH, Guo W, Hu Z, Liang S, Fan J, Liu H (2015) A review on the sustainability of constructed wetlands for wastewater treatment: design and operation. Bioresour Technol 175:594–601
Wu S, Kuschk P, Brix H, Vymazal J, Dong R (2014) Development of constructed wetlands in performance intensifications for wastewater treatment: a nitrogen and organic matter targeted review. Water Res 57:40–55

Development of Future Bioformulations for Sustainable Agriculture

19

Muhammad Ijaz, Qasim Ali, Sana Ashraf, Muhammad Kamran, and Abdul Rehman

Abstract

Intensive use of chemical fertilizers and pesticides for increased food production has resulted in many health hazards to humans and animals. The incessant application of these hazardous chemicals is also degrading agroecosystems. The beneficial role of soil microbes in sustainable agriculture has provided insights for decreasing the reliance on pesticides and use of chemicals for food production. In recent years, development of inoculants for sustainable agriculture has provided an alternative. However, application of these bioformulations has many hindrances and has been met with social reluctance, especially in developing countries. Because of the high specificity of bioformulations to crop and soil types, this cost-effective and green strategy faces many hurdles in comparison with chemical fertilizers and pesticides. Moreover, the viability and effectiveness of inoculants relies on the carrier material and preservation conditions. For their success in sustainable agriculture, careful selection of microbe types and extensive field evaluations are needed. This chapter critically reviews the different types and different aspects of bioformulation development for sustainable agriculture.

Keywords

Agriculture · Bioformulation · Bacteria · Fungi · Viruses · Sustainable

---

M. Ijaz · Q. Ali (✉) · A. Rehman
College of Agriculture, Bahauddin Zakariya University, Layyah, Pakistan
e-mail: ali_qasim@bzu.edu.pk

S. Ashraf
Institute of Soil and Environmental Sciences, University of Agriculture, Faisalabad, Pakistan

M. Kamran
Department of Agronomy, University College of Agriculture, University of Sargodha, Sargodha, Pakistan

V. Kumar et al. (eds.), *Microbiome in Plant Health and Disease*,
https://doi.org/10.1007/978-981-13-8495-0_19

19.1 Introduction

Bioformulations are biologically active materials containing single or multiple beneficial microbes or their metabolites, immersed in cost-effective carrier materials, which are applied to stimulate plant development and fertility, and to overcome phytopathogens (Arora et al. 2010). Burges and Jones (1998) noted that bioformulations contain aids to protect the microbes, to transfer the microbes to their final locations, and to boost the functions of the microbes. An operative definition of a bioformulation should include an active component, a carrier substance, and an additive material. The active component is typically a viable organism; it can be a functional microbe or spore, and its subsistence during the period of preservation is indispensable for efficient formulation development (Hynes and Boyetchko 2006). Bioagents added for bioformulation development provide an active and potential microbial consortium in the zone around the roots, influencing plant growth by multiple processes (Lugtenberg and Kamilova 2009). Mycorrhizal fungi and microbes belonging to the *Azospirillum*, *Azotobacter*, *Bacillus*, *Bradyrhizobium*, *Mesorhizobium*, *Pseudomonas*, *Rhizobium*, and *Trichoderma* genera are frequently used in bioformulation development. Besides *Bacillus thuringiensis* (Bt), plant growth–promoting bacteria (PGPBs) are employed frequently in cell-based formulations. The production volume of bioformulations is low in comparison with agrochemicals; their contribution is below 5% of the total items available for application as agrochemicals (Arora 2015).

An appropriate carrier substance, which is inactive in nature, assists the active components (viable cells). It guarantees that the cells are safely delivered to the vicinity of the plant or within the plant body, and that they work efficiently for promoting plant growth or destroying specific pests. Carrier substances also prolong the shelf life of the formulation (Burges and Jones 1998). The carrier is a nonliving material that must supply the cells with a favorable environment. Diverse substances can be employed as carriers, such as soils, useless plant biomass, inert substances (polymers, vermiculite, perlite), or liquids (Bashan et al. 2014). Additives such as gums, silica gel, methyl cellulose, and starch are also added to enhance the physicochemical and nutritive characteristics of bioformulations (Schisler et al. 2004). According to Jeyarajan and Nakkeeran (2000), the features of an ideal formulation are as follows:

1. It must have an extended shelf life.
2. It must not be hazardous to the growing crop.
3. It must dissolve effectively in H_2O and must deliver the bacteria.
4. It must withstand unfavorable environmental situations.
5. It must be inexpensive and must provide consistent control of plant disease.
6. It must be compatible with other agrochemicals.
7. It must be cost effective and freely accessible for formulation development.

The research conducted in the area for advancement of bioformulations has resulted in:

1. Development of carrier characteristics
2. Exploration of microorganisms with improved characteristics for plant growth
3. Enhancement of the metabolic state of the cells and their potential to employ intracellular accumulated substances for subsistence inside the carrier material (Kadouri et al. 2005)

In bioformulations, bacterial cells should be able to tolerate many unfavorable situations, such as desiccation and possibly hot circumstances. The bacteria should sustain high survival rates and have the capability to enhance plant growth during prolonged time periods. For survival, bacteria use diverse approaches such as formation and storage of osmolytes or polyhydroxyl alkanoates (PHAs). Osmomodified cells that retain osmolytes, such as trehalose or glycine betaine, exhibit a much stronger capacity to bear dehydration than non-osmomodified microbes, significantly enhancing their plant development potential (Bonaterra et al. 2005). Microbes with excessive PHA levels can survive better than those with diminished levels, as PHAs offer the cells the capability to tolerate multiple adverse physicochemical stresses (Morel et al. 2012). The ideal microbe–plant mutualistic association includes a diazotrophic microbial relationship with the growing plant. Diazotrophs convert atmospheric nitrogen to ammonia. Certain diazotrophs and other PGPBs (*Pseudomonas* and *Bacillus*) also yield phytohormones, siderophores, and phosphate-solubilizing molecules, among other complexes (Morel and Castro-Sowinski 2013).

The existing literature shows that combined application of different helpful microbes with diverse plant growth–promoting characteristics has additional or synergistic consequences for plant development and productivity. New findings also suggest that application of microbial or plant material–based metabolites for development of bioproducts can enhance crop yield and growth (Morel et al. 2015). In this chapter, the current knowledge regarding the addition of microorganisms and secondary metabolites to bioformulations that enhance crop yield is discussed.

19.2 The Worldwide Scenario

Globally, products containing viable microbes for plant growth promotion and suppression of plant pathogens are being applied (Gasic and Tanovic 2013). However, there has been no collective analysis of the available literature. One possible reason is the inconsistency of the terminology that is used. In many developing countries, the term "biofertilizer" is used; however, in other countries, the term "bioinoculant" is used for such agents that promote plant growth. In both cases, either microbe-based materials or microbes themselves are implemented (Chen et al. 2006) to improve the bioavailability of essential minerals that enhance soil fertility and subsequent plant uptake of nutrients. Most cultivators worldwide now regularly use agrochemicals for different crops. The most developed biofertilizer market in the world is in Europe, and it expanded from about US$2.5664 billion in 2012 to US$4.5822 billion in 2017 (PRWeb 2014). In North America, the biofertilizer industry started in 2012 and was predicted to expand by 14.4% between 2013 and 2018 (MicroMarket Monitor 2015). Both North America and Europe contributed more than 50% to the worldwide market income. China is also endorsing biofertilizer application for cultivation of crops. India has 151 different biofertilizer development industries. Among all, nitrogen-fixing biofertilizers were the most extensively implemented, accounting for over 78% of the global demand in 2012 (AgroNews 2014). In the area of biological control, the most efficacious biopesticides are those that contain Bt, and they account for 95% of all microbes employed. Worldwide, 322 Bt-containing formulations are being produced, generating US$210 million in revenue each year (CAB International Centre 2010). The field application of other available biopesticides is also growing. In the recent past, market surveys have been conducted by different organizations to gather data on biopesticides, but the accuracy of those reports is unclear. One of the major reasons is that the criteria used in market surveys may differ, as many companies and agroindustries include subcategories such as pheromones, essential oils, insect growth regulators, plant-induced protectants, plant growth promoters, biochemicals, and microbes in biopesticides, while others produce only microbe-based products. Thakore (2006) reported that the revenue for biopesticides in 2005 was US$672 million, but there was no explanation of the classes involved. Business organizations are strongly engaged in direct marketing surveys; thus, they have possibly amassed large quantities of consistent data, and it has been established that at the global level the biopesticide industry is growing at a rate of 10% each year.

19.3 Types of Bioformulation

Two main categories of bioformulations are available: solids and liquids (Burges and Jones 1998); however, various subcategories of bioformulation are now available and are being used all over the world (Fig. 19.1).

19.3.1 Solid Formulations

These include granules, microgranules, wettable powders (WPs), wettable granules (WGs)/water-dispersible granules (WDGs), and dusts. They are developed by mixing a binder, a dispersant, a wetting agent, etc. (Knowles 2008).

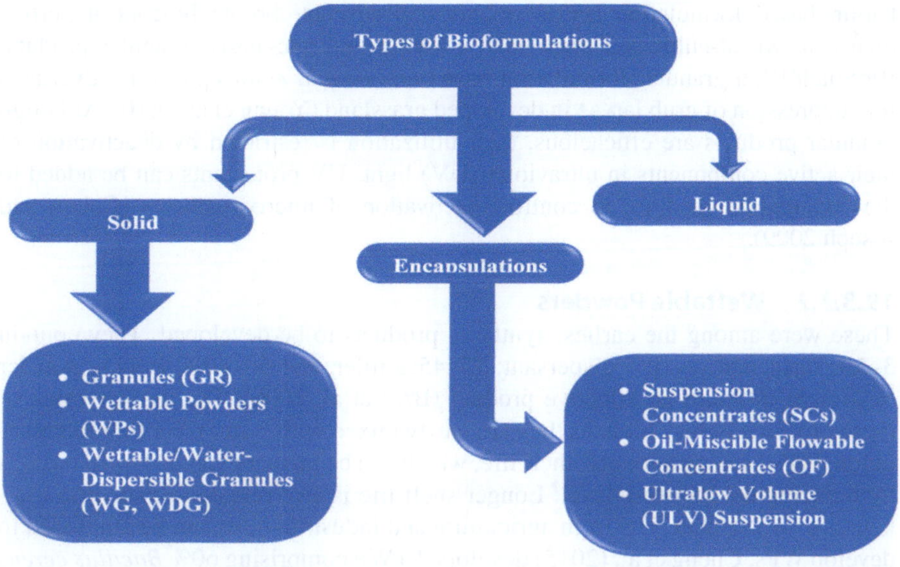

Fig. 19.1 Different types of bioformulations

19.3.1.1 Granules

These are dry substances and comprise a carrier material, a binder, and an active component. The proportion of active components in granules ranges between 5% and 20% (Brar et al. 2006). Depending on the volume of the particles, granules are grouped into macrosubstances (measuring up to 1000 μm) and microgranules (up to 600 μm). The granules must be nondusty and noncaking, and they must flow easily and break down in the cultivating soil to deliver the active component. They are typically harmless, with no danger of inhalation, and are generally employed for soil reclamation. Preservation and an extended shelf life are the most important factors for granular formulations (Callaghan and Gerard 2005). Generally employed granules include gluten, cottonseed flour and sugars, cornmeal baits, wheatmeal fragments, diatomaceous earth, semolina (durum) wheat flour, gelatin or acacia gum, and sodium alginate. MET52®, a granular bioformulation of *Metarhizium anisopliae* var. *anisopliae* strain F52, is extensively employed for biological control of black vine weevil larvae in pulpy fruits and nonedible plants (Ansari and Butt 2012). Germ-free rice is employed as a carrier material, while alginate prill is implemented in SoilGard®. This formulation consists of *Trichoderma virens* as the active component and is sold by Certis LLC for extermination of soilborne disease caused by *Pythium* and *Rhizoctonia*. The choice of different carrier materials may influence the performance of the active components in the field. In one study, Mejri et al. (2013) evaluated the bioherbicidal potential of *Pseudomonas trivialis* X33d by applying two different granular products and observed that a semolina and kaolin–based formulation (Pesta) manifested stronger brome biocontrol performance in wheat-cultivated areas than a kaolin and talc–based granular product. BioShield™, a granular formulation based on *Serratia entomophila*, is advertised for suppression of grub larvae in developed grassland (Young et al. 2010). Although granular products are efficacious, their utilization is restricted by deactivation of their active components in ultraviolet (UV) light. UV protectants can be added to the formulation medium to control inactivation of microorganisms (Cohen and Joseph 2009).

19.3.1.2 Wettable Powders

These were among the earliest synthetic products to be developed. They contain 3–5% surfactant, 1–10% dispersant, 15–45% filler, and 50–80% specific powder by weight to attain an effective product (Brar et al. 2006). The market value of these products is very high, as they are easily mixed with water prior to implementation. WPs have a prolonged shelf life, which can be further extended to 18 months by reducing the moisture level. Longer shelf life is also related to their firm marketplace. Waste materials from agriculture and industrial units can be employed to develop WPs. Cheng et al. (2015) developed a WP comprising 60% *Bacillus cereus* powder, 28.9% diatomite as a carrier material, 6% alkyl naphthalene sulfonate as a wetting element, 1% K_2HPO_4 as a stabilizer, 4% sodium lignin sulfonate as a disperser, and 0.1% β-cyclodextrin as a UV protectant in an initial experiment, and they observed that the product was beneficial in biological control of postharvest disorder in comparison with synthetic chemical application. Woo et al. (2014) surveyed the use of *Trichoderma*-based formulations in crop cultivation and observed that about 55.3% of *Trichoderma* products in the market are available as WPs.

19.3.1.3 Wettable/Water-Dispersible Granules

These formulations are also known as dry flowables. They are formulated to form WPs that are easily applicable, eco-friendly, and readily soluble in H_2O. They consist of wetting and dispersing compounds just like those used in WPs; however, the content of the dispersing compound is normally higher. Like WPs, WDGs have a prolonged shelf life. WDGs have extensive potential for nematode suppression. An antagonistic fungus is utilized to suppress powdery mildew produced by different types of pathogen in different fruits and vegetables, and has been developed as a WDG (Falk et al. 1995). Chumthong et al. (2008) processed H_2O-soluble granules comprising *Bacillus megaterium* for bioelimination of rice sheath blight and indicated that these bioproducts displayed excellent physical properties such as good H_2O dissolution and optimal viscosity, making them appropriate for spray distribution.

19.3.1.4 Dusts

These are among the oldest formulations and generally contain a 10% concentration of the active component with a particle size of 50–100 μm. They have been used for a long time. However, handling and application difficulties are associated with dusts

(Harris and Dent 2000). Dust-based beauverial protein essence is being applied in biological control.

19.3.2 Liquid Formulations

These products are also known as liquid suspensions and comprise biomass suspensions in H_2O, oils, or mixtures of both. A standard liquid product contains 10–40% microbes, 35–65% carrier liquid (oil or H_2O), 3–8% surfactant, 1–5% dispersant, and 1–3% suspender component (Brar et al. 2006). Liquid formulation may be of the following types.

19.3.2.1 Suspension Concentrates

Suspension concentrates (SCs) are manufactured by combining solid active component(s) with a low dissolving potential in H_2O and reasonable stability to hydrolysis (Tadros 2013). SCs are added to H_2O prior to application. Their preservation and dissolution can be enhanced by inclusion of surfactants and different additives. Cultivators normally utilize more SCs than WPs because they are non-dusty and are simple to quantify and transfer into the spray container.

19.3.2.2 Oil-Miscible Flowable Concentrates

Oil-miscible flowable concentrates (OFs) are stabilized suspensions of active component(s) in a liquid for dispersion in an organic solvent prior to application (Singh and Merchant 2012).

19.3.2.3 Ultralow-Volume Suspensions

Ultralow-volume (ULV) suspensions are ready for application by a ULV apparatus, which produces a very light spray (Singh and Merchant 2012).

19.3.2.4 Oil Dispersions

Oil dispersions (ODs) are safe suspensions of active component(s) in an H_2O-immiscible solution or oil. ODs have validated a growing importance over the past decade. Mbarga et al. (2014) formulated a soybean oil–based *Trichoderma asperellum* formulation and found that it had good capability for suppression of cacao black pod disease, with a longer conidia half-life than that seen in a liquid suspension. Specific defensive strategies are a prerequisite for use of fungus-containing ODs. During long-term storage the active component (conidia) may be precipitated at the base of the tank. Very few *Trichoderma*-containing liquid products are employed for biological control; examples are Trichojet, Enpro-Derma, and Trichorich-L (Woo et al. 2014). Oil-containing products are considered best for foliar application and have been assessed as effective in improving the performance of entomopathogens (Feng et al. 2004). Oil evaporation is low, thus the formulation retains its efficacy for a long period and can be used as an emulsion (oil in H_2O).

19.4 Encapsulation

Encapsulation involves developing a film or holding microbial cells inside a polymeric substance to form beads that are penetrable by minerals, gases, and metabolites for sustaining cellular activity inside the beads (John et al. 2011). Depending on the mass of the beads that are formed, two kinds of methods—macroencapsulation (involving beads that are millimeters to centimeters in size) and microencapsulation (involving beads that are 1–1000 μm in size) are adopted (Nordstierna et al. 2010). Macroencapsulation methods are more fruitful than microencapsulation methods. Encapsulation is done for protection of active components from unfavorable environmental conditions. Presently, gelatin, starch, cellulose, and some other polymers are employed for encapsulation of active components (Cheze-Lange et al. 2002). Preservation can be improved by coating the capsules with dyes. Both solid and liquid formulations have been extensively applied in agricultural systems; dry products are normally favored over wet products, as they offer a prolonged shelf life and are simple to preserve and transport (Burges and Jones 1998). The production of a bioformulation is a challenging procedure, and the previous research in this area has not been adequate. The growing pressure to compose new products to be used in place of agrochemicals has attracted attention from entrepreneurs in this area, and they are financing different projects to fabricate cost-effective and efficacious technology. Some technological improvements to develop Bt-containing products have provided considerable assistance for their commercial production. For example, micellar-enhanced ultrafiltration is a method being implemented to isolate soluble organic components, such as thuringiensin, from an aqueous efflux (Tzeng et al. 1999). Likewise, in situ product removal includes biochemical material subtraction through a fermentation procedure and is effectively implemented in subtraction of Bt toxin proteins (Agrawal and Burns 1996).

19.5 Available Bioformulations

Bioformulations are being extensively applied in agriculture. They are mainly valued for their potential contributions in the areas of biological control and biofertilization.

19.5.1 Formulations for Nutrient Uptake

In recent years, application of microbes has become recognized as a productive way to provide growth nutrients to plants, as it can considerably minimize the application of chemical fertilizers; thus, production of biofertilizers for different crops at commercial levels is increasing (Trabelsi and Mhamdi 2013). Microbe-based bioformulations for enhancing the bioavailability of nutrients are explained in Sects. 19.5.1.1, 19.5.1.2, 19.5.1.3 and 19.5.1.4.

19.5.1.1 Nitrogen

Nitrogen (N) is a vital plant macronutrient needed in high concentrations (1–3% on the basis of dry weight), but only a small proportion of nitrogen fertilizers applied to farming soils is taken up by plants (Kraiser et al. 2011). According to a survey, about 50% is taken up by crops, 25% is released into the lower atmosphere, 20% is discharged into water systems, and only 5% is deposited in the soil pool (Garnett et al. 2009). This is the main reason why the total amount of synthetic N applied to agricultural crops worldwide has increased dramatically from 12 to 104 Tg/year (i.e., from 12 to 104 million tonnes per year) in the past. Biological nitrogen fixation (BNF) is a natural process converting elemental nitrogen into plant-available nitrogen and has substantial ecological and economic benefits (Gothwal et al. 2009). However, the proficiency of N_2 fixation is restricted and is absolutely constrained to the majority of the phyla of bacteria and methanogenic archaea. Symbiotic N_2 fixation inside the nodules of vascular growing plants is conducted by two main classes of bacteria: rhizobia and *Frankia* (Franche et al. 2009). Legumes constitute the third largest family of flowering plants accounting for approximately 27% of the world's crop production including important crop legumes: soybean (*Glycine max*), peanut (*Arachis hypogaea*), mung bean (*Vigna radiata*), chickpea (*Cicer arietinum*), lentil (*Lens culinaris*), common bean (*Phaseolusvulgaris*), pea (*Pisum sativum*), and alfalfa (*Medicago sativa*). BNF yields approximately 200 million tonnes of nitrogen per year (Peoples et al. 2009), reducing the costs of crop fertilization for farmers. Different researchers have documented the function of rhizobia in viable crop cultivation and have concluded that farming systems applying rhizobial inoculum, rather than N fertilizers, can supply sufficient N to legumes (Arora et al. 2010). Both leguminous seeds and soil can be supplied with legume inoculants containing active rhizobia. Legume inoculants may contain one or multiple types of bacteria that are beneficial for the specific host and are commercially applied for development of powder/granular and liquid formulations (Lupwayi et al. 2006). Peat is frequently used as a carrier substance in legume inoculation development. The cell count is reliant on the environmental circumstances and the rhizobial species applied. Although direct utilization of rhizobia has been very effective, the current implementation of Nod factors or lipochitooligosaccharides (LCOs) has also had considerable influences on crop production in soils containing limited rhizobia (Kidaj et al. 2012). A liquid product of free-living N_2 fixers such as *Azospirillum*- and *Azotobacter*-containing cyanobacteria has been also commercialized in different countries and has resulted in a substantial intensification of the agricultural yield. Different endophytic bacteria such as *Achromobacter*, *Azoarcus*, and *Burkholderia* can also fix N_2 (Franche et al. 2009).

19.5.1.2 Phosphate

Growing plants take up phosphorus (P) as phosphate ions from the growth medium—that is, the soil solution. Phosphate is possibly one of the least accessible plant minerals present in the root zone, as a consequence of its inorganic fixation and development of organic complexes. The P level in normal soils is approximately 0.05% (w/w), of which only 0.1% is accessible to growing crops. Approximately

80% of applied P may not be available to growing plants. Globally, a total area of nearly 5.7 billion hectares of soil has been documented as being deficient in phosphate (Vassilev and Vassileva 2006). Biological processes in the soil, such as microbial functions, achieve transformation of insoluble forms of P to a plant-available form (orthophosphate), which is an essential quality of arbuscular mycorrhizal fungi (AMFs) and phosphate-solubilizing bacteria (PSBs) (Khan et al. 2007). In recent years, agriculturalists have extensively applied microbial inoculum containing phosphate-solubilizing microbes (PSMs). Field application of PSMs has increased the production of soybean, maize, wheat, mung bean, and chickpea. The phosphate-solubilizing potential of PSMs was discovered much earlier, but it was not possible to make them commercially successful as a bioformulation. Quality assurance is indispensable for production of consistent and pollution-free bioproducts; however, in the field, proficiency is dependent on environmental factors such as salinity, pH, moisture, temperature, and other soil conditions (Khan et al. 2009). Among PSB-containing biofertilizers, products containing *Aspergillus* sp., *Bacillus* sp., *Pseudomonas* sp., and *Penicillium* sp. are preferred (Sharma et al. 2013). Phosphobacterin is among the oldest *Bacillus megatherium*–containing biofertilizers, and roughly ten million hectares were reclaimed with this product in Russia in 1958. In India, P Sol B® has been extensively used to inoculate cultivated soil and contains *Pseudomonas striata* (NCIM 2847). Fosfosol® is a phosphatic biofertilizer containing *Penicillium janthinellum* and is utilized extensively in Colombia (Moreno-Sarmiento et al. 2007).

19.5.1.3 Potassium

For optimum plant growth, bioavailability of potassium (K) is as important as bioavailability of N and P. Many important functions of plants are based on K availability. K (a macronutrient) is essential for enzymatic functions of numerous physiological reactions (including protein synthesis, starch synthesis, and photosynthesis) and also promotes resistance to infections and insects (Rehm and Schmitt 2002). The lithosphere contains about 2.5% potassium, but its concentration varies widely in the soil, ranging from 0.04% to 3%. Soil K is accessible to plants in four diverse reservoirs (Syers 1998):

1. The soil solution
2. Exchangeable potassium
3. Fixed potassium
4. Lattice potassium

Among these, the soil solution and exchangeable K are continuously accessible for plant utilization, but for rapid crop growth with sufficient K, acquirement only from these sources is not enough and an external source (potassic fertilizer) is needed. Globally, India is the fourth largest user of K fertilizers after the USA, China, and Brazil. The fixed K pool in soils is dissolved by discharge of organic acids from bacteria that enhance the level of K^+ in the soil medium (Meena et al. 2014). Their potential to solubilize minerals enriched with potassium—such as orthoclase, micas, and illite—is of great concern in composition of bioinoculants with the capacity to deliver soluble K to growing plants (Sheng and Lin 2006). In a few countries, particularly South Korea and China, K biofertilizers have been evaluated widely. For the formulation of K biofertilizers, mostly those PSBs that can also dissolve potassium-containing ores are selected (Ahmed and El-Araby 2012). *Frateuria aurantia* has been shown to be a highly potent K-solubilizing bacterium and is employed in the formulations of Symbion-K, Biosol-K, and K Sol B® bioproducts.

19.5.1.4 Iron

Almost all life forms on earth need iron (Fe) in the form of several proteins and pigments. In soil, its level is between 7 and 500 g kg^{-1}, mostly in the insoluble Fe (III) form, which easily hydrolyzes to give $Fe(OH)_2^+$, $Fe(OH)_3^+$, and Fe $(OH)_4^+$. In soil, the concentration of Fe (III) is high, but plants uptake ferrous (II) iron and the accessibility of each form of Fe is determined by the pH and O_2 level in the soil (Fageria et al. 1990). It has been recognized that microbes living near developing roots drop their redox potential, increasing the level of Fe (II) ions for plant utilization. This microbe-mediated Fe uptake is enhanced by iron chelators called siderophores. Various siderophores are found in the rhizosphere, but those released by pseudomonads are considered to have a higher affinity for chelating ferric ions (Meyer 2000). A high level of siderophore production by PGPBs provides them with competitive benefits in comparison with other microbes. Microbe-assisted Fe uptake has been recognized as a promising way to provide effective mineral uptake to plants, and many study findings have shown that application of microbial inoculants at a low level with the potential to chelate iron promotes plant growth. One of the essential characteristics of efficient siderophore-releasing microbes is that they may control soil fungal pathogens by chelating accessible iron and making it inaccessible to other organisms (Beneduzi et al. 2012). Several study reports have stated that application of microbes enhances Fe uptake by growing plants (Saha et al. 2015), but the availability of bioformulations for Fe uptake is low. In India, Fe Sol B®, a product of Agri Life Bio Solutions, is an iron-mobilizing biofertilizer currently used for several edible crops.

19.6 Plant Growth–Promoting Bioformulations

Plant growth–promoting rhizobacteria (PGPRs) are useful for both plant development and decreasing insect pest attacks. One of the communal ways of utilizing bacterial inoculants in soil is in the form of bioformulations. The sustainability of the inoculum in an applicable formulation for a definite time period is important for marketing of the technology (Bashan 1998). According to the existing literature, *Bacillus* bioformulations can persist for up to 1 year (El-Hassan and Gowen 2006). Carriers containing preparations of two PGPRs such as *Bacillus subtilis* and *Pseudomonas corrugata* developed into formulations have also been assessed for growth enhancement, rhizosphere colonization, and their sustainability during storage (Trivedi et al. 2005).

Viswanathan and Samiyappan (2008) reported that *Pseudomonas* spp. has the capability to control *Colletotrichum falcatum*, which causes systemic infection in sugarcane stalks in field environments. These experiments also showed that *Pseudomonas* spp. enhanced cane and sugar productivity. Many sugar producers in India have shown keen interesting in using this novel procedure to control *C. falcatum* in sugarcane.

Chakravarty and Kalita (2011) showed suppression of bacterial wilt with a concomitant enhancement in the productivity of bioformulation-treated crops compared with inoculated controls, reinforcing the utility of *Pseudomonas fluorescens* as a biocontrol agent for bacterial wilt in brinjal, as well as PGPRs. Still, rigorous screening of native populations of *P. fluorescens*, development of better carriers, and large-scale field experiments in different climatic circumstances are required to develop formulations with improved disease suppression performance in the field. Chuaboon and Prathuangwong (2007) discussed future prospects for PGPR bioformulations to be employed for enhancement of growth and health in economic crops. A humic acid and *P. fluorescens*–based bioformulation has been established as a suitable replacement for synthetic fertilizers. This bioformulation is in a liquid form; possesses a long shelf life with no contamination or without carriers; is easy to handle, store, and transport; and is easy to use with irrigation. This bioformulation can be utilized for dual purposes such as crop protection and production. The fungus *Fusarium oxysporum*, which causes wilt in tomato, was selected as a specimen to assess the efficacy, viability, and inhibitory characteristics of the liquid bioformulation. Two crop varieties—radish and tomato—were selected for testing of this liquid bioformulation in field trials (Agrawal et al. 2014).

Chakraborty et al. (2013) showed that three isolates—*Serratia marcescens*, *Bacillus amyloliquefaciens*, and *B. pumilus*—had capability as plant growth developers to enhance the growth of tea plants in field experiments. The intensification in growth was attributed to phosphate solubilization, defense enzymes, and augmented buildup of phenolics. The sustainability of the isolates in bioformulations of talc, sawdust, and rice husks was also inspected. In comparison with *S. marcescens*, bioformulations containing *B. amyloliquefaciens* and *B. pumilus* were more valuable for field application because of the formation of endospores by these bacilli.

19.7 Biopesticides: Biocontrol and Formulation

Hazardous use of synthetic chemicals for managing pests creates problems not only in plants but also in humans and animals by polluting the surrounding environment. That is why scientists and researchers seek to develop novel, inexpensive, and eco-friendly chemicals from natural sources, called biochemicals. Globally, according to a rough estimation, 1400 natural chemicals are being traded, and this trend is gradually increasing.

Generally, microbes (bacteria, fungi, and viruses) are being used for formulation of useful biopesticides. The type and name of each biopesticide depends upon the type of organisms used in the formulation of the pesticide. Selection of biological agents depends upon the viability, strength, and potential of the inoculum for effectiveness in the field (Ash 2010). A number of scientists and researchers are working on formulation of natural biocontrol chemicals and have published reports discussing production methodologies and biopesticide formulations (Ehlers and Shapiro-Ilan 2005).

19.7.1 Bacteria

Bacteria-based natural pesticides are considered safe and economical products, and are attracting more attention. *B. thuringiensis* (Bt) is commonly used as a natural, Gram-positive biopesticide in soil (Bravo et al. 2011). Generally, Bt is considered a first-generation biopesticide, with a mixture of spores and crystals from native strains being used in commercial products (Rosas–Garcia 2009). However, advancements in molecular genetic modification have allowed engineering of Bt (Cerda and Maurizio 2004) for use in second-generation products containing various insecticidal crystal proteins. Biopesticide formulations containing *P. fluorescens* have been found to be beneficial because of their selectivity in nature for target pests and are categorized as third-generation products (Young et al. 2008). Biopesticides containing *P. fluorescens* cells engineered to produce Bt Cry delta-endotoxin are cultured and then chemically treated for toxin fixation within the cells. Nanocapsules are prepared to minimize degradation of the toxin and to stabilize the formulation for treatment of plant leaves. Hence, the storage life of the product has been enhanced by adoption of this advanced technology. Developed countries are adopting semisolid and liquid-state fermentation technologies to produce Bt commercially and at an industrial level, respectively, whereas in developing states, semisolid and solid-state fermentation techniques are most commonly used to develop Bt on a small scale (Devi et al. 2005). Various phytopathogens are naturally controlled by use of biopesticides containing *Pseudomonas* strains (Tewari and Arora 2014). For instance, various diseases can be suppressed by using diverse *P. fluorescens* strains, as they are well known to yield a range of antibiotics or antifungal metabolites (Weller 2007). Various *Pseudomonas*-based products such as Bio Save, BlightBan, Cedomon, Biocoat, and Victus are used to control various diseases in a natural way. Moreover, to manage soilborne seedling diseases and pathogens in fruits and vegetables, *Agrobacterium radiobacter*, *Burkholderia cepacia*, and *Streptomyces griseoviridis* have been used (Leonard and Julius 2000). Generally, entomopathogenic bacteria can be simply generated in an in vitro setup, except for *Bacillus popilliae* and its close relatives, which can be generated only in their natural hosts.

19.7.2 Fungi

Scientists and researchers have repeatedly attempted to develop commercial formulations of biopesticides based on fungi such as *M. anisopliae* and *Beauveria bassiana*; however, these attempts have been less successful than bacterial formulations

(McCoy 1990). Internationally, a commercial mycoinsecticide is available that is formulated from *B. bassiana*, and it alone accounts for about 34% of the total commercial fungal product market (De Faria and Wraight 2007). Approximately 750 types of fungal species have been identified as entomopathogenic in nature; however, very few have been found to be effective as biopesticides to manage insect pests (Copping 2009). Various reasons are responsible for these discrepancies, but mainly the shelf life of fungal formulations is short and mass production costs are high. The delicate nature of conidia and hyphae is an additional reason for poor preservation of the active components beyond a certain time limit. Various formulations contain *Trichoderma harzianum*, *Trichoderma asperellum*, *Trichoderma gamsii*, *Coniothyrium minitans*, *Aspergillus flavus*, and *Chondrostereum purpureum* (Auld 2002). Mostly, *Trichoderma* spp. are being used in research studies to quantify their potential for biocontrol, and progress in commercialization of these formulations is being made with the passage of time by adoption of innovative techniques for mass production of these fungi (Vinale et al. 2008). Genetic modifications have also been utilized to enhance fungal action against target pests. Fungal transformations have been made using Ca^{2+} and polyethylene glycol–mediated protoplast transformation, electroporation, and particle bombardment techniques (Gielesen and van den Berg 2013). In 2014, the famous mycologist Jaronski published a report on solid-state fermentation (also known as solid substrate fermentation) and liquid-state fermentation (also known as submerged culture fermentation) for mass production of entomopathogenic fungi. Entomopathogenic fungi can be mass produced by using various organic substrates such as broken rice, cassava chips, coconut and cotton cake, kodo and finger millet, rice husks, and wheat bran. Recently, inorganic substances such as diatomite (calcined diatomaceous earth) and clay granules with open pores have also been utilized for fungal formulation (Jaronski 2014). Use of industrial effluent and sludge without water as high-mineral inputs for development of fungi has also been described (Verma et al. 2007).

19.7.3 Viruses

Viruses cause severe diseases not only in humans but also in animals and insects. Seven virus families—Baculoviridae, Iridoviridae, Parvoviridae, Picornaviridae, Poxviridae, Reoviridae, and Rhabdoviridae—are commonly known for causing infections in pests, especially in insects, making them suitable for use as biocontrol agents. These viruses infect insects and form occlusion bodies which confirm their role in biocontrol; however, Baculoviridae are mostly used in bioformulations (Kalawate 2014). The Baculoviridae family has been categorized into four genera: *Alphabaculovirus*, *Betabaculovirus*, *Gammabaculovirus*, and *Deltabaculovirus* (Reid et al. 2014). According to an estimate, 30 products of different types made from baculoviruses, including over different 20 species, have been made commercially available after registration as biopesticides (Rao et al. 2015). The world's largest viral insecticide producer is China, which produces over 32 registered viral biopesticides (Sun 2015). Europe and the USA also have sizable markets for viral biopesticides, which are sold in those markets under the trade names Madex 3 (Andermatt Biocontrol), Granupom (AgrEvo), Carpovirusine (NPP-Calliope), Carposin (Agrichem), Virin-Gyap (NPO Vector), and CYD-X (Thermo Trilogy).

Viral insecticides can be developed in vitro and in vivo. Generally, an original host is used in in vivo conditions, and advanced methodology for in vivo production in insects has been described by several scientists (van Beek and Davis 2007). However, during in vivo production processes, various difficulties such as bacterial contamination and virus degeneration can occur, so scientists prefer in vitro production of viruses using controlled methods and have recommended it as being better than in vivo production (Nguyen et al. 2011). Viral insecticides have been developed in the form of dense WPs rather than liquid or granular formulations.

19.7.4 Consortium-Based Inoculants

Commercial bioformulations containing single strains are available on the market. Bioformulation using a mixed culture or coinoculation with microbes has been found to be more effective in improving plant growth and development. Coinoculation of mycorrhizal microbes with rhizobia has achieved better results in leguminous crops, and farmers are adopting this practice in their fields to improve nutritional levels in leguminous plants and to enhance drought and osmotic tolerance in lucerne, soybean, broad bean, chickpea, and pigeon pea. Plant growth is stimulated by utilization of a combination of rhizobia and PSBs in legumes (Messele and Pant 2012). As phosphate and potassium are vital for optimal and sustainable crop production, an integrated approach using PSBs with a K-solubilizing bacterial inoculum can be applied to achieve the required yield targets (Hu et al. 2006). Combined application of fungi (AMFs) with a bacterial inoculum increases accumulation of nitrogen and phosphorus in common bean shoots in comparison with a single inoculum (Tajini et al. 2012). Azotobacter, cyanobacteria, and microalgae consortia can be utilized as biofertilizers and biostimulators in various crops, as recommended by Zayadan et al. (2014). In Vietnam a consortium-based biofertilizer, sold under the commercial name BioGro™, is being marketed that contains two bacilli, *P. fluorescens* strains, and a soil yeast (Cong et al. 2009).

19.8 Constraints in Manufacturing and Sale of Bioformulations

Since biopesticides are products manufactured through use of living microorganisms, every step in the production process, from the start to the end of the process, is crucial to sustain the microbial load and vigor. Biopesticide production technology requires appropriate quality assurance and the aid of sophisticated equipment to confirm the availability of these valuable formulations on the market. Quality inoculant production is of great importance for supplying essential nutrients to crop plants and making cropping systems sustainable (Kabi 1997). The richest source of microorganisms on the earth is soil. These microorganisms, inhabiting the soil rhizosphere, are beneficial to plants and their ecosystems. Anthropogenic activities, including indiscriminate use of synthetic chemicals such as fertilizers and pesticides, are affecting soil fertility and productivity, disturbing the ecosystem and harming animals as well as human life. Ultimately, soils are becoming unfertile and arid, and are losing valuable microbes (Seneviratne and Kulasooriya 2013). However, environmentally friendly bioformulations are not attracting much attention in the agromarket, because of various limitations associated with them. The chief constraints associated with operative bioformulation development are discussed in Sects. 19.8.1, 19.8.2 and 19.8.3

19.8.1 High Costs of Production

Bioformulation production is expensive and requires advanced hi-tech instruments and controlled sterile conditions. These are the reasons why bioformulations are not attracting more attention. Nonsterile carrier inoculants are widely used in the formulation of biopesticides, causing bacterial contamination. It is worth noting, however, that, after a decade of their use, no health issues due to use of nonsterile carrier materials have been reported. The production of bacterial inoculant formulations is claimed to be inexpensive in comparison with production of agrochemicals, although screening of bacterial strains on a large scale is still needed to explore their biological activities (Bashan 1998). Biopesticides are temperature-sensitive and environment-sensitive products; thus, enormous caution is needed during every step from the beginning to the end of their production, transportation/distribution, and application. They need to be properly packaged and stored, and suitable carrier materials need to be used, which require a lot of investment (Arora et al. 2001). Generally, companies with large production facilities invest more and understand the marketing process much better than the others. The process for registration of a new product is also a great hindrance, as it is often costly and time consuming in the development of new products (Ehlers 2006).

19.8.2 Shelf Life

Appropriate skill and special facilities are required for bioformulation storage, and most farmers, producers, and retailers lack these facilities. Bioformulation production methods, packaging, storage, and transportation need to sustain the formulation's shelf life. To avoid problems, air-dried and lyophilized preparations of bioformulations have been adopted (Kosanke et al. 1992). The shelf life of a bioinoculant depends upon its water content during storage. The lower the water content is, the longer the shelf life will be, and the product will remain viable for a long period. Thus, the bacterial load in the formulation will remain dormant, uncontaminated, environmentally safe, and fertilizer responsive during application. Low water content during formulation processes is very critical, especially for bacterial (non-spore-forming) formulations (Shah-Smith and Burns 1997). Numerous variables influence the survival of bacteria during the formulation process, such as the bacterial cultivation culture, the time of harvesting from the medium, the physiological state of the bacteria, and the dehydration rate and technology (Paul et al. 1993).

19.8.3 Unpredictable Performance: Fate of Inoculants Introduced into the Soil

The performance of bioformulations in the field can be inconsistent, and this is considered a major marketing constraint for their commercialization. Prompt decay of the population of active cells is another major reason for failure of bioformulations to achieve their objectives. Soil is heterogeneous in nature and contains a mixture of microorganisms, which are interlinked with each other, so the action of an applied biopesticide is unpredictable under inconsistent indigenous circumstances. Microorganisms and the roots of growing plants gain mutual benefits by helping each other. The action of the microorganisms may be promotional or inhibitory depending upon the bioformulation concentrations in the soil. This growth/survival–inhibitory influence of soil has been termed soil microbiostasis. It has been recognized that poor availability of mineral sources to microbes in the soil can be linked to numerous unfavorable abiotic and biotic aspects. The activity and fate of inoculant organisms in the soil can be determined by the physiological characteristics of the organisms. These physiological characteristics play a vital role in the capability of the inoculant bacteria to survive and colonize the soil; however, often these characteristics are unknown. Hence, the responses of different species can vary in terms of their persistence and their actions in the soil rhizosphere. Therefore, to obtain efficient and effective inoculants, a systematic assortment technique is mandatory. Moreover, the inherent physiological traits of the organisms and both abiotic and biotic soil elements play significant roles. Abiotic aspects of the soil—such as its texture, type, pH, temperature, and moisture—exert their influence on the inoculant inhabitants of the soil by imposing numerous different types of stress on the inoculant cells (Evans et al. 1993).

19.9 Roles of Metabolites in Future Bioformulations

Additives and metabolites play vital roles in making bioformulations more reliable and effective. Flavonoids are preliminary biomolecules used in combination with rhizobial inoculants to promote nodulation. Addition of flavonoids to rhizobial inoculants improves nodulation, N_2 fixation, and vigor to combat abiotic stresses. Nodulation occurs through secretion of single LCO compounds by root-nodulating bacteria called rhizobia. The association between rhizobia and legume roots is known to be symbiotic in nature, and LCO biomolecules are vital in this association, influencing the crop yield positively (Oldroyd 2013). Various PGPRs, including rhizobia and pseudomonads, also secrete crucial exopolysaccharide (EPS) metabolites, which improve nodule formation, root establishment, and formation of biofilms, protecting introduced cells under stress conditions, counteract contaminants, and provide a carbon reservoir (Tewari and Arora 2014). Protection of nitrogenase enzymes in rhizobia is also provided by EPS metabolites that are prolific developers of biopolymers. Production of EPSs is economical, and they are easy to produce extensively. EPS-containing bioformulations not only shield cells but also support colonization of the rhizosphere by the bioinoculant. EPSs can also help to shield plant roots under abiotic stresses. As shown in Fig. 19.2, inclusion of LCOs, flavonoids, and/or phytohormones increases the performance of bioformulations, resulting in enhanced commercial products.

The activity and effectiveness of the microbial load can be enhanced by combined application of these biomolecules with plant macro- and micronutrients such as rock P, K, S, or Zi, along with solubilizing and sulfur-oxidizing bacteria. Sources of these substances are readily available and inexpensive worldwide in comparison with usage of soluble types of P and K (Abou-el-Seoud and Abdel-Megeed 2012). Sulfuric acid is produced by sulfur-oxidizing bacteria during phosphate and sulfur solubilization processes that increase the bioavailability of iron and zinc (micronutrients) (Stamford et al. 2008).

Various research studies have shown the supplementary importance of cellulose, amino acids, starch, molasses, and wastewater, along with bioinoculants, in a bioformulation (Arora 2015). These supplementary substances act as continuous nutrient reservoirs, releasing nutrients in a slow manner so that microbial populations can be maintained in the bioformulation during storage. Inclusion of microcrystalline cellulose achieves worthwhile results in this regard. Protection of microorganisms from UV radiation can be achieved by utilizing molasses. Therefore, the shelf life and survivability of plant growth–promoting microorganisms in a bioformulation can be enhanced by making these amendments (Brar et al. 2006). PGPRs produce biosurfactants that exhibit a mixture of properties (such as antimicrobial, antiviral, and anti-insecticidal activities) and act as emulsifiers, wetting, and dispersing agents (Thavasi et al. 2015). It is proposed to augment these substances in bioformulations and utilize them as plant growth–promoting microorganism carriers and dispersal agents. The importance of biosurfactants is greater in liquid inoculants for foliar spraying of plants. Phagostimulants and attractants such as sucrose, molasses, edible oil, glutamate, and pheromones can be added to bioformulations particularly biopesticides to attract pathogens towards the antagonistic inoculant microbe and trap them (Farrar and Ridgway 1995). Advanced formulations should contain metabolites that not only increase the efficiency and shelf life of biopesticides but also are acceptable and have a wider application range internationally. Metabolites include protectants, adjuvants, attractants, stimulants, antimicrobials, and precursors of biological origin. Microbial metabolites are being implemented in different industries globally. Such bioformulations will be more effective in varying climatic and geographic conditions. Formulations comprising bacterial cells and metabolites will be a consistent technology.

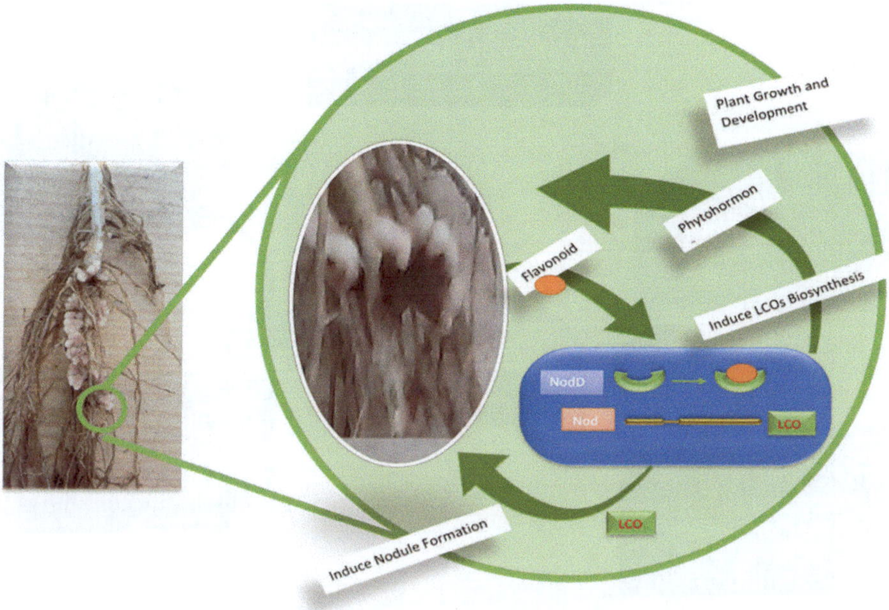

Fig. 19.2 Effects of secondary metabolites on bioformulations

19.10 Research Areas for Further Development of Bioformulations

Plant diseases can be easily managed by developing and utilizing biocontrol techniques. Bioformulation is one of these techniques and is cost effective and environmentally friendly (Heydari and Gharedaghli 2007). Bioformulations from antagonistic bacteria and other biocontrol agents are gaining great importance and attention internationally, particularly in countries where soilborne diseases pose major problems.

The optimal formulation and effectiveness of biocontrol agents are significant aspects of their efficacy. Formulations that have been developed and analyzed can be helpful for suppressing plant diseases and possibly other plant–pathogen combinations. These bioformulations have great potential as natural pesticides and can replace chemical fungicides. In sustainable agriculture, bioformulations should be used in integrated pest management (IPM) (Ardakani et al. 2010). A bioformulation can boost product stability, shield bacteria against different environmental conditions, and provide an initial food source. Application of PGPRs either to promote crop health or to control plant diseases depends on production of commercial formulations with appropriate carriers that maintain the viability of the bacteria for a substantial time period. It is important to estimate the survival of the immobilized bacteria in different carriers and also their capability to retain the qualities required for plant growth promotion (Aeron et al. 2011).

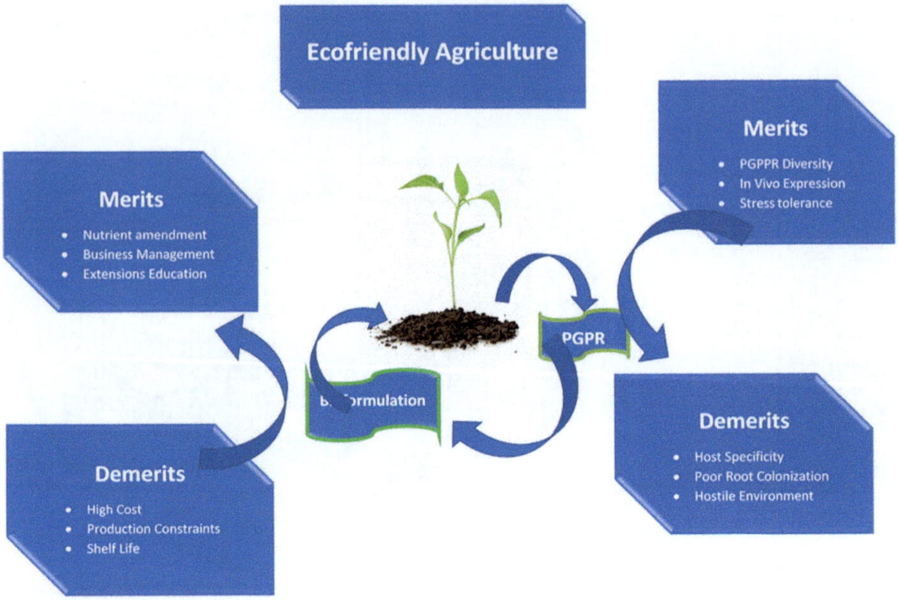

Fig. 19.3 Research and development strategy for bioformulation technology

Research on N fixation and phosphate solubilization by PGPRs is progressing, but little research has been conducted on solubilization of potassium, which is the third main essential macronutrient for plant development. Research in this area will not only promote the use of bioinoculants but also create confidence in their utilization among cultivators. Apart from that, future research focusing on optimizing the growth conditions and prolonging the shelf life of PGPR products that are not phytotoxic to crop plants and that tolerate unfavorable environmental situations, achieve higher productivity, and are cost effective for farmers to use will also be beneficial (Gupta et al. 2015). A talc-based bioformulation of a *P. fluorescens* RRb-11 isolate showed maximum shelf life and survivability in the rhizosphere, reducing the disease intensity of bacterial blight of rice and thereby increasing the yield when employed as a seed treatment, seedling root dip, and soil drench in combination (Jambhulkar and Sharma 2014). Strong technical and research-based relationships among researchers, agriculturists, microbiologists, biotechnologists, industrialists, and farmers are vital needs in this regard (Fig. 19.3).

19.11 Conclusion

Use of consortia containing multitrait plant growth–promoting microbes may be useful in formulation of novel bioinoculants that can offer inexpensive, reasonable, and appealing substitutes for costly agrochemicals. The inoculant industry is facing various challenges to make better-quality formulations that offer a long shelf life and more viable and resistive cells in rhizosphere surroundings, and that are easy to use and economical for farmers. More research to explore the practical aspects of mass production and formulation are needed in order to develop effective, stable, safer, more economical, and novel bioformulations.

References

Abou-el-Seoud II, Abdel-Megeed A (2012) Impact of rock materials and biofertilizations on P and K availability for maize (*Zea maize*) under calcareous soil conditions. Saudi J Biol Sci 19:55–63

Aeron A, Dubey RC, Maheshwari DK, Pandey P, Bajpai VK, Kang SC (2011) Multifarious activity of bioformulated *Pseudomonas fluorescens* PS1 and biocontrol of *Sclerotinia sclerotiorum* in Indian rapeseed (*Brassica campestris* L.). Eur J Plant Pathol 131:81–93

Agrawal A, Burns MA (1996) Application of membrane-based preferential transport to whole broth processing. Biotechnol Bioeng 55:581–591

Agrawal P, Pandey SC, Manjunatha RAH (2014) Development of liquid formulation for the dual purpose of crop protection and production. J Environ Res Develop 8(3):378–383

AgroNews (2014) Biofertilizers market—global industry analysis, size, share, growth, trends and forecast, 2013–2019. http://report.agropages.com/ReportDetail-1292.htm. Accessed 12 Nov 2018

Ahmed HFS, El-Araby MMI (2012) Evaluation of the influence of nitrogen fixing, phosphate solubilizing and potash mobilizing biofertilizers on growth, yield, and fatty acid constituents of oil in peanut and sunflower. Afr J Biotechnol 11:10079–10088

Ansari MA, Butt TM (2012) Evaluation of entomopathogenic fungi and a nematode against the soil-dwelling stages of the crane fly *Tipula paludosa*. Pest Manag Sci 68:1337–1344

Ardakani SS, Heydari A, Khorasani N, Arjmandi R (2010) Development of new bioformulations of *Pseudomonas fluorescens* and evaluation of these products against damping-off cotton seedlings. J Plant Pathol 92(1):83–88

Arora NK (2015) Plant microbe symbiosis: applied facets. Springer, New Delhi, p 383

Arora NK, Kumar V, Maheshwari DK (2001) Constraints, development and future of the inoculants with special reference to rhizobial inoculants. In: Maheshwari DK, Dubey RC (eds) Innovative approaches in microbiology. Singh and Singh, Dehradun, pp 241–245

Arora NK, Khare E, Maheshwari DK (2010) Plant growth promoting rhizobacteria: constraints in bioformulation, commercialization and future strategies. In: Maheshwari DK (ed) Bacteria and plant health. Springer, Berlin, pp 97–116

Ash GJ (2010) The science, art and business of successful bioherbicides. Biol Control 52:230–240

Auld B (2002) Fungi as biocontrol agents: progress, problems and potential. Plant Pathol 51:518–518

Bashan Y (1998) Inoculants of plant growth promoting bacteria use in agriculture. Biotechnol Adv 6:729–770

Bashan Y, de-Bashan LE, Prabhu SR, Hernandez JP (2014) Advances in plant growth promoting bacterial inoculant technology: formulations and practical perspectives (1998–2013). Plant Soil 378:1–33

Beneduzi A, Ambrosini A, Pasagila LMP (2012) Plant growth promoting rhizobacteria (PGPR): their potential as antagonistic and biocontrol agents. Genet Mol Biol 35:1044–1051

Bonaterra A, Camps J, Montesinos E (2005) Osmotically induced trehalose and glycine betaine accumulation improves tolerance to desiccation, survival and efficacy of the postharvest biocontrol agent *Pantoea agglomerans* EPS125. FEMS Microbiol Lett 250:1–8

Brar SK, Verma M, Tyagi RD, Valero JR (2006) Recent advances in downstream processing and formulations of *Bacillus thuringiensis* based biopesticides. Process Biochem 41:323–342

Bravo A, Likitvivatanavong S, Gill SS, Sobero NM (2011) *Bacillus thuringiensis*: a story of a successful bioinsecticide. Insect Biochem Mol Biol 41:423–431

Burges HD, Jones KA (1998) Formulation of microbial biopesticides: beneficial microorganisms, nematodes and seed treatments. Kluwer Academic, Dordrecht, p 411

CAB International Centre (2010) The 2010 world wide biopesticides market summary. CAB International Centre, Wallingford

Callaghan MO, Gerard FM (2005) Establishment of *Serratia entomophila* in soil from a granular formulation. N Z Plant Prot 58:122–125

Cerda H, Maurizio GP (2004) Genetic engineering with *Bacillus thuringiensis* and conventional approaches for insect resistance in crops. Crit Rev Plant Sci 23:317–323

Chakraborty U, Chakraborty BN, Chakraborty AP, Sunar K, Dey PL (2013) Plant growth promoting rhizobacteria mediated improvement of health status of tea plants. Indian J Biotechnol 12(1):20–31

Chakravarty G, Kalita MC (2011) Management of bacterial wilt of brinjal by *Pseudomonas fluorescens* based bioformulation. ARPN J Agric Biol Sci 6(3):1–11

Chen YP, Rekha PD, Arun AB, Shen FT, Lal WA, Young CC (2006) Phosphate solubilizing bacteria from subtropical soil and their tricalcium phosphate solubilizing abilities. Appl Soil Ecol 34:33–41

Cheng H, Li L, Hua J, Yuan H, Cheng S (2015) A preliminary preparation of endophytic bacteria CE3 wettable powder for biological control of postharvest diseases. Not Bot Hortic Agrobot 43:159–164

Cheze-Lange H, Beunard D, Dhulster P, Guillochon D, Caze A-M, Morcellet M, Saude N, Junter GA (2002) Production of microbial alginate in a membrane bioreactor. Enzym Microb Technol 30:656–661

Chuaboon W, Prathuangwong S (2007) Biological control of cauliflower soft rot using bacterial antagonist and its risk assessment. J Thai Phytopathol 21:63–48

Chumthong A, Kanjanamaneesathian M, Pengnoo A, Wiwattanapatapee R (2008) Water-soluble granules containing *Bacillus megaterium* for biological control of rice sheath blight: formulation, bacterial viability and efficacy testing. World J Microbiol Biotechnol 24:2499–2507

Cohen E, Joseph T (2009) Photostabilization of *Beauveria bassiana* conidia using anionic dyes. Appl Clay Sci 42:569–574

Cong PT, Dung TD, Hien TM (2009) Inoculant plant growth–promoting microorganisms enhance utilisation of urea–N and grain yield of paddy rice in southern Vietnam. Eur J Soil Biol 45:52–61

Copping LG (2009) The manual of biocontrol agents: a world compendium. British Crop Protection Council, Alton, p 851

De Faria MR, Wraight SP (2007) Mycoinsecticides and mycoararicides: a comprehensive list with worldwide coverage and international classification of formulation types. Biol Control 43:237–256

Devi PSV, Ravinder T, Jaidev C (2005) Cost-effective production of *Bacillus thuringiensis* by solid-state fermentation. J Invertebr Pathol 88:163–168

Ehlers RU (2006) Use of biotechnology in biological plant protection. Pflanzenschutz Schr Dtsch Phytomed Ges 8:17–31

Ehlers RU, Shapiro-Ilan DI (2005) Mass production. In: Grewal PS, Ehlers RU, Shapiro-Ilan DI (eds) Nematode biocontrol agents. CAB International Centre, Wallingford, pp 65–78

El-Hassan SA, Gowen SR (2006) Formulation and delivery of the bacterial antagonist *Bacillus subtilis* for management of lentil vascular wilt caused by *Fusarium oxysporum* f. sp. *lentis*. J Phytopathol 154:148–155

Evans J, Wallace C, Dobrowolski N (1993) Interaction of soil type and temperature on the survival of *Rhizobium leguminosarum* bv. *viciae*. Soil Biol Biochem 25:1153–1160

Fageria NK, Baligar VC, Edwards DG (1990) Soil–plant nutrient relationships at low pH stress. In: Baligar VC, Duncan RR (eds) Crops as enhancers of nutrient use. Academic, San Diego, pp 475–507

Falk SP, Gadoury DM, Cortesi P, Pearson RC, Seem RC (1995) Partial control of grape powdery mildew by the mycoparasite *Ampelomyces quisqualis*. Plant Dis 79:483–490

Farrar JRR, Ridgway RL (1995) Enhancement of activity of *Bacillus thuringiensis* Berliner against four lepidopterous insect pests by nutrient based phagostimulants. J Entomol Sci 30:29–42
Feng MG, Pu XY, Ying SH, Wang YG (2004) Field trials of an oil-based emulsifiable formulation of *Beauveria bassiana* conidia and low application rates of imidacloprid for control of false-eye leaf hopper *Empoasca vitis* on tea in southern China. Crop Prot 23:489–496
Franche C, Lindstrom K, Elmerich C (2009) Nitrogen-fixing bacteria associated with leguminous and non-leguminous plants. Plant Soil 321:35–59
Garnett T, Conn V, Kaiser BN (2009) Root based approaches to improving nitrogen use efficiency in plants. Plant Cell Environ 32:1272–1283
Gasic S, Tanovic B (2013) Biopesticide formulations, possibility of application and future trends. Pestic Phytomed 28:97–102
Gielesen B, van den Berg M (2013) Transformation of filamentous fungi in microtiter plate. In: Gupta VK, Tuohy MG, Ayyachamy M, Turner KM, O'Donovan A (eds) Laboratory protocols in fungal biology: current methods in fungal biology. Springer, New York, pp 343–347. http://sci-hub.tw/10.1007/978-1-4614-2356-0_30
Gothwal R, Nigam V, Mohan M, Sasmal D, Ghosh P (2009) Screening of nitrogen fixers from rhizospheric bacterial isolates associated with important desert plants. Appl Ecol Environ Res 6:101–109
Gupta G, Parihar SS, Ahirwar NK, Snehi SK, Singh V (2015) Plant growth promoting rhizobacteria (PGPR): current and future prospects for development of sustainable agriculture. J Microbial Biochem Technol 7:96–102
Harris J, Dent D (2000) Priorities in biopesticide research and development in developing countries. CAB International Centre, Wallingford
Heydari A, Gharedaghli A (2007) Integrated pest management on cotton in Asia and North Africa. In: INCANA, Tehran
Hu X, Chen J, Guo J (2006) Two phosphate and potassium solubilizing bacteria isolated from Tianmu Mountain, Zhejiang, China. World J Microbiol Biotechnol 22:983–990
Hynes RK, Boyetchko SM (2006) Research initiatives in the art and science of biopesticide formulations. Soil Biol Biochem 38:845–849
Jambhulkar PP, Sharma P (2014) Development of bioformulation and delivery system of *Pseudomonas fluorescens* against bacterial leaf blight of rice *Xanthomonas oryzae* pv. *oryzae*. J Environ Biol 35:843–849
Jaronski ST (2014) Mass production of entomopathogenic fungi: state of the art. In: Morales-Ramos JA, Rojas MG, Shapiro-Ilan DI (eds) Mass production of beneficial organisms: invertebrates and entomopathogens. Elsevier, London, pp 357–414
Jeyarajan R, Nakkeeran S (2000) Exploitation of microorganisms and viruses as biocontrol agents for crop disease management. In: RK Upadhyay et al. (eds) Biocontrol potential and their exploitation in sustainable agriculture. Kluwer Academic/Plenum, New York, pp 95–116
John RP, Tyagi RD, Brar SK, Surampalli RY, Prevost D (2011) Bio-encapsulation of microbial cells for targeted agricultural delivery. Crit Rev Biotechnol 31:211–226
Kabi MC (1997) Impact of biofertilizer on rural development. In: Proceedings of National Conference on Impact of Biotechnology and Modern Horticulture in Rural Development, Jadavpur
Kadouri D, Jurkevitch E, Okon Y, Castro-Sowinski S (2005) Ecological and agricultural significance of bacterial polyhydroxyalkanoates. Crit Rev Microbiol 31:55–67
Kalawate A (2014) Microbial viral insecticides. In: Sahayaraj K (ed) Basic and applied aspects of biopesticides. Springer, New Delhi, pp 47–68
Khan MS, Zaidi A, Wani PA (2007) Role of phosphate-solubilizing microorganisms in sustainable agriculture—a review. Agron Sustain Dev 27:29–43
Khan MS, Zaidi A, Wani PA (2009) Role of phosphate solubilizing microorganisms in sustainable agriculture. In: Lictfouse E et al (eds) Sustainable agriculture. Springer, Dordrecht, pp 551–570
Kidaj D, Wielbo J, Skorupska A (2012) Nod factors stimulate seed germination and promote growth and nodulation of pea and vetch under competitive conditions. Microbiol Res 167:144–150

Knowles A (2008) Recent developments of safer formulations of agrochemicals. Environmentalist 28:35–44

Kosanke JW, Osburn RM, Shuppe GI, Smith RS (1992) Slow rehydration improves the recovery of dried bacterial populations. Can J Microbiol 38:520–525

Kraiser T, Gras DE, Gutierrez AG, Gonzalez B, Gutierrez RA (2011) A holistic view of nitrogen acquisition in plants. J Exp Bot 62:1455–1466

Leonard GC, Julius JM (2000) Review biopesticides: a review of their action, applications and efficacy. Pest Manag Sci 56:651–676

Lugtenberg B, Kamilova F (2009) Plant-growth-promoting rhizobacteria. Annu Rev Microbiol 63:541–556

Lupwayi NZ, Clayton GW, Rice WA (2006) Rhizobial inoculants for legume crops. J Crop Improv 15:289–321

Mbarga JB, Begoude BAD, Ambang Z, Meboma M, Kuate J, Schiffers B, Hoopen GMT (2014) A new oil-based formulation of *Trichoderma asperellum* for the biological control of cacao black pod disease caused by *Phytophthora megakarya*. Biol Control 77:15–22

McCoy CW (1990) Entomogenous fungi as microbial pesticides. In: Baker RR, Dunn PE (eds) New direction in biological control: alternatives for suppressing agricultural pests and diseases. Alan R. Liss, New York, pp 139–159

Meena VS, Maurya BR, Verma JP (2014) Does a rhizospheric microorganism enhance K^+ availability in agricultural soils? Microbiol Res 169:337–347

Mejri D, Gamalero E, Souissi T (2013) Formulation development of the deleterious rhizobacterium *Pseudomonas trivialis* X33d for biocontrol of brome (*Bromus diandrus*) in durum wheat. J Appl Microbiol 114:219–228

Messele B, Pant LM (2012) Effects of inoculation of *Sinorhizobium ciceri* and phosphate solubilizing bacteria on nodulation, yield and nitrogen and phosphorus uptake of chickpea (*Cicer arietinum* L.) in Shoa Robit area. J Biofertil Biopestic 3:1–6

Meyer JM (2000) Pyoverdines: pigments, siderophores and potential taxonomic markers of fluorescent *Pseudomonas* species. Arch Microbiol 174:135–142

MicroMarket Monitor (2015) North America biofertilizer market by application (cereals & grains, fruits & vegetables, pulses & oilseeds), by type (nitrogen fixing biofertilizers, phosphate solubilizing biofertilizers, potash mobilizing biofertilizers), by source, by geography—analysis and forecast to 2019. http://www.micromarketmonitor.com/market/north-america-bio-fertilizer-5250154124.html. Accessed

Morel MA, Castro-Sowinski S (2013) The complex molecular signaling network in microbe–plant interaction. In: Arora NK (ed) Plant microbe symbiosis: fundamentals and advances. Springer, New Delhi, pp 169–199

Morel MA, Brana V, Castro-Sowinski S (2012) Legume crops, importance and use of bacterial inoculation to increase the production. In: Goyal A (ed) Crop plant. InTech, Rijeka, pp 217–240

Morel MA, Cagide C, Minteguiaga MA, Dardanelli MS, Castro-Sowinski S (2015) The pattern of secreted molecules during the co-inoculation of alfalfa plants with *Sinorhizobium meliloti* and *Delftia* sp. strain JD2: an interaction that improves plant yield. Mol Plant-Microbe Interact 28:134–142

Moreno-Sarmiento N, Moreno-Rodriguez LF, Uribe D (2007) Biofertilizantes para la agricultura en Colombia. In: Izaguirre-Mayoral ML, Labandera C, Sanjuan J (eds) Biofertilizantes en Iberoamerica: vision tecnica, cientifica y empresarial. Denad Internacional, Montevideo, pp 38–45

Nguyen Q, Qi YM, Wu Y, Chan LC, Nielsen LK, Reid S (2011) In vitro production of *Helico verpa* baculovirus biopesticides—automated selection of insect cell clones for manufacturing and systems biology studies. J Virol Methods 175:197–205

Nordstierna L, Abdalla AA, Nordin M, Nyden M (2010) Comparison of release behavior from microcapsules and microspheres. Prog Org Coat 69:49–51

Oldroyd GE (2013) Speak, friend, and enter: signaling systems that promote beneficial symbiotic associations in plants. Nat Rev Microbiol 11:252–263

Paul E, Fages J, Blanc P, Goma G, Pareilleux A (1993) Survival of alginate-entrapped cells of *Azospirillum lipoferum* during dehydration and storage in relation to water properties. Appl Microbiol Biotechnol 40:34–39

Peoples MB, Brockwell J, Herridge DF, Rochester IJ, Alves BJ, Urquiaga S, Boddey RM, Dakora FD, Bhattarai S, Maskey SL, Sampet C (2009) The contributions of nitrogen-fixing crop legumes to the productivity of agricultural systems. Symbiosis 48(1–3):1–7

PRWeb (2014) Europe bio fertilizer market is expected to reach $4,582.2 million in 2017 new report by MicroMarket Monitor. https://www.prweb.com/releases/bio-fertilizer/market/prweb12040386.htm. Accessed 18 Sept 2019

Rao GVR, Kumar CS, Sireesha K, Kumar PL (2015) Role of nucleopolyhedroviruses (NPVs) in the management of lepidopteran pests in Asia. In: Sree KS, Varma A (eds) Biocontrol of lepidopteran pests: use of soil microbes and their metabolites. Springer, Cham, 11–52

Rehm G, Schmitt M (2002) Potassium for crop production. Regents of the University of Minnesota, Saint Paul. https://extension.umn.edu/phosphorus-and-potassium/potassium-crop-production Accessed 14 June 2019

Reid S, Chan L, Van Oers M (2014) Production of entomopathogenic viruses. In: Morales-Ramos JA, Rojas MG, Shapiro-Ilan DI (eds) Mass production of beneficial organisms. Elsevier, Amsterdam, pp 437–482

Rosas-Garcia NM (2009) Biopesticide production from *Bacillus thuringiensis*: an environmentally friendly alternative. Recent Pat Biotechnol 3:28–36

Saha M, Sarkar S, Sarkar B, Sharma BK, Bhattacharjee S, Tribedi P (2015) Microbial siderophores and their potential applications: a review. Environ Sci Pollut Res 23:3984

Schisler DA, Slininger PJ, Behle RW, Jackson MA (2004) Formulation of *Bacillus* sp. for biological control of plant diseases. Phytopathology 94:1267–1271

Seneviratne G, Kulasooriya SA (2013) Reinstating soil microbial diversity in agroecosystems: the need of the hour for sustainability and health. Agric Ecosyst Environ 164:181–182

Shah-Smith DA, Burns RG (1997) Shelf-life of a biocontrol *Pseudomonas putida* applied to sugarbeet seeds using commercial coating. Biocontrol Sci Tech 7:65–74

Sharma SB, Sayyed RZ, Trivedi MH, Gobi TA (2013) Phosphate solubilizing microbes: sustainable approach for managing phosphorus deficiency in agricultural soils. SpringerPlus 2:587

Sheng XF, Lin YH (2006) Solubilization of potassium-bearing minerals by a wild-type strain of *Bacillus edaphicus* and its mutants and increased potassium uptake by wheat. Can J Microbiol 52(1):66–72

Singh KN, Merchant K (2012) The agrochemical industry. In: Kent JA (ed) Handbook of industrial chemistry and biotechnology. Springer, New York, pp 643–699

Stamford NP, Lima RA, Lira MA Jr, Santos CERS (2008) Effectiveness of phosphate and potash rocks with *Acidithio bacillus* on sugarcane yield and their effects on soil chemical attributes. World J Microbiol Biotechnol 24:2061–2066

Sun X (2015) History and current status of development and use of viral insecticides in China. Viruses 7:306–319

Syers JK (1998) Soil and plant potassium in agriculture. International Fertiliser Society, Newcastle upon Tyne

Tadros T (2013) Suspension concentrates. In: Tadros T (ed) Encyclopedia of colloid and interface science. Springer, Berlin/Heidelberg, 1334–1334

Tajini F, Trabelsi M, Drevon JJ (2012) Combined inoculation with *Glomus intraradices* and *Rhizobium tropici* CIAT899 increases phosphorus use efficiency for symbiotic nitrogen fixation in common bean (*Phaseolus vulgaris* L.). Saudi J Biol Sci 19:157–163

Tewari S, Arora NK (2014) Multifunctional exopolysaccharides from *Pseudomonas aeruginosa* pf23 involved in plant growth stimulation, biocontrol and stress amelioration in sunflower under saline conditions. Curr Microbiol 69:484–494

Thakore Y (2006) The biopesticide market for global agricultural use. Ind Biotechnol 2:194–208

Thavasi R, Marchant R, Banat IM (2015) Biosurfactant applications in agriculture. In: Naim KN, Vardar-Sukan F (eds) Biosurfactants production and utilization-processes, technologies, and economics. Taylor & Francis, Boca Raton, pp 313–327

Trabelsi D, Mhamdi R (2013) Microbial inoculants and their impact on soil microbial communities: a review. Bio Med Res Int 86:32–40

Trivedi P, Pandey A, Palni LMS (2005) Carrier-based preparations of plant growth promoting bacterial inoculants suitable for use in cooler regions. World J Microbiol Biotechnol 21:941–945

Tzeng YM, Tsun HY, Chang YN (1999) Recovery of thuringiensin with cetylpyridinium chloride using micellar-enhanced ultrafiltration process. Biotechnol Prog 15:580–586

Van Beek N, Davis DC (2007) Baculovirus insecticide production in insect larvae. Methods Mol Biol 388:367–378

Vassilev SV, Vassileva CG (2006) A new approach for the classification of coal fly ashes based on their origin, composition, properties, and behavior. Fuel 86:1490–1512

Verma M, Brar S, Tyagi R, Surampalli R, Valero J (2007) Industrial wastewaters and dewatered sludge: rich nutrient source for production and formulation of biocontrol agent, *Trichoderma viride*. World J Microbiol Biotechnol 23:1695–1703

Vinale F, Sivasithamparam K, Ghisalberti EL, Marra R, Woo SL, Lorito M (2008) Trichoderma–plant–pathogen interactions. Soil Biol Biochem 40:1–10

Viswanathan R, Samiyappan R (2008) Bio-formulation of fluorescent *Pseudomonas* spp. induces systemic resistance against red rot disease and enhances commercial sugar yield in sugarcane. Arch Phytopathol Plant Protect 41(5):377–388

Weller DM (2007) Pseudomonas biocontrol agents of soil borne pathogens: looking back over 30 years. Phytopathology 97:250–256

Woo SL, Ruocco M, Vinale F, Nigro M, Marra R, Lombardi N, Pascale A, Lanzuise S, Manganiello G, Lorito M (2014) Trichoderma-based products and their widespread use in agriculture. Open Mycol J 8:71–126

Young J, Choi J, Gringorten L, Belanger L, Morel L, Bourque D, Masson L, Groleau D, Mıguez CB (2008) Production of an insecticidal crystal protein from *Bacillus thuringiensis* by the methylotroph *Methylobacterium extorquens*. Appl Environ Microbiol 74:5178–5182

Young SD, Townsend RJ, Swaminathan J, Callaghan MO (2010) *Serratia entomophila*–coated seed to improve ryegrass establishment in the presence of grass grubs New Zealand. Plant Prot 63:229–234

Zayadan BK, Matorin DN, Baimakhanova GB, Bolathan K, Oraz GD, Sadanov AK (2014) Promising microbial consortia for producing biofertilizers for rice fields. Microbiology 83:391–397

Exploring Diversity of Bacterial Endophyte Communities Using Advanced Sequencing Technology

20

Inga Tamošiūnė, Elena Andriūnaitė, Vidmantas Stanys, and Danas Baniulis

Abstract

The endophytic bacteria are universal occupants of the host plant tissues, and microbial endophytic population play an important and distinctive role in the agro-ecosystems functioning. Exploration of the vast diversity of plant-associated and endophytic microbiome has been largely accelerated by the introduction of high-throughput sequencing technologies. Microbial DNA direct amplification from samples of plant tissue and implementation of the massive parallel sequencing technology generates remarkably extensive data that provides an expansive insight into the composition of the plant-associated microbial communities. This review presents an overview of the research experience built up during the decade of application of the next-generation sequencing (NGS) in the metagenomic analysis of the endophytic bacterial communities. We outline the methods and sequencing approaches used in the metagenomic analysis of the plant-associated microbiome and discuss methodological challenges associated with the application of the high-throughput sequencing technology. The metagenomic studies have provided new knowledge about plant genotype- and tissue-specific complexity of the indigenous endophytic communities as well as their dynamics during plant development. Of particular interest are the new insights into the effect of agricultural practices and environmental factors on the endophytic microbiome composition that pave the way to practical implications and development of efficient and sustainable agriculture.

Keywords

Bacteria · Diversity · Endophyte · Metagenomics · Sequencing

I. Tamošiūnė · E. Andriūnaitė · V. Stanys (✉) · D. Baniulis
Institute of Horticulture, Lithuanian Research Centre for Agriculture and Forestry, Babtai, Kaunas reg., Lithuania
e-mail: v.stanys@lsdi.lt

V. Kumar et al. (eds.), *Microbiome in Plant Health and Disease*,
https://doi.org/10.1007/978-981-13-8495-0_20

20.1 Introduction

All plants develop close interaction with microorganisms that is a key and main basis of plant productivity and health (Andreote et al. 2009; Berg et al. 2014; Turner et al. 2013). The bacterial endophytes colonize intercellular spaces of the plant cell walls, xylem vessels of plant leaves, stems and roots; the endophytes are also observed in flower tissues (Compant et al. 2010), in fruits (de Melo Pereira et al. 2012) and in seeds (Cankar et al. 2005; Johnston-Monje and Raizada 2011; Trognitz et al. 2014).

The roots of a plant have been considered and recognized as the main access point for prospective endophytes from the environment (Compant et al. 2010); therefore, high endophytic population density is characteristic to roots and other underground plant tissues as compared to phyllosphere or seeds (Hallmann et al. 2001). Endophytic infection takes place at young tissues of the root elongation and differentiation zone (Hurek et al. 1994), at cracks, occurring at the emergence points of lateral roots (Reinhold and Hurek 1988) or the wounds inflicted by pathogens or mechanical damage (Compant et al. 2010). On the other hand, root nodule-forming bacteria have specific mechanisms for active root system penetration (Hardoim et al. 2008). Once bacteria relocate to the root conductive tissues, it is followed by ascending migration of bacterial endophytes from roots to shoots and leaves (Chi et al. 2005; Compant et al. 2010). In addition, colonization of the phyllosphere occurs at wounds as a result of herbivore or other mechanical damage (Compant et al. 2010). Certain indigenous endophytic bacteria that form an intimate relationship with the host plant could be transmitted through seeds (Cankar et al. 2005; Johnston-Monje and Raizada 2011; Shahzad et al. 2018; Trognitz et al. 2014).

Mainly due to its simple application, the most common methodology used to study bacterial endophytes is based on the characterization of isolates that live in the internal plant tissues or seeds obtained by disinfection of plant surfaces and cultivation on microbial growth media (Miche and Balandreau 2001; Ryan et al. 2008). Due to long culture periods and elaborate culture techniques, culture-dependent procedures are time-consuming (Miche and Balandreau 2001). Besides, regardless of the efforts focused on optimization of endophytic bacteria isolation and culturing, such as media composition, incubation conditions, and various supplements (Eevers et al. 2015; Hamaki et al. 2005), the accepted view is that the culture-dependent approach recovers only a small proportion of all plant-associated bacteria and has a limited capability to reflect the diversity of the endophytome (Eevers et al. 2015). However, despite the limitations, this approach provides the bacterial isolates that are useful for plant colonization and functional studies (Lebeis 2014).

During the last decades, a number of culture-independent approaches focusing on the application of restriction enzymes and polymerase chain reaction (PCR) for analysis of the diversity of the bacterial communities were developed (reviewed by Andreote et al. (2009)). The DNA sequence-based methods gathered an extensive amount of data regarding the endophytic microbial diversity using various metagenomic sequencing techniques and proved invaluable in exploring evolutionary conservation and phylogeny of plant-associated microorganisms as they identify the complete genetic sequence. Sequencing of hypervariable regions of the small subunit ribosomal RNA gene (16S rRNA) allow accurate taxonomic identification (Turner et al. 2013). Extensive studies of plant-associated microorganisms using Sanger sequencing of metagenomic libraries brought to light new aspects of endophytic diversity, including identification of new unculturable species and dynamics of endophyte population, and provided hints about their physiological and ecological significance in the complex interaction between the plant-host and bacterial endophytes (Xia et al. 2015). Culture-dependent and DNA sequencing-based methods have certain advantages and disadvantages; therefore, it was suggested that polyphasic approach that combines the two different approaches should be used in order to more efficiently reflect the richness of bacterial communities interacting with plants (Lebeis 2014).

Massive parallel sequencing, also called second-generation sequencing (SGS) or next-generation sequencing (NGS), technology uses largely parallelized platforms for DNA amplification followed by sequencing of millions to billions of short reads in a single step. Complementing the traditional methods by the novel high-throughput sequencing technology allows characterization of bacterial endophytome in unprecedented detail. These techniques are capable of revealing the complete diversity of the microbiome in a single run and largely replaced the Sanger sequencing-based analysis of metagenomic libraries (Lebeis 2014). The major advantage of the next-generation sequencing technologies, such as pyrosequencing (e.g., 454), reverse dye terminator (e.g., Illumina), sequencing by oligonucleotide ligation and detection (e.g., SOLiD) and proton detection (e.g., Ion Torrent), is their ability to generate an enormous volume of data for the reasonable cost (Metzker 2010). Application of such techniques allows a detection of a considerable fraction of the uncultivable and rare bacterial species that are missed using the cultivation-based approach (Kisand and Wikner 2003). The application of the second-generation sequencing requires preparation of amplified sequencing libraries before the sequencing; therefore, it should be kept in mind that precision of these powerful tools can be biased by the efficacy of DNA extraction, amplification of the region of interest using PCR, and also sequencing reactions (Andreote et al. 2009; Kchouk et al. 2017).

The aim of this review is to summarize the advances in understanding the diversity and functioning of bacterial endophytic communities by the introduction of novel massive parallel sequences techniques for metagenomic analysis. An outline of the methods and sequencing approaches used in metagenomic analysis of microbiome and overview of methodological challenges associated with the application of the methods is provided, followed by a summary of new insights into the dynamics and complexity of endophytic populations and how it is modulated by agricultural practices and environmental factors.

20.2 Application of Novel Sequencing Methods in Metagenomic Analysis of Endophytome

20.2.1 Sequencing Platforms and Methods Used in Microbial Metagenome Analysis

The introduction of massively parallel sequencing technologies opens up new capabilities to explore composition and function of microbial community in situ. Culture-independent methodologies, based on analysis of DNA extracted directly from plant samples, are the key tool for studying the functional diversity of plant-associated endophytes (Beckers et al. 2016). Initial studies of plant-associated microorganisms using next-generation sequencing were carried out using the 454 pyrosequencing technology as it was the first commercially available NGS instrument introduced by Roche (Knief 2014). Studies on endophytic communities of potato (Manter et al. 2010) and cottonwood (Gottel et al. 2011) were among the first examples of NGS technology and pyrosequencing applications that shed light on the diversity of plant endophytome.

Several years later, first 16S rRNA-based metagenomic studies of plant endophytome using the Illumina sequencing technology were published and were dedicated to sugar beet, *Aloe vera* ((L.) Burm.f.), *Sorghastrum nutans ((L.) Nash) and Spartina alterniflora* (Loisel.) plants (Akinsanya et al. 2015; Arenz et al. 2015; Hong et al. 2015; Shi et al. 2014). Although currently 454 pyrosequencing continues to be useful in the analysis of microbial communities (Correa-Galeote et al. 2018; Gadhave et al. 2018; Gonzalez-Escobedo et al. 2018; Hakim et al. 2018), the Illumina platform became the most popular technique as illustrated by the large number of studies published last year (Chen et al. 2018; Dijkhuizen et al. 2018; Du et al. 2018; Fitzpatrick et al. 2018; Gaby et al. 2018; Gao and Shi 2018; Huang 2018; Huang et al. 2018; Kunda et al. 2018; Montanari-Coelho et al. 2018; Passera et al. 2018; Ren et al. 2018; Saminathan et al. 2018; Sun et al. 2018; Szymanska et al. 2018; Thiem et al. 2018; Wang et al. 2018).

The Ion Torrent semiconductor sequencing platform provides similar sequencing quality and faster sequencing time as compared to other next-generation sequencing strategies (Kchouk et al. 2017), and its pertinence for microbiome analysis has been substantiated by the studies on environmental samples, such as compost (Blaya et al. 2016), soil or rhizospheric microbiomes (Bell et al. 2015; Li et al. 2017a), as well as diversity of fungal endophytes (Kemler et al. 2013); however, this technology has been rarely used in bacterial endophytome analysis. The semiconductor sequencing technology was used to study the diversity of the bacterial endophytic community of willow trees growing in petroleum hydrocarbons-contaminated soil (Tardif et al. 2016), and recently, a study on bacterial diversity of rice rhizosphere and endorhizosphere has been published by Moronta-Barrios et al. (2018).

The future focus of the microbial metagenomic analysis is on the emerging third-generation sequencing technologies such as the real-time sequencing of single molecule (also called PacBio) by Pacific Biosciences or nanopore sequencing by Oxford Nanopore and their potential in the characterization of microbial diversity. These technologies are referred as single molecule sequencing as they do not need a PCR amplification step and distinguished from the NGS by longer reads generated in a relatively short time (Kchouk et al. 2017). Longer sequence reads ensure high accuracy in OTU assignment that ideally represents a better phylogenetic analysis of microbial communities (Knief 2014). For example, analysis of microbial community structure of lake water samples using the PacBio full-length 16S rRNA sequencing resulted in less ambiguous taxonomic classification and better phylogenetic resolution as compared to the results obtained using the Illumina sequencing of the V4 domain of the 16S rRNA gene (Singer et al. 2016). In case of bacterial endophytes, third-generation sequencing technology was mainly used for bacterial genome assembly (Lumactud et al. 2017; Passera et al. 2018; Utturkar et al. 2017), and no metagenomic analysis results have been published so far.

Amplicon sequencing and whole-genome shotgun sequencing are the two sequencing approaches used with current NGS platforms for the metagenomic investigation of microbiome populations. The amplicon sequencing technique involves the extraction of genomic DNA, conserved genomic region amplification, and sequencing. As compared with other rRNAs, the 16S rRNA is the most conserved (Rajendhran and Gunasekaran 2011; Vetrovsky and Baldrian 2013); therefore, the amplicon sequencing strategy that includes selection of the conserved hypervariable domains of 16S rRNA gene has been universally employed for fast taxonomic identification and phylogenetic profiling of the microbial communities (Bartram et al. 2011; Tian and Zhang 2017). The 16S rRNA amplicon sequencing technique has been extensively employed to characterize the biodiversity of plant-associated microbiota (Diwan et al. 2018; Zuniga et al. 2017). Alternatively, several investigations have highlighted that the 16S rRNA amplicon sequencing approach could lead to incorrect interpretation of biological information mainly due to the limitations of DNA extraction procedures, selected hypervariable regions or distinct PCR primers (Beckers et al. 2016; Hong et al. 2009; Sharpton 2014).

Next-generation sequencing techniques are progressively exploited for amplicon sequencing of fungal and bacterial marker genes, other than 16S rRNA gene, in order to characterize the specific taxonomic composition of communities in the phyllosphere and rhizosphere. For example, the amplicon sequencing of a dinitrogenase reductase gene (*nifH*) is commonly used for characterization of diazotroph community associated with plants (Gaby and Buckley 2014). The conserved nature of the *nifH* gene has made it a popular molecular marker to investigate the total and functional diversity of diazotrophic bacteria associated with various plants, such as maize, poplar, rice, sugarcane, sorghum and switchgrass (Bahulikar et al. 2014; Gaby et al. 2018; Ji et al. 2014; Kifle and Laing 2015; Knoth et al. 2014; Mareque et al. 2018; Prayitno and Rolfe 2010).

Since the 16S rRNA gene has comparatively low sequence divergence among related bacterial taxa, Barret et al. (2015) used molecular marker based on gyr B, a gene which encodes the DNA gyrase β subunit, which is commonly used as a phylogenetic marker for several genera of bacteria (Watanabe et al. 2001). Application of this approach in metagenomic analysis provided insights into the taxonomic composition of the endophytic community at the genus or species level (Barret et al. 2015).

An alternative approach to the amplicon sequencing method is the whole-genome shotgun sequencing that is targeting the entire genomic content of the sample. This technology is based on DNA extraction followed by its trimming into smaller fragments that are sequenced randomly, and subsequently representative DNA units are constructed from the smaller overlapping DNA fragments (Quince et al. 2017; Sharpton 2014). Randomly sequenced reads from taxonomically informative genomic loci or coding sequences aligned to various genomic locations enable to investigate the variety and function of overall bacterial societies in a single analysis (Sharpton 2014). Furthermore, the taxa could be defined more accurately at the species level using the shotgun sequencing. For example, shotgun metagenomics has been used to study the taxonomic and functional diversities of microbiome populations in the rhizosphere of soybean and *Lotus japonicus* (L.) plants (Mendes et al. 2014; Unno and Shinano 2013). However, shotgun sequencing is expensive and requires more information for biologically meaningful interpretation of the data; therefore, currently, most of the metagenomic studies on plant endophyte communities use simpler and more cost-efficient amplicon sequencing approach.

20.2.2 Methodological Considerations for Sample Preparation and Analysis

Without a doubt, the massive parallel sequencing technologies have revolutionized microbial biodiversity research, however, the benefits came with distinct methodical problems in the plant-associated microbiome studies. A major barrier for detecting endophytic bacterial communities using metagenomic approaches is the overwhelming ratio of plant to bacterial DNA in samples with low abundance of the plant-associated bacteria (Beckers et al. 2016). This procedural shortcoming is related to the poor separation of bacterial endophytic sequences of homologous sequences from plant compartments, such as plant nucleus and plastids (Govindasamy et al. 2014). Bacterial DNA enrichment methods could provide a solution to overcome this limitation in the preparation of the metagenomic libraries (Beckers et al. 2016). There are few DNA or bacterial cell enrichment strategies that could improve plant to bacterial DNA ratio in the metagenomic library and consequently increase the depth of the analysis.

Several studies were focusing on the enrichment of the plant-associated bacterial cells or DNR prior to PCR amplification. Ikeda et al. (2009) employed the sequence of differential-centrifugation steps followed by density-gradient centrifugation and efficiently enriched endophytic bacteria from leaves and stems of soybean and rice. Mora-Ruiz et al. (2015) applied differential centrifugation for endophytic bacteria enrichment before DNA isolation. Later, Utturkar et al. (2016) described a modified density gradient and differential centrifugation-based approach for endophytic bacteria separation from *Populus* (L.) roots that significantly reduced plant DNA contamination. Hydrolysis of the plant cell walls and differential centrifugation were used to enrich target genes from tissues of *Maytenus hookeri* (Loes.) (Jiao et al. 2006). Wang et al. (2008) applied enzymatic treatment for bacterial DNA enrichment from stems of *Mallotus nudiflorus* ((L.) Kulju & Welzen). For bacterial DNA ratio enrichment, Nikolic et al. (2011) applied overnight shaking of the small pieces of potato tubers in sodium chloride solution. Maropola et al. (2015) compared the results of metagenomic analysis of sorghum-associated endophytic bacterial diversity of the samples prepared using different DNA extraction procedures and concluded that the SDS-based DNA extraction protocols resulted in similar composition of the agriculturally important bacterial genera.

In addition, incomplete extraction of microbial DNA could affect the results of metagenomic analysis. Due to the simple application, most plant microbiome studies use commercial DNA extraction kits that rely on mechanical or enzymatic lysis of bacterial cells to release DNA (Zielinska et al. 2018). The DNA extraction procedure must guarantee that bacterial cell lysis provides sufficient genomic material and eliminates plant-derived enzymes and phytochemicals. The extraction step is often critical as the bacterial species might require different conditions for cell lysis and no single DNA extraction method exists that is efficient to obtain information about all present phylogenetic groups (de Bruin and Birnboim 2016). Further, comparison of different DNA preparation procedures has revealed that not only plant DNA extraction protocols, but also DNA yield and purity affects the results of metagenomic analysis of endophytic bacteria diversity (Maropola et al. 2015). Therefore, the selection of optimal DNA preparation conditions is crucial to ensure quality of the analysis.

Inaccuracies in the results of metagenomic analysis using NGS technologies could be also related to PCR primer design, DNA amplification procedure and sequencing artifacts (Brooks et al. 2015; Pinto and Raskin 2012). The PCR amplification step is of particular importance, especially in amplicon sequencing, where it is applied at least twice – at the initial sample preparation step and during sequencing library amplification (Pinto and Raskin 2012). Therefore, the selection of optimal primer pairs that cover the region of interest and ensure correct biological conclusions is essential (Thijs et al. 2017).

During the last decade, a large number of findings on microbial ecology and composition of bacterial communities have been established using high-throughput amplicon sequencing of 16S rRNA gene (Beckers et al. 2016). Short (<500 bp) hypervariable regions of the 16S rRNA locus are amplified using PCR and analyzed using NGS technology (Diwan et al. 2018). In order to accurately detect plant-associated communities of bacteria, taxonomic studies rely on the use of the most informative PCR primers to amplify all bacterial phylotypes present in the sample, and insufficient resolution of community profiling could be related to the choice of primers (Brooks et al. 2015; Thijs et al. 2017). The homology between bacterial 16S rRNA, chloroplast and mitochondrial DNA raises further challenges in the selection of the primers suitable for the assessment of the endophytic bacteria (Beckers et al. 2016). The regions of 16S rRNA gene used by different studies for the metagenomic analysis of plant endophytome are illustrated in Fig. 20.1, and the primer sets, plant tissues, and plants used in the analyses with references are provided in Table 20.1.

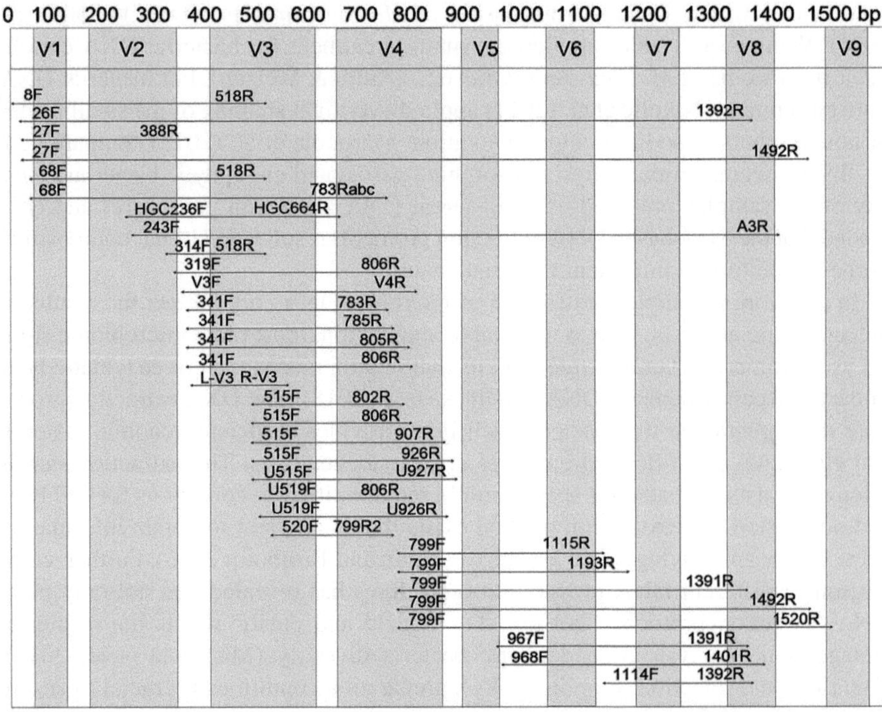

Fig. 20.1 Segments used for the amplicon-based metagenomic analysis of plant endophytome mapped to the structure of the 16S rRNA gene that includes hypervariable regions V1–V9

There are several reports of taxon-specific primer sets successfully applied to reduce the interference of plant DNA (Chelius and Triplett 2001; Wemheuer and Wemheuer 2017). One of the universal bacterial primers 799F designed by Chelius and Triplett (2001) has two base pair mismatch on the 3′ end of the primer to eliminate chloroplastic 16S rRNA sequences. This primer and its modifications (799F2, 783R, 783Rabc) have been applied for the metagenomic analysis by a number of studies with varying results (Bodenhausen et al. 2013; Bulgarelli et al. 2012; Hanshew et al. 2013; Jones et al. 2013; Leveau and Tech 2011; Lundberg et al. 2012; Rastogi et al. 2012; Sagaram et al. 2009; Shade et al. 2013; Sun et al. 2018). Other universal primers were successfully used to amplify bacteria DNA from leaves, stems, and roots of plants, such as *Arabidopsis*, wheat, spinach, rice, banana, maize, and grapevine (Barret et al. 2015; Bodenhausen et al. 2013; Ding and Melcher 2016; Du et al. 2018; Kembel et al. 2014; Mashiane et al. 2017; Moyes et al. 2016; Rastogi et al. 2012; Rua et al. 2016; Sun et al. 2018; Wang et al. 2016a; Wemheuer and Wemheuer 2017). Jackson et al. (2013) used a specific combination of primer set that targets only DNA of bacteria without amplifying residual DNA of chloroplast. Mitochondrial DNA of plant is co-amplified, but significantly larger fragments are produced that are separated on agarose gel (Jackson et al. 2013).

Table 20.1 16S rRNA gene-specific primer sets used in the metagenomic analysis of plant endophytome

Primer pair	Plants and tissues	References
8F – 518R	Roots and stems of sorghum (*Sorghum bicolor* (L.) Moench)	Maropola et al. (2015)
26F – 1392R	Pumpkin plants (*Cucurbita pepo* L.)	Eevers et al. (2016b)
27F – 388R	Roots of potato (*Solanum tuberosum* L.)	Manter et al. (2010)
27F – 1492R[a]	Roots of poplar (*Populus deltoides* W. Bartram ex Marshall)	Utturkar et al. (2016)
	Roots of caliph medic (*Medicago truncatula* Gaertn.)	Yaish et al. (2016)
	Shoot tips of banana (*Musa* sp. cv. Grand Naine)	Thomas and Sekhar (2017)
68F – 518R	Roots, stems, and leaves of poplar tree (*Populus tremula L. × Populus alba L.)*	Beckers et al. (2016)
68F – 783Rabc	Roots, stems, and leaves of poplar tree (*Populus tremula L. × Populus alba L.)*	Beckers et al. (2016)
HGC236F – HGC664R	Shoot tips of banana plants (*Musa* sp., AAA, Giant Cavendish cv. Baxi)	Du et al. (2018)
243F – A3R	Stem, root and grain of rice (*Oryza sativa* (L.) cv. Wusimi)	Wang et al. (2016a)
314F – 518R	Shoot tips of banana (*Musa* sp. cv. Grand Naine)	Thomas and Sekhar (2017)
319F – 806R	Seeds of Danshen (*Salvia miltiorrhiza* Bge)	Chen et al. (2018)
V3F – V4R	Roots of Indian rice (*Oryza sativa* L.)	Sengupta et al. (2017)
	Whole moso bamboo plant (*Phyllostachys edulis* Carrière J.Houz.)	Liu et al. (2017)
341F – 783R[a]	Roots, stems, and leaves of poplar tree (*Populus tremula L. × Populus alba L.)*	Beckers et al. (2016)
	Roots of Zucchini (*Cucurbita pepo* ssp. Pepo)	Eevers et al. (2016a)
341F – 785R[a]	Roots, stems, and leaves of poplar tree (*Populus tremula L. × Populus alba L.)*	Beckers et al. (2016)
	Roots of C3 *(Triticum aestivum L., T. Monococcum L., T. Turgidum Desf., Secale cereale L., Hordeum vulgare L., Avena sativa L., Festuca arundinaceae Schreb., Brachypodium distachyon (L.) P.Beauv.)* and C4 (*Sorghum bicolor L. Moench, S. Laxiflorum F.M.Bailey, Sorghastrum nutans (L.) Nash, Miscanthus sinensis Andersson, Bothriochloa bladhii (Retz.) S.T.Blake, Zea mays L., Pennisetum americanum (L.) R.Br., Eragrostis tef (Zucc.) Trotter*) grasses	Naylor et al. (2017)
	Stem, root and grain of rice (*Oryza sativa* (L.) cv. Wusimi)	Wang et al. (2016a)
	Shoot tips of banana plants (*Musa* sp., AAA, Giant Cavendish cv. Baxi)	Du et al. (2018)
	Leaves, stems, and roots of aloe (*Aloe vera* (L.) Burm.f.)	Akinsanya et al. (2015)
	Roots of halophyte (*Salicornia europaea* L.)	Szymanska et al. (2018)
	Seeds of wheat (*Triticum aestivum* (L.) cv. Heixiaomai)	Huang et al. (2016)

(continued)

Table 20.1 (continued)

Primer pair	Plants and tissues	References
341F – 805R[a]	Whole chinese leek plant (*Allium tuberosum* Rottler ex Sprengel cv. Dajingou)	Huang (2018)
	Leaves of soybean (*Glycine max* (L.) Merrill)	Montanari-Coelho et al. (2018)
341F – 806R	Roots of ginseng (*Panax notoginseng* (Burkill) F. H. Chen ex C. Y. Wu & K. M. Feng)	Tan et al. (2017)
	Roots of rice (*Oryza sativa* L.)	Kunda et al. (2018)
L-V3 – R-V3	Sugar beet (*Beta vulgaris* (L.) cv. Xintian and cv. Beta 580)	Shi et al. (2014)
515F – 802R	Roots of Venezuelan rice (*Oryza* spp. cv. Pionero 2010 FL, cv. DANAC SD20A)	Moronta-Barrios et al. (2018)
515F – 806R[a]	Seeds of cabbage, cauliflower, turnip, broccoli, canola (*Brassicaceae* sp.), radish (*Raphanus sativus* var. *Sativus* L.), garden rocket (*Diplotaxis tenuifolia* (L.) DC.), tomato (*Solanum lycopersicum* L.), carrot (*Daucus carota* L.), bean (*Phaseolus vulgaris* L.), thale cress (*Arabidopsis thaliana* (L.) Heynh.), barrel clover (*Medicago truncatula* Gaertn.)	Barret et al. (2015)
	Fruits of watermelon (*Citrullus* spp. cv. with red and yellow flesh)	Saminathan et al. (2018)
	Tissues of *Arabis alpina* L., *Dysphania ambrosioides* (L.) Mosyakin & Clemants and *Veronica ciliata* L.	Sun et al. (2018)
	Roots of Venezuelan rice (*Oryza* spp. cv. Pionero 2010 FL, cv. DANAC SD20A)	Moronta-Barrios et al. (2018)
	Roots, stalks, leaves, and young shoots of sugarcane plants (*Saccharum officinarum* L.)	de Souza et al. (2016)
515F – 907R	Leaves of japonica rice (*Oryza* spp. cv. Wuxiangjing 14)	Ren et al. (2015a)
515F – 926R	Leaves of *Stevia rebaudiana* (Bertoni) plants	Yu et al. (2015)
	Roots and seeds of common reed (*Phragmites australis* (Cav.) Trin. ex Steud.) and narrowleaf cattail (*Typha angustifolia* L.)	Gao and Shi (2018)
	Roots of soya bean and alfalfa (*Medicago sativa* L.)	Xiao et al. (2017)
U515F – U927R	Roots of broccoli (*Brassica* sp. cv. Green sprouting broccoli)	Gadhave et al. (2018)
U519F – 806R	Fruits of watermelon (*Citrullus* spp. cv.with red and yellow flesh)	Saminathan et al. (2018)
U519F – U926R	Roots of maize (*Zea mays* L.)	Correa-Galeote et al. (2018)
520F – 799R2	Stems and roots of willow cultivars (*Salix purpurea* cv. Fish Creek and *Salix miyabeana* cv. SX67)	Tardif et al. (2016)

(continued)

Table 20.1 (continued)

Primer pair	Plants and tissues	References
799F – 1115R	Leaves, stems, tassels, and seeds of transgenic Bt maize (*Zea* sp. cv. MON 810) and its isogenic parental line (*Zea* sp. cv. non-Bt)	Mashiane et al. (2017)
	Needles of pine (*Pinus flexilis* E.James)	Moyes et al. (2016)
	Needles of pine (*Pinus radiata* D.Don)	Rua et al. (2016)
	Needles of pines (*Pinus flexilis* E.James and *Pinus engelmannii* Carr.)	Carrell and Frank (2014)
799F – 1193R[a]	Roots, stems, and leaves of poplar tree (*Populus tremula L. × Populus alba L.)*	Beckers et al. (2016)
	Roots and leaves of *Arabidopsis thaliana* (L.) Heynh.	Bodenhausen et al. (2013)
	Seeds of *Arabidopsis thaliana* (L.) Heynh.	Truyens et al. (2016a)
	Roots of *Arabidopsis thaliana* (L.) Heynh.	Bulgarelli et al. (2012)
	Leaves, stems, and roots of halophyte (*Messerschmidia sibirica* L.)	Tian and Zhang (2017)
	Stems of fynbos plants (*Erepsia anceps* (Haw.) Schwantes, *Phaenocoma prolifera* (L.) D.Don, *Leucadendron laureolum* Fourc.)	Miyambo et al. (2016)
799F – 1391R	Roots, stems, and leaves of poplar tree (*Populus tremula L. × Populus alba L.)*	Beckers et al. (2016)
799F – 1492R[a]	Leafy green vegetables (*Lactuca* spp., iceberg lettuce, romaine lettuce, red leaf lettuce, green leaf lettuce, and baby spinach)	Jackson et al. (2013)
	Needles of pine (*Pinus flexilis* E.James)	Moyes et al. (2016)
	Needles of pine (*Pinus radiata* D.Don)	Rua et al. (2016)
	Roots of Salicornia (*Salicornia europaea* L.)	Zhao et al. (2016)
	Roots of caliph medic roots (*Medicago truncatula* Gaertn.)	Yaish et al. (2016)
	Romaine lettuce (*Lactuca sativa* L.)	Rastogi et al. (2012)
799F – 1520R	Branches of grapevine (*Vitis vinifera* L.)	Campisano et al. (2014)
967F – 1391R	Roots, stems, and leaves of poplar tree (*Populus tremula L. × Populus alba L.)*	Beckers et al. (2016)
968F – 1401R	Roots of durum wheat (*Triticum turgidum* var. durum)	Yang et al. (2012)
1114F – 1392R	Roots of *Arabidopsis thaliana* (L.) Heynh.	Lundberg et al. (2012)

[a]Different studies used primer pairs corresponding to the same region, but with a modified primer sequence

Selective primers target only a single section of the hypervariable domains; therefore, this representation of the bacterial diversity could be incomplete due to the limited coverage of 16S rRNA gene sequencing data (Starke et al. 2014; Thijs et al. 2017). Further, a lack of primer specificity has the potential to introduce bias into the representation of the microbial diversity. Beckers et al. (2016) evaluated a series of commonly used primer pairs for the analysis of bacterial communities of poplar trees (*Populus alba* L.; *Populus tremula* L.), and the study revealed that application of suboptimal primer pairs not only consumes sequencing resources for amplification of non-target sequences but also results in biased representation of certain taxa (Beckers et al. 2016; Ghyselinck et al. 2013). Al-Awadhi et al. (2013) noted that certain primers fail to amplify the target 16S rRNA gene domains when DNA mixture is used. Further, bacteria-specific primers could fail to reveal actinobacterial sequences that are rich in GC contents; therefore, to overcome this problem, actinobacteria-specific primers were used by Du et al. (2018) to study endophytic actinobacterial community of the banana (*Musa* sp.) shoot tips.

Peptide nucleic acid (PNA) clamps is a promising technique for elimination of host organellar sequences (plastid and/or mitochondrial) during PCR amplification of the 16S rRNA fragments. Arenz et al. (2015) described the application of non-target host DNA blocking primers for endophytic bacteria analysis in plants. In this report, the application of two chloroplast and mitochondrial rRNA-specific PNA clamps increased efficiency of bacterial DNA amplification approximately 300-folds in comparison to typical PCR procedure (Arenz et al. 2015). Later, Fitzpatrick et al. (2018) examined the efficiency of the PNA clamps for the analysis of the root microbial communities from 32 plant species and demonstrated that PNA clamps did not introduce bias in detection of individual bacterial taxa and could be a useful tool to reduce host contamination during amplification. However, the sequence of PNA clamps should be carefully selected based on host species as even single mismatch of nucleotide could result in severe reduction in its efficiency (Fitzpatrick et al. 2018).

Another PCR amplification-related bias is a chimera formation when the forward primer mistakenly shifts its position to the reverse primer. This results in amplification of two or more unrelated DNA fragments at once (Pinto and Raskin 2012). To avoid the chimera formation, a lower number of PCR cycles are used.

Thus, in order to efficiently characterize the plant-associated microbiome, at least two enrichment methods could be used to provide an independent estimate on the structure of microbial communities. Furthermore, the efficacy of the genetic material extraction procedures should be verified. Lastly, an appropriate representation of the bacterial diversity requires selection of the optimal primer pairs that cover a region of 16S rRNA or other locus that properly represents the taxonomic composition of the microbiome, and host-specific blocking primers could be used to avoid wasteful co-amplification of plant-derived sequences.

20.2.3 Quality of Metagenomic Data Analysis

Next-generation sequencing technologies enable taxonomic profiling of plant-associated microbial communities at high resolution and depth; however, the success of analysis depends on several steps, including data analysis. The biological and functional information of plant-associated microbial communities cannot be meaningfully interpreted without effective taxonomic classification. Nevertheless, even with well-characterized environmental samples, it is challenging to refer taxonomy structure through the short 16S rRNA amplicon reads (Mizrahi-Man et al. 2013).

An assessment of the operational taxonomic units (OTUs) is the crucial step that can affect final interpretation of microbial community structure. OTUs could be identified using alignment-, tree-, or phylogeny-based methods (Holovachov et al. 2017). The most often used bacterial taxonomic assignments are based on identification of representative sequences by sequence comparison with the reference database (Schloss et al. 2011). However, plant microbiome is composed of a large variety of bacterial species that includes rare bacterial taxa underrepresented in available databases. New groups of taxonomy not present in the reference taxonomy could be erroneously assigned to known reference groups; therefore, such classification could be detrimental to microbiome investigation as some bacterial taxa are overestimated while other remains overlooked (Murali et al. 2018). Recently, Murali et al. (2018) developed a new method IDTAXA that could be promising approach for plant endophytome studies. This new method uses machine learning principles to lessen over classification mistakes and has been shown to have higher accuracy than other popular classifiers.

In the future, current limitations associated with partial 16S rRNA gene sequencing could be also resolved through the introduction of sequencing technologies that are capable to provide larger segments or entire sequence of the 16S rRNA gene, such as third-generation sequencing platforms. Singer et al. (2016) demonstrated that application of the PacBio sequencing platform revealed underestimated specific microbial genera in ecological samples. Potentially, such technology could be successfully adopted for the plant-associated microbial community profiling.

Currently, there are many NGS sequencing projects dedicated to analysis of plant-associated microbial community structure; however, comparative analysis of independently produced metagenomics data is rather intricate due to different approaches used for sample preparation, PCR amplification, and data analysis. Furthermore, the content and quality of metagenomics date depend on multiple factors that need to be taken into account. Kim et al. (2017) summarized experimental design suggestions for human microbiome research that could provide important insights for environmental samples as well. A proper sample collection and storing is stressed as an important factor in high quality metagenomic library preparation, as well as introduction of positive and negative controls into analysis could reflect on the efficacy of microbial community representation (Kim et al. 2017).

20.3 Novel Insights into Complexity and Dynamics of Bacterial Endophyte Community

The start of the next-generation sequencing era in metagenomic analysis of plant endophytome has been marked by the application of pyrosequencing to examine the bacterial communities associated with plants, which colonize roots of diverse plants such as potato (*Solanum tuberosum* L.) (Manter et al. 2010) and cottonwood (*Populus deltoides* Bartram ex Marshall) (Gottel et al. 2011) During the following years, endophytomes of a large variety of agricultural and plant species were characterized using next-generation sequencing. Metagenomic profiling using NGS technology has emerged as a sensitive and efficient tool capable to reflect a complex diversity of the endophytic microbiome (Eevers et al. 2016a).

Despite the additional taxonomic diversity identified using a metagenomic analysis as compared to culture-dependent approach, the majority of endophytic bacterial communities are distributed among five bacterial phyla – *Actinobacteria*, *Proteobacteria*, *Firmicutes*, *Acidobacteria*, and *Bacteroidetes*. These entire bacterial phyla could be found across the plant microbiome but differ in abundance and distribution (Bulgarelli et al. 2012). Endophytic bacteria have been isolated from various plants and their organs, which shows that these communities do not represent random assemblies of microorganisms, and few dominant phyla and other subgroups can be predictable in specific tissues or organs. Significant differences in plant endophytome composition have been observed in studies performed with different plant organs (Akinsanya et al. 2015; Bodenhausen et al. 2013; Li et al. 2017a; Maropola et al. 2015; Romero et al. 2014; Sarria-Guzman et al. 2016; Tian and Zhang 2017; Yang et al. 2017), host plant species and developmental stage (Barret et al. 2015; Liu et al. 2017; Miyambo et al. 2016; Pinto et al. 2014; Shi et al. 2014; Xiao et al. 2017), growing season (Bulgarelli et al. 2012; Shen and Fulthorpe 2015), growing conditions (Carrell and Frank 2014; Carrell and Frank 2015; Chen et al. 2018; Moyes et al. 2016; Ren et al. 2015a), soil type (Bulgarelli et al. 2012; Carrell and Frank 2014; Kunda et al. 2018; Lundberg et al. 2012; Tardif et al. 2016; Truyens et al. 2016a; Yaish et al. 2016), cultivation practice (Campisano et al. 2014; Perez-Jaramillo et al. 2018; Tan et al. 2017; Tian and Zhang 2017; Yang et al. 2012) or a combination of observed factors.

Endophytic and rhizospheric bacterial populations exhibit diverse abundance of main groups at the phylum level. Significantly higher bacterial colonization and diversity has been found in the plant rhizosphere than in roots and other plant tissues (Akinsanya et al. 2015; Gottel et al. 2011; Maropola et al. 2015; Romero et al. 2014; Tian and Zhang 2017). In contrast, the bacterial diversity of the endophytic compartment of leaves was higher than the bacterial diversity associated with phyllosphere communities (Bodenhausen et al. 2013; Tian and Zhang 2017). The majority of studies found that in the rhizosphere, the bacterial communities were dominated by *Proteobacteria* (mostly α-*Proteobacteria* subclass), and this was observed in cottonwood trees (Gottel et al. 2011), *Arabidopsis thaliana* ((L.) Heynh.) (Bodenhausen et al. 2013; Lundberg et al. 2012), tomato (*Solanum lycopersicum L.*) (Romero et al. 2014), *Anthurium andraeanum* (Linden ex André) (Sarria-Guzman et al. 2016), and halophyte (*Messerschmidia sibirica* L.) (Tian and Zhang 2017). A later study comparing bacteria communities in different compartments (leaves, roots, and stems) of *A. thaliana* showed that different parts of the plant had been colonized by the same bacterial genera that differed in abundance (Bodenhausen et al. 2013).

Sequences of γ-*Proteobacteria*, mostly *Pseudomonas* sp., were abundant in all plant compartments of *A. thaliana* (Bodenhausen et al. 2013). *Pseudomonas* dominated in stem samples of *Aloe vera* (L. Burm.f.) (Akinsanya et al. 2015), and γ-*Proteobacteria* in general was most abundant class of the endophytic community in tomato leaves (*Solanum lycopersicum L.*) (Romero et al. 2014), tree peony

(*Paeonia Sect. Moutan*) (Yang et al. 2017), sorghum (*Sorghum bicolor (L.) Moench*) (Maropola et al. 2015) and orchid (*Dendrobium catenatum* Lindl.) (Li et al. 2017b) samples.

The most obvious difference between bacterial communities has been observed for different plant genotype or plant development stages. Microbiome communities primarily determined by plant species has been described in studies with soya bean (*Glycine max* (L.) Merr.), alfalfa (*Medicago sativa* L.) (Xiao et al. 2017) and three fynbos plants (*Erepsia anceps* (Haw.), *Leucadendron laureolum* Lam. Fourc. and *Phaenocoma prolifera* L. (Miyambo et al. 2016). These findings were similar to the results described for bacterial communities inhabiting phyllosphere of tree leaves where higher variability in bacterial communities was observed across different tree species than within individual trees of the same species (Redford et al. 2010). However, an extensive study, including 28 plant genotypes belonging to different varieties and species (affiliated mostly to the *Brassicaceae* family) reveled that in some cases, variation of endophytic bacterial community structure during plant growth stages was more prominent than differences observed among different plant genotypes (Barret et al. 2015). Bacterial community changes between different development stages were observed in the microbiome of *Arabidopsis* roots (Lundberg et al. 2012) and rice leaves (Ren et al. 2015a). A variation of the endophytic bacteria community composition during the growth of sugar beet (*Beta vulgaris* L.) was described in the study by Shi et al. (2014) where α-*Proteobacteria* population increased during the growth period and γ-*Proteobacteria* decreased. As indicated by Liu et al. (2017), the complexity of bacterial communities in moso bamboo (*Phyllostachys edulis* Carrière J. Houz) tissues gradually increased during growth of the seedlings. On the other hand, in grapevine (*Vitis vinifera* L.) fruits, a sharp decline in eukaryotic biodiversity during ripening was observed (Pinto et al. 2014).

Several studies revealed a seasonal variation of composition of the endophytic community in perennial trees and *Arabidopsis* plants. Shen and Fulthorpe (2015) sampled endophytome of Manitoba maple (*Acer negundo* L.), Chinese elm (*Ulmus parvifolia* Jacq.) and Siberian elm (*Ulmus pumila* L.), over three seasons, and culture-independent analysis revealed significant differences of endophytic community structure. The genus of *Sanguibacter* (*Actinobacteria*) and *Erwinia* (γ-*Proteobacteria*) were dominant in all plant species during summer, meanwhile β-*Proteobacteria* with the majority of these OTUs corresponding to the genus of *Ralstonia* was more abundant in samples collected during winter and fall (Shen and Fulthorpe 2015). In another study, approximately tenfold difference in the relative abundance of *Actinobacteria* in *Arabidopsis* root samples collected during the fall or spring seasons was observed (Bulgarelli et al. 2012).

A number of studies are dedicated to transfer of endophytic bacteria in plant tissues. The seeds represent a source of vertical transmission of endophytic bacteria (Cankar et al. 2005; Johnston-Monje and Raizada 2011; Shahzad et al. 2018; Trognitz et al. 2014). Recent analysis of the core microbiome of Danshen (*Salvia miltiorrhiza* Bge) seeds from seven diverse cultivation areas showed that the main bacterial composition was consistent through all samples (Chen et al. 2018). This suggested that the seed-associated microbiome of *S. miltiorrhiza* consists of a core set of microbial taxa and represents a distinct source for endophytic bacterial colonization of the host plant across different environments.

A primary study using metagenomic 16S rRNA and gyrB sequencing provided hints about vertically transmitted endophytes was published by Barret et al. (2015). The extensive study included 28 plant genotypes evaluated during three different physiological stages (seeds, germinating seeds, and seedlings) and revealed a strong variation in bacterial composition among the seed samples. However, several bacterial OTUs assigned to species of *Pantoea*, *Pseudomonas,* and *Xanthomonas* based on 16S rRNA gene sequencing and *Pantoea agglomerans* detected using gyrB marker were detected systematically in all seed samples. Further, major decrease in bacterial diversity was observed in the transition period from germinating seeds into seedlings, which perhaps indicates a strong force of selection applied on seed-borne microorganisms by young plant. *Pantoea* and *Pseudomonas* were common among all the samples of seeds and seedlings, but in seeds less prevalent genera *Bacillus* and *Massilia,* were significantly enriched in seedlings (Barret et al. 2015).

Truyens et al. (2016b) showed that endophytic assemblage in the radicle of *Arabidopsis* resembled that of the seed despite the distinct bacterial composition of the substrate used for plant growth. *Pseudomonas* and *Rhizobium* genera were dominant in the radicle emerging from germinating seeds harvested from plants growing on sand; meanwhile, the radicles of plants growing in potting soil also included *Stenotrophomonas* spp. as the main constituent of bacterial assemblage. This would suggest that vertical endophyte transfer is the main source of endophytic bacteria at the early growth stage. However, endophytic composition of the leaf samples was different compared to the radicle samples, and detected bacteria were mainly derived from the environment and not from the seeds (Truyens et al. 2016b). In another study, a striking distinction between the seed, rhizospheric and root bacterial endophytes populations of wetland plant common reed (*Phragmites australis* (Cav.) Trin. ex Steud.), and narrow leaf cat tail (*Typha angustifolia* L.) was described (Gao and Shi 2018). Although the three bacterial communities were closely related for the two plants they appear as distinct assemblages rather than a subgroup of root endophytic bacterial communities or rhizobacterial communities, suggesting adaptation of the bacteria to the distinct habitats. The dominant genera in the seeds were *Desulfobacter*, *Geobacter*, *Thiobacillus*, *Sulfurimonas*, *Methyloversatilis*, and *Dechloromonas* that were absent or scarcely represented in root samples (Gao and Shi 2018).

Diversity and vertical transmission routes of rice endophytic bacteria were assessed using 16S rRNA sequencing (Wang et al. 2016a; Wang et al. 2016b). *Sphingomonas* and *Pseudomonas* spp. were proposed to represent the seed-borne indigenous bacteriome that is transmitted from seeds to seedlings and mature plant stems and roots during the plant development (Wang et al. 2016b). Meanwhile, genera *Brevundimonas*, *Petrobacter*, *Sphingobium*, *Cetobacterium*, *Methylobacterium*, and *Devosia* were not found in seedlings, suggesting that they were more likely to colonize rice roots from the rhizosphere. The results of sequencing of actinobacteria-enriched 16S rRNA libraries prepared from grains, stems, and roots of rice suggested that *Streptomycess* spp. was transmitted in rice plants vertically. Instead, the sequences corresponding to genera *Pseudonocardia* and *Dietzia* were isolated from tissues of root, stem of rice and were most likely transmitted from roots (Wang et al. 2016a).

A recent study with maize revealed putative vertically transmitted endophytes assigned to genera of *Bacillus*, *Halomonas*, and *Shewanella* that were dominant in germs and were transmitted to sprouts (Wang et al. 2018). Another study revealed that endophytic bacteria assigned to genera of *Erwinia* and *Cupriavidus* could be transmitted from seeds into sprouts of wheat (Huang et al. 2016). These genera were not dominant in seeds or sprout tissues, however, bacteria from *Erwinia* genera had been described as disease suppressing agents in wheat (Kempf 1989), and the vertically transmitted endophytes may play a significant role in wheat plants growth (Huang et al. 2016).

In peanut (*Arachis hypogaea* L.), bacterial genera *Halothiobacillus* and *Synechococcus* were dominant in germs, cotyledons, and sprouts (Huang et al. 2017). Similar results were also described for soybean (*Glycine max*) where *Halothiobacillus* was dominant genus in seeds and sprouts (Huang et al. 2018). This might imply a similar association with indigenous vertically transmitted endophytes in the two related plants of the *Fabaceae* family. However, the role of *Halothiobacillus* sp. remains elusive as plant development-encouraging characters of the genus have not been described so far.

20.4 Endophytome Is Shaped by Agricultural Practices and Environmental Factors

20.4.1 Effect of Agricultural Practices

Land management practices of agriculture, for example, irrigation and tillage, alter soil characteristics that may result in reduction of diversity of soil microflora owing to desiccation, compaction of soil, mechanical damage, pore size reduction, and food resource access disruption (Garcia-Orenes et al. 2013; Jangid et al. 2008). On the other hand, use of fertilizers and chemical pesticides can result in significant changes in composition and metabolic activity of the soil microfloral community (Pampulha and Oliveira 2006; Zhong et al. 2010). Many bacterial endophytes originate from the plant-associated microbial population in the rhizospheric zone (Hardoim et al. 2008), and microbiomeal diversity of host plant rhizospheric zone itself is defined by overall composition of the microbial pool of soil and further refined by specific plant-microbial interactions (Sorensen and Sessitsch 2006). Therefore the effect of the agronomic practices on the overall soil microbial community could be expected to reflect differences in endophyte populations of agricultural crop plants. Before use of sophisticated sequencing technology in metagenomics analysis, the research was aimed to elicit an effect of agricultural practices on composition of the endophytic bacteria populations is limited to several studies which demonstrated that the endophyte community was susceptible to different nitrogen fertilization levels in sugarcane and rice (Fuentes-Ramirez et al. 1999; Sasaki et al. 2013) or application mineral versus organic fertilizer and herbicides in maize (Seghers et al. 2004; Xia et al. 2015). Xia et al. (2015) assessed the outcome of organic versus traditional practices on endophyte diversity in corn, tomato, melon, and pepper. Sequencing analysis of cloned *nifH* gene libraries was used to establish the effect of flooding on rice endophytic diazotrophic communities (Ferrando and Fernandez 2015).

The application of next-generation sequencing of metagenomic 16S rRNA gene libraries conveyed new information on the effect on agricultural practices on endophytic bacterial diversity. Study by Perez-Jaramillo et al. (2018) showed that constant availability of macronutrients and water caused physiological differences between cultivated and wild plants, such as a shallower root system of the domesticated crops and less complex sugar exudates. These changes also reflected in plant endophytic root communities, where the most common taxonomic shift was decrease in *Bacteroidetes* abundance and increased abundances of the *Actinobacteria* and *Proteobacteria* in cultivated crop plants (Perez-Jaramillo et al. 2018).

In the context of climate change, it is important to know how plant bacterial endophytes respond to elevated temperature and CO_2 levels. The complex treatment of elevated temperature and CO_2 levels and altered nitrogen fertilization affected the relative abundance of the leaf endophytic bacteria assigned to *Moraxellaceae, Enterobacteriaceae*, *Comamonadaceae Xanthomonadaceae*, *Microbacteriaceae*, and *Sphingobacteriaceae* families at different growth stages of rice (Ren et al. 2015a, b). Elevated CO_2 revealed a significant effect on the microbial composition at the tillering stage independent of the nitrogen fertilization levels, and at the filling stage the effect was observed under high nitrogen conditions (Ren et al. 2015a). Also, the richness and diversity of leaf endophytic community were significantly affected by elevated CO_2 levels in combination with elevated temperature (Ren et al. 2015a).

One of the common problems in agriculture is the continuous crop cultivation. Rhizospheric soil bacterial diversity diminished with extended period of constant cultivation of *Panax notoginseng* ((Burkill) F.H.Chen) but had no significant effect on endophytic community (Tan et al. 2017). The prolonged cultivation was also associated with establishment of pathogenic bacteria or fungi and development of root-rot disease, where bacterial diversity was higher in rhizospheric soils and healthy roots of *P. notoginseng* as compared to diseased one. The main reasons for changes in the bacterial communities were total pH of soil, the amount of phosphorus, and organic matter. On the other hand, prolonged cultivation of maize leads to increased richness of the soil and endophytic bacterial communities (Correa-Galeote et al. 2018). The root-dominant genera *Bradyrhizobium, Sphingomonas*, *Methylophilus*, and *Herbaspirillum* were more abundant in roots of plants grown in soil that were cultivated with maize crop for at least 5 years compared with plants that were grown in fallow soils.

Pest management practices could also effect plant-associated microbial community. Campisano et al. (2014) assessed differences in the endophytic population composition in grapevine plantation cultivated using integrated pest management (IPM) and organic conditions. The study revealed that *Staphylococcus*, *Mesorhizobium*, and *Caulobacter* genera were relatively more abundant in plants from organic vineyards, while *Ralstonia*, *Burkholderia*, and *Stenotrophomonas* were more abundant in grapevines from the integrated pest management vineyards. Intriguingly, the bacteria belonging to the *Mesorhizobium* genus that promotes plant growth through nodule formation and nitrogen fixation was more abundant in plants grown without using any chemical fertilizer (Bloemberg and Lugtenberg 2001). *Mesorhizobia*, a nodule-forming bacteria, are well known for their capability to associate symbiotically with root of plants in wide species range.

Nitrogen is often the growth-limiting plant nutrient. Therefore, several studies explored the role of endophytes in plant nitrogen metabolism. The primary study on rice endophyte metagenome carried out using shotgun library sequencing approach revealed that endophytes might be involved in the entire process of nitrogen cycle, including N_2 fixation and denitrification (Sessitsch et al. 2011). The later NGS-based metagenomic analyses were focused mainly on conifer species and supported the role of endophytic bacteria in N_2 fixation. Composition analysis of endophytic bacteria in *Pinus flexilis* and *P. engelmannii* revealed that generally the endophytic populations clustered according to the host species, however the same phylotype related to *Gluconacetobacter diazotrophicus* and other N_2-fixing acetic acid bacterial endophytes consistently dominated in both of the conifers (Carrell and Frank 2014). Further, the association between limber pine (*Pinus flexilis* E.James) and prospective N_2-fixing acetic acid bacteria was demonstrated by Moyes et al. (2016), suggesting that endophytic bacteria inside conifer needle microbiota represent an important source of nitrogen for subalpine forests. Only low abundance of the acetic acid bacteria was detected in foliar endophyte communities of coast red wood (*Sequoia sempervirens* (D.Don) Endl.) and giant sequoia (*Sequoiadendron giganteum* (Lindl.) J.Buchholz) (Carrell and Frank 2015) which might reflect the effect of the different grow environment. However, the endophytome of coast redwood included N_2-fixing lichen-associated lineage of Rhizobiales-1 and might be a source of nitrogen for redwood trees as well.

In agriculture, it is becoming increasingly popular to use bacterial bioinoculants containing plant growth-promoting endophytes as means to complement indigenous seed or soil bacterial community with strains capable to enhance plant fitness and growth. Gadhave et al. (2018) investigated the effect of seed inoculation and soil treatment with universal soil bacteria, *Bacillus subtilis*, *Bacillus amyloliquefaciens*, and *Bacillus cereus*, on endophytic bacterial community's diversity in roots of autumn broccoli. The study showed that *Bacillus* failed to establsih as endophyte, which could be attributed as soil type or plant species-specific effect, as *Pseudomonas* sp. is more common than *Bacillus* sp. in *Brasicca napus* plants (Gadhave et al. 2018; Graner et al. 2006). Despite that, the *Bacillus* supplementation had species specific as well as generic effect on endosphere composition. The most prominent effect was induced by *Bacillus amyloliquefaciens* and resulted in a large decline in dominant *Pseudomonas* as well as *Rhizobium bacteria* and an increase in relative abundance of other genera, mostly *Sphingomonas, Tahibacter*, *Variovorax, and*

Dyadobacter (Gadhave et al. 2018). This was attributed to antagonistic activities among the soil bacteria within the plant rhizosphere or endosphere.

The study by Yang et al. (2012) provided evidence that chickpea and pea crop and their termination time influence the composition of bacterial communities inhabiting the durum wheat roots that are sown after rotation, and it could have a significant influence on wheat crop production. In durum wheat, higher grain yield was correlated with higher presence of endophytic *Acidobacteria* and *Actinobacteria* and depletion of *Firmicutes*, whereas higher abundance of *Actinobacteria* and *Proteobacteria* in the endosphere was observed when the crop was sown after harvesting early pulses crop as compared to late harvested pulses crop.

In addition, the importance of agricultural practices that help to maintain natural diversity of plant endophytic bacteria is highlighted by observations that crop plants might become a niche for human pathogens and a source for outbreaks of food-borne diseases (Brandl 2006). Significantly higher diversity of culturable endophytic bacteria in different tissues of maize, melon, pepper and tomato was observed when cultivated using organically compared to traditional procedures (Xia et al. 2015), and a decline of species antagonistic to bacterial pathogens in endosphere and soil was connected with potential plant colonization by human pathogens (Latz et al. 2012). However, NGS-based metagenomic analysis of packaged leafy salad vegetables at the point of consumption did not show significant differences in endophytic bacteria composition between organic versus conventional practices or surface-sterilized versus non-sterilized preparations (Jackson et al. 2013).

20.4.2 Adaptation to Soil Salinity

There is a growing interest in microorganisms found in association with plants growing in harsh environmental conditions that help plants to gain tolerance to abiotic stresses. Several studies were focused on endophytes associated with salt-tolerant and halophyte plant species or agricultural plants grown in saline soils and which provide evidence that associated microbiota might be important for adaptation to the adverse conditions.

Mora-Ruiz et al. (2015) studied endophytes of two plants of the subfamily *Salicornioideae*, *Allenrolfea vaginata* ((Griseb.) Kuntze) from Chile and *Arthrocnemum macrostachyum* ((Moric.) K.Koch) from four locations in Chile and Spain. The most abundant and common species identified by metagenomic analysis were *Alkalibacillus salilacus*, *Chromohalobacter canadensis*, *Halomonas* spp. and *Kushneria* spp. These halophilic bacteria have been found associated with salt-tolerant plants from different locations. Meanwhile, sequences assigned to *Pseudomonas* sp. were common only in the Chilean samples. Later study by Zhao et al. (2016) revealed a noticeable dissimilarity in the bacterial endophytic populations from different life stages of the related halophyte plant *Salicornia europaea* (L.) roots. The richest endophytic bacteria diversity was detected at the seedling stage, and thereafter, variety of endophytic bacteria declined during flower and fruit setting stage. Genera *Serpens*, *Halomonas*, *Pseudomonas*, *Azomonas*, and

Pantoea were observed during all growth phases. Recently, the study by Szymanska et al. (2018) on endophytic bacteria composition in *S. europaea* (L.) roots from two trial locations in Poland with diverse origins and soil salinity levels showed that higher salinity did not have an adverse outcome on biodiversity of the bacteria. However, a distinct taxonomic composition was observed for the two sites and was attributed to the distinct adaptation of halotolerant and halophytic microorganisms.

Another coastal halophyte, *Messerschmidia sibirica* (L.) that belongs to *Boraginaceae* family, was studied by Tian and Zhang (2017). Bacterial communities of plants collected in Shandong Peninsula of China varied across the different plant tissues, and the roots had a higher diversity of bacteria than leaves or stems. Unlike for the *Salicornioideae* plants, dominant genera identified in tissues of the *M. sibirica* did not include halophilic bacteria and were assigned as *Pseudomonas*, *Bacillus*, *Sphingomonas*, *Streptomyces*, *Microbacterium*, *Rhizobium*, and *Nocardioides*.

Further, studies with common agricultural or model plants reported that endophytic bacteria might help the host plant in coping with soil salinity stress. Kunda et al. (2018) studied rice samples collected from the seaside salinity area in West Bengal, India. Several of the identified bacterial genera were new for rice or for other plants growing in saline environments and included *Hydrogenoanerobacterium*, *Ruminiclostridium*, *Aerinimonas*, *Arcobacter*, *Luteibacter*, *Chitinophaga*, *Lactobacillus*, *Lutispora*, *Legionella*, *Arcobacter*, *Propionivibrio*, *Hydrogenispora*, *Sulfospirillum*, *Pseudolabrys*, *Oxobacter*, and *Acetobacterium*.

Yaish et al. (2016) studied the endophyte composition of Caliph medic (*Medicago truncatula* Gaertn.), a forage crop, and model legume plant. In this study, soil was collected from fields used to grow different *Medicago* species and the stress was induced by increased salt concentration. Although the pairwise overall variation analysis did not show statistically significant effect of soil salinity on the root endophytic communities, out of 41 identified OTUs, 29 were enriched differentially when the plants were exposed to salinity stress. Several differentially enriched species of *Marinobacter*, *Streptomyces* and *Halomonas* genera identified in this study were halophilic bacteria characteristic to marine and salt-polluted environments. Also, it was proposed that the differences in the community partially could be due to a presence of opportunistic phytopathogens that might infect plants during stress periods, such as *Pseudomonas aeruginosa*, which was enriched in the communities isolated from the plant roots treated with salinity, as well as the appearance of the beneficial microbes antagonistic to fungal pathogens, such as *P. stutzeri* (Yaish et al. 2016).

Further, Thiem et al. (2018) showed that bacterial community diversity within the roots of Black alder (*Alnus glutinosa* Gaertn.) was reduced in saline soils. Sequences affiliated with halotolerant/halophilic bacteria *Sphingomonas*, *Granulicella*, and *Rhodanobacter* dominated the saline site; on the other hand, *Rhizobium* and *Bradyrhizobium* were more abundant at the nonsaline location.

20.4.3 Effect of Soil Contaminants

Endophytic bacteria could help plants to cope with phytotoxicity of heavy metals, organic pollutants and to enhance phytoremediation effectiveness (Ma et al. 2016; Taghavi et al. 2005). Metagenomic analysis of bacterial endophytic population of two hyper accumulators (*Arabis alpina* L., *Dysphania ambrosioides* (L.) Mosyakin & Clemants) from Pb-Zn-contaminated sites revealed that in roots and shoots of above two plants, *Pseudomonas* was the most prevailing genus (Sun et al. 2018). Although it is known that certain strains of *Pseudomonas* sp. are able to stand heavy metal higher concentrations (AL-Saleh and Akbar 2015; Chen et al. 2014), their beneficial role for plant heavy metal tolerance and accumulation remains to be elucidated. Meanwhile, Sun et al. (2018) demonstrated that another endophytic bacterium of *Microbacterium* sp., isolated from *A. alpine* (L.), had a plant growth-promoting effect under heavy metal stress.

The microbiome of willow trees (*Salix purpurea* cv. Fish Creek, and *Salix miyabeana* cv. SX67) grown in petroleum hydrocarbon-contaminated soil was significantly influenced by contamination level (Tardif et al. 2016). However, the effect was less noticeable in rhizospheric zone, root, and stem parts as compared to the nonrhizospheric soil, which suggests that environment of plant might act as a defensive buffer region against increasing contamination levels (Tardif et al. 2016). The increased relative abundance of *Pseudomonas* sp. in the plant tissues was the most prominent difference associated with increasing contamination. Also, genera *Dickeya*, *Steroidobacter*, *Sinorhizobium*, *Sphingobium*, and *Rhizobium* were more abundant in plants growing at higher contamination levels.

Diversity and composition of zucchini (*Cucurbita pepo* L.) root endophytic community was modified by the addition of 2,2-bis(p-chlorophenyl)-1,1-dichloroethylene (DDE) to the plant nutrient medium where dominant *Rhizobium* sp. is surpassed by *Stenotrophomonas* sp. upon exposure to the pollutant (Eevers et al. 2016b). Less prominent effect on endophyte diversity was observed in shoots that were dominated by *Pseudomonas* sp. and is likely due to lower accumulation levels of the DDE. Pesticide degrading strains of *Stenotrophomonas maltophilia* have been previously described (Lu et al. 2009; Mukherjee and Roy 2013), therefore it was proposed that the bacterial endophytes associated with the zucchini roots might play a role in remediation of organic pollutants such as pesticide 2,2-bis(p-chlorophenyl)-1,1,1-trichloroethane (DDT) and its degradation product DDE (Eevers et al. 2016b).

20.4.4 Endophytome of Genetically Modified (GM) Plants

Several studies aimed to investigate the effect of the plant genetic modification on soil or plant-associated microorganisms were focused on the *Bacillus thuringiensis* (Bt) corn expressing Cry1Ab toxin, and different results about diversity of bacteria in the Bt maize rizosphere or soil microbiome depending on plant growth stage, soil type or environmental factors were reported (e.g., Castaldini et al. 2005; Cotta et al.

2013; Icoz et al. 2008; Tan et al. 2010). Application of conventional molecular analysis techniques, such as PCR-DGGE, terminal restriction fragment length polymorphism (T-RFLP), and sequencing of 16S rRNA gene clone libraries, conveyed that endophyte communities in maize roots were not perturbed by the expression of Bt-toxin in transgenic maize compared to isogenic cultivars (da Silva et al. 2014; Prischl et al. 2012). On the contrary, recent NGS-based metagenomic study targeting V5–V6 region of the 16S rRNA gene revealed shifts in abundance and diversity of the endophytic bacterial population in phyllosphere of the Bt maize cultivar (MON810) as compared to the isogenic parental line (Mashiane et al. 2017). Differences in the abundance of some bacterial genera, including *Brachybacterium*, *Burkholderia*, *Enterobacter*, *Rhodococcus*, and *Acidovorax*, species of these are known as useful endophytes, were observed between cultivars (Mashiane et al. 2017). Diversity indices suggested higher diversity and more even distribution of bacterial species in the phyllosphere of nontransgenic maize as compared to Bt maize.

Another study on leaf endophytic bacteria diversity in GM soybean plants (cultivar BR 16 over-expressing *AtAREB1* transcription factor conferring drought tolerance (event 1Ea2939)) and conventional soybean cultivar BR 16 revealed that genetic modification had less prominent effect on the diversity of microbial community (Montanari-Coelho et al. 2018). The main differences in microbial community composition were represented by the orders *Bacillales* and *Pseudomonadales*, prevailing in transgenic plants and the order *Clostridiales*, prevailing in the conventional cultivar. In addition, genus *Rummeliibacillus* was the more frequent in transgenic cultivar meanwhile *Geobacter*, *Lysinibacillus*, and *Paludibacter* were more frequent in the conventional cultivar. Thus, the differences detected in transgenic maize and soybean plants imply that the endophytic community is sensitive to subtle physiological changes in plant induced by genetic modification, and the effect appears species- and/or genetic modification-specific.

20.5 Concluding Remarks

The massive parallel sequencing is a remarkable technology that encompasses extensive metagenomic data and is capable to directly reflect on genetic diversity of the microbiome. Although current NGS technology, based on sequencing of relatively short DNA reads, provides limited depth of taxonomic identification that could barely reach beyond the family level, the extensive data provides valuable insights into the composition of the plant-associated microbiome communities.

The decade of application of the NGS in the metagenomic analysis of plant-associated microbial communities have largely complemented an understanding about the immense taxonomic diversity of endophytic bacteria and in particular among the unculturable taxa. The plant endophytic microbiome is dominated by *Actinobacteria* and *Proteobacteria*, and to the smaller degree, it includes *Acidobacteria, Bacteroidetes*, and *Firmicutes* phyla that varies in abundance within different plant compartments and during development stages. The endophytes have been frequently reported as α-, β-, and γ-subgroups of *Proteobacteria* phylum, the γ-subgroup being the most varied and dominated group that comprises common endophytic and soil bacterial species.

The metagenomic studies expanded the evidence that the endophytic microbiome composition is determined by environmental factors as well as plant genotype. Further insights into the role of agricultural practices and environmental factors in shaping the structure of the endophytic microbiome have been revealed, stressing the need of farming systems that would maintain the natural plant endophytic bacterial diversity and benefit from the stress-reducing and plant growth promoting properties of the endophytes. Thus, the metagenomic studies provide important knowledge about the diversity of bacterial endophyte communities that could be a useful tool in development of balanced and viable agriculture.

Acknowledgments This research was funded by the long-term research program "Genetics and directed genotype development of agricultural and forestry plants" implemented by the Lithuanian Research Centre for Agriculture and Forestry.

References

Akinsanya MA, Goh JK, Lim SP, Ting AS (2015) Metagenomics study of endophytic bacteria in *Aloe vera* using next-generation technology. Genom Data 6:159–163. https://doi.org/10.1016/j.gdata.2015.09.004S

Al-Awadhi H, Dashti N, Khanafer M, Al-Mailem D, Ali N, Radwan S (2013) Bias problems in culture-independent analysis of environmental bacterial communities: a representative study on hydrocarbonoclastic bacteria. SpringerPlus 2:369. https://doi.org/10.1186/2193-1801-2-369

Al-Saleh E, Akbar A (2015) Occurrence of *Pseudomonas aeruginosa* in Kuwait soil. Chemosphere 120:100–107. https://doi.org/10.1016/j.chemosphere.2014.06.031

Andreote FD, Azevedo JL, Araujo WL (2009) Assessing the diversity of bacterial communities associated with plants. Braz J Microbiol 40:417–432. https://doi.org/10.1590/S1517-83822009000300001

Arenz BE, Schlatter DC, Bradeen JM, Kinkel LL (2015) Blocking primers reduce co-amplification of plant DNA when studying bacterial endophyte communities. J Microbiol Methods 117:1–3. https://doi.org/10.1016/j.mimet.2015.07.003

Bahulikar RA, Torres-Jerez I, Worley E, Craven K, Udvardi MK (2014) Diversity of nitrogen-fixing bacteria associated with switchgrass in the native tallgrass prairie of northern Oklahoma. Appl Environ Microbiol 80:5636–5643. https://doi.org/10.1128/AEM.02091-14

Barret M, Briand M, Bonneau S, Preveaux A, Valiere S, Bouchez O, Hunault G, Simoneau P, Jacquesa MA (2015) Emergence shapes the structure of the seed microbiota. Appl Environ Microbiol 81:1257–1266. https://doi.org/10.1128/AEM.03722-14

Bartram AK, Lynch MD, Stearns JC, Moreno-Hagelsieb G, Neufeld JD (2011) Generation of multimillion-sequence 16S rRNA gene libraries from complex microbial communities by assembling paired-end illumina reads. Appl Environ Microbiol 77:3846–3852. https://doi.org/10.1128/AEM.02772-10

Beckers B, Op De BM, Thijs S, Truyens S, Weyens N, Boerjan W, Vangronsveld J (2016) Performance of 16s rDNA primer pairs in the study of rhizosphere and endosphere bacterial microbiomes in metabarcoding studies. Front Microbiol 7:650. https://doi.org/10.3389/fmicb.2016.00650

Bell TH, Cloutier-Hurteau B, Al-Otaibi F, Turmel MC, Yergeau E, Courchesne F, St-Arnaud M (2015) Early rhizosphere microbiome composition is related to the growth and Zn uptake of willows introduced to a former landfill. Environ Microbiol 17:3025–3038. https://doi.org/10.1111/1462-2920.12900

Berg G, Grube M, Schloter M, Smalla K (2014) Unraveling the plant microbiome: looking back and future perspectives. Front Microbiol 5:148. https://doi.org/10.3389/fmicb.2014.00148

Blaya J, Marhuenda FC, Pascual JA, Ros M (2016) Microbiota characterization of compost using omics approaches opens new perspectives for Phytophthora root rot control. PLoS One 11:e0158048. https://doi.org/10.1371/journal.pone.0158048

Bloemberg GV, Lugtenberg BJ (2001) Molecular basis of plant growth promotion and biocontrol by rhizobacteria. Curr Opin Plant Biol 4:343–350. https://doi.org/10.1016/S1369-5266(00)00183-7

Bodenhausen N, Horton MW, Bergelson J (2013) Bacterial communities associated with the leaves and the roots of *Arabidopsis thaliana*. PLoS One 8:e56329. https://doi.org/10.1371/journal.pone.0056329

Brandl MT (2006) Fitness of human enteric pathogens on plants and implications for food safety. Annu Rev Phytopathol 44:367–392. https://doi.org/10.1146/annurev.phyto.44.070505.143359

Brooks JP, Edwards DJ, Harwich MD Jr, Rivera MC, Fettweis JM, Serrano MG, Reris RA, Sheth NU, Huang B, Girerd P, Strauss JF III, Jefferson KK, Buck GA (2015) The truth about metagenomics: quantifying and counteracting bias in 16S rRNA studies. BMC Microbiol 15:66. https://doi.org/10.1186/s12866-015-0351-6

Bulgarelli D, Rott M, Schlaeppi K, Ver Loren van TE, Ahmadinejad N, Assenza F, Rauf P, Huettel B, Reinhardt R, Schmelzer E, Peplies J, Gloeckner FO, Amann R, Eickhorst T, Schulze-Lefert P (2012) Revealing structure and assembly cues for *Arabidopsis* root-inhabiting bacterial microbiota. Nature 488:91–95. https://doi.org/10.1038/nature11336

Campisano A, Antonielli L, Pancher M, Yousaf S, Pindo M, Pertot I (2014) Bacterial endophytic communities in the grapevine depend on pest management. PLoS One 9:e112763. https://doi.org/10.1371/journal.pone.0112763

Cankar K, Kraigher H, Ravnikar M, Rupnik M (2005) Bacterial endophytes from seeds of Norway spruce (*Picea abies* L. Karst). FEMS Microbiol Lett 244:341–345. https://doi.org/10.1016/j.femsle.2005.02.008

Carrell AA, Frank AC (2014) *Pinus flexilis* and *Picea engelmannii* share a simple and consistent needle endophyte microbiota with a potential role in nitrogen fixation. Front Microbiol 5:333. https://doi.org/10.3389/fmicb.2014.00333

Carrell AA, Frank AC (2015) Bacterial endophyte communities in the foliage of coast redwood and giant sequoia. Front Microbiol 6:1008. https://doi.org/10.3389/fmicb.2015.01008

Castaldini M, Turrini A, Sbrana C, Benedetti A, Marchionni M, Mocali S, Fabiani A, Landi S, Santomassimo F, Pietrangeli B, Nuti MP, Miclaus N, Giovannetti M (2005) Impact of Bt corn on rhizospheric and soil eubacterial communities and on beneficial mycorrhizal symbiosis in experimental microcosms. Appl Environ Microbiol 71:6719–6729. https://doi.org/10.1128/AEM.71.11.6719-6729.2005

Chelius MK, Triplett EW (2001) The diversity of archaea and bacteria in association with the roots of *Zea mays* L. Microb Ecol 41:252–263. https://doi.org/10.1007/s002480000087

Chen L, Luo S, Li X, Wan Y, Chen J, Liu C (2014) Interaction of Cd-hyperaccumulator *Solanum nigrum* L. and functional endophyte *Pseudomonas* sp. Lk9 on soil heavy metals uptake. Soil Biol Biochem 68:300–308. https://doi.org/10.1016/j.soilbio.2013.10.021

Chen H, Wu H, Yan B, Zhao H, Liu F, Zhang H, Sheng Q, Miao F, Liang Z (2018) Core microbiome of medicinal plant *Salvia miltiorrhiza* seed: a rich reservoir of beneficial microbes for secondary metabolism? Int J Mol Sci 19. https://doi.org/10.3390/ijms19030672

Chi F, Shen SH, Cheng HP, Jing YX, Yanni YG, Dazzo FB (2005) Ascending migration of endophytic rhizobia, from roots to leaves, inside rice plants and assessment of benefits to rice growth physiology. Appl Environ Microbiol 71:7271–7278. https://doi.org/10.1128/AEM.71.11.7271-7278.2005

Compant S, Clement C, Sessitsch A (2010) Plant growth-promoting bacteria in the rhizo- and endosphere of plants: their role, colonization, mechanisms involved and prospects for utilization. Soil Biol Biochem 42:669–678. https://doi.org/10.1016/j.soilbio.2009.11.024

Correa-Galeote D, Bedmar EJ, Arone GJ (2018) Maize endophytic bacterial diversity as affected by soil cultivation history. Front Microbiol 9:484. https://doi.org/10.3389/fmicb.2018.00484

Cotta SR, Dias ACF, Marriel IE, Gomes EA, van Elsas JD, Seldin L (2013) Temporal dynamics of microbial communities in the rhizosphere of two genetically modified (GM) maize hybrids in tropical agrosystems. Antonie van Leeuwenhoek 103:589–601. https://doi.org/10.1007/s10482-012-9843-7

da Silva DA, Cotta SR, Vollu RE, Jurelevicius DA, Marques JM, Marriel IE, Seldin L (2014) Endophytic microbial community in two transgenic maize genotypes and in their near-isogenic non-transgenic maize genotype. BMC Microbiol 14:332. https://doi.org/10.1186/s12866-014-0332-1

de Bruin OM, Birnboim HC (2016) A method for assessing efficiency of bacterial cell disruption and DNA release. BMC Microbiol 16:197. https://doi.org/10.1186/s12866-016-0815-3

de Melo Pereira GV, Magalhaes KT, Lorenzetii ER, Souza TP, Schwan RF (2012) A multiphasic approach for the identification of endophytic bacterial in strawberry fruit and their potential for plant growth promotion. Microb Ecol 63:405–417. https://doi.org/10.1007/s00248-011-9919-3

de Souza RS, Okura VK, Armanhi JS, Jorrin B, Lozano N, da Silva MJ, Gonzalez-Guerrero M, de Araujo LM, Verza NC, Bagheri HC, Imperial J, Arruda P (2016) Unlocking the bacterial and fungal communities assemblages of sugarcane microbiome. Sci Rep 6:28774. https://doi.org/10.1038/srep28774

Dijkhuizen LW, Brouwer P, Bolhuis H, Reichart GJ, Koppers N, Huettel B, Bolger AM, Li FW, Cheng S, Liu X, Wong GK, Pryer K, Weber A, Brautigam A, Schluepmann H (2018) Is there foul play in the leaf pocket? The metagenome of floating fern *Azolla* reveals endophytes that do not fix N_2 but may denitrify. New Phytol 217:453–466. https://doi.org/10.1111/nph.14843

Ding T, Melcher U (2016) Influences of plant species, season and location on leaf endophytic bacterial communities of non-cultivated plants. PLoS One 11:e0150895. https://doi.org/10.1371/journal.pone.0150895

Diwan V, Albrechtsen HJ, Smets BF, Dechesne A (2018) Does universal 16S rRNA gene amplicon sequencing of environmental communities provide an accurate description of nitrifying guilds? J Microbiol Methods 151:28–34. https://doi.org/10.1016/j.mimet.2018.05.025

Du X, Zhai Y, Deng Q, Tan H, Cao L (2018) Illumina-based sequencing analysis directed selection for actinobacterial probiotic candidates for banana plants. Probiotics Antimicrob Proteins 10:284–292. https://doi.org/10.1007/s12602-017-9293-7

Eevers N, Gielen M, Sanchez-Lopez A, Jaspers S, White JC, Vangronsveld J, Weyens N (2015) Optimization of isolation and cultivation of bacterial endophytes through addition of plant extract to nutrient media. Microb Biotechnol 8:707–715. https://doi.org/10.1111/1751-7915.12291

Eevers N, Beckers B, Op De BM, White JC, Vangronsveld J, Weyens N (2016a) Comparison between cultivated and total bacterial communities associated with *Cucurbita pepo* using cultivation-dependent techniques and 454 pyrosequencing. Syst Appl Microbiol 39:58–66. https://doi.org/10.1016/j.syapm.2015.11.001

Eevers N, Hawthorne JR, White JC, Vangronsveld J, Weyens N (2016b) Exposure of *Cucurbita pepo* to DDE-contamination alters the endophytic community: a cultivation dependent vs a cultivation independent approach. Environ Pollut 209:147–154. https://doi.org/10.1016/j.envpol.2015.11.038

Ferrando L, Fernandez SA (2015) Strong shift in the diazotrophic endophytic bacterial community inhabiting rice (*Oryza sativa*) plants after flooding. FEMS Microbiol Ecol 91:fiv104. https://doi.org/10.1093/femsec/fiv104

Fitzpatrick CR, Lu-Irving P, Copeland J, Guttman DS, Wang PW, Baltrus DA, Dlugosch KM, Johnson MTJ (2018) Chloroplast sequence variation and the efficacy of peptide nucleic acids for blocking host amplification in plant microbiome studies. Microbiome 6:144. https://doi.org/10.1186/s40168-018-0534-0

Fuentes-Ramirez LE, Caballero-Mellado J, Sepulveda J, Martinez-Romero E (1999) Colonization of sugarcane by *Acetobacter diazotrophicus* is inhibited by high N-fertilization. FEMS Microbiol Ecol 29:117–128

Gaby JC, Buckley DH (2014) A comprehensive aligned nifH gene database: a multipurpose tool for studies of nitrogen-fixing bacteria. Database (Oxford) 2014:bau001. doi. https://doi.org/10.1093/database/bau001

Gaby JC, Rishishwar L, Valderrama-Aguirre LC, Green SJ, Valderrama-Aguirre A, Jordan IK, Kostka JE (2018) Diazotroph community characterization via a high-throughput nifH amplicon sequencing and analysis pipeline. Appl Environ Microbiol 84. https://doi.org/10.1128/AEM.01512-17

Gadhave KR, Devlin PF, Ebertz A, Ross A, Gange AC (2018) Soil inoculation with *Bacillus* spp. modifies root endophytic bacterial diversity, evenness, and community composition in a context-specific manner. Microb Ecol 76:741–750. https://doi.org/10.1007/s00248-018-1160-x

Gao T, Shi XY (2018) Taxonomic structure and function of seed-inhabiting bacterial microbiota from common reed (*Phragmites australis*) and narrowleaf cattail (*Typha angustifolia* L.). Arch Microbiol 200:869–876. https://doi.org/10.1007/s00203-018-1493-3

Garcia-Orenes F, Morugan-Coronado A, Zornoza R, Scow K (2013) Changes in soil microbial community structure influenced by agricultural management practices in a mediterranean agro-ecosystem. PLoS One 8:e80522. https://doi.org/10.1371/journal.pone.0080522

Ghyselinck J, Pfeiffer S, Heylen K, Sessitsch A, De VP (2013) The effect of primer choice and short read sequences on the outcome of 16S rRNA gene based diversity studies. PLoS One 8:e71360. https://doi.org/10.1371/journal.pone.0071360

Gonzalez-Escobedo R, Briones-Roblero CI, Pineda-Mendoza RM, Rivera-Orduna FN, Zuniga G (2018) Bacteriome from *Pinus arizonica* and *P. durangensis*: diversity, comparison of assemblages, and overlapping degree with the gut bacterial community of a bark beetle that kills pines. Front Microbiol 9:77. https://doi.org/10.3389/fmicb.2018.00077

Gottel NR, Castro HF, Kerley M, Yang Z, Pelletier DA, Podar M, Karpinets T, Uberbacher E, Tuskan GA, Vilgalys R, Doktycz MJ, Schadt CW (2011) Distinct microbial communities within the endosphere and rhizosphere of *Populus deltoides* roots across contrasting soil types. Appl Environ Microbiol 77:5934–5944. https://doi.org/10.1128/AEM.05255-11

Govindasamy V, Franco CMM, Gupta VVSR (2014) Endophytic actinobacteria: diversity and ecology. In: Verma VC, Gange AC (eds) Advances in endophytic research. Springer, New Delhi, pp 27–59

Graner G, Persson P, Meijer J, Alstrom S (2006) A study on microbial diversity in different cultivars of *Brassica napus* in relation to its wilt pathogen, *Verticillium longisporum*. FEMS Microbiol Lett 224:269–276. https://doi.org/10.1016/S0378-1097(03)00449-X

Hakim S, Mirza BS, Zaheer A, Mclean JE, Imran A, Yasmin S, Sajjad MM (2018) Retrieved 16S rRNA and nifH sequences reveal co-dominance of *Bradyrhizobium* and Ensifer (*Sinorhizobium*) strains in field-collected root nodules of the promiscuous host *Vigna radiata* (L.) R. Wilczek. Appl Microbiol Biotechnol 102:485–497. https://doi.org/10.1007/s00253-017-8609-6

Hallmann J, Quadt-Hallmann A, Miller WG, Sikora RA, Lindow SE (2001) Endophytic colonization of plants by the biocontrol agent *Rhizobium etli* G12 in relation to *Meloidogyne incognita* Infection. Phytopathology 91:415–422. https://doi.org/10.1094/PHYTO.2001.91.4.415

Hamaki T, Suzuki M, Fudou R, Jojima Y, Kajiura T, Tabuchi A, Sen K, Shibai H (2005) Isolation of novel bacteria and actinomycetes using soil-extract agar medium. J Biosci Bioeng 99:485–492. https://doi.org/10.1263/jbb.99.485

Hanshew AS, Mason CJ, Raffa KF, Currie CR (2013) Minimization of chloroplast contamination in 16S rRNA gene pyrosequencing of insect herbivore bacterial communities. J Microbiol Methods 95:149–155. https://doi.org/10.1016/j.mimet.2013.08.007

Hardoim PR, van Overbeek LS, Elsas JD (2008) Properties of bacterial endophytes and their proposed role in plant growth. Trends Microbiol 16:463–471. https://doi.org/10.1016/j.tim.2008.07.008

Holovachov O, Haenel Q, Bourlat SJ, Jondelius U (2017) Taxonomy assignment approach determines the efficiency of identification of OTUs in marine nematodes. R Soc Open Sci 4:170315. https://doi.org/10.1098/rsos.170315

Hong S, Bunge J, Leslin C, Jeon S, Epstein SS (2009) Polymerase chain reaction primers miss half of rRNA microbial diversity. ISME J 3:1365–1373. https://doi.org/10.1038/ismej.2009.89

Hong Y, Liao D, Hu A, Wang H, Chen J, Khan S, Su J, Li H (2015) Diversity of endophytic and rhizoplane bacterial communities associated with exotic *Spartina alterniflora* and native mangrove using Illumina amplicon sequencing. Can J Microbiol 61:723–733. https://doi.org/10.1139/cjm-2015-0079

Huang Y (2018) Comparison of rhizosphere and endophytic microbial communities of Chinese leek through high-throughput 16S rRNA gene Illumina sequencing. J Integr Agric 17:359–367. https://doi.org/10.1016/S2095-3119(17)61731-3

Huang Y, Kuang Z, Wang W, Cao L (2016) Exploring potential bacterial and fungal biocontrol agents transmitted from seeds to sprouts of wheat. Biological Control 98:27–33. https://doi.org/10.1016/j.biocontrol.2016.02.013

Huang Y, Kuang Z, Deng Z, Zhang R, Cao L (2017) Endophytic bacterial and fungal communities transmitted from cotyledons and germs in peanut (*Arachis hypogaea* L.) sprouts. Environ Sci Pollut Res Int 24:16458–16464. https://doi.org/10.1007/s11356-017-9254-4

Huang Y, Zhang M, Deng Z, Cao L (2018) Evaluation of probiotic diversity from soybean (*Glycine max*) seeds and sprouts using Illumina-based sequencing method. Probiotics Antimicrob Proteins 10:293–298. https://doi.org/10.1007/s12602-017-9305-7

Hurek T, Reinhold-Hurek B, Van Montagu M, Kellenberger E (1994) Root colonization and systemic spreading of *Azoarcus* sp. strain BH72 in grasses. J Bacteriol 176:1913–1923

Icoz I, Saxena D, Andow DA, Zwahlen C, Stotzky G (2008) Microbial populations and enzyme activities in soil in situ under transgenic corn expressing cry proteins from *Bacillus thuringiensis*. J Environ Qual 37:647–662

Ikeda S, Kaneko T, Okubo T, Rallos LE, Eda S, Mitsui H, Sato S, Nakamura Y, Tabata S, Minamisawa K (2009) Development of a bacterial cell enrichment method and its application to the community analysis in soybean stems. Microb Ecol 58:703–714. https://doi.org/10.1007/s00248-009-9566-0

Jackson CR, Randolph KC, Osborn SL, Tyler HL (2013) Culture dependent and independent analysis of bacterial communities associated with commercial salad leaf vegetables. BMC Microbiol 13:274. https://doi.org/10.1186/1471-2180-13-274

Jangid K, Williams MA, Franzluebbers AJ, Sanderlin JS, Reeves JH, Jenkins MB, Endale DM, Coleman DC, Whitman WB (2008) Relative impacts of land-use, management intensity and fertilization upon soil microbial community structure in agricultural systems. Soil Biol Biochem 40:2843–2853. https://doi.org/10.1016/j.soilbio.2008.07.030

Ji SH, Gururani MA, Chun SC (2014) Isolation and characterization of plant growth promoting endophytic diazotrophic bacteria from Korean rice cultivars. Microbiol Res 169:83–98. https://doi.org/10.1016/j.micres.2013.06.003

Jiao JY, Wang HX, Zeng Y, Shen YM (2006) Enrichment for microbes living in association with plant tissues. J Appl Microbiol 100:830–837. https://doi.org/10.1111/j.1365-2672.2006.02830.x

Johnston-Monje D, Raizada MN (2011) Conservation and diversity of seed associated endophytes in *Zea* across boundaries of evolution, ethnography and ecology. PLoS One 6:e20396. https://doi.org/10.1371/journal.pone.0020396

Jones RT, Sanchez LG, Fierer N (2013) A cross-taxon analysis of insect-associated bacterial diversity. PLoS One 8:e61218. https://doi.org/10.1371/journal.pone.0061218

Kchouk M, Gibrat JF, Elloumi M (2017) Generations of sequencing technologies: from first to next generation. Biol Med (Aligarh) 9:3. https://doi.org/10.4172/0974-8369.1000395

Kembel SW, O'Connor TK, Arnold HK, Hubbell SP, Wright SJ, Green JL (2014) Relationships between phyllosphere bacterial communities and plant functional traits in a neotropical forest. Proc Natl Acad Sci U S A 111:13715–13720. https://doi.org/10.1073/pnas.1216057111

Kemler M, Garnas J, Wingfield MJ, Gryzenhout M, Pillay KA, Slippers B (2013) Ion Torrent PGM as tool for fungal community analysis: a case study of endophytes in *Eucalyptus grandis* reveals high taxonomic diversity. PLoS One 8:e81718. https://doi.org/10.1371/journal.pone.0081718

Kempf HJ (1989) *Erwinia herbicola* as a biocontrol agent of *Fusarium culmorum* and *Puccinia recondita* f. sp. *tritici* on wheat. Phytopathol 79:990–994

Kifle MH, Laing MD (2015) Isolation and screening of bacteria for their diazotrophic potential and their influence on growth promotion of maize seedlings in greenhouses. Front Plant Sci 6:1225. https://doi.org/10.3389/fpls.2015.01225

Kim D, Hofstaedter CE, Zhao C, Mattei L, Tanes C, Clarke E, Lauder A, Sherrill-Mix S, Chehoud C, Kelsen J, Conrad M, Collman RG, Baldassano R, Bushman FD, Bittinger K (2017) Optimizing methods and dodging pitfalls in microbiome research. Microbiome 5:52. https://doi.org/10.1186/s40168-017-0267-5

Kisand V, Wikner J (2003) Combining culture-dependent and -independent methodologies for estimation of richness of estuarine bacterioplankton consuming riverine dissolved organic matter. Appl Environ Microbiol 69:3607–3616. https://doi.org/10.1128/AEM.69.6.3607-3616.2003

Knief C (2014) Analysis of plant microbe interactions in the era of next generation sequencing technologies. Front Plant Sci 5:216. https://doi.org/10.3389/fpls.2014.00216

Knoth JL, Kim SH, Ettl GJ, Doty SL (2014) Biological nitrogen fixation and biomass accumulation within poplar clones as a result of inoculations with diazotrophic endophyte consortia. New Phytol 201:599–609. https://doi.org/10.1111/nph.12536

Kunda P, Dhal PK, Mukherjee A (2018) Endophytic bacterial community of rice (*Oryza sativa* L.) from coastal saline zone of West Bengal: 16S rRNA gene based metagenomics approach. Meta Gene 18:79–86. https://doi.org/10.1016/j.mgene.2018.08.004

Latz E, Eisenhauer N, Rall BC, Allan E, Roscher C, Scheu S, Jousset A (2012) Plant diversity improves protection against soil-borne pathogens by fostering antagonistic bacterial communities. J Ecol 100:597–604. https://doi.org/10.1111/j.1365-2745.2011.01940.x

Lebeis SL (2014) The potential for give and take in plant-microbiome relationships. Front Plant Sci 5:287. https://doi.org/10.3389/fpls.2014.00287

Leveau JHJ, Tech JJ (2011) Grapevine microbiomics: bacterial diversity on grape leaves and berries revealed by high-throughput sequence analysis of 16S rRNA amplicons. Acta Hortic (905):31–42. https://doi.org/10.17660/ActaHortic.2011.905.2

Li L, Fan F, Song A, Yin C, Cui P, Li Z, Liang Y (2017a) Microbial composition and diversity are associated with plant performance: a case study on long-term fertilization effect on wheat growth in an Ultisol. Appl Microbiol Biotechnol 101:4669–4681. https://doi.org/10.1007/s00253-017-8147-2

Li O, Xiao R, Sun L, Guan C, Kong D, Hu X (2017b) Bacterial and diazotrophic diversities of endophytes in *Dendrobium catenatum* determined through barcoded pyrosequencing. PLoS One 12:e0184717. https://doi.org/10.1371/journal.pone.0184717

Liu F, Yuan Z, Zhang X, Zhang G, Xie B (2017) Characteristics and diversity of endophytic bacteria in moso bamboo (*Phyllostachys edulis*) based on 16S rDNA sequencing. Arch Microbiol 199:1259–1266. https://doi.org/10.1007/s00203-017-1397-7

Lu Z, Sang L, Li Z, Min H (2009) Catalase and superoxide dismutase activities in a *Stenotrophomonas maltophilia* WZ2 resistant to herbicide pollution. Ecotoxicol Environ Saf 72:136–143. https://doi.org/10.1016/j.ecoenv.2008.01.009

Lumactud R, Fulthorpe R, Sentchilo V, van der Meer JR (2017) Draft genome sequence of *Microbacterium foliorum* strain 122 isolated from a plant growing in a chronically hydrocarbon-contaminated site. Genome Announc 5:e00434–e00417. https://doi.org/10.1128/genomeA.00434-17

Lundberg DS, Lebeis SL, Paredes SH, Yourstone S, Gehring J, Malfatti S, Tremblay J, Engelbrektson A, Kunin V, Del Rio TG, Edgar RC, Eickhorst T, Ley RE, Hugenholtz P, Tringe SG, Dangl JL (2012) Defining the core *Arabidopsis thaliana* root microbiome. Nature 488:86–90. https://doi.org/10.1038/nature11237

Ma Y, Rajkumar M, Zhang C, Freitas H (2016) Beneficial role of bacterial endophytes in heavy metal phytoremediation. J Environ Manage 174:14–25. https://doi.org/10.1016/j.jenvman.2016.02.047

Manter DK, Delgado JA, Holm DG, Stong RA (2010) Pyrosequencing reveals a highly diverse and cultivar-specific bacterial endophyte community in potato roots. Microb Ecol 60:157–166. https://doi.org/10.1007/s00248-010-9658-x

Mareque C, da Silva TF, Vollu RE, Beracochea M, Seldin L, Battistoni F (2018) The endophytic bacterial microbiota associated with sweet sorghum (*Sorghum bicolor*) is modulated by the application of chemical N fertilizer to the field. Int J Genomics 2018:7403670. https://doi.org/10.1155/2018/7403670

Maropola MK, Ramond JB, Trindade M (2015) Impact of metagenomic DNA extraction procedures on the identifiable endophytic bacterial diversity in *Sorghum bicolor* (L. Moench). J Microbiol Methods 112:104–117. https://doi.org/10.1016/j.mimet.2015.03.012

Mashiane RA, Ezeokoli OT, Adeleke RA, Bezuidenhout CC (2017) Metagenomic analyses of bacterial endophytes associated with the phyllosphere of a Bt maize cultivar and its isogenic parental line from South Africa. World J Microbiol Biotechnol 33:80. https://doi.org/10.1007/s11274-017-2249-y

Mendes LW, Kuramae EE, Navarrete AA, van Veen JA, Tsai SM (2014) Taxonomical and functional microbial community selection in soybean rhizosphere. ISME J 8:1577–1587. https://doi.org/10.1038/ismej.2014.17

Metzker ML (2010) Sequencing technologies - the next generation. Nat Rev Genet 11:31–46. https://doi.org/10.1038/nrg2626

Miche L, Balandreau J (2001) Effects of rice seed surface sterilization with hypochlorite on inoculated *Burkholderia vietnamiensis*. Appl Environ Microbiol 67:3046–3052. https://doi.org/10.1128/AEM.67.7.3046-3052.2001

Miyambo T, Makhalanyane TP, Cowan DA, Valverde A (2016) Plants of the fynbos biome harbour host species-specific bacterial communities. FEMS Microbiol Lett 363:fnw122. https://doi.org/10.1093/femsle/fnw122

Mizrahi-Man O, Davenport ER, Gilad Y (2013) Taxonomic classification of bacterial 16S rRNA genes using short sequencing reads: evaluation of effective study designs. PLoS One 8:e53608. https://doi.org/10.1371/journal.pone.0053608

Montanari-Coelho KK, Costa AT, Polonio JC, Azevedo JL, Marin SRR, Fuganti-Pagliarini R, Fujita Y, Yamaguchi-Shinozaki K, Nakashima K, Pamphile JA, Nepomuceno AL (2018) Endophytic bacterial microbiome associated with leaves of genetically modified (AtAREB1) and conventional (BR 16) soybean plants. World J Microbiol Biotechnol 34:56. https://doi.org/10.1007/s11274-018-2439-2

Mora-Ruiz MR, Font-Verdera F, Diaz-Gil C, Urdiain M, Rodriguez-Valdecantos G, Gonzalez B, Orfila A, Rossello-Mora R (2015) Moderate halophilic bacteria colonizing the phylloplane of halophytes of the subfamily *Salicornioideae* (*Amaranthaceae*). Syst Appl Microbiol 38:406–416. https://doi.org/10.1016/j.syapm.2015.05.004

Moronta-Barrios F, Gionechetti F, Pallavicini A, Marys E, Venturi V (2018) Bacterial microbiota of rice roots: 16S-based taxonomic profiling of endophytic and rhizospheric diversity, endophytes isolation and simplified endophytic community. Microorganisms 6:14. https://doi.org/10.3390/microorganisms6010014

Moyes AB, Kueppers LM, Pett-Ridge J, Carper DL, Vandehey N, O'Neil J, Frank AC (2016) Evidence for foliar endophytic nitrogen fixation in a widely distributed subalpine conifer. New Phytol 210:657–668. https://doi.org/10.1111/nph.13850

Mukherjee P, Roy P (2013) Persistent organic pollutants induced protein expression and immunocrossreactivity by *Stenotrophomonas maltophilia* PM102: a prospective bioremediating candidate. Biomed Res Int 2013:714232. doi:https://doi.org/10.1155/2013/714232

Murali A, Bhargava A, Wright ES (2018) IDTAXA: a novel approach for accurate taxonomic classification of microbiome sequences. Microbiome 6:140. https://doi.org/10.1186/s40168-018-0521-5

Naylor D, DeGraaf S, Purdom E, Coleman-Derr D (2017) Drought and host selection influence bacterial community dynamics in the grass root microbiome. ISME J 11:2691–2704. https://doi.org/10.1038/ismej.2017.118

Nikolic B, Schwab H, Sessitsch A (2011) Metagenomic analysis of the 1-aminocycloprop ane-1-carboxylate deaminase gene (acdS) operon of an uncultured bacterial endophyte colonizing *Solanum tuberosum* L. Arch Microbiol 193:665–676. https://doi.org/10.1007/s00203-011-0703-z

Pampulha ME, Oliveira A (2006) Impact of an herbicide combination of bromoxynil and prosulfuron on soil microorganisms. Curr Microbiol 53:238–243. https://doi.org/10.1007/s00284-006-0116-4

Passera A, Marcolungo L, Casati P, Brasca M, Quaglino F, Cantaloni C, Delledonne M (2018) Hybrid genome assembly and annotation of *Paenibacillus pasadenensis* strain R16 reveals insights on endophytic life style and antifungal activity. PLoS One 13:e0189993. https://doi.org/10.1371/journal.pone.0189993

Perez-Jaramillo JE, Carrion VJ, de Hollander M, Raaijmakers JM (2018) The wild side of plant microbiomes. Microbiome 6:143. https://doi.org/10.1186/s40168-018-0519-z

Pinto AJ, Raskin L (2012) PCR biases distort bacterial and archaeal community structure in pyrosequencing datasets. PLoS One 7:e43093. https://doi.org/10.1371/journal.pone.0043093

Pinto C, Pinho D, Sousa S, Pinheiro M, Egas C, Gomes AC (2014) Unravelling the diversity of grapevine microbiome. PLoS One 9:e85622. https://doi.org/10.1371/journal.pone.0085622

Prayitno J, Rolfe B (2010) Characterization of endophytic diazotroph bacteria isolated from rice. HAYATI J Biosci 17:73–78. https://doi.org/10.4308/hjb.17.2.73

Prischl M, Hackl E, Pastar M, Pfeiffer S, Sessitsch A (2012) Genetically modified Bt maize lines containing cry3Bb1, cry1A105 or cry1Ab2 do not affect the structure and functioning of root-associated endophyte communities. Appl Soil Ecol 54:39–48. https://doi.org/10.1016/j.apsoil.2011.12.005

Quince C, Walker AW, Simpson JT, Loman NJ, Segata N (2017) Shotgun metagenomics, from sampling to analysis. Nat Biotechnol 35:833–844. https://doi.org/10.1038/nbt.3935

Rajendhran J, Gunasekaran P (2011) Microbial phylogeny and diversity: small subunit ribosomal RNA sequence analysis and beyond. Microbiol Res 166:99–110. https://doi.org/10.1016/j.micres.2010.02.003

Rastogi G, Sbodio A, Tech JJ, Suslow TV, Coaker GL, Leveau JH (2012) Leaf microbiota in an agroecosystem: spatiotemporal variation in bacterial community composition on field-grown lettuce. ISME J 6:1812–1822. https://doi.org/10.1038/ismej.2012.32

Redford AJ, Bowers RM, Knight R, Linhart Y, Fierer N (2010) The ecology of the phyllosphere: geographic and phylogenetic variability in the distribution of bacteria on tree leaves. Environ Microbiol 12:2885–2893. https://doi.org/10.1111/j.1462-2920.2010.02258.x

Reinhold B, Hurek T (1988) Location of diazotrophs in the root interior with special attention to the kallar grass association. Plant Soil 110:259–268

Ren G, Zhang H, Lin X, Zhu J, Jia Z (2015a) Response of leaf endophytic bacterial community to elevated CO_2 at different growth stages of rice plant. Front Microbiol 6:855. https://doi.org/10.3389/fmicb.2015.00855

Ren G, Zhu C, Alam MS, Tokida T, Sakai H, Nakamura H, Usui Y, Zhu J, Hasegawa T, Jia Z (2015b) Response of soil, leaf endosphere and phyllosphere bacterial communities to elevated CO2 and soil temperature in a rice paddy. Plant Soil 392:27–44. https://doi.org/10.1007/s11104-015-2503-8

Ren Z, Tang S, Jiang Y, Jiang M, Zheng S, Liu W, Yang Z, Sang S, Chen Z, Xia T, Yin M (2018) High-throughput sequencing analysis of endophytic bacteria diversity in fruits of white and red pitayas from three different origins. Pol J Microbiol 67:27–35. https://doi.org/10.5604/01.3001.0011.6139

Romero FM, Marina M, Pieckenstain FL (2014) The communities of tomato (*Solanum lycopersicum* L.) leaf endophytic bacteria, analyzed by 16S-ribosomal RNA gene pyrosequencing. FEMS Microbiol Lett 351:187–194. https://doi.org/10.1111/1574-6968.12377

Rua MA, Wilson EC, Steele S, Munters AR, Hoeksema JD, Frank AC (2016) Associations between ectomycorrhizal fungi and bacterial needle endophytes in *Pinus radiata*: implications for biotic selection of microbial communities. Front Microbiol 7:399. https://doi.org/10.3389/fmicb.2016.00399

Ryan RP, Germaine K, Franks A, Ryan DJ, Dowling DN (2008) Bacterial endophytes: recent developments and applications. FEMS Microbiol Lett 278:1–9. https://doi.org/10.1111/j.1574-6968.2007.00918.x

Sagaram US, DeAngelis KM, Trivedi P, Andersen GL, Lu SE, Wang N (2009) Bacterial diversity analysis of Huanglongbing pathogen-infected citrus, using PhyloChip arrays and 16S rRNA gene clone library sequencing. Appl Environ Microbiol 75:1566–1574. https://doi.org/10.1128/AEM.02404-08

Saminathan T, GarcÃa M, Ghimire B, Lopez C, Bodunrin A, Nimmakayala P, Abburi VL, Levi A, Balagurusamy N, Reddy UK (2018) Metagenomic and metatranscriptomic analyses of diverse watermelon cultivars reveal the role of fruit associated microbiome in carbohydrate metabolism and ripening of mature fruits. Front Plant Sci 9:4. https://doi.org/10.3389/fpls.2018.00004

Sarria-Guzman Y, Chavez-Romero Y, Gomez-Acata S, Montes-Molina JA, Morales-Salazar E, Dendooven L, Navarro-Noya YE (2016) Bacterial communities associated with different *Anthurium andraeanum* L. plant tissues. Microbes Environ 31:321–328. https://doi.org/10.1264/jsme2.ME16099

Sasaki K, Ikeda S, Ohkubo T, Kisara C, Sato T, Minamisawa K (2013) Effects of plant genotype and nitrogen level on bacterial communities in rice shoots and roots. Microbes Environ 28:391–395. https://doi.org/10.1264/jsme2.ME12212

Schloss PD, Gevers D, Westcott SL (2011) Reducing the effects of PCR amplification and sequencing artifacts on 16S rRNA-based studies. PLoS One 6:e27310. https://doi.org/10.1371/journal.pone.0027310

Seghers D, Wittebolle L, Top EM, Verstraete W, Siciliano SD (2004) Impact of agricultural practices on the *Zea mays* L. endophytic community. Appl Environ Microbiol 70:1475–1482. https://doi.org/10.1128/AEM.70.3.1475-1482.2004

Sengupta S, Ganguli S, Singh PK (2017) Metagenome analysis of the root endophytic microbial community of Indian rice (*O. sativa* L.). Genom Data 12:41–43. https://doi.org/10.1016/j.gdata.2017.02.010

Sessitsch A, Hardoim P, Doring J, Weilharter A, Krause A, Woyke T, Mitter B, Hauberg-Lotte L, Friedrich F, Rahalkar M, Hurek T, Sarkar A, Bodrossy L, van Overbeek L, Brar D, van Elsas JD, Reinhold-Hurek B (2011) Functional characteristics of an endophyte community colonizing rice roots as revealed by metagenomic analysis. Mol Plant Microbe Interact 25:28–36. https://doi.org/10.1094/MPMI-08-11-0204

Shade A, Caporaso JG, Handelsman J, Knight R, Fierer N (2013) A meta-analysis of changes in bacterial and archaeal communities with time. ISME J 7:1493–1506. https://doi.org/10.1038/ismej.2013.54

Shahzad R, Khan AL, Bilal S, Asaf S, Lee IJ (2018) What is there in seeds? Vertically transmitted endophytic resources for sustainable improvement in plant growth. Front Plant Sci 9:24. https://doi.org/10.3389/fpls.2018.00024

Sharpton TJ (2014) An introduction to the analysis of shotgun metagenomic data. Front Plant Sci 5:209. https://doi.org/10.3389/fpls.2014.00209

Shen SY, Fulthorpe R (2015) Seasonal variation of bacterial endophytes in urban trees. Front Microbiol 6:427. https://doi.org/10.3389/fmicb.2015.00427

Shi Y, Yang H, Zhang T, Sun J, Lou K (2014) Illumina-based analysis of endophytic bacterial diversity and space-time dynamics in sugar beet on the north slope of Tianshan mountain. Appl Microbiol Biotechnol 98:6375–6385. https://doi.org/10.1007/s00253-014-5720-9

Singer E, Bushnell B, Coleman-Derr D, Bowman B, Bowers RM, Levy A, Gies EA, Cheng JF, Copeland A, Klenk HP, Hallam SJ, Hugenholtz P, Tringe SG, Woyke T (2016) High-resolution phylogenetic microbial community profiling. ISME J 10:2020–2032. https://doi.org/10.1038/ismej.2015.249

Sorensen J, Sessitsch A (2006) Plant-associated bacteria lifestyle and molecular interactions. In: van Elsas JD, Jansson JK, Trevors JT (eds) Modern soil microbiology. CRC Press, Boca Raton, pp 211–236

Starke IC, Vahjen W, Pieper R, Zentek J (2014) The influence of DNA extraction procedure and primer set on the bacterial community analysis by pyrosequencing of barcoded 16S rRNA gene amplicons. Mol Biol Int 2014:548683. https://doi.org/10.1155/2014/548683

Sun W, Xiong Z, Chu L, Li W, Soares MA, White JF, Jr., Li H (2018) Bacterial communities of three plant species from Pb-Zn contaminated sites and plant-growth promotional benefits of endophytic *Microbacterium* sp. (strain BXGe71). J Hazard Mater 370:225–231. doi:https://doi.org/10.1016/j.jhazmat.2018.02.003

Szymanska S, Borruso L, Brusetti L, Hulisz P, Furtado B, Hrynkiewicz K (2018) Bacterial microbiome of root-associated endophytes of *Salicornia europaea* in correspondence to different levels of salinity. Environ Sci Pollut Res Int 25:25420–25431. https://doi.org/10.1007/s11356-018-2530-0

Taghavi S, Barac T, Greenberg B, Borremans B, Vangronsveld J, van der Lelie D (2005) Horizontal gene transfer to endogenous endophytic bacteria from poplar improves phytoremediation of toluene. Appl Environ Microbiol 71:8500. https://doi.org/10.1128/AEM.71.12.8500-8505.2005

Tan F, Wang J, Feng Y, Chi G, Kong H, Qiu H, Wei S (2010) Bt corn plants and their straw have no apparent impact on soil microbial communities. Plant Soil 329:349–364. https://doi.org/10.1007/s11104-009-0163-2

Tan Y, Cui Y, Li H, Kuang A, Li X, Wei Y, Ji X (2017) Diversity and composition of rhizospheric soil and root endogenous bacteria in *Panax notoginseng* during continuous cropping practices. J Basic Microbiol 57:337–344. https://doi.org/10.1002/jobm.201600464

Tardif S, Yergeau E, Tremblay J, Legendre P, Whyte LG, Greer CW (2016) The willow microbiome is influenced by soil petroleum-hydrocarbon concentration with plant compartment-specific effects. Front Microbiol 7:1363. https://doi.org/10.3389/fmicb.2016.01363

Thiem D, Golebiewski M, Hulisz P, Piernik A, Hrynkiewicz K (2018) How does salinity shape bacterial and fungal microbiomes of *Alnus glutinosa* roots? Front Microbiol 9:651. https://doi.org/10.3389/fmicb.2018.00651

Thijs S, Op De BM, Beckers B, Truyens S, Stevens V, Van Hamme JD, Weyens N, Vangronsveld J (2017) Comparative evaluation of four bacteria-specific primer pairs for 16S rRNA gene surveys. Front Microbiol 8:494. https://doi.org/10.3389/fmicb.2017.00494

Thomas P, Sekhar AC (2017) Cultivation versus molecular analysis of banana (*Musa* sp.) shoot-tip tissue reveals enormous diversity of normally uncultivable endophytic bacteria. Microb Ecol 73:885–899. https://doi.org/10.1007/s00248-016-0877-7

Tian XY, Zhang CS (2017) Illumina-based analysis of endophytic and rhizosphere bacterial diversity of the coastal halophyte *Messerschmidia sibirica*. Front Microbiol 8:2288. https://doi.org/10.3389/fmicb.2017.02288

Trognitz F, Piller K, Nagel M, Borner A, Bacher C-F, Rechlik M, Mayrhofer H, Sessitsch A (2014) Isolation and characterization of endophytes isolated from seeds of different plants and the application to increase juvenile development. In: Brandstetter A, Geppner M, Grausgruber H, Buchgraber K (eds) Tagung Zukünftiges Saatgut – Produktion, Vermarktung, Nutzung und Konzervierung (Future seed – production, marketing, use and conservation), Hohere Bundeslehr und Forschungsanstalt. Raumberg-Gumpenstein, Austria, pp 25–28

Truyens S, Beckers B, Thijs S, Weyens N, Cuypers A, Vangronsveld J (2016a) Cadmium-induced and trans-generational changes in the cultivable and total seed endophytic community of *Arabidopsis thaliana*. Plant Biol (Stuttg) 18:376–381. https://doi.org/10.1111/plb.12415

Truyens S, Beckers B, Thijs S, Weyens N, Cuypers A, Vangronsveld J (2016b) The effects of the growth substrate on cultivable and total endophytic assemblages of *Arabidopsis thaliana*. Plant Soil 405:325–336. https://doi.org/10.1007/s11104-015-2761-5

Turner TR, James EK, Poole PS (2013) The plant microbiome. Genome Biol 14:209. https://doi.org/10.1186/gb-2013-14-6-209

Unno Y, Shinano T (2013) Metagenomic analysis of the rhizosphere soil microbiome with respect to phytic acid utilization. Microbes Environ 28:120–127. https://doi.org/10.1264/jsme2.ME12181

Utturkar SM, Cude WN, Robeson MS Jr, Yang ZK, Klingeman DM, Land ML, Allman SL, Lu TY, Brown SD, Schadt CW, Podar M, Doktycz MJ, Pelletier DA (2016) Enrichment of root endophytic bacteria from *Populus deltoides* and single-cell-genomics analysis. Appl Environ Microbiol 82:5698–5708. https://doi.org/10.1128/AEM.01285-16

Utturkar SM, Klingeman DM, Hurt RA Jr, Brown SD (2017) A case study into microbial genome assembly gap sequences and finishing strategies. Front Microbiol 8:1272. https://doi.org/10.3389/fmicb.2017.01272

Vetrovsky T, Baldrian P (2013) The variability of the 16S rRNA gene in bacterial genomes and its consequences for bacterial community analyses. PLoS One 8:e57923. https://doi.org/10.1371/journal.pone.0057923

Wang HX, Geng ZL, Zeng Y, Shen YM (2008) Enriching plant microbiota for a metagenomic library construction. Environ Microbiol 10:2684–2691. https://doi.org/10.1111/j.1462-2920.2008.01689.x

Wang W, Zhai Y, Cao L, Tan H, Zhang R (2016a) Endophytic bacterial and fungal microbiota in sprouts, roots and stems of rice (*Oryza sativa* L.). Microbiol Res 188-189:1–8. https://doi.org/10.1016/j.micres.2016.04.009

Wang W, Zhai Y, Cao L, Tan H, Zhang R (2016b) Illumina-based analysis of core actinobacteriome in roots, stems, and grains of rice. Microbiol Res 190:12–18. https://doi.org/10.1016/j.micres.2016.05.003

Wang W, Li Y, Qin W, Sun C, Tan H, Cao L (2018) Using Illumina-based sequence analysis to guide probiotic candidate selection and isolation. Probiotics Antimicrob Proteins 10:478–484. https://doi.org/10.1007/s12602-017-9298-2

Watanabe K, Nelson J, Harayama S, Kasai H (2001) ICB database: the gyrB database for identification and classification of bacteria. Nucleic Acids Res 29:344–345

Wemheuer B, Wemheuer F (2017) Assessing bacterial and fungal diversity in the plant endosphere. Methods Mol Biol 1539:75–84. https://doi.org/10.1007/978-1-4939-6691-2_6

Xia Y, DeBolt S, Dreyer J, Scott D, Williams MA (2015) Characterization of culturable bacterial endophytes and their capacity to promote plant growth from plants grown using organic or conventional practices. Front Plant Sci 6:490. https://doi.org/10.3389/fpls.2015.00490

Xiao X, Chen W, Zong L, Yang J, Jiao S, Lin Y, Wang E, Wei G (2017) Two cultivated legume plants reveal the enrichment process of the microbiome in the rhizocompartments. Mol Ecol 26:1641–1651. https://doi.org/10.1111/mec.14027

Yaish MW, Al-Lawati A, Jana GA, Vishwas PH, Glick BR (2016) Impact of soil salinity on the structure of the bacterial endophytic community identified from the roots of Caliph medic (*Medicago truncatula*). PLoS One 11:e0159007. https://doi.org/10.1371/journal.pone.0159007

Yang C, Hamel C, Gan Y, Vujanovic V (2012) Bacterial endophytes mediate positive feedback effects of early legume termination times on the yield of subsequent durum wheat crops. Can J Microbiol 58:1368–1377. https://doi.org/10.1139/w2012-114

Yang R, Liu P, Ye W (2017) Illumina-based analysis of endophytic bacterial diversity of tree peony (*Paeonia* Sect. *Moutan*) roots and leaves. Braz J Microbiol 48:695–705. https://doi.org/10.1016/j.bjm.2017.02.009

Yu X, Yang J, Wang E, Li B, Yuan H (2015) Effects of growth stage and fulvic acid on the diversity and dynamics of endophytic bacterial community in *Stevia rebaudiana* Bertoni leaves. Front Microbiol 6:867. https://doi.org/10.3389/fmicb.2015.00867

Zhao S, Zhou N, Zhao ZY, Zhang K, Tian CY (2016) High-throughput sequencing analysis of the endophytic bacterial diversity and dynamics in roots of the halophyte *Salicornia europaea*. Curr Microbiol 72:557–562. https://doi.org/10.1007/s00284-016-0990-3

Zhong W, Gu T, Wang W, Zhang B, Lin X, Huang Q, Shen W (2010) The effects of mineral fertilizer and organic manure on soil microbial community and diversity. Plant Soil 326:511–522. https://doi.org/10.1007/s11104-009-9988-y

Zielinska S, Radkowski P, Blendowska A, Ludwig-Galezowska A, Los JM, Los M (2018) The choice of the DNA extraction method may influence the outcome of the soil microbial community structure analysis. MicrobiologyOpen 6:e00453. https://doi.org/10.1002/mbo3.453
Zuniga C, Zaramela L, Zengler K (2017) Elucidation of complexity and prediction of interactions in microbial communities. Microb Biotechnol 10:1500–1522. https://doi.org/10.1111/1751-7915.12855

Perceptions of Microbe–Microbe and Plant–Microbiome Interfaces: The Metagenomic Maneuver

21

Nikhi Verma and Vivek Kumar

Abstract

The interactive relationship between plant and microbe starts on the surface of the plant, such as on the roots, leaves, or other parts. Various different types of microbes thrive on different parts of the plant. Microbes may also enter plant cells and are then known as endophytes; this relationship consists of an intricate interaction between plant and microbe. This intricate interaction occurs also at a genomic level. The plant–microbiome interaction may be beneficial or harmful for both, and sometimes it may be neutral. The presence of another microbe in the vicinity also changes the relationship between a microbe and its host. Now we understand that this type of communication or interface is very complicated, and, to understand this scenario, we need the help of modern techniques such as a metagenomic approach or next-generation sequencing. The soil and plant microbiome community plays a significant role in providing essential nutrients to the host plant and also in recycling nutrients and carbon in the environment. On the other hand, we do not know much about novel or nonisolated soil microbes; therefore, their functions are unknown to us. Using a metagenomic approach we can reveal the identity of an unknown soil microbe along with its functional gene information. Most of these associate microbes that enter into plant tissues communicate with their host very closely. This interaction influences metabolic aspects, nutritional uptake potential and transport, signaling of hormones, and stress mitigation, ultimately resulting in plant growth and development. In this respect, a metagenomic approach can be proficiently linked to other omics techniques to offer a multifarious picture of in-progress occurrences that exploit the communication between the plant and its total microbiome. The

N. Verma
Department of Medical Mycology, VP Chest Institute, University of Delhi, Delhi, India

V. Kumar (✉)
Himalayan School of Biosciences, Swami Rama Himalayan University, Dehradun, Uttarakhand, India

V. Kumar et al. (eds.), *Microbiome in Plant Health and Disease*,
https://doi.org/10.1007/978-981-13-8495-0_21

applied and practical purposes of these metagenomic studies, apart from providing a clearer vision of ecological biological diversity and ecological aspects of microbes, is to provide detailed and valuable tactics for enhancing crop production and protection against host pathogens.

Keywords
Microbe · Interaction · Metagenomics · Plant · Sequencing

21.1 Introduction

Microbes are present ubiquitously in the environment, from hydrothermal vents to the human intestine. They are tiny invaders that can work wonders and, at the same time, harm us. These astounding adventurers are the fundamental unit of life on the globe, not only supporting life but also maintaining it. Although they are present everywhere in the world, only a diminutive portion of them is known. Enormous numbers of microorganisms have been isolated, but the functions of many of them are still not known. To decipher microbes in detail, isolation of microbes is followed by their identification using molecular technology. With advancements in molecular techniques such as polymerase chain reaction (PCR) and metagenomics, this has become straightforward to some extent. Hence, the microbiome can be studied (Gilbert et al. 2010; Turnbaugh et al. 2007).

From the human body to plants, all are inhabited by microbes comprising viruses, fungi, bacteria, etc. Their association with the host plays a fundamental part in their health, growth, and development. Two types of plant–microbiome alliance benefit plants. First, highly specific interactions indicate an alliance of significant precision, developing in a symbiotic environment. The second type is commensalism, an alliance between two organisms or microorganisms where the pathogen gains a benefit while the host neither gains anything nor is harmed by the interaction. Secretion of nutrients from plants by microbes (fungi and bacteria) during their growth in concurrence with the roots imparts no observable advantage to the plant. The microbiome corresponding to a plant is considered its second genome. Each single environment related to the plant (the endosphere, phyllosphere, and rhizosphere) exhibits a particular microbial association with a particular function. Molecular interpretation by implementation of technologies such as next-generation sequencing (NGS) reveals that with use of current methods, only a miniscule portion—namely, 5%—of microbes have been cultured, while the vast majority of microbes and their functions remain unknown (Mendes et al. 2013).

The earth is rich with an enormous variety of microbes and plants, and is greatly influenced by their interactions. The plant–microbiome interaction is diverse and can be good, bad, or neutral. A good plant–microbiome interaction results in a symbiotic relationship, whereas a bad interaction can lead to negative consequences. The interaction between plants and microbes is a key determinant of plant fitness and yield, and has attracted considerable attention lately. Highly compacted soil is usually colonized by a massive number of microorganisms, which can be advantageous or malignant (Berendsen et al. 2012).

Researchers have found that the population of microorganisms is much higher than that of the plant cells. Soil contains many beneficial mycorrhizal microorganisms, which have a symbiotic relationship with the roots of plants by exchange of nutrients and, furthermore, nitrogen fixation. In contrast to mycorrhizal microorganisms, there are also enormous numbers of pathogens that affect plant machinery. Therefore, to counterattack, plants have developed defense mechanisms to manage such exposure. Molecular evidence indicates that nearly 700 million years ago, the alliance of green algae with mycorrhizal fungi was crucial in the development of terrestrial plants. With the exceptions of *Arabidopsis thaliana* and various members of the Brassicaceae family, the majority of plants have retained this symbiotic relationship, which facilitates uptake of mineral nutrients (phosphate, etc.) via the roots. Microbes colonizing plants also play a significant role in biogeochemical cycles (Philippot et al. 2009).

In the rhizosphere region, around 5–20% of photosynthates (products of photosynthesis) is liberated. Furthermore, every year, each plant liberates 500 g of isoprene and 100 g of methanol into the surroundings. In the case of methanol, this represents between 0.016% and 0.14% of photosynthetic products (photosynthates), mainly based on the type of plant. For microbes, both are prospective reservoirs of carbon and energy. Notably, plants in agricultural soils trigger microbial denitrification and methane formation, thus promoting release of nitrogen oxide (N_2O) and methane (CH_4). These gases contribute to the greenhouse effect (Wrage et al. 2001; Conrad et al. 2006).

The plant–microbiome interaction is a systemic interaction. In soil, plants excrete massive amounts of substances (exudates)—gums, saps, etc. Thus, the first step in the interaction is identification of such exudates by microbes. There is an assumption that plants are capable of acquiring microbes through plant exudates such as carbohydrates and amino acids. These plant exudates can differ according to the plant type and its biotic or physical factors. Berg et al. (2016) reported that different plants have different and specific microbial profiles. In a comparison of the rhizosphere inhabitation of two therapeutic plants—babuna (*Matricaria chamomilla*, commonly known as chamomile) and nightshade (*Solanum distichum*)—it was observed that despite being cultivated under the same conditions, they had nonidentical structural and functional microbial profiles. Furthermore, exudates from the same plant differ according to plant developmental phases determining specific microbial agglomerations (Chaparro et al. 2013).

So far, a few compounds (plant exudates) responsible for particular interactions, such as flavonoids in pea rhizobia and strigolactones as signaling molecules for fungi—namely, arbuscular mycorrhizal fungi (AMFs)—have been identified by scientists. A model for microbial colonization was introduced by Reinhold-Hurek et al. (2015). The community of microbes occupies a wide range and is affected by the type of soil and other environmental aspects. Nearer the rhizosphere (plant roots), the community is more specialized with fewer species. Only a handful of species are able to penetrate the plant roots and become established inside the plant. Moreover, the microbial community differs between the different regions of the plant after invading it (Akiyama et al. 2005).

21.2 Strategies for Deciphering the Plant Microbiome

Typical microbiology includes isolation and culture of microorganisms from nature by utilizing different culture (nutrient) media and conditions for growth based on the selected organisms. However, for particular research on physiology and genetics, an axenic (pure) culture of a specific organism is needed; techniques based on culture miss the huge amount of microbial diversification that is present in an environment. In microbial ecology, a variety of non-culture-dependent molecular approaches are used. For deciphering prokaryotes, the universal 16S ribosomal RNA (rRNA) gene PCR amplification is generally used. Sequencing of the mutable segments of this gene permits accurate identification of taxa. The use of advanced high-capacity sequencing methods has been extensively endorsed, as they permit identification of large numbers of sequences—i.e., thousands to millions in a sample—revealing a profusion of even infrequently occurring species of microbes. However, for deciphering microorganisms such as fungi (which are eukaryotic), the corresponding 18S rRNA gene probably cannot adequately elucidate taxonomic differences, so a hypervariable region (HVR) in the internal transcribed spacer is often used (Bentley et al. 2008; Margulies et al. 2005). Figure 21.1 describes a few metagenomic approaches to bacterial identification.

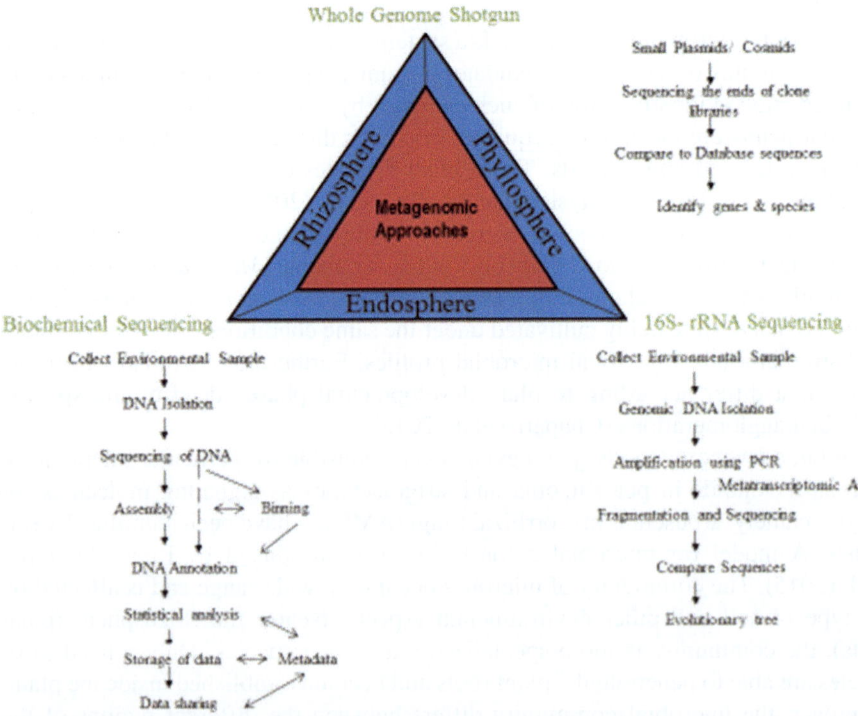

Fig. 21.1 Metagenomic approaches to phylogenetic studies

The drawback of this is that amplification of genomic DNA (gDNA) via PCR is indelibly unfair in the design of a primer and usually identifies only the organisms of interest. Complicated environments are colonized by creatures from all of life's kingdoms. Eukaryotic organisms (fungi, nematodes, protozoa, etc.) are present in soils worldwide. Some of these organisms can be crucial phytopathogens, while the rest are grazers of bacteria, whereas archaea execute vital biochemical reactions specifically in agricultural soils, such as oxidation of ammonia (NH_3) and formation of methane (methanogenesis). Microorganisms present in a community intercommunicate as well as interacting with the host plant, so it is crucial to capture the microbiome diversity to the fullest extent possible. For this purpose, application of universal examinations such as metagenomics, metaproteomics, and metatranscriptomics permits synchronous evaluation and collation of microbial populations over every single domain of life. Metagenomics can reveal the functional capacity of a microbiome, while metaproteomics and metatranscriptomics provide systematic depictions of protein richness and community-wide gene expression. Metatranscriptomics has revealed alterations at the kingdom level in the framework of crop plant rhizosphere microbiomes (Hong et al. 2009; Pinto and Raskin 2012).

21.3 The Rhizosphere Environs

The soil has a biologically active region surrounding the roots of the plant, which contains soilborne microorganisms (including fungi and bacteria), and this is referred to as the rhizosphere, where the roots are affected by biotic and abiotic characteristics. This region of the soil is well known to provide an ecological niche that is appropriate for many helpful microbes in the soil. The prolific activity of microbes in this zone aids various biotic and ecological processes that are crucial for the health of the plant. Research on intercommunications between plants and the soil microbiome within the rhizospheric region is essential for comprehension of a wide spectrum of fundamental processes—namely, the working of the ecosystem, the cycling of nutrients, and the isolation of carbon. An enormous challenge facing ecologists is to associate the variety of microbes that exist in the rhizosphere with the roles they play in the natural ecosystem. However, the most daunting task encountered by microbiologists and phytologists while deciphering these intercommunications is that most microorganisms that reside in this rhizospheric region are unable to be cultured in an artificial environment. In fact, over 99% of microbe species that exist in the soil cannot be cultured in an artificial environment. Analyses of communities of bacteria isolated from numerous environments have discovered that the proportion of culturable cells is not indicative of the richness or microbial community diversification existing in the environment; generally, it is noted that the direct microscopical enumeration exceeds the viable cell enumeration by various orders of magnitude. However, current approaches in genomics and molecular methods do offer opportunities to connect structural diversification to activities taking place in the rhizosphere (Hiltner 1904).

Various rhizospheric microbes are plant growth regulators, encouraging growth and development of seeds, while mycorrhizal fungi may offer vegetation an enhanced capacity for uptake of nutrients, higher production, and drought tolerance, and may stimulate diversification of the plant. Since microbes associated with the rhizosphere perform various metabolic abilities and play fundamental parts in the health of a plant, understanding of their community framework is crucial for comprehension of the roles played by them individually, and metagenomics has the potential to answer various significant questions about the uncultivated proportion of the rhizospheric community (Dakora 2003).

The rhizosphere zone is affected by the roots of plants by means of rhizodeposition of scraped cells, mucilage, and excretions (exudates). Several compounds are carried by exudates from the roots—mainly sugars and organic acids—but they also contain various fatty acids, hormones, amino acids, antimicrobial compounds, vitamins, and factors required for growth. The chief factors of the rhizosphere microbiome framework are root exudates. The composition of root exudates can differ among particular species of plant and cultivars and also with the growth phase and the age of the plant. Moreover, root exudates are affected by the microbiome, as plants propagated under axenic conditions have noticeably distinctive compositions. For example, *A. thaliana* was shown to have a distinct composition of root exudates and likewise specific communities of rhizospheric bacteria, while the rhizospheric bacterial communities associated with other successions demonstrated great similarity (Bertin et al. 2003).

Rhizodeposition consists of several different components apart from root exudates. Mucilage release and shedding of root cells results in accumulation of a huge quantity of substances in the rhizosphere, including polymers of plant cell walls such as pectin and cellulose. Degradation of cellulose by microbial inhabitants of the soil is extensive in soils that contain excessive quantities of organic matter. Methanol is released as a result of pectin decomposition and can be utilized by microorganisms as a source of carbon. Within the rhizosphere, active metabolism of methanol has been noted. Apart from making a carbon source available to the inhabitants of the rhizosphere, the roots of the plant also provide a substrate to anchor microorganisms. Therefore, there is an important overlap between bacterial attachment to roots and to a static structure of wood (Stursova et al. 2012; Haichar et al. 2007)

Research on microbiomes in the rhizospheric region has revealed phenomenal identical division of microbial phyla, while in comparison of strains and species of microorganisms, the differences between different cultivars of plants become pronounced. The samples, especially those belong to the alpha (α) and beta (β) classes, are generally dominated by Proteobacteria, while other substantial groups include Bacteroidetes, Actinobacteria, Acidobacteria, Firmicutes, Verrucomicrobia, and Planctomycetes (Inceoglu et al. 2011; Teixeira et al. 2010).

Rhizobacteria that stimulate plant growth, which function by means of several mechanisms, are of special importance in the environment of the rhizosphere. Endosymbiotic bacteria such as *Rhizobium* spp. and bacteria that fix nitrogen (N_2), including aerobic and free-living *Azotobacter* spp., offer a fixed source of nitrogen for the plant, whereas minerals containing phosphorus can be solubilized by a number of different bacteria, enhancing its bioavailability. Manipulation of phytohormones by microbes—especially gibberellins, ethylene, and auxins—may also promote growth or mitigate drought stress. Several rhizobacteria that promote growth in a plant function in opposition to phytopathogens by generating antimicrobial compounds or by interfering with virulence factors through effectors provided by type 3 secretion systems (T3SSs). In particular, actinomycetes are said to generate a vast variety of compounds that have antiviral, antibacterial, insecticidal, antifungal, and nematicidal properties. In the soil and the rhizosphere, actinomycetes are usually present as one of the most copious classes of bacteria, and they are particularly refined in communities of endophytes (Rezzonico et al. 2005).

The close relationships between root exudates and the rhizospheric composition of microbes is well established (Broeckling et al. 2008; Badri et al. 2009, 2013; Micallef and Shiaris 2009; Chaparro et al. 2012, 2013). Root exudates contain numerous chemicals that can behave as signaling molecules, substrates, etc., to coordinate alterations in the composition of microbes. Lately, it has been revealed that the framework of root exudates of *Arabidopsis* alter a gradient in development of the plant (Chaparro et al. 2013). The released quantities of sugars and sugar alcohols were much greater at first but were then reduced during the growth of the plant. On the other hand, the released quantities of phenolics and amino acids became amplified over time. It was therefore concluded that during the initial phases of development, sugars are released by root seedlings as a substrate for a vast variety of microbes, but as the plant grows, it secretes certain substrates and probably antimicrobial compounds in an attempt to select specific microbes to reside in the rhizosphere. In the region of the rhizosphere, this prospective selection of microbes as the plant grows might be connected with the possibilities of advantageous microbes inhibiting pathogenic microbes (Mendes et al. 2011), activating induced systemic tolerance (IST) to control abiotic stress (Selvakumar et al. 2012), amplifying the innate immunity of the plant (Zamioudis and Pieterse 2012), improving mineral nutrition (Bolan 1991; van der Heijden et al. 2008), and optimizing the overall health of the plant (Berendsen et al. 2012; Chaparro et al. 2012).

21.4 The Phyllosphere Environs

The region of the plant above the ground that is occupied by microbes is known as the phyllosphere, which acts as a common niche for cooperation between microorganisms and the plant. The leaf blade has been referred to as the phylloplane. Leaves are usually exposed to the airstream and dust stream, leading to establishment of particular flora with the help of waxes, appendages, and cuticles, superficially, which further aid in the enlacement of microbes. The microorganisms present in the phyllosphere may exist or multiply on leaves as determined by the area of influence of substances in leaf exudates. These leaf exudates carry fundamental nutrient components ($C_6H_{12}O_6$ (glucose), $C_{12}H_{22}O_{11}$ (sucrose), amino acids, etc.), and these particular microbial habitats may provide a niche for fixing of nitrogen and release of compounds that are able to promote plant growth. Moreover, the microorganisms that exist in the phyllospheric region may play a fundamental role in preventing plant diseases by controlling airborne pathogens. Microorganisms present on the surface of the leaf are referred to as extremophiles, as they can survive an extreme range of temperatures (5–55 °C) and exposure to ultraviolet radiation. Various microorganisms (*Pseudomonas*, *Pantoea*, *Diplococcus*, *Azotobacter*, *Xanthomonas*, *Bacillus*, etc.) have been observed in various crop plant phyllospheres (Dobrovolskya et al. 2017).

The plant's aerial surface is presumed to be quite low in nutrients in comparison with the rhizosphere. The microorganisms do not colonize the leaves uniformly; they are influenced by the structures of the leaf, such as stomata, veins, and hairs. Approximately 10^7 microorganisms per square centimeter colonize the leaf surface. The phyllosphere is a very potent environment in comparison with the rhizosphere, with the inhabitant microorganisms being influenced by large variations in temperature, radiation, and humidity throughout the day and night. Abiotic factors such as these influence the microbiome of the phyllosphere through alterations in the metabolism of the plant. In particular, precipitation and air currents are considered to cause terrestrial commutation among microorganisms inhabiting the phyllosphere. In one study, the metabolite profile of the *A. thaliana* leaf was changed interestingly by transportation of soil microorganisms to the roots; enhanced concentrations of various amino acids in the metabolome of the leaflet were matched by enhanced herbivory from insects, suggesting cross-talk between the aboveground and belowground plant parts (Lindow and Brandl 2003).

Using PCR, rRNA genes have been amplified to profile communities of bacteria and fungi present in the phyllospheric regions of different plants. It was observed that the abundance of microbes was very high in warmer climates and in humid climates as compared with temperate climates. Correspondingly, the phylum of bacteria found to be dominant was Proteobacteria (classes alpha and beta), generally with Actinobacteria and Bacteroidetes also being present. During summer, various plants' phyllospheric zones were noted to be dominated by lactic acid bacteria (LABs), e.g., firmicutes in the Mediterranean. The mode of their metabolism was suggested to permit LABs to withstand warm and parched weather conditions. However, this was not the case in other seasons. At higher levels of taxonomy of microorganisms, the microbes of several different plants' phyllospheres can be found to be identical, but clear differences are visible at the strain and species levels, considering the finely modulated metabolic adaptations needed to exist in the aforementioned environments. However, the microbiomes of the rhizospheric region are homological to the soil; some resemblance has been observed between the microbiomes of the phyllosphere and of the air (Vokou et al. 2012).

Proteogenomic analysis of several microbiomes of the phyllosphere has revealed species that absorb and digest amino acids, carbohydrates, and ammonium derived from plants, meaning that these compounds are principal sources of nitrogen (N) and carbon (C) in the phyllospheric region. Researchers have also discovered that *Methylobacterium* spp. and other methylotrophs are very prolific phyllospheric microorganisms, which dynamically absorb, digest, and metabolize methyl alcohol (CH_3OH) extracted from the pectin of the plant (Galbally and Kirstine 2002).

21.5 The Endosphere Environs

Endophytes are microorganisms that reside in internal regions of the plant (the leaf, root, stem, etc.) without affecting the host plant. The term "endophyte" is derived from Greek; "endo-" means "within" and "-phyte" means "plant." Endophytes are usually presumed to be nonpathogenic. They cause no noticeable symptoms; however, they include dormant (latent) pathogens that can cause infection in certain environmental situations and/or host genotypes. Endophytic microbes are considered a subpopulation of the rhizospheric microbiome; however, their features are different from those of rhizospheric bacteria, suggesting that not all rhizospheric bacteria can invade plants. If they enter the host, they can alter their own metabolism and thus become adjusted to the host's internal environment. Although it is usually reckoned that microbes (bacteria) isolated from plant tissues after sterilization of the surface are "endophytic," the case is different for aerial parts and root surfaces, as there are several niches on them where microbes may persist even after treatment with the chemicals that are commonly used for sterilization of surfaces. To confirm whether specific bacteria are truly endophytic or not, techniques such as confocal microscopy, transmission electron microscopy (TEM), and embedding of samples in resin are used (Compant et al. 2010; Monteiro et al. 2012; James 2000). The early researchers used 'sonication technique' to eliminate the surface layers of plant tissue, and the tissue that remained was used to characterize the endophytic microbiome. Research such as this has revealed that endophytes mainly inhabit killed or moribund cells and the intercellular apoplast; endophytes were not found to inhabit cells that were alive, unlike true symbioses such as the arrangement between rhizobia and legumes. Generally, they exist in the vessels of the xylem, where they can be translocated from the roots to other parts of the plant above the ground (Lundberg et al. 2012; Bulgarelli et al. 2012).

But this is the question: How do endophytic bacteria invade their hosts originally? The foremost evidence indicates that they invade most probably through cracks that occur naturally. Endophytic microorganisms mainly invade the host (plant) through punctures, which happen naturally as a consequence of plant development or via root hairs and epidermal juxtapositions. Endophytic microbes can be spread mainly via two pathways: horizontal or vertical. The endophytic microbiome may be altered by factors such as different environmental influences, the plant development phase, and the physicochemical soil structure (Lian et al. 2008; Mitter et al. 2017). The chief route taken by endophytic bacteria for colonization appears to be the rhizospheric environment. Endophytes arrive in the rhizospheric region via hemotaxis in the direction of elements of root exudates, proceeding by attachment. The elements that have been shown to play roles in this attachment are exopolysaccharides and lipopolysaccharides; they help to attach endophytic bacteria to the tissues of the plant. The favored attachment site following entry is the root zone (apical) with a layer of (thin-walled) surface root—for instance, the zone of cell elongation and the zone of root hairs, with little cracks produced by the development of lateral roots. Furthermore, the zone of differentiation and the intercellular spaces in the epidermis and the root regions have also been suggested to be favored locations for colonization by microbes. Wounds, crevices in roots caused by arthropods, and sites for development of lateral roots are usually presumed to be major entry points for penetration by microbes. Cellulytic enzymes such as endopolygalacturonases and endoglucanases need to be produced by the bacteria to hydrolyze exothermal walls for penetration (Suman et al. 2016).

The age of the plant is inversely proportional to its bacterial concentration, meaning that younger plants have higher concentrations of bacteria than more mature plants. In addition, the endophytic bacterial concentration is lower than the epiphytic concentration. Analysis of the metagenome (Sessitsch et al. 2012) and transcripts of *nifH* and 16S rRNA analysis (culture-independent approaches) have shown enormous endophytic diversification in staple food crops such as rice and sugarcane. Lately, 16S rRNA high-throughput sequencing has been used to describe the vital endophytic microbiome of *A. thaliana* (Fischer et al. 2012).

21.6 Metatranscriptomic Analysis

Given the trends in ongoing research, it appears probable that metagenomic data sets will continue to increase quickly and will soon dominate the data sets of complete genome sequences obtained from cultured microorganisms. However, such data sets provide information only on genomic matter; there is no apparent hint of dynamic expression or expression of genes. Although techniques such as quantitative PCR (qPCR) may be used on natural samples for quantification of gene expression, these are finite and are normally used to quantify small numbers of known genes. With the accomplishment of more than 134 metagenomic sequences, the examination of universal alterations in expression of genes, known as transcriptomics, is a progressively interesting mechanism for analyzing the molecular motives of metabolic and ecological features (Liu and Zhu 2005). The techniques used for metagenome gene expression analysis are discussed in Sects. 6.1–6.5.

21.6.1 Differential Display Polymerase Chain Reaction

Differential display PCR (DD-PCR), or differential display reverse transcription PCR (DDRT-PCR), is a technique, fully based on PCR, that permits comparison of several samples of RNA at the same time and further aids identification of both induced and suppressed genes. This technique involves two fundamental steps: (1) construction of a complementary DNA (cDNA) library for every single sample of RNA isolated from different communities, along with a degenerated set, requiring reverse transcription by anchoring of oligodeoxythymidine nucleotides to the end; and (2) amplification by PCR of partial sequences chosen randomly from the library of cDNA with an authentic anchored deoxythymidine (dT) primer and an arbitrary primer (upstream). DDRT-PCR is carried out using the same sets of primers on different cell populations (Liang and Pardee 1992). The basis of this approach is to compare the pools of messenger RNA (mRNA) isolated from microbes grown under different conditions, with subsequent reverse transcription and amplification by PCR at random sites, following by sequencing. This methodology was previously employed for enrichment of genes with a preferred microbe, observing their induced expression when microorganisms were exposed to controlled conditions, then it started being used on total RNA straightforwardly extracted from samples collected from the environment (Fleming et al. 1998; Aneja et al. 2004; Sharma et al. 2004). In this metagenomic sphere, the current examples involve the invention of a novel operon for degradation of 2,4-dinitrophenol (Walters et al. 2001), and, in mixed cultures, genes for cyclohexanone monooxygenase (Brzostowicz et al. 2003). Therefore, DDRT offers an effective strategy for deciphering the expression of genes in microbes that are present in the environment, independently from sequence understanding and without culturing. The major limitation of this approach originates from the information that there is no transcript signal present globally in bacteria that permits homogeneous amplification of total mRNA (Vieites et al. 2009).

21.6.2 cDNA Amplified Fragment Length Polymorphism

cDNA amplified fragment length polymorphism (cDNA-AFLP) is another valuable advanced PCR technique in which primers (random hexamers) are used to synthesize cDNA from total RNA (Egert et al. 2006). Two restriction endonucleases are used to digest the fragments obtained; generally a 4- or 6-bp-long cutter is used, and then the ends of the fragment adaptors are ligated. The amplified products are separated by electrophoresis, and the lengths of the fragments obtained are approximately 100–400 bp. Band intensity differences can be visualized and thus allow evaluation of comparable differences in degrees of gene expression. Identification of the corresponding whole-length cDNA is generally required for further evaluation of particular transcripts of interest. Although this technique has the ability to connect the microbial encoding capacity with a function in the environment, its relevance to an approach such as metagenomics is finite so far, as the rRNA stability is low and the few examples have been confined basically to intestinal samples (Egert et al. 2006).

21.6.3 Suppression Subtractive Hybridization

Suppression subtractive hybridization (SSH) is a technique used extensively for DNA molecule separation, differentiating two samples of DNA that are closely related. There are two prime applications of SSH: (1) subtraction of cDNA and (2) subtraction of genomic DNA (Rebrikov et al. 2004). In practice, for production of either subtracted cDNA or gDNA libraries, SSH is a highly effective and accepted technique. This technique is based on a PCR suppression effect and combines normalization and subtraction in a single process. This combination works in the following manner: the normalization step equalizes the concentrations of fragments of

DNA within the selected population, while the subtraction step scoops out the repeated sequences that exist in the compared populations. This dramatically increases the chances of obtaining less abundant differentially manifested cDNA or genomic DNA fragments, and it makes analysis of the subtracted library easier (Rebrikov et al. 2004). In an ingenious study, researchers employed this method amalgamated with a metagenomic strategy to discover unexpectedly large differences in the community structures of archaea in the rumen microorganism communities of two bulls fed similar diets and accommodated together, which would have been quite difficult to identify by applying other standard techniques (Galbraith et al. 2004).

Subtracted libraries of cDNA to identify genes expressed differentially among samples collected from the environment can be produced by applying the SSH technique. This strategy will lead to the separation of exclusive novel niches and pathways of active metabolism (Rebrikov et al. 2004). The following steps are used to create subtractive libraries: (1) isolation of mRNA from various comparable samples, (2) generation of cDNA, and (3) subtraction of cDNA populations. Preliminary examination disclosed that metagenomic data of 1–2 giga base pairs of polluted versus ancient sites are transformed into 30–200 SSH clones of approximately 20,000 bp each, i.e., 0.001% subtractive clones. The escaping DNA fragments are subtracted and may be cloned to compose short libraries of SSH, supplying a surplus of gene targets active in opposition to pollutants in a manner completely independent from coinciding roles at ancient and pollutant sites. Thereby, cDNA is prepared, for instance, for further subtraction to isolate snippets parallel to genes whose level of expression is enhanced. Here, the samples collected from the polluted sites were referred to as the "tester" while the ancient samples were referred to as the "driver" (Vieites et al. 2009).

21.6.4 Catalyzed Reporter Deposition–Fluorescence In Situ Hybridization

Catalyzed reporter deposition–fluorescence in situ hybridization (CARD-FISH) is another powerful technique for qualitative evaluation of gene activity in vivo. However, this methodology is applicable only for quantification of transcripts of genes that have already been studied. The actual sequence of the gene must be known to construct the probe for this particular tool. Therefore, this limits the usage of this methodology to the study of metagenomes based on activity, as in most cases, one must work with unfamiliar genes and should make efforts to discover new roles and activities rather than working on already studied genes (Vieites et al. 2009).

21.6.5 DNA Microarray

The DNA microarray is one of the most powerful technologies and has an immense capacity to elucidate the meaning of microbial systems. It is a technology developed by Stephen Fodor in the late 1980s and is also known as the biochip or DNA chip. Genomic technology based on the biochip is a robust tool for observing a large number of gene expressions in a single experiment simultaneously (Hoheisel 2006). At the beginning, this technology was intended for characterizing an individual species transcriptionally, but it has been applied to environmental usage in recent years (Zhou and Thompson 2002, 2004; Adamczyk et al. 2003; El Fantroussi et al. 2003; Taroncher-Oldenburg et al. 2003; Zhou 2003; Loy et al. 2004; Tiquia et al. 2004; Bodrossy et al. 2006; An and Parsek 2007). One of the biggest challenges in employing the DNA chip for examining samples collected from the environment is the low detection sensitivity of hybridization based on the microarray, in combination with the small amount of biomass frequently present in environmental setting samples. DNA chips for expression characterization can be split into two broad groups: (1) those based on deposition of precompiled DNA probes and (2) those based on oligonucleotide probes synthesized in situ. An example of an oligonucleotide probe is the Affymetrix array. A lot of applications use DNA microarrays involving, for instance, profiling of microbe communities isolated from environmental samples such as water and soil (Zhou 2003; Eyers et al. 2004), detection of pathogens in clinical samples and those isolated from the field (Bodrossy and Sessitsch 2004), and checking of food and water for contamination by bacteria (Lemarchand et al. 2004). To decipher the diversity of microbes in different environments, several varieties of DNA microarray have been employed—for instance, those involving oligonucleotides made up of 20–70 bp (Ward et al. 2007), fragments of DNA (cDNA) amplified by PCR (Wu et al. 2004), and complete genome DNA. To date, meta DNA chip research has been used to observe gene expression worldwide in more than 20 distinct environments, covering a massive area of research on diversity (Bae et al. 2005).

The microarrays used to outline the libraries of the metagenome may offer a constructive proposal for quick characterization of numerous clones. For instance, a fosmid library was procured and further arranged on a glass slide (Sebat et al. 2003). This process is known as a metagenome microarray (MGA). In this particular format, the notions of the "probe" and "target" are the inverse of those in common cDNA and oligonucleotide microarrays. Here, the fosmid clones referred to as "targets" are found on the slide and a particular gene probe is tagged and employed for hybridization. This biochip format may offer a worthwhile approach to screening of metagenomes for identifying clones quickly from libraries of metagenomes without the necessity for use of tedious methods for screening of several target genes. Researchers (Sebat et al. 2003; Park et al. 2008) have employed this biochip program to screen metagenome libraries with complete genomes of microbes and genomes of communities. To assess eukaryotic soil microbe communities' functional diversification, an experimental strategy was evaluated by Bailly et al. (2007) on the basis of building and screening a library of cDNA from a metatranscriptome by utilizing forest soil–extracted polyadenylated mRNA. The variety of organisms was analyzed by sequencing of a segment of rRNA genes (18S) and cDNA. The evaluation of the metatranscriptome revealed that the taxonomic division did not match; it was estimated that the soil samples might contain more than 180 species and 70% of the sequences were connected to protists and fungi. DNA-based biochip identification strategies integrated with the complete community of an amplified genome have been used to examine the structures of low-biomass groundwater microbial communities (Wu et al. 2006). However, this strategy could not be adopted for conventional use, and is used for activity examination based on mRNA. One problem in detection of mRNAs isolated from environmental samples by using biochip hybridization is getting an adequate quantity of mRNAs for evaluation. Prior to hybridization, a few types of amplification signal are required. Nevertheless, amplification based on random PCR is not a suitable option, because of amplification bias and consequent loss of quantitative data (Nygaard and Hovig 2006). Furthermore, the gene-after-gene feature of conventional PCR strongly limits the benefits of functional gene analysis by microarray use. To resolve this issue, a brand new technique called whole-community RNA amplification (WCRA) has been devised for amplifying a complete community of RNAs randomly to provide an adequate quantity of mRNAs isolated from environmental samples for microarray analysis (Gao et al. 2006).

The mRNA half-life is short, which leads to one of the major complications associated with the microarray (Selinger et al. 2003; Andersson et al. 2006), and in archaea and bacteria the mRNA represents only a small part of the complete RNA. Lately, various methods have been evolved for solving these challenges. It is quite a daunting task to decipher the expression of genes using a chip of a DNA sample isolated from an environment. First of all, in cDNA microarrays based on PCR, sensitivity may sometimes be an issue, as only genes from populations contributing more than 5% of the DNA community can be detected. Secondly, the samples may carry several contaminants from the environment that alter the RNA quality and hybridization of the DNA (Zhou and Thompson 2002); hence, extraction of undegraded mRNA becomes quite difficult (Bürgmann et al. 2003). The specificity of the extraction procedure plays a fundamental part and should differ according to the sampling location, as there needs to be enough differentiation between the probes. In addition, annotation and extensive protein functional characterization remain difficult and error-prone procedures, as systems microbiology depends greatly on the overall knowledge of gene product functions (Morrison et al. 2006).

21.7 Host Effect on the Plant Microbiome

The communications between a plant and the microbes surrounding it are extremely powerful and complicated. Remarkably, the plant's immune system is thought to make a significant contribution in characterizing the microbiome structure of the plant. *A. thaliana* mutants lacking an innate immune response called systemic acquired resistance (SAR) manifested variations in the formation of the bacterial community in the rhizospheric region in comparison with the wild type, while SAR activated chemically did not effect a notable change in the bacterial community of the rhizospheric region. Furthermore, in the phyllospheric region of *A. thaliana*, the variety of endophytes was reduced by induction of salicylic acid intermediary resistance, while plants lacking defense mediated via jasmonate revealed greater epiphytic variety. The authors of this study suggested that the outcomes of plant resistance procedures in the microbiome were inconsistent and for restraining a few bacterial communities, SAR was responsive (Kniskern et al. 2007).

Among various plant-related bacteria, especially rhizobia, the production of phytohormones such as indole-3-acetic acid (IAA) occurs worldwide, while other phytohormone gibberellins can be produced only by some species of *Bacillus*. Interference with the signaling of jasmonate and ethylene by hormone analogs produced by *Pseudomonas syringae* results in the opening of stomata and entry of pathogens. It has been reported that these bacteria can degrade hormones, as well as their precursors. For instance, plant ethylene signaling can be inhibited by microbial deamination of 1-aminocyclopropane-1-carboxylic acid (ACC), resulting in high plant tolerance of environmental stress (Glick 2005).

A few chemical signals released by plants promote particular interactions, the majority of which are identified from variant organisms. For example, flavonoids activate multiple reactions in root pathogens, mycorrhizae, rhizobia, and different plants. Furthermore, strigolactones stimulate branching of hyphae in the case of mycorrhizal fungi and foster germination of seed in parasitic plants. However, not many genes of plants and pathways contribute to the formation of multiple interactions with distinct microbes; examples include the evolutionary pathways that are divided between mycorrhizal symbiosis and infection caused by oomycetes and rhizobial symbiosis and infection caused by nematodes. It is still unknown how these pathways are communicated with other members of the microbiome and also whether they are able to interact or not (Damiani et al. 2012).

An extensive variety of antimicrobial compounds is produced by plants both constitutively and in response to disease-causing microbes or viruses. The plant kingdom has a variety of compounds—such as alkaloids, phenolics, and terpenoids—that are present worldwide, while the rest are limited to specific groups; glucosinolates, for instance, are produced by members of the Brassicales order and by *Arabidopsis*. An exogenous glucosinolate produced by transgenic *Arabidopsis* further changes the communities of fungi and bacteria in the rhizospheric region and root tissue. The *Avena strigosa* species, whose seeds are consumable and are commonly known as oats, produces avenacins, which are triterpenoid saponins that protect the plants against fungal pathogens. Mutants of oat deficient in avenacins are more sensitive to fungal pathogens and have distinct communities of culturable fungi colonizing their roots in comparison with wild-type oats that have the same genotypes. Unexpectedly, though, a recent universal study of the microbes colonizing the rhizosphere of the above two genotypes observed small differences between the fungal communities. Amoebozoa and Alveolata, which belong to the Eukaryota domain, were severely affected in the mutant by the deficiency of avenacins, while bacterial communities remained unaffected. This shows how a minor alteration in the genotype of a plant can have complicated and unnoticed impacts on the plant microbiome. Other research found no remarkable variations in microbes colonizing the rhizospheric region between normal (wild-type) maize and maize modified genetically for production of an insecticidal toxin by a bacteria known as *Bacillus thuringiensis* (Bt for short; thus, the toxin is called Bt toxin), although the fact that it was insecticidal could have been the reason for the lack of significant differences.

Moreover, in the case of wheat, when the gene *pm3b*is was introduced in the rhizospheric region, it conferred resistance to molds but had only negligible impacts on pseudomonads and mycorrhizal fungal colonies. Resistance to disease, involving production of antimicrobial compounds, is a characteristic introduced as an outcome of genetic manipulation or molecular breeding in an attempt to manage disease. These changes may or may not influence the inhabitants of the microbiome, possibility with unnoticed impacts on the plant, and should be evaluated individually. It should be noted that the yields of disease resistance genes are usually unspecified (Meyer et al. 2013).

21.8 Interplay of Microbial Complexity and Metagenomics

Evolution results in microbiome complexity, as well as acquisition of some beneficial characteristics. The universal consequence of microbial metabolic approaches is an amalgamation of interactions with universal importance on very miniscule scales. To maintain life on this planet, two types of interactions are necessary to attain biogeochemical cycles. The first type is microbiological interaction and the second type is chemical interdependence. Associations of microbes (i.e., microbial communities that interact and operate in an alliance) achieve more than the same organisms can achieve individually (Lozupone and knight 2008). Coming to the second type of interaction—chemical interdependence—a sequence of interactions varying from obligate to nominal is considered to occur between representatives of microbial communities. In any instance, microbial communities where the representatives communicate are different from microbial assemblages where the representatives merely coexist. If we see the issue of massive diversification of worldwide microbial species, the issue was raised nearly 100 years ago as to whether particular microorganism species are found worldwide or are restricted to a few geographical areas. The current understanding suggests that although a vast proportion of microbes are not culturable, they are confined to a particular habitation and geographical position, although a few completely cosmopolitan organisms do exist, such as deep-sea marine group I archaea (DeLong 2006) and the marine obligate, hydrocarbon-degrading Gammaproteobacteria, e.g., Alcanivorax. Advanced technologies have become accessible; moreover, studies may reveal that more microorganisms are present worldwide than previously realized (Yakimov et al. 2007).

It is important to note that the proportional richness of a few groups of microbes is not linked on a mandatory basis to the significance of that group in the operation of that community. In a group, ordinary organisms may not inevitably perform a crucial function despite being present in large numbers, while organisms that constitute only 0.1% of the group (such as nitrogen fixers) may be of central importance (Dinsdale et al. 2008). Efficient characterization of this central diversification will impart a novel perception of metabolic functions and mutuality (dependency on each other) underlying microbial existence, and the function of each and every organism present in ecosystems. In this situation, it is important to study microbial and enzymatic complements in various niches and how they negotiate the functioning of the community. Additionally, ongoing and subsequent systems microbiology approaches can impart a perspective allowing us to comprehend the complicated characteristics of microbial communities, their dynamics, and their influences on naturalistic channels. Systems approaches to microbial profiles could resolve basic queries regarding environmental microbiology, such as which organisms are present and what their activities are. For such an evaluation, there are few steps to follow. First of all, it is important to identify the community members under investigation and also their relationships with their host. Nevertheless, as previously stated, only a few microbes are easily culturable. To decipher such microbial communities without culturing every single microbe in spite of their involution and commutability, an approach has been developed with advancements in molecular techniques, now popularly known as "metagenomics" or "environmental genomics" (Ferrer et al. 2008). The word "metagenomics" describes the study of DNA or genomic material recovered from the environment (Tringe et al. 2005). The fundamental purposes of sequence-based metagenomics are to reconstruct the metabolism of the life-forms (i.e., organisms) forming the community and to envisage their functional contributions to the biological community (ecosystem). A comprehensive analysis of genomic statistics may be concatenated with analysis of the expression of genes, usually known as a transcriptome, to identify the genes and their characteristics by utilizing a set of DNA sequences (Cavalieri and Grosu 2004; Ferrer et al. 2008) or for probing a particular bacterium. Nevertheless, despite belonging to similar species, organisms have sequence commutability of genes, and incompleteness of genomic data, due to the subtlety of communities, is common in application of this technique. Advances in the analytical technique of mass spectrometry have played a fundamental role in solution of this problem, enabling extensive developments in protein study (proteomics) and metabolomics/interactomics (Urisman et al. 2005).

21.9 Conclusions and Future Prospects

The associate host microbiome can be considered a secondary genome or an extension of the host genome. The host microbiome associated with the phyllosphere (above ground), the rhizosphere (below ground), and the endosphere (internal) tissues of the same host plant are very dissimilar. Interestingly, the microbiome that occupies the same place (niche) in a variety of plants can also differ, especially when we view the microbiome from different taxonomic angles such as the genus, species, and different strains. This is where specific metabolic capabilities are required to use host-derived carbon sources and tolerate host defenses. Abiotic conditions such as the temperature, moisture, and pH have broad effects on the microbiome directly and indirectly through the host.

Our insights into plant–microbe interactions will be enhanced by understanding of how numerous microbiomes synergize with each other and also with their host plants. Next-generation sequencing (NGS) has the capabilities to carry out such research work, but it requires improvement in computational technology to speed up the process of discovery. The progress and advancements in manifold strategies to generate such progress indicate that NGS will become the foremost tool for exploration of phytomicrobiomes. These strategies include increased computing speed, condensing of NGS sequence data sets, enrichment of microbial sequences, data simplification of known sequences, and changes in Basic Local Alignment Search Tool (BLAST) searches. The use of e-probes in BLAST searches creates possibilities to help examine or explore multiple microbe interactions with each other and also with the host plant.

Therefore, it is clear that understanding of molecular details and insights into multimicrobe communications with host plants will necessitate experiments on model systems with identified microbial combinations in greenhouses and plant growth chambers. Conversely, understanding of which multimicrobe–plant synergism requires examination and exploration can be expedited using metagenomic tactics that compare explicit sets of microbiomes with developmental and physiological phenotypes in naturally growing plants or crops grown under field conditions.

References

Adamczyk J, Hesselsoe M, Iversen N, Horn M, Lehner A, Nielsen PH, Schloter M, Roslev P, Wagner M (2003) The isotope array, a new tool that employs substrate-mediated labelling of rRNA for determination of microbial community structure and function. Appl Environ Microbiol 69:6875–6887

Akiyama K, Matsuzaki K, Hayashi H (2005) Plant sesquiterpenes induce hyphal branching in arbuscular mycorrhizal fungi. Nature 435:824–827

An D, Parsek MR (2007) The promise and peril of transcriptional profiling in biofilm communities. Curr Opin Microbiol 3:292–296

Andersson AF, Lundgren M, Eriksson S, Rosenlund M, Bernander R, Nilsson P (2006) Global analysis of mRNA stability in the archaeon *Sulfolobus*. Genome Biol 7(10):R99. https://doi.org/10.1186/gb-2006-7-10-r99

Aneja MK, Sharma S, Mayer J, Schloter M, Munch JC (2004) RNA fingerprinting—a new method to screen for differences in plant litter degrading microbial communities. J Microbiol Methods 59:223–231

Badri DV, Quintana N, El Kassis EG, Kim HK, Choi YH, Sugiyama A, Verpoorte R, Martinoia E, Manter DK, Vivanco JM (2009) An ABC transporter mutation alters root exudation of phytochemicals that provoke an overhaul of natural soil microbiota. Plant Physiol 151:2006–2017

Badri DV, Chaparro JM, Zhang R, Shen Q, Vivanco JM (2013) Application of natural blends of phytochemicals derived from the root exudates of *Arabidopsis* to the soil reveal that phenolic-related compounds predominantly modulate the soil microbiome. J Biol Chem 288:4502–4512

Bae JW, Rhee SK, Park JR, Chung WH, Nam YD, Lee I, Kim H, Park YH (2005) Development and evaluation of genome probing microarrays for monitoring lactic acid bacteria. Appl Environ Microbiol 71:8825–8835

Bailly J, Fraissinet-Tachet L, Verner MC, Debaud JC, Lemaire M, Wésolowski-Louvel M, Marmeisse R (2007) Soil eukaryotic functional diversity, a metatranscriptomic approach. ISME J 7:632–642

Bentley DR, Balasubramanian S, Swerdlow HP, Smith GP, Milton J, Brown CG, Hall KP, Evers DJ, Barnes CL, Bignell HR, Boutell JM, Bryant J, Carter RJ, Keira Cheetham R, Cox AJ, Ellis DJ, Flatbush MR, Gormley NA, Humphray SJ, Irving LJ, Karbelashvili MS, Kirk SM, Li H, Liu X, Maisinger KS, Murray LJ, Obradovic B, Ost T, Parkinson ML, Pratt MR, Rasolonjatovo IM, Reed MT, Rigatti R, Rodighiero C, Ross MT, Sabot A, Sankar SV, Scally A, Schroth GP, Smith ME, Smith VP, Spiridou A, Torrance PE, Tzonev SS, Vermaas EH, Walter K, Wu X, Zhang L, Alam MD, Anastasi C, Aniebo IC, Bailey DM, Bancarz IR, Banerjee S, Barbour SG,

Baybayan PA, Benoit VA, Benson KF, Bevis C, Black PJ, Boodhun A, Brennan JS, Bridgham JA, Brown RC, Brown AA, Buermann DH, Bundu AA, Burrows JC, Carter NP, Castillo N, Chiara E Catenazzi M, Chang S, Neil Cooley R, Crake NR, Dada OO, Diakoumakos KD, Dominguez-Fernandez B, Earnshaw DJ, Egbujor UC, Elmore DW, Etchin SS, Ewan MR, Fedurco M, Fraser LJ, Fuentes Fajardo KV, Scott Furey W, George D, Gietzen KJ, Goddard CP, Golda GS, Granieri PA, Green DE, Gustafson DL, Hansen NF, Harnish K, Haudenschild CD, Heyer NI, Hims MM, Ho JT, Horgan AM, Hoschler K, Hurwitz S, Ivanov DV, Johnson MQ, James T, Huw Jones TA, Kang GD, Kerelska TH, Kersey AD, Khrebtukova I, Kindwall AP, Kingsbury Z, Kokko-Gonzales PI, Kumar A, Laurent MA, Lawley CT, Lee SE, Lee X, Liao AK, Loch JA, Lok M, Luo S, Mammen RM, Martin JW, McCauley PG, McNitt P, Mehta P, Moon KW, Mullens JW, Newington T, Ning Z, Ling Ng B, Novo SM, O'Neill MJ, Osborne MA, Osnowski A, Ostadan O, Paraschos LL, Pickering L, Pike AC, Pike AC, Chris Pinkard D, Pliskin DP, Podhasky J, Quijano VJ, Raczy C, Rae VH, Rawlings SR, Chiva Rodriguez A, Roe PM, Rogers J, Rogert Bacigalupo MC, Romanov N, Romieu A, Roth RK, Rourke NJ, Ruediger ST, Rusman E, Sanches-Kuiper RM, Schenker MR, Seoane JM, Shaw RJ, Shiver MK, Short SW, Sizto NL, Sluis JP, Smith MA, Ernest Sohna Sohna J, Spence EJ, Stevens K, Sutton N, Szajkowski L, Tregidgo CL, Turcatti G, Vandevondele S, Verhovsky Y, Virk SM, Wakelin S, Walcott GC, Wang J, Worsley GJ, Yan J, Yau L, Zuerlein M, Rogers J, Mullikin JC, Hurles ME, McCooke NJ, West JS, Oaks FL, Lundberg PL, Klenerman D, Durbin R, Smith AJ (2008) Accurate whole human genome sequencing using reversible terminator chemistry. Nature 456:53–59

Berendsen RL, Pieterse CMJ, Bakker P (2012) The rhizosphere microbiome and plant health. Trends Plant Sci 17:478–486

Berg G, Rybakova D, Grube M, Koberl M (2016) The plant microbiome explored: implications for experimental botany. J Exp Bot 67:995–1002

Bertin C, Yang XH, Weston LA (2003) The role of root exudates and allelochemicals in the rhizosphere. Plant Soil 256:67–83

Bodrossy L, Sessitsch A (2004) Oligonucleotide microarrays in microbial diagnostics. Curr Opin Microbiol 7:245–254

Bodrossy L, Stralis-Pavese N, Konrad-Köszler M, Weilharter A, Reichenauer TG, Schöfer D, Sessitsch A (2006) mRNA-based parallel detection of active methanotroph populations by use of a diagnostic microarray. Appl Environ Microbiol 72:1672–1676

Bolan NS (1991) A critical review on the role of mycorrhizal fungi in the uptake of phosphorus by plants. Plant Soil 134:189–207

Broeckling CD, Broz AK, Bergelson J, Manter DK, Vivanco JM (2008) Root exudates regulate soil fungal community composition and diversity. Appl Environ Microbiol 74:738–744

Brzostowicz PC, Walters DM, Thomas SM, Nagarajan V, Rouvière PE (2003) mRNA differential display in a microbial enrichment culture: simultaneous identification of three cyclohexanone monooxygenases from three species. Appl Environ Microbiol 69:334–342

Bulgarelli D, Rott M, Schlaeppi K, Ver Loren van Themaat E, Ahmadinejad N, Assenza F, Rauf P, Huettel B, Reinhardt R, Schmelzer E, Peplies J, Gloeckner FO, Amann R, Eickhorst T, Schulze-Lefert P (2012) Revealing structure and assembly cues for *Arabidopsis* root–inhabiting bacterial microbiota. Nature 488:91–95

Bürgmann H, Widmer F, Sigler WV, Zeyer J (2003) mRNA extraction and reverse transcription-PCR protocol for detection of nifH gene expression by *Azotobacter vinelandii* in soil. Appl Environ Microbiol 69:1928–1935

Cavalieri D, Grosu P (2004) Integrating whole-genome expression results into metabolic networks with pathway processor. Curr Protoc Bioinformatics. Chapter 7: Unit 7.6. https://doi.org/10.1002/0471250953.bi0706s05

Chaparro JM, Sheflin AM, Manter DK, Vivanco JM (2012) Manipulating the soil microbiome to increase soil health and plant fertility. Biol Fertil Soils 48:489–499

Chaparro JM, Badri DV, Bakker MG, Sugiyama A, Manter DK, Vivanco JM (2013) Root exudation of phytochemicals in *Arabidopsis* follows specific patterns that are developmentally programmed and correlate with soil microbial functions. PLoS One 8:10

Compant S, Clement C, Sessitsch A (2010) Plant growth–promoting bacteria in the rhizo and endosphere of plants: their role, colonization, mechanisms involved and prospects for utilization. Soil Biol Biochem 42:669–678

Conrad R, Erkel C, Liesack W (2006) Rice cluster I methanogens, an important group of Archaea producing greenhouse gas in soil. Curr Opin Biotechnol 17:262–267

Dakora FD (2003) Defining new roles for plant and rhizobial molecules in sole and mixed plant cultures involving symbiotic legumes. New Phytol 158:39–49

Damiani I, Baldacci-Cresp F, Hopkins J, Andrio E, Balzergue S, Lecomte P, Puppo A, Abad P, Favery B, Herouart D (2012) Plant genes involved in harbouring symbiotic rhizobia or pathogenic nematodes. New Phytol 194:511–522

DeLong EF (2006) Archaeal mysteries of the deep revealed. Proc Natl Acad Sci U S A 103:6417–6418

Dinsdale EA, Edwards RA, Hall D, Angly F, Breitbart M, Brulc JM, Furlan M, Desnues C, Haynes M, Li L, McDaniel L, Moran MA, Nelson KE, Nilsson C, Olson R, Paul J, Brito BR, Ruan Y, Swan BK, Stevens R, Valentine DL, Thurber RV, Wegley L, White BA, Rohwer F (2008) Functional metagenomic profiling of nine biomes. Nature 452:629–632

Dobrovolskaya T, Khusnetdinova K, Manucharova N, Golovchenko A (2017) Structure of epiphytic bacterial communities of weeds. Microbiology 86(2):257–263

Egert M, de Graaf AA, Smidt H, de Vos WM, Venema K (2006) Beyond diversity: functional microbiomics of the human colon. Trends Microbiol 14:86–91

El Fantroussi S, Urakawa H, Bernhard AE, Kelly JJ, Noble PA, Smidt H, Yershov GM, Stahl DA (2003) Direct profiling of environmental microbial populations by thermal dissociation analysis of native rRNAs hybridized to oligonucleotide microarrays. Appl Environ Microbiol 69:2377–2382

Eyers L, George I, Schuler L, Stenuit B, Agathos SN, Fantroussi SEI (2004) Environmental genomics: exploring the unmined richness of microbes to degrade xenobiotics. Appl Environ Microbiol 66:123–130

Ferrer M, Beloqui A, Timmis KN, Golyshin PN (2008) Metagenomics for mining new genetic resources of microbial communities. J Mol Microbiol Biotechnol 16:109–123

Fischer D, Pfitzner B, Schmid M, Simões-Araújo JL, Reis VM, Pereira W, Ormeño-Orrillo E, Hai B, Hofmann A, Schloter M, Martinez-Romero E, Baldani JI, Hartmann A (2012) Molecular characterisation of the diazotrophic bacterial community in uninoculated and inoculated field-grown sugarcane (*Saccharum* sp.). Plant Soil 356:83–99

Fleming JT, Yao WH, Sayler GS (1998) Optimization of differential display of prokaryotic mRNA: application to pure culture and soil microcosms. Appl Environ Microbiol 64:3698–3706

Galbally IE, Kirstine W (2002) The production of methanol by flowering plants and the global cycle of methanol. J Atmos Chem 43:195–229

Galbraith EA, Antonopoulos DA, White BA (2004) Suppressive subtractive hybridization as a tool for identifying genetic diversity in an environmental metagenome: the rumen as a model. Environ Microbiol 6:928–937

Gao H, Yang ZK, Gentry TJ, Wu L, Schadt CW, Zhou J (2006) Microarray-based analysis of microbial community RNAs by whole-community RNA amplification. Appl Environ Microbiol 2:563–571

Gilbert JA, Meyer F, Jansson J, Gordon J, Pace N, Tiedje J, Ley R, Fierer N, Field D, Kyrpides N, Glöckner FO, Klenk HP, Wommack KE, Glass E, Docherty K, Gallery R, Stevens R, Knight R (2010) The Earth Microbiome Project: meeting report of the "1st EMP Meeting on Sample Selection and Acquisition" at Argonne National Laboratory October 6 2010. Stand Genomic Sci 3:249–253

Glick BR (2005) Modulation of plant ethylene levels by the bacterial enzyme ACC deaminase. FEMS Microbiol Lett 251:1–7

Haichar FEZ, Achouak W, Christen R, Heulin T, Marol C, Marais MF, Mougel C, Ranjard L, Balesdent J, Berge O (2007) Identification of cellulolytic bacteria in soil by stable isotope probing. Environ Microbiol 9:625–634

Hiltner L (1904) Über neuere Erfahrungen und Probleme auf dem Gebiet der Bodenbakteriologie und unter besonderer Berücksichtigung der Grüdungen und Brache. Arbeiten der Deutschen Landwirtschafts-Gesellschaft H 98:59–78

Hoheisel JD (2006) Microarray technology: beyond transcript profiling and genotype analysis. Nat Rev Genet 7:200–210

Hong SH, Bunge J, Leslin C, Jeon S, Epstein SS (2009) Polymerase chain reaction primers miss half of rRNA microbial diversity. ISME J 3:1365–1373

Inceoglu O, Abu Al-Soud W, Salles JF, Semenov AV, van Elsas JD (2011) Comparative analysis of bacterial communities in a potato field as determined by pyrosequencing. PLoS One 6:11

James EK (2000) Nitrogen fixation in endophytic and associative symbiosis. Field Crop Res 65:197–209

Kniskern JM, Traw MB, Bergelson J (2007) Salicylic acid and jasmonic acid signalling defense pathways reduce natural bacterial diversity on *Arabidopsis thaliana*. Mol Plant-Microbe Interact 20:1512–1522

Lemarchand K, Masson L, Brousseau R (2004) Molecular biology and DNA microarray technology for microbial quality monitoring of water. Crit Rev Microbiol 30:145–172

Lian J, Wang Z, Zhou S (2008) Response of endophytic bacterial communities in banana tissue culture plantlets to *Fusarium* wilt pathogen infection. J Gen Appl Microbiol 54(2):83–92

Liang P, Pardee AB (1992) Differential display of eukaryotic messenger RNA by means of the polymerase chain reaction. Science 257:971–976

Lindow SE, Brandl MT (2003) Microbiology of the phyllosphere. Appl Environ Microbiol 69:1875–1883

Liu WT, Zhu L (2005) Environmental microbiology-on-a-chip and its future impacts. Trends Biotechnol 23:174–179

Loy A, Kusel K, Lehner A, Drake HL, Wagner M (2004) Microarray and functional gene analyses of sulfate-reducing prokaryotes in low-sulfate, acidic fens reveal co-occurrence of recognized genera and novel lineages. Appl Environ Microbiol 70:6998–7009

Lozupone CA, Knight R (2008) Species divergence and the measurement of microbial diversity. FEMS Microbiol Rev 32:557–578

Lundberg DS, Lebeis SL, Paredes SH, Yourstone S, Gehring J, Malfatti S, Tremblay J, Engelbrektson A, Kunin V, del Rio TG, Edgar RC, Eickhorst T, Ley RE, Hugenholtz P, Tringe SG, Dangl JL (2012) Defining the core *Arabidopsis thaliana* root microbiome. Nature 488:86–90

Margulies M, Egholm M, Altman WE, Attiya S, Bader JS, Bemben LA, Berka J, Braverman MS, Chen YJ, Chen Z, Dewell SB, Du L, Fierro JM, Gomes XV, Godwin BC, He W, Helgesen S, Ho CH, Irzyk GP, Jando SC, Alenquer ML, Jarvie TP, Jirage KB, Kim JB, Knight JR, Lanza JR, Leamon JH, Lefkowitz SM, Lei M, Li J, Lohman KL, Lu H, Makhijani VB, McDade KE, McKenna MP, Myers EW, Nickerson E, Nobile JR, Plant R, Puc BP, Ronan MT, Roth GT, Sarkis GJ, Simons JF, Simpson JW, Srinivasan M, Tartaro KR, Tomasz A, Vogt KA, Volkmer GA, Wang SH, Wang Y, Weiner MP, Yu P, Begley RF, Rothberg JM (2005) Genome sequencing in microfabricated high-density picolitre reactors. Nature 437:376–380

Mendes R, Kruijt M, De Bruijn I, Dekkers E, van der Voort M, Schneider JH Piceno YM, DeSantis TZ, Andersen GL, Bakker PA, Raaijmakers JM (2011) Deciphering the rhizosphere microbiome for disease-suppressive bacteria. Science 332:1097–1100

Mendes R, Garbeva P, Raaijmakers JM (2013) The rhizosphere microbiome: significance of plant beneficial, plant pathogenic, and human pathogenic microorganisms. FEMS Microbiol Rev 37:634–663

Meyer JB, Song-Wilson Y, Foetzki A, Luginbuhl C, Winzeler M, Kneubühler Y, Matasci C, Mascher-Frutschi F, Kalinina O, Boller T, Keel C, Maurhofer M (2013) Does wheat genetically modified for disease resistance affect root-colonizing pseudomonads and arbuscular mycorrhizal fungi*?* PLoS One 8:12

Micallef SA, Shiaris MPC-CA (2009) Influence of *Arabidopsis thaliana* accessions on rhizobacterial communities and natural variation in root exudates. J Exp Bot 60:1729–1742

Mitter B, Pfaffenbichler N, Flavell R, Compant S, Antonielli L (2017) A new approach to modify plant microbiomes and traits by introducing beneficial bacteria at flowering into progeny seeds. Front Microbiol 8:11. https://doi.org/10.3389/fmicb.2017.00011

Monteiro RA, Balsanelli E, Wassem R, Marin AM, Brusamarello-Santos LCC, Schmidt MA, Tadra-Sfeir MZ, Pankievicz VCS, Cruz LM, Chubatsu LS, Pedrosa FO, Souza EM (2012) Herbaspirillum–plant interactions: microscopical, histological and molecular aspects. Plant Soil 356:175–196

Morrison N, Wood AJ, Hancock D, Shah S, Hakes L, Gray T, Tiwari B, Kille P, Cossins A, Hegarty M, Allen MJ, Wilson WH, Olive P, Last K, Kramer C, Bailhache T, Reeves J, Pallett D, Warne J, Nashar K, Parkinson H, Sansone SA, Serra PR, Stevens R, Snape J, Brass A, Field D (2006) Annotation of environmental OMICS data: application to the transcriptomics domain. OMICS 2:172–178

Nygaard V, Hovig E (2006) Options available for profiling small samples: a review of sample amplification technology when combined with microarray profiling. Nucleic Acids Res 34:996–1014

Park SJ, Kang CH, Chae JC, Rhee SK (2008) Metagenome microarray for screening of fosmid clones containing specific genes. FEMS Microbiol Lett 284:28–34

Philippot L, Hallin S, Borjesson G, Baggs EM (2009) Biochemical cycling in the rhizosphere having an impact on global change. Plant Soil 321:61–81

Pinto AJ, Raskin L (2012) PCR biases distort bacterial and archaeal community structure in pyrosequencing datasets. PLoS One 7:16

Rebrikov DV, Desai SM, Siebert PD, Lukyanov SA (2004) Suppression subtractive hybridization. Methods Mol Biol 258:107–134

Reinhold-Hurek B, Bunger W, Burbano CS, Sabale M, Hurek T (2015) Roots shaping their microbiome: global hotspots for microbial activity. Annu Rev Phytopathol 53:403–424

Rezzonico F, Binder C, Defago G, Moenne-Loccoz Y (2005) The type III secretion system of biocontrol *Pseudomonas fluorescens* KD targets the phytopathogenic chromista *Pythium ultimum* and promotes cucumber protection. Mol Plant-Microbe Interact 18:991–1001

Sebat JL, Colwell FS, Crawford RL (2003) Metagenomic profiling: microarray analysis of an environmental genomic library. Appl Environ Microbiol 8:4927–4934

Selinger DW, Saxena RM, Cheung KJ, Church GM, Rosenow C (2003) Global RNA half-life analysis in *Escherichia coli* reveals positional patterns of transcript degradation. Genome Res 13:216–223

Selvakumar G, Panneerselvam P, Ganeshamurthy AN, Maheshwari DK (2012) Bacterial mediated alleviation of abiotic stress in crops. In: Maheshwari DK (ed) Bacteria in agrobiology: stress management. Springer, New York, pp 205–224

Sessitsch A, Hardoim P, Doring J, Weilharter A, Krause A, Woyke T, Mitter B, Hauberg-Lotte L, Friedrich F, Rahalkar M, Hurek T, Sarkar A, Bodrossy L, vanOverbeek L, Brar D, van Elsas JD, Reinhold-Hurek B (2012) Functional characteristics of an endophyte community colonizing rice roots as revealed by metagenomic analysis. Mol Plant-Microbe Interact 25:28–36

Sharma S, Aneja MK, Mayer J, Schloter M, Munch JC (2004) RNA fingerprinting of microbial community in the rhizosphere soil of grain legumes. FEMS Microbiol Lett 240:181–186

Stursova M, Zifcakova L, Leigh MB, Burgess R, Baldrian P (2012) Cellulose utilization in forest litter and soil: identification of bacterial and fungal decomposers. FEMS Microbiol Ecol 80:735–746

Suman A, Yadav AN, Verma P (2016) Endophytic microbes in crops: diversity and beneficial impact for sustainable agriculture. In: Singh DP, Singh HB, Prabha R (eds) Microbial inoculants in sustainable agricultural productivity. Vol 1: research perspectives. Springer, Dordrecht, pp 117–143

Taroncher-Oldenburg G, Griner EM, Francis CA, Ward BB (2003) Oligonucleotide microarray for the study of functional gene diversity in the nitrogen cycle in the environment. Appl Environ Microbiol 69:1159–1171

Teixeira L, Peixoto RS, Cury JC, Sul WJ, Pellizari VH, Tiedje J, Rosado AS (2010) Bacterial diversity in rhizosphere soil from Antarctic vascular plants of Admiralty Bay, maritime Antarctica. ISME J 4:989–1001

Tiquia SM, Wu L, Chong SC, Passovets S, Xu D, Xu Y, Zhou J (2004) Evaluation of 50-mer oligonucleotide arrays for detecting microbial populations in environmental samples. BioTechniques 36:664–675

Tringe SG, von Mering C, Kobayashi A, Salamov AA, Chen K, Chang HW, Podar M, Short JM, Mathur EJ, Detter JC, Bork P, Hugenholtz P, Rubin EM (2005) Comparative metagenomics of microbial communities. Science 308:554–557

Turnbaugh PJ, Ley RE, Hamady M, Fraser-Liggett CM, Knight R, Gordon JI (2007) The human microbiome project. Nature 449:804–810

Urisman A, Fischer KF, Chiu CY, Kistler AL, Beck S, Wang D, DeRisi JL (2005) E-predict: a computational strategy for species identification based on observed DNA microarray hybridization patterns. Genome Biol 9:R78

van der Heijden MGA, Bardgett RD, Van Straalen NM (2008) The unseen majority: soil microbes as drivers of plant diversity and productivity in terrestrial ecosystems. Ecol Lett 11:296–310

Vieites JM, Guazzaroni ME, Beloqui A, Golyshin PN, Ferrer M (2009) Metagenomic approaches in systems microbiology. FEMS Microbiol Rev 33(1):236–255

Vokou D, Vareli K, Zarali E, Karamanoli K, Constantinidou HIA, Monokrousos N, Halley JM, Sainis I (2012) Exploring biodiversity in the bacterial community of the Mediterranean phyllosphere and its relationship with airborne bacteria. Microb Ecol 64:714–724

Walters DM, Russ R, Knackmuss H, Pouviere PE (2001) High-density sampling of a bacterial operon using mRNA differential display. Gene 273:305–315

Ward BB, Eveillard D, Kirshtein JD, Nelson JD, Voytek MA, Jackson GA (2007) Ammonia oxidizing bacterial community composition in estuarine and oceanic environments assessed using a functional gene microarray. Environ Microbiol 9:2522–2538

Wrage N, Velthof GL, van Beusichem ML, Oenema O (2001) Role of nitrifier denitrification in the production of nitrous oxide. Soil Biol Biochem 33:1723–1732

Wu L, Thompson DK, Liu X, Fields MW, Bagwell CE, Tiedje JM, Zhou J (2004) Development and evaluation of microarray-based whole-genome hybridization for detection of microorganisms within the context of environmental applications. Environ Sci Technol 38:6775–6782

Wu L, Liu X, Schadt CW, Zhou J (2006) Microarray-based analysis of subnanogram quantities of microbial community DNAs by using whole-community genome amplification. Appl Environ Microbiol 72 4931–4941

Yakimov MM, Timmis KN, Golyshin PN (2007) Obligate oil degrading marine bacteria. Curr Opin Biotechnol 18:257–266

Zamioudis C, Pieterse CM (2012) Modulation of host immunity by beneficial microbes. Mol Plant-Microbe Interact 25:139–150

Zhou J, Thompson DK (2002) Challenges in applying microarrays to environmental studies. Curr Opin Biotechnol 13:204–207

Zhou J (2003) Microarrays for bacterial detection and microbial community analysis. Curr Opin Microbiol 6:288–294

Zhou J, Thompson DK (2004) Microarray technology and applications in environmental microbiology. Adv Agron 82:183–270

Metagenomic Approach in Relation to Microbe–Microbe and Plant–Microbiome Interactions

22

Abdul Rehman, Muhammad Ijaz, Komal Mazhar, Sami Ul-Allah, and Qasim Ali

Abstract

Metagenomics is the study of the genetic material of microbes in their natural living environment, which involves complex microbial communities. This study helps to identify the sequences of uncultured microbes present in the plant rhizosphere and their diversity in the environment through quicker and less expensive methods. This is a new challenge for technicians and bioinformaticians to gain knowledge about microbes through millions of genomes. From culture, only single-clone data are obtained, which are then further used in sequence formation, whereas in metagenomics, usually data on more than 10,000 species in microbial communities are studied. From these samples, new genes and their functions have been observed. In the future, this technique will be used in the same way as 16S ribosomal RNA gene sequencing methods are used to describe profiles of microbial communities. Metagenomics is new to science and is a novel technique for handling of genomic data by scientists, used in the last few years.

Keywords

Ecosystem · Plant · Microbe · Metagenomics · Interaction · Sequencing

A. Rehman · M. Ijaz (✉) · S. Ul-Allah · Q. Ali
College of Agriculture, Bahauddin Zakariya University, Layyah, Pakistan
e-mail: abdul-pbg@bzu.edu.pk; muhammad.ijaz@bzu.edu.pk

K. Mazhar
Department of Forestry and Range Management, University of Agriculture Faisalabad, Faisalabad, Pakistan

V. Kumar et al. (eds.), *Microbiome in Plant Health and Disease*,
https://doi.org/10.1007/978-981-13-8495-0_22

22.1 Introduction

Microbes are involved in various vital and beneficial processes such as plant protection and plant growth promotion. In the rhizospheric zone the microbial community also faces internal competition, and this competition also depends upon the type of plant root exudates. They contain many genes and proteins, which aid plant growth and development (Abd-Elsalam et al. 2010; Lakshmanan et al. 2014). Rhizospheric microbes have been described as plant developers, since they have so many beneficial functions such as nitrogen fixation, solubilization of phosphorus, phytohormone production, biocontrol of diseases, and stress mitigation (Tsurumaru et al. 2015; Majeed et al. 2015; Massart et al. 2015). Communities of microbes and their effects on plant health greatly depend upon the genotype, the plant growth stage, and the soil type and texture (Broeckling et al. 2008). These microbes are involved in the nutrient cycle, water cycle, and production of various metabolites (Buscot and Varma 2005). Apart from those microbes that give benefits to plants, there are also many dangerous microbes that can cause disease and nutrient shortages. They can break protective shields in plants and disrupt protective mechanisms (Sharma et al. 2011; Mendes et al. 2013).

Researchers have revealed that soil contains 4×10^6 different microbial taxa, while others have reported that more than one million different microbes are present in 1 g of rhizospheric soil (Gans et al. 2005). The numbers and diversity of microbes present in the rhizosphere are huge, but we do not have detailed information about their diversity (Singh et al. 2010). Scientists estimate that more than 99% of species present in the rhizosphere are beneficial to plants (Dinsdale et al. 2008; Kumar et al. 2015). This hypothesis has also been validated by culture-independent approaches. So, there are more opportunities for technicians to utilize microbes for better development of crops and their tolerance of their environment (Biteen et al. 2016). Although under laboratory conditions, only some microbes are culturable, metagenomic studies have enabled a lot of information on unculturable microflora to be obtained.

22.2 Metagenomics

Metagenomics is the study of any group of different genomes of microbes taken from any niche, and describes their origin, ecology, and diversity. This collective study shows the structures of microbe communities and their interactions. This technique allows researchers to study complex systems mediated by microbes in the rhizosphere. The availability of nucleotide sequencing and high-throughput techniques helps in sequencing of large amounts of microbial DNA (Metzker 2010) and allows us to see the rhizospheric community. It also provides baseline information on taxonomic composition (Lagos et al. 2015) and tells us about fungal microbes in the rhizosphere (LeBlanc et al. 2015).

22.3 Approaches to Metagenomics

There are two types of metagenomic approaches found through culture-independent analysis: physical based and sequence based (Rabausch et al. 2013). The conventional sequence-based technique, 16S ribosomal RNA (rRNA), has been found to be more reliable for phylogenetic study of rhizospheric microbes; however, this technique does not provide information about metabolic and dynamic aspects of microbes. Those aspects can be studied by functional metagenomics (Soni et al. 2012; Aßhauer et al. 2015); thus, functional metagenomics is used for analysis of novel genes and their characteristics in a family (Nacke et al. 2011; Illeghems et al. 2015).

For development of a genomic library based on metagenomics, the first main step is isolation of DNA from the rhizosphere and the environment. There are many methods to isolate pure DNA and intact DNA (Bertrand et al. 2005). Assessment of pure DNA is difficult because of the presence of humic acid and polyphenolic compounds, which mingle with DNA samples and are precipitated (Streit and Schmitz 2004). These compounds are the source of hindrances in the restriction and digestion of DNA, the process of ligation, and the activity of the Taq polymerase enzyme in polymerase chain reaction (PCR) (Sharma et al. 2007). New technology has developed methods of DNA extraction that isolate DNA without these restricting compounds (Tanveer et al. 2016). DNA extraction using different methods reveals differences in the purity of the DNA (Niemi et al. 2001).

22.4 Metagenomic Techniques

A wide range of technologies in molecular study have allowed researchers to characterize bacterial diversity in populations in an environment. There are numerous molecular techniques that are helpful in microbiology, such as PCR, cloning and sequencing of genes (especially ribosomal genes), restriction and terminal restriction fragment length polymorphism, fluorescent in situ hybridization, and denaturing gradient gel electrophoresis. In microbial diversity, 16S rRNA is used as a marker gene as it is stored in microbes through several years of evolution. Thus, it allows the study of bacteria and archaea, illuminating the distribution of taxonomy and evolutionary association of microbes.

Various next-generation techniques for identification and characterization of organisms have been developed by engineers. Amplicon gene sequencing is used for investigation of the diversity of bacteria and fungi through 16S rDNA and internal transcribed spacers, respectively. Metagenome sequencing illuminates the diversity and physiological abilities of microbes from a specific environment. Sequencing of complementary DNA (cDNA) helps in the study of expression of genes and their functions, to analyze their potential in metabolic activities. It is also called metatranscriptomic analysis. In metaproteomics, protein sequencing and characteristics of a community of plant and microbes can be studied.

Table 22.1 Metagenomic techniques used for assessment of microbial communities

Techniques	Aim of the study	References
Amplicon gene sequencing of conserved marker genes, 16S ribosomal RNA	Rhizobacterial population of *Arachis hypogaea*	Haldar and Sengupta (2015)
	Bacterial and fungal rhizospheric communities in hydrocarbon-contaminated soils	Bell et al. (2014)
Metagenome sequencing	Soybean rhizosphere	Mendes et al. (2013)
	454 pyrosequencing to analyze rhizosphere fungal communities during soybean growth	Sugiyama et al. (2014)
	Grassland plant community richness and soil edaphics	LeBlanc et al. (2015)
Metatranscriptome sequencing	Root surface microbiome	Ofek-Lalzar et al. (2014)
	Rhizospheric microbiome assemblage affected by plant development	Chaparro et al. (2014)
Metaproteomic profiling	Phyllosphere and rhizosphere of rice	Knief et al. (2012)
	Sugarcane rhizosphere	Lin et al. (2013)
Metabolomic profiling	Tomato root mycorrhizae	Tschaplinski et al. (2014) and Rivero et al. (2015)

Mass spectrometry is used to analyze extracted protein quantities and their molecules, with their involvement in metabolic activities. In recent technological developments, combined use of these techniques has provided more accurate results, i.e., mass spectrometry with gas chromatography and mass spectrometry with liquid chromatography. They give more reliable results both quantitively and qualitatively in plants and their tissues, and also in the rhizospheric niche (Zhang et al. 2012).

With regard to plant microbes, microorganisms in the rhizosphere are especially important. Metagenomics helps in broadening the view for analysis of the rhizosphere. Thus, novel discoveries greatly depend upon new methods and technologies (Table 22.1).

22.5 Bioinformatic Tools

22.5.1 Software for Metagenomic Analysis

Metagenomics is very helpful for providing expressive results of nucleotide sequences, because a lot of nucleotide sequences are generated by metagenomic sequencing. On the other hand, bioinformatic software is also essential for DNA sequencing as devised by Sanger. There are many sequencing platforms for next-generation sequencing, such as Illumina, PacBio, and 454 pyrosequencing. Illumina and PacBio software have been developed for the reading of short and long sequencing of metagenomes. Mothur (https://www.mothur.org) is open-source software that is used for Sanger sequencing and amplicon analysis. Similarly, two other software packages—CARMA and MEGAN (a metagenome analyzer)—are also beneficial in the field of metagenome data sets (Gerlach et al. 2009; Caporaso et al. 2014; Gerlach and Stoye 2011; Huson and Weber 2013). PICRUST software connects taxonomic classifications from metaprofiling results to metabolic information (Langille et al. 2013).

22.5.2 Platforms for Metagenomic Analysis

There are many metagenomic stages that give information about microbial multiplicity analysis. Nowadays, there are big challenges in the analysis of environmental sequences and ways to analyze different types of data such as both taxonomic and functional data. To overcome all of these types of problems, community-enabling cloud compatibility platforms are available, including IMG/M, Web CARMA, and CAMERA software (Gerlach et al. 2009).

22.6 Rhizosphere Metagenomics

First of all, it is very important to understand the interactions between rhizospheric microorganisms and plants because the rhizosphere provides a more favorable environment than bulk soil (Valentine 2007). So, assessment of the microbial community structure in the soil is founded mainly on the use of culture-dependent and culture-independent methods, including soil metagenomics (Daniel 2005). In the field of agriculture, rice crops have been exploited by using metagenomics. In these crops, various traits such as protein secretions, nitrogen fixation, and quorum sensing (QS) were forecast by metagenomics (Knief et al. 2012). Recently, by use of a culture-independent approach, the influence of wild and cultivated rice genotypes on the rhizospheric bacterial community was established (Shenton et al. 2016). For identification of microbiome traits directly from barley rhizospheric soil that had received no phosphate fertilizer for the previous 15 years, a functional metagenomic approach was applied. The study revealed that phosphorus solubilization was mainly associated with the nonculturable microbiome present in the rhizospheric soil (Chhabra et al. 2013). Furthermore, in red mangroves from the Red Sea, targeted metagenomic approaches were applied to identify the variety of soil fungal microorganisms (Simoes et al. 2015; Alzubaidy et al. 2016). Both culture-dependent and culture-independent methods are very beneficial for understanding the interactions in the microbiome and enhance our understanding of how we can get more benefit from the microflora that remain in rhizospheric soil (Naz et al. 2014). So, combination of a metagenomic approach with other rapid molecular techniques provides a better way to understand the microbial wealth of the rhizosphere (Knief et al. 2011; Unno and Shinano 2013). A recent report by Jin et al. (2016) revealed that metagenomic analysis of rhizospheric soil can provide an overview of the functional regions of a protein domain and can be developed for protein optimization and functional characterization. Recently, for plant growth, metagenomic study has shown that a small "core" rhizospheric bacterial community may interact synergistically with an arbuscular mycorrhizal fungus (AMF) and other presumably beneficial bioinoculants (Valverde et al. 2016).

22.7 Major Microbial Groups in the Rhizospheric Soil Metagenome

Traditionally, the study of microbial diversity has focused on isolation of microbes on diverse culture media and understanding of their metabolic variations. Under laboratory circumstances, growth media are used to culture selective microbes, although only certain subpopulations of microbes present in environmental samples can be cultured, while others cannot be grown in different media or under such conditions. With use of typical cultivation methods, diversity of complex bacterial communities is inevitable. Moreover, many of the "unculturable" bacteria characterize new phylotypes, families, and divisions in the archaeal and bacterial domains (Sharma et al. 2005). It was previously estimated that of 61 different phyla, 31 were unculturable (Hugenholtz and Tyson 2008). Furthermore, research using sequence-based metagenomics has shown that the rice rhizosphere is dominated by various microbes such as Proteobacteria, Acidobacteria, Planctomycetes, Firmicutes, Actinomycetes, and Firmicutes (Arjun and Harikrishnan 2011; Knief et al. 2011; Mahyarudin and Rusmana 2015; Bhattacharyya et al. 2016). Similarly, the wheat rhizosphere shows associations between the Planctomycetes and Chloroflexi phyla (Naz et al. 2014), which are also linked with Actinobacteria, Firmicutes, archaea, fungi, viruses, and unclassified taxa (Hernandez-Leon et al. 2012).

Moreover, Actinobacteria, γ-Proteobacteria, and ascomycetous divisions dominated the rhizosphere of soybean (Bresolin et al. 2010). Among All fungi, ascomycota is predominantly found in bulk soil as well as in rhizoshpere of soyabean (Li et al. 2010). A predominance of Proteobacteria and Bacteroidetes was identified by metagenomic analysis of a vegetable rhizosphere (Jackson et al. 2013). So, it is concluded that Proteobacteria are the main unculturable bacteria present in soil rhizospheres (Hernandez-Leon et al. 2012; Unno and Shinano 2013; Shenton et al. 2016; Bhattacharyya et al. 2016). Common rhizospheric microbes are listed in Table 22.2.

22.8 Application of Next-Generation Sequencing Technologies in Revealing Plant–Microbe Interactions

Microbes present in the rhizosphere are associated with different plant parts and are known as polyspheric, endospheric, and rhizospheric microbes (Vorholt 2012). They affect plant health and growth through beneficial, lethal, or neutral modes of action (Newton et al. 2010). Most microbes present in the soil do no harm to plants, but some have specific functions that do cause trouble for plants (Zhan and Sun

Table 22.2 Common rhizospheric microbes associated with some plants/crops

Rhizosphere	Common microbes	References
Wheat (*Triticum aestivum*)	*Achromobacter, Bacillus, Cellulomonas, Clostridia, Gallionella, Herbaspirillum, Nocardia, Microbacterium, Mycobacterium,* uncultured bacteria	Valverde et al. (2016)
	Azoarcus, Balneimonas, Bradyrhizobium, Gemmatimonas, *Rhodoplanes, Rubellimicrobium, Skermanella*	Naz et al. (2014)
Rice	*Actinokineospora, Actinoplanes, Geodermatophilus, Kocuria, Streptomyces*	Mahyarudin and Rusmana (2015)
	Acidovorax, Anaeromyxobacter, Azospirillum, Bradyrhizobium, Dechloromonas, Desulfovibrio, Geobacter	Knief et al. (2011)
Sugarcane	*Azospirillum, Bacillus, Belnapia, Bradyrhizobium, Chitinophaga, Chryseobacterium, Cohnella, Rhizobium, Streptomyces, Terrimonas, Tumebacillus*	Pisa et al. (2011)
Lettuce	*Alkanindiges, Burkholderia, Novosphingobium, Sphingobium, Sphingomonas*	Schreiter et al. (2014)
Tobacco (*Nicotiana tabacum*)	*Bordetella, Burkholderia, Flavobacterium*	Brinkmann and Tebbe (2007)
Soybean	*Bacillus, Bradyrhizobium, Rhizobium, Stenotrophomonas, Streptomyces*	Sugiyama et al. (2014)
Maize	*Arthrobacter, Bacillus, Burkholderia, Kitasatospora,* uncultured bacteria	Oliveira et al. (2009)
Arabidopsis thaliana	*Arthrobacter, Flavobacterium, Kineosporiaceae, Massilia*	Bodenhausen et al. (2013)
Para grass (*Urochloa mutica*)	*Anaerosinus, Bacillus, Caldilinea,* Chloroflexi, *Clostridium, Microcoleus*	Mukhtar et al. (2016)

2012; Vacheron et al. 2013). Many microorganisms are well known for nitrogen fixation and aid plant growth and development. It has also been found that through quorum sensing, bacterial communities communicate with plants (Ferluga and Venturi 2009; Subramoni and Venturi 2009). Multiple studies have revealed that homologous receptors of the QS LuxR system are involved in the rhizosphere and in the development of roots and their structure (Bai et al. 2012).

Microbes express some genes related to phytic acid such as citrate synthase and alkaline phosphate genes (Unno and Shinano 2013). Mycorrhizal microbes also have genes related to metabolism of phosphorus, nitrogen, potassium, and iron (Mendes et al. 2014). They have potential for enhancing host development and nutrition uptake. These genes are also used in modulating the health of the plant, suppressing disease effects, and simulating hormonal cycles. They also have the ability to modify plant mechanisms in stress conditions (Mendes et al. 2011; Zolla et al. 2013). To determine the relation between metagenomics and meta-proteomics, rice and its associated microbial activity were studied. This showed the complexity of the community in the host plant rhizosphere (Knief et al. 2012). Both metagenomic and metaproteomic methods are powerful tools and provide a detailed picture of active and functional microbes. Various transcripts have been revealed after metatranscriptomic analysis in *Arabidopsis* at different growth stages, such as genes related to disease, which are active at the flowering and bolting stages (Chaparro et al. 2014).

The study of metabolite sets in microbial communities is called microbial metabolomics. Detailed observation of phenomena occurring during interaction of the host and its microbial communities can provide a clear picture of the physiological condition of the microorganisms (Venturi and Fuqua 2013). In addition, QS signaling also describes the host and microbial relationship and modes of signaling in plant growth and development (Hartmann et al. 2014). LuxR solo receptors have been found to be involved in the association between the plant and its related microbes (Patel et al. 2013; Subramoni et al. 2015). Bacterial LuxR solos have been studied in pathogen zones. They are involved in the production of low molecular weight substances which brings pathogenicity and its related genes into the host. Microbes also exhibit symbiotic phenomena with the host plant and provide them with benefits (Venturi and Fuqua 2013).

This is the initial stage of interkingdom communion based on chemicals within plants and microbes. So, research related to bacteria can reveal this communication behavior to prevent the impacts of disease and promote plant growth. Next-generation sequencing has given us a better way to understand microbial communities at low cost with high output. However, the function behind the microbes' metabolism is still under research. So, different information collected from different modern technologies should give way to new research to explore the potential of microbes. This information will be helpful in the future for modulation of the microbiome to minimize the incidence of disease and enhance gross plant productivity.

22.9 Plant–Microbe Interactions

In the last decade, utilization of microbes for disease reduction, states of diseases, immunity, and plant health have been focused on by researchers. This work has shown the significant importance of microbes in the plant rhizosphere. Many human disease conditions are caused by the interaction of microorganisms with the host's genetic system and the environment. Observations at the genus level or below it have shown that every human being has a special microbiome. At the phylum and group levels, the case is different; there are representative drifts that have importance, which distinguish between disease-state microbiomes and healthy-state microbiomes, such as irritable bowel syndrome and obesity (Greenblum et al. 2012) and atherosclerosis (Koeth et al. 2013). This concept was referred to as a "supraorganism" in the Human Genome Project, where it was described by Turnbaugh et al. (2007) that if humans are thought of as a combination of microbial and human cells, in which the human genetic landscape acts as an accumulation of the genes in the human genome and the microbiome, with human metabolic features as a merging of human and microbial traits, then the final image that appears is one of a human known as a supraorganism.

In accordance with that first project on the human microbiome (Turnbaugh et al. 2007), the idea of this hypothesis can be extended in plants instead of humans. Thus, in a plant within a given habitat, there is a set of genes that are contained by the plant microbiome and can be determined by the plant's phonology, phylogeny, etc. This raises a query about the habitat, which is determined by a range—a place containing both biotic factors (plants and microorganisms) and abiotic factors (soil and exudates). Within the soil, the interaction of plants with microbes and the effects of roots through their exudates are called the rhizosphere. And it is not only the soil in contact with the roots that helps in constructing the belowground environment; also, belowground plant microbes are successfully expanded through exudates. Plants microbes are further classified at the time of interaction (Bais et al. 2004) at the time of interface where microbes and plants alternate their information. The rhizosphere and parts of plants (from above ground to the underground structures) are included as the habitat where microbes can cooperate in an internal (endophytic) or adherent (epiphytic) manner.

The interactions between plants, soil, and microbes are not new, but this perception of the plant microbiome involves "microbe–soil–microbe–plant–microbe" interactions rather than merely the "soil–microbe–plant" interface. It has been shown that plants contain a smaller genome than the microbiome community and that interactions of a microbial nature enable the plant genome to expand and create a "second genome" for the plant (Bernedsen et al. 2012).

If we study the belowground soil microbiome, endophytes, the rhizoplane, the rhizosphere, bulk soil, and epiphytes are addressed. The range of the soil that is not penetrated by roots is called bulk soil. Root exudations are not present in bulk soil, so it is not affected by chemotaxis. The concentrations of organic compounds are greater in the rhizosphere than in bulk soil, according to research. It has been reported that a wide range of microbe variation is present in the rhizosphere (Egamberdiyeva et al. 2008; Mendes et al. 2011), and the soil acts as a medium for the transmission of exudates between the plant and microbes through plant roots and attached soil. The effects of climate change on the plant rhizosphere ultimately affect the plants and microbes present in the rhizosphere and their efficiency (Bais et al. 2004). If we study the plant closely, the rhizoplane is present, which is involved in the interaction of plant tissues with the soil. Those microbes that are attached to plant tissues are known as epiphytes, and those that are present inside the body of the plant are known as endophytes (Bulgarelli et al. 2013). The microbial lifestyle is very important to understand, because it is very complex, and it is important to understand the basic phenomena of microbial interactions with plants. It is also very important to recognize the types of microbes, which may be epiphytes or endophytes, and their functions in stress conditions. Their lifestyle is much more complex than is explored here. A dynamic environment exists between the rhizosphere and the rhizoplane. The rhizosphere greatly influences the activities of microbes and the microbes influence the rhizosphere according to the nature of the soil. Ions that can move can be restored more easily than nonmotile ions (which are depleted quickly), i.e., ammonium, potassium, and phosphorus (Neumann and Romheld 2002). The pH of the rhizosphere may differ from that of bulk soil by 2–3 pH units as a direct result of different biological activities. These activities affect the relative solubilities of major nutrients; for example, in insoluble inorganic soil, more phosphorus is present and thus is solubilized by the interactive effects of plants and microbes (Neumann and Romheld 2002).

Interpolation of generalities from recent data measurements of the microbial rhizosphere by using structural sorting is difficult in view of biogeographical or temporal fluxes. This shows general variations and adaptation of the environment by the microbes in the rhizosphere, changing the nutrient cycle and creating negative impacts on the ecosystem (Rout and Callaway 2009, 2012). This implication shows the influence of the plant microbiome on the ecosystem and also the enhanced availability of nutrients to plants, i.e., the availability of nitrogen and carbon in their utilized forms.

The services of the ecosystem are interlinked with the functional traits of the plants and microbes that form the soil, its composition, organic decomposition, mineralization of nutrients, and plant products (de Bello et al. 2010). Role of microbe in plant production has not escaped those that are fimiliar with crops, where Soybean has been manipulated by these nitrogen fixing bacteria for high yield by various microbes in rhizosphere (Harris et al. 1985). Many examples of plants that have been modified by plant growth–promoting bacteria (PGPBs) have been described in the literature. Many PGPB activities help plants to perform better—for example, phosphate solubilization, nitrogen fixing, and enhanced hormonal production (James 2000; Martinez-Romero 2006; Hardoim et al. 2008).

It is clear that the interactions of microbes with the plant genome, in the form of hormonal signaling, greatly influence genes and their functions. Plant microbial changes in the soil ecosystem result in novel characteristics and functions that may be related closely to the macroevolution of plants, e.g., polyploidization. From either the micro- or macroperspective, microbes exert their influence on expression of traits in plants. For better understanding of the dynamics of the microbiome in an ecosystem, expression of traits, regulation, and functions should be known, and changes occur through microbial impacts. Plant microbes are those that use the currency of exudates in the rhizosphere and have complex forms of communication within the belowground ecosystem, with significant implications for plants.

22.9.1 The Currency of the Microbiome: Exudates

In the root zone, the exudates that are released are plant chemicals or may come from microbes. Through exudates, plants can communicate with microorganisms to obtain help for adaptation to stress conditions such as drought, pathogenic spores, and toxicity from metal ions. On the other hand, microbes also receive benefits from plant exudates in the form of nutrients. In that case, the plant microbiome is the source from which the plant genome may be spread to other regions. Sections 22.9.2

and 22.9.3 provide a detailed discussion of plant and microbial exudate release and uptake, and their effects on plants function.

22.9.2 Plant Uptake and Release

The secretions that plants release may be of various types in terms of both their structures and their constituents, and they also vary from plant to plant and from species to species. They may be low molecular weight compounds (e.g., amino acids and organic molecules) or high molecular weight compounds (e.g., carbohydrates, sugars, lipids, and many proteins) (Badri and Vivanco 2009). Plant exudates are mostly sugars and amino acid molecules (Jaeger et al. 1999), which provide the basis of many other functions in cells, such as defense mechanisms against pathogens and other disease-containing molecules. These exudates also provide energy and other aids for the metabolic activities of microorganisms. A well-known example is the grass *Sorghum halepense*, which excretes sorgoleone, a very rich source of allelopathic properties (Czarnota et al. 2001), from its root hairs (Czarnota et al. 2003). Moreover, recent research has shown that microbes present in the rhizosphere have the ability to use allelopathic molecules as a carbon ion source (Gimsing et al. 2009). Thus, various benefits of these exudates make them important resources for microbes. Research has revealed that plants and microbes have good chemistry in production of plants, making them a good source of nutrients for the microbes.

Nutrients and other chemical molecules are taken up through the roots and then stored in free spaces in the roots. These act as a chemical currency for plants, as many significant processes involve these components, i.e., creation of protective layers and nutrient substrates, and other benefits in the ecosystem. Plant chemicals are related to specific characteristics present in the genotype (Lesuffleur et al. 2007), and their concentration and quality also depend on their composition (Carvalhais et al. 2011; Matilla et al. 2010). The antimicrobial defense system of the plant includes terpenoids, isoflavonoids, and flavonoids (Hardoim et al. 2008). Among them, isoprenoids are a wide range of metabolic compounds both functionally and structurally. Usually, primary metabolites activate many processes in cells; for example, molecules resulting from isoprenoids are the main factor in seed emergence (gibberellin 3 (GA3) and IAA) and photosynthesis (phytopigments). On the other hand, secondary metabolites help defend the plant against pathogens and expand the niche via allelopathic components.

There are different factors that influence the chemical composition of roots secretions: the concentration and level of carbon dioxide, drought stress, and shortage of nutrients. These factors greatly influence the phytochemistry of exudates; for example, higher concentrations of CO_2 have a major effect on the composition of exudates and large amounts of CO_2 are stored in the root zone. This concentration also varies in different plant species (Cheng and Gershenson 2007; Phillips et al. 2006), affecting the yield and biomass of plants both positively and negatively. In one reported study, CO_2 had a positive effect on biomass in clover and rye but had a negative effect on maize (Phillips et al. 2006), although it increased the amino acid content of the maize exudates. This was possibly a result of the photosynthetic pathway of C4 plants, which results in a high growth rate with higher concentrations of CO_2, although the raised quantity of amino acids could also have been due to microbial interaction in the rhizosphere of the maize plants (Bever 1994; Klironomos 2002). However, the phenomenon behind this is currently unclear; thus, scientists need to identify the main processes involved.

It is not necessary for all types of environmental stresses to elicit equal responses from the rhizosphere area and roots; for example, during shortages of nitrogen and phosphorus, maize plants release smaller amounts of amino acids and other organic acids (Carvalhais et al. 2011). In addition, microbes also play a major role in the quantity and quality of these plant roots exudates as they are also helpful in protecting plants from external damage. Less study has been done on plant exudate–microbe interactions, but their participation in plant growth is now more evident (Boller and He 2009; Doornbos et al. 2012; Reading and Sperandio 2006).

22.9.3 Microbial Uptake and Release

Plant exudates are the main basis of communities of microbes, which are formed as a result of them. In addition to being a currency for plants, exudates also help microbes in the forms of CO_2 and other nutrients. The rhizosphere contains large quantities of sugars and amino acids (Jones 1998) and a wide range of microbes (Bernedsen et al. 2012). Many researchers have tried to find ways to measure the utilization of those exudates through respiration and other assays, e.g., ECO microplates. Small molecules such as organic compounds accelerate the respiration rate with the help of microbes (van Hees et al. 2005), which may increase or decrease the quantity of nutrients involved in energy metabolism in microbe communities. The phenomenon behind this has been proved by research on soil amendments, which has shown that microbes present in rhizosphere can utilize these nutrients before the plant (Kielland 1994; Owen and Jones 2001). Some prokaryotes use a specific process for utilization of amino acids instead of sugar; this process is found in *Pseudomonas putida* KT2440. It was shown that in this PGPB, the same protein was used for uptake of amino acids as was previously used for sugar utilization (Moreno et al. 2009).

Secretion or exudation releases by microbes in the rhizosphere ecosystem contain various types of chemicals and their ions, which help in different processes such as protection from disease molecules, uptake of nutrients, and promotion of plant growth. The cycle of transformation of nutrients in the soil is complex. Thus, microbes are involved in biogeochemical cycling and their exudates act as catalysts in these transformation cycles. In these cycles, many plant nutrients are present that are essential for plant growth and development, such as nitrogen, phosphorus, alkaline metalloids, and other micronutrients such as zinc, iron, and cobalt. (Stevenson and Cole 1999).

Nitrogen is the main element for plant growth and other key stages and factors in the survival of plants, but this phenomenon is highly reliant on the role of the microbiome present in the rhizosphere. Nitrogen-fixing prokaryotes convert nitrogen gas into ammonia, which is an available form for primary producers to utilize. This process enters the nitrogen element into the transformation cycle of the ecosystem. Conversion of nitrogen gas takes place through the enzyme nitrogenase (Howard and Rees 1996). Genes related to nitrogenase have been found in both anaerobic and aerobic habitats (Zehr et al. 2003). Thus, cyclic activities are the results of microbial interactions and exudations in the soil ecosystem.

Sometime, microbial exudates contain significant excretions that are helpful in plant growth and increase levels of hormones that are essential for the plant, such as IAA and 1-aminocyclopropane-1-carboxylate (ACC). Inoculation with bacteria that are responsible for ACC deaminase in the rhizosphere of wheat enhanced the nutrient uptake efficiency and root development (Shaharoona et al. 2008). This enzyme suppressed the production of ethylene in the roots, which helped to decrease the effects of various environmental stresses (Honma and Shimomura 1978; Glick et al. 2007; Hardoim et al. 2008).

Microbial exudates contain the basis of various antifungal and antibiotic compounds that are used nowadays. Exudation of antimicrobial substances by prokaryotes and eukaryotes is considered a common process. However, only very small fractions of them are usable (Piel 2011). The reason behind this may be the complex interactions of those microbes within soil ecosystems (Buée et al. 2009; Curtis et al. 2002; Torsvik et al. 1990, 2002a, b). As a result, the microbes help to maintain plant diversity in the ecosystem (Czaran et al. 2002). Additionally, it was found that some exudates of microbes act as plant pathogens and can cause disease in plants; these are low molecular weight substances (Boller and He 2009). So, it is concluded that microbial exudates can trigger other microorganisms to react against the plant immune system or to suppress the plant immune system's ability to combat pathogenic substances, or they may accelerate the plant exudation process. Through detection of essential microbial exudates, many other functions of these exudates have been found, which is not surprising for soil microbial communities, as they also contain some functional redundancy.

22.10 Impacts on Plant Functions

Microbes present in the soil and in the human gut perform the same functions and activities for their hosts (Bernedsen et al. 2012)—for example, uptake of nutrients (Van et al. 2008) and pathogen protection (Doornbos et al. 2012)—and play roles in the immune system of the host (Neal et al. 2012; Neal and Ton 2013; Van et al. 2009). For exchange of nutrients and exudate release, plants need energy, which varies from species to species and from plant to plant, and also depends on the weight of the molecule being taken up or utilized. An active transport mechanism is required for high molecular weight transportation (Badri and Vivanco 2009). The concentrations and composition of root exudates are highly dependent on adenosine triphosphate (ATP)–binding cassette transporters (Badri et al. 2009). A diffusion process is used to transport low molecular weight substances through membranes (Badri and Vivanco 2009; Jones and Dangle 2006).

Plants are the basic consumers of microbial exudates, and study results have revealed that plants themselves create their own rhizosphere for microbial activities (Bernedsen et al. 2012; Friesen et al. 2011). For example, fluorescent pseudomonads, which can produce the antimicrobial compound 2,4-diacetylphloroglucinol (2,4-DAPG), are widely spread over a number of plant species (Mavrodi et al. 2011). This is a wide-spectrum antimicrobial and can protect plants against a wide range of pathogens, most of which are fungal (Raaijmakers et al. 2009). Control of pathogens is an activity performed by soil microbial organisms/communities through their exudates. Formation of 2,4-DAPG in root ecosystems is observed in a wide range of plant varieties, directly related to uptake of wheat (Mendes et al. 2011; Raaijmakers et al. 2009; Weller et al. 2002).

Along with protection from pathogens, these microbial exudates also have effects on plant traits. Microbes intercede in plants' biochemical reactions and functions; thus, they perform novel functions for cells by altering their pathways. Microbes perform a variety of metabolic activities through which they secrete chemicals into the plant, which are of benefit to it. All known hormones are produced by the activities of these microbes (Friesen et al. 2011). This novel ability alters the physiological mechanism of plants by increasing or suppressing phytohormones; for example, IAA has been found in 80% of bacteria present in plant root zones (Loper and Schroth 1986). IAA has the ability to slow down or accelerate plant growth through increases or decreases in its concentration (Glick 1999; Patton and Glick 1996; Sarwar and Kremer 1995). The concentration effect on the growth of plants was first reported in a microbial community where many prokaryotes were present and produced large amounts of IAA, whereas a community containing fewer prokaryotes produced only low concentrations of IAA (Rout et al. 2013a, b). Variations in the concentration of IAA are also affected by climate stress and the phenology of the plant, but these hormones increase plant robustness and stress tolerance (Friesen et al. 2011; Kaplan et al. 2013). PGPBs that contain a variety of genes and express them where necessary, especially for plant growth mechanisms, are called "competent" (Hardoim et al. 2008). When PGPBs are effective in dual plant growth processes such as ACC deaminase and phosphate solubilization, they are considered to have "dual traits" (Baig et al. 2012).

22.11 Impacts on Bacterial Functions

In the root zone, the primary advantage is the availability of amino acids and organic substances from microbial activities (Nelson 2004). Through root exudates, various phenotypic functions are carried out—for example, chemical effects, stress tolerance, gene modulation of sporulation and competence, formation of biofilms on roots, and degradation of polychlorinated biphenyl (Amador et al. 2010; Toussaint et al. 2012; Mader et al. 2002; Rudrappa et al. 2008). Biofilm formation and disassembly by rhizospheric bacteria are carried out by the actions of root exudates

(Kolodkin-Gal et al. 2010). By formation of biofilm on roots, plants are protected from many microbial activities and other fungal attacks. These exudates may be terpenoids, isoflavonoids, or flavonoids (Hardoim et al. 2008). These microbes act as symbionts for plants and protect them from pathogens, i.e., as epiphytes or endophytes, as well as aiding plant growth.

22.12 Ecology of the Microbiome

In this review, only plant-related microbes such as rhizospheric endophytes and epiphytes are discussed. They are plant-related organisms that directly or indirectly affect the traits of plants and soil ecosystems. Although their effects on aboveground seeds and other habitats are often ignored, protection against herbivory, seed pollination, and protection against predation and pathogen attack of pathogen are among their effects above ground (Friesen et al. 2011). These interactions deserve to be expanded upon so that plant genome spread over allover the globe and variation should be exhibited. In the ecosystem, various factors make up the microbial cycle structure that benefits the plant; these factors are the microorganisms themselves, the plant, and the soil, all of which form the complete structure and its cycles.

22.12.1 The Rhizosphere and Rhizoplane

It is very difficult to define a rhizospheric community, and research done using third- and fourth-generation techniques has shown that plants are highly influenced by these organisms (Hiltner 1904). Through use of modern technologies, the interactions of microbes with plants are now being revealed and many genes that are involved have now been identified. Many biotic and abiotic stresses in soils are due to climate stress, which affects microbial activities in the soil. These changes in soils and microorganism activities determine the difference between microbial communities and their zones. This happens in arid areas, where abundant microbial communities are present in soils, showing great diversity (Aguirre-Garrido et al. 2012; Ben-David et al. 2011; Kaplan et al. 2013; Yu et al. 2012). Because of these microbes, the soil gains a high water-holding capacity along with more nutrients and a more diverse microbial population (Schade and Hobbie 2005).

Like gut function, the state of the root environment is informative about the health of the plant and the state of the microbial community when it is affected by disease, and a healthy rhizosphere can be detected from its microbiome (Burdon and Thrall 2009). For better quality of plants, production, and sustainability, the rhizosphere and rhizoplane are studied in different aspects. The mechanisms behind plant interactions and microbial activity should be clear to understand the positive and negative effects of those activities on the soil ecosystem. For example, PGPBs have been shown to enhance induced systemic resistance, detected from jasmonic acid priming in plant leaves and broad resistance to pathogens (Van et al. 2008; Pineda et al. 2010; Vander Ent et al. 2009; Van Oosten et al. 2008). Successively, plants show a great range of defense signaling in the upper layer of the soil as well as below it (Ahmad et al. 2011; Neal et al. 2012; Neal and Ton 2013). In the rhizosphere, development of a web network is started by the plant and microbes, and when the variety of plant reservoirs is known, it can be easy to find solutions for different problems in the soil ecosystem; such problems include climate change, low yields, low soil fertility, pathogen attacks, and bioremediation (Bernedsen et al. 2012; Curtis et al. 2002).

22.12.2 Epiphytes and Endophytes

Communities present as root endophytes and epiphytes may change completely, as in the rhizosphere, with evolution and symbiotic effects (Boller and He 2009; Compant et al. 2010; Compant et al. 2005) promoting a good relationship with the plant. Microbial colonization of plant tissues in either an epiphytic or endophytic manner is based on the molecular methods that are utilized. Through exudate communications, a colonization process occurs between the microbes and the plant (Deakin and Broughton 2009; Elasri et al. 2001; Hardoim et al. 2008). Microbes use a specific approach to colonize their host partner for exchange of substances between them. For example, maize exudates released to induce systemic resistance are helpful in recruitment of the PGPB *P. putida* (Neal et al. 2012). Underlying this, a chemotaxis gene has been found in varieties of microbes living in the rhizosphere, and its expression is necessary for improvement of plant traits and soil treatment (Hardoim et al. 2008). In plants, different chemical exudates that are beneficial to PGPBs are produced as part of plant responses; an organic acid in tomato is an example of a chemotactic substance (de Weert et al. 2002) as is an amino acid in rice (Bacilio-Jimenez et al. 2003). Genes have been found in bacteria that aid establishment of colonies on roots, with type IV pilli enabling movement, efflux of isoflavonoids, and rearrangement of DNA, further affecting microbial colonies (Bohm et al. 2007; Palumbo et al. 1998; Dekkers et al. 1998).

Microbes present in plant tissue show endophytic lifestyles (Bulgarelli et al. 2013). Many domesticated crops contain different endophytic microorganisms, while many wild-type species contain invasive microbes (Compant et al. 2008; Hallmann et al. 1997; James et al. 2002; Rout and Chrzanowski 2009). The importance of endosymbiont interactions with plants is largely unknown except for a few of them, such as nodule bacteria in leguminous crops and fungal mycorrhizae. For decades, the importance of nitrogen-fixing bacteria has been known to be significant in sugarcane and *Sorghum bicolor* (Baldani et al. 1996; James et al. 1997; Kirchof et al. 2001).

In agricultural biotechnology, genes related to microbes that influence plant growth and yield have attracted more interest. For example, *Bacillus* strains show ACC deaminase production along with phosphate solubilization in plants, functioning as a growth promoter with dual traits (Baig et al. 2012). However, other phosphate-solubilizing bacteria merely enhance plant growth without producing ACC deaminase and thus do not show dual traits (Zaidi and Khan 2005).

Transmission of plant epiphytes and endophytes occurs horizontally (Friesen et al. 2011). This causes a struggle between symbionts for expressions of traits in the plant, which may be expressed as alterations in the ecosystem (Bever et al. 2009; Kiers and Denison 2008). By this coevolutionary mechanism, evolutionary shifts are caused by symbiotic and mutualistic effects. Endophytes such as fungi are transmitted vertically into host plants, depending on host fitness (Clay and Schardl 2002). As these symbionts are transferred sexually to the host, chances of increased evolution exist between them. Vertical transmission is more beneficial for the plant than horizontal transmission (Clay and Schardl 2002; Sachs et al. 2004).

Transmission can be applicable in a wide range of environments and influences host evolutionary mechanisms, as in the case of *Sorghum halepense*, where it only transmitted N_2-fixing endophytic organism to soil (Rout et al. 2013a, b). This shows the capability of invaders to regulate the pathogen responses consequents by those endosymbionts (N_2-fixing endophytic organism). It is hypothesized that when the host is very close to the microbes, there is a high degree of transmission of organisms (Rudgers et al. 2009). It has been stated that plant pathogens increase their density when in a monoculture form (Tilman 2000), so endophytes of *Sorghum* are transmitted horizontally in plant rhizomes. This mechanism is common in invader plants, as the endophytes reproduce sexually in the plant rhizosphere and a wide range of endophytes are present in the host roots (Baldani et al. 1996; James et al. 1997; Kirchof et al. 2001; Rout and Chrzanowski 2009). Comprehensive study of plants and microbes has determined the activity and interactions of microbes with their host plants and has shown their genetic and environmental effects on their ecology.

22.13 Importance of the Microbiome to Plant Genomics

Genomic study of plants has shown the mechanism behind the interaction of plants with microbes and has helped to identify the factors responsible for better plant performance. The condition and structure of the macrobiotic community are greatly influenced by plant traits; for example, roots exhibit characteristics of the microbiome (Morris and Djordjevic 2006; Spaepen et al. 2008). Changes in root structure show overall plant health; by an increase in border cells, resistance to pathogens is also increased (Chen et al. 2012) because border cells provide a defensive layer at the root tip (Curlango-Rivera et al. 2013). In agribusiness, enhancement of root growth systems for control of pathogenic effects is being actively considered, as many invasive plants spread below ground and thereby expand their niche. Thus, these invasive plants increase their numbers and can sometimes become very problematic invaders (Rout et al. 2013a, b). By suppression of these invading plants in the soil, their growth is also retarded in comparison with the growth of unsuppressed plants. Thus, the specific role of microbes in plant development is understandable in molecular techniques, particularly use of RNA-Seq. The impacts of microbes on plants are very broad in general, affecting the growth of roots, shoots, leaves, stems, and even flowers. Microbial activity has also been identified in invasive species, and their physiological mechanisms, shoot and leaf allocation, growth and fitness, and total biomass are also the result of microbial activities (Friesen et al. 2011; van Kleunen et al. 2010).

Different microbes can cause differences in the plant microbiome because of their various behaviors and can influence the plant genome and phenotypic traits according to their structures. Plant microbes also cause changes in the ecosystem by their direct effects. Changes created by these plant microbes have effects at a group level, such as pathogen attacks on wheat and rye, as reported earlier (Burdon and Thrall 2009). Moreover, by supporting the plants through an enhanced root system for expansion of their niche, those microbes also aid plant ecology by influencing plant-to-plant competition, herbivory, defense, and pollination (Klironomos 2002; Cahill et al. 2008; Friesen et al. 2011). Limitations in nutrients and traits for acquisition of these resources greatly affect the primary productivity of plants (Lambers et al. 2008), as well as that of the microbes involved in biogeochemical cycles.

In microbial communities, structure and function are of great significance and warrant more attention from researchers. For persistence of plants, adaptive traits are dictated by ecology. Severe environments such as deserts, drought, and salt stresses are areas where microbiomes are studied for their mechanisms related to drought stress (Kaplan et al. 2013). Microbes plays major roles in the functions and performance of plant metabolism. A few years ago, the roles of microbiomes were studied, and they are known to be essential aspects of research on the genetics and metabolism of host plants. Modern technologies—i.e., third- and fourth-generation sequencing, if used—will be helpful in finding trait loci for construction of genomic libraries. Several known quantitative loci traits have been used for identification of genomes of interest (Hu et al. 2003; Jang et al. 2008; Paterson et al. 1995). By identification of phenotypic variations in plants, caused by gene signaling molecules, the microbial contributions to those functions can be studied. Extreme conditions may be biotic or abiotic; both are influenced by regulation of plant cascade hormones and pathogen attacks. Microbiome selection by specific host plants is very complicated. Plants use environmental factors to increase their ability to select microbes, but these are not completely adequate for the purpose; it is use of exudates that gives plants greater control in selecting their microbial community (Doornbos et al. 2012).

22.14 Conclusion

Environmental evidence has shown the significance of microbe interactions with plants and their growth (Friesen et al. 2011). The plant microbiome is a major aspect of expansion of the genome of plants in an environment through feedback mechanisms. These feedback mechanisms are important in plant–soil–microbial interactions (Bever 1994). The types of impacts that soil microbes have on plants may vary over time to become parasitic or mutualistic (Callaway and Rout 2011). Irrelevant growth of microbial parasites shows negative impacts on plants, whereas beneficial microbes give positive feedback to the plants (Klironomos 2002). These plant–soil–microbial interactions and shifts in biodiversity affect the structure and functions of the ecosystem and become more complex in determining plant growth and gene expression. These gene patterns are present in specific plant species. The occurrence of horizontal transfer of genes and rapid evolution of microbes facilitate more phenomena of genetic differences and variations in local plants (Rout and Callaway 2012). New technology has provided better understanding of plant microbiomes, ecology, and transcriptomes in plants. This process is done by microbes in rhizosphere (Bernedsen et al. 2012; Curtis et al. 2002) and are linked with idle genes and is necessary for the nutrient transmission as involved in nitrogen-fixing bacteria (Zehr et al. 2003). Dispersal favors plants that can transfer their microbes and motivate them to persist in different soils (de Bello et al. 2010). Previous studies have shown that variations in plants influence the composition of the microbiome (Bernedsen et al. 2012).

Study of the complexity of these mechanisms has helped us to understand the selection and quantification of various microbial functions and exudates of plants, and their impacts on genome functioning. Understanding of the functions and mechanisms of an organism provides greater understanding of the whole organism itself. For all plants, the presence and activities of microbes are of the utmost importance. The influences of microbes and their functions on plants vary between species and between plants, and they also vary depending on genetic and environmental factors. Scientists should focus on structural changes and behavioral changes of microbes within soil ecosystems, as well as functional changes, so that changes related to these microbes in plants in various habitats can be understood. In addition, functional screening using metagenomics and metatranscriptomics will help to determine plant phonological traits developed by microbiomes and will also assess the function of the "second genome" in plants.

References

Abd-Elsalam KA, Almohimeed I, Moslem MA, Bahkali AH (2010) M13-microsatellite PCR and rDNA sequence markers for identification of *Trichoderma* (Hypocreaceae) species in Saudi Arabian soil. Genet Mol Res 9:2016–2024

Aguirre-Garrido JF, Montiel-Lugo D, Hernandez-Rodriguez F, Torres-Cortes G, Millan V, Toro N (2012) Bacterial community structure in the rhizosphere of three cactus species from semi-arid highlands in Central Mexico. Antonie Van Leeuwenhoek 101:891–904

Ahmad S, Veyrat N, Gordon-Weeks R, Zhang Y, Martin J, Smart L (2011) Benzoxazinoid metabolites regulate innate immunity against aphids and fungi in maize. Plant Physiol 157:317–327

Alzubaidy H, Essack M, Malas TB, Bokhari A, Motwalli O, Kamanu FK, Jamhor SA, Okhtar NA, Antunes A, Simoes MF, Alam I, Bougouffa S, Lafi FF, Bajic VB, Archer JAC (2016) Rhizosphere microbiome metagenomics of gray mangroves (*Avicennia marina*) in the Red Sea. Gene 576:626–636

Amador CI, Canosa I, Govantes F, Santero G (2010) Lack of CbrB in *Pseudomonas putida* affects not only amino acids metabolism but also different stress responses and biofilm development. Environ Microbiol 12:1748–1761

Arjun JK, Harikrishnan K (2011) Metagenomic analysis of bacterial diversity in the rice rhizosphere soil microbiome. Biotechnol Bioinformatics Bioeng 1:361–367

Aßhauer KP, Wemheuer B, Daniel R, Meinicke P (2015) Tax4Fun: predicting functional profiles from metagenomic 16S rRNA data. Bioinformatics 31:2882–2884

Bacilio-Jimenez M, Aguilar-Flores S, Ventura-Zapata E, Perez-Campos E, Bouquelet S, Zenteno E (2003) Chemical characterization of root exudates from rice (*Oryza sativa*) and their effects on the chemotactic response of endophytic bacteria. Plant Soil 249:271–277

Badri DV, Vivanco JM (2009) Regulation and function of root exudates. Plant Cell Environ 32:666–681

Badri DV, Quintana N, El Kassis EG, Kim K, Choi YH (2009) An ABC transporter mutation alters root exudation of phytochemicals that provoke an overhaul of natural soil microbiota. Plant Physiol 151:2006–2017

Bai XG, Todd CD, Desikan R, Yang YP, Hu XY (2012) N-3-oxo-decanoyl-lhomoserine-lactone activates auxin-induced adventitious root formation via hydrogen peroxide– and nitric oxide–dependent cyclic GMP signaling in mung bean. Plant Physiol 158:725–736

Baig KS, Arshad M, Shaharoona B, Khalid A, Ahmed I (2012) Comparative effectiveness of *Bacillus* spp. possessing either dual or single growth-promoting traits for improving phosphorus uptake, growth and yield of wheat (*Triticum aestivum* L.). Ann Microbiol 62:1109–1119

Bais HP, Park SW, Weir TL, Callaway RM, Vivanco JM (2004) How plants communicate using the underground information superhighway. Trends Plant Sci 9:26–32

Baldani JI, Pot B, Kirchof G, Falsen E, Baldani VLD, Olivares FL (1996) Emended description of *Herbaspirillum*; inclusion of [*Pseudomonas*] *rubrisubalbicans*, a mild plant pathogen, as *Herbaspirillum rubrisubalbicans* comb. nov.; and classification of a group of clinical isolates (Ef group 1) as *Herbaspirillum* species 3. Int J Syst Bacteriol 46:802–810

Bell TH, El-Din Hassan S, Lauron-Moreau A, Al-Otaibi F, Hijri M, Yergeau E, St-Arnaud M (2014) Linkage between bacterial and fungal rhizosphere communities in hydrocarbon-contaminated soils is related to plant phylogeny. ISME J 8:331–343. https://doi.org/10.1038/ismej.2013.149

Ben-David EA, Zaady E, Sher Y, Nejidat A (2011) Assessment of the spatial distribution of soil microbial communities in patchy arid and semi-arid landscapes of the Negev Desert using combined PLFA and DGGE analyses. FEMS Microbiol Ecol 76:492–503

Bernedsen RL, Pieterse CMJ, Bakker AHM (2012) The rhizosphere microbiome and plant health. Trends Plant Sci 17:478–486

Bertrand H, Poly F, Van VT, Lombard N, Nalin R, Vogel TM, Simonet P (2005) High molecular weight DNA recovery from soils prerequisite for biotechnological metagenomic library construction. J Microbiol Methods 62:1–11

Bever JD (1994) Feedback between plants and their soil communities in an old field community. Ecology 75:1965–1977

Bever JD, Richardson SC, Lawrence BM, Holmes J, Watson M (2009) Preferential allocation to beneficial symbiont with spatial structure maintains mycorrhizal mutualism. Ecol Lett 12:13–21

Bhattacharyya PKS, Roy M, Dasa S, Raya D, Balachandar S, Karthikeyan AK, Mohapatra NT (2016) Elucidation of rice rhizosphere metagenome in relation to methane and nitrogen metabolism under elevated carbon dioxide and temperature using whole genome metagenomic approach. Sci Total Environ 542:886–898

Biteen JS, Paul C, Blainey CZG, Chun M, George M (2016) Church tools for the microbiome: nano and beyond. ACS Nano 10:6–37

Bodenhausen N, Horton MW, Bergelson J (2013) Bacterial communities associated with the leaves and the roots of *Arabidopsis thaliana*. PLoS One 8:e56329. https://doi.org/10.1371/journal.pone.0056329

Bohm M, Hurek T, Reinhold-Hurek B (2007) Twitching motility is essential for endophytic rice colonization by the N_2-fixing endophyte *Azoarcus* sp. strain BH72. Mol Plant-Microbe Interact 20:526–533

Boller T, He SY (2009) Innate immunity in plants: an arms race between pattern recognition receptors in plants and effectors in microbial pathogens. Science 324:742–744

Bresolin JD, Bustamante MMC, Kruger RH, Silva MRSS, Perez KS (2010) Structure and composition of bacterial and fungal community in soil under soybean monoculture in the Brazilian Cerrado. Braz J Microbiol 41:391–403

Brinkmann N, Tebbe C (2007) Differences in the rhizosphere bacterial community of a transplastomic tobacco plant compared to its non-engineered counterpart. Environ Biosaf Res 6:113–119

Broeckling CD, Broz AK, Bergelson J, Manter DK, Vivanco JM (2008) Root exudates regulate soil fungal community composition and diversity. Appl Environ Microbiol 74:738–744

Buée M, Reich M, Murat C, Morin E, Nilsson RH, Uroz S (2009) 454 pyrosequencing analyses of forest soils reveal an unexpectedly high fungal diversity. New Phytol 184:844–856

Bulgarelli D, Schlaeppi K, Spaepen S, van Themaat EVL, Schulze-Lefert P (2013) Structure and functions of the bacterial microbiota of plants. Annu Rev Plant Biol 64:807–838

Burdon JJ, Thrall PH (2009) Coevolution of plants and their pathogens in natural habitats. Science 324:755–756

Buscot F, Varma A (eds) (2005) Microorganism in soils: roles in genesis and functions. Springer, Berlin/Heidelberg, pp 139–153

Cahill JF, Elle E, Smith GR, Shore BH (2008) Disruption of a belowground mutualism alters interactions between plants and their floral visitors. Ecology 89:791–801

Callaway RM, Rout ME (2011) Soil biota and plant invasions: biogeographic effects on plant–microbe interactions. In: Richardson DH (ed) Fifty years of invasion ecology: the legacy of Charles Elton. Wiley-Blackwell, West Sussex, pp 131–142

Caporaso JG, Kuczynski J, Stombaugh J, Bittinger K, Bushman FD, Costello EK, Fierer N, Peña AG, Goodrich JK, Gordon JI, Huttley GA, Kelley ST, Knights D, Koenig JE, Ley RE, Chaparro JM, Badri DV, Vivanco JM (2014) Rhizosphere microbiome assemblage is affected by plant development. ISME J 8:790–803

Carvalhais LC, Dennis PG, Fedoseyenko D, Hajirezaei MR, Borriss R, von Wiren N (2011) Root exudation of sugars, amino acids, and organic acids by maize as affected by nitrogen, phosphorus, potassium, and iron deficiency. J Plant Nutr Soil Sci 174:3–11

Chaparro JM, Badri DV, Vivanco JM (2014) Rhizosphere microbiome assemblage is affected by plant development. ISME J 8:790–803

Chen AH, Chen LJ, Wu ZJ (2012) Relationships among persistence of *Bacillus thuringiensis* and cowpea trypsin inhibitor proteins, microbial properties and enzymatic activities in the rhizosphere soil after repeated cultivation with transgenic cotton. Appl Soil Ecol 53:23–30

Cheng W, Gershenson A (2007) Carbon fluxes in the rhizosphere. In: Cardon ZG, Whitbeck JL (eds) The rhizosphere—an ecological perspective. Academic, San Diego, pp 31–56

Chhabra S, Brazil D, Morrissey J, Burke JI, Gara FO, Dowling DN (2013) Characterization of mineral phosphate solubilization traits from a barley rhizosphere soil functional metagenome. MicrobiologyOpen 2:717–724

Clay K, Schardl C (2002) Evolutionary origins and ecological consequences of endophyte symbiosis with grasses. Am Nat 160:99–127

Compant S, Duffy B, Nowak J, Clement C, Barka EA (2005) Use of plant growth–promoting bacteria for biocontrol of plant diseases: principles, mechanisms of action, and future prospects. Appl Environ Microbiol 42:669–678

Compant S, Kaplan H, Sessitsch A, Nowak J, Barka EA, Clement C (2008) Endophytic colonization of *Vitis vinifera* L. by *Burkholderia phytofirmans* strain PsJN: from the rhizosphere to inflorescence tissues. FEMS Microbiol Ecol 63:84–93

Compant S, Clement C, Sessitsch A (2010) Plant growth–promoting bacteria in the rhizo- and endosphere of plants: their role, colonization, mechanisms involved and prospects for utilization. Soil Biol Biochem 71:4951–4959

Curlango-Rivera G, Pew T, Vanetten HD, Zhongguo X, Yu N, Hawes MC (2013) Measuring root disease suppression in response to a compost water extract. Phytopathology 103:255–260

Curtis TP, Sloan WT, Scannell JC (2002) Estimating prokaryotic diversity and its limits. Proc Natl Acad Sci U S A 99:10494–10499

Czaran TL, Hoekstra RF, Pagie L (2002) Chemical warfare between microbes promotes biodiversity. Proc Natl Acad Sci U S A 99:786–790

Czarnota MA, Paul RN, Dayan FE, Nimbal CI, Weston LA (2001) Mode of action, localization of production, chemical nature, and activity of sorgoleone: a potent PSII inhibitor in *Sorghum* spp. root exudates. Weed Technol 15:813–825

Czarnota MA, Rimando AM, Weston LA (2003) Evaluation of root exudates of seven sorghum accessions. J Chem Ecol 29:2073–2083
Daniel R (2005) The metagenomics of soil. Nat Rev Microbiol 3:470–478
de Bello F, Lavorel S, Diaz S, Harrington R, Cornelissen JHC (2010) Towards an assessment of multiple ecosystem processes and services via functional traits. Biodivers Conserv 19:2873–2893
de Weert S, Vermeiren H, Mulders IHM, Kuiper I, Hendrickx N, Bloemberg GV (2002) Flagella-driven chemotaxis towards exudate components is an important trait for tomato root colonization by *Pseudomonas fluorescens*. Mol Plant-Microbe Interact 15:1173–1180
Deakin WJ, Broughton WJ (2009) Symbiotic use of pathogenic strategies: rhizobial protein secretion systems. Nat Rev 7:312–320
Dekkers LC, Phoelich CC, van der Fits L, Lugtenberg JJ (1998) A site-specific recombinase is required for competitive root colonization by *Pseudomonas fluorescens* WCS365. Proc Natl Acad Sci U S A 95:7051–7056
Dinsdale EA, Edwards RA, Hall D, Angly F, Breitbart M, Brulc JM (2008) Functional metagenomic profiling of nine biomes. Nature 452:629–628
Doornbos R, van Loon L, Bakker P (2012) Impact of root exudates and plant defense signaling on bacterial communities in the rhizosphere. Agron Sustain Dev 32:227–234
Egamberdiyeva D, Kamilova F, Validov S, Gafurova L, Kucharova Z, Lugtenberg B (2008) High incidence of plant growth-stimulating bacteria associated with the rhizosphere of wheat grown in salinated soil in Uzbekistan. Environ Microbiol 10:1–9
Elasri M, Delorme S, Lemanceau P, Stewart G, Laue B, Glickman E (2001) Acyl-homoserine lactone production is more common among plant-associated *Pseudomonas* spp. than among soilborne *Pseudomonas* spp. Appl Environ Microbiol 67:1198–1209
Ferluga S, Venturi V (2009) OryR is a LuxR-family protein involved in interkingdom signaling between pathogenic *Xanthomonas oryzae* pv. *oryzae* and rice. J Bacteriol 191:890–897
Friesen ML, Porter SS, Stark SC, von Wettberg EJ, Sachs JL, Martinez Romero E (2011) Microbially mediated plant functional traits. Annu Rev Ecol Evol Syst 42:23–46
Gans J, Wolinsky M, Dunbar J (2005) Computational improvements reveal great bacterial diversity and high metal toxicity in soil. Science 309:1387–1390
Gerlach W, Stoye J (2011) Taxonomic classification of metagenomic shotgun sequences with CARMA3. Nucleic Acids Res 39:e91
Gerlach W, Jünemann S, Tille F, Goesmann A, Stoye J (2009) WebCARMA: a web application for the functional and taxonomic classification of unassembled metagenomic reads. BMC Bioinformatics 10:430. https://doi.org/10.1186/1471-2105-10-430
Gimsing AL, Blum J, Dyan FE, Locke MA, Sejer LH, Jacobsen CS (2009) Mineralization of the allelochemical sorgoleone in soil. Chemosphere 76:1041–1047
Glick BR (1999) The enhancement of plant growth by free-living bacteria. Can J Microbiol 41:109–117
Glick BR, Todorovic B, Czarny J, Cheng Z, Duan J, McConkey B (2007) Promotion of plant growth by bacterial ACC deaminase. Crit Rev Plant Sci 26:227–242
Greenblum S, Turnbaugh PJ, Borenstein E (2012) Metagenomic systems biology of the human gut microbiome reveals topological shifts associated with obesity and inflammatory bowel disease. Proc Natl Acad Sci U S A 109:594–599
Haldar S, Sengupta S (2015) Impact of plant development on the rhizobacterial population of *Arachis hypogaea*: a multifactorial analysis. J Basic Microbiol 55:922–928
Hallmann L, Quadt-Hallmann A, Mahaffee WF, Kloepper JW (1997) Bacterial endophytes in agricultural crops. Can J Microbiol 43:895–914
Hardoim PR, van Overbeek LS, van Elsas JD (2008) Properties of bacterial endophytes and their proposed role in plant growth. Trends Microbiol 16:463–471
Harris D, Pacovsky RS, Paul EA (1985) Carbon economy of soybean–*Rhizobium*–*Glomus* associations. New Phytol 101:427–440
Hartmann A, Rothballer M, Hense BA, Schröder P. Bacterial (2014) quorum sensing compounds are important modulators of microbe-plant interactions. Frontiers in plant science. 5:131

Hernandez-Leon R, Martinez-Trujillo M, Valencia-Cantero E (2012) Construction and characterization of a metagenomic DNA library from the rhizosphere of wheat (*Triticum aestivum*). Phyton Int J Exp Bot 81:12

Hiltner L (1904) Über neuere Erfahrungen und Probleme auf dem Gebiet der Bodenbakteriologie und unter besonderer Berucksichtigung der Grundungung und Brache. Arbeiten der Deutschen Landwirtschaftlichen Gesellschaft 98:59–78

Honma M, Shimomura T (1978) Metabolism of 1-aminocyclopropane-1-carboxylic acid. Agric Biol Chem 42:1825–1831

Howard JB, Rees DC (1996) Structural basis of biological nitrogen fixation. Chem Rev 96:2965–2982

Hu FY, Tao DY, Sacks E, Xu P, Li J (2003) Convergent evolution of perenniality in grasses. Proc Natl Acad Sci U S A 100:4050–4054

Hugenholtz P, Tyson GW (2008) Microbiology: metagenomics. Nature 455:481–483

Huson DH, Weber N (2013) Microbial community analysis using MEGAN. Methods Enzymol 531:465–485. https://doi.org/10.1016/B978-0-12-407863-5.00021-6

Illeghems K, Weckx S, De Vuyst L (2015) Applying meta-pathway analyses through metagenomics to identify the functional properties of the major bacterial communities of a single spontaneous cocoa bean fermentation process sample. Food Microbiol 50:54–63

Jackson CR, Randolph KC, Osborn SL, Tyler HL (2013) Culture dependent and independent analysis of bacterial communities associated with commercial salad leaf vegetables. BMC Microbiol 13:274

Jaeger CH III, Lindow SE, Miller W, Clark E, Firestone MK (1999) Mapping of sugar and amino acid availability in soil around roots with bacterial sensors of sucrose and tryptophan. Appl Environ Microbiol 65:2685–2690

James E (2000) Nitrogen fixation in endophytic and associative symbiosis. Field Crop Res 65:197–209

James EK, Olivares FL, Baldani JI, Dobreiner J (1997) *Herbaspirillum*, an endophytic diazotroph colonizing vascular tissue in leaves of *Sorghum bicolor*. J Exp Bot 48:785–797

James EK, Gyaneshwar P, Mathan N, Barraquio WL, Reddy PM, Iannetta PP (2002) Infection and colonization of rice seedlings by the plant growth–promoting bacterium *Herbaspirillum seropedicae* Z67. Mol Plant-Microbe Interact 15:894–906

Jang CS, Lemke C, Tang H, Bowers JE, Paterson AH (2008) Evolutionary fate of rhizome specific genes in a non-rhizomatous *Sorghum* genotype. Heredity 102:266–273

Jin Z, Di Rienzi SC, Janzon A, Werner JJ, Angenent LT, Dangl JL, Fowler DM, Ley RE (2016) Novel rhizosphere soil alleles for the enzyme 1- aminocyclopropane-1-carboxylate deaminase queried for function with an in vivo competition assay. Appl Environ Microbiol 82:1050–1059

Jones DL (1998) Organic acids in the rhizosphere—a critical review. Plant Soil 1:25–44

Jones JDG, Dangle JL (2006) The plant immune system. Nature 444:323–329

Kaplan D, Maymon M, Agapakis CM, Lee A, Wang A, Prigge BA (2013) A survey of the microbial community in the rhizosphere of two dominant shrubs of the Negev Desert highlands, *Zygophyllum dumosum* Boiss. and *Atriplex halimus*, using cultivation-dependent and -independent methods. Am J Bot 100:1713–1725

Kielland K (1994) Amino acid absorption by arctic plants: implications for plant nutrition and nitrogen cycling. Ecology 75:2373–2383

Kiers ET, Denison RF (2008) Sanctions, cooperation, and the stability of plant rhizosphere mutualisms. Annu Rev Ecol Evol Syst 39:215–236

Kirchof G, Eckert B, Stoffels M, Baldani JI, Reis V, Hartman A (2001) *Herbaspirillum frisingense* sp. nov., a new nitrogen-fixing bacterial species that occurs in C4-fibre plants. Int J Syst Evol Microbiol 51:157–168

Klironomos JN (2002) Feedback with soil biota contributes to plant rarity and invasiveness in communities. Nature 417:67–70

Knief C, Delmotte N, Chaffron S, Stark M, Innerebner G, Wassmann R, von Mering C, Vorholt A (2011) Metaproteogenomic analysis of microbial communities in the phyllosphere and rhizosphere of rice. ISME J:1–13

Knief C, Delmotte N, Chaffron S, Stark M, Innerebner G, Wassmann R, von Mering C, Vorholt JA (2012) Metaproteogenomic analysis of microbial communities in the phyllosphere and rhizosphere of rice. ISME J 6:1378–1390

Koeth RA, Wang Z, Levison BS, Buffa JA, Org E, Sheehy BT (2013) Intestinal microbiota metabolism of L-carnitine, a nutrient in red meat, promotes atherosclerosis. Nat Med 19:576–585. https://doi.org/10.1038/nm.3145

Kolodkin-Gal I, Romero D, Cao S, Clardy J, Kolter R, Losick R (2010) D-amino acids trigger biofilm disassembly. Science 328:627–629

Kumar S, Krishnani KK, Bhushan B, Brahmane MP (2015) Metagenomics: retrospect and prospects in high throughput age. Biotechnol Res Int 2015:121735. https://doi.org/10.1155/2015/121735

Lagos ML, Maruyama F, Nannipieri P, Mora ML, Ogram A, Jorquera MA (2015) Current overview on the study of bacteria in the rhizosphere by modern molecular techniques: a mini review. J Soil Sci Plant Nutr 15:04–523

Lakshmanan V, Selvaraj G, Bais HP (2014) Functional soil microbiome: belowground solutions to an aboveground problem. Plant Physiol 166:689–700

Lambers H, Raven JA, Shaver GR, Smith SE (2008) Plant nutrient-acquisition strategies change with soil age. Trends Ecol Evol 23:95–103

Langille MGI, Zaneveld J, Caporaso JG, McDonald D, Knights D, Reyes JA (2013) Predictive functional profiling of microbial communities using 16S rRNA marker gene sequences. Nat Biotechnol 31:814–821

LeBlanc N, Kinkel LL, Kistler HC (2015) Soil fungal communities respond to grassland plant community richness and soil edaphics. Microb Ecol 70:188–195

Lesuffleur F, Paynel F, Bataille MP, Cliquet JB (2007) Root amino acid exudation: measurement of high efflux rates of glycine and serine from six different plant species. Plant Soil 294:235–246

Li C, Weidong XL, Ying K, Wang WJ (2010) Effect of monoculture soybean on soil microbial community in the northeast China. Plant Soil 330:423

Lin W, Wu L, Lin S, Zhang A, Zhou M, Lin R, Wang H, Chen J, Zhang Z, Lin R (2013) Metaproteomic analysis of ratoon sugarcane rhizospheric soil. BMC Microbiol 13:135

Loper JE, Schroth MN (1986) Influence of bacterial source of indole-3-acetic acid on root elongation of sugar beet. Phytopathology 76:386–389

Mader U, Antelmann H, Buder T, Dahl MK, Hecker M, Homuth G (2002) *Bacillus subtilis* functional genomics: genome-wide analysis of the DegS-DegU regulon by transcriptomics and proteomics. Mol Gen Genomics 268:455–467

Mahyarudin I, Rusmana YL (2015) Metagenomic of actinomycetes based on 16S rRNA and nifH genes in soil and roots of four Indonesian rice cultivars using PCR-DGGE. Hayati J Biosci 22:113–121

Majeed A, Abbasi KM, Hameed S, Imran A, Rahim N (2015) Isolation and characterization of plant growth–promoting rhizobacteria from wheat rhizosphere and their effect on plant growth promotion. Front Microbiol 6:198

Martinez-Romero E (2006) Dinitrogen-fixing prokaryotes. In: Dworkin M, Falkow S, Rosenberg E, Schleifer KH, Stackebrandt E (eds) The prokaryotes. Springer, New York, pp 793–817

Massart S, Martinez-Medina M, Jijakli MH (2015) Biological control in the microbiome era: challenges and opportunities. Biol Control 89:108

Matilla MA, Ramos JL, Bakker P, Doornbos R, Badri DV, Vivanco JM (2010) *Pseudomonas putida* KT2440 caused induced systemic resistance and changes in *Arabidopsis* root exudation. Environ Microbiol Rep 2:381–388

Mavrodi DV, Mavrodi OV, Parejko JA, Weller DM, Thomashow LS (2011) The role of 2,4-diacetylphloroglucinol– and phenazine-1-carboxylic acid–producing *Pseudomonas* spp. In: Maheshwari DK (ed) Natural protection of wheat from soilborne pathogens. Bacteria in agrobiology: plant nutrient management. Springer, Berlin, pp 60–63

Mendes R, Kruijt M, de Bruijn I, Dekkers E, van der Voort M, Schneider JH, Piceno YM, DeSantis TZ, Andersen GL, Bakker PA, Raaijmakers JM (2011) Deciphering the rhizosphere microbiome for disease-suppressive bacteria. Science 332:1097–1100

Mendes R, Garbeva P, Raaijmakers JM (2013) The rhizosphere microbiome: significance of plant beneficial, plant pathogenic, and human pathogenic microorganisms. FEMS Microbiol Rev 37:634–663. https://doi.org/10.1111/1574-6976.12028

Mendes LW, Kuramae EE, Navarrete AA, Van Veen JA, Tsai SM (2014) Taxonomical and functional microbial community selection in soybean rhizosphere. ISME J 8:1577–1587

Metzker ML (2010) Sequencing technologies—the next generation. Nat Rev Genet 11:31–46. https://doi.org/10.1038/nrg2626

Moreno R, Martines-Gomariz M, Yuste L, Gil C, Rojo F (2009) The *Pseudomonas putida* Crc global regulator controls the hierarchical assimilation of amino acids in a complete medium: evidence from proteomic and genomic analyses. Proteomics 9:2910–2928

Morris AC, Djordjevic MA (2006) The *Rhizobium leguminosarum* biovar *trifolii* ANU794 includes novel developmental responses on the subterranean clover cultivar Woogenellup. Mol Plant-Microbe Interact 19:471–479

Mukhtar S, Mirza MS, Awan HA, Asma M, Mehnaz S, Malik K (2016) Microbial diversity and metagenomic analysis of the rhizosphere of para grass (*Urochloa mutica*) growing under saline conditions. Pak J Bot 48:779–791

Nacke H, Will C, Herzog S, Nowka B, Engelhaupt M, Daniel R (2011) Identification of novel lipolytic genes and gene families by screening of metagenomic libraries derived from soil samples of the German biodiversity exploratories. FEMS Microbiol Ecol 78:188–201

Naz I, Mirza MS, Bano A (2014) Molecular characterization of rhizosphere bacterial communities associated with wheat (*Triticum aestivum* L.) cultivars at flowering stage. J Anim Plant Sci 24:1123–1134

Neal AL, Ton J (2013) Systemic defense priming by *Pseudomonas putida* KT2440 in maize depends on benzoxazinoid exudate from roots. Plant Signal Behav 8:e22655. https://doi.org/10.4161/psb.22655

Neal AL, Ahmad S, Gordon-Weeks R, Ton J (2012) Benzoxazinoids in root exudates of maize attract *Pseudomonas putida* to the rhizosphere. PLoS One 7:35498. https://doi.org/10.1371/journal.pone.0035498

Nelson EB (2004) Microbial dynamics and interactions in the spermosphere. Annu Rev Phytopathol 42:271–309

Neumann G, Romheld V (2002) Root-induced changes in the availability of nutrients in the rhizosphere. In: Waisel Y, Eshram E, Beeckman T (eds) Plant roots: the hidden half. Marcel Dekker, New York, pp 617–649

Newton AC, Fitt BD, Atkins SD, Walters DR, Daniell TJ (2010) Pathogenesis, parasitism and mutualism in the trophic space of microbe–plant interactions. Trends Microbiol 18:365–373

Niemi RM, Heiskanen I, Wallenius K, Lindstrom K (2001) Extraction and purification of DNA in rhizosphere soil samples for PCR-DGGE analysis of bacterial consortia. J Microbiol Methods 45:155–165

Ofek-Lalzar M, Sela N, Goldman-Voronov M, Green SJ, Hadar Y, Minz D (2014) Niche and host-associated functional signatures of the root surface microbiome. Nat Commun 5:4950

Oliveira CA, Alves VMC, Marriel IE, Gomes EA, Scotti MR, Carneiro NP, Guimarães CT, Schaffert RE, Sá NMH (2009) Phosphate solubilizing microorganisms isolated from rhizosphere of maize cultivated in an oxisol of the Brazilian Cerrado biome. Soil Biol Biochem 41:1782–1787

Owen AG, Jones DL (2001) Competition for amino acids between wheat roots and rhizosphere microorganisms and the role of amino acids in plant N acquisition. Soil Biol Biochem 33:651–657

Palumbo JD, Kado CI, Phillips DA (1998) An isoflavonoid-inducible efflux pump in *Agrobacterium tumefaciens* is involved in competitive colonization of roots. J Bacteriol 180:3107–3113

Patel HK, Suárez-Moreno ZR, Degrassi G, Subramoni S, González JF (2013) Bacterial LuxR solos have evolved to respond to different molecules including signals from plants. Front Plant Sci 4:447

Paterson AH, Schertz KF, Lin YR, Liu SC, Chang YL (1995) The weediness of wild plants: molecular analysis of genes influencing dispersal and persistence in Johnson grass, *Sorghum halepense* (L.) Pers. Proc Natl Acad Sci U S A 92:6127–6131

Patton CL, Glick BR (1996) Bacterial biosynthesis of indole-3-acetic acid. Can J Microbiol 42:207–220

Phillips DA, Fow TC, Six J (2006) Root exudation (net efflux of amino acids) may increase rhizodeposition under elevated CO_2. Glob Chang Biol 12:561–567

Piel J (2011) Approaches to capturing and designing biologically active small molecules produced by uncultured microbes. Annu Rev Microbiol 65:431–453

Pineda A, Zheng S, van Loon JJA, Pieterse CMJ, Dicke M (2010) Helping plants to deal with insects: the role of beneficial soil-borne microbes. Trends Plant Sci 15:507–514

Pisa G, Magnani GS, Weber H, Souza EM, Faoro H, Monteiro RA, Daros E, Baura V, Bespalhok JP, Pedrosa FO, Cruz LM (2011) Diversity of 16S rRNA genes from bacteria of sugarcane rhizosphere soil. Braz J Med Biol Res 44:1215–1221

Raaijmakers JM, Paulitz TC, Steinberg C, Alabouvette C, Moenne-Loccoz Y (2009) The rhizosphere: a playground and battlefield for soilborne pathogens and beneficial microorganisms. Plant Soil 321:341–361

Rabausch U, Juergensen J, Ilmberger N, Böhnke S, Fischer S, Schubach B (2013) Functional screening of metagenome and genome libraries for detection of novel flavonoid-modifying enzymes. Appl Environ Microbiol 79:4551–4563

Reading NC, Sperandio V (2006) Quorum sensing: the many languages of bacteria. FEMS Microbiol Lett 254:1–11

Rivero J, Gamir J, Aroca R, Pozo MJ, Flors V (2015) Metabolic transition in mycorrhizal tomato roots. Front Microbiol 6:598

Rout ME, Callaway RM (2009) An invasive plant paradox. Science 324:724–725

Rout ME, Callaway RM (2012) Interactions between exotic invasive plants and soil microbes in the rhizosphere suggest 'everything is not everywhere'. Ann Bot 110:213–222

Rout ME, Chrzanowski TH (2009) The invasive *Sorghum halepense* harbors endophytic N_2-fixing bacteria and alters soil biogeochemistry. Plant Soil 315:163–172

Rout ME, Chrzanowski TH, DeLuca TH, Westlie TK, Callaway RM, Holben WE (2013a) Bacterial endophytes enhance invasive plant competition. Am J Bot 100:1726–1737

Rout ME, Chrzanowski TH, Smith WK, Gough L (2013b) Ecological impacts of the invasive grass *Sorghum halepense* on native tallgrass prairie. Biol Invasions 15:327–339

Rudgers JA, Afkhami MA, Rua MA, Davitt SH, Hammer S, Huguet VM (2009) A fungus among us: broad pattern of endophyte distribution in the grasses. Ecology 90:1531–1539

Rudrappa T, Czymmek KJ, Pare PW, Bais HP (2008) Root-secreted malic acid recruits beneficial soil bacteria. Plant Physiol 148:1547–1556

Sachs JL, Mueller UG, Wilcox TP, Bull JJ (2004) The evolution of cooperation. Q Rev Biol 79:135–160

Sarwar M, Kremer RJ (1995) Enhanced suppression of plant growth through production of L-tryptophan-derived compounds by deleterious rhizobacteria. Plant Soil 172:261–269

Schade J, Hobbie SE (2005) Spatial and temporal variation in the islands of fertility in the Sonoran Desert. Biogeochemistry 73:541–553

Schreiter S, Ding G, Heuer H, Neumann G, Sandmann M, Grosch R et al (2014) Effect of the soil type on the microbiome in the rhizosphere of field-grown lettuce. Front Microbiol 5:144

Shaharoona B, Naveed M, Arshad M, Zahir ZA (2008) Fertilizer-dependent efficiency of pseudomonads for improving growth, yield and nutrient use efficiency of wheat (*Triticum aestivum* L.). Appl Microbiol Biotechnol 79:147–155

Sharma R, Ranjan R, Kapardar RK, Grover A (2005) Unculturable bacterial diversity: an untapped resource. Curr Sci 89:72–77

Sharma PK, Capalash N, Kaur J (2007) An improved method for single step purification of metagenomic DNA. J Mol Biotechnol 36:61

Sharma SK, Johri BN, Ramesh A, Joshi OP, Prasad SV (2011) Selection of plant growth–promoting *Pseudomonas* spp. that enhanced productivity of soybean wheat cropping system in central India. J Microbiol Biotechnol 21:1127–1142

Shenton M, Iwamoto C, Kurata N, Ikeo K (2016) Effect of wild and cultivated rice genotypes on rhizosphere bacterial community composition. Rice 9:42

Simoes MF, Antunes A, Ottoni CA, Amini MS, Alam I, Alzubaidy H, NA Mokhtar JA, Archer C, Bajic VB (2015) Soil and rhizosphere associated fungi in gray mangroves (*Avicennia marina*) from the Red Sea—a metagenomic approach. Genomics Proteomics Bioinformatics 13:310–320

Singh C, Soni R, Jain S, Roy S, Goel R (2010) Diversification of nitrogen fixing bacterial community using nifH gene as a biomarker in different geographical soils of western Indian Himalayas. J Environ Biol 31:553–556

Soni R, Acahrya C, Primalatha K (2012) Metagenomics technology. In: Kumar A, Pareek A, Gupta SM (eds) Biotechnology in medicine and agriculture principles and practice. IK International, New Delhi, p 835

Spaepen S, Dobbelaere S, Croonenborghs A, Vanderleyden J (2008) Effects of *Azospirillum brasilense* indole-3-acetic acid production on inoculated wheat plants. Plant Soil 312:15–23

Stevenson FJ, Cole MA (1999) Cycles of soil: carbon, nitrogen phosphorus, sulphur and micronutrients, 2nd edn. Wiley, New York

Streit WR, Schmitz RA (2004) Metagenomics—the key to the uncultured microbes. Curr Opin Microbiol 7:492–498

Subramoni S, Venturi V (2009) PpoR is a conserved unpaired LuxR solo of *Pseudomonas putida* which binds N-acyl homoserine lactones. BMC Microbiol 9:125

Subramoni S, Florez Salcedo DV, Suarez-Moreno ZR (2015) A bioinformatic survey of distribution, conservation, and probable functions of LuxR solo regulators in bacteria. Front Cell Infect Microbiol 5:16

Sugiyama A, Ueda Y, Zushi T, Takase H, Yazaki K (2014) Changes in the bacterial community of soybean rhizospheres during growth in the field. PLoS One 9:e100709

Tanveer A, Yadav S, Yadav D (2016) Comparative assessment of methods for metagenomic DNA isolation from soils of different crop growing fields. 3 Biotech 6(2):220

Tilman D (2000) Global environmental impacts of agricultural expansions: the need for sustainable and efficient practices. Proc Natl Acad Sci U S A 96:5995–6000

Torsvik V, Goksøyr J, Daae FL (1990) Comparison of phenotypic diversity and DNA heterogeneity in a population of soil bacteria. Appl Environ Microbiol 56:776–781

Torsvik V, Daae FL, Sandaa RA, Øvreås L (2002a) Microbial diversity and function in soil: from genes to ecosystems. Curr Opin Microbiol 5:240–245

Torsvik V, Øvreås L, Thingstad TF (2002b) Prokaryotic diversity—magnitude, dynamics, and controlling factors. Science 296:1064–1066

Toussaint JP, Pham TTM, Barriault D, Sylvestre M (2012) Plant exudates promote PCB degradation by a rhodococcal rhizobacteria. Appl Microbiol Biotechnol 95:1589–1603

Tschaplinski TJ, Plett JM, Engle NL, Deveau A, Cushman KC, Martin MZ, Doktycz MJ, Tuskan GA, Brun A, Kohler A, Martin F (2014) *Populus trichocarpa* and *Populus deltoides* exhibit different metabolomic responses to colonization by the symbiotic fungus *Laccaria bicolor*. Mol Plant-Microbe Interact 27:546–556

Tsurumaru H, Okubo T, Okazaki K, Hashimoto M, Kakizaki K, Hanzawa E, Hiroyuki T, Noriyuki A, Fukuyo T, Yasuyo S, Seishi I, Minamisawa K (2015) Metagenomic analysis of the bacterial community associated with the taproot of sugar beet. Microbes Environ 30:63–69

Turnbaugh PJ, Ley RE, Hamady M, Fraiser-Liggett CM, Knight R, Gordon JI (2007) The Human Microbiome Project: a strategy to understand the microbial components of the human genetic and metabolic landscape and how they contribute to normal physiology and predisposition to disease. Nature 449:804–810

Unno Y, Shinano T (2013) Metagenomic analysis of the rhizosphere soil microbiome with respect to phytic acid utilization. Microbes Environ 28:120–127

Vacheron J, Desbrosses G, Bouffaud ML, Touraine B, Moënne-Loccoz Y (2013) Plant growth–promoting rhizobacteria and root system functioning. Front Plant Sci 4:356

Valentine DL (2007) Adaptations to energy stress dictate the ecology and evolution of the Archaea. Nat Rev Microbiol 5:316–323

Valverde A, De-Maayer P, Oberholster T, Henschel J, Louw MK, Cowan D (2016) Specific microbial communities associate with the rhizosphere of *Welwitschia mirabilis*, a living fossil. PLoS One 11:e0153353

van der Heijden MGA (2008) The unseen majority: Soil microbes as drivers of plant diversity and productivity in terrestrial ecosystems Ecol Lett 11:651-651

van der Heijden MGA, and Horton TR (2009) Socialism in soil? The importance of 536 mycorrhizal fungal networks for facilitation in natural ecosystems. J Ecol 97:1139-1150

van Hees PAW, Jones DL, Nyberg L, Holstrom SJM, Godbold DL, Lundstrom US (2005) Modeling low molecular weight organic acid dynamics in forest soils. Soil Biol Biochem 37:517–531

van Kleunen M, Weber E, Fischer M (2010) A meta-analysis of trait differences between invasive and non-invasive plant species. Ecol Lett 13:235–245

Van Oosten VR, Bodenhausen N, Reymond P, Van Pelt JA, Van Loon LC, Dicke M (2008) Differential effectiveness against herbivorous insects in *Arabidopsis*. Mol Plant-Microbe Interact 21:919–930

Vander Ent S, VanHulten M, Pozo MJ, Czechowski T, Udvardi MK, Pieterse CMJ (2009) Priming of plant innate immunity by rhizobacteria and b-aminobutyric acid: differences and similarities in regulation. New Phytol 183:419–431

Venturi V, Fuqua C (2013) Chemical signaling between plants and plant pathogenic bacteria. Annu Rev Phytopathol 51:17–37

Vorholt JA (2012) Microbial life in the phyllosphere. Nat Rev Microbiol 10:828840

Weller DM, Raaijmakers JM, Gardener BBM, Thomashow LS (2002) Microbial populations responsible for specific soil suppressiveness to plant pathogens. Annu Rev Phytopathol 40:309–348

Yu J, Grishkan I, Steinbrener Y (2012) Microfungal community diversity in *Zygophyllum dumosum* and *Hammada scoparia* root zones in the northern Negev Desert. J Basic Microbiol 52:1–12

Zaidi A, Khan MS (2005) Interactive effect of rhizospheric microorganisms on growth, yield and nutrient uptake of wheat. J Plant Nutr 28:2079–2092

Zehr JP, Jenkins BD, Short SM, Steward GF (2003) Nitrogenase gene diversity and microbial community structure: a cross-system comparison. Environ Microbiol 5:539–554

Zhan J, Sun Q (2012) Diversity of free-living nitrogen-fixing microorganisms in the rhizosphere and non-rhizosphere of pioneer plants growing on wastelands of copper mine tailings. Microbiol Res 167:157–165

Zhang AH, Sun H, Wang P, Han Y, Wang XJ (2012) Modern analytical techniques in metabolomics analysis. Analyst 137:293–300

Zolla G, Badri DV, Bakker MG, Manter DK, Viyanco JM (2013) Soil microbiomes vary in their ability to confer drought tolerance to *Arabidopsis*. Appl Soil Ecol 68:1–9

MIX
Papier aus verantwortungsvollen Quellen
Paper from responsible sources
FSC® C105338

If you have any concerns about our products,
you can contact us on
ProductSafety@springernature.com

In case Publisher is established outside the EU,
the EU authorized representative is:
Springer Nature Customer Service Center GmbH
Europaplatz 3, 69115 Heidelberg, Germany

Printed by Libri Plureos GmbH
in Hamburg, Germany